NEW SIGNAL PROCESSING RESEARCH

NEW SIGNAL PROCESSING RESEARCH

TAKUMI MAEDA
EDITOR

Nova Science Publishers, Inc.
New York

For permission to use material from this book please contact us:
Telephone 631-231-7269; Fax 631-231-8175
Web Site: http://www.novapublishers.com

LIBRARY OF CONGRESS CATALOGING-IN-PUBLICATION DATA

New signal processing research / Takumi Maeda, editor.
p. cm.
ISBN 978-1-60456-479-2 (hardcover)
1. Signal processing--Congresses. I. Maeda, Takumi.
TK5102.5.N4864 2008
621.382'2--dc22 2008008021

Published by Nova Science Publishers, Inc. ✢ New York

CONTENTS

PREFACE

Signal processing is the analysis, interpretation, and manipulation of signals. Signals of interest include sound, images, and biological signals such as ECG, radar signals, and many others. Processing of such signals includes storage and reconstruction, separation of information from noise (for example, aircraft identification by radar), compression (for example, image compression), and feature extraction (for example, speech-to-text conversion). This new book presents the latest research in the field from around the world.

Chapter 1 gives a detailed review on symmetrical wavelets including the second-order Mexican-hat, Morlet and hyperbolic, and discusses various applications of the Mexican-hat wavelet. Higher-order symmetrical wavelets based on the second-order ones are also discussed in detail. Admissibility conditions of high-order wavelets are quantitatively elaborated with new results. Frequency-domain studies on these wavelets are also given. Chronological developments of symmetrical wavelets are thoroughly given with additional insight on key features. Differences between orthogonality and non-orthogonality using multi-resolution analysis are also discussed. Applications of symmetrical wavelets in various scientific disciplines are highlighted. It will be shown that there exists a strong link among different scientific disciplines which has not been explored and documented. Future work on second- and higher-order symmetrical wavelets is indicated. Applications of the Morlet wavelet, which is commonly used to study seismic signals, will be given in an upcoming publication.

The rather recent advent of wireless sensor networks (WSNs) has posed new challenges to the scientific community. The perspective taken in Chapter 2 is that WSNs may provide an appropriate testbed for some new (or at least less classical) basic principles that are emerging, often just under the driving force of this applicative scenario. Accordingly, the authors select some of the emerging paradigms and discuss their impact on the analysis and design of WSNs. A few case studies are provided, which are borrowed from the authors' recent works on the subject.

DNA microarray images consist of thousands of spots with weak intensity arranged in a matrix whose rows correspond to genes and columns correspond to experiments. The DNA microarray technology has emerged as a powerful and cost-effective tool for understanding of gene expression, regulation and interaction through a simultaneous study of thousands of genes. A successful microarray technology depends on seven main stages: experiment design, microarray fabrication, microarray image processing, data mining, biological illustration, data publication and data management. The focus of Chapter 3 is thus on a survey of recent advances in image analysis and data mining for DNA microarray technology. In image analysis, the authors discuss techniques for grid recognition, spot segmentation and quantification. In data mining, they discuss techniques for missing data imputation and classifications. The authors thoroughly survey recent literature on these tasks and briefly describe the fundamentals behind recent developments. They also present simulation results to illustrate the applications of some typical techniques.

In Chapter 4, the authors present the relevant mathematical background to address two well defined signal and image processing problems, namely, the problem of structured noise filtering and the problem of interpolation of missing data. The former is addressed by recourse to oblique projection based techniques whilst the latter, which can be considered equivalent to impulsive noise filtering, is tackled by appropriate interpolation methods.

In many applications of digital signal processing, such as audio and image processing, it is found that the filter bank with the changed channel number and filter coefficients can provide better performance than the time-invariant filter bank. This observation motivated us to generally study the time-varying filter bank. It is known that the theory of time-invariant filter banks is quite complete and has arrived a very high level. In comparison with time-invariant filter banks the research of time-varying filter banks is relatively new. In the authors' research on this topic, a series of theories are established to analyze the time-varying filter bank and multirate system. In Chapter 5, the authors provide a summarization of the developed theories and introduce also some new research results on this topic.

A filter is a simple system that can modify the components at certain frequencies, for example, the magnitude, phase, or group delay, for the input signals. Digital filters played an important role in the applications while using the digital signal processing. In the design of frequency-selective filters, the desired filter characteristics are specified in the frequency domain in terms of the desired magnitude and phase response of the filter. Many design algorithms have been proposed to meet the desired filter specifications. In general, linear algebra based methods have received considerable attention due to the powerful computation of digital computers. However, the typical linear algebra based methods require computationally intensive matrix inversion, and so these algorithms cannot meet the real-time requirement. Applying neural network to achieve the design of digital filter work is a new area of research. In Chapter 6, the Hopfield neural network (HNN) based approaches are introduced to carry out the finite impulse response (FIR) filter design. The proposed HNN based approach uses an improved feedback neural network to implement the design task, and is comparatively more effective and practical than the previous neural-based methods presented by Bhattacharya. The proposed structure is unified, regular and applicable for the designs of differentiators, Hilbert transformers, all-pass phase equalizers, FIR Nyquist filters, and quadrature mirror filter, etc., with significant reduction of computational complexity and hardware cost. Simulation results indicate that the proposed technique can achieve better or comparable results as compared to existing methods. If the hardware implementation is considered, for example, in a single IC chip, the proposed approach would provide a solution with real-time processing.

Adaptive filtering has become a more and more intense field of research for the last forty years. One of the starting points in almost all of this activity has been based on assuming Gaussian random signals that are then processed by the adaptive systems. Nonetheless, during the last few years, a huge number of theoretical advances and practical applications have been researched and developed without assuming the signals to be Gaussian. Some examples are underwater acoustic noise, low frequency atmospheric noise and many types of man-made noise. In addition, the presence of outliers may cause the adaptive algorithm to become unstable if the used algorithms are based on the Gaussian assumption.

The adaptive processing field has also experienced a change in its usual approaches the last few years since many new algorithms have been proposed in a non-Gaussian framework. One of the most relevant characteristics of these new algorithms is their robustness to impulsive (non-Gaussian) noise; Chapter 7 reviews these algorithms.

Gene expression profiles were shown to be useful in genomic signal processing when discriminating between cancer and normal (healthy) examples and/or between different types of cancer. K-nearest neighbors (k-NN) is one of the classification algorithms that demonstrated good performance for gene expression based cancer classification. Given that distance metric is fixed, the conventional k-NN has a single parameter (k - the number of nearest neighbors for each example) to set, which makes k-NN a very attractive choice in addition to the fact that it does not need training.

Classification performance of any classifier, including a k-NN, is typically characterized by classification error achieved on independent examples, which are often unavailable for the considered task. Thus, unbiased and low-variance error estimation is of ultimate importance in this case. The authors found that bolstered error satisfies these requirements and it was therefore chosen for their study. Bolstered error estimation is built on random sampling in the neighborhood of each example (with example-dependent neighborhood radius) and computing the number of errors made on such artificially created data. Because of random sampling, all examples can be employed in assessing the error, unlike cross-validation or bootstrap procedures.

In Chapter 8, the authors investigate the link between k-NN bolstered error and dataset complexity characterizing how difficult to classify a certain dataset. Their measure for the dataset complexity is the normalized Wilcoxon rank sum statistic. Through extensive simulation coupled with the copula method for analysis of association in bivariate data, the authors show that dataset complexity and bolstered error are related in terms of several dependence types such as positive quadrant dependence, tail monotonicity, and stochastic monotonicity.

As a result, the authors propose a new scheme for generating ensembles of k-NN classifiers, which is based on the selection of low complexity feature subsets for k-NNs in the ensemble, which constitutes to choosing accurate k-NNs according to the found dependence relation. The candidate subsets are randomly sampled from the whole set of the original features in order to make predictions of individual k-NNs diverse.

Experiments carried out on eight gene expression datasets containing different types of cancer demonstrate that the authors' ensemble generating scheme is superior (in terms of bolstered resubstitution error) to a single best classifier in the ensemble and to the traditional ensemble construction scheme that is ignorant of dataset complexity. It also outperforms the redundancy-based filter, especially designed to remove irrelevant genes.

In recent years, signal processing with distributed multiple-sensor systems has attracted the attention of several investigators and has been gaining importance. This interest has been sparked by the requirement of military surveillance systems to be more reliable and immune to electronic attack than single sensor systems. In Chapter 9, signal processing with distributed multiple-sensor detection systems with data fusion is considered. The distributed sensors monitor a common volume and send their local decisions to a fusion center, which further combines them into a final global decision. There are two main approaches for signal processing in distributed multiple-sensor detection systems. The first is centralized signal processing, where all sensor observations are transmitted to a central processor that performs optimal detection of targets based on sensors observations. The second approach is decentralized signal processing, where the signal processing is distributed among the sensors and a fusion center. The local sensors monitor the same object scene and pass their local decisions about the same hypothesis to the fusion center that combines them into a final global decision. The contribution of this chapter is divided into two main parts. In the first part, the problem of obtaining and comparing the global performance improvement in centralized and decentralized approaches is considered and analyzed. The needed number of sensors in the decentralized approach, to achieve a performance non inferior to the centralized approach, for the same signal to noise ratio, is determined. The dependence of global performance improvement on the desired false alarm probability, the number of sensors, and the signal to noise ratio at each individual sensor, is also addressed. In the second part of the contribution, the authors propose a simple algorithm for optimizing multiple sensor detection system with data fusion, in terms of both the sensors and the fusion center, according to Neyman-Pearson strategy. The proposed algorithm determines, for a given false alarm probability, the corresponding optimum setting of the thresholds both at the fusion center and at the sensors. It scans all possible solutions and picks the solution that gives the largest detection probability. The proposed algorithm is found to be simple, accurate and fast. The performance analysis is provided using Monte Carlo simulation in case of Gaussian and Rayleigh distributed observations.

Chapter 10 proposes a gray-level Gaussian transformation for image enhancement in the spatial domain. Comparisons with a power-law gray-level transformation are made. Image histograms of both transformations are estimated. A new formula for the useful energy in images is proposed to automate the performance evaluation process of image enhancement. Relative merits of the new transformation are discussed.

In multi-path mobile environments, signal transmission between a base station (BS) and user equipment (UE) is usually complicated because the signals are blocked, reflected or refracted by random scatterers. These scatterers can be modelled by using a suitable distribution. An angle-of-arrival (AoA) probability density function (pdf) based on the scatterer distribution is used to describe features of signal transmission between the BS and the UE. The Gaussian AoA pdf was shown to be effective in modelling the scatterers. The AoA pdf in which the scatterer distribution obeys a hyperbolic sech function is derived in this paper. Approximate hyperbolic and Gaussian AoA pdfs under small and large AoA conditions are also given. Power azimuth spectra (PAS) and entropy of the approximate pdfs as functions of the angular spread are computed. Relative merits of each pdf are discussed in Chapetr 11. Possible further work is also outlined.

Noncoherent receivers are used in digital communication systems when carrier phase estimation is difficult or impossible. It is known that on the additive white Gaussian noise (AWGN) channel, optimal noncoherent receiver structure is usually implemented using matched filters followed by square law detectors. Square law detection results in non-Gaussian, signal dependent noise distribution at detector's output which significantly complicates channel estimation and data detection tasks especially if no training sequence is used (blind estimation). In Chapter 12, the authors present a new approach for optimum noncoherent receiver design. The proposed receiver performs blind channel estimation and data detection on channels characterized by intersymbol interference (ISI). They model the received signal as hidden Markov model (HMM) where the observation probability has a non-central chi-square distribution (NCCS). The proposed algorithm proceeds in 4 steps. The first step is based on a recently developed EM algorithm for the NCCS distribution and its finite mixtures. This novel algorithm involves only closed form expressions and can be easily implemented. The second step is a standard Baum-Welch algorithm. At the third step, the correspondence between the channel states and the transmitted bits is established. At the last step, the data is detected by symbol by symbol maximum a posteriori probability (MAP) algorithm or Viterbi algorithm.

Simulation results are presented for three communication channels: wireless communication with on-off keying (OOK) modulation on AWGN channel, wireless communication with OOK modulation on frequency selective fading channel and direct detection optical channel with ISI and OOK modulation. The proposed approach can also be used in noncoherent receivers for signal to noise ratio (SNR) estimation. Performance of EM based SNR estimation algorithm is presented.

In research and engineering applications, it is usually necessary to solve the underdetermined system for its sparse solution, whereas most of current methods can not solve this problem effectively. In Chapter 13, a novel algorithm is proposed to solve underdetermined system sparsely. In this algorithm, the solution space is coded with some particles firstly, and then the solution space is compressed in this particle-coded space by the evolution way of particle swarm optimization method to convert the underdetermined system to a non-underdetermined one as possible. Finally, the underdetermined system is solved in the compressed sub-space robustly after it is converted to a non-underdetermined one. During the solving procedure, a new l_0 constrained fitness function is introduced to select the global optimal sparse solution for the underdetermined system. The proposed method is compared with Minimum Norm (MN) and Focal Underdetermined system solver (FOCUSS) by solving two different underdetermined systems: one was a random system, the other was an underdetermined system derived from a 3-shell sphere head model, which is usually adopted in EEG source imaging. The current results confirmed a good sparse solution for the underdetermined system could be achieved with the proposed method in this chapter.

Space-time adaptive processing (STAP) for bistatic airborne radar offers several advantages, such as the higher possibility of detecting stealth targets. Moreover, the ability to passively provide surveillance data without alerting the target under observation has implications for peace keeping operations and counter-terrorism activities. One way to achieve these capabilities is to deploy reconfigurable bistatic sensor systems where passive receiver(s), such as unmanned aerial vehicles (UAVs), are not colocated with the active transmitter(s). These transmitters can be placed in secure and defensible locations remote from the surveillance area, thus decreasing the possibility of detection by other systems.

However, bistatic radar operation does have its disadvantages. The major problem with bistatic radar systems is the range dependent nature of the clutter echoes. Due to the motion of the two platforms, the ground clutter echoes will exhibit a Doppler shift that is proportional to the velocity of both platforms. The physical geometry of the environment causes the spatio-temporal clutter spectrum to change between different ranges. This results in training range gates that are statistically different from the test range gate, which broadens the clutter suppression filter notch and consequently degrades the processor performance. This region of research has not been emphasized enough in the current literature.

The major objective of Chapter 14 is twofold: i) discuss the concept of STAP; ii) present training methods to address the range dependency problem for bistatic radar operation in a dense environment of ground-moving targets. The work in this chapter is directed towards what may be called 'small STAP', where the number of spatial channels is small and the array is non-uniform. The work is motivated by a desire to minimize the amount of navigational data associated with both the transmitter and receiver. Furthermore, it is directed towards environments where all range gates may contain targets.

A novel technique is proposed to predict the range-dependent inverse covariance matrix, which is used to compute the STAP filter weights, by utilizing linear prediction theory. An extension of the technique to non-linear prediction is also presented. The two proposed techniques give an improvement in clutter suppression performance over conventional STAP techniques. They also provide mitigation against the additional clutter notches resulting from range and Doppler ambiguities. These techniques eliminate the requirement for a uniform linear array and can be applied to arrays of arbitrary configuration. No navigational data or parameters estimation are necessary as only the clutter data is required.

Nonnegative Matrix Factorization (NMF) finds nonnegative and usually sparse lower-rank representations of the input data. Basically, the factorization is achieved by alternating minimization of a given cost function subject to nonnegativity constraints. Unfortunately, the underlying optimization problem in NMF is generally non-unique, and a variety of assumptions and computational strategies has been proposed to select suitable approximations to the true nonnegative factors. The problems of non-uniqueness in NMF have been studied by many researchers, e.g., D. Donoho and V. Stodden; F. Theis, K. Stadlthanner and T. Tanaka; and M. Heiler and C. Schnorr, who emphasize that additional constraints (aside from nonnegativity) considerably relax the non-uniqueness. A common approach is to use sparsity and/or orthogonality constraints. In the authors' approach, other constraints are also pointed out, e.g., smoothness, binarity or unimodality. Additionally, the optimization algorithm crucially affects the estimated factors since the cost function is non-convex, independent of the additional constraints. The gradient descent techniques that are often used for solving this problem do not guarantee a unique solution. Depending on initialization, NMF may even provide spurious or misleading factors. In Chapter 15, the authors discuss a possibility of using other (morthological) constraints in NMF, and some optimization algorithms with iterative steering to the global solution.

The Empirical Mode Decomposition is a technique to decompose a given signal into a set of elemental signals called Intrinsic Mode Functions. The Empirical Mode Decomposition is the base of the so-called "Hilbert-Huang Transform" that comprises also a Hilbert Spectral Analysis and an instantaneous frequency computation. A modified improved algorithm for the Empirical Mode Decomposition is described in Chapter 16. Its interpretation in terms of AM/FM modulation is done. Numerical simulations are carried out to empirically evaluate the algorithm.

A fast computation method of acquiring a high resolution (HR) image from its blurred and downsampled low resolution (LR) image by using the discrete cosine transform (DCT) is presented in Chapter 17. The blurring and downsampling process is mathematically modeled in matrix form and then the inverse problem of the process is solved by using the regularization approach. Since the blurring and downsampling matrix is very large, a large computation time is needed to directly solve the problem. The DCT makes the blurring and downsampling matrix be a sparse matrix, which leads to the analytical solution in scalar form in the DCT domain. Therefore, the HR image can be quickly generated from the LR image by the DCT fast computation method. Computer simulation has shown that the proposed method is superior to the cubic spline interpolation in the HR image restoration performance.

Chapter 18 describes some robust techniques for the localization of distributed sources. For a distributed source, the system receives a spatially dispersed signal consisting of several arriving rays spread around some nominal direction. In this study, the authors distinguish two types of distributed sources: coherently distributed (CD) and incoherently distributed (ID) sources. They present, for these two types of sources, some robust methods for decoupled estimation of their angular parameters (DOA and angular spread). Compared with existing methods, the proposed methods improve the robustness to the mismodeling of the spatial distribution of source, and can work even in the case where the different sources have different angular distribution shapes.

For tracking maneuvering target in a cluttered environment, a novel tracking algorithm based on maximum entropy fuzzy clustering is proposed in Chapter 19. First, the interacting multiple model (IMM) approach is used to solve the maneuver problem, and the false alarms generated by clutter are accommodated through a maximum entropy fuzzy probabilistic data association filter (MEF-PDAF). Second, to avoid the unobservability problem of passive target tracking, a nonlinear measurement model of multiple passive sensors is formulated. Moreover, to reduce the computational load, the characteristic of the discrimination factor is analyzed, and the influence of the clutter density on it is considered, which enables the algorithm eliminate invalid measurements. Finally, simulation results show that the proposed algorithm has advantages over the conventional IMM-PDAF algorithm in terms of simplicity and efficiency.

In: New Signal Processing Research
Editor: Takumi Maeda, pp. 1-54
ISBN: 978-1-60456-479-2

Chapter 1

CONTINUOUS SYMMETRICAL WAVELETS AND THEIR APPLICATIONS — PART I: THE MEXICAN-HAT WAVELET

Khoa N. Le[1]
Griffith School of Engineering
Centre for Wireless Monitoring and Applications
Griffith University, Gold Coast campus, Parklands Drive,
Southport, QLD4215, Australia

Abstract

This chapter gives a detailed review on symmetrical wavelets including the second-order Mexican-hat, Morlet and hyperbolic, and discusses various applications of the Mexican-hat wavelet. Higher-order symmetrical wavelets based on the second-order ones are also discussed in detail. Admissibility conditions of high-order wavelets are quantitatively elaborated with new results. Frequency-domain studies on these wavelets are also given. Chronological developments of symmetrical wavelets are thoroughly given with additional insight on key features. Differences between orthogonality and non-orthogonality using multiresolution analysis are also discussed. Applications of symmetrical wavelets in various scientific disciplines are highlighted. It will be shown that there exists a strong link among different scientific disciplines which has not been explored and documented. Future work on second- and higher-order symmetrical wavelets is indicated. Applications of the Morlet wavelet, which is commonly used to study seismic signals, will be given in an upcoming publication.

Key words: symmetrical wavelets, non-orthogonality, correlation, continuous wavelet transform, Mexican-hat wavelet, Morlet wavelet, hyperbolic wavelet

---

[1] E-mail address: K.Le@griffith.edu.au. Tel. (+617) 5552 9175, Fax. (+617) 5552 8065.

1. Introduction

1.1. Overview

The wavelet transform has become very popular and found many applications in various areas of signal processing including image processing, digital water-marking, denoising, image compression and filter banks. The history on wavelets dated back to the beginning of the twentieth century, to 1910, when Haar first proposed a wavelet, Haar wavelet (HaW), which later was shown to be the simplest orthogonal wavelet [1]. In 1982, a "Mexican-hat" wavelet (MHW), whose name implies its shape, which is the negative second-order derivative or Laplacian operator ∇^2 of a Gaussian pulse $\exp(-x^2/\sigma)$, was used by Marr in image processing and vision studies, of which wavelet zero-crossings were employed for edge detection [2]. It was also shown by Marr that the second-derivative function was the most satisfactory tool for detecting intensity changes and edges in images. It should be noted that in his book published in 1982, Marr did not mention the word "wavelet", nor refer to the Haar "wavelet". In other words, he independently used the "wavelet" tool for vision analyses by means of the ∇^2 operator of the Gaussian pulse, which leads to studies of high-order wavelets using high-order derivatives.

A few years later, another function which is a modulated Gaussian pulse, was numerically used by the French geophysicist, J. Morlet, in the field of seismology for processing seismic data. Morlet realised the effectiveness of the pulse because of its finite-time compact and coined the name "wavelet" meaning "short pulse", however, he did not theoretically show how the method worked. Having learned about the sound results from Morlet's work, another French physicist, A. Grossmann, theoretically studied Morlet's work and together they showed that square integrable functions, whose Fourier transforms vanished on a half plane belonged to a Hardy subspace H^2 which is part of the real-number space $L^2(\mathrm{R})$, could be analysed by using square integrable functions from a different subspace, whose Fourier transforms did not vanish on a half plane, by employing scaling (dilating) and translating (shifting), which formed the foundation for the continuous transform [3]. This technique, and Gabor's work on filtering the Fourier transform with a window function to study non-stationary signals proposed in 1946 [4] leading to work on the Short-Time Fourier Transform (STFT), correspond to "coherent states" in theoretical physics — i.e. a family of square integrable functions generated from one basis function, called the mother function, by expanding itself via translating and scaling — of which the former is a special case of a Weyl-Heisenberg group and the latter of an affine "$cx + d$" group [5]. It was also Daubechies, in her paper published in 1990 [5], who coined the term "phase space", which was borrowed from quantum mechanics, to describe the 2-D expansion via translation (time) and dilation (frequency) of wavelets.

The work of Grossmann and Morlet published in 1984, after the proposal of the STFT by Gabor in 1946 [4], was recognized widely and opened up new research directions on wavelets whose roles have been central in signal processing and found many applications in a wide range of different disciplines. In view of the work by Gabor, Grossmann's and Morlet's work showed that there were different ways signal energy could be viewed using different tools such as the Fourier transform and the wavelet transform. In other words, not only could signals be represented by a sum of infinite "basis" functions such as sinusoids and co-sinusoids proposed by Fourier, but also by a sum of finite pulses with finite time support, i.e. the pulses decaying to zero at infinity, which could expand themselves by dilating (using a scale index a) and translating or shifting (using a time index b), which can be explicitly represented as [3]

$$\psi(t) = \frac{1}{\sqrt{a}} \cdot \psi\left(\frac{t-b}{a}\right), \tag{1}$$

where $\psi(t)$ is the mother wavelet defined in the time domain t.

The time index b can be used to shift the wavelet to a specific time location that requires to be explored. By varying the scale a, the wavelet can be expanded (for large a) yielding a coarse resolution, or compressed (for small a) yielding a finer resolution [6]. The wavelet transform is mathematically given by Eq. (2) [3, 6]

$$\hat{\psi}(a,b) = \int_{-\infty}^{+\infty} f(t) \cdot \frac{1}{\sqrt{a}} \cdot \psi\left(\frac{t-b}{a}\right) dt, \tag{2}$$

where $\hat{\psi}(a,b)$ is the wavelet transform of the input signal $f(t)$.

To be an admissible wavelet, a function must satisfy the following admissibility constraint given by Eq. (3) [3]

$$0 < C_{\psi} = \int_{-\infty}^{+\infty} \frac{[\hat{\psi}_F(\omega)]^2}{\omega} d\omega < \infty, \tag{3}$$

where $\hat{\psi}_F(\omega) = \int_{-\infty}^{+\infty} \psi(t) \cdot e^{-j\omega t} dt$ is the Fourier transform of $\psi(t)$, which imposes that the wavelet function must be zero mean, because if the signal frequency component at $\omega = 0$ (DC condition) is different from zero, i.e. $\hat{\psi}_F(\omega) \neq 0$, then Eq. (3) approaches infinity which makes it undefined. To achieve the condition imposed by Eq. (3), we must have [6]

$$\int_{-\infty}^{+\infty} \psi(t)\, dt = 0, \text{ and } \psi(t) = 0 \text{ for } t \to \infty, \tag{4}$$

which implies that $\psi(t)$ must be compactly support or finite.

After taking the wavelet transform of the signal $f(t)$, the signal can be reconstructed by using the following formula [3]

$$f(t)=\frac{1}{C_{\psi}}\cdot\int_{-\infty}^{+\infty}\int_{-\infty}^{+\infty}\hat{\psi}(a,b)\cdot\frac{1}{\sqrt{a}}\cdot\psi\left(\frac{t-b}{a}\right)\cdot\frac{1}{a^{2}}\,da\,db \tag{5}$$

which is clear that if C_{ψ} is infinity or zero then Eq. (5) is undefined and the wavelet is not admissible. Thus, the zero-mean constraint imposed on continuous wavelets ensures the signal reconstruction.

Continuing on Grossmann's and Morlet's work, Daubechies and Meyer further developed discrete wavelet transforms by confining the discrete scale a and time b in a lattice model [7, 8], i.e. expansions of a and b are perpendicular or orthogonal in 2-D space. This work leads to proposals of orthogonal and orthonomal wavelet bases with significant contributions from Daubechies, Grossmann, Meyer, Coifman, Jansen, Mallat, Morlet, Tchamitchian and others [8]. However, a major breakthrough, proposed in 1989 by Mallat and Meyer, was the definition of multi-resolution yielding important concepts such as subband coding, filter banks and discrete wavelet transforms [9] which have attracted wide attention. The introduction of multiresolution requires a scaling function, $\varphi(t)$, together with the wavelet function, to completely represent signals in the L^2(R) space if they are orthogonal to each other. One of the main purposes of the wavelet and Fourier expansions is to approximately represent signals in the L^2(R) space. For the Fourier expansion, to cover the L^2(R) space, its only basis function is $e^{jx} = \cos(x) + j\sin(x)$, which requires no expansions because of the infinite time support of $\cos(x)$ and $\sin(x)$. For the wavelet expansion, since the wavelet function possesses finite-time support, it must be expanded [10] to cover the L^2(R) space. The basis function is called "mother wavelet" without the involvements of the scale a and translation b.

Suppose that there exists a scaling function $\varphi(t)$, which is orthogonal to a given wavelet function $\psi(t)$, and $\varphi(t)$ spans a closed reference subspace S_0 in L^2(R). A family of scaling functions expanding in L^2(R) is defined as [1, 10]

$$\varphi_{a,b}(t)=2^{a/2}\cdot\varphi(2^{a}t-b), \tag{6}$$

where a and b are positive integer scale and time indices respectively

As the scale function expands, the number of subspaces created by dilating and translating the scaling function increases, and eventually these subspaces will cover part of L^2(**R**). It should be noted that since the expansion process is linear and generated from only one basis function $\varphi(t)$, subsequent subspaces from the same family do not overlap, in fact, it was shown by Mallat [1, 9] that they were nested subspaces

$$S_{-\infty}\subset\ldots\subset S_{-3}\subset S_{-2}\subset S_{-1}\subset S_0\subset S_1\subset S_2\subset\ldots\subset S_{\infty}, \tag{7}$$

where $S_{-\infty} = \{0\}$ and $S_{\infty} = L^2$(R) by definition. The higher is the subspace number over the right-hand side of Eq. (7), the finer is its resolution. Similarly, the further away from the reference subspace S_0 a subspace is to the left in Eq. (7), the finer is its resolution. Moreover, the scale function can be explicitly expressed as [1]

$$\varphi(t) = \sum_{n=0}^{+\infty} h(n) \cdot \sqrt{2} \cdot \varphi(2t - n) \tag{8}$$

where by definition, the discrete sequence $h(n)$ consists of the scaling coefficients.

Because scaling subspaces are nested, they clearly cannot cover the entire $L^2(\mathrm{R})$ space. In fact, the scaling function provides the coarse framework for signal approximation in the $L^2(\mathrm{R})$ space. To cover the difference subspaces which are gaps among adjacent scaling subspaces, the wavelet function and its expansions are required. The wavelet expansion is given as

$$\psi_{a,b_1}(t) = 2^{a/2} \cdot \psi(2^a t - b_1), \tag{9}$$

where a and b_1 are positive integer scale and time indices respectively.

With the addition of wavelet spaces, and from Eq. (7), the space $L^2(\mathrm{R})$ is now given as [1]

$$L^2(\mathrm{R}) = S_0 \oplus W_0 \oplus W_1 \oplus W_2 \oplus \ldots \oplus W_\infty, \tag{10}$$

where the symbol $\oplus$ indicates direct subspace addition.

A graphical representation of Eq. (7) is given in Figure 1. The wavelet function can also be expressed in terms of the scaling function as given by Eq. (11) [1]

$$\psi(t) = \sum_{n=0}^{+\infty} h_1(n) \cdot \sqrt{2} \cdot \varphi(2t - n), \text{ where } h_1(n) \text{ is a discrete sequence} \tag{11}$$

By using $\varphi(t)$ and $\psi(t)$, the signal can be represented as [1, 9]

$$f(t) = \sum_{k=-\infty}^{+\infty} c(k) \cdot \varphi_k(t) + \sum_{m=0}^{+\infty} \sum_{k=-\infty}^{+\infty} d(m,k) \cdot \psi_{m,k}(t), \tag{12}$$

where $c(k)$ and $d(m, k)$ are the scale and wavelet coefficients respectively. It should be noted that the former coefficients can be considered as giving coarse details of $f(t)$, and the latter give finer details of $f(t)$ based on $c(k)$.

For $\varphi(t)$ to be orthogonal to $\psi(t)$ it is required that

$$\int_{-\infty}^{+\infty} \varphi_{a,b}(t) \cdot \psi_{a,b_1}(t)\, dt = 0. \tag{13}$$

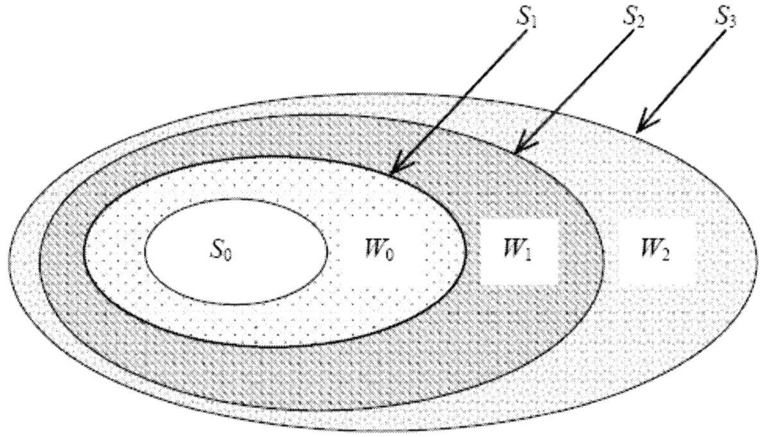

Figure 1. Graphical representation of the multiresolution concept given by Eqs. (7) and (10); note $\varphi(t)$ is orthogonal to $\psi(t)$.

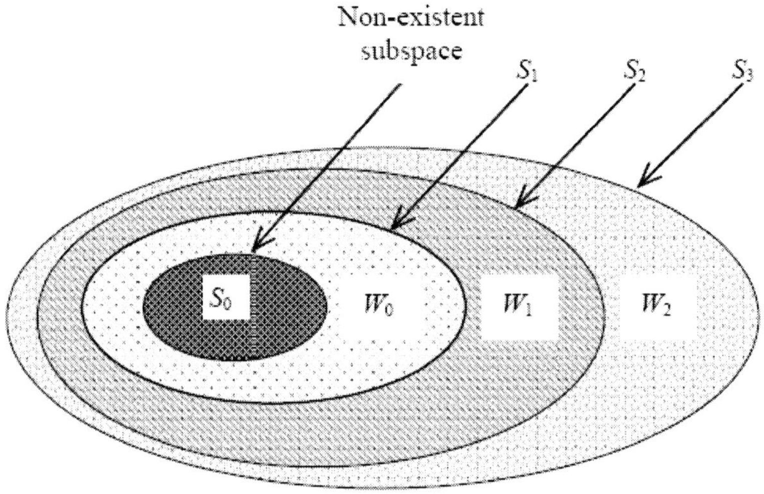

Figure 2. Multi-resolution analysis of symmetrical wavelets whose scaling function $\varphi(t)$ and its space S_0 do not exist.

From Eq. (11), given a wavelet function $\psi(t)$, its scaling function $\varphi(t)$ only exists if the sequence $h_1(n)$ exists, which is not always the case in practice [1]. There are many orthogonal wavelets whose scaling functions exist such as Daubechies [11], Meyer [12], and non-orthogonal wavelets whose scaling functions do not exist such as Mexican-hat [12] and Morlet [3, 12]. Wavelets are generally divided into four major groups [12]

1. Group 1 (orthogonal with *FIR* filtering): This group includes the Daubechies, Coiflets and Symlets wavelets;
2. Group 2 (bi-orthogonal with *FIR* filtering): The Bior-Splines wavelet belongs to this group;
3. Group 3 (orthogonal with scale function): The wavelet and its scale function exist, but its *FIR* filter does not. The Meyer wavelet is a typical member of this group;

4. Group 4 (*FIR* filter and scaling function do not exist): This has been labelled "crude" wavelet group because symmetrical wavelets in this group can be mathematically described by their explicit expressions. The Mexican-hat, hyperbolic and Morlet wavelets belong to this group.

Groups 1 to 3, which include many proposed orthogonal and bi-orthogonal wavelets, have been extensively studied and developed as explained earlier. Group 4 which consists of symmetrical wavelets whose explicit expressions exist, have often been overlooked [13], because their scaling functions do not exist, which make discrete reconstruction using these wavelets difficult. As a result, their applications in data compression, estimation and approximation are somewhat limited. However, the main advantage of these symmetrical wavelets is that they are regular with a number of vanishing moments, compact with finite-time support, and more importantly, can be used to detect intensity changes and edges in images as shown by Marr [2], who effectively used cross correlation and convolution functions of these wavelets on the input image. Thus, it is evident that symmetrical wavelets have found useful applications where their symmetry was explored, including astrophysics, weather prediction, signal processing, astronomy, image processing, signal analysis of ECG and EEG, fractal, chaos, water-marking, denoising and fault detection. These applications exploit the symmetry of the wavelets as discussed in detail later. A number of review papers and books on wavelets and related topics can be found in [1, 5, 6, 8, 9, 11, 14-17]. It is clear that for the Mexican-hat and Morlet wavelets, their scaling functions with their subspaces do not exist, therefore according to the multiresolution analysis, they do not cover the entire $L^2(\mathrm{R})$ space which makes signal approximation using these wavelets difficult. However, reconstruction in difference subspaces spanned by the wavelet expansion, except S_0, is possible even though they do not cover the entire $L^2(\mathrm{R})$ space. One way to overcome this difficulty is to imagine that the first subspace V_0 starts from infinity and subsequent subspace numbers are reduced until the real V_0 is reached [18]. It should therefore be stressed that the main focus of using these wavelets is therefore not to fully expand the $L^2(\mathrm{R})$ space, but to expand the difference subspaces where their symmetrical features via correlation functions can be optimally employed, which is the key idea in many applications.

The two most well-known symmetrical wavelets in the literature are Mexican-hat and Morlet which have been extensively studied and their band-peak frequency, minimum number of sampling points, scale resolution, maximum number of scales have been explicitly given as well as numerically estimated [19-22]. The main aim of this paper is to review useful applications of the Mexican-hat wavelet (MHW) that have been independently employed in different disciplines, and also to briefly introduce a new symmetrical hyperbolic wavelet (HyW), which is generated from the hyperbolic kernel (HyK) used in Cohen time-frequency power spectrum class (C-TFPSC). Background on the Morlet wavelet is discussed in this paper, however, its applications will be reported in a future publication. Possible connections among various applications are also given where appropriate. For each application, detailed interpretations and possible future work will be discussed.

Useful applications of the Mexican-hat and Morlet wavelets motivate further research in applying higher-order hyperbolic and Mexican-hat wavelets to existing and new applications.

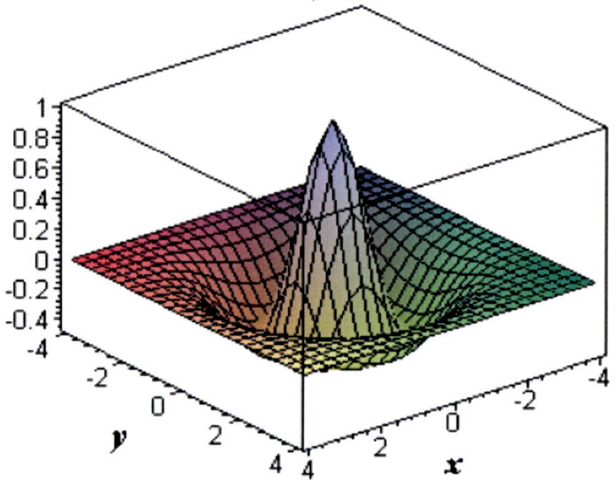

Figure 3. Magnitude of the 2-D MHW [Eq. (17)] for $\sigma = 2$ in the spatial domain.

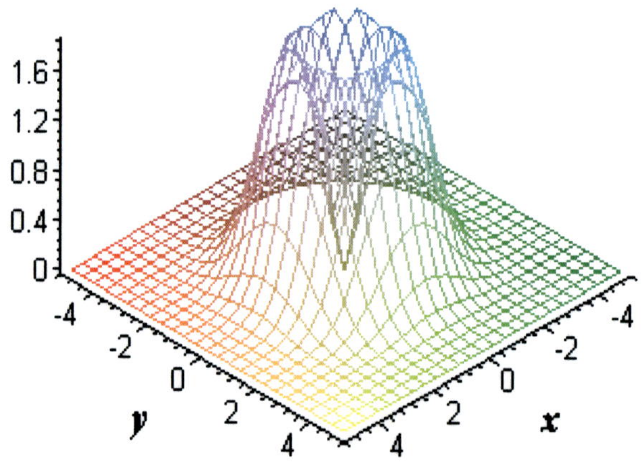

Figure 4. 2-D frequency response magnitude of the MHW [Eq.(18)], for $\sigma = 2$, the vertical axis shows the magnitude of the response.

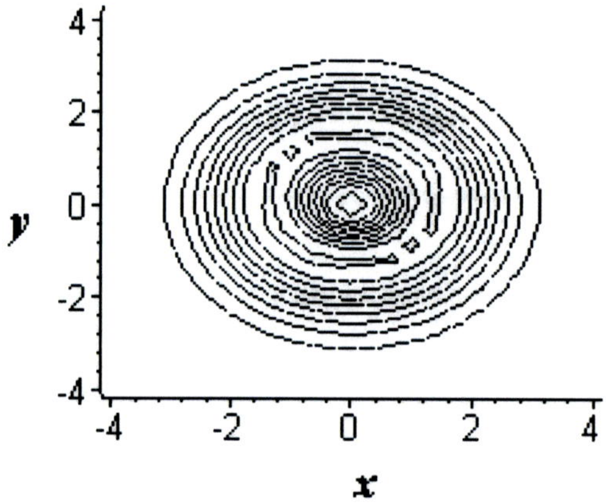

Figure 5. Contour plot of the magnitude of frequency response shown in Figure 4 for the 2-D MHW.

The eighth-order MHW was also shown by Grossmann to be effective for edge detection in image processing [5]. Higher-order MHW's and HyW's, which have also been shown to be useful with finer scale resolutions compared to the second-order wavelets [21], will be briefly discussed.

1.2. Organisation

The paper is organized as follows. Section 2.1 provides background on the second-order MHW, MoW and HyW. Section 2.2 discusses high-order MHW and HyW which are high-order derivatives of the corresponding CWK and HyK. Section 3 outlines applications of the MHW that have been independently carried out across different disciplines. In addition, interpretation and possible connections among these applications are discussed. Section 4 concludes the main findings of this review paper.

1.3. Notation

A number of notations are used in this paper to avoid unnecessary text repetitions and to provide clarity.

Mexican-hat wavelet	MHW
Morlet wavelet	MoW
Second-order hyperbolic wavelet	HyW
Haar wavelet	HaW
Continuous wavelet transform	CWT
Inverse continuous wavelet transform	ICWT
Multi-resolution analysis	MRA
Choi-Williams kernel	CWK
Hyperbolic kernel	HyK
Cohen's time-frequency power spectrum class	C-TFPSC
Short-Time Fourier Transform	STFT
Fourier Power Spectrum	FPS
Multi-reference matched-filter	MRMF
Finite impulse response	FIR
Signal-to-noise ratio	SNR
Point-spread function	PSF
Artificial neural network	ANN
Automatic phase recognition	APR
All-electron density-functional	AE-DF
VanderLugt correlator	VLC
Joint transform correlator	JTC
Wavelet power spectrum	WPS
Mars obiter laser altimeter	MOLA
Logarithmic harmonic filter	LHF
Cosmic microwave background	CMB

Maximum entropy method	MEM
Image-guided radiotherapy	IGRT
Amorphous silicon electronic portal imaging device	a-Si EPID
Phase only CHF	POCHF
Circular harmonic filter	CHF

2. Background

Symmetrical wavelets have found many useful applications where their symmetry is exploited by using the concept of correlation, which measures the similarities between two different signals (cross correlation) and of the signal with itself (auto correlation), as a lag parameter τ shifts to different locations on the time axis. More explicitly, the cross-correlation function $R_{fg}(\tau)$ between two functions $f(t)$ and $g(t)$ is given by [23]

$$R_{fg}(\tau) = \int_{-\infty}^{+\infty} f(t) \cdot g(t \pm \tau)\, dt \tag{14}$$

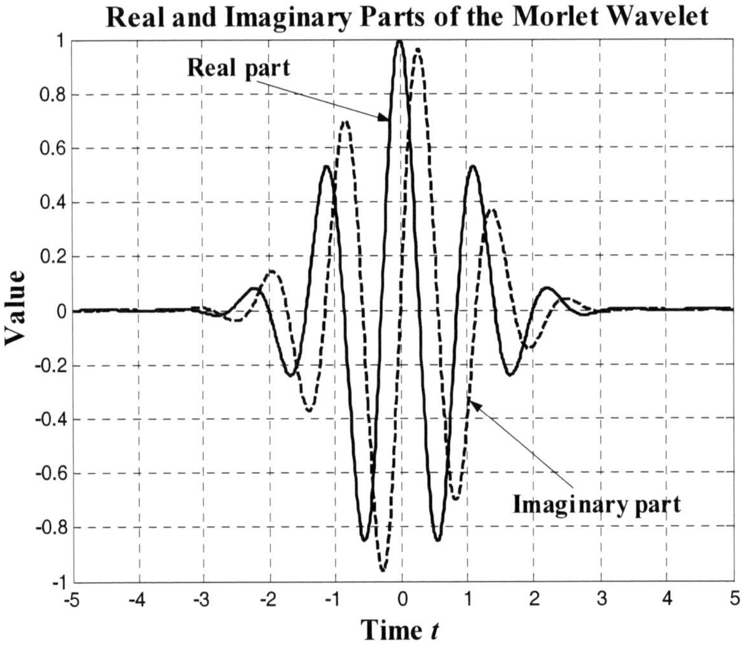

Figure 6. Real and imaginary parts of the MoW [Eq.)], for $\sigma = 2$, and $\omega_\psi = 5.5$.

It should be noted that the smaller is the lag parameter τ, the larger is the correlation function, which implies that correlation functions attain their peaks at $\tau = 0$ or zero shift. This idea is clearly understood in the case of the auto-correlation function where $f = g$ from Eq. (14). It should also be noted that the selection of a plus sign or minus sign in Eq. (14) does not alter values of the correlation function. Since a correlation function measures the similarities between two signals, it is clear that for signals that possess a high degree of similarity to the wavelet function, their cross-correlation function will be larger than when correlating with signals with less similarity. This is the key idea that has been widely employed by researchers in various disciplines to extract features from an input signal/image by convolving it with a wavelet of similar features. For example, to detect Gaussian signals, one can effectively use the MHW since it is the Laplacian of a Gaussian pulse.

Symmetrical wavelets can generally be divided into two types: second-order and high-order [20, 21]. The former type is based on the Laplacian operator ∇^2 which is equivalent to the second-order derivative of an admissible kernel used for C-TFPSC. The Mexican-hat or Choi-Williams wavelet, named after its corresponding kernel CWK used in C-TFPSC, MoW and HyW are second-order wavelets. Higher-order wavelets can be obtained by taking high-order derivatives of the corresponding kernels as reported in [20, 21]. High-order wavelets, such as the fourth-order and in particular eighth-order MHW's, were effectively used by Grossmann for edge detection as reported in [5]. Second-order wavelets have been successfully employed in various applications and therefore will be studied first. Higher-order wavelets will be discussed in Section 2.2.

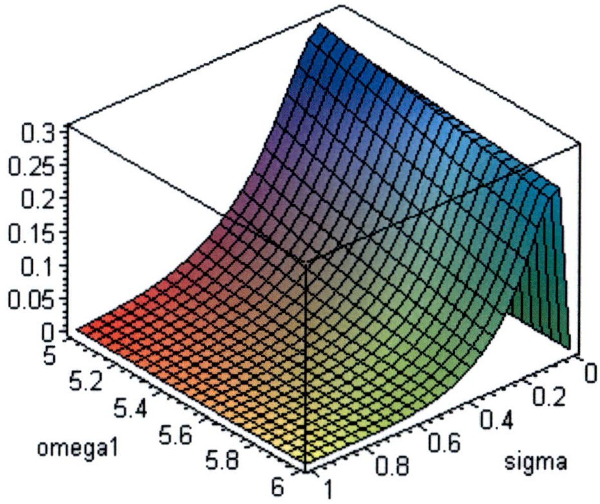

Figure 7. Magnitude frequency response of the MoW under DC conditions [$\omega = 0$ from Eq. (22)], plotted against ω_ψ (omega1 on graph) and σ, the vertical axis shows the magnitude of the response. The smaller σ becomes, the more difficult it is for wavelet reconstruction using the MoW as its magnitude response when $\omega = 0$ increases.

2.1. Second-Order Wavelets

2.1.1. Mexican-Hat

The 1-D MHW is given by Eqs. (15) and (16) [20, 21] in time and frequency domains respectively

$$\psi_{MHW}(t) = \frac{2}{\sigma} \cdot \left(1 - \frac{2t^2}{\sigma}\right) \cdot \exp\left(-\frac{t^2}{\sigma}\right), \tag{15}$$

where σ is the kernel control parameters, and

$$\hat{\psi}_{MHW}(\omega) = \sqrt{\pi\sigma} \cdot \omega^2 \cdot \exp\left(-\frac{\sigma\omega^2}{4}\right). \tag{16}$$

From Eq. (16), it is clear that the reconstruction or inverse continuous wavelet transform can be achieved by using the MHW, because its frequency response vanishes at $\omega = 0$. The 2-D MHW is given by Eq. (17)

$$\psi_{MHW2D}(x, y) = \frac{2}{\sigma} \cdot \left(1 - \frac{2(x^2 + y^2)}{\sigma}\right) \cdot \exp\left(-\frac{x^2 + y^2}{\sigma}\right) \tag{17}$$

and plotted in Figure 10, from which it is clear that the wavelet is local around the origin.

The 2-D Fourier transform of the MHW is given by Eq. (18)

$$\hat{\psi}_{MHW2D}(\omega_x, \omega_y) = \sqrt{\pi\sigma} \cdot (\omega_x^2 + \omega_y^2) \cdot \exp\left(-\frac{\sigma(\omega_x^2 + \omega_y^2)}{4}\right), \tag{18}$$

and plotted in Figure 11 for $\sigma = 2$.

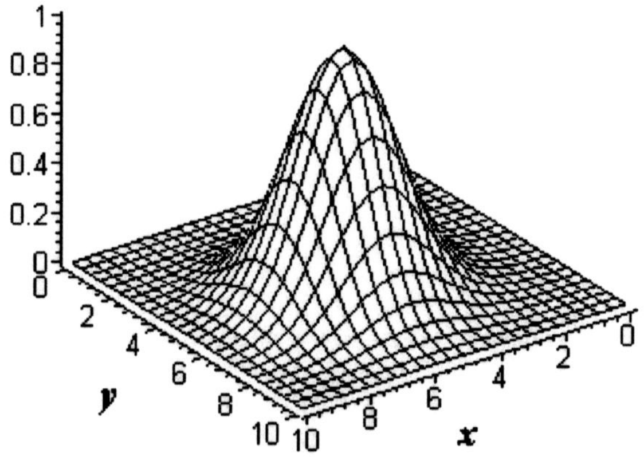

Figure 8. Magnitude of frequency response of the 2-D MoW [Eq. (23)], for $\omega_\psi = 5.5$ and $\sigma = 2$.

From Figure 11, it can be seen that MHW frequency response has a sharp rise for ω_x and ω_y in the immediate neighbourhood away from the origin which enables wavelet reconstruction. This feature is also useful for zero-crossing detection in images [2, 18]. It should be noted that the MHW is optimally effective if the input image possesses circular symmetry. If the image salient features are located in one particular direction, then the MoW can be more effective than the MHW as will be seen shortly. From Figure 12, it appears that the 2-D MHW frequency response has two symmetrical and distinctive peaks or "notches", located at (−1, 1) and (1, −1), which can split if the wavelet order increases as it is the case for the HyW as will be seen later in Section 2.1.3. For $\sigma = 2$, the frequency response of the 1-D MHW is compared with those of the MoW and HyW and is plotted in Figure 10.

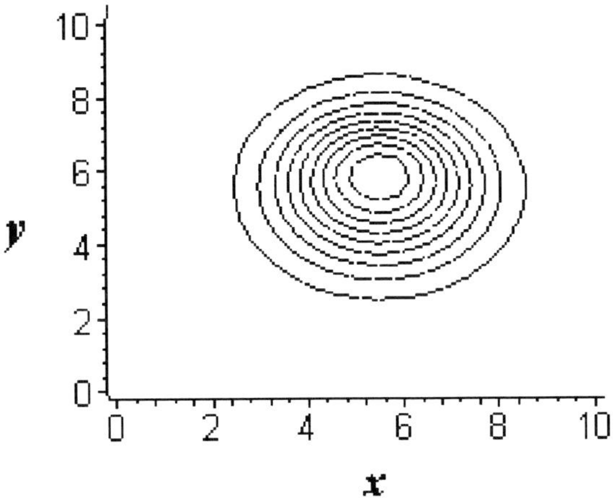

Figure 9. Contour plot of the frequency response of the 2-D MoW shown in Figure 8.

2.1.2. Morlet

The generalised 1-D MoW with the wavelet parameter σ is given by Eq. (19) [19-21]

$$\psi_{MoW}(t) = \exp(j\omega_\psi t)\cdot\exp(-t^2/\sigma), \tag{19}$$

where $5 \le \omega_\psi \le 6$ so that the wavelet is admissible.

The 1-D and 2-D real MoW's are given by Eqs. (20) and (21) respectively

$$\psi_{MoW}(t) = \cos(\omega_\psi t)\cdot\exp(-t^2/\sigma), \tag{20}$$

$$\psi_{MoW}(x, y) = [\cos(\omega_\psi x) + \cos(\omega_\psi y)]\cdot\exp[-(x^2 + y^2)/\sigma] \tag{21}$$

The MoW can be considered as a modulated Gaussian pulse, whose real and imaginary parts are plotted in Figure 13. Its Fourier transform is given by Eq. (22) [19, 21, 22]

$$\hat{\psi}_{MoW}(\omega) = \begin{cases} \sqrt{\pi\sigma}\cdot\exp\left(-\dfrac{(\omega-\omega_\psi)^2}{4\sigma}\right), & \text{for } \omega > 0, \\ 0, & \text{for } \omega \le 0, \text{ where } 5 \le \omega_\psi \le 6, \text{ and } \sigma \ne 0. \end{cases} \tag{22}$$

From Eqs. (5) and (22), it is clear that reconstruction is not possible for the MoW if $\hat{\psi}_{MoW}(\omega)$ is too large at $\omega = 0$. To examine the wavelet reconstruction properties of the MoW, the magnitude of its frequency response given by Eq. (22) when $\omega = 0$ is plotted in Figure 14 against σ and ω_ψ, from which it is seen to be strongly dependent on σ. As ω_ψ increases to 6, its magnitude slowly decreases but is still $O(10^{-1})$ which is not sufficiently small to enable reconstruction or ICWT capability. For $\sigma \geq 1$, it is evident that the wavelet frequency response is approximately zero and the Morlet ICWT can be obtained. For σ infinitely small, reconstruction cannot be achieved. Thus, for the 1-D MoW, the conditions under which its ICWT exists are: $5 \leq \omega_\psi \leq 6$ and $\sigma \geq 1$. In the literature, typical values of $\sigma = 2$ and $\omega_\psi = 5.5$ are usually employed [19]. For $\sigma = 2$ and $\omega_\psi = 5.5$, the MoW frequency response is compared with those of the MHW and HyW, and is plotted in Figure 10.

The 2-D frequency response of the MoW is given by Eq. (23), and plotted in Figure 8.

$$\hat{\psi}_{MoW2D}(\omega_x, \omega_y) = \begin{cases} \sqrt{\pi\sigma} \cdot \exp\left(-\dfrac{(\omega_x - \omega_\psi)^2 + (\omega_y - \omega_\psi)^2}{4\sigma} \right), & \text{for } \omega_x, \omega_y > 0, \\ 0, & \text{for } \omega_x, \omega_y \leq 0, \text{ where } 5 \leq \omega_\psi \leq 6, \text{ and } \sigma \neq 0. \end{cases} \tag{23}$$

From Figure 8, it is evident that the MoW is directional (in terms of its peak's location in 2-D space) and therefore can be used to detect salient features which are located in specific directions in an image [24, 25]. From Figure 9, it should be noted that the MoW has a near-perfect circular symmetry. However, its peak is not located at the origin and therefore it is not rotationally invariant, but this drawback can be overcome by employing four MoW's in the four quadrants [24, 25] of the coordinate so that the combined wavelet is symmetrical in 2-D space.

2.1.3. Hyperbolic

The 1-D HyW, which is generated from the first-order HyK [26], and its Fourier transform are given by Eqs. (24) and (25) respectively [21, 22]. Eq. (25) is plotted in Figure 10.

$$\psi_{Hy1}(t) = 2\beta^2 \cdot \text{sech}(\beta t) \cdot \{1 - 2 \cdot [\text{sech}(\beta t)]^2\}, \tag{24}$$

$$\hat{\psi}_{Hy1}(\omega) = \frac{\pi\omega^2}{\beta} \cdot \text{sech}\left(\frac{\pi\omega}{2\beta}\right), \tag{25}$$

where β is the wavelet control parameter.

From Eq. (25), it is clear that reconstruction can be analytically achieved by using the HyW since its frequency response at $\omega = 0$ is zero. Because of the presence of a hole at the origin in the frequency domain, which enables wavelet reconstruction, the HyW and MHW are called "notch" band-pass filters because of their finite time support properties. In digital image processing [18], these filters are useful for image enhancement and restoration.

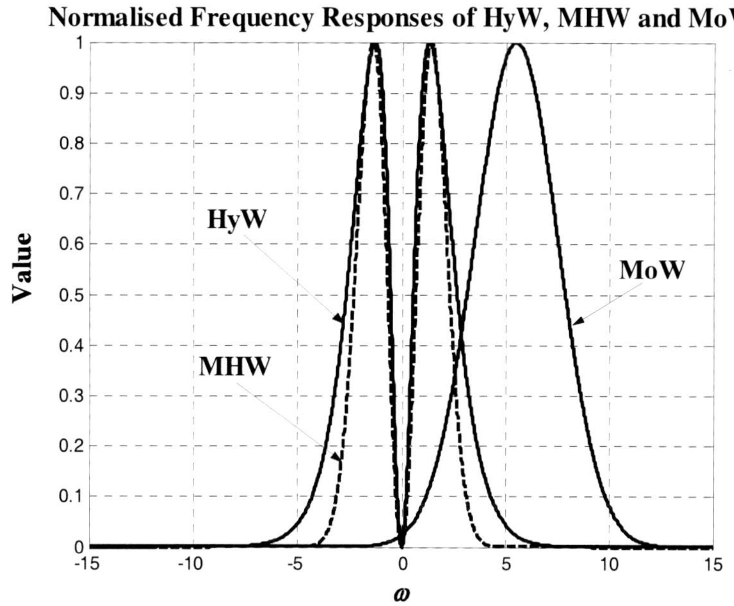

Figure 10. Normalised 1-D magnitude frequency responses of the HyW [Eq. (25)] for $\beta = 1$, MHW [Eq. (16)] for $\sigma = 2$, and MoW [Eq. (22)] for $\omega_\psi = 5.5$ and $\sigma = 2$.

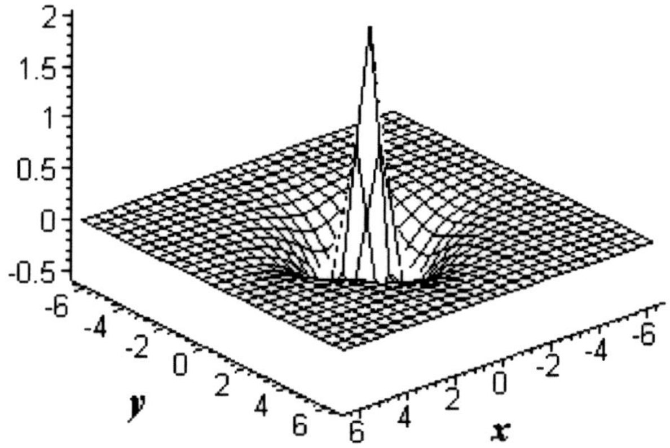

Figure 11. Magnitude of the 2-D spatial-domain HyW [Eq. (26)] for $\beta = 1$.

From Figure 10, and Eqs. (16), (22) and (25), it is clear that the MHW and HyW frequency responses are even functions of ω in the frequency domain, while the frequency response of the MoW has a small, but non-zero, value at $\omega = 0$ which makes it analytically unsuitable for wavelet reconstruction as discussed in Section 2.1.2. The MHW frequency response has a steeper rising edge and roll-off than the HyW which suggests that the former is more local and more effective than the latter for edge detection. However, by having a longer roll-off in the frequency domain, the HyW is more suitable to smoothen rough edges in an image and therefore can be effectively used for circular-particle detection in the human brain [27] as will be discussed shortly. The frequency responses of the MHW and HyW are evidently more local than that of the MoW. The MoW frequency response is asymmetrical and undefined for $\omega < 0$, causing difficulties when studying quadratic non-linear interactions as will be discussed in [28]. The 2-D time and frequency responses of the HyW are given by Eqs. (26) and (27), and plotted in Figure 11 and Figure 12 respectively for $\beta = 1$.

$$\psi_{Hy1\_2D}(x,y) = 2\beta^2 \cdot \text{sech}(\beta x) \cdot \text{sech}(\beta y) \cdot \{1 - 2 \cdot [\text{sech}(\beta x) \cdot \text{sech}(\beta y)]^2\}, \quad (26)$$

$$\hat{\psi}_{Hy1\_2D}(\omega_x, \omega_y) = \frac{\pi(\omega_x^2 + \omega_y^2)}{\beta} \cdot \left[\text{sech}\left(\frac{\pi\omega_x}{2\beta}\right) \cdot \text{sech}\left(\frac{\pi\omega_y}{2\beta}\right)\right] \quad (27)$$

As can be seen in Figure 11, the HyW is rotationally invariant which is somewhat similar to the MHW. From Figure 3 and Figure 10, it can be seen that the MHW has a less steep rising-edge close to the origin compared with the HyW, which suggests that the latter is more effective than the former for edge detection, which is the most useful application the MHW has found in the literature. From Figure 12 for $\beta = 1$, it is evident that the HyW has sharp edges which are distributed symmetrically, but not circularly, in the frequency domain, which is useful in detecting image features that are located close to the origin. As ω_x (x on graph) and ω_y (y on graph) increase, the HyW is rotationally invariant and can be considered as being similar to the MHW. From Figure 4, for $\sigma = 2$, the MHW has a circular distribution for these edges and there are no distinctive edges present in its frequency response which indicates that the MHW is a genuine circular-symmetrical wavelet, which is useful for images with perfect symmetry.

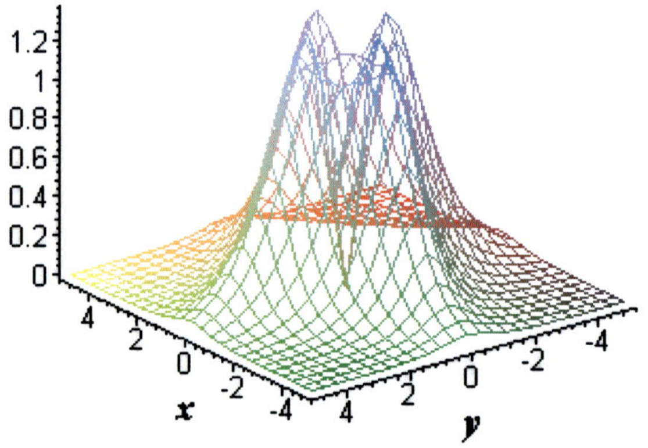

Figure 12. Magnitude of the 2-D frequency response of the 2nd-order HyW, for $\beta = 1$ [Eq. (27)].

From Figure 12 for $\beta = 1$, the HyW is symmetrical about the vertical axis. As with the MHW, for ω_x and ω_y close to the origin, the frequency response vanishes which means that reconstruction is possible. For the MoW, to be effective for image processing, four MoW's are used to cover the entire four quadrants [29]. From Figure 13 and Figure 15, it can be seen that the HyW has four symmetrical and distinctive peaks in the frequency domain, approximately located at (−1, 0), (0, 1), (1, 0) and (0, −1), which can be effectively used to detect possible symmetry in images, or to focus only on salient features with other features suppressed. As a result, the fast rising and falling edges of the HyW (Figure 12) enable more detailed studies on images than using the MHW or MoW. As the wavelet order m increases to eight, these peaks will split and become independent peaks as can be seen in Figure 17. For $m > 8$, the number of peaks in the frequency domain remains unchanged, which can be suggested that high-order HyW's reach equilibrium for $m = 8$.

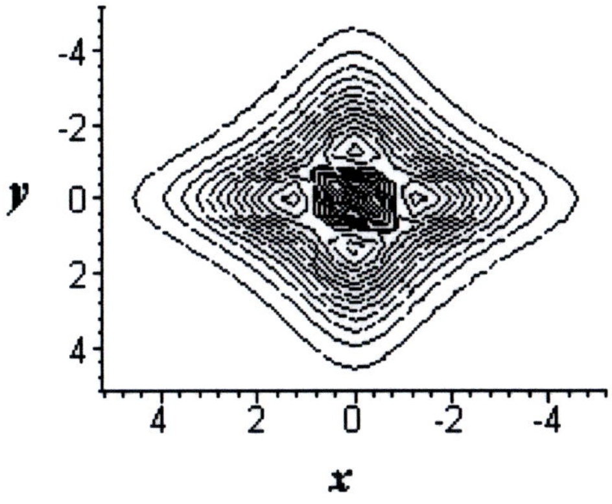

Figure 13. Contour plot of the 2nd-order 2-D HyW frequency response shown in Figure 12.

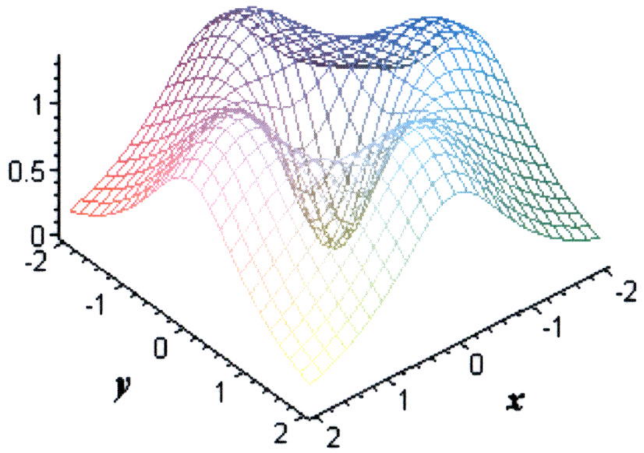

Figure 14. Magnified version of the magnitude of the 2-D frequency response of the 2nd-order HyW, for $\beta = 1$ [Eq. (27)] shown in Figure 12.

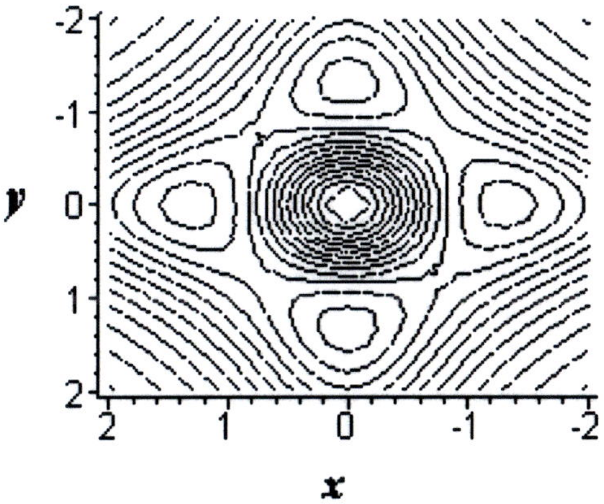

Figure 15. Contour plot of the magnified magnitude 2-D frequency response of the 2nd-order HyW, for $\beta = 1$ shown in Figure 14.

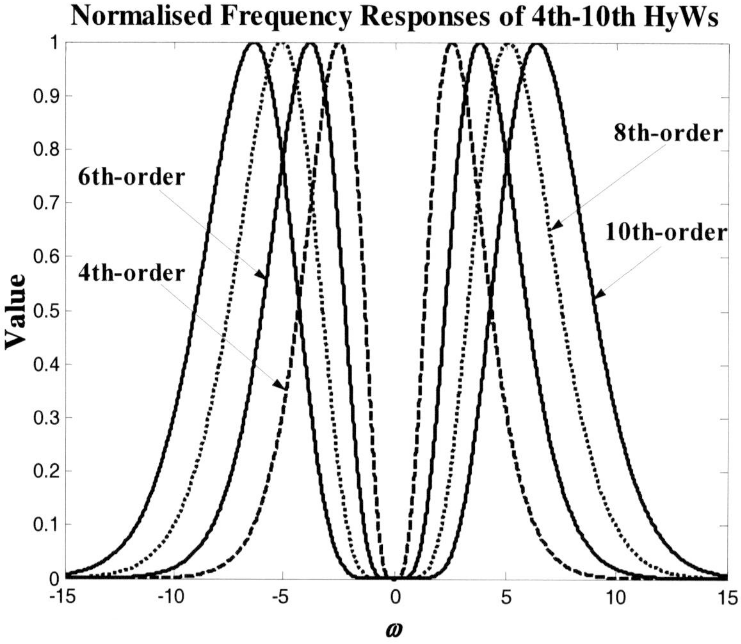

Figure 16. 1-D magnitude frequency responses of 4th-, 6th-, 8th- and 10th-order HyW's, for $\beta = 1$ [Eq. (30)].

Moreover, from Figure 17 and Figure 18, it can be seen that the 8th-order HyW was more stable than its lower-order members because it completely vanishes at the origin. As the wavelet order m increases, the empty region becomes wider and suppresses some part(s) of an image causing blurring. This establishes a trade-off between the wavelet stability and image blurriness which can be linked to cross-term suppression effectiveness of the HyK [26]. In [20], it was noted that the 10th-order HyW was the optimal wavelet in the entire HyW high-order wavelet family as its number of oscillatory side lobes remains unchanged when m increases beyond 8.

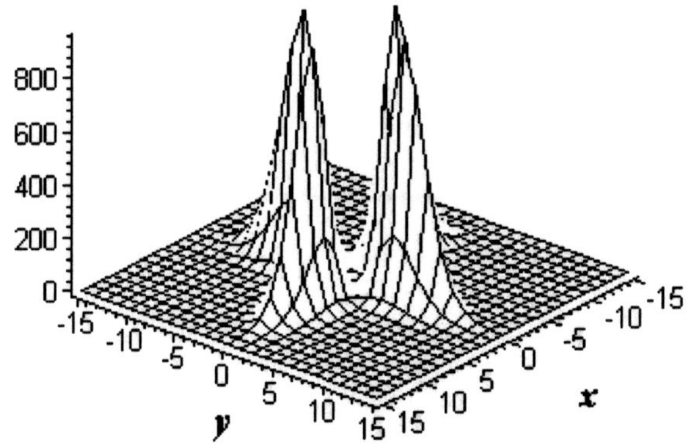

Figure 17. The 8th-order HyW 2-D magnitude frequency response [Eq. (31)], for $\beta = 1$, the vertical axis shows the magnitude of the response.

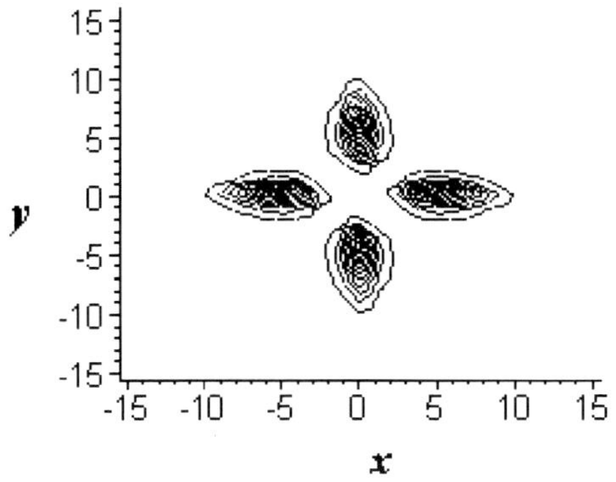

Figure 18. Contour plot of the 8th-order 2-D HyW magnitude frequency response as shown in Figure 17. Note that there is no energy smearing in the wavelet's stop-band at the origin formed by its four peaks which suggests that the wavelet can be effective in suppressing low-frequency and periodic noise sources in 2-D images [18].

Proposals of new wavelets, such as the HyW and its family, play an important role in diversifying time-frequency and wavelet analyses to other scientific disciplines such as medicine, astronomy, physics, fluid mechanics and structures, where more sophisticated signal processing tools are required to provide more insight into new and existing applications. As a result, new research directions are opened up in which the HyW will be compared with the MHW and MoW by applying it to various applications in which the former two wavelets have been found useful. From that, its true performance and effectiveness can scientifically be justified.

The regularity of a wavelet is measured by its number of vanishing moments. By definition, a wavelet which possesses nth-order vanishing moment has up to $(n - 1)$ vanishing moments [1, 10, 11], which means

$$\int_{-\infty}^{+\infty} \psi(t) \cdot t^{n-1} = 0 \tag{28}$$

It is clear from Eq. (28) that the higher is the number of vanishing moments, the higher is the polynomial order that the wavelet can suppress. This is particularly useful for detecting features that are embedded in constant or linear background-noise sources since these will be completely suppressed by the wavelet as will be discussed later in various applications in this paper. Eq. (28) is explicitly satisfied by the MHW, MoW and HyW with an identical number of vanishing moments of one (second-order moment), because

$$\int_{-\infty}^{+\infty} \psi(t) \cdot t = 0 \tag{29}$$

where $\psi(t)$ can be replaced by $\psi_{MHW}(t)$ [Eq. (15)], $\psi_{MoW}(t)$ [Eq. (19)], and $\psi_{Hy1}(t)$ [Eq. (24)].

It should be noted that even though the MHW, MoW and HyW have the second-order vanishing moment, linear trends and background which are odd-order polynomials, i.e. 3rd, 5th, 7th, ..., can also be suppressed since these wavelets are even functions.

2.2. High-Order Hyperbolic and Mexican-Hat Wavelets

For completeness, high-order wavelets are generated by taking even high-order derivatives of an admissible kernel used in C-TFPSC [20]. In [20], it was shown that odd high-order derivatives are not suitable wavelets, and more importantly, they cannot be considered as admissible kernels for Cohen's class. Further analyses can be found in [20], in which numerical and theoretical normalisation constants and wavelet parameters such as band-peak frequency and scale resolution were studied in detail.

2.2.1. Hyperbolic Wavelets

For the HyK, the 1-D and 2-D frequency responses of its even *m*th-order wavelets are given by Eqs. (30) and (31) respectively [20]

$$\hat{\psi}_{Hy\_m}(\omega) = \frac{\pi\omega^m}{\beta} \cdot \text{sech}\left(\frac{\pi\omega}{2\beta}\right), \tag{30}$$

$$\hat{\psi}_{Hy\_m2D}(\omega_x, \omega_y) = \frac{\pi(\omega_x^m + \omega_y^m)}{\beta} \cdot \left[\text{sech}\left(\frac{\pi\omega_x}{2\beta}\right) \cdot \text{sech}\left(\frac{\pi\omega_y}{2\beta}\right)\right] \tag{31}$$

From Eq. (30), it is clear that the *m*th-order HyW's are reconstructable since they vanish at $\omega = 0$, which makes the family useful. Eq. (30) is plotted against ω for $m = 4$, 6, 8 and 10 in Figure 16, from which it can be seen that the higher is the wavelet order m, the longer is the roll-off, and the less local are the wavelets. However, the higher is the wavelet order m, the sharper are its frequency response's peaks with very large magnitude, which is useful for amplification purposes and focusing on one particular part of an image, as can be seen in Figure 17, for $m = 8$ from Eq. (31).

2.2.2. Mexican-Hat Wavelets

The high-order MHW's were obtained by taking even, high-order derivatives (4th, 6th, 8th and 10th) of the CWK, and they were given in [20]. Their corresponding frequency responses are given by

$$\hat{\psi}_{CW4}(\omega) = \frac{\sqrt{\pi\sigma} \cdot \exp\left(\frac{-\sigma\omega^2}{4}\right)}{\sigma^2} \cdot \left(\sigma^2\omega^2 + 12\sigma\omega - 2\sigma + 12\right), \tag{32}$$

$$\hat{\psi}_{CW6}(\omega) = \frac{\sqrt{\pi\sigma}\cdot\exp\left(\frac{-\sigma\omega^2}{4}\right)}{\sigma^3}\cdot\left(\sigma^3\omega^3 + 30\sigma^2\omega^2 - 6\sigma\omega(\sigma-30) - 60\sigma + 120\right), \quad (33)$$

$$\hat{\psi}_{CW8}(\omega) = \frac{\sqrt{\pi\sigma}\cdot\exp\left(\frac{-\sigma\omega^2}{4}\right)}{\sigma^4}\cdot\left(\begin{array}{l}\sigma^4\omega^4 + 56\sigma^3\omega^3 - 12\sigma^2\omega^2(\sigma-70)\\ -336\sigma\omega(\sigma-10) + 12\sigma^2 - 1680\sigma + 1680\end{array}\right), \quad (34)$$

$$\hat{\psi}_{CW10}(\omega) = \frac{\sqrt{\pi\sigma}\cdot\exp\left(\frac{-\sigma\omega^2}{4}\right)}{\sigma^5}\cdot\left(\begin{array}{l}\sigma^5\omega^5 + 90\sigma^4\omega^4 - 20\sigma^3\omega^3(\sigma-126) - 360\sigma^2\omega^2(3\sigma-70)\\ +60\sigma\omega(\sigma^2 - 252\sigma + 1{,}260) + 1{,}080\sigma^2 - 50{,}400\sigma + 30{,}240\end{array}\right). \quad (35)$$

It is clear from Eqs. (32)–(35) that high-order MHW's are not reconstructable for general values of σ as they are non-zero when $\omega = 0$. However, for some specific values of σ, reconstruction can be achieved. From Eq. (32), the ICWT of the 4th-order MHW can be obtained by setting $\sigma = 6$. Similarly, from Eqs. (33)–(35), to achieve reconstruction, approximate values of σ for the 6th-, 8th- and 10th-order MHW's are: 2 (exact), 139 and 1, 46 and 0.6 respectively. This suggests that the HyW high-order wavelets are more robust, effective (with a finer scale resolution [20-22]) and stable than the MHW high-order wavelets since the former wavelets can be analytically reconstructed for all values of β. From this, it is clear that the MHW can be replaced by the HyW in applications where the former wavelet has been found useful. It should be noted that the analysis of high-order MHW's was shown to be difficult with explicit parameter expressions, such as band-peak frequency and scale resolution, and those of the 6th-order MHW did not exist [20], which suggests that high-order MHW's are less practical than the second-order MHW. Moreover, to be effective with a fine scale resolution and a reasonably large total number of scales, the control parameter σ of the MHW must be set to specific values or ranges which can not include the values required for reconstruction of the wavelet. This establishes a trade-off in the wavelet performance and its reconstruction. For the HyW, this problem does not exist since all high-order HyW's are always reconstructable (see Eq. (30)), and therefore β can be set to specific values to achieve a finer scale resolution or a larger/smaller total number of scales, depending on the application. The effectiveness of the HyW, however, must be carefully examined by applying it in specific applications.

Another interesting point to note is that the further is the filter's pass-band from the origin, the more effective it is in terms of smoothing edges and sharp transitions over high-frequency regions. In addition, the longer low-frequency stop-band of the high-order filters compared with lower-order filters can be effectively used to suppress periodic noise [18] as can be seen in Figure 16 and Figure 18. From this point of view, high-order MHW's and HyW's are more effective than their lower-order wavelets as reported in [20, 21]. However, as the filter's band-pass position is too high in the frequency domain, more high-frequency detailes of the image will be suppressed, which establishes a trade-off between image blurring over low frequencies and image smoothing over high frequencies. More details on this topic will be discussed in a future publication.

3. Applications and Discussion

The MHW has found many useful applications, which are classified in five major fields: image processing, medicine, optics, astronomy and astrophysics, and fluid mechanics. These applications are discussed in chronological order which explains the origin and main motivations of each application. From that, it is possible to identify possible developments using the MHW for later applications based on previous ones. Although applications are discussed chronologically, two exceptions are present: use of the MHW in neural networks separately published in 1996 [30] and 2002 [31], and a series of seven papers in astronomy published from 1999 to 2002 [32-38]. Section 3.3 briefly reviews applications of the MHW which was considered as Laplacian-of-Gaussian (LoG) from 1988 to present. Section 3.4 outlines a number of applications, starting from 1980 which was earlier or concurrent to Marr's work, in which the Mexican-hat shape was observed and no modelling analyses on the wavelet were carried out. Because of a very wide range of applications in different disciplines, details of each application are not given in the interest of brevity and as it is not the main focus of the paper. Instead, general motivation and main ideas are discussed as a guide, which can be used to create new developments and applications. The effectiveness and methods of using the MHW in each application are emphasised.

3.1. Early Applications of the MHW from 1984 to 1996

Marr's work of applying the Laplace operator to the Gaussian mask showed that the MHW was rotationally invariant or isotropic, i.e. it possesses circular symmetry in the 2-D spatial or time domain, which was the main reason why the MHW has found many useful applications. In the frequency domain, the wavelet is called a "notch" filter with a hole at $\omega = 0$ because wavelet reconstruction is possible as explained in Section 2.1.1.

In 1984, Reiboeck and Altmann [39] reported the use of the MHW in the first processing stage of retinal mapping onto striate cortex of mammals. In this work, the MHW was used as a rotational-invariant filter whose radius increased linearly with retinal eccentricity. The MHW was considered similar to a receptive field of retinal ganglion cells. A logarithmic polar transformation was used to convert scales and translations in the original image to its centre, yielding the modified image $g(u, v)$, which was then convolved with the MHW to remove edges and of distortion, creating another image $h(u, v)$. This image was then transformed using an R-transform, which can be realised by using simple neural networks, to complete the retinal mapping. In this application, centralising scales and translations in the original image was the most important feature before taking the logarithmic polar transform, and therefore only the rotationally-invariant MHW and HyW were suitable for this task since the MoW is directional. However, the MoW can be used to smoothen the $g(u, v)$ image before taking the R-transform, because of its sharp-rising peaks which can effectively suppress rough edges in the image. As the retinal mapping size increases, so does the number of neural networks, which motivates the use of parallel computing so that higher efficiency can be achieved.

It should be noted that symmetrical wavelets and their transforms have been employed to study turbulence and chaos extensively. In 1992, Brasseur and Wang used the 2-D MHW to study transitions from Gaussian conditions to turbulence [40]. In particular, nonlinearities and intermittent vortical structures were examined by using the wavelet bispectrum in 4-D. Because of graphical difficulties in displaying 4-D images, their 2-D contour plots were displayed instead, which is the disadvantage of the method. However, the wavelet bispectrum yielded much more insight on how intermittent vortical structures and nonlinearities were formed during turbulence. From that, it was possible to study nonlinear interactions among different wavelet scales during turbulence. This method was later developed further by Milligen in 1995 [41, 42] to study turbulence at random time interval. This particular application showed the effectiveness of the wavelet bispectrum, as similar to its Fourier bispectrum counterpart, in studying turbulence. However, a large amount of computing power is required so that the wavelet bispectrum calculations can be efficiently carried out, which is the main limitation of the method. The principle of correlation was still applicable in this case in which wavelet bispectral peaks were clearly identified by examining the 2-D contour graphs. As clearly stated in [40], the wavelet bispectra of other wavelets can be computed, and for each wavelet, the basic features of intermittent vortical structures can be successfully identified, which was because of the involvement of two independent scales in the wavelet bispectrum which lowers the degree of dependence on the wavelet shape for this particular application. In other words, employment of other wavelets could yield further improvements.

In 1993, the 1-D MHW and its CWT were successfully used to estimate scaling exponents in histograms of turbulent data as reported in [43]. It was shown by Vergassola *et al.* that if a local increment in turbulent data $|v(x+p) - v(x)|$ obeys the power law p^h, then its corresponding CWT coefficients can be described by a^h, where p is the lag parameter in the data, h the scaling exponent, $v(x)$ the input turbulent data, and a the wavelet scale. The scaling exponent h was then estimated by applying a least-square fit method to the wavelet coefficients. It was observed from the experiment that the datum histograms were noisy because of a finite number of inertial range effects. To remedy the problem, they found a complex wavelet and suggested a method to design an optimal wavelet by decomposing the analysed wavelet (which is the MHW in this case) into a number of wavelets using a Hermite function of order p, H_p, given as follows

$$g(x,\{c_p\}) = \sum_{p=1}^{M} c_p \cdot H_p \cdot \exp(-x^2/2), \tag{36}$$

where c_p is a set of random coefficients with $\sum_{p=1}^{M} c_p = 1$, and M the maximum scaling exponent. The decomposition of the Mexican-hat CWT, given by Eq. (36), is

$$T_g(a,\{c_p\}) = \sum_{p=1}^{M} c_p \cdot T_{H_p}(a), \tag{37}$$

where T_g and $T_{H_p}(a)$ are the wavelet transforms of g and H_p respectively.

Eqs. (36) and (37) were then tested on Monte Carlo simulation data. For each set of $\{c_p\}$, the correlation coefficients $K\{c_p\}$ of $\log|T(a, \{c_p\})|$ versus $\log(a)$ were estimated and the optimal wavelet g_{opt} is the wavelet that has the largest set of $K\{c_p\}$. This work opens up new research directions using wavelets in fractal and chaos where the MoW and HyW can be employed. Higher-order of the MHW, MoW and HyW can also find this application useful.

The 2-D MHW, given by Eq. (17) with $\sigma = 2$, was used for edge detection in an image by calculating their auto-correlation function as explained in Section 1.4. Rotational-invariant features of the 2-D MHW were effectively used [44]. An optimal scale for edge detection was estimated as $a \approx 0.4$. For $a > 0.4$, the image became distorted and unrecognisable. It was also found that increasing a did not improve the effectiveness of the wavelet, instead, decreasing a, i.e. stretching the wavelet which results in coarser frequency resolutions, enhanced edge-detection effectiveness. The authors combined the original circular harmonic filter (CHF) and its modified form, phase only CHF (POCHF), with the CWT to form the MHW CWT CHF (WT-CHF) and CWT POCHF (WT-POCHF) correlators, whose performance was studied by calculating the ratio of the correlation peak intensity to the largest side lobe with different orientation angles. They concluded that the WT-CHF had the largest ratio and therefore was more effective than the CHF and POCHF correlators. Both the WT-CHF and WT-POCHF correlators yielded sharper correlation peaks with larger *SNR*s than the CHF and POCHF correlators, which suggested that the MHW CWT is effective when employing with the CHF correlator to estimate correlation peaks in images. It should be stressed that image orientation was employed in this case which brings up an idea that "directional" wavelets such as the MoW [29] is more effective than the MHW even though the latter was shown to perform very effectively in this particular application. From [20, 26], it was shown that the HyK and HyW were more noise robust than the CWK and MHW respectively, which suggested that the HyW is more effective than the MHW in edge detection. Performance of the MHW, MoW and HyW can then be compared and analysed so that the optimal wavelet for edge detection using the correlation method can be identified. It should also be stressed that higher-order HyWs were explicitly shown to be reconstructable [with vanished frequency responses at $\omega = 0$ as shown by Eq. (30)] and also more effective with finer scale resolutions than the second-order HyW and MoW, which suggests that high-order HyW's are useful in this case. The MHW was then later extended to be directional with the introduction of two independent scales a and a_1, and an orientation angle θ, as will be discussed in [45].

The MHW was used in medicine to analyse tissue reaction to immunocytochemistry antibody and in-situ hybridisation in human brains by extracting immunoreaction products or counting the number of auto-radiographic grains, i.e. circular "gold" particles created by chemical processes between the human brain and antibodies, in images captured by using a very high resolution scanner [27]. The wavelet was particularly useful for extracting "gold" particles from grey particles because the former have a higher image density, which was also known as enhancing gold grains in an image with a constant background in extracting antibody products. The symmetrical shape of the MHW was effectively utilised by using the fact that particles located close to its centre would have a much larger weighting factor than those away from the centre, thus yielding smoother edges in the image. This concept is similar to the weighting concept employed in cross-term suppression in C-TFPSC for the CWK and HyK [26, 46], whose weighting factor is smaller than that of the CWK, and therefore is more effective in suppressing cross terms. A 9×9 mask, given by Eq. (38) and plotted in Figure 20, of the 2-D MHW image used in this case, was employed to detect gold particles

$$\begin{bmatrix} 0 & 0 & 0 & -1 & -1 & -1 & 0 & 0 & 0 \\ 0 & -1 & -1 & -3 & -3 & -3 & -1 & -1 & 0 \\ 0 & -1 & -3 & -3 & -1 & -3 & -3 & -1 & 0 \\ -1 & -3 & -3 & 6 & 13 & 6 & -3 & -3 & -1 \\ -1 & -3 & -1 & 13 & 24 & 13 & -1 & -3 & -1 \\ -1 & -3 & -3 & 6 & 13 & 6 & -3 & -3 & -1 \\ 0 & -1 & -1 & -3 & -3 & -3 & -1 & -1 & 0 \\ 0 & -1 & -1 & -3 & -3 & -3 & -1 & -1 & 0 \\ 0 & 0 & 0 & -1 & -1 & -1 & 0 & 0 & 0 \end{bmatrix}. \tag{38}$$

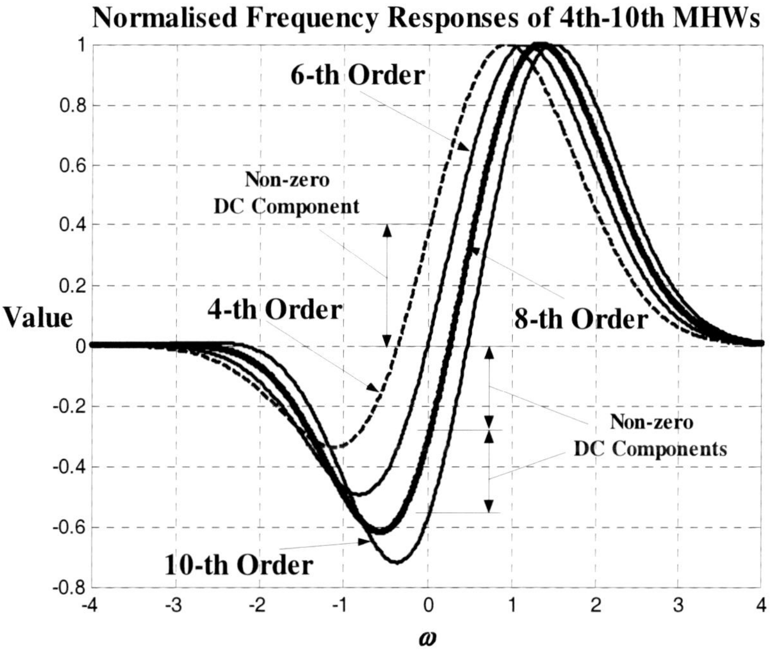

Figure 19. Magnitude frequency responses of the 4th- to 10th-order MHW's, for $\sigma = 2$, under which wavelet reconstruction can only be achieved for the 6th-order MHW with a zero DC component. The approximate values of σ that would enable wavelet reconstruction under DC conditions of the 4th-, 8th- and 10th-order MHW's are 6 (exact), 140 or 1, and 46 or 0.6 respectively, by substituting $\omega = 0$ to Eqs. (32), (34) and (35). Specific values of σ can then be chosen to suit a specific application, because the larger is σ, the narrower is the filter's pass-band [18] in the frequency domain which can compromise its effectiveness.

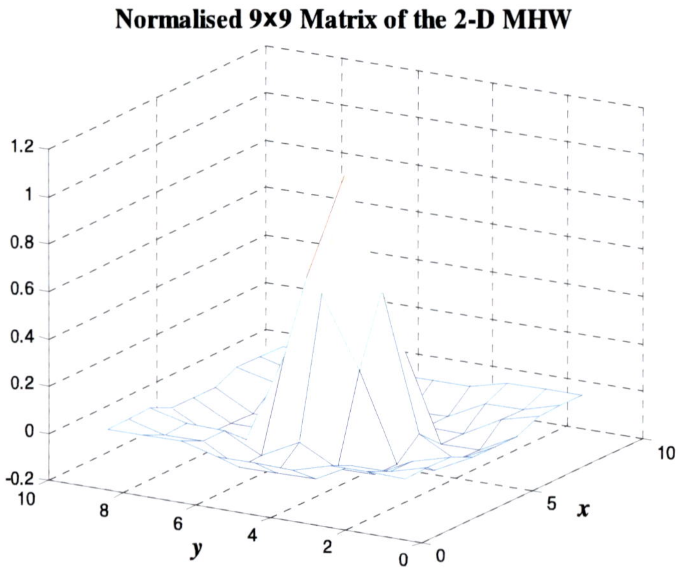

Figure 20. The normalised 9×9 2-D mask of the MHW for antibody-product detection in human brains.

The MHW mask given by Eq. (38) was reported to be effective even though the image background was highly contrast because of the presence of uranyl acetate and lead citrate counterstains.

This suggests that the energy distribution of human antibody-products are of Gaussian type and rotationally invariant because the correlation of the MHW, which is the second-order Laplacian operator of the Gaussian pulse, possessed high peaks yielding effective grain detection under noisy conditions. Recall that the MHW has been known to be most effective in detecting edges and linear trends present in a constant background [45], not in a high-contrast background. This yields a number of large peaks which make the detection of the right objects more difficult.

This work, thus, provides another useful application of the MHW which can be effectively employed to detect salient features in images with a high-contrast background. From this, further investigations, such as finding the brightness level and designing the filter image coefficients for optimum detection of gold particles, can be carried out. The work reported in [27] can also be extended to other living tissues in the human body which can provide fruitful interactions between image processing and medicine. Since the HyW is similar to the MHW, it can also be employed so that further detection improvements can be identified.

The 1-D MHW was used to briefly study large eddy structures, with a finite bandwidth [28], in velocity signals of a plane turbulent jet [47], by convolving it with turbulent data so that large eddy structures can be successfully revealed. The wavelet power spectrum (WPS) and Fourier power spectrum (FPS) were also used to study the energy distribution at different wavelet scales. It was found that the largest and smallest wavelet transform coefficients corresponded to the largest eddy structure and intermittency respectively, which suggested that eddy structures can be effectively studied by using the cross correlation function between the MHW and the data. Moreover, this fact also suggests that the velocity signal of a plane jet is of Gaussian type and therefore can be more effectively studied by using the bispectrum which has been employed for Gaussianity tests in various signals.

The 1-D MHW and its transform were briefly used to study spatial heterogeneity of geologic structures and spatial changes in water permeability at Radioactive Waste Management Site Area 5 in Nevada [48]. In this work, the authors used the wavelet variance, which is the average square sum of the signal wavelet coefficients, to represent the signal energy distribution because the energy is conserved by the wavelet transform. Changes in the signal energy patterns using the wavelet variance were then examined to study the correlation among different wavelet scales used to successfully reveal the signal behaviour. They also showed that the wavelet variance was useful for fractal and chaotic studies which were extensively studied in [41, 42, 49, 50].

Following on the work reported by Marr in 1980 [2], in 1995, Saloma [51] found that optical mirrors and lenses, whose point-spread functions (PSFs) are similar to the MHW and Fourier transform of the HaW, were effective in measuring their similarities with an input object in optical analysers. The circular symmetry of the MHW made it easier to profile the mirror, and it was found that Huygens-Fresnel and Fraunhofer diffraction phenomina corresponded to the wavelet transform process. However the work reported by Saloma was by no means completed and more work was required so that a systematic profiling of optical mirrors could be obtained as promised by the author [51].

In 1995, Milligen *et al.* [41, 42] used the Morlet wavelet bispectrum and bicoherence as useful tools to study turbulence and showed that they were very effective in identifying linear interactions. Following on Milligen *et al.*'s successful works, in 1996, Jha and Saxena [28] used the MHW to identify non-linear interactions in ADITYA tokamak edge turbulence. The wavelet power spectrum and wavelet bicoherence were successfully employed, as similar to the bispectrum and bicoherence, to separate low- and high-frequency components. In this work, Jha and Saxena, having learned more about previous works by Milligen *et al*, reasoned that the MoW was not a perfect wavelet to study turbulence and quadratic interactions because it does not strictly meet the admissibility constraint of zero-mean value, and its frequency response is not symmetrical. These requirements can be met by using the MHW which is an admissible wavelet whose frequency response is symmetrical and an even function, which is more suitable for non-linear quadratic interactions. Moreover, the MHW is a real wavelet which can also be used to estimate kurtosis of a time series as a function of frequency. The wavelet bicoherence, $b_W(a_1, a_2)$, is given as

$$[b_W(a_1,a_2)]^2 = \frac{[B_W(a_1,a_2)]^2}{\left(\int_0^T [\hat{\psi}(a_1,\tau)\cdot\hat{\psi}(a_2,\tau)]^2\, d\tau\right)\cdot \int_0^T [\hat{\psi}(a,\tau)]^2\, d\tau} \tag{39}$$

where a_1, a_2 and a are wavelet scales with $1/a = 1/a_1 + 1/a_2$, $B_W(a_1, a_2) = \int_0^T \hat{\psi}(a_1,\tau)\cdot\hat{\psi}(a_2,\tau)\cdot\hat{\psi}(a,\tau)\, d\tau$ is the wavelet bispectrum, and T finite duration of the input data. The frequency f_a at a particular scale a can be estimated as

$$f_a = \frac{f_s}{a\pi\sqrt{2}}, \tag{40}$$

where f_s is the sampling frequency of the input data.

Because of symmetry, the following frequency regions of the bicoherence are plotted: $0 < f_1 < f_{max}$ and $-f_{max} < f_2 < f_{max}/2$, requiring the existence of "negative" frequencies which are undefined if using the MoW. Because the MHW has been successfully used to study turbulence, it can be suggested that the HyW can find these applications useful since both wavelets possess similar properties in the time and frequency domains. Moreover, the HyW was successfully employed to study ECG, music and speech signals in [49].

The uniqueness in shape of symmetrical wavelets has been mainly employed for correlation and convolution purposes where salient features in 1-D data or 2-D images can be extracted. This key advantage has also been employed to statistically study triggered earthquakes, which are believed to be triggered by previous earthquakes within a 1,000 km-radius circle. A triggering hypothesis was proposed by Lomnitz [52] and tested on a 90-year period earthquake data set using a Pearson-based good-of-fit test, which is given as

$$\chi^2 = \sum \frac{(f_{predicted} - f_{observed})^2}{f_{observed}}, \tag{41}$$

where the summation is over nine intervals from 100 km to 1,000 km.

It was found that the Mexican-hat distribution was accepted with $\chi \approx 98\%$, which is higher than that of a null hypothesis of 90%. The usefulness of symmetrical wavelets in seismic data was not a surprise because Morlet proposed his wavelet in the 1980's and numerically applied it to seismic data processing with great success. Because of the even symmetry and localisation in both time and frequency domains, the effectiveness of the MHW and HyW in processing seismic data is ought to be sound. This can be considered as analogous to the case where the successful work by Milligen *et al.*, using the MoW to study quadratic interactions, was later improved by Jha and Saxena [28] by using the MHW.

The CWT of symmetrical wavelets was employed in 2-D optical correlators with Dammann's gratings using VanderLugt's configuration [53]. The 2-D MHW was translated continuously and scaled discretely using separate discrete optical channels or lenses as can be seen in Figure 21. There were two main stages which are separated by the middle plane. In the first stage, an input 2-D image was filtered by Dammann's grating, which is a 2-D equally-spaced square matrix generating different diffraction orders. This process "split" the image into different parts, which are simultaneously processed by independent optical lenses or "channels", which can be considered as different dilating and shifting versions of a mother wavelet. The role of the multi-reference matched-filter (MRMF) was to store the Fourier transforms of expanded versions of the mother wavelets. The filter was designed so that it matched the location of the modified image (after multiplying it with Dammann grating) with the location of each Fourier spectrum produced by lens 1. Lens 2 performed the 2-D CWT of the spectral wavelet coefficients from the first MRMF. The second stage performed the ICWT to reconstruct the image, of which lens 3 performed the Fourier transform of the 2-D wavelet transform coefficients produced from the first stage of the system, and stored these spectral wavelet coefficients in the second MRMF. Lens 4 performed the 2-D ICWT on these coefficients and the image was reconstructed on the output plane. For correlation purposes, the MHW was arbitrarily chosen as the mother wavelet because of its effectiveness in edge detection [2]. It is also interesting to implement the MoW and HyW on this system since these wavelets have been shown to be effective for edge detection and for salient-feature extraction in time series [42, 49]. It should be noted that the system can be expanded further

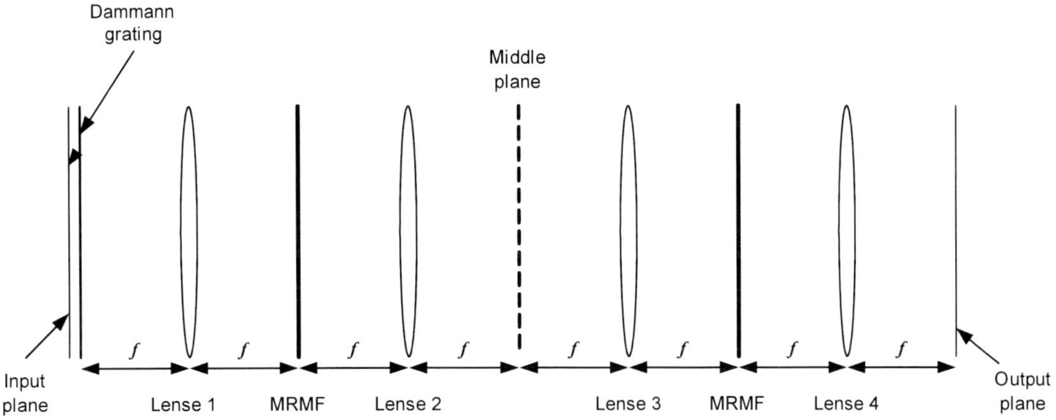

Figure 21. Wavelet optical processor using VanderLugt's configuration with Dammann's grating.

The MHW and the CWT have been used extensively in artificial neural networks (ANNs) and automatic pattern recognition (APR) for image compression, feature extraction from clutters and speech-formant detection. Because of its simplicity, there have been proposals where the CWT was calculated in real-time applications using parallel computing. More details can also be found in [30]. In particular, Linsker's neural network model on oriented receptive fields was mathematically modelled by using the MHW [31]. It was found that different layers in Linsker's model possessed Gaussian characteristics in which the Gaussian pulse was effectively employed to study them. However, according to Yamazaki's work [31], the MHW was also effective in modelling these layers whose arbor function, determining the effective area of synaptic connections to postsynaptic cells, was approximated as the MHW, called Gaussian arbour function. The correlation of the arbour function was computed, and it was suggested that the function could be used to control spatial frequency of different receptive fields.

In 1997, Chen *et. al.* [54] used the 1-D Mexican-hat CWT to measure "ramp" frequency and microfront times of coherent turbulent eddies. It was reported that the air and temperature directly above a turbulent surface obey a ramp pattern of which a slow decrease or increase is followed by a rapid change back to its previous point or the equilibrium point. An optimal scale of the MHW was estimated by realising that a fast change in the input data series, which corresponds to a coherent structure, yields a change in sign or zero crossing in the wavelet transform. The optimal scale a_{opt} was estimated by using the wavelet variance, whose first-order derivative changes its sign as the scale a increases from zero to infinity. This also implies that at a_{opt}, the input data and the wavelet are optimally correlated, hence, ramp structures in the data can be successfully detected. Ramp structures, with their frequency and microfront times, were measured on three different surfaces: straw mulch, Douglas-fir forest and bare soil. It was reported that the MHW could not identify consecutive and close coherent structures. However, the results obtained by using the MHW and a new ramp model proposed by the authors were consistent. In their measurements, it was also noted that the largest ramp coherent structure was accurately detected with smaller structures being suppressed. This is mainly because the optimal scale a_{opt} was employed, which will identify the largest peak in a cross correlation function between the data and the wavelet. Therefore, it can be suggested that the wavelet variance should not be estimated at the optimal scale but over the whole scale range so that distinctive peaks can be identified. From that, consecutive ramp coherent structures can be detected. From the results of this work, it can also be suggested that ramp coherent structures are of Gaussian types and therefore can be effectively detected by using the MoW which is the modulation of the Gaussian pulse. It is also worth trying the HyW in this case with strong amplification peaks, and therefore it can be used to detect the most dominant ramp coherent structures in turbulent data.

The MHW was also used to study wetness variability in six different regions in eastern China using historic data collected in the last 1,033 years by Jiang, Zhang and Fraedrich [55]. In this study, wavelet zero crossings, locations at which the wavelet peaks change their sign, were used to detect a change in condition between wetness and dryness with positive peaks in the wavelet coefficients indicating severe dryness such as drought, and negative peaks indicating severe wetness such as flood or storm. This technique is similar to the technique employed in [54] by Chen, Novak and Black, in which ramp structures were successfully detected by locating the change in sign of the wavelet variance function. In this study, the wavelet transform coefficients were used, and it was shown that the MHW was very effective in detecting correlation peaks and local variations of the coefficients which provide insight into the behaviour of weather patterns in a particular year. The MHW was used in this study because it is effective for turbulence studies as successfully reported in [28, 50, 56-58], with a finer resolution compared to orthogonal wavelets [59]. This suggests that turbulent structures such as ramp coherent are of Gaussian type. However, in [55], there was not sufficient evidence indicating that the weather pattern in eastern China was of Gaussian type, implying that symmetrical wavelets such as MoW and HyW can be effectively employed in this particular application as suggested by Jiang, Zhang and Fraedrich [55]. Since the MoW is directional, it can be more effective than the MHW by providing more accurate locations and magnitude of local peaks in some specific directions. The HyW was shown to possess a finer scale resolution than the MHW [20-22] which can yield a better understanding on weather local behaviour over a time period. From that, weather predictions can also be possible.

3.2. The MHW as a "Mature" Wavelet from 1997 to Present

Up to this point, the MHW has been successfully employed in various disciplines including medicine, optics, neural networks and fluid mechanics. In particular, the MHW was used to detect gold particles in the human brains by using a high-resolution camera to capture digital images. The time period from 1997 till present can be considered as the "mature" period of the MHW whose new forms and developments are proposed and effectively utilised with more work focusing on fluid mechanics, image processing and optics were carried out compared to the time period prior to 1997 as reported in Section 3.1. In addition, astronomy, nuclear physics and topography have been useful applications of the MHW.

The MHW not only has been employed as an isotropic wavelet, but also as an anisotropic wavelet. In [45], it was successfully used by Malik and Subramanian in their paper published in 1997 as a matched filter to detect salient features in astronomical data from the first CfA survey. The 2-D MHW, different from its normal 2-D circular structure, was elliptical with two independent scales a and a_1, and was mathematically given as

$$\psi(x, y, a, a_1) = \left[1 - \left(\frac{x^2}{a^2} + \frac{y^2}{(a_1)^2}\right)\right] \cdot \exp\left(-\frac{x^2}{2a^2} - \frac{y^2}{2(a_1)^2}\right), \text{ with } \sigma = 2. \quad (42)$$

The main advantage of this wavelet over the 2-D isotropic and circular anisotropic wavelets is that it can focus on a specific direction by varying an orientation angle θ of the ellipse's axes to the horizontal axis, to extract useful features in an image. This makes the MHW wavelet a "directional" wavelet in a way and therefore it can be used in applications where rotationally-invariant wavelets cannot be employed. It was shown in [45] that the elliptical 2-D MHW was very effective in detecting large underlying structures and hidden curves. The wavelet was also shown to be insensitive to constant backgrounds and linear trends because it is a second-order moment wavelet as shown in Section 2.1.3. An important statement made by Malik and Subramanian was that symmetrical wavelets including the Mexican-hat, Morlet and hyperbolic, even though do not completely cover the L^2(R) space, could be effectively employed to detect important features in 1-D data series and 2-D images by using their unique shape as the key advantage. By correlating the wavelet with a 2-D image, their similarities can be extracted, thus, by knowing the wavelet properties, one can determine those of the image. The MHW has a peak and a number of slow oscillatory side lobes which clearly do not "react" with a constant background and a linear trend since these features are not the wavelet features and they do not vary significantly with time. The wavelet's smooth rising peak clearly indicates that it can be effectively used to detect intensity changes in 1-D data series or 2-D images in fine detail [45]. In [27], it was found that the MHW could be used to detect salient features in 2-D images possessing a high-contrast background which further strengthens the idea of using the wavelet for detection under the presence of constant backgrounds and linear trends. The work in [27] also yielded further thoughts on the use of the elliptical MHW for grain detection in human brains because the grains could be distributed in a particular direction in the image. By fine tuning the orientation angle θ and choosing the correct interaction between two independent scales a and a_1 to enhance noise robustness [60], effective detection can be achieved.

In 1999, Kailas and Narasimha used the 2-D MHW to study small-scale motions existing in large-scale coherent structures in turbulent flows [61]. Two mixing layers were studied: laminar, which consists of small-scale coherent structures, and turbulent which consists of a mixture of different motions. The interfaces among large- and small-scale motions in each layer were then examined. The MHW was employed because of its effectiveness in detecting sharp gradients in images as reported by Marr [2], which could be used to effectively identify coherent and non-coherent structures. For laminar mixing layers, it was found that small wavelet scales, corresponding to high frequencies, were required to successfully detect small-scale coherent structures in turbulent flows. The finest region in small-scale coherent structure of the flow was detected with the smallest possible scale of the wavelet at $a = 0.02$, which corresponds to the peak of the cross correlation function between the turbulent image and the wavelet. For turbulent mixing layers, small wavelet scales were not used to detect large-scale coherent structures. It was found that for $0.02 \leq a \leq 0.16$, both small- and large-scale coherent structures were more clearly revealed by using the 2-D Mexican-hat CWT, and the higher was the scale, the clearer was large-scale coherent structures being revealed and vice versa. Overall, it was satisfactorily reported by Kailas and Narasimha that the 2-D MHW was effective in revealing small- and large-scale coherent structures in turbulent flows. It was also noted by the authors that the use of orthonormal wavelets, even though could yield "optimal" performance, was optimally unsuitable for turbulent flows because of their coarse resolutions. Non-orthogonal wavelets, even though are redundant, they are flexible in choosing suitable scales for a particular set of data or images, which yield more insight to local behaviour of the signal/image energy concentration. Moreover, these wavelets are symmetrical which are useful for symmetry and edge detection in images [2]. Motivated from this work, other symmetrical wavelets such as Morlet and hyperbolic can also be used to study turbulent flows so that detailed interactions among coherent structures in each layer of a turbulent flow can be gained.

Han, Cho and Ihm employed the MHW to develop an all-electron density-functional (AE-DF) program [62]. The MHW was chosen because it is a smooth wavelet which is suitable for expanding Hamiltonian and plane waves, whereas an orthogonal wavelet such as Daubechies could not because of its singularity in shape. This motivates the use of the MoW and HyW so that atomic parameters of different molecules can be estimated by using different methods. From that, the best wavelet can be identified. It should also be noted that because smooth symmetrical wavelets (MHW, MoW and HyW), which have often been overlooked and only there exists a small number in the literature [13], are particularly useful in atomic calculations, it is important to develop possible links among different disciplines in signal processing such as time-frequency and wavelet analyses [20, 21], so that more symmetrical wavelets can be proposed and effectively used. Although the MHW is an over-complete basis whose scaling function does not exist, it was reported that [62] the wavelet did not cause any instability in the electron calculations. In particular, it was used to compute local density approximations and generalised gradient approximations of hydrogen (H), oxygen (O), carbon (C), manganese (Mg) and silicon (Si) atoms including cut-off energy and atom support radius. Moreover, vibrational frequencies of CO, H_2O and H_2 atoms were numerically estimated. From that, atomic peaks, which are located at the atom nucleus, were identified which showed the effectiveness of the MHW for detecting sharp-rising structures. The only disadvantage of the MHW in this application was that a large amount computing power was required for atomic calculations, which motivates the use of parallel computing.

In 2000, Zhang *et al* [63, 64] reported interesting results on using the 2-D circular anisotropic MHW to extract corners and edges in optical images by using a joint transform correlator (JTC), which employs a bismuth silicon oxide $Bi_{12}SiO_{20}$ photorefractive crystal as a recording material. They showed that the JTC performed better than the VanderLugt correlator (VLC) with a higher *SNR*, and also it was advantageous in real-time applications. The performance of a number of wavelets including the Mexican-hat, Haar and Robert's gradient was compared, and it was found that the Robert's gradient was the most effective. In their second work [65], a BSO-crystal JTC combining the 2-D MHW with circular harmonic filters (CHFs) up to the second order was reported to be more effective than the first case in edge and corner detection. The BSO-crystal JTC was reported to suffer from a low *SNR*, but effective in edge detection. The main idea in Zhang *et al.*'s work was to use the MHW to simultaneously improve the correlator's *SNR*, and retaining its effectiveness in detecting image edges and corners as carried out in their previous work [63, 64]. They concluded that the MHW could improve the *SNR* (compared to the earlier work [63, 64]) for large scales or high frequencies. It should be noted that this work did not use the elliptical 2-D MHW used in [45] which could yield further improvements in the correlator's *SNR*. A number of questions arise which may ignite further significant work on optical JTC: Why can only large scales improve the correlator's *SNR*? Can higher-order CHFs be employed so that the *SNR* or edge-detection effectiveness can be further improved? The works by Zhang *et al.* [63-65] also open up new research directions in optical image processing in which the MoW and HyW can also be employed.

In 2000, Alam and Chain successfully utilised the 2-D MHW to improve multiple-target detection techniques [66] by using a joint transform correlator. A reference image was used to recognise different input images, and to form joint images. The 2-D Fourier transforms of the joint images were filtered by using the 2-D MHW wavelet in the frequency domain, obtained by dilating and translating its frequency response by using different scale and time indices. The process of multiplying the Fourier transform of joint images with the wavelet filter in frequency domain was equivalent to correlating these images in the time domain and yielding correlation peaks at the output. It is useful to note that in experimental optics, wavelet filters were realised by using a spatial light modulator (SLM). The outcome of modifying the joint image Fourier transforms with different wavelet filters in the frequency domain yielded three different correlation outputs: autocorrelation of the reference image and input images, cross correlation between the reference image and input images, and cross correlation between the input images, of which the first autocorrelation should be retained to successfully detect the target input images, whereas the remaining correlation functions should be suppressed. The MHW has been used successfully for edge detection and was shown to be effective in this case by improving the magnitude ratio between the autocorrelation of the reference and input images to the cross-term artefacts, which are always present in a multi-component image. This concept was clearly demonstrated in time-frequency signal processing context using Cohen's time-frequency power spectrum class with the CWK [26, 46, 67, 68]. This work opens up new application(s) for the HyW because the HyK was shown to be more effective than the CWK [26] by having a larger cross-term weighting function resulting in better cross-term suppression in the time-frequency plane. This suggests that the HyW, which is shown to be a circular and rotationally invariant wavelet, can be effectively employed in this application to improve the magnitude ratio of the autocorrelation to cross-term artefacts in optical correlators for multi-target detection.

Following on the work by Alam and Chain [66], Cheng and Chen [69] used the 2-D MHW, in combination with a logarithmic harmonic filter (LHF), to detect objects at different angles. The LHF was known to be effective in detecting images at different projection angles. However, if the angle is large or approaches 90^0, then the LHF method is not effective. The MHW was specifically used in this case since it is rotationally invariant, yielding effective edge detection independent of the object size and projection angle, which is useful for practical automatic guidance systems such as road-sign recognition. A 2-D input image was first 2-D wavelet-transformed using the MHW. This process gave wavelet coefficients of the image describing its distinctive differences to other images. The wavelet-transformed coefficients were then correlated with the LHF so that the image could be realised. From [69], it was shown that images were recognised with the projection angle in the range from -83^0 to

83^0, compared with -62^0 to 62^0 by using only the LHF method. This showed that the 2-D Mexican-hat CWT coefficients could be effectively used to distinguish images at different projection angles. The MHW 2-D wavelet transform was also used as optical filters, which could be considered as "transparence" located at the input plane, for edge detection in the spatial domain [70] and spatial-frequency domain by varying the scale and time indices so that local changes in the input image could be tracked. The configuration of the system in this work was similar to Figure 21, however, only two lenses and two planes (input and output) were employed in the spatial domain. In the spatial-frequency domain, an extra lens was used to obtain the image frequency response. It is clear that the work reported in this application can be carried out by using other wavelets such as the HyW and MoW.

Spectral studies of topography tracks at the northern caps on the planet Mars were carried out by using the variance of the Mexican-hat wavelet transform [71]. Two-dimensional data were collected by the Mars Obiter Laser Altimeter (MOLA) over a 2-year period from 1999 till 2001. The MHW was chosen in this application because of its effectiveness in edge detection, which could be used to identify important features of the polar caps on Mars and yielding more insight to their geographical properties. The wavelet transforms of different tracks were computed and their variances were estimated as a function of the scale a. From that, the power law of the variance as a function of a was numerically estimated. It was noted in this work that symmetrical wavelets, especially the MHW, was particularly useful for topography because of finer resolutions could be obtained, compared with using orthogonal wavelets. As a result, a sharp change in the ice thickness was detected 24 km from the North pole on Mars. The HyW has been shown to possess similar properties to the MHW and therefore it is also suitable for this application. Moreover, the HyW was shown to possess a finer scale resolution than the MHW [20-22] which further strengthens its suitability for this application. It was also noted in this work that the wavelet transforms of different tracks at specific directions were carried out which motivates the use of directional wavelets such as MoW and elliptical MHW [45]. Another approach to topography studies on Mars is to use the MHW and HyW to scan through all 2-D images that have been collected by MOLA. From that, distinctive features of each image can be readily identified which would further enhance understanding on the Red Planet.

The MHW was successfully used to study Poisson data using the software "WAVDETECT" [72] in the field of astrophysics for source detection. The ideas in this work — including wavelet-data correlation, background estimation, source-detection-threshold computation, variance estimation and source characterisation — were then successfully used by Vielva, Herranz, Sanz and Cayon in astronomy to study point sources embedded in a cosmic microwave background as will be discussed shortly. The MHW was mainly chosen in this work because of threefold: (1) it has two vanishing moments which can be effectively used in suppressing a constant background, (2) it is limited both in the time and frequency domains which can effectively suppress aliasing, (3) its explicit expressions exist which reduce computation time. It should be noted that the higher is the number of vanishing moments, the better is the wavelet in suppressing a constant background or linear trends, which suggests that high-order wavelets, studied in Section 2.2, are more effective than the second-order Mexican-hat, Morlet and hyperbolic wavelets. Since the HyW and MHW are both circular wavelets and possess similar characteristics, the former can also be used in this case to study Poisson data. It was shown [72] that the MHW was effective in source detection using Poisson-sampling fluctuation data in a constant background which suggested that the data are of Gaussian type. From the works by Vielva, Herranz, Sanz and Cayon, it was shown that point-source data were in fact of Gaussian type, which explained the effectiveness of the MHW.

The MHW was also successfully used in detecting defects in structural beams [73] in the field of civil engineering which widens the application range of the MHW. The idea behind this work was that the singular property of a wavelet was effectively utilised to detect minute singularities in a damaged beam by calculating its Lipschitz exponent α using the vibrational modes. A function $f(x)$ possesses a Lipschitz exponent $\alpha \geq 0$ at the displacement $x = v$, if there exist a constant K and a polynomial P_v of order m, where m is the largest integer such that $m \leq \alpha$, so that

$$f(x) = P_v(x) + \varepsilon_v(x), \text{ and } \varepsilon_v(x) \leq K \cdot | x - v |^{\alpha}. \tag{43}$$

From Eq. (43), it is clear that if $0 < \alpha < 1$, then the polynomial P_v does not exist since the maximum integer order m cannot be found. Under this condition, $f(x)$ can be approximated by its "error function" $\varepsilon(x)$ and α can be numerically estimated. However, if $\alpha \geq 1$, it is required that $P_v(x)$ to be eliminated so that α can be numerically estimated, which motivates the use of wavelet vanishing moments to numerically estimate α. It should be noted that under this condition $|x - v|$ and $|x - v|^{\alpha}$ are small. Consider a wavelet whose vanishing-moment order $n \geq \alpha$, then the CWT of $f(x)$, given by Eq. (43), is given as

$$\hat{\psi}_f(a,b) = \hat{\psi}_{\varepsilon_v}(a,b), \tag{44}$$

with $\hat{\psi}_P(a,b) = 0$, because $\int_{-\infty}^{+\infty} \psi(a,b) \cdot P_v(x) = 0$, which allows one to estimate α.

From [73], it was found that the Lipschitz exponent for a damaged beam was of $1 < \alpha < 2$ using Euler's beam theory, which means that the wavelet must have at least two vanishing moments. The MHW was chosen because it has two vanishing moments and was shown to be effective in edge detection, which is useful to locate defect(s) in damaged beams. In their study, they also considered high-order wavelets of the Gaussian pulse, and suggested that there was a trade-off between the number of vanishing moments and support sizes. Finally, they concluded that the second-order MHW (with $\sigma = 1$) was used because only the second-order vanishing moment was required and therefore better support sizes were achieved. Moreover, the MHW is a compact-support wavelet which could be useful in detecting fine local energy changes in vibrational modes of the beam. To estimate α, the CWT was applied to the beam's vibrational modes, from which, the defect location was identified because its CWT was continuous for all scales in the semi-logarithmic plot. From this, it could be suggested that defects in beam's vibrational modes behave chaotically [41, 42, 49, 50] with a smooth CWT representing a broad energy distribution, instead of a discrete CWT possessed by harmonic vibrational modes. The decay along the CWT was then plotted against the scale index and the slope of the straight line was numerically estimated, yielding $\alpha \approx 1.31$ as expected. This particular application was one of the first in the field of civil engineering using a circular wavelet such as the MHW. It is clear that the HyW, as shown in Section 2.1.3, is also a circular wavelet with two vanishing moments and can be employed to estimate the Lipschitz exponent of damaged beams. For circular wavelets which are also rotationally invariant, it is advantageous since these wavelets are uniform and therefore defects can be detected without biasing in any specific directions. However, it should also be noted that the main focus of this application was employing wavelets with at least two vanishing moments, which suggested that the MoW could also be useful, even though it is a directional wavelet. As a result, improvements on the accuracy of the Lipschitz's exponent estimation process can be achieved. In fact, the MoW was reported to be effective in studying sonic echo pile data in which locations of pile toes were successfully identified [13]. Further work can be carried out to detect defects in different types of concrete beams, whose material properties are significantly different. From this, it is possible to classify defect detection benchmarks in different classes of concrete beams, which establishes new standards in the construction of buildings. Further work in this area will be investigated in the near future by using the MHW, HyW and MoW.

The 2-D isotropic and anisotropic MHW's have also found useful applications in astronomy in a series of papers by Sanz *et al.* [32] on denoising in cosmic microwave background (CMB) sky maps, by Vielva *et al.* [33, 34] on Planck point-source extraction from CMB sky maps, by Herranz *et al.* [35] on salient-feature extraction of simulated Planck point source time series, by Cayon *et al.* [36, 37] on Gaussian detection in *COBE*-DMR maps which are full-sky cosmic microwave data, and by Martinez-Gonzalez *et al.* [38] on Gaussian detection in CMB sky maps using the 2-D anisotropic MHW. In [32], Sanz *et al.* showed that the MHW was better than the Haar wavelet by having a larger wavelet-coefficient variance. They also used the Daubechies fourth-order wavelet to denoise 2-D CMB sky maps by using MRA. It should be noted that because the MHW scaling function does not exist, it cannot be used for a complete signal approximation in the $L^2(\mathrm{R})$ space as explained in Section 1. However, the contribution of this work started a series of successfully works which will be discussed shortly.

In [33, 34], Vielva *et al.* showed that CMB data were of Gaussian type so that the isotropic MHW would yield an optimal cross-correlation function over certain optimum scale ranges. They formulated an explicit expression of the wavelet coefficients as a function of the scale a, $W(a)$ which is given by Eq. (45), in terms of the point source position x_0, its magnitude B, and area A which is mainly used for normalisation, under noisy conditions of variance σ_a

$$\frac{W(a)}{a} = 2\sqrt{2\pi} \cdot \frac{B}{A} \cdot \frac{(a/\sigma_a)^2}{[1+(a/\sigma_a)^2]^2} \tag{45}$$

From Eq. (45), they showed that for certain scale ranges, the wavelet coefficients attained maximum values relative to their dispersion due to the presence of CMB data: $\sigma_{wave}(a) = 2\pi a^2 \cdot \int_{-\infty}^{+\infty} P(f) \cdot f \cdot |\hat{\psi}(af)|^2 \, df$, where $P(f)$ is the data power spectrum, f the frequency variable, and $\hat{\psi}(af)$ the Fourier transform of the MHW. The optimal scale range is explicitly given as

$$\frac{W(a_{opt})}{\sigma_{wave}(a)} > \frac{B}{A\sigma}, \tag{46}$$

where σ is the dispersion of the real data in space.

From Eqs. (45) and (46), the optimal scale range can be obtained which enables establishment of an "optimal" MHW to study Planck point sources in a CMB. They also worked out the amplification factor as

$$A_{amp} = \frac{W(a_{opt}) \cdot \sigma}{(B/A) \cdot \sigma_{wave}(a)}. \tag{47}$$

The wavelet covariance was also used to study the correlation among different scales. They analysed CMB datum maps at 875 GHz, and concluded that the MHW performed better than the Gaussian filter for well-chosen wavelet scale ranges over high frequencies. It was also found that for cases where the CMB data possessed similar variations to the point sources, the MHW was not effective in point-source detection because the cross-correlation function did not provide dominant correlation peaks. However, for cases where the CMB data possessed higher emission than that of the point sources, the wavelet performed better with a larger number of point sources detected than the previous case. The authors also outlined a number of possibilities to improve the overall performance of their work, including pre-noising and 2-D elliptical MHW employment (as reported in [45]), and pointed out that elliptical asymmetry was not useful for point-source detection.

A maximum entropy method (MEM) was employed in combination with the MHW to further improve point-source detection [33]. The main motivation of the work reported in [33] was to overcome the ineffectiveness of the MWH when point sources possessed similar variation with the wavelet, as reported in their previous work [34]. In this work, they showed that the MEM and MHW were complementary to each other of which the former was very effective for low-to-immediate flux detection, and the latter for bright-source detection. It should be noted that bright sources present stand-out features with irregularities and non-uniformness of an image, which can be detected by using correlation. The MHW was able to estimate the magnitude of strong bright sources with acceptable accuracy. The main idea of this work was to separate bright sources from low-to-immediate fluxes. First, the data map was independently analysed using the MHW to remove bright sources, yielding a modified map consisting of only low-to-immediate sources, which were then used as the input to the MEM for reconstruction. An error or residual data map, which mostly consisted of noise, was then produced by subtracting the estimated point sources with the original data map. The error map was then analysed to check whether there are any remaining point sources embedded in noise using the MHW. From [33] and their previous work [34], Vielva *et al.* concluded that the MHW was a robust and effective tool for point-source detection in CMB maps. Moreover, the combination of the MHW and MEM provided better results — in terms of bright-source subtraction and contamination prevention of which the significance of using the MHW was to successfully estimate bright-source magnitude — than separately using each method.

Continuing on from the work by Vielva *et al.*, Cayon *et al.* employed the 2-D anisotropic sphere MHW, which was a rotationally-invariant wavelet, and the HaW, for Gaussian detection in full-sky cosmological data *COBE*-DMR over a period of four years [36, 37], by realizing that the datum distribution was on a sphere, not on a plane as was carried out in the literature. This work was the first paper in a series of two papers dealing with the subject. By utilising the fact that the MHW was the second-order derivative of the Gaussian pulse, the cross-correlation function between the data and wavelet was formed, leading to the conclusion that the data were of Gaussian type. They also found that there were not significant differences between the 2-D MHW and 2-D HaW even though the former was rotationally invariant, and the latter focused on certain directions in space which could be useful for data whose symmetry is significantly different from it. The second paper by Martinez-Gonzalez *et al.* [38], studied in detail the performance of the spherical MHW and compared it with the spherical HaW, which is an orthogonal wavelet, for Gaussian and non-Gaussian detection. They also showed how a wavelet was projected on to the spherical coordinate. It was shown that the MHW could detect skewness and kurtosis at levels that are much smaller in magnitude than the orthogonal HaW. This is mainly because the former is a regularly-smooth wavelet with two vanishing moments, localized in frequency and symmetrical even though it is non-orthogonal.

This two-paper series opens up new research work and possible connections between astronomy and signal processing in a sense that up to this point, there have been only two wavelets that were projected on to the spherical coordinate, even though there are other symmetrical wavelets literature, such as Morlet and hyperbolic. It is thus interesting to conduct further studies on which new wavelets can be employed so that detailed conclusions on the roles and effectiveness of symmetrical wavelets in processing cosmological data can be reached. In particular, the use of the MoW, a modulated Gaussian wavelet, and HyW, a Gaussian-like wavelet, as shown in [20-22]. Another point to make is that the elliptical 2-D anisotropic MHW wavelet, reported in [45], can find this application useful since the wavelet direction can be controlled by varying its orientation angle θ. Moreover, it is also worth to examine the effectiveness of orthogonal wavelets for this application even though the main focus is to extract salient features of point sources embedded in CMB maps. All of these aspects can be implemented so that further improvements in studying full-sky cosmological data can be achieved.

The final paper in studying point sources in astronomy was contributed by Herranz *et al.* [35] after some early successful works by Vielva *et al.* [33, 34]. In their work, Herranz *et al.* studied full-sky cosmological data using 1-D time-ordered series, with no artifacts caused by correlation between the data and CMB noise as the main advantage, even though these series suffer from low *SNR*s compared with studies using 2-D images. They also proposed necessary criteria for a 1-D optimal adaptive filter, and explicitly obtained its expressions. The performance of the optimal filter, MHW and Gaussian filter was compared, and it was found that the former two filters were more effective than the latter. The optimal filter appeared to be more effective in discriminating low-magnitude noisy components, which were uniformly distributed in the full sky, than the MHW and Gaussian filters. The MHW was shown to be most effective for noise components distributed on the Galactic plane.

The 2-D MHW also found useful applications in digital water-marking by effectively identifying intensity changes in an image, yielding detection of salient features which were considered as reference points [60], where digital watermarks were then embedded. The key idea behind this application utilised the fact that the MHW is a smooth and regular wavelet with symmetrical properties. In their work, Tang and Hang identified that the wavelet was rotationally invariant because its 2-D image was finely concentrated at the origin, yielding uniformness in all directions in 2-D space as can be seen in Figure 3. As a result, normalised-image detection was identical to original-image detection which significantly simplified the detection process. Moreover, the wavelet was shown to be noise robust by correctly choosing two orthogonal scales in 2-D space. The idea of using the local peaks of the cross correlation function between the wavelet and the image utilised the fact that the correlation of the MHW with an object (1-D or 2-D) could be used as to approximately represent the input signal whose details can be revealed by the wavelet expansion in the difference subspaces as explained in Section 1.1. The work reported in [60] opens a number of new research directions in which rotationally-invariant wavelets such as the HyW can be effectively utilised. It was shown in [60] that there was a trade-off between image visibility and noise robustness which motivates further investigations using other wavelets such as Morlet, hyperbolic, or the elliptical Mexican-hat [45]. By using the elliptical MHW, it is possible to concentrate the energy of the watermark on one random direction in the image, and thus can avoid severe external attacks. It should be noted that the MoW is directional in 2-D space and could be used in a similar way to the elliptical MHW for watermark hiding in 2-D images.

The 2-D MHW was used to automatically detect fiducial tungsten markers, whose shape is spherical, in 2-D medical portal images [74]. Tungsten markers were implanted into living organs, usually on the skin, whose movements could be visualised for adaptive image-guided radiotherapy (IGRT). The markers in portal images were captured by using an amorphous silicon electronic portal imaging device (a-Si EPID) at a very high resolution. Because marker implantation played a vital role in guiding surgeons to achieving accurate operation on the patient, marker detection was of significant importance for a successful operation. Since the shape of markers was similar to the MHW, convolution of portal images with the 2-D MHW gave doctors distinctive peaks whose locations approximately identified the markers. The convolution process could be considered as cross-correlating the image with the 2-D MHW, thus yielding locations of the similarities features in the image to those of the wavelet. The MHW was employed in this case for unsupervised marker detection, i.e. automatic detection, because it is a circular and rotationally-invariant wavelet, which was used to scan through portal images without prior knowledge of where the markers were. For this reason, the MoW cannot be employed in this case because it is directional, however, the HyW can be utilised. After the convolution process, a scanner was used to measure the grayness level in the modified image, then regions whose grayness level was higher than the predefined threshold were examined to locate the markers. Because the MHW has two vanishing moments, constant and linear background noise sources were almost eliminated. However, marker detection still suffered from background images of gas pockets and bone edges, and it was found by Buck *et al.* [74] that the larger was the marker, the better was the efficiency of the detection process. The elliptical MHW [45] was also mentioned as a useful tool to detect cylindrical markers. Because of its similar characteristics to the MHW of circular and rotationally-invariant nature, the HyW can be employed for marker detection. The elliptical HyW can also be derived so that possible improvements can be achieved.

The 1-D CWT of the MHW was successfully used to estimate the scaling exponent of turbulent data by decomposing the wavelet into a number of smaller wavelets using the Hermite function up to a certain order. The MHW and its transform were used by Chen and Hu [56] to decompose turbulent data with large-scale vortices, such as wind velocity and temperature collected at Huaihe river basin in China, into coherent structures. By definition, coherent structures are distinctively large-scale oscillation patterns existing in turbulent data, which were shown to exist in large-scale motions. Chen and Hu showed that the discrete-orthogornal wavelet transform was not suitable in this particular case because of its coarse resolution. The wavelet energy of the coherent structures, which contained most of the turbulent energy, were then studied and their correlation coefficients were numerically estimated. Coherent structures in turbulent data were then successfully reconstructed by using the Mexican-hat ICWT [Eq. (5)]. Dimensionless durations of the wind velocity in 2-D were estimated by using the CWT, and it was found that there were strong correlations among them. It was also shown that the asymmetric first-order derivative of the Gaussian pulse was also effective in estimating the time duration of coherent structures. From this work, it could be suggested that coherent structures in turbulent data are Gaussian type structures since they can be successfully detected and reconstructed using the MHW. The asymmetric first-order derivative of the Gaussian pulse was shown to be effective in this respect which further strengthens this point. Further work in this direction can also be possible with the use of the MoW, HyW and its higher-order family members.

3.3. The Mexican-Hat Wavelet as Laplacian-of-Gaussian (LoG)

It appears in the literature that the MHW and LoG are two different concepts even though they are identical as explained earlier. The main difference is that they have been applied and employed in different areas and context. Thus, for completeness, this subsection aims to briefly explore a number of typical applications of the LoG as an edge detector being applied in various different fields.

The LoG has been well-known in image processing, in particular edge detection, by convolved with images and then zero-crossings are detected as edges. A large amount of theoretical work on the LoG for edge detection [75-95] has been reported in the literature. The LoG has been applied in various different fields such as optics, speech, stereopsis, object recognition and movement tracking. The following applications mainly used the LoG as an efficient edge detector. Unlike the works presented in [75-95], the following works use the LoG as the main tool for particular applications with minimal theoretical development.

In 1988, Rangathan and Shah [96] proposed a VLSI single-chip design of the LoG in 2-D to compute scale space at a higher speed. The main advantages of this method are two fold: (1) Its algorithm does not approximate the LoG in 2-D, instead, four 1-D convolutions were used to generate one 2-D convolution; (2) The design was more advanced by using VLSI technology than other methods proposed in the literature. In addition, it was based on basic properties of the LoG, namely scaling, separability and symmetry. However, at the time of publication, the design was not built. Jordan III and Bovik [97] reported two matching algorithms in stereo images, namely *Intensity Matching* using the LoG edge detector, and *Color Matching* using chromatic gradients which characterise the intensity of zero crossings. The authors noted that the use of the LoG edge detector even though was effective, but not efficient due to long processing time. This is consistent with the work of Rangathan and Shah [96] published in the same year. Jordan III and Bovik then proposed a new method using chromatic gradients because edges do not have chromatic information to improve the efficiency of stereo-image matching with promising results.

In 1989, Kaneko [98] briefly reported a method using the LoG with a coarse-to-fine matching strategy to detect disparity in trinocular stereo images. Disparity in stereo images is represented by the brightness of the image. A zero-crossing map of an image was created first by using the LoG then a disparity map was generated. Kaneko found that the LoG was effective with its scale σ of 2, 6, 8 and 16. In 1990, Vernon and Tistarelli [99] used optical flow using a camera, which was mounted on robot's end effectors, to determine depth maps. The LoG was used for zero-crossing detection from filtered images. From the obtained zero-crossing images, original images were reconstructed and optical flow was determined by matching contours in LoG reconstructed images with original images. In 1991, Raman, Sarkar and Boyer [100] used the LoG for edge detection to study tissue boundary refinement in magnetic resonance images (MRI). The LoG was extensively used in their work as it was considered an optimal edge detector. In the same paper, the authors investigated a trade-off between reducing error and increasing scale of the LoG filter. They came up with a contour-based technique for edge detection at coarse scales, which was shown to be precise and efficient for medical applications.

Boyer, Sotak, Jr. and Schenk [101] constructed a graph-theoretic abstraction of images by using the LoG in stereopsis to solve stereo correspondence problems. The LoG was particularly used to extract edge contours in an image, which are then used to reconstruct the original image. Comparisons were then made between the reconstructed and the original images. In this work, the authors clearly indicated that the LoG was one of many edge detectors which was used in experiments. This opens up further work on applying the MoW and HyW for stereopsis as the LoG is mathematically equivalent to MHW. Wuescher and Boyer [102] also developed techniques on how to partition LoG contours in stereo images to basic primitives using a constant curvature criterion to determine depth maps.

In 1995, Godbole and Amin [88] used a mathematical morphology method, LoG filter, Derich's and Shen's algorithm for edge and overlap detection in medical images. Performance of each method was compared and it was found that mathematical morphology was the most suitable method for medical images, especially in detecting overlapped human lungs. In 1997, Yagi, Ohshima and Funahashi [103] modelled a neuronal circuit of the outer retina consisting of cones, horizontal cells and bipolar cells. From the model, they found that there exists a quantitative relationship between bipolar cells' receptive fields and the LoG. Bipolar cells have two basic operations: (1) smoothing operation which can be considered as similar to a Gaussian-like weighting function, (2) contrast enhancement operation which is mathematically similar to the LoG. Chhabra [104] estimated optical flow in real time along contours of significant intensity change in images by using zero-crossings given by the LoG edge detector. The main difference of this work from the work presented by Lai in [105] is that the present work employs output of the LoG edge detector to estimate optical flow in real time, whereas Lai used the LoG as an operator to propose a new regularisation formula. This shows that the LoG, or the MHW, is useful across various different fields.

In 1998, Raj, Dunn and Cannon [106] found that the LoG was useful and insensitive to edges in scanning directions of a field emission scanning electron microscopy device, which is used to study particle orientation in ceramic green bodies. This was then used to explain quantitatively the non-uniform sintering shrinkages in tape cast ceramic microstructure. Girouard *et al.* [107] attempted to automatically setup a deviation measurement system using an electronic portal imaging device (EPID) in pelvic fields. Bone edges were detected by using the LoG. By varying the scale parameter σ, bone edges in different bone images taken by the EPID were effectively detected, i.e. large values of σ was used for poor-quality images and vice versa. The authors also used the FFT to reduce computational time of image convolution in edge detection. In addition, for large σ, the LoG was found to be more robust than for cases of smaller σ.

Jarvis and Macalpine [108] convolved the LoG with Hubble-Deep-Field images to detect possible active high-redshift galactic nuclei. By using the LoG, it was possible to remove background noise and flat galactic background, leaving point-like components at the centre of the image. This is mainly because the LoG is isotropic and rotationally invariant. This work was performed before a series of seven papers published in the *Monthly Notices of the Royal Astronomy Society* by Vielva [33, 34], Herranz [35], Cayon [36, 37], Sanz [32] and Martinez-Gonzalez [38] between 1999 and 2002 as discussed in Section 3.2. It is worthy to note that these works [32-38, 108] were consistent with each other.

In 1999, Dai and Khorram [109] proposed an improved LoG edge detector for image segmentation. Edges can be considered as zero crossings in the resultant convolved image with the LoG by detecting sign changes in the image's pixels. The main problem is that the pixels can have low values at weak edge points, which are considered as noisy pixels. To remedy this problem, a threshold is used to filter out pixels with lower or zero values. However, this in turn causes discontinuity at weak edge points. A new and improved LoG edge detector called "Thin and Robust Zero-Crossing" was proposed with a two-step operation: (1) edge selection and point marking; (2) edge refinement and sorting. The new LoG edge detector was also shown to be effective in region detection.

In 2000, Souza, Ruiz and Cruz [110] studied palpebral fissure, which is the area formed by the upper and lower eyelids, by extracting its significant features using the LoG with promising results. The authors noted that the LoG was more effective than other edge detectors because it can detect palpebral contours. This work could further be improved by using the "improved" LoG detector proposed by Dai and Khorram [109] in 1999. Gakkai D. J. T and Gakkai D. T. [111] briefly reported a method using the LoG to locate underground pipes by analysing pulse radar images, which can be generated by shifting the sending and receiving antennas in one direction over the ground and hence creating 2-D images. The LoG was used to detect edges and Hough transform was also employed to detect the depth of the pipes. The method was also shown to be able to detect deeply buried pipes. Banarse, France and Duller [112] employed Gabor, Gaussian and LoG filters for face and hand-written digit recognition. Performance of the three filters was compared and it was found that the Gabor filter outperformed the LoG and Gaussian filters, whereas the Gaussian filter was outperformed by the LoG filter. Chen *et al.* [113] utilised the LoG to analyse spatial changes of craniofacial structures for orthodontic treatment, which is dependent on cephalometric tracing manually applied on to the patient. The manual superimposition of cephalometric causes errors which create side effects on patients. With the use of the LoG, images taken by orthodontists were filtered by the LoG to detect zero crossings, and hence the accurate cephalogram superimposition applied on the patient. This process was then automated, which takes approximately 6 minutes, to significantly reduce the errors with the current accuracy up to the order of 0.05 millimetres.

In 2001, Lu *et al.* [114] used the wavelet transform to extract mountain areas from satellite images. In this work, they also found that the LoG was particularly useful for ridge-line detection as terrain slope was effectively detected and estimated. In 2002, Shingo, Satoko and Masato [115] used the LoG to improve local visibility in digital colour images because of incompatibility among different input (digital camera) and output devices (printer). Images are divided into smaller regions using the LoG whose optimal conversion curve was then constructed using adaptive histogram equalisation. The image was then recovered by combining sub-regions at the end with correct adjustment on their brightness and darkness.

In 2004, Lai [105] proposed a modified regularization formulation to compute optical flow in image motion analysis. The new formula employed the LoG operator instead of the traditional Laplacian operator to smooth the image and to eliminate noise. As a result, a new data constraint was established with the following weighting function

$$w(x, y, t) = \frac{1}{\sqrt{F_x^2(x, y, t) + F_y^2(x, y, t) + c_c}}, \tag{48}$$

where $F_x(x, y, t) \cdot u(x, y, t) + F_y(x, y, t) \cdot v(x, y, t) + F_t(x, y, t) = 0$, with $u(x, y, t)$ and $v(x, y, t)$ are optical flow vectors, and c_c is the weighting factor which can be used to control error amplification. The main advantage of using the LoG in this case is that noise can be effectively controlled by varying the weighting factor c_c.

Cheng, Wang and Shi [116] recently used the LoG to detect and to estimate sizes of microcalcification clusters in human mammograms. Because of the fuzziness of microcalcification clusters in mammograms, fuzzification was also employed. Mammogram images were fuzzified then filtered using the LoG. After the filtering process, local maxima of the image are identified as microcalcification clusters which will be removed. The LoG is specifically useful in this case because of its ability to differentiate between fine microcalcification clusters and dense tissues in mammograms which are usually mistaken as microcalcifications. The algorithm is graphically given in Figure 22.

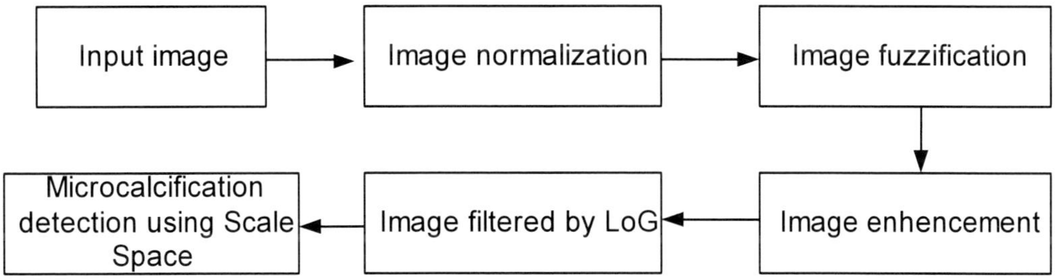

Figure 22. Microcalcification detection algorithm.

It should be noted that applications using the MHW as the LoG should be applicable to the HyW and MoW because of similarities among the three wavelets as explained in Section 2.1. However, a lot more work need to be carried out to access the usefulness of the HyW and MoW in these applications.

3.4. The "Mexican-Hat" Concept from a Chemical Physics Point of View

This subsection briefly discusses related applications in which the Mexican-hat concept was widely and effectively utilised. A large number of papers in the field of chemical physics on studying potential energy surfaces were carried out using Jahn-Teller's coupling formula. These works concluded that potential-energy surfaces possessed a "Mexican-hat" shape, starting with the early work by Van Ouwerkerk [117] on retina response in 1980, when the MHW was first proposed by Marr [2]. Although the Mexican-hat shape was successfully observed in these theoretical works, it has not been theoretically studied which can lead to possible new research directions in modelling potential-energy surfaces using the hyperbolic and Morlet wavelets whose properties are similar to those of the Mexican-hat wavelet. Existence of the Mexican-hat shape has been identified and observed in various theoretical aspects and scientific applications as follows:

- Spatial response of cats and humans on a small-light stimulus [117];
- Jahn-Teller coupling of the $Cu(Im)_6^{2+}$ ions in crystal $Zn(Im)_6Cl_2{\cdot}4H_2O$ [118];
- Potential surfaces of the Cu^{2+} ions doped in K_2ZnF_4 [119];
- Electronic structures of the $YBa_2Cu_3O_7$ high conducting-temperature superconductor [120];
- Potential surfaces of the centres of *trans*-$Cu(H_2O)_2(Cl_4)^{2-}$, *trans*-$Cu(NH_3)_2Cl_4^{2-}$ and *trans*-$Cu(H_2O)(NH_3)(Cl_4)^{2-}$ [121];
- Optical modelling of the elastic scattering of α particles on Ca at 1.37 GeV by using the Woods-Saxon form [122];
- Iron-sulphur cluster in reduced rubredoxin [123];
- Exponentially decreasing synaptic-coupling correlations in 1-D and 2-D in neural networks [124];
- Cosmological consequences of the Peccei-Quinn field during inflation under vacuum fluctuation conditions [125, 126];
- The mechanism of chiral-symmetry restoration in high-temperature quantum chromo-dynamics (QCD) was statistically simulated by using the MHW [127];
- The Fulton-Gouterman transformation applied to the *E-e* Jahn-Teller problem creating "exotic" states [128];
- Distribution theory of neocortex [129];
- Visual data in area 17 of the mammalian cortex [130];
- Synaptic coupling functions of intra-cortical distances among cortical cells [131];
- One of the exact solutions to Einstein's equation describing the evolution of inflationary universe models [132];
- Linsker and Miller neural models [133];
- Quasi-topological solitons potential surfaces in 2-D [134];
- Variation of hybrid inflation [135];
- Quasi 2-D GaAs of the 3-D CuCl radiative renormalisations [136];
- Spatial-frequency transformation of cortical neurons [137];
- The electron double resonance spectra of hydrated neutral homo-nuclear metal clusters such as $^7Li_3$, $^6Li\ ^7Li_2$ and $^7Li_3(H_2O)$ [138];

- Introduction of the Mexican-hat potential and compared it with the standard $|\psi^4|$ field theory [139];
- Ginzburg-Landau potential surfaces [140];
- Investigations of the Nambu-Jona-Lasinio model using the Dirac sea which can be considered as the manifestation of the MHW [141];
- Orientation distributions in human vision using neural networks [142];
- Coupling of integrate-and-fire neurons in the 1-D ring topology [143];
- The MHW was used to transform the n-length hydrophobic free energy sequences $H_{i = 1, 2, \ldots, n}$ of proteins [144];
- Diffusion of store elastic energy at an infinite contact surface of two materials of different elastic constants [145];
- Potential surfaces of the Jahn-Teller coupling between the Mn e_g electron and the oxygen octahedral distortions in $LaMnO_3$ [146];
- Molecular picture of excited states and fragmentation paths of the Na_5F_4 cluster [147];
- Off-centre displacement and rotation of the Li^+ ion in KCl crystal [148];
- Higgs potential surfaces and coupling [149];
- Coupled rotation-vibration of hydrogen molecules in a confining cylindrical potential [150].

It should be noted that the above-mentioned works present various potential research directions in which the MHW in one hand, and symmetrical wavelets in the other hand, could be useful. For example, the MoW or HyW could be used to model potential-energy surfaces of a particular process in chemical physics. It should also be noted that the HyW is quite similar to the MHW which suggests that the former wavelet can find the latter's applications useful. However, a lot more work is required to complete the big picture in which symmetrical wavelets are thoroughly studied and applied to existing and new applications. With the advancement of new technology, in particular the use of high-speed cameras with very fine resolutions yielding 2-D images of high quality, there is always room for new applications and further improvements.

Conclusions

This paper has reviewed symmetrical wavelets and focused on useful applications of the Mexican-hat (Choi-Williams) wavelet over the last 20 years since 1980, when it was successfully studied by Marr. Based on the multiresolution analysis proposed by Mallat, it was shown that symmetrical wavelets could not span the entire $L^2(\mathrm{R})$ space because their scaling functions do not exist. However, this is not the main focus in using these wavelets, which have shown to be useful in edge detection, brightness detection, particle detection, and noise suppression. Symmetrical wavelets are useful since they can be effectively employed to extract important features in 2-D images based on the principle of correlation.

The hyperbolic wavelet was introduced and studied and it was found that this wavelet is rotationally invariant and similar to the Mexican-hat wavelet, which is a desirable property for image processing and edge detection. The Morlet wavelet was also discussed in which it was found that the wavelet is not a circular wavelet. Instead, it is directional and has also found useful applications in processing seismic data and in other fields. The 1-D and 2-D Mexican-hat wavelet, Morlet wavelet and hyperbolic wavelet were studied in the time and frequency domains. Useful applications of the Mexican-hat wavelet can be classified into five major fields: image processing, medicine, optics, astronomy and astrophysics, and fluid mechanics (turbulence).

In addition, a significant application of the wavelet was found in the field of civil engineering by the work of Hong *et al.* in 2002 where locations of defects in concrete beams were identified. The work of Malik and Subramanian published in 1997 significantly showed that the Mexican-hat wavelet could be made to be anisotropic by realising the elliptic symmetry instead of circular symmetry in the time domain with the use of two orthogonal scales and a directional angle θ. A series of seven papers published from 1999 to 2002 by Sanz, Vielva, Herranz, Cayon and Martinez-Gonzales successfully showed that the wavelet can be effectively utilised in the field of astronomy.

The Mexican-hat wavelet was also very useful as evidenced by Mize's work published in 1994. In image processing, it was successfully used for watermark injection into digital images by the work of Tang and Hang published in 2003. In optics, Zhang's work published in 2000 showed that the Mexican-hat wavelet was more effective than the joint transform correlator. A number of significant practical works on studying turbulence and coherent structures utilising the wavelet were successfully reported between 1997 and 2003 by Chen *et al.*, Kailas *et al.* and Jiang *et al.* It should be noted that work on turbulence started in 1994 with a breakthrough paper by Milligen *et al.* in 1995, and later with a review paper by Farge in 1996. Further research work on modelling the Mexican-hat shape in chemical physics also provides interesting and challenging tasks. The Mexican-hat wavelet has also been used as Laplacian-of-Gaussian in many different fields as briefly outlined in Section 3.3.

Because of a large number of publications in the literature, applications using the Mexican-hat wavelet in various disciplines are difficult to keep track of, which can undermine its potential. Its usefulness can open up new research directions for the Morlet wavelet and hyperbolic wavelet, which has been shown to possess similar characteristics to the Mexican-hat wavelet. Moreover, higher-order hyperbolic wavelets have been shown to be more effective than the second-order Mexican-hat and hyperbolic wavelets by having finer scale resolutions. This can promise useful and challenging work in the near future when high-order hyperbolic wavelets are applied to existing and new applications.

Research on the Mexican-hat wavelet has constantly been carried out in all fields of science. The author has been aware that on submission of this paper, papers reported on the Mexican-hat wavelet applications have continuously been published in the literature. The reader therefore should incorporate these additional papers to this paper so that a more comprehensive review can be obtained.

Acknowledgements

The author would like to thank Dr. K. P. Dabke for carefully proof-reading the manuscript, and for his constructive suggestions which greatly improved the quality of the paper.

References

[1] Burrus, C.S., Gopinath, R.A., and Guo, H., *Introduction to Wavelets and Wavelet Transform: A Primer*, 1998, New Jersey: Prentice Hall, Upper Saddle River, 1-100.

[2] Marr, D., *Vision: A Computational Investigation into the Human Representation and Processing of Visual Information*, 1982, San Francisco: W. H. Freeman and Company, 54-67.

[3] Grossmann, A. and Morlet, J., "Decomposition of Hardy functions into square integrable wavelets of constant shape," *SIAM Journal on Mathematical Analysis*, July 1984, 15(4), pp. 723-36.

[4] Gabor, D., "Theory of communication," *Journal of the Institute of Electrical Engineers (London)*, 1946, 93(III), pp. 429-57.

[5] Daubechies, I., "The wavelet transform, time-frequency localization and signal analysis," *IEEE Transactions on Information Theory*, Sept. 1990, 36(5), pp. 961-1005.

[6] Rioul, O. and Vetterli, M., "Wavelets and signal processing," *IEEE Signal Processing Magazine*, Oct. 1991, 8(4), pp. 14-38.

[7] Daubechies, I., Grossmann, A., and Meyer, Y., "Painless non-orthogonal expansions," *Journal of Mathematical Physics*, May 1986, 27(5), pp. 1271-83.

[8] Heil, C.E. and Walnut, D.F., "Continuous and discrete wavelet transforms," *SIAM Review*, Dec. 1989, 31(4), pp. 628-66.

[9] Mallat, S., *A Wavelet Tour of Signal Processing*, 1998, New York: Academic Press, 485-532.

[10] Chui, C.K., *An Introduction to Wavelets*, 1992, San Diego: Academic Press, 1-23, 119-150.

[11] Daubechies, I., *Ten Lectures on Wavelets*, 1992, Philadelphia: SIAM, 20-100.

[12] Misiti, M., et al., *Wavelet Toolbox for Use with MATLAB--A User's Guide*, 1997, MA: The MathWorks, Inc., 6.2-6.114.

[13] Addison, P.S., Watson, J.N., and Feng, T., "Low-oscillation complex wavelets," *Journal of Sound and Vibration*, 2002, 254(4), pp. 733-62.

[14] Strang, G., "Wavelets and dilation equations: A brief introduction," *SIAM Review*, Dec. 1989, 31(4), pp. 614-27.

[15] Vetterli, M. and Herley, C., "Wavelets and filter banks: theory and design," *IEEE Transactions on Signal Processing*, Sept. 1992, 40(9), pp. 2207-32.

[16] Jawerth, B. and Sweldens, W., "An overview of wavelet based multiresolution analyses," *SIAM Review*, Sept. 1994, 36(3), pp. 377-412.

[17] Mallat, S., "Wavelets for vision," *Proceedings of the IEEE*, April 1996, 84(4), pp. 604-14.

[18] Gonzalez, R.C. and Woods, R.E., *Digital Image Processing*, 2002, New Jersey: Prentice Hall, Upper Saddle River, 567-600.

[19] Jordan, D., Miksad, R.W., and Powers, E.J., "Implementation of the continuous wavelet transform for digital time series analysis," *Review of Scientific Instruments*, Mar. 1997, 68(3), pp. 1484-94.

[20] Le, K.N., "High-order symmetrical wavelets," *Journal of Sound and Vibration*, 2007, 304(1-2), pp. 297-325.

[21] Le, K.N., Dabke, K.P., and Egan, G.K., "Hyperbolic wavelet family," *Review of Scientific Instruments*, 2004, 75(11), pp. 4678-93.
[22] Le, K.N., Dabke, K.P., and Egan, G.K. *The hyperbolic wavelet function*, in *SPIE Aerosense Proceedings: Wavelet Applications VIII*, April 2001, Orlando, Florida, USA: SPIE.
[23] Lathi, B.P., *Modern Digital and Analog Communication Systems*, 1998, NewYork: Oxford University Press, 30-44.
[24] Antoine, J.-P., et al., "Image analysis with two-dimensional continuous wavelet transform," *Signal Processing*, April 1993, 31(3), pp. 241-72.
[25] Antoine, J.-P. and Vandergheynst, P., "Two-dimensional directional wavelets in image processing," *International Journal of Imaging Systems and Technology*, 1996, 7(3), pp. 152-65.
[26] Le, K.N., Dabke, K.P., and Egan, G.K., "Hyperbolic kernel for time-frequency power spectrum," *Optical Engineering*, 2003, 42(8), pp. 2400-15.
[27] Mize, R.R., "Quantitative image analysis for immunocytochemistry and in situ hybridization," *Journal of Neuroscience Methods*, Oct. 1994, 54(2), pp. 219-37.
[28] Jha, R. and Saxena, Y.C., "Wavelet analysis of ADITYA edge turbulence: Evidence of non-linear interaction," *Physics of Plasmas*, Aug. 1996, 3(8), pp. 2979-89.
[29] Antoine, J.-P., Vandergheynst, P., and Murenzi, R., "Two-dimensional direction wavelets in image processing," *International Journal of Imaging Systems and Technology*, 1996, 7(3), pp. 152-65.
[30] Szu, H.H., Telfer, B., and Garcia, J., "Wavelet transforms and neural networks for compression and recognition," *Neural Networks*, 1996, 9(4), pp. 695-708.
[31] Yamazaki, T., "A mathematical analysis of the development of oriented receptive fields in Linsker's model," *Neural Networks*, 2002, 15(2), pp. 201-7.
[32] Sanz, J.L., et al., "Wavelets applied to cosmic microwave background maps: a multiresolution analysis for denoising," *Mon. Notices of the R. Astron. Soc.*, 1999, 309(3), pp. 672-80.
[33] Vielva, P., et al., "Combining maximum-entropy and the Mexican-hat wavelet to reconstruct the microwave sky," *Mon. Notices of the R. Astron. Soc.*, 2001, 328(1), pp. 1-16.
[34] Vielva, P., et al., "Predicted Plank extragalactic point-source catalogue," *Mon. Notices of the R. Astron. Soc.*, 2001, 326(1), pp. 181-91.
[35] Herranz, D., et al., "Point source detection and extraction from simulated Plank time-ordered data using optimal adaptive filters," *Mon. Notices of the R. Astron. Soc.*, Aug. 2002, 334(3), pp. 533-41.
[36] Cayon, L., et al., "Spherical Mexican-hat wavelet: an application to detect non-Gaussianity in the *COBE*-DMR maps," *Mon. Notices of the R. Astron. Soc.*, 2001, 326(4), pp. 1243-8.
[37] Cayon, L., et al., "Isotropic wavelets: a powerful tool to extract point sources from cosmic microwave background maps," *Mon. Notices of the R. Astron. Soc.*, 2000, 315(4), pp. 757-61.
[38] Martinez-Gonzalez, E., et al., "The performance of spherical wavelets to detect non-Gaussianity in the cosmic microwave background sky," *Mon. Notices of the R. Astron. Soc.*, 2002, 336(1), pp. 22-32.

[39] Reiboeck, H.J. and Altmann, J., "A model for size- and rotation-invariant," *Biological Cybernetics*, 1984, 51(2), pp. 113-21.
[40] Brasseur, F.G. and Wang, Q., "Structural evolution of intermittency and anisotropy at different scales analysed using three-dimensional wavelet transforms," *Physics of Fluids A: Fluids Dynamics*, Nov. 1992, 4(11), pp. 2538-54.
[41] van Milligen, B.P., Hidalgo, C., and Sanchez, E., "Nonlinear phenomena and intermittency in plasma and turbulence," *Physical Review Letters*, Jan. 1995, 74(3), pp. 395-8.
[42] van Milligen, B.P., et al., "Wavelet bicoherence: a new turbulence analysis tool," *Physics of Plasmas*, Aug. 1995, 2(8), pp. 3017-32.
[43] Vergassola, M., et al., "Wavelet analysis of a Gaussian Kolmogorov signal," *J. Phys. A: Math. Gen.*, Nov. 1993, 26(22), pp. 6093-9.
[44] Cheng, H.-S. and Liang, T.-C., "Rotational invariant pattern recognition using a composite circular harmonic and 2-D isotropic Mexican-hat wavelet filter," *Optics Communications*, Nov. 1994, 112(1-2), pp. 9-15.
[45] Malik, R.K. and Subramanian, K., "Matched-filter analysis of large scale structures in the 1st CfA slice using an anisotropic wavelet basis," *Astronomy and Astrophysics*, 1997, 317(2), pp. 318-27.
[46] Choi, H.I. and Williams, W.J., "Improved time-frequency representation of multi-component signals using exponential kernels," *IEEE Transactions on Signal Processing*, June 1989, 37(6), pp. 862-71.
[47] Li, H. and Nozaki, T., "Wavelet analysis for the plane turbulent jet (analysis of large eddy structure)," *JSME International Journal Series B: Fluids and Thermal Engineering*, Nov. 1995, 38(4), pp. 525-31.
[48] Li, B.-L. and Loehle, C., "Wavelet analysis of multiscale permeabilities in the subsurface," *Geophysical Research Letters*, Dec. 1995, 22(23), pp. 3123-6.
[49] Le, K.N., Dabke, K.P., and Egan, G.K., "Hyperbolic wavelet power spectra of non-stationary signals," *Optical Engineering*, 2003, 42(10), pp. 3017-37.
[50] Farge, M., et al., "Wavelets and turbulence," *Proceedings of the IEEE*, April 1996, 84(4), pp. 639-69.
[51] Saloma, C., "Mirrors with point-spread functions that exhibit wavelet-related characteristics," *Optics Letters*, Oct. 1995, 20(19), pp. 1943-5.
[52] Lomnitz, C., "Search of a worldwide catalog for earthquakes triggered at intermediate distances," *Bulletin of the Seismological Society of America*, April 1996, 86(2), pp. 193-8.
[53] Ouzieli, I. and Mendlovic, D., "Two-dimensional wavelet processor," *Applied Optics*, Oct. 1996, 35(29), pp. 5839-46.
[54] Chen, W., Novak, M.D., and Black, T.A., "Coherent eddies and temperature structure functions for three contrasting surfaces. Part I: Ramp model with finite microfront time," *Boundary-Layer Meteorology*, 1997, 84(1), pp. 99-123.
[55] Jiang, J., Zhang, D.e., and Fraedrich, K., "Historic climate variability of wetness in East China (960-1992): A wavelet analysis," *International Journal of Climatology*, July 1997, 17(9), pp. 969-81.
[56] Chen, J. and Hu, F., "Coherent structures detected in atmospheric boundary-layer turbulence using wavelet transforms at Huaihe river basin, China," *Boundary-Layer Meteorology*, 2003, 107(2), pp. 429-44.

[57] Jha, R., Mattoo, S.K., and Saxena, Y.C., "Structures, wavelet and intermittency in tokamak edge turbulence," *Phys. Plasmas*, 1997, 4(8), pp. 2982-8.

[58] Kevlahan, N.K.-R. and Vassilicos, J.L., "The space and scale dependencies of the self-similar structure of turbulence," *Proceedings of the Royal Society of London Series A*, Nov. 1994, 447(1930), pp. 341-63.

[59] Shensa, M.J., "Discrete inverses for non-orthogonal wavelet transforms," *IEEE Transactions on Signal Processing*, April 1996, 44(4), pp. 798-807.

[60] Tang, C.-W. and Hang, H.-M., "A feature-based robust digital image watermarking scheme," *IEEE Transactions on Signal Processing*, April 2003, 51(4), pp. 950-9.

[61] Kailas, S.V. and Narasimha, R., "The eduction of structures from flow imagery using wavelets. Part I: The mixing layer," *Experiments in Fluids*, July 1999, 27(2), pp. 167-74.

[62] Han, S., Cho, K., and Ihm, J., "Wavelets in all-electron density-functional calculations," *Physical Review B*, July 1999, 60(3), pp. 1437-40.

[63] Zhang, H., et al., "Optical implementation of a photorefractive joint transform correlator with wavelet," *Optics Communications*, 2000, 181(4-6), pp. 223-30.

[64] Zhang, H., et al., "Image feature extraction with various wavelet functions in a photorefractive joint transform correlator," *Optics Communications*, Nov. 2000, 185(4-6), pp. 277-84.

[65] Zhang, H., et al., "Rotationally invariant pattern recognition in a photorefractive joint transform correlator using circular harmonic filters and the wavelet transform," *Optics Communications*, Aug. 2001, 195(5-6), pp. 383-91.

[66] Alam, M.S. and Chain, D., "Efficient multiple target recognition using a joint wavelet transform processor," *Optical Engineering*, May 2000, 39(5), pp. 1203-10.

[67] Cohen, L., *Time-Frequency Analysis*, 1995, New York: Prentice Hall, EngleWoods Cliffs, 136-289.

[68] Cohen, L., "Time-frequency distribution--A review," *IEEE Proceedings*, July 1989, 77(7), pp. 941-81.

[69] Cheng, Y.-S. and Chen, H.-C., "Projection-invariant pattern recognition with logarithmic harmonic function and wavelet transform," *Applied Optics*, Sept. 2001, 40(26), pp. 4661-6.

[70] Tan, L., et al., "Filtering theory and application of wavelet optics at the spatial-frequency domain," *Applied Optics*, Jan. 2001, 40(2), pp. 257-60.

[71] Malamud, B.D. and Turcotte, D.L., "Wavelet analyses of Mars polar topography," *Journal of Geophysical Research*, Aug. 2001, 106(E8), pp. 17497-504.

[72] Freeman, P.E., et al., "A wavelet-based algorithm for the spatial analysis of Poisson data," *Astrophysical Journal Supplement Series*, Jan. 2002, 138(1), pp. 185-218.

[73] Hong, J.-C., et al., "Damage detection using the Lipschitz exponent estimated by the wavelet transform: applications to vibration modes of a beam," *International Journal of Solids and Structures*, 2002, 39(7), pp. 1803-16.

[74] Buck, D., Alber, M., and Nusslin, F., "Potential and limitations of the automatic detection of fiducial markers using an amorphous silicon flat-panel imager," *Physics in Medicine and Biology*, 2003, 48(6), pp. 763-74.

[75] Conradsen, K. and Nilsson, G., "Data dependent filters for edge enhancement of Landsat images," *Computer Vision, Graphics, and Image Processing*, 1987, 38(2), pp. 101-21.

[76] Chen, J.S. and Medioni, G., "Detection, localization, and estimation of edges," *IEEE Transactions on Pattern Analysis and Machine Intelligence*, 1989, 11(2), pp. 191-8.
[77] Lu, Y. and Jain, R.C., "Behavior of edges in scale space," *IEEE Transactions on Pattern Analysis and Machine Intelligence*, 1989, 11(4), pp. 337-56.
[78] Shann, R.T. and Oakley, J.P., "Novel approach to boundary finding," *Image and Vision Computing*, 1990, 8(1), pp. 32-6.
[79] Alter-Gartenberg, R., Huck, F.O., and Narayanswamy, R., "Image recovery from edge primitives," *Journal of Optical Society of America A*, 1990, 7(5), pp. 898-911.
[80] Defee, I. and Neuvo, Y., "Median-based zero-crossing edge detectors for closely spaced edges," *CVGIP: Graphical Models and Image Processing*, 1991, 53(2), pp. 196-203.
[81] Feng, J., Lin, W.-C., and Chen, C.-T., "Epicardial boundary detection using fuzzy reasoning," *IEEE Transactions on Medical Imaging*, 1991, 10(2), pp. 187-99.
[82] Zhu, Q.M., et al., "Edge extraction by active defocusing," *Spatial Vision*, 1991, 5(4), pp. 253-67.
[83] Goshtasby, A., "On edge focusing," *Image and Vision Computing*, 1994, 12(4), pp. 247-56.
[84] Goshtasby, A. and Shyu, H.-L., "Edge detection by curve fitting," *Image and Vision Computing*, 1995, 13(3), pp. 169-77.
[85] Mehrotra, R. and Zhan, S., "A computational approach to zero-crossing-based two-dimensional edge detection," *Graphical Models and Image Processing*, 1996, 58(1), pp. 1-17.
[86] Van Vliet, L.J. and Verbeek, P.W., "Edge localization by MoG filters: multiple-of-Gaussians," *Pattern Recognition Letters*, 1994, 15(5), pp. 485-96.
[87] Masoud, A.A. and Bayoumi, M.M., "Using local structure for the reliable removal of noise from the output of the LoG edge detector," *IEEE Transactions on Systems, Man, and Cybernetics*, 1995, 25(2), pp. 328-37.
[88] Godbole, S. and Amin, A., "Mathematical morphology for edge and overlap detection for medical images," *Real-Time Imaging*, 1995, 1(3), pp. 191-201.
[89] Qian, R.J. and Huang, T.S., "Optial edge detection in two-dimensional images," *IEEE Transactions on Image Processing*, 1996, 5(7), pp. 1215-20.
[90] Cortelazzo, G., Della Guistina, D., and Mian, G.A., "Frequency domain design of FIR and IIR Laplacian of Gaussian filters for edge detection," *Circuits, Systems, and Signal Processing*, 1996, 15(5), pp. 555-72.
[91] Shen, B. and Sethi, I.K., "Convolution-based edge detection for image/video in blcok DCT domain," *Journal of Visual Communication and Image Representation*, 1996, 7(4), pp. 411-23.
[92] Kennedy, L.M. and Basu, M., "A Gaussian derivative operator for authentic edge detection and accurate edge localization," *International Journal of Pattern Recognition and Artificial Intelligence*, 1999, 13(3), pp. 367-80.
[93] Gunn, S.R., "On the discrete representation of the Laplacian of Gaussian," *Pattern Recognition*, 1999, 32(8), pp. 1463-72.
[94] Di Sciasio, E. and Guaragnella, C., "Simple nonlinear dual-window operator for edge detection," *Optical Engineering*, 2002, 41(9), pp. 2181-5.
[95] Zhang, L. and Bao, P., "Edge detection by scale multiplication in wavelet domain," *Pattern Recognition Letters*, 2002, 23(14), pp. 1771-84.

[96] Ranganathan, N. and Shah, M., "A VLSI architecture for computing scale space," *Computer Vision, Graphics, and Image Processing*, 1988, 43(2), pp. 178-204.

[97] Jordan III, J.R. and Bovik, A.C., "Computational stereo vision using color," *IEEE Control Systems Magazine*, 1988, 8(3), pp. 31-6.

[98] Kaneko, T., "Disparity detection from trinocular stereo images," *The Transactions of the IEICE*, 1989, E72(1), pp. 13-15.

[99] Vernon, D. and Tistarelli, M., "Using camera motion to estimate range for robotic parts manipulation," *IEEE Transactions on Robotics and Automation*, 1990, 6(5), pp. 509-21.

[100] Raman, S.V., Sarkar, S., and Boyer, K.L., "Tissue boundary refinement in magnetic resonance images using contour-based scale space matching," *IEEE Transactions on Medical Imaging*, 1991, 10(2), pp. 109-21.

[101] Boyer, K.L., Sotak Jr, G.E., and Schenk, A.F., "Structural stereopsis: potential for automatic stereo camera calibration," *Optical Engineering*, 1991, 30(3), pp. 288-99.

[102] Wuescher, D.M. and Boyer, K.L., "Robust contour decomposition using a constant cuvature criterion," *IEEE Transactions on Pattern Analysis and Machine Intelligence*, 1991, 13(1), pp. 41-51.

[103] Yagi, T., Ohshima, S., and Funahashi, Y., "The role of retinal bipolar cell in early vision: an implication with analogue networks and regularization theory," *Biological Cybernetics*, 1997, 77(3), pp. 163-71.

[104] Chhabra, A.K., "Real-time computation of optical flow along contours of significant intensity change," *Real-Time Imaging*, 1997, 3(2), pp. 87-99.

[105] Lai, S.-H., "Computation of optical flow under non-uniform brightness variations," *Pattern Recognition Letters*, 2004, 25(8), pp. 885-92.

[106] Raj, M., Dunn, S.M., and Cannon, W.R., "Measurement of particle orientation in tape cast," *Journal of Computer-Assisted Microscopy*, 1998, 10(1), pp. 33-51.

[107] Girouard, L.M., et al., "Automatic setup deviation measurement with electronic portal images for pelvic fields," *Medical Physics*, 1998, 25(7), pp. 1180-5.

[108] Jarvis, R.M. and Macalpine, G.M., "Possible high-redshift, low-luminosity active galactic nuclei in the Hubble deep field," *The Astronomical Journal*, 1998, 116(6), pp. 2624-31.

[109] Dai, X. and Khorram, S., "A feature-based image registration algorithm using improved Chain-Code representation combined with invariant moments," *IEEE Transactions on Geoscience and Remote Sensing*, 1999, 37(5), pp. 2351-62.

[110] Souza, A.D.A., Ruiz, E.E.S., and Cruz, A.A.V., "Palpebral fissure morphology segmentation and measurement using image processing," *IEEE Engineering in Medicine and Biology*, 2000, 19(1), pp. 114-9.

[111] Gakkai, D.J.T. and Gakkai, D.T., "Underground pipe signal extraction using LoG filter," *IEICE Transactions on Information and Systems*, 2000, E83-D(1), pp. 112-5.

[112] Banarse, D.S., France, I., and Duller, A.W.G., "Analysis and application of a self-organising image recognition neural network," *Advances in Engineering Software*, 2000, 31(12), pp. 937-44.

[113] Chen, Y.-T., et al., "Feature curves based cephalometric superimposition," *Journal of Medical and Biological Engineering*, 2000, 20(4), pp. 211-8.

[114] Lu, T., et al., "Mountain areas extraction by introduction of wavelet transform and its application to ridge lines detection," *The Journal of the IIEEJ*, 2001, 30(3), pp. 293-7.

[115] Shingo, A., Satoko, I., and Masato, N., "Automation of local visibility improvement for color digital image," *Transactions of the Institute of Electrical Engineers of Japan, Part C*, 2002, 122-C(10), pp. 1762-71.
[116] Cheng, H.D., Wang, J., and Shi, X., "Microcalcification detection using fuzzy logic and scale space approaches," *Pattern Recognition*, 2004, 37(2), pp. 363-75.
[117] Van Ouwerkerk, H.J., et al., "Instabilities in a continuous medium model for the retina," *Biological Cybernetics*, 1980, 39(1), pp. 11-14.
[118] Keijzers, C.P., et al., "The nature of the Jahn-Teller effect in the copper-doped Hexaimidazole Zinc Dichloride Tetrahydrate complex," *Journal of Magnetic Resonance*, 1983, 52(2), pp. 211-20.
[119] Riley, M.J., Hitchman, M.A., and Reinen, D., "Effects of vibronic coupling on the EPR spectra of copper(II) doped K_2ZnF_4," *Chemical Physics*, Feb. 1986, 102(1-2), pp. 11-28.
[120] Keller, J., "A two step approach to high T_c superconductivity," *Revista Mexicana de Faisica*, July 1988, 34(3), pp. 415-9.
[121] Riley, M.J., et al., "Temperature dependence of the EPR spectra of the centers $Cu(H_2O)_2Cl_4^{2-}$, $Cu(NH_3)_2Cl_4^{2-}$, and $Cu(H_2O)(NH_3)_2Cl_4^{2-}$) in Cu^{2+}-doped NH_4Cl: An interpretation using a model of dynamic vibronic coupling," *Inorganic Chemistry*, 1988, 27(11), pp. 1924-34.
[122] Nakano, M., et al., "Optical potentials for alpha particles on Ca isotopes at 1.37 GeV," *Physical Review C*, Sept. 1989, 40(3), pp. 1323-34.
[123] Siddhanta, A.K. and Bhattacharyya, B.D., ""Mexican-Hat" parameters of a non-heme protein," *Journal of Molecular Structure (Theochem)*, 1989, 184(1-2), pp. 39-47.
[124] Monasson, R., "Properties of neural networks storing spatially correlated patterns," *J. Phys. A: Math. Gen.*, 1992, 25(13), pp. 3701-20.
[125] Lyth, D.H., "Axions and inflation: Vacuum fluctuations," *Physical Review D*, May 1992, 45(10), pp. 3394-3404.
[126] Lyth, D.H. and Stewart, E.D., "Axions and inflation: String formation during inflation," *Physical Review D*, July 1992, 46(2), pp. 532-8.
[127] DeTar, C. and Dong, S.-J., "Toward an effective chiral model of high-temperature QCD," *Physical Review D*, June 1992, 45(12), pp. 4681-94.
[128] Eiermann, H. and Wagner, M., "Exotic states in the E-e Jahn-Teller problem," *Journal of Chemical Physics*, Mar. 1992, 96(6), pp. 4509-17.
[129] Reggia, J.A., et al., "A competitive distribution theory of neocortical dynamics," *Neural Computation*, May 1992, 4(3), pp. 287-317.
[130] Stetter, M., Lang, E.W., and Muller, A., "Emergence of orientation selective simple cells simulated in deterministic and stochastic neural networks," *Biological Cybernetics*, 1993, 68(5), pp. 465-76.
[131] Worgotter, F. and Niebur, E., "Cortical column design: a link between the maps of preferred orientation and orientation tuning strength?," *Biological Cybernetics*, 1993, 70(1), pp. 1-13.
[132] Barrow, J.D., "Exact inflationary universes with potential minima," *Physical Review D*, Mar. 1994, 49(6), pp. 3055-8.
[133] Soodak, R.E., "Simulation of visual cortex development under lid-suture conditions: enhancement of response specificity by a reverse-Hebb rule in the absence of spatially patterned input," *Biological Cybernetics*, 1994, 70(4), pp. 303-9.

[134] Bachas, C. and Tomaras, T.N., "Ribbons around Mexican hats," *Nuclear Physics B*, 1994, B428(1-2), pp. 209-20.

[135] Roberts, D., Liddle, A.R., and Lyth, D.H., "False vacuum inflation with a quartic potential," *Physical Review D*, April 1995, 51(8), pp. 4122-8.

[136] Ivanov, A.L. and Haug, H., "Radiative renormalizations for excitonic molecules," *Physical Review Letters*, Jan. 1995, 74(3), pp. 438-41.

[137] Levi, R., et al., "Frequency-spatial transformation: a proposal for parsimonious intra-cortical communication," *International Journal of Neural Systems*, Nov. 1996, 7(5), pp. 591-8.

[138] Mile, B., Sillman, P.D., and Yacob, A.R., "The first ENDOR spectrum of naked metal clusters: $^{7}Li_3$, ^{6}Li, $^{7}Li_2$ and $^{7}Li_3\ (H_2O)_x$ in an adamantane matrix," *Chemical Physics Letters*, Dec. 1996, 263(5), pp. 710-4.

[139] Theodorakis, S., "Analytic vortex solutions in an unusual Mexican-hat potential," *Physical Review E*, Oct. 1997, 56(4), pp. 4809-12.

[140] Anglin, J., "Cold, dilute, trapped bosons as an open quantum system," *Physical Review Letters*, July 1997, 79(1), pp. 6-9.

[141] Fiolhais, M., et al., "Soliton formation in sigma models," *Physical Review C*, Dec. 1997, 56(6), pp. 3311-9.

[142] Ringach, D.L., "Tuning of orientation detectors in human vision," *Vision Research*, 1998, 38(7), pp. 963-72.

[143] Lin, J.K., et al., "Irregular synchronous activity in stochastically-coupled networks of integrate-and-fire neurons," *Network-Computation in Neural Systems*, 1998, 9(3), pp. 333-44.

[144] Selz, K.A., Mandell, A.J., and Shlesinger, M.F., "Hydrophobic free energy eigenfunctions of pore, channel, and transporter proteins contain beta-burst patterns," *Biophysical Journal*, Nov. 1998, 75(5), pp. 2332-42.

[145] Colin, J., Grilhe, J., and Junqua, N., "Transformation of a cubic precipitate to a butterfly shape due to localized instabilities," *Acta Mater.*, 1998, 46(4), pp. 1249-55.

[146] Popovic, Z. and Satpathy, S., "Cooperative Jahn-Teller coupling in the manganites," *Physical Review Letters*, Feb. 2000, 84(7), pp. 1603-6.

[147] Durand, G. and Spiegelmann, F., "Molecular picture of excited states and fragmentation paths of the Na/sub 5/F/sub 4/cluster," *European Physical Journal D*, Feb. 2001, 13(2), pp. 237-43.

[148] Georgiev, M., et al., "Off-center Li^{+} ions: solid state rotators described by Mathieu's nonlinear-oscillator equation," *The European Physical Journal B*, Sept. 2002, 29(2), pp. 273-7.

[149] Bjorken, J.D., Pakvasa, S., and Tuan, S.F., "Yet another extension of the standard model: Oases in the desert?," *Physical Review D*, Sept. 2002, 66(5), pp. 053008/1-15.

[150] Yildirim, T. and Harris, A.B., "Quantum dynamics of a hydrogen molecule confined in a cylindrical potential," *Physical Review B*, 2003, 67(24), pp. 245413/1-15.

In: New Signal Processing Research
Editor: Takumi Maeda, pp. 55-87
ISBN 978-1-60456-479-2

Chapter 2

WIRELESS SENSOR NETWORKS: A TESTBED FOR EMERGING MILESTONES IN COMMUNICATION THEORY

Stefano Marano [*] ***and Vincenzo Matta*** [†]
Department of Information and Electrical Engineering (DIIIE)
University of Salerno, via Don Melillo, 84084, Fisciano (SA), Italy

Abstract

The rather recent advent of wireless sensor networks (WSNs) has posed new challenges to the scientific community. The perspective taken in this chapter is that WSNs may provide an appropriate testbed for some new (or at least less classical) basic principles that are emerging, often just under the driving force of this applicative scenario. Accordingly, we select some of the emerging paradigms and discuss their impact on the analysis and design of WSNs. A few case studies are provided, which are borrowed form the authors' recent works on the subject.

1. Prologue

Scientists and researchers of scientific disciplines undoubtedly enjoy some remarkable advantages with respect to many people employed in other professions. The most important is that they usually love their work, and perhaps second in order of relevance is that their work may be surprisingly easy. Indeed, good research is the result of a quite simple recipe: Let you mind run away unbounded, provided that your reasoning does not conflict with the few basic principles of your discipline. It is not an accident that in many successful research teams there exists a strong interaction between a certain number of young researchers and one, or more, team leaders. The latter exercise a centripetal force in the attempt of anchoring the scientific approach to some well-known prime principles, while the former instinctively apply a centrifugal force that leads to unexplored landscapes. The equilibrium point provides scientific advances.

[*] E-mail address: marano@unisa.it
[†] E-mail address: vmatta@unisa.it

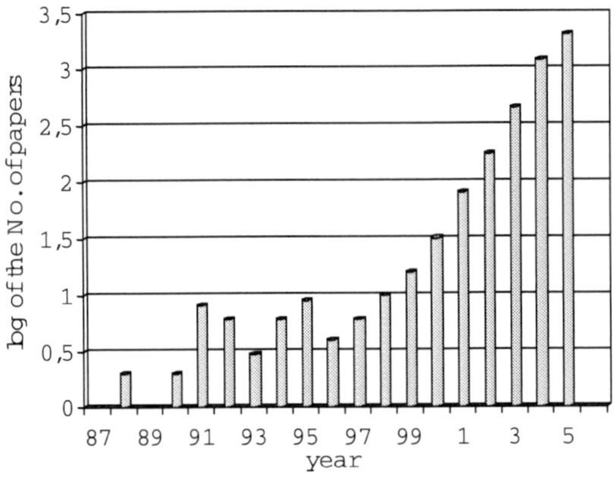

Figure 1. The exponential growth of the number of technical papers related to *sensor networks*, over an interval of ten years, as results from querying the site www.ieeexplore.ieee.org.

The attitude to explore new landscapes is the quintessential of the scientific investigation and can be preserved only by instilling a solid scientific culture. On the other hand, prime principles are not immutable. As the human knowledge evolves, new phenomena are investigated or known phenomena are investigated from a different perspective. Sometimes this is related to the emergence of *new* prime principles. An abused example of such evolution is the advent of quantum mechanics in which the tools and the principles of probability theory became the fundamental guidelines for investigating the behavior of atomic particles. Einstein's reluctance in accepting such an a uncertain (probabilistic) description of the nature is the proof of the difficulties of (finding and) digesting new basic principles. Nonetheless, this difficult process is vital for the success of the scientific investigation.

The above provoking statement about the ease of the scientific investigation can now be rephrased as follows. Research progresses are easy if an overwhelming creativity and curiosity is guided by the *most suitable* basic principles of that discipline.

In this chapter we explore some emerging milestones (*i.e.*, prime principles) that are assuming increasing relevance in designing and analyzing wireless sensor networks (WSNs). The organization is as follows. Section 2. presents the main themes of this chapter and provides a general overview. The selected topics are then reconsidered more in depth in separate subsections of Sect. 3., while examples of applications of the general principles are provided in Sect. 4.. We do not encourage the reader to go directly to the conclusions in Sect. 5. since, in fact, no conclusions are reported.

2. Introduction

A couple of years ago we queried a popular WEB search engine with the keyword "sensor networks" and something like one million of documents were found. To restrict the field, we then limited the search only to technical documents such as those collected in the site of the *IEEE* – `www.ieeexplore.ieee.org` – and approximately 6000 documents were found with the same query. What is really impressive is the exponential growth of the number of documents related to the sensor networks, in the years from 1996 to 2006, as depicted in Fig. 1. This should be enough to believe that, in effect, a strong trend from the classical centralized systems to distributed architectures made of a multitude of simple entities was, and currently still is, really in force [1–7]. It is not a case that sensor networks are generally credited as a key technology for the 21st century. Such technology has driven a renewed interest in fields like collaborative signal and information processing, distributed information fusion, multiuser information theory, cross-layer system design, decentralized inference, network security, and so on. Indeed, the analysis and design of WSNs require methodologies and tools from different fields, and the forthcoming discussion focuses on two such fields: communications and statistics. In both of them, and especially in their interplay, WSNs are experiencing the emergence of new reference principles and milestones.

A WSN is typically made of a large number of elementary units – sensors or nodes – deployed over a certain area to be surveyed. These nodes are typically miniaturized, cheap, lightweight devices, powered by on-board batteries of quite limited capacity [8–14]. A classical system architecture prescribes that the nodes sense the environment and collect relevant data to be delivered (in single or multi-hop fashions) to a common fusion center (FC) whose role is that of data fusion. Not uncommon, however, are some alternative architectures such as, for instance, the *fully decentralized* scheme, whose peculiarity is the absence of a central unit like the FC. The nodes of a fully decentralized WSN exchange data with their neighbors with the typical goal of cooperatively reaching agreement, or consensus, about a common value [15–18].

Historically, the development of WSNs was motivated by military applications, such as battlefield surveillance, target tracking, detection of intrusions, contaminants, radiations, and so forth. However, the advent of WSNs is expected to have a quite pervasive impact also in our every-day life, and in fact many civilian applications can be conceived, including environment and habitat monitoring, healthcare applications, home automation, traffic control, video surveillance, telemedicine, industrial and manufacturing automation, distributed robotics, physical security, building and structures monitoring, and the list could continue.

The design of WSNs obeys some generally accepted rules. Three such rules are of particular relevance to us and, reduced to their simplest and basic forms, can summarized in the following *postulates*:

1. Sensor networks are application specific.

2. Design metrics and criteria can be unconventional.

3. Sensor networks are severely energy constrained.

In the first, the peculiar nature of each network is emphasized. This may represents a marked difference from, for instance, classical digital communication systems that are designed regardless of the nature of the information source and, in this sense, are application independent. The second postulate is closely connected to the main theme of this chapter since it reveals that WSN design may be related to new prime principles. To elaborate on the third, we refer to [19]. One expression of the many Moore's Laws tells us that DSP devices require power that drops by a factor of 1.6 per year. On the other hand, Shannon's theory dictates how much SNR (signal-to-noise ratio) is needed at the receiver (*e.g.*, the fusion center) to reliably recover the delivered messages (*e.g.*, that from the nodes), and Maxwell's theory rules the propagation losses incurred by the electromagnetic signal. Since these two theories do not modify with time, the bottleneck is the energy that the sensors spend for communications, more than that consumed in the local computation [19]. Therefore, the last postulate can be modified as follows.

3′. Sensor networks are severely constrained in terms of *communication* energy.

As evidenced by the above list of WSNs applications, the typical task of a sensor network is that of providing some form of *inference*. Thus, we are interested in an *inference* task, subject to *communication* constraints. In addition, it should be stressed that the sensor network is, by its own nature, a multiterminal environment where many senders deliver data to one (or more) fusion center. Roughly speaking, these two last observations are the essential reasons why WSNs are a suitable testbed for the new paradigms emerging on the evanescent border between communication and statistics.

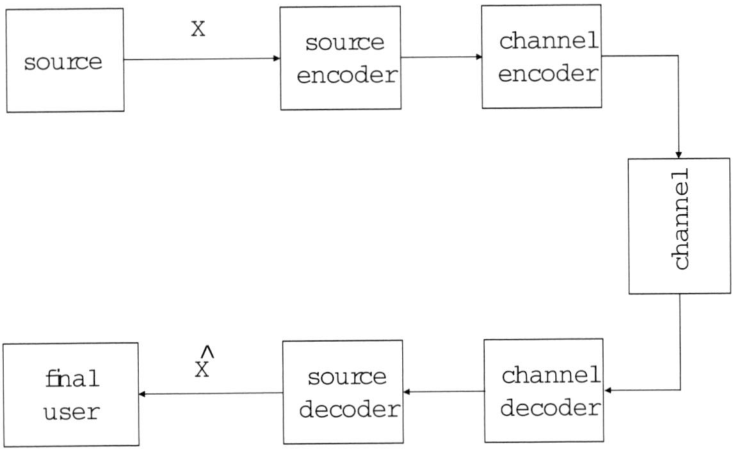

Figure 2. The classic block scheme of a digital communication system.

To elaborate, let us refer to the point-to-point (*i.e.*, single-terminal) block scheme in Figure 2. Virtually all textbooks of digital communications present a similar conceptual scheme, so that the meaning of each block should be clear. For our purposes, to the same level of abstraction, a WSNs can be modeled by the block scheme in Figure 3. Here the goal is *not* to recover the source data as reliably as possible, but rather to make inference, from the data available to the receiver, about a parameter that rules the sensors' observations. We are moving form the classical *reconstruction* scenario to an inference scenario over a communication system. This requires a joint perspective (and a cross fertilization) between

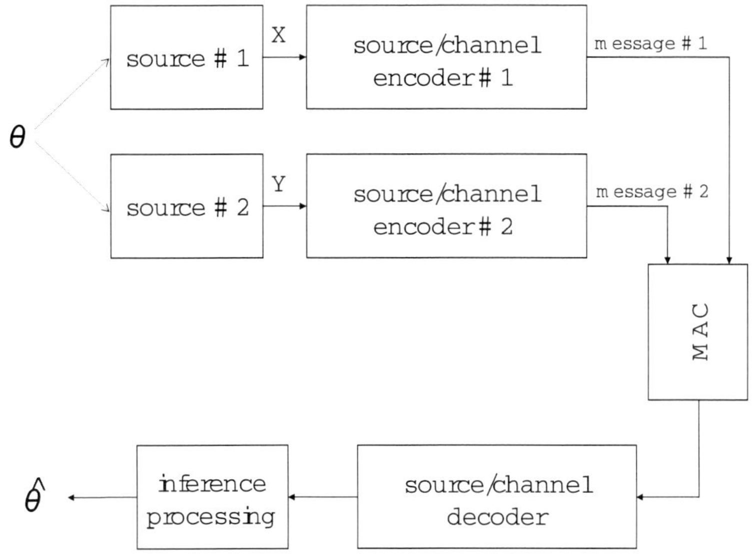

Figure 3. The block scheme relevant to WSNs. With respect to the scheme in Figure 2 two differences are easily noted: (i) Here we have many senders (two in the figure) and one receiver, and (ii) the goal is not to recover the observations (X, Y) but to infer the state of the nature θ that rules their distribution $p_\theta(x, y)$. MAC stems for Multiple Access Channel.

communications and statistics, namely, broadly speaking, this asks for a cross-layer approach[1]. This is the first basic fact. The second basic fact is the multiterminal nature of the system: There are many senders (two, in the figure) and one receiver. The impact of these two modifications to the well known and understood scheme in Figure 2 may be dramatic and unexpected.

For example, a little thought reveals that using known techniques about the data reconstruction may be wasteful in the scheme of Figure 3, as now the actual matter is still to compress, but with the goal of preserving not the accuracy of the observations (X, Y), but that of an embedded parameter θ [21]. Furthermore, it is well known that in multiterminal scenarios, Shannon's separation principle usually fails to be true [22]. As a consequence, the bit may fail to be the universal *measure unit* of sources and channels [23], and a joint perspective of these two entities is needed [24]. To complicate further the matter, one should recall that, as our second postulate implies, well-known practical design recipes for a classical communication scheme may fail to work with WSNs. For instance, complicated channel codes may be impossible to be implemented in a sensor network.

Furthermore, let us consider the classic concept of channel capacity. It describes how

[1]To avoid a possible misunderstanding, we would like to stress that the term "cross-layer" is used in the literature with different meanings, the emerging school of thought being, perhaps, the viewpoint taken in [20]. In this chapter, however, the term cross-layer is used in a much broader sense: We simply refer to the fact that a system design/analysis requires *joint* perspectives of seemingly different fields.

many bits of information can be reliably transferred from the input to the output, per single channel use. An underlying key assumption is that the *time* (or degrees of freedom) available for the communication plays a central role, since it is rather obvious that if an unbounded amount of time is available to transmit any single input symbol, such a delivering can be made as reliable as desired (for instance, by transmitting an unbounded number of times that symbol). Now, in WSNs the point is that the time constraint may be not so relevant and the concept of capacity should be accordingly modified to face with other constraints. Otherwise stated, in many instances, the transmitting nodes are supplied with quite limited energy and this leads us to explore alternative concepts of the channel capacity, such as the capacity per unit cost (see [25]), where this latter is just the energy spent for delivering the input symbols.

Finally, it is important to stress that, due to their own nature, WSNs are particularly vulnerable to many kind of attacks, such as unauthorized intrusions, message interception, node impairments, eavesdroppers, and so forth. The general issue of *security* is certainly a key topic in WSN applications.

The design and analysis of WSNs involve all the above issues and, in this sense, WSNs seem a natural testbed for the associated paradigms, formalizations, and tools. The following section provides a more in-depth overview of these topics. Then, in the section devoted to the examples, we touch upon our research activity of the last few years, which is presented here as case studies. It should be understood, and we want to make this point very explicit, that we have no pretence of generality or exhaustiveness or completeness, since the selected examples simply mirror the authors' (biased) lines of investigations. In any case, entry points to the general literature are also provided.

3. Selected Paradigms for WSNs

3.1. Inference versus Data Reconstruction *(Recovering the Messages may be Wasteful)*

In 1979 the newborn multiterminal inference theory moved its first steps in the scientific community. It was Toby Berger who introduced it at the *IEEE Shannon Theory Workshop* held in New York [26]. As perfectly elucidated in [27], *"This was the first attempt to combine two seemingly different kinds of problems which had been separately investigated in the void between statistics and information theory."* The basic scenario of multiterminal estimation/detection problems can be again summarized as in Figure 3 where we focus, for the time being, on *source* coding aspects (that is, there is an ideal channel whose output allows us to recover exactly all its inputs). We are interested in making inference about a certain phenomenon, say θ, which can be monitored by several spatially-separate sensors (in Figure 3 we consider two observers, for the sake of clarity, and denote them as "sources"). Sensors collect their observations X and Y, which are statistically related to θ via the distribution $p_\theta(x, y)$, and then send their data toward a fusion center, which is demanded to produce $\widehat{\theta}$, the final inference about θ. In the absence of any constraint on the transmission rate, the problem would be definitely trivial, being solvable by standard tools from statistics. However, intriguing (and also realistic) problems arise when meaningful constraints come into scene. In this case there exists a tradeoff between the rate at which the information may travel across the channel, and the goodness of the inference produced at the fusion center.

The question can be regarded as an extension of the classical rate/distortion tradeoff [28]. There are, however, some complications which make the issue rather tricky: What is the information truly relevant to estimation/detection purposes? What about the role played by the statistical correlation among sensors?

There are two widely accepted paradigms for dealing with multiterminal inference in a meaningful way. The first one was pioneered by Zhang and Berger [29], and successively considered by Han and Amari [27]. They consider a *deterministic* parameter θ, with n independent pair of observations $\{(X_i, Y_i)\}_{i=1}^n$, and with the interpretation that the sequence $\{X_i\}_{i=1}^n$ is taken at the first sensor and $\{Y_i\}_{i=1}^n$ at the second. The observations are drawn from[2] $\prod_{i=1}^{n} p_\theta(x_i, y_i)$, and the final inference about θ is accomplished by the fusion center, which receives finite-rate encoded versions of the observations.

The second paradigm, which can be regarded as the Bayesian counterpart of the former, was initially adopted, among others, by Yamamoto [30], Flynn and Gray, [31], and recently reconsidered by Berger [32]. Here we have n successive (independent) realizations $(\theta_1, \theta_2, \ldots, \theta_n)$ of a *random* parameter θ, which has to be estimated from the encoded versions of the sensors' measurements $\{(X_i, Y_i)\}_{i=1}^n$. It is worth noting that, in this case, the k-th pair of observations (X_k, Y_k) contains information *only* about the corresponding k-th realization of the random source.

The common question is: How can the local data be efficiently compressed, *in view of* the estimation task carried out at the fusion center? At first glance, one is encouraged to tackle the general multiterminal problem with the standard compression tools available for the point-to-point scenario, valid even for inference problems, see, *e.g.*, [33, 34]. Unfortunately, the decentralized structure of the network (enforcing the constraint of no data-sharing among the remote sensors) makes the matter harder. The basic difference between single and multiterminal environments is that, in the latter, it is not possible to construct the optimal estimator $\widehat{\theta}(X, Y)$ at the remote sensors.

A related issue is that of finding the optimal rate/distortion region for multiterminal estimation, a topic for which there exists an abundant literature, and we have no pretence at all to review the subject. One of the most recent advances, to whom we refer the interested reader, can be found in [35]. Of specific interest is the CEO problem which will play a role in the following discussion. The CEO (Chief Executive Officer) problem was introduced by Berger, Zhang and Viswanathan in [32], and deals with a distributed estimation system where a team of spatially-separated agents observe conditionally independent versions of the same source, and have to communicate with a fusion center (the CEO) for estimation purposes. The agents cannot convene, and this is a distributed inference problem. In its earlier version, the CEO was solved for an infinite number of agents [32], and for discrete alphabet random variables. The analogous (infinite agents) for quadratic distortion and Gaussian variables has been fully characterized by Oohama in [36]. The *exact* sum-rate/distortion bound, for a quadratic Gaussian CEO, in the case of a finite number of agents, has been only recently discovered in [37].

Finding the ultimate limits of the rate/distortion region is challenging, and there is no

[2]Obviously, correlation is admitted among sensors' observations. Moreover, the solution for the no-correlation case is trivial, and can be shown to entail a zero-rate compression scheme, see [27].

doubt that this is of paramount importance for our distributed systems. However, the spreading of applications in sensor networks also created a strong need of practical recipes for designing good distributed estimation systems. The most important existing techniques can be categorized in two big families: (i) cyclic optimization; (ii) Slepian-Wolf encoding. The cyclic optimization procedure, first introduced in [31], stems from the celebrated Lloyd & Max algorithm for designing optimal quantizers [38, 39]. The approach amounts to first select an appropriate metric for inference purposes as the objective function of the algorithm. Then, a classical Lloyd & Max procedure is applied cyclically to each sensor's quantizer, the other sensors' codebooks being fixed. In this way, at each step, the quantizers' regions are updated, and the algorithms can be shown to converge. The concept of [31] was generalized to hypothesis testing, and reprised in [40], where scalar quantizers with entropy coded output are employed. In [40], the problem of selecting a suitable decision metric to be used by the algorithm is addressed, and several solutions are applied in connection to different applicative scenarios.

In [41], the cyclic procedure borrowed from [31, 40] is considered in the context of parameter estimation under the Bayesian paradigm, as well as under the Fisher one. It is shown how different distortion metrics can lead to different computational problems, depending upon the joint distribution of the observations and/or the source.

Another research field which plays a fundamental role in the context of distributed source coding, is that of *practical* Slepian-Wolf coding. As pointed out by Verdú [42], "*Despite the existence of potential applications, the conceptual importance of SlepianWolf coding has not been mirrored in practical data compression.*" Perhaps stimulated by this sentence, quite recently there have been enormous advances in the field of distributed source coding, which are briefly summarized below.

In [43], the idea of applying channel coding techniques for distributed compression was first proposed. As clearly elucidated in [5], it is possible to consider a fictitious channel connecting the observations X and Y, such that standard channel coding techniques can be used to the aim of distributed compression. Authors in [43] propose a novel algorithm named DISCUS (Distributed Source Coding Using Syndromes), which is the first effective attempt to approach the envisioned Slepian-Wolf bounds. In [44, 45], LDPC (Low Density Parity Check) codes are used as a way to approach the distributed compression issue. We remark on passing the curious story of LDPC codes: They were invented by Gallager in 1962 [46, 47], and only recently have been rediscovered by the channel coding community as being capacity-achieving [48, 49].

3.2. Channel Capacity per Unit Cost *(Time May Not Be an Issue)*

In the previous section we implicitly focus on a source-only perspective. Accordingly, the communication constraint was set in terms of bit-rate, thus dealing with ideally noiseless bit-pipes. The other side of the story is that one is usually faced with the problem of noisy channels, for which physical constraints in terms of, *e.g.*, the energy of the signals entering the channel, are usually imposed. These constraints, along with the effect of noise, give rise to limitations in terms of the amount of information which can be reliably recovered at the destination. Shannon's famous theorem on channel capacity provides us with a simple measure, *i.e.*, that single number called channel capacity C, which gives the achievable upper bound on the number of *bits per channel use* which can be reliably transferred through the channel.

However, in its basic form, the channel capacity theorem is formulated for the classic point-to-point scenario and refers to a standard communication model where the focus in on the number of channel uses (*i.e.*, time). How does the perspective change when dealing with WSNs? First, it is well-known that the definition of capacity can be extended to consider multiuser scenarios such as those of WSNn (see classical textbooks like [22] and therein references). Furthermore, a large amount of literature deals with *alternative* definitions of capacity that better account for the peculiarities of the sensor network where, for instance, the energy consumption can be a more critical issue with respect to the number of channel uses. Examples of alternative definitions include the throughput and transport capacity [50], or the bits-per-Joule capacity that represents the maximum number of bits that the network can deliver per Joule of energy [51]. Actually, investigations in this direction (*i.e.*, when energy is a more severe constraint than time) are certainly not new and can be traced back to [52] and then to [53]. However, credit is due to Verdú [25] (who elaborates on [54]) for popularizing the concept of channel capacity per unit cost, where the relevant *rate* of transmission R is defined as the ratio between the logarithm of the number of messages and the cost to be spent for sending the associated codewords (in the standard definition the ratio is with respect to the length of the codewords, namely with respect to the number of channel uses for sending the message).

Among the many findings in [25], Verdú shows that, notably, when the channel has a zero-cost input symbol (think, for instance, to the no-signal symbol in an on/off modulation scheme), the capacity per unit cost is obtained by working at an arbitrarily low cost. This may have a strong impact on a WSN design. Indeed, the traditional network resource allocation focuses on maximizing the speed at which the information can flow through the channel. New guidelines in the WSN design, instead, may push the system in the direction of achieving the capacity per unit cost. Although it may seems that this process represents a tradeoff between fast communications and energy efficient communications, it is actually true that the energy efficiency is not necessarily paid in the coin of slow communication, as we learn by reading the first few paragraphs of [25].

Before ending this section, let us mention that the above viewpoint about the channel capacity can be generalized to other metrics that traditionally measure the performance of a system with respect to some time-like index (see also the final section in [25]). In this respect, in Sect. 4.2., we discuss an instance of such perspective, when the problem is a detection one, and the relevant metric is the Kullback-Leibler divergence. To briefly anticipate the setup, in Sect. 4.2. we consider independent and identically distributed observations, so that the Kullback-Leibler divergence is additive, and one can consider the divergence corresponding to a single observation (sensor) as the best achievable error exponent, or, put another way, a detection metric *per single sensor*. It makes sense, in our WSN environment, to judge the efficiency of the detection scheme with respect to constraints other than the number of sensors. We accordingly shall work in terms of *divergence per unit cost*, where the cost will be suitably chosen for the energy-constrained networks that we are interested in.

3.3. *(Un)*Separation Principle

Up to now, we considered the problem of conveying information toward a final destination in a genuinely distributed system like a WSN definitely is. Basically, two roads were traced. On the one hand, one has to judiciously select the quality and amount of information to be transferred, depending on the final performance metric, which is in turn related to the specific application that the WSN is deployed for (estimation, detection, and so on). On the other hand, information delivering is not for free, and a suitable metric for the cost is needed (classical rate, capacity per unit cost, transmission/activation probability, and so forth).

This evident bifurcation of the system optimization into two separate stages, however, has been implicitly assumed. We have not claimed (and we cannot) that a layered design perspective would be *globally* optimal. Were we too naive/optimistic in reasoning as before? Are there some relevant applications wherein this separate optimization is optimal? Is a separate approach totally useless?

First of all, the reason for posing the question in such a way is probably a heritage of well-celebrated results of information theory. In fact, consider for a while the " $(E = mc^2)$-equivalent" result in information theory, that is, reliable information transfer of a source over a noisy channel is possible if and only if

$$H < C, \tag{1}$$

where H is the source entropy, and C the channel capacity. This is known as the "*source-channel coding*" , or "*binary interface*", or "*separation*" theorem, and basically joins the well-known results by Shannon about source and channel coding. The first name given to the theorem is quite natural, and perhaps deserves no further comments.

The name "*binary interface*" tells something more. As a matter of fact, the theorem states that one can know whether reliable transmission is possible by simply comparing two numbers, the first, H, characterizing the quantity of information conveyed by a source, and the other, C, measuring the quantity of information that a channel can reliably convey. The key point is that the measure unit for H and C is the *bit*: Conceptually, the binary-digit (digital) era is simply a consequence of the binary interface theorem.

As to the third name, "*separation theorem*", this due to the fact that the proof of the theorem actually embodies the following stronger property: If reliable transmission is possible, there always exists a way to get it by *separately* designing the source and channel coding stages.

The above considerations explain why an ideal bifurcation took place since the milestone Shannon's theorem was introduced, and why two markedly different research communities (source coding and channel coding) were born.

Let us go one step ahead. The classic guidelines and milestones described above hold true for the single-terminal, point-to-point scenario, *and* when the focus is on reconstruction. What about our WSN environment? In fact, in general, the separation theorem does *not* hold true in multiterminal settings.

Let us first consider the transmission of a distributed source of information (remote sensors, that cannot talk each other), toward a common fusion center (decoder), through a common multiple access channel (MAC). On the source side, the extension of the Shannon source-coding theorem to the multiterminal case was provided by the (already quoted) milestone Slepian-Wolf theorem [55], which shows that, despite the physical separation of the sources (sensors), reliable source representation is still possible if the sum-rate is $R = R_X + R_Y > H(X, Y)$, with the additional constraint that the single rates cope with $R_X > H(X|Y)$ and $R_Y > H(Y|X)$.

On the channel side, a definition of the MAC capacity was seen to embody some subtleties. Indeed, the channel capacity region of the MAC [22] looks like an easy extension of its single-terminal counterpart. However, a famous counterexample (see [22]) considers a distributed source with a specific correlation structure and a specific kind of MAC such that the entropy-region of the source does not intersect the channel capacity region (that is, the multiterminal equivalent of $H > C$), while a*zero-error* source-recovery is possible. It is possible and also easy: It can be achieved with a fully *uncoded* transmission, namely, by simply putting the uncoded source-output as channel-input. Put another way, in our multiterminal setting, there might be the need of crossing the layers of source and channel coding. This tells that the potential statistical dependence between the channel inputs do not allow, in general, to proceed as in the single-terminal case. Otherwise stated, it is no longer true that a simple comparison between the information content of the distributed source (say, the joint entropy $H(X, Y)$), and a number characterizing the channel (sum-capacity C), is sufficient to infer whether reliable transmission is possible or not.

According to the above arguments, the lack of separation seems to be ascribed to the potential dependence among the remote sensors' observations. However, this is not the only case. Consider, for instance, the following simple example where the focus is not on data reconstruction. Assume that each sensor of the WSN observes a noisy version X_i of a Gaussian random variable θ to be estimated. Under the assumption of additive Gaussian noise N_i, and selecting the MSE metric for judging estimation performances, this is usually referred to as the quadratic Gaussian CEO problem [32, 36, 37]. We have

$$X_i = \theta + N_i. \tag{2}$$

Furthermore, let us assume that the information travels toward the final fusion center through a Gaussian MAC, whose output is

$$Y = \sum_{i=1}^{n} X_i + W, \tag{3}$$

where W is the channel noise term, and n is the number of sensors. A first attempt in designing the system could be that of joining the rate-distortion results available for this case, with the channel capacity results pertaining to the Gaussian MAC. Following the outstanding viewpoint of [56–58], let us look instead to the following mind-provoking scheme: What if we simply transmit the *uncoded* X_i over the MAC, without any coding? Neither source codes nor channel codes are required, therefore no large block sizes are involved; no delay is imposed, and the system complexity is amazingly low.

To answer, let us scale the received signal Y by the number of sensors n. It is easily seen that, for large n, the channel output will converge towards the sample mean of the data, which is, loosely speaking, an optimal estimator of θ. In [58], it is shown how this scheme *exponentially* outperforms the separate approach. How can it be? The explanation lies in the fact that, in the separation philosophy, our system first attempts to reliably recover all the X_is, and then to fuse the reconstructed data for the final estimation task. Oppositely, in the proposed uncoded scheme, it is recognized that we are ultimately interested in a *summary* of the data (their arithmetic mean) and, as a matter of the fact, that the (additive) channel is automatically doing the job for us! Again, in a sense, we are crossing the layers of communications and statistics.

Summarizing, we can definitely say that a cross-layer perspective is expected to be rewarding in WSNs, especially when, as it can be grasped from the above discussion, different layers are somewhat *matched*, and one can capitalize on such matching. This viewpoint is taken in [24, 59], where general results are provided. However, we must also keep in mind that if the separation theorem is not valid in general, the same goes for the cross-layer approach. Indeed, there are a lot of situations where the matching between different layers is either absent or difficult to exploit (and a separate design may be *practical*, *convenient* and, in some case, *optimal*).

3.4. *(Crossing the Layers of)* Security and Inference

The literature on secure networking for general sensor networks is abundant, and we do not attempt to provide a review; the interested reader is referred to [60] and therein references. Also, needless to say, there is quite a lot of literature in decentralized detection, entry-points can be found in *e.g.* [4, 8–12, 61–66]. Oppositely, the literature on distributed detection where security aspects play a prominent role is very scarce, see [67–70]. This is particularly surprisingly since, due to their own nature, the inference-oriented WSNs that we are dealing with, are very vulnerable to many different types of attacks.

Let us start from a single-user scenario and let us focus, for the time being, on a pure communication problem (no inference). From an information theoretical perspective, needless to say, this story begins with a paper by Shannon [71] where a private-key communication is addressed. Somehow disappointingly, it turned out that perfect secrecy would require a secret key whose entropy is at least as large as that of the message itself. Much later, Wyner considered the different setup of reliable communication in the presence of an intruder (the wire-tapper) which listens the messages through a *wire-tap channel* [72]. The intruder perfectly knows the encoding and decoding strategies used for transmissions, and the goal is to achieve the best information rate compatible with a prescribed level of secrecy, which refers to the capability of the intruder of decoding the messages. The approach involves quantities like channel capacity and equivocation; the main result is the optimal region (rate versus equivocation) for secure communication, and the definition of *secrecy capacity* as the maximum rate achievable by the main user, compatible with the maximum of wire-tap confusion (maximum equivocation). Further wire-tap scenarios are addressed in [73].

Csiszár and Körner [74] generalized the secure communication problem to a broadcast scenario where the messages are sent to both the main user and the wire-tapper, but there are confidential messages that must be received only by the main user. Further extensions include [75, 76], which address the case of the relay channel with confidential messages, and [77,78] where secure communications over the MAC with confidential messages is studied. Also relevant are [79], the MIMO secure communication problem addressed in [80], and the cooperative approach in [81]. Fading channels has also been recently considered, see *e.g.*, [82–84].

The communication problem studied in [85] focuses specifically on a sensor network, whose nodes reach agreement (network *orientation*) about a common message to be successively delivered to a mobile agent that travels across the network. This scheme is referred to as Cooperative SEnsor Networks with Mobile Access, or C-SENMA for short. Assuming that a certain fraction of the nodes are misinformed, the aim in [85] is to compute the communication capacity of the system, *i.e.* the supremum of the achievable rates, defined, as usual, as the logarithm of the number of messages in the codebook divided by the length of a codeword. It is shown that the same capacity of the absence of misinformation can be achieved, provided that individual nodes can be polled by the mobile agent. Conversely, if such polling is not allowed, and sensors transmit according to a preassigned scheduling, the reduction of the capacity amounts exactly to the fraction of misinformed sensors. The case of energy constrained network that, as we know, is very relevant to practical applications, is also addressed in the same paper. In [86] further analysis involving LDPC codes is offered. For our purposes it is worth noting that, despite the original C-SENMA formulation considers misinformed nodes, the setup is somewhat equivalent to assuming the presence of an intruder that plagues the network, and that *infects* a fraction of the nodes. The infected sensors are forced to deliver fictitious messages randomly chosen among the possible message set. Accordingly, the approach pursued in [85] is useful to model sensor networks subject to certain specific kind of attacks.

Among the most insidious types of attack, hereafter we concentrate our attention on that called Byzantine. The Byzantine paradigm was introduced in the computer science community by Lamport, Shostak and Pease in [87], see also [88]. Further developments can be found in [89], this latter being focused on information theoretical aspects. More recently, Byzantine models have been used in the context of network security in [90]. The original Byzantine generals problem consists in a clique of traitorous generals that conspire to prevent loyal generals to form consensus on attack or retreat. The classical Byzantine general problem is solvable if and only if less than one third of the generals are traitorous. That is to say, consensus in the presence of a Byzantine attack is possible if and only if less than $1/3$ of the generals are traitorous.

The capacity of a sensor network under Byzantine attack is addressed in [91]. There, it is assumed that the Byzantine sensors have full knowledge of the system and there is no restriction on what they can transmit. Unlike the misinformed scenario of [85], where the fictitious messages are randomly chosen, in [91] the Byzantine sensors are sometimes malicious, but they can also behave as honest some other times, in order not to allow sensor identification (honest/Byzantine) by the fusion center.

As said at the beginning of this section, the literature on security in distributed detection is very scarce. However, one version of the Byzantine generals problem specifically related to distributed detection in WSNs can be found in [92, 93]. This offers another key example of cross-layer approach where the layers involved are that of inference (detection) and that of security prescriptions. The basic scenario addressed in [92, 93] is described later, in Sect. 4.3..

3.5. *(Crossing the Layers of)* Network Architecture and Inference Paradigms

In this section we first present very briefly a specific network architecture, known as SENMA, in which a mobile agent plays the role of fusion center. Then, we describe how the inference (detection, in our case) procedure should be tailored to the moving (sequential) nature of the agent, thus motivating the adoption of a dated detection approach, perhaps still not popular enough: Wald's sequential paradigm.

3.5.1. SENMA

The SENMA (Sensor Networks with Mobile Agents) architecture was initially proposed in [94] and has attracted much interest, in view of its nice properties of scalability, robustness and energy saving [19, 94–96]. With the SENMA scheme, the network is endowed with one or more mobile agents that travel across the sensor area and query the nodes falling inside their fields of view (namely, sufficient close to the agent). Oppositely to the earlier concept of software mobile agents [97], in the SENMA the mobile units are hardware devices such as manned or unmanned aerial or terrestrial vehicles. Therefore, it is evident how the communications between nodes and the central unit are facilitated by the geographic proximity, and the limited resources of the sensors can be suitably balanced by the powerful characteristics of the mobile agent. In fact, this latter may perform many critical network operations much more efficiently, and the burden of sophisticated processing can be thus shifted away from the sensors.

In addition, the mobile agents need not always be present in the area where the network is deployed. They can work only when it is *necessary*. The stages of data collection (sensors) and data fusion (agents) can be more easily managed and separated, if needed. This provides further desirable characteristics and properties. Finally, the mobile agents can work only when it is *safe* to do it. The advantages in terms of data protection, interception avoidance, sensor masking, attack countermeasures, and in general of network security should be self-evident.

3.5.2. Wald's Sequential Detection

Sequential detection was introduced by Wald in 1947 [98] and, since then, it has been investigated by many authors, see *e.g.*, [65, 99–103]. In a sequential test the number of samples is not fixed in advance so that the sequential paradigm is particularly suited for applications where the number of data can be virtually infinite. Thus, the question arises of how many samples are to be processed in order to ensure a certain performance level. Indeed, exceeding this number would imply an improvement of the performances over that prescribed at the design state, an improvement which we are not interested in, and larger delays for making the final decision, since observations are typically collected sequentially in time. Wald showed that there exists a sequential strategy, called SPRT (Sequential Probability Ratio Test), which performs optimally in the class of all the sequential strategies. As this latter includes as special case the usual fixed-sample-size (FSS) tests, it follows that SPRT performs better than the classical likelihood ratio test of the Neyman-Pearson framework. To be more precise, let us consider testing two simple statistical hypotheses $\mathcal{H}_1$ against $\mathcal{H}_0$, by means of iid (independent identically distributed) data. Given a set of observations $\boldsymbol{r}^m$, the well-known Neyman-Pearson approach of maximizing the detection probability given a constraint over the false alarm probability, leads to a test based on the log-likelihood ratio $L(\boldsymbol{r}^m) = \sum_{i=1}^{m} l(r_i)$, where $l(r_i)$ is the log-likelihood ratio per sample. That is

$$L(\boldsymbol{r}^m) = \begin{cases} \geq \gamma & \Rightarrow \text{decide } \mathcal{H}_1, \\ < \gamma & \Rightarrow \text{decide } \mathcal{H}_0. \end{cases} \tag{4}$$

Instead, Wald proposed the following alternative [98]:

$$L(\boldsymbol{r}^m) = \begin{cases} \geq \gamma_1 & \Rightarrow \text{decide } \mathcal{H}_1, \\ \leq \gamma_0 & \Rightarrow \text{decide } \mathcal{H}_0, \\ \in]\gamma_0, \gamma_1[& \Rightarrow \text{take another sample}. \end{cases} \tag{5}$$

Here we have two thresholds, and the number of samples to make a decision is a random variable, say N. Now, given a pair (P_d, P_f) of detection and false alarm probabilities, it has been shown by Wald that the SPRT achieves the minimum $E[N]$ in the class of all sequential tests (FSS are included therein): The SPRT is optimum in the sense that it minimizes the average sample number for given performances, amongst all fixed- and variable-length tests.

For the SPRT the following approximations hold

$$\gamma_1 \approx \ln\left(\frac{P_d}{P_f}\right), \qquad \gamma_0 \approx \ln\left(\frac{1-P_d}{1-P_f}\right), \tag{6}$$

$$\mathrm{E}[N|\mathcal{H}_1] \approx \frac{D_b(P_d||P_f)}{D(f_1||f_0)}, \qquad \mathrm{E}[N|\mathcal{H}_0] \approx \frac{D_b(P_f||P_d)}{D(f_0||f_1)}, \tag{7}$$

where $D(\cdot\,||\cdot\,)$ is the Kullback–Leibler divergence, $D_b(x, y)$ denotes the Kullback–Leibler divergence between the binary pmfs $(x, 1-x)$ and $(y, 1-y)$ [22], and f_0 and f_1 stem for the pdf of the data $\mathcal{H}_0$ and $\mathcal{H}_1$, respectively.

In a WSN designed for a detection task, the conceptual relationships between the SENMA architecture and the sequential test should be evident: Since the mobile unit of the SENMA sequentially polls the network to collect data from the nodes, the sequential paradigm exploited in SPRTs appears as a very natural framework. An example of application is presented in Sect. 4.4..

4. Examples

The previous sections present a set of guidelines, milestones, reference principles, and design recipes that should be kept in mind in the design and analysis of WSNs. The aim of the present section is to offer some example of applications where the above guidelines and tools are exploited. We would like to stress three points.

First, the presented examples consider rather idealized system models whose immediate practical applicability is not the main issue for us. On the other hand, the examples do deal with practical problems, so that the approach *is* practically-oriented.

Second, the reader will easily discover how the theory illustrated in the previous sections represents the common background and guideline, even though hereafter we usually avoid to further stress the underlying general principles.

Third, we reiterate that the examples are selected form the authors' recent works, or in any case are strictly related to the authors' lines of research. As a consequence, they only give an overview of some of *our* research lines, and nothing more than that.

4.1. Inference-Based Multiple Access [104, 105]

Consider a classic estimation problem, where a nonrandom parameter θ has to be guessed from a set of independent measurements $x_1, x_2, \ldots, x_n$, which are collected at the remote nodes, and whose statistical distribution depends upon θ. The final estimation routine is left to a fusion center, which receives data sent by the sensors through a multiple access channel.

Motivated by the lesson learned from section 3.3., we argue that the access policy to the channel must be carefully designed in view of the final estimation task. The already discussed example of Gaussian CEO over a Gaussian MAC [58], suggests to exploit the additive nature of the channel for estimation purposes. However, (i) the quoted example refers to a collection of estimation problems, *e.g.*, infinitely many emissions of a *random* source θ; and (ii) the results hold for the specific Gaussian case, the optimal estimator being additive.

As a matter of fact, the general case of a deterministic parameter, and arbitrarily distributed observations is solved in [104, 106] (for a detection-oriented approach, see [107, 108]). Specifically, consider discrete (*e.g.*, quantized) iid observations $q_1, q_2, \ldots, q_n$, with $q_i \in [1, 2, \ldots, k]$, and with a common probability mass function (pmf) $p_\theta(q)$. An observation is key to devise a smart access scheme: The *optimal* estimator corresponding to the absence of any channel, is solely a function of the empirical pmf, that is, the *type* of observations. Accordingly, let us elaborate as follows. Define a set of k orthonormal waveforms $u_1, u_2, \ldots, u_k$. The i-th sensor transmits over the channel the waveform

$$s_i = \sqrt{E} u_{q_i}, \quad i = 1, 2, \ldots, n, \tag{8}$$

where E is the available energy. At the receiver side, one gets

$$r = \sqrt{E} \sum_{i=1}^{n} u_{q_i} + w = \sqrt{E} \sum_{j=1}^{k} N_j u_j + w, \tag{9}$$

where N_j is just the number of occurrences of symbol j in the whole data-vector. Scaling by the number of sensors n, the first term is just the empirical pmf. On the other hand, as n becomes large, the noise term goes to zero so that asymptotically perfect recovering of the type is possible.

The described access scheme is known as type based multiple access (TBMA) and, of course, admits a rigorous mathematical formalization. In [104], an estimator based upon this approach is properly designed, and its asymptotic properties are derived. Specifically, defined $I_q(\theta)$ the (per-sample) Fisher information of q, the TBMA estimator $\widehat{\theta}_{TBMA}$ is

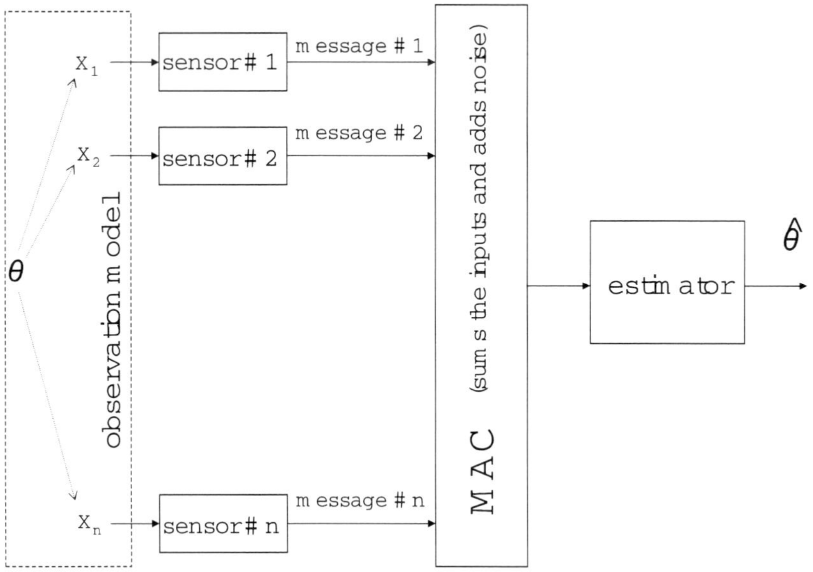

Figure 4. Block scheme valid for both the TBMA and the LBMA estimators. In the TBMA the sensors' observations are quantized versions of the original continuous one, and the messages are based on pulse-position-modulation orthogonal signaling. For the LBMA the observations are continuous, and the messages entering the MAC are the (derivative of the) local likelihoods functions regarded as function of time (*i.e.*, $\theta \leftrightarrow t$). When n goes to infinity, the noise can be beaten and asymptotically efficient estimators can be built from the MAC output.

proved to be asymptotically efficient, that is

$$\sqrt{n}(\widehat{\theta}_{TBMA} - \theta) \xrightarrow{d} \mathcal{N}\left(0, \frac{1}{I_q(\theta)}\right), \tag{10}$$

where $\xrightarrow{d}$ stems for convergence in distribution when n goes to infinity, and $\mathcal{N}(a, b)$ is our shortcut for a Gaussian distribution with mean a and variance b. Thus, for n sufficiently large, the optimal performances corresponding to the absence of noise are preserved, despite the presence of the MAC. How can it be?

The explanation relies upon the fact that one capitalizes on the additive nature of the channel and adds *coherently* the waveforms corresponding to the same observed symbol. Thus, the coherent "work" of many sensors (recall that n is allowed to increase without bound) asymptotically beats the effect of the noise, whose power does not increase with n. Note that the key-element for the above scheme to work is the match between the additive nature of the sufficient statistic, and the additive nature of the multiple access channel: We are interested in a summary of the data for inference purposes, and the channel automatically computes it for us. Furthermore, in [104], the presence of a zero-mean fading is also addressed and the performance degradation is studied, showing that optimal performances are irremediably lost, due to the lack of coherence.

The above scheme assumes discrete observations. Very often in practice, this actually corresponds to the quantization of some continuous observations. However, no optimal recipe seems available for designing such quantizers. This motivates the analysis in [105] where an approach directly working on continuous observations x_i is proposed. The rationale behind is as follows. Let $f_\theta(x)$ be the common pdf of the continuous iid observations $\{x_i\}_{i=1}^n$, and define the score function

$$s(x,\theta) = \frac{\partial \log f(x,\theta)}{\partial \theta}. \tag{11}$$

Each sensor just sends over the channel the (score-)waveform $A \cdot s(x_i, t)$, where the amplitude A is tuned just to cope with an energy constraint, and we have identified the parameter space with the observation time interval, that is, $\theta \leftrightarrow t$. The received signal at the output of an additive MAC will be the superposition of all the score waveforms sent by the remote nodes, plus the noise term:

$$r(t) = A \sum_{i=1}^{n} s(x_i, t) + w(t) = A s(\boldsymbol{x}, t) + w(t), \tag{12}$$

where $s(\boldsymbol{x}, t)$ is the aggregate score corresponding to the vector of observations $\boldsymbol{x} = [x_1, x_2, \ldots, x_n]$. As it can be seen, this access scheme is essentially based upon the very basic properties of the log-likelihood function, and is accordingly referred to as likelihood based multiple access (LBMA). Following the same lines as in the previous reasoning, one is expected, asymptotically with the number of sensors, to beat the effect of the noise, such that the uncorrupted score function is retrieved. This latter is sufficient for building an (asymptotically) optimal maximum likelihood estimator.

Again, the above intuitive arguments admit a formal justification. With some technical assumptions about the degrees of freedom available to the communication system, different estimators are proposed in [105], which share the common asymptotic optimality property: *Let T be the observation time, and W the available channel bandwidth. If W scales as n^α, $1/2 < \alpha < 1$, the LBMA estimator $\widehat{\theta}_{LBMA}$ is such that*

$$\sqrt{n}(\widehat{\theta}_{LBMA} - \theta) \xrightarrow{d} \mathcal{N}\left(0, \frac{1}{I(\theta)}\right), \tag{13}$$

where $I(\theta)$ is the (per sample) Fisher information achievable from the continuous observations x_i.

Again, we have used the fact that, for independent observations, the overall log-likelihood is additive, *and* such is the multiple access channel. This matches the estimation problem to the communication one, and, one more time, "breaking the wall" between statistics and communications leads to asymptotically optimum performances.

4.2. Perfectly Secret Distributed Detection

Let us consider an energy constrained WSN designed for a binary detection task, say $\mathcal{H}_1$ against $\mathcal{H}_0$. The relevant energy constraint (spent by a sensor for its communication tasks) is here assumed proportional to the transmission probability of the sensors, complying with the viewpoint in, *e.g.*, [66]: Due to the packet overhead for communication protocols, encryption, synchronization, authentication, and so forth, the cost of a transmission may be better modeled in terms of the number of (*i.e.*, probability of) communication tasks activated by the sensor, rather than in terms of how many bits of informations are sent. Thus, we may limit the number of communication tasks, by using a censoring strategy: The data collected at the sensors are really delivered to the fusion center only if these data are believed sufficiently *informative* for the detection. In judging the amount of information content of the data, each sensors computes the local likelihood and, as we know from the many studies available [62, 66, 109–111], a suitable singly-connected *no-send* region is defined. The transmission is activated only if the local likelihood is outside the no-send region and, in this case, the local likelihood is transmitted to the fusion center; otherwise, the sensor stays silent for saving energy.

Let $p_i(y), i = 0, 1$ be the distributions of the local likelihood under the two hypotheses, and assume that these density functions have no point mass and common support. Also, let β_1 and β_0 the sensors' transmission probabilities (the same for all the sensors: the homogeneous network is made of identical sensors sensing iid data) under $\mathcal{H}_1$ and $\mathcal{H}_0$, respectively. We have

$$\beta_i = 1 - \int_{\tau_l}^{\tau_u} p_i(y)\, dy, \quad i = 0, 1. \tag{14}$$

The cost constraint is fulfilled by enforcing the condition $\max\{\beta_1, \beta_0\} \leq \beta$, for a prescribed $\beta \in (0, 1)$.

Following *the* milestone of modern communications [112], a source of information is suitably characterized by its Shannon's entropy. In our inference-oriented system, however, the source outputs are not relevant *per se*. What is relevant is how much a source output compels with one or the other state of the nature (statistical hypothesis), and this concept can be formalized via the Kullback–Leibler divergence [113, 114], that measures the *distance* between the two hypotheses which is embedded in the source outputs. In this regard, we are relying upon Stein's lemma [115] for binary detection problems involving iid observations and completely characterized (simple, in the statistical terminology) hypotheses. This is as follows. The best (minimum achievable) probability of a missed detection $1 - P_d$ (true $\mathcal{H}_1$, decide $\mathcal{H}_0$), compatible with any prescribed false-alarm probability P_f (true $\mathcal{H}_0$, decide $\mathcal{H}_1$), verifies:

$$\lim_{n\to\infty} \frac{\log(1 - P_d)}{n} = -D(\mathcal{H}_0 || \mathcal{H}_1), \tag{15}$$

where n is the number of iid samples available, and $D(\mathcal{H}_0||\mathcal{H}_1)$ is the Kullback–Leibler distance between the (marginal) distributions of the samples under the two hypotheses.

Thus, let D be the divergence of the source, *i.e.*, of the sensors outputs (local likelihoods computed by the sensors). On accounting for the censoring strategy, this can be written as

$$D = \int_{y_l}^{\tau_l} p_0(y) \log \frac{p_0(y)}{p_1(y)}\, dy + \int_{\tau_u}^{y_u} p_0(y) \log \frac{p_0(y)}{p_1(y)}\, dy + (1 - \beta_0) \log\left(\frac{1 - \beta_0}{1 - \beta_1}\right), \tag{16}$$

where $y_l \geq 0$ and $y_u \leq \infty$ are the extreme points of the likelihood-pdf support.

Let us now discuss the very peculiar aspect of this application example. The WSNs are vulnerable to many types of attacks, and we assume that there is one attacker that senses the communication channel between sensors and fusion center. The messages from the sensors are encrypted so that their content is unaccessible to the intruder, but this is not enough for a system that we want to be *perfectly secret*. Indeed, the attacker, by monitoring the channel, *sees* if this latter is in 'idle' or in 'busy' state and, using this information, the intruder can make its inference about the state of the nature. If we want to design a perfectly (detection) secret system in this setting, the best countermeasure is to make the information available to the intruder completely useless for the detection task. The divergence perceived by the attacker using its binary (idle/busy) data is

$$\beta_0 \log\left(\frac{\beta_0}{\beta_1}\right) + (1-\beta_0)\log\left(\frac{1-\beta_0}{1-\beta_1}\right), \tag{17}$$

and we conclude that the property of perfect secrecy is obtained by enforcing the condition $\beta_0 = \beta_1$.

Now, we want to maximize the divergence at the fusion center with the above two constraints: that $\max\{\beta_1, \beta_0\} \leq \beta$ and that $\beta_1 = \beta_0$. Thus, according to eq. (17), we want to maximize:

$$D(\beta) = \int_{y_l}^{\tau_l} p_0(y)\log\frac{p_0(y)}{p_1(y)}\,dy + \int_{\tau_u}^{y_u} p_0(y)\log\frac{p_0(y)}{p_1(y)}\,dy, \tag{18}$$

subject to

$$\beta_1 = \beta_0 \leq \beta, \tag{19}$$

where $D(\beta)$ is our definition of the *divergence-cost* function, and the notation emphasizes the dependence upon the maximum allowable cost β.

The following result can be proven. *The divergence-cost function $D(\beta)$ that characterizes the system performance is strictly increasing and strictly concave* [3] $\cap$ *in the interior of $\beta \in (0,1)$. In addition, there exists a unique pair of censoring thresholds $\tau_l(\beta) \leq 1$ and $\tau_u(\beta) \geq 1$, which are monotonic and, in the limit of vanishingly small β, converge to the extremes of the likelihood-pdf support.*

Exactly as the capacity per unit cost defined through the capacity-cost function (see the seminal paper by Verdú [25]), we can define the *divergence per unit cost*:

$$\mathbf{D} = \sup_{\beta}\frac{D(\beta)}{\beta} = \lim_{\beta\to 0}\frac{C(\beta}{\beta}, \tag{20}$$

where the last equality is a consequence of the above properties of[4] $D(\beta)$. The operational meaning of $\mathbf{D}$ is as follows. The system detection performances for a network of n sensors is $nD(\beta)$. Now, assume that $\gamma = n\beta$ is the overall energy spent by the network. For a prescribed γ (large enough), among two systems working at different per-sensor costs, say β' and β'', the more convenient is that with a lower value of $D(\beta)/\beta$. Put another way, we are not comparing the two systems for the same number of sensors, but rather for the same energy expense. Therefore, in our distributed network with perfect secrecy, the most economic way to convey information suited for the detection task amounts to use more and more sensors, with vanishing energy expense per each sensor.

[3]Similar properties, as well known, hold for the capacity-cost function [116].

[4]The capacity per unit cost studied by Verdú [25] shares the same properties, when the channel has a free input symbol.

4.3. Detection in WSNs under Byzantine Attack [92,93]

Here we consider a WSN engaged in a detection task, with the remote nodes collecting iid discrete observations about the binary state of the nature. Thus these observations are drawn iid from pmf p, if $\mathcal{H}_1$ is in force, or from pmf q, if $\mathcal{H}_0$ is in force. However, this is actually true only for the "honest" sensors, as the system is subject to a Byzantine attack: A fraction of the network' nodes are reprogrammed by an intruder in such a way that these "Byzantine" sensors emit iid accordingly to x under $\mathcal{H}_1$ and to y under $\mathcal{H}_0$. The attack is particularly insidious since the intruder knows the state of the nature that the network is aimed to discover, since the fusion center does not know which sensor is Byzantine, and since the Byzantines cooperatively design their attacking distributions x and y.

More specifically, let us assume that n, the total number of sensors in the network, is so large that large deviation results can be applied, and let $\alpha\, n$ be the average number of Byzantine sensors (α also referred to as the attacking power), which is known to the system designer. For α greater than or equal to a certain blinding threshold, say α_b, it turns out that the performance of the system can be nullified: The data collected at the fusion center give no chance to make a nontrivial guess about the state of the nature. The threshold value, which is always less than or equal to $1/2$, is only a function of the two hypotheses and is

$$\alpha_b = \frac{\sum\limits_{k\in\mathcal{K}} (q_k - p_k)^+}{1 + \sum\limits_{k\in\mathcal{K}} (q_k - p_k)^+}, \tag{21}$$

where $(c)^+ = c$ when $c \geq 0$ and $(c)^+ = 0$ otherwise, and $\mathcal{K}$ is the common alphabet of the observations under the two hypotheses.

Conversely, for $\alpha < \alpha_b$ the detection capability of the network cannot be brought to zero and the best test exponent $\Delta(\alpha)$ (in the Stein's lemma sense [115]) of the network remains strictly positive. Such an exponent corresponds to the best for the network once that the Byzantine sensors have chosen their best attacking distributions:

$$\Delta(\alpha) \stackrel{def}{=} \min_{x,y} D(z||w), \tag{22}$$

where z and w are the pmfs of the data collected at the fusion center (sent by honest and Byzantine sensors in a proportion of $(1-\alpha)$ and α, respectively), and $D(\cdot||\cdot)$ is the Kullback–Leibler divergence between pmfs [22, 113, 114].

The optimal attacking distributions (optimal for the intruder, of course) can be found in the form

$$\begin{cases} x_k = \dfrac{1-\alpha}{\alpha}\,(\gamma_x q_k - p_k)^+, \\[2ex] y_k = \dfrac{1-\alpha}{\alpha}\,(\gamma_y p_k - q_k)^+, \end{cases} \tag{23}$$

where $0 < \gamma_x, \gamma_y \leq 1$ are constants to be set in order to fulfill $\sum\limits_{k\in\mathcal{K}} x_k = \sum\limits_{k\in\mathcal{K}} y_k = 1$, by mens of a procedure known as "water-filling".

As a matter of fact, the optimal test employed at the fusion center can be recognized to be a censored version of the original test between p and q, in that eqs. (23) yield the least favorable distributions well-known in the context of robust detection [117], see

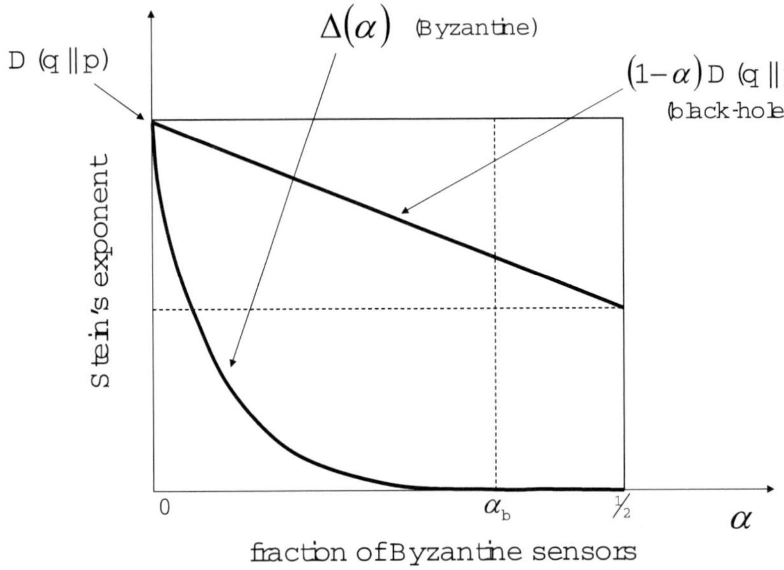

Figure 5. The optimal text exponent resulting from an intruder that selects the best distributions (x, y) to launch its Byzantine attack, and the fusion center that, being aware of the presence of the attacker, selects its best Neyman-Pearson test designed over the α-contaminated distributions $w = (1-\alpha)p + \alpha x$, $z = (1-\alpha)q + \alpha y$, where (p, q) are the original distributions representing the two possible states of the nature. Also shown is the test performance of a wilder attacker that simply destroys a fraction α of the nodes, which is called *black-hole* attacker.

also [118, 119]. This strong analogy with robust detection theory implies that a Byzantine change of the attacking distributions when the network is not aware of this (*i.e.*, when it still implements the censored test), cannot provide any advantage from the intruder's viewpoint. In addition, if only an upper bound $\bar{\alpha}$ for the attacking power is known at the fusion center, as it may be often the case in practice, the censored test designed with $\bar{\alpha}$ achieves an exponent no smaller than $\Delta(\bar{\alpha})$.

It can be also shown, finally, that the optimal test exponent $\Delta(\alpha)$ is continuous, decreasing, and convex $\cup$ over the interval $\alpha \in (0, \alpha_b)$, with $\Delta(0) = D(q||p)$ (which we assume > 0) and $\lim_{\alpha \to \alpha_b} \Delta(\alpha) = 0$. The typical behavior is shown in Figure 5 along with the exponent of a black-hole attack. This latter simply amounts to destroy the fraction α of the sensors under the intruder's control and, as it can be seen in the figure, is by far less effective than the more sophisticated Byzantine attack.

4.4. Sequential Detection in SENMA Architectures [110]

Consider a network of sensors polled by a mobile agent according to the SENMA architecture. Each sensor implements a sequential test based on its own locally gathered data. As soon as a new sensor falls inside the rover's field of view, it is queried from the rover for delivering its local decision, if any. Clearly, if the local test is still in progress at that sensor, no local decision is available and no data are sent to the rover. Thus, before making its decision, the node is in *sleep* mode, and the rover's polling simply *turns off* that sensor which, thus, will never reveal itself to the rover. Conversely, once a local decision is made, a polling signal switches it from sleep to transmit mode: the local decision (say ± 1) is delivered to the rover and then the sensor *turns off* (further delivering is inhibited).

The mobile agent collects data (local decisions) from the sensors that terminated their task when queried. These data are used to implement another sequential test, now at the rover level. This latter test, when eventually concluded, provides the final decision about the state of the nature.It is worth noting that the querying/responding policy avoids any collision concerns over the common communication channel.

The observation model assumes Brownian motions with and without a known drift, for $\mathcal{H}_1$ and $\mathcal{H}_0$, respectively. That is

$$\mathcal{H}_i : \; x_t^{(s)} = w_t^{(s)} + i\theta t, \quad t \in (0, \infty), \quad s = 1, 2, \ldots, \tag{24}$$

where $i = 0, 1$, represents the state of the nature, s is the sensor index, and w_t is a Brownian motion with variance $\sigma^2 t$, that evolves over the continuous time axis t. The known constant θ rules the signal-to-noise ratio per unit time $\rho = \theta^2/\sigma^2$ (measured in sec^{-1}).The likelihood ratio, obtained via Cameron-Martin formula (see e.g., [120]), results in $l_t^{(s)} = \theta\, x_t^{(s)} - \theta^2 t/2$, and is used to run the local SPRT at sensor s, namely

$$l_t^{(s)} \begin{cases} \geq \gamma_1 & \Rightarrow \text{ sensor } s \text{ decide } \mathcal{H}_1, \\ \leq \gamma_0 & \Rightarrow \text{ sensor } s \text{ decide } \mathcal{H}_0, \\ \in]\gamma_0, \gamma_1[& \Rightarrow \text{ sensor } s \text{ needs more data.} \end{cases} \tag{25}$$

Assume that, for simplifying the analysis, the local miss and false alarm probability are equal, and let p_e such an error probability. We know that $\gamma_0 = -\gamma_1 = \log\left[p_e/(1-p_e)\right]$ [98].

The rover travels across the surveyed area with constant velocity and the sensors are disseminated completely at random, so that the total number of polled sensors at time t is a Poisson counting process with intensity per unit of time λ (sec^{-1}). Consequently, also the data collected by the rover can be modeled as Poisson processes. However, since as time elapses the probability of taking a decision at the remote sensors increases, such processes are inhomogeneous. Specifically, let X_t and Y_t be the total number of $+1$ and -1, respectively, received by the rover up to time t. These are independent inhomogeneous processes whose log-likelihood ratio can be easily found; the resulting SPRT at the rover is

$$L_t = (X_t - Y_t) \log \frac{1 - p_e}{p_e} \quad \begin{cases} \geq \Gamma_1 & \Rightarrow \text{ decide } \mathcal{H}_1, \\ \leq \Gamma_0 & \Rightarrow \text{ decide } \mathcal{H}_0, \\ \in]\Gamma_0, \Gamma_1[& \Rightarrow \text{ more data are needed.} \end{cases} \tag{26}$$

It is clear that a tradeoff exists between the *amount* and the *reliability* of the information that the rover may collect: Imposing p_e large (namely close to $1/2$) implies that many sensors end their task shortly so that the rover can potentially collect a large amount of data. The other side of the story is that these data are poorly informative in the sense that the decisions delivered to the mobile agent are subject to a large probability of being wrong.

The reverse is true with very small values of p_e which imply reliable local decisions but, at the same time, very long decision time at the sensor level and, in turn, low probability of collecting data at the rover. Note that the fewer the transmissions, the larger the energy saving form the sensors. With these ideas in mind, we choose as parameter to be optimized over the local sensors' performance p_e, and as objective the minimization of the average time $E[T]$ for making the final decision, where T is the stopping time for the test (26). For the details of the optimization we refer the reader to [110]. Here we only report the final performances which result in amazingly simple and physical-insighting formulas (see *e.g.*, eqs. (27) and (28) below).

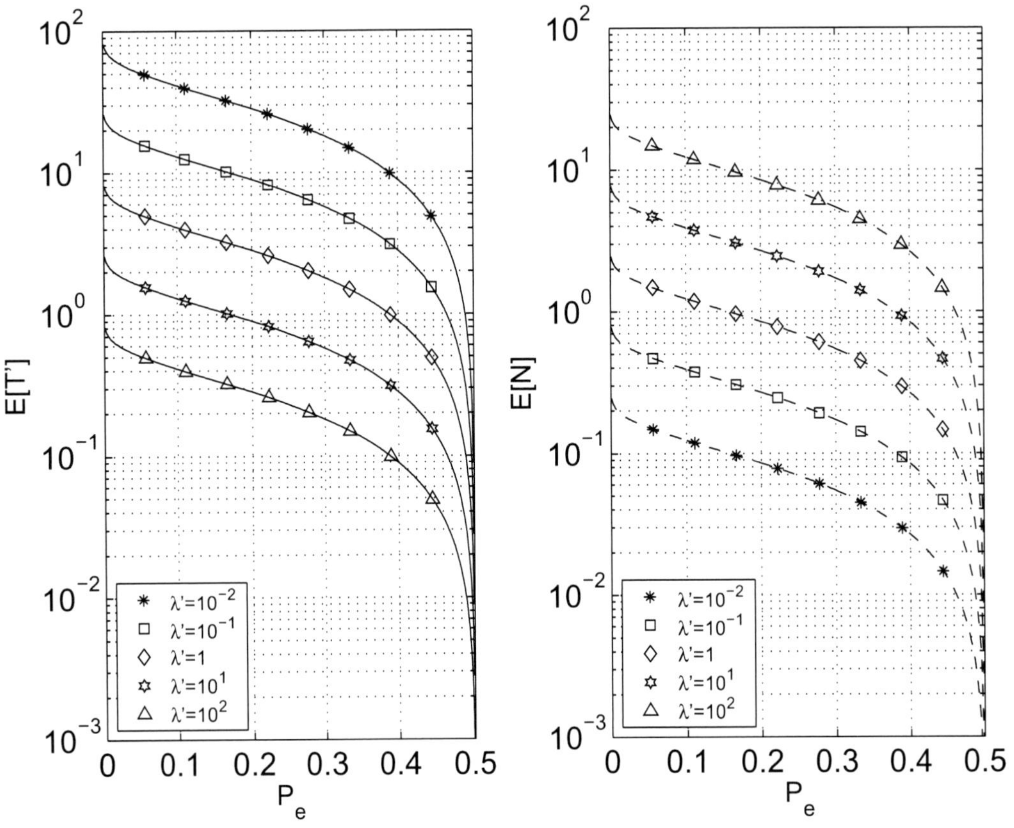

Figure 6. Normalized average time $E[T']$ needed to end the test *(left)*, and average number of transmitting sensors (which is a proxy of the total energy consumption) $E[N]$ *(right)*, as function of the error probability P_e of the test. The plots are parametrized in λ', the normalized density of sensors.

First, let N be the number of symbols globally emitted by the ensemble of sensors when the rover's detection process ends, which is also a proxy of the total amount of energy spent by the network; let P_e be the global error probability (miss detection and false alarm) at the rover; let $D_b(x)$ be the Kullback-Leibler divergence [22] between the binary pmfs $(x, 1-x)$ and $(1-x, x)$; finally, let η be the ratio between the average number of *transmitting* sensors and the average number of *encountered* sensors. As shown in [110], η is in the order of 0.3, which means that $\approx 30\%$ of the encountered sensors actually communicate their decisions to the rover. Its inverse, $1/\eta$ represents, in a sense, the energy saving with respect to a system in which *all* polled sensors send their local decisions. We have the following approximate relationships, which are valid for symmetry both under $\mathcal{H}_0$ and under $\mathcal{H}_1$:

$$E[T] = \sqrt{\frac{2D_b(P_e)}{\eta\lambda\,\rho}}, \tag{27}$$

$$E[N] = \sqrt{\frac{2D_b(P_e)\,\eta\lambda}{\rho}}. \tag{28}$$

The above equations deserve some comments.

We note that the average time to take the final decision is proportional to $(\eta\lambda\rho)^{-1/2}$: by doubling either the SNR ρ or the *talking-node density* $\eta\lambda$ drops by the decision time of a factor ≈ 0.7. As one expects, lowering the P_e implies larger processing times and this dependence is regulated by the square root of the binary divergence function $\sqrt{D_b(P_e)}$. While similar comments apply to the number of transmissions $E[N]$, the role of the product $\eta\lambda$ is opposite: the larger is the density of the talking nodes, the larger is $E[N]$. As a consequence, the ratio between the processing time and the average number of symbols received by the rover is only ruled by the talking-node density, *i.e.*, $E[T]/E[N] = (\eta\lambda)^{-1}$.

It may be convenient to introduce the following normalized dimensionless quantities:

$$\lambda' \stackrel{def}{=} \lambda/\rho, \tag{29}$$

$$E[T'] \stackrel{def}{=} E[T]\rho \tag{30}$$

so that the system performances can be even easier summarized as

$$E[T'] = \sqrt{\frac{2D_b(P_e)}{\eta\lambda'}}, \tag{31}$$

$$E[N] = \sqrt{2D_b(P_e)\,\eta\lambda'}, \tag{32}$$

where it is worth noting that $E[T']\,E[N] = 2D_b(P_e)$. Equations (31) and (32) are depicted in Figure 6.

5. Conclusion

We have tried to draw what we expect to be the road ahead for an exciting travel across the different "layers" of statistics, communications, security, their interplay, and the related emerging milestones in a WSN applicative scenario. We are, however, not in the position of offering conclusions and, instead of summarizing, we opt for leaving this section empty; the certainly large number of researchers that will take part to this travel will write, at due time, the first (unavoidably partial) conclusions.

References

[1] C. Chong and S. Kumar, Sensor networks: Evolution, opportunities, and challenges. *Proc. IEEE*, vol. 91, no. 8, pp. 1247–1256, Aug. 2003.

[2] H. Gharavi and S. Kumar, Scanning the issue – special issue on sensor network and applications. *Proc. IEEE*, vol. 91, no. 8, pp. 1151–1153, Aug. 2003.

[3] A. Krasnopeev, J. Xiao, and Z. Luo, Minimum energy decentralized estimation in a wireless sensor network with correlated sensor noises. *EURASIP Journal on Wireless Communications and Networking*, no. 4, pp. 473–482, 2005.

[4] R. Niu and P. K. Varshney, Distributed detection and fusion in a large wireless sensor network of random size. *EURASIP Journal on Wireless Communications and Networking*, no. 4, pp. 462–472, 2005.

[5] S. S. Pradhan, J. Kusuma, and K. Ramchandran, Distributed compression in a dense microsensor network. *IEEE Signal Processing Mag.*, vol. 19, pp. 51–60, Mar. 2002.

[6] Z. Yang, M. Dong, L. Tong, and B. M. Sadler, MAC protocols for optimal information retrieval pattern in sensor networks with mobile access. *EURASIP Journal on Wireless Communications and Networking*, no. 4, pp. 493–504, 2005.

[7] F. Zhao, J. Liu, J. Liu, L. Guibas, and J. Reich, Collaborative signal and information processing: An information-directed approach. *Proc. IEEE*, vol. 91, no. 8, pp. 1199–1209, Aug. 2003.

[8] R. Viswanathan and P. K. Varshney, Distributed detection with multiple sensors: Part I – fundamentals. *Proc. IEEE*, vol. 85, no. 1, pp. 54–63, Jan. 1997.

[9] R. S. Blum, A. Kassam, and H. V. Poor, Distributed detection with multiple sensors: Part II – advanced topics. *Proc. IEEE*, vol. 85, no. 1, pp. 64–79, Jan. 1997.

[10] J.-F. Chamberland and V. V. Veeravalli, Decentralized detection in sensor networks. *IEEE Trans. Signal Processing*, vol. 51, no. 2, pp. 407–416, Feb. 2003.

[11] R. R. Tenney and N. R. Sandell Jr., Detection with distributed sensors. *IEEE Trans. Aerosp. Electron. Syst.*, vol. AES-17, pp. 501–510, 1981.

[12] J. N. Tsitsiklis, Decentralized detection. in *Advances in Signal Processing*, H. V. Poor and J. B. Thomas, Eds. JAI Press, 1993, pp. 297–344.

[13] A. Sayeed, D. Estrin, G. Pottie, K. Ramchandran, and M. Pursley, Guest editorial self-organizing distributed collaborative sensor networks. *IEEE J. Select. Areas Commun.*, vol. 23, no. 4, pp. 689–692, Apr. 2005.

[14] S. D. Servetto, R. Knopp, A. Ephremides, S. Verdú, and S. B. Wicker, Guest editorial fundamental performance limits of wireless sensor networks. *IEEE J. Select. Areas Commun.*, vol. 22, no. 6, pp. 961–965, Aug. 2004.

[15] R. Olfati-Saber, A. Fax, and R. M. Murray, Consensus and cooperation in networked multi-agent systems. *Proc. IEEE*, vol. 95, no. 1, pp. 215–233, Jan. 2007.

[16] J. Tsitsiklis and M. Athans, Convergence and asymptotic agreement in distributed decision problems. *IEEE Trans. Automat. Contr.*, vol. 29, pp. 42–50, Jan. 1984.

[17] S. Boyd, A. Ghosh, B. Prabhakar, and D. Shah, Randomized gossip algorithms. *IEEE Trans. Inform. Theory*, vol. 52, no. 6, pp. 2508–2530, June 2006.

[18] S. Barbarossa and G. Scutari, Bio-inspired sensor network design. *IEEE Signal Processing Mag.*, vol. 24, no. 3, pp. 26–35, May 2007.

[19] B. M. Sadler, Fundamentals of energy-constrained sensor network systems. *IEEE Aerosp. Electron. Syst. Mag*, vol. 20, no. 8, Aug. 2005.

[20] M. Chiang, S. H. Low, A. R. Calderbank, and J. C. Doyle, Layering as optimization decomposition: A mathematical theory of network architectures. *Proc. IEEE*, vol. 95, no. 1, pp. 255–312, Jan. 2007.

[21] S. Marano, V. Matta, and P. Willett, Asymptotic design of quantizers for decentralized MMSE estimation. *IEEE Trans. Signal Processing*, in print.

[22] T. Cover and J. Thomas, *Elements of Information Theory*. New York: John Wiley & Sons, 1991.

[23] S. Verdú, Current trends in communication theory. lecture at University of Naples, Italy (*slides of presentation*), May 26, 2006.

[24] M. Gastpar, B. Rimoldi, and M. Vetterli, To code, or not to code: lossy source-channel communication revisited. *IEEE Trans. Inform. Theory*, vol. 49, no. 5, pp. 1147–1158, May 2003.

[25] S. Verdú, On channel capacity per unit cost. *IEEE Trans. Inform. Theory*, vol. 36, no. 5, pp. 1019–1030, Sept. 1990.

[26] T. Berger, Decentralized estimation and decision theory. in *IEEE 7th Spring Workshop on Information Theory*, Mt. Kisco, NY, USA, Sep. 1979.

[27] T. S. Han and S. Amari, Statistical inference under multiterminal data compression. *IEEE Trans. Inform. Theory*, vol. IT-44, no. 6, pp. 2300–2324, Oct. 1998.

[28] C. E. Shannon, Coding theorems for a discrete source with a fidelity criterion. *IRE National Convention Record, Part 4*, pp. 142–163, 1959.

[29] Z. Zhang and T. Berger, Estimation via compressed information. *IEEE Trans. Inform. Theory*, vol. 34, no. 2, pp. 198–211, Mar. 1988.

[30] H. Yamamoto and K. Itoh, Source coding theory for multiterminal communication systems with a remote source. *The Transactions of the IECE of Japan*, vol. E-63, no. 10, pp. 700–706, Oct. 1980.

[31] T. J. Flynn and R. M. Gray, Encoding of correlated observations. *IEEE Trans. Inform. Theory*, vol. 33, no. 6, pp. 773–787, Nov. 1987.

[32] T. Berger, Z. Zhang, and H. Viswanathan, The CEO problem. *IEEE Trans. Inform. Theory*, vol. 42, no. 3, pp. 887–902, May 1996.

[33] R. L. Dobrushin and B. S. Tsybakov, Information transmission with additional noise. *IRE Transactions on Information Theory*, vol. 8, no. 5, pp. 293–304, July 1962.

[34] T. Fine, Optimal mean square quantization of a noisy input. *IEEE Trans. Inform. Theory*, vol. 11, no. 2, pp. 293–294, 1965.

[35] S. D. Servetto, The sum-rate of multiterminal source coding with two encoders. *IEEE Trans. Inform. Theory*, Nov. 2006 *(submitted)*.

[36] Y. Oohama, The rate-distortion function for the quadratic Gaussian CEO problem. *IEEE Trans. Inform. Theory*, vol. 44, no. 3, May 1988.

[37] J. Chen, X. Zhang, T. Berger, and S. Wickler, An upper bound on the sum-rate distortion function and its corresponding rate allocation schemes for the CEO problem. *IEEE J. Select. Areas Commun.*, vol. 22, no. 6, pp. 977–987, Aug. 2004.

[38] S. P. Lloyd, Least squares quantization in pcm. *IEEE Trans. Inform. Theory*, vol. 28, no. 2, pp. 129–136, Mar. 1982 (previously unpublished Bell Laboratories Technical Note, 1957).

[39] J. Max, Quantizing for minimum distortion. *IRE Transactions on Information Theory*, vol. 6, no. 1, pp. 7–12, Mar. 1960.

[40] M. Longo, T. Lookabaugh, and R. Gray, Quantization for decentralized hypothesis testing under communication constraints. *IEEE Trans. Inform. Theory*, vol. 36, no. 2, pp. 241–255, Mar. 1990.

[41] W. M. Lam and A. R. Reibman, Design of quantizers for decentralized estimation systems. *IEEE Trans. Commun.*, vol. 41, no. 11, pp. 1602–1605, Nov. 1993.

[42] S. Verdú, Fifty years of shannon theory. *IEEE Trans. Inform. Theory*, vol. 44, no. 6, pp. 2057–2078, Oct. 1998.

[43] S. S. Pradhan and K. Ramchandran, Distributed source coding using syndromes (discus): design and construction. *IEEE Trans. Inform. Theory*, vol. 49, no. 3, pp. 626–643, Mar. 2003.

[44] A. D. Liveris, Z. Xiong, and C. N. Georghiades, Compression of binary sources with side information at the decoder using ldpc codes. *IEEE Commun. Lett.*, vol. 6, no. 10, pp. 440–442, Oct. 2002.

[45] Z. Xiong, A. D. Liveris, and S. Cheng, Distributed source coding for sensor networks. *IEEE Signal Processing Mag.*, vol. 21, no. 5, pp. 80–94, Sept. 2004.

[46] R. G. Gallager, Low density parity check codes. *IRE Transactions on Information Theory*, vol. 8, pp. 21–28, 1962.

[47] ——, Low density parity check codes. in *no. 21 in Research Monograph Series*. Cambridge, MA: MIT Press, 1963.

[48] D. J. C. MacKay and R. M. Neal, Good codes based on very sparse matrices. in *Cryptography and Coding. 5th IMA Conference in lecture notes in computer science, Colin Boyd Edition*, vol. 1025, Dec. 18-20 1995, pp. 100–111.

[49] D. J. C. MacKay, Good error-correcting codes based on very sparse matrices. *IEEE Trans. Inform. Theory*, vol. 45, no. 2, pp. 399–431, Mar. 1999.

[50] P. Gupta and P. R. Kumar, The capacity of wireless networks. *IEEE Trans. Inform. Theory*, vol. 46, no. 2, pp. 388–404, Mar. 2000.

[51] V. Rodoplu and T. Meng, Bits-per-joule capacity of energy-limited wireless networks. *IEEE Trans. Wireless Commun.*, vol. 6, no. 3, pp. 857–865, Mar. 2007.

[52] M. J. E. Golay, Note on the theoretical efficiency of information reception with PPM. *Proc. IRE*, vol. 37, 1949.

[53] V. Rodoplu and T. Meng, Channel capacity in bits per joule. *IEEE J. Oceanic Eng.*, vol. 11, no. 1, pp. 97–99, Jan. 1986.

[54] R. G. Gallager, Energy limited channels: Coding, multiaccess and spread spectrum. MIT, MA, Tech. Rep., Nov. 1987.

[55] D. Slepian and J. K. Wolf, Noiseless coding of correlated information sources. *IEEE Trans. Inform. Theory*, vol. 19, no. 4, pp. 471–480, July 1973.

[56] M. Gastpar, M. Vetterli, and P. L. Dragotti, Sensing reality and communicating bits: a dangerous liaison. *IEEE Signal Processing Mag.*, vol. 23, no. 4, pp. 70–83, July 2006.

[57] M. Gastpar, To code or not to code. Ph.D. dissertation, Ecole Polytechnique Fédérale (EPFL), Lausanne, Switzerland, 2002.

[58] M. Gastpar and M. Vetterli, On the capacity of large Gaussian relay networks. *IEEE Trans. Inform. Theory*, vol. IT-51, no. 3, Mar. 2005.

[59] S. Vembu, S. Verdú, and Y. Steinberg, The source-channel separation theorem revisited. *IEEE Trans. Inform. Theory*, vol. 41, no. 1, pp. 44–54, Jan. 1995.

[60] E. Shi and A. Perrig, Designing secure sensor networks. *IEEE Trans. Wireless Commun.*, vol. 11, no. 6, pp. 38–41, Dec. 2004.

[61] P. K. Varshney, *Distributed Detection and Data Fusion*. New York, NY: Springer, 1997.

[62] S. Appadwedula, V. V. Veeravalli, and D. Jones, Energy efficient detection in sensor networks. *IEEE J. Select. Areas Commun.*, vol. 23, pp. 639–702, Apr. 2005.

[63] C. K. Sestok, M. R. Said, and A. V. Oppenheim, Randomized data selection in detection with applications to distributed signal processing. *Proc. IEEE*, vol. 91, no. 8, pp. 1184–1198, Aug. 2003.

[64] P. F. Swaszek and P. Willett, Parley as an approach to distributed detection. *IEEE Trans. Aerosp. Electron. Syst.*, vol. 31, no. 1, pp. 447–457, Jan. 1995.

[65] V. Veeravalli, T. Basar, and H. V. Poor, Decentralized sequential detection with a fusion center performing the sequential test. *IEEE Trans. Inform. Theory*, vol. 39, no. 2, pp. 433–442, Mar. 1993.

[66] C. Rago, P. Willett, and Y. Bar-Shalom, Censoring sensors: a low-communication-rate scheme for distributed detection. *IEEE Trans. Aerosp. Electron. Syst.*, vol. 32, no. 2, pp. 554–568, Apr. 1996.

[67] F. Ye, H. Luo, S. Lu, and L. Zhang, Statistical en-route filtering of injected false data in sensor networks. *IEEE J. Select. Areas Commun.*, vol. 23, no. 4, pp. 839–850, Apr. 2005.

[68] X. Luo, M. Dong, and Y. Huang, On distributed fault-tolerant detection in wireless sensor networks. *IEEE Trans. Comput.*, vol. 55, no. 1, pp. 58–70, Jan. 2006.

[69] T. Clouqueur, K. K. Saluja, and P. Ramanathan, Fault tolerance in collaborative sensor networks for target detection. *IEEE Trans. Comput.*, vol. 53, no. 3, pp. 320–333, Mar. 2004.

[70] W. Du, J. Deng, Y. Han, and P. Varshney, A witness-based approach for data fusion assurance in wireless sensor networks. in *Proc. GLOBECOM,*, 2003, pp. 1435–1439.

[71] C. E. Shannon, Communication theory of secrecy systems. *Bell System Technical Journal*, vol. 28, pp. 656–715, Oct. 1949.

[72] A. D. Wyner, The wire-tap channel. *AT & T Bell Labs Tech. J.*, vol. 54, no. 8, pp. 1355–1387, Oct. 1975.

[73] L. H. Ozarow and A. D. Wyner, Wire-tap channel II. *AT & T Bell Labs Tech. J.*, vol. 63, pp. 2135–2157, 1984.

[74] I. Csiszár and J. Körner, Broadcast channels with confidential messages. *IEEE Trans. Inform. Theory*, vol. IT-24, no. 3, pp. 339–348, May 1978.

[75] Y. Oohama, Coding for relay channels with confidential messages. in *Proc. of IEEE Information Theory Workshop*, Cairns, Australia, Sept. 2-7 2001, pp. 87–89.

[76] Y. Oohama, Relay channels with confidential messages. *IEEE Trans. Inform. Theory*, Nov. 2007 *(submitted)*.

[77] R. Liu, I. Maric, R. D. Yates, and P. Spasojevic, The discrete memoryless multiple access channel with confidential messages. in *Proc. of the International Symposium on Information Theory (ISIT)*, Seattle, WA, July 9-14 2006.

[78] Y. Liang and H. V. Poor, Generalized multiple access channels with confidential messages. *IEEE Trans. Inform. Theory*, Nov. 2006 *(submitted)*.

[79] E. Tekin and A. Yener, The Gaussian multiple access wire-tap channel. *IEEE Trans. Inform. Theory*, Nov. 2006 *(submitted)*.

[80] A. O. Hero, Secure space-time communication. *IEEE Trans. Inform. Theory*, vol. 49, pp. 1–16, Dec. 2003.

[81] L. Lai and H. El Gamal, The relay-eavesdropper channel: Cooperation for secrecy. *IEEE Trans. Inform. Theory*, Dec. 2006 *(submitted)*.

[82] P. Parada and R. Blauth, Secrecy capacity of SIMO and slow fading channels. in *Proc. of the International Symposium on Information Theory (ISIT)*, Chicago, IL, USA, Sept. 2005.

[83] J. Barros and M. R. D. Rodrigues, Secrecy capacity of wireless channels. in *Proc. of the International Symposium on Information Theory (ISIT)*, Seattle, WA, July 9-14 2006.

[84] P. K. Gopala, L. Lai, and H. El Gamal, On the secrecy capacity of fading channels. *IEEE Trans. Inform. Theory*, Dec. 2006 *(submitted)*.

[85] Z. Yang and L. Tong, Cooperative sensor networks with misinformed nodes. *IEEE Trans. Inform. Theory*, vol. IT-51, no. 12, pp. 4118–4133, Dec. 2005.

[86] Z. Yang and L. Tong, On the error exponent and the use of ldpc codes for cooperative sensor networks with misinformed nodes. *IEEE Trans. Inform. Theory*, vol. 53, no. 9, pp. 3265–3274, Sept. 2007.

[87] L. Lamport, R. Shostak, and M. Pease, The Byzantine generals problem. *ACM Transactions on Programming Languages and Systems*, vol. 4, pp. 382–401, July 1982.

[88] D. Dolev, The Byzantine generals strike again. *Journal of Algorithms*, vol. 3, no. 1, pp. 14–30, 1982.

[89] B. Pfitzmann and M. Waidner, Information theoretic pseudosignatures and Byzantine agreement for $t \geq n/3$. IBM Research Report, Tech. Rep. RZ2882, Nov 1996.

[90] T. Ho, B. Leong, R. Koetter, M. Médard, M. Effrons, and D. Karger, Byzantine modification detection in multicast networks using randomized network coding. in *IEEE Proc. Intl. Sym. Inform. Theory*, June 27-July 2 2004, p. 143.

[91] O. Kosut and L. Tong. (July 14, 2006) Capacity of cooperative fusion in the presence of byzantine sensors. [Online]. Available: `http://arxiv.org/PS_cache/cs/pdf/0607/0607076v1.pdf`

[92] S. Marano, V. Matta, and L. Tong, Distributed detection in the presence of byzantine attack in large wireless sensor networks. in *Military Communications Conference (MILCOM 2006)*, Washington, D.C., USA, Oct. 2006.

[93] S. Marano, V. Matta, and L. Tong, Distributed inference in the presence of byzantine sensors. in *Fortieth Annual Asilomar Conference on Signals, Systems, and Computers (ASILOMAR 2006)*, Asilomar Conference Centre, Pacific Grove, CA, USA, Oct. 29 – Nov. 1 2006.

[94] L. Tong, Q. Zhao, and S. Adireddy, Sensor networks with mobile agents. in *Proceedings of MILCOM 2003*, Boston MA, Oct. 2003.

[95] S. Appadwedula, V. Veeravalli, and D. Jones, Energy-efficient detection in sensor networks. *IEEE J. Select. Areas Commun.*, vol. 23, no. 4, pp. 693–702, Apr. 2005.

[96] S. Appadwedula, V. Veeravalli, and D. Jones, Energy-efficient detection in sensor networks. in *Proceedings of Intern. Conf. on Information Fusion (FUSION 2003)*, Cairns, Queensland, Australia, July 2003, pp. 764–771.

[97] D. Kotz and R. S. Gray, Mobile agents and the future of the internet. *IEEE Trans. Automat. Contr.*, vol. 28, pp. 1081–1090, Dec. 1983.

[98] A. Wald, *Sequential Analysis*. New York: Dover, 1947.

[99] D. Siegmund, *Sequential Analysis: Tests and Confidence Intervals*. New York: Springer-Verlag, 1985.

[100] M. Woodroofe, *Nonlinear Renewal Theory in Sequential Analysis*. Philadelphia, PA: SIAM, 1982.

[101] A. G. Tartakovsky and I. A. Ivanova, Approximations in sequential rules for testing composite hypotheses and their accuracy in the problem of signal detection from post-detector data. *Prob. Inform. Transm.*, vol. 28, no. 1, pp. 55–64, Jan.-Mar. 1992.

[102] S. Tantaratana, Some recent results on sequential detection. in *Advances in Statistical Signal Processing*. New York: JAI, 1993, vol. 2, pp. 265–296.

[103] S. Tantaratana and V. Poor, Asymptotic efficiencies of truncated sequential tests. *IEEE Trans. Inform. Theory*, vol. IT-28, pp. 911–923, Nov. 1982.

[104] G. Mergen and L. Tong, Type based estimation over multiaccess channels. *IEEE Trans. Signal Processing*, vol. 54, no. 2, pp. 613–626, Feb. 2006.

[105] S. Marano, V. Matta, L. Tong, and P. Willett, A likelihood based multiple access for estimation in sensor networks. *IEEE Trans. Signal Processing*, vol. 55, no. 11, pp. 5155–5166, Nov. 2007.

[106] K. Liu, H. El Gamal, and A. Sayeed, Decentralized inference over multiple-access channels. *IEEE Trans. Signal Processing*, vol. 55, pp. 3445–3455, July 2007.

[107] K. Liu and A. M. Sayeed, Type-based decentralized detection in wireless sensor networks. *IEEE Trans. Signal Processing*, vol. 55, pp. 1899–1910, May 2007.

[108] G. Mergen, V. Naware, and L. Tong, Asymptotic detection performance of type-based multiple access over multiaccess fading channels. *IEEE Trans. Signal Processing*, vol. 55, pp. 1081–1092, Mar. 2007.

[109] S. Appadwedula, V. V. Veeravalli, and D. Jones, Decentralized detection with censoring sensors. *IEEE Trans. Signal Processing*, to appear.

[110] S. Marano, V. Matta, P. Willett, and L. Tong, Cross-layer design of sequential detectors in sensor networks. *IEEE Trans. Signal Processing*, vol. 54, no. 11, pp. 4105–4117, Nov. 2006.

[111] W. P. Tay, J. N. Tsitsiklis, and M. Z. Win, Asymptotic performance of a censoring sensor network. *IEEE Trans. Inform. Theory*, vol. 53, pp. 4191–4209, Nov. 2007.

[112] C. E. Shannon, A mathematical theory of communication. *Bell Syst. Tech. Journ.*, vol. 27, pp. 379–423, 623–656, July-Oct. 1948.

[113] S. Kullback and R. A. Leibler, On information and sufficiency. *Ann. Math. Statist.*, vol. 22, pp. 79–86, 1951.

[114] S. Kullback, *Information Theory and Statistics*. New York: Dover, 1968.

[115] H. Chernoff, A measure of asymptotic efficiency for tests of a hypothesis based on a sum of observations. *Annals Math. Statist*, vol. 23, pp. 493–507, 1952.

[116] R. J. McEliece, *The Theory of Information and Coding*. London, UK: Addison-Wesley, 1997.

[117] P. J. Huber, A robust version of the probability ratio test. *Ann. Math. Statist.*, vol. 36, no. 6, pp. 1753–1758, Dec. 1965.

[118] S. A. Kassam and H. V. Poor, Robust techniques for signal processing: A survey. *Proc. IEEE*, vol. 73, no. 3, pp. 433–481, Mar. 1985.

[119] P. J. Huber, *Robust Statistics*. New York: John Wiley & Sons, 1981.

[120] H. V. Poor, *An Introduction to Signal Detection and Estimation*. New York: Springer-Verlag, 1988.

In: New Signal Processing Research
Editor: Takumi Maeda, pp. 89-126
ISBN 978-1-60456-479-2

Chapter 3

Recent Advances in Image Analysis and Data Mining for DNA Microarray Processing

Ying Shan[1]*, ***Guang Deng***[1†] ***and Ju Jia Zou***[2‡]
[1] Department of Electronic Engineering
La Trobe University Bundoora
Victoria 3086, Australia
[2] School of Engineering
The University of Western Sydney
NSW 1797, Australia

Abstract

DNA microarray images consist of thousands of spots with weak intensity arranged in a matrix whose rows correspond to genes and columns correspond to experiments. The DNA microarray technology has emerged as a powerful and cost-effective tool for understanding of gene expression, regulation and interaction through a simultaneous study of thousands of genes. A successful microarray technology depends on seven main stages: experiment design, microarray fabrication, microarray image processing, data mining, biological illustration, data publication and data management. The focus of this paper is thus on a survey of recent advances in image analysis and data mining for DNA microarray technology. In image analysis, we discuss techniques for grid recognition, spot segmentation and quantification. In data mining, we discuss techniques for missing data imputation and classifications. We thoroughly survey recent literature on these tasks and briefly describe the fundamentals behind recent developments. We also present simulation results to illustrate the applications of some typical techniques.

1. Introduction

The DNA microarray technology [86] has become an increasingly active branch of bioinformatics research. It has emerged as a powerful and cost-effective tool for the understanding

*E-mail address: y.shan@latrobe.edu.au
†E-mail address: d.deng@latrobe.edu.au. (Corresponding author).
‡E-mail address: j.zou@uws.edu.au

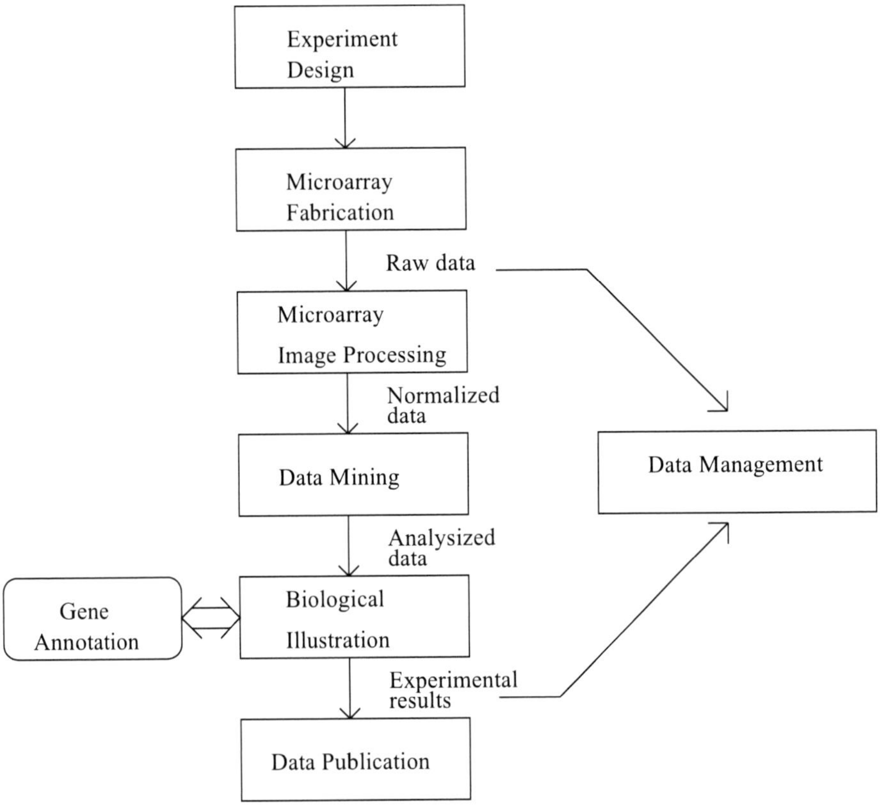

Figure 1. Diagram for work flow of microarray technology, which is composed of seven main stages, i.e., experiment design, microarray fabrication, microarray image processing, data mining, biological illustration, data publication and data management.

of gene expression, regulation and interaction through a simultaneous study of thousands of genes. The DNA microarray technology allows the study and classification of genes in a much shorter time than ever before. Thus functional applications of microarrays are used extensively in clinical research, agricultural development and drug discovery [1], and many other areas. The goal in clinical research is to develop new techniques for the diagnose and treatment of human disease. Challenges in drug discovery motivate the application of microarray technology. In agricultural applications, the aim is to identify plant pathogens to improve plant protection.

In general, the microarray technology consists of seven stages: experiment design, microarray fabrication, microarray image processing, data mining, biological illustration, data publication and data management. The work flow is shown in Figure 1. It has been recognized that, in the whole work flow, microarray image analysis and data mining have prominent impacts on the identification of differentially expressed genes.

The DNA microarray images are made up of thousands of spots with weak intensity arranged in a compact matrix whose rows correspond to genes and columns correspond to various experiments. Since they are different from images of natural scenes and medical images, specialized techniques have been developed for the processing and extraction of information. In image analysis, there are three main tasks: gird recognition, spot segmentation and quantification. In data mining, theories and algorithms for statistical analysis are becoming indispensable tools for missing value estimation and data analysis based on supervised/unsupervised learning.

The aim of this paper is to provide an extensive survey and evaluation of recent advances in microarray image and data analysis. We also discuss some interesting research problems. The rest of this paper is organized as follows. In Section 2., we discuss advances on microarray image processing analysis. After a brief introduction of microarray experiment, we describe techniques used in grid recognition, spot segmentation and quantification. In Section 3., we discuss the missing data imputation methods including methods based on classical tools such as the k-nearest neighbours and least squares and methods based on Bayesian, projection onto convex sets and meta-data. We also present simulation results. In Section 4., we discuss several data classification techniques, including unsupervised, supervised and Bayesian classification. Finally, Section 5. summarizes techniques in image analysis and data mining techniques for DNA microarray research.

2. Microarray Image Processing and Analysis

2.1. Introduction

A typical microarray slide usually contains several rectangular or square blocks of intensity spots immobilized in a lattice or grid. Often a microarray slide is a 2×4 cm^2 membrane or a microscope glass or a silicon substrate. After scanned by lasers with two specific wavelengths, the microarray slide is converted into a pair of microarray images in the TIFF format using 16 bits to represent the intensity of each pixel. An illustration of microarray images is shown in Figure 2. It can be seen that the region of the microarray image carrying features is composed of thousands of the microscopic spots arranged in a tiny and compact matrix. Each image appears as an $M \times N$ gene expression matrix whose rows correspond to M genes, and columns correspond to N experiments. Main difficulties arising in image analysis are originated from the microarray images to be processed, which are called "raw data" for the whole work flow of microarray technology. Raw data is a particular type of images comprising a large number of spots with weak intensity. They are different from either natural scene images or medical images. Since the microarray experiment is a complex electric-optimal-chemical procedure, noise is frequently found in the raw data. Noise filtering is usually performed to ensure high quality biological measurements are obtained.

The aim of the image analysis step is to reduce an image of spots into a data table which represents gene expression. The performance of image processing has a significant influence on further data analysis steps in the complete microarray analysis workflow. We divide this step into three main stages [111]: grid recognition, spot segmentation and quantification. We will discuss techniques for these three stages in Sections 2.3., 2.4. and 2.5., respectively.

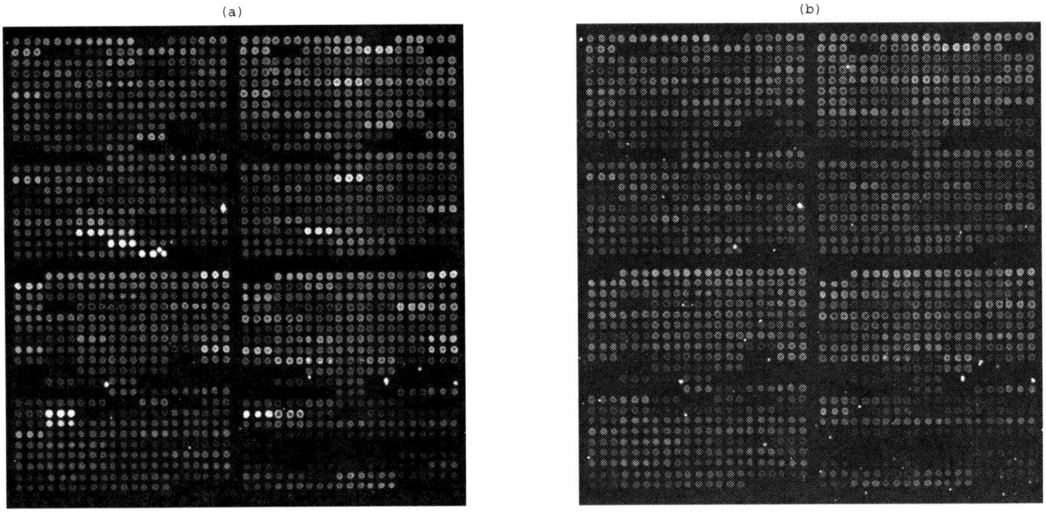

Figure 2. A pair of microarray images (2 × 2 blocks each) in TIFF format 16 bits. Spots are packed in blocks with weak intensity. (a) the microarray image from Cy3 channel, (b) the microarray image from Cy5 channel.

2.2. Generation of Microarray Images

We briefly review the fundamentals in microarray experiment and describe the generation of microarray images, both of which are helpful in fully understanding DNA microarray methodology. We then outline typical procedures in microarray image analysis.

2.2.1. Microarray Experiments

The target of a microarray experiment is to generate a pair of microarray images, namely *raw data*, representing the abundance of mRNA from a reference sample and a target sample, i.e., in practice usually a normal sample and an abnormal sample.

A microarray experiment starts with a specific experiment design which determines what can be inferred from the data and which is essential to the successful use of microarrays. Under the careful experiment setup, the DNA molecules containing interested genes, will be de-helixed under suitable chemical or heat conditions. DNA strands, called probes, taken from time-points during a biological process, for example, are immobilized onto a glass or silicon substrate at a known position in the array. This is in practice an automatic dipping procedure, completed by a high-speed robot. On the other hand, a target sample (e.g. from tumor cells) and a reference sample (e.g. from normal cells) are labeled with red color and green color, respectively, by fluorescent dyes Cy5 and Cy3 for distinguishing reason. The choice of Cy5 (635nm, red) and Cy3 (532nm, green) is for the reason that they are brighter, more photo stable and give less background than most other fluorophors. Two samples are usually mixed with equivalent volume in a tube. The mixture is then allowed to hybridize with the probes on the array simultaneously. An illustration of the microarray experiment is shown in Figure 3. After hybridization, the array is scanned by laser with

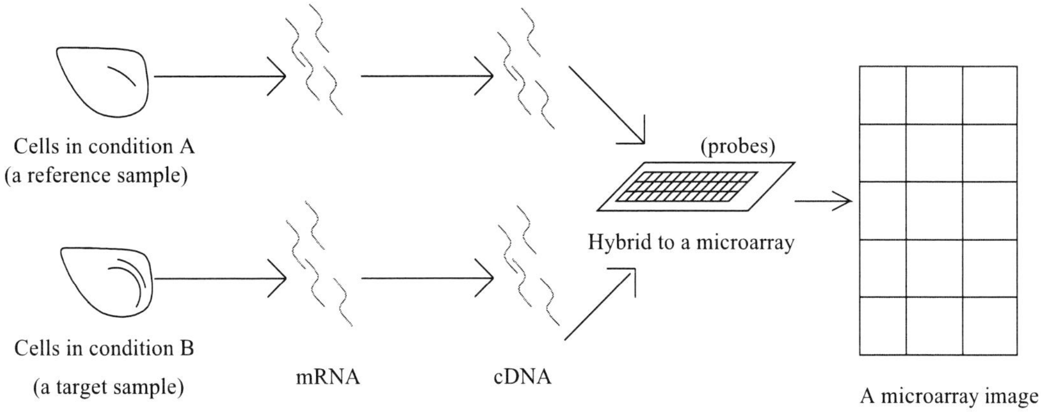

Figure 3. An illustration diagram of the fabrication procedure of cDNA microarrays.

two wavelengths to excite each dye. Two gray scale microarray images are generated. The intensity of each pixel is represented by 16 bits. As seen in Figure 2, the images correspond to Cy3 channel and Cy5 channel, respectively. Each image, typically over 30 megabytes in size, appears as an $(M \times N)$ matrix whose rows stand for M genes, and columns correspond to N experiments. Usually the number of genes is greater than the number of experiment conditions.

In practice, images are converted to 8 bits/pixel for visualization and analysis. We also perform this conversion in our simulation experiments in this section. After the microarray experiment, the two remaining important tasks are analysis and storage. We focus on microarray analysis in this paper, including image analysis and data mining techniques. The management, storage and reporting of microarray data have been standardized [74]. A widely accepted international standard is MIAME (Minimum Information About a Microarray Experiment) [76]. Most of microarray manufacturers and software producers have MIAME compliant products. A detailed review on storage and transmission of microarray images can be found in [74].

2.2.2. Procedures in Microarray Images Analysis

To successfully implement image processing on microarray images, it is important to understand the characteristics of microarray images. Since a microarray experiment is a chemical-electrical-optical process, numerous variations could happen. Sources of variations [10] include

- Grid geometry: Any rotational offset of a slide caused by a rotated 2D grid in a microarray image with respect to the image edge.

- Spot morphology: A spot may not always be round, and its size may not always be fixed. Spots may have different diameters. They may come in all shapes, such as donut, oval or pear, etc, caused by distortion of dipping pin, dirt on the slide or other treatment.

- Pixel classification: Though ideally all pixels on a microarray image will be categorized into two groups: foreground and background, one may find the artifacts as the third group of pixels, because of the presence of dust or dirt, or the hybridization error in the experiment.

- Normalization: This refers to the variations in foreground/background contrast. Incorrect high background and/or low foreground would impair the accuracy in extracting foreground and background intensity. The variations of foreground and background intensities may be caused by uneven slide material.

The task of gene expression is to obtain a data table containing the ratio of intensity of foreground and background. Microarray image analysis can be divided into three main steps [111]: (1) grid recognition, (2) spot separation and (3) quantification. Grid recognition or alignment is a procedure of assigning coordinates to the spots. Spot separation or segmentation allows determining the borders of each spot. Quantification or intensity extraction aims at generating a gene expression matrix by calculating the foreground intensity and background intensity. We discuss methods and challenges in each step below.

2.3. Grid Recognition

There are two types of grid alignment approaches: template-based (i.e. manual or semi automated approaches) and data-driven approaches (i.e. automatic approaches) [11]. The former is employed in some software packages, while the latter, which allows high throughput analysis, is used in recent algorithms.

2.3.1. Manual Griding

Using manual griding, the user can adjust spot size, spot spacing and grid location based on grid templates. Currently available software packages that have incorporated template-matching steps include GenePix Pro by Axon Instruments [8], and ScanAlyze by Eisen [32]. Advantages of manual griding include:

- Independent of parameters once gridding starts.

- Simple procedure.

- Not prone to be intervened by spurious or missing spots.

Therefore, template-based approach is appropriate for cases in which the measured grid geometry does not distort much from the expected grid model. Some latest gridding methods can employ this non-parametric procedure as the initial step [37]. Nevertheless, one obvious disadvantage is that it is less effective if distortions or irregularities are not defined by the template in advance.

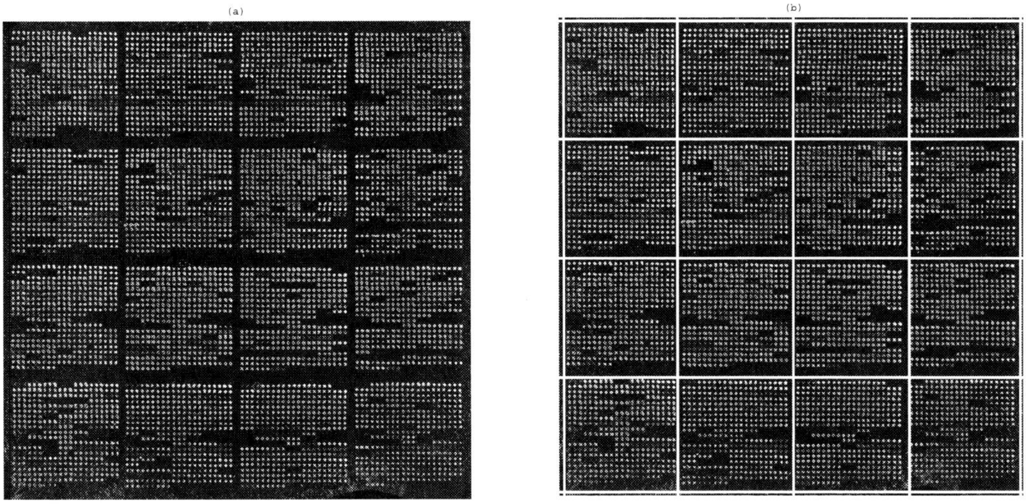

Figure 4. An illustration of block alignment result. (a) the original microarray image consists of 4×4 blocks, (b) the block alignment result.

2.3.2. Automatic Gridding

It is able to automatically identify or index each block in the whole slide as well as each individual spot in a block. Automatic gridding is a data-driven method capable of finding irregular grids and depends on parameters and statistic analysis. Automatic grid recognition consists of two essential components: 1) block alignment which detects and corrects rotated blocks; 2) grid matching which generates grid coordinates. A detailed description of these two components is presented below.

- Block alignment

There are two steps: a block localization and a rotation detection. For the block localization, the basic dogma is to treat the profiles (e.g. projections) along x-axis and y-axis, respectively. For an ideal image without rotation, we can extract the horizontal and vertical profile and then locate the most substantial valleys in the two profile curves [33]. Figure 4 shows a diagram for the block alignment result.

For a rotated block, we need to identify the block rotation by carefully examining the intensity projection profiles along the x-axis and y-axis. If the size of the block is greater than average, then the block is identified as being rotated. The direction of rotation can be determined by comparing pixel intensities in the upper half and lower half of the image. To calculate the rotation angle, the distorted region is iteratively rotated by a small degree until its projection is closely matched to a normal profile [105]. This is based on the assumption that there are only a few blocks rotated by a small degree.

- Grid matching

Grid matching finds the address of each spot in a block. Methods based on mathematical morphology have been studied [41], [17]. The R and G intensities are added along the

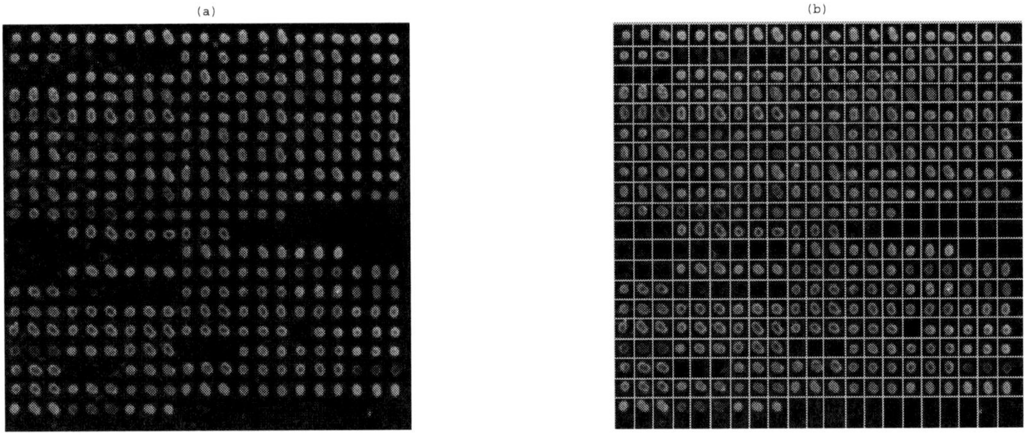

Figure 5. Grid matching by profiles in 2D. (a) the original image, (b) the grid matching result.

vertical and horizontal directions in an array matrix. The obtained signals have multiple peaks. The location of each peak is regarded as the coordinate of a spot center and the average of two successive peak values is taken to define the global boundary of a spot mask. A simulation result from this strategy can be seen in Figure 5.

An efficient method [7] using the Radon Transform (RT) can find both the spot rotation and coordinates simultaneously. The method of addressing coordinates of each spot is described as follows,

$$x_{ij} = x_0 + i \triangle y \sin\beta + j \triangle x \sin\alpha \tag{1}$$

and

$$y_{ij} = y_0 + i \triangle y \cos\beta + j \triangle x \cos\alpha, \tag{2}$$

where (i, j) are the positions of a spot in the array, (x_0, y_0) are the coordinates of the upper left spot in one block, $(\triangle x, \triangle y)$ are the grid spacing and (α, β) are the angles between the grid directions and the x axis.

There are software packages which provide functions for determining the rotation angles and coordinates of spots. For example, Gridline is able to detect other primitive shapes than circular shape and has a build-in speed versus accuracy trade-off mechanism to accommodate user's requirements [9]. Gridline has been embedded in a set of components of the software *Image to Knowledge* (I2K) which is available from the website http://alg.ncsa.uiuc.edu/tools/docs/i2k/manual/index.html.

2.4. Spot Segmentation

Segmentation is a key step in microarray image analysis. Recent advances in segmentation can be categorized into three classes: (a) methods based on spatial features, (b) methods based on intensity information, and (c) hybrid methods which use both spatial and intensity information. A summary of these methods is listed in Table 1.

2.4.1. Segmentation Based on Spatial Features

Since round spots are the primitive shape in microarray images, the circular constraint is adopted in many software packages, such as ScanAlyze [32] and GenePix [8]. The former identifies all the objectives with a fixed radius, whereas the latter is able to detect the foreground spots with an adaptive span. Mathematical morphology, which is widely used in binary and gray-scale image processing, has also found applications in microarray image processing. Two typical examples are the watershed transform and seeded region growing (SRG) [111], [110]. The SRG algorithm has been successfully intergrated into two software packages: Spot [23] and SLOCO [45]. Edge detection [82] is also used in segmentation. There are many edge detection algorithms such as Canny, Sobel, Laplacian and zero-crossing in the second derivative. Results of using typical edge detection algorithms are shown in Figure 6.

2.4.2. Segmentation Based on Intensity Information

This type of segmentation of pixels is performed within a spot mask and is based entirely on the intensity information. The method implemented in the QuantArray software uses a threshold for classification. The threshold is calculated from the histogram of pixel values in a square mask around a spot. Chen et al. [27] introduced a method using a circular target mask to cover all foreground pixels and iteratively computing a Mann-Whitney test as the threshold for classification. An improvement [42] is achieved by using a modified Mann-Whitney test and carefully choosing the start point and the search strategy.

Applications of K-means clustering and Gaussian mixture models are two representative developments in microarray image segmentation.

- K-means clustering [18]

The aim is to cluster the pixels into the foreground and the background. Mean values for the foreground and background pixels are initialized as the highest intensity and the lowest intensity in the image, respectively. Then each pixel in a spot mask will be classified depending on its distance to the two mean values. After all pixels are processed, two new mean values are calculated for the two groups of pixels. This process is continued until the convergence condition is satisfied. Two extensions are the fuzzy C-means clustering (FCM) [106] and the Partitioning Around Medoids (PAM) [77]. In FCM the weighted sum of the Euclidean distance between all observations to both cluster centers is minimized. In this sense, the FCM uses a "softer" clustering strategy than K-means clustering. PAM uses the Manhattan distance which is insensitive to outliers and is appropriate for the particular task of microarray image clustering.

- Gaussian mixture models (GMM) [17]

Methods using Gaussian mixture models rely on the assumption that the intensity x within a spot is a random variable following a GMM of n components. Parameters to be estimated

Table 1. Summary of advances in segmentation algorithms for microarray image analysis

(a) methods based on spatial features	description
(a.1) Using shape constraints	
Fixed circles: ScanAlyze [32]	Circle spot with fixed radius
Adaptive circles: GenePix [8]	Circle spot with adaptive radius
(a.2) Using mathematical morphology	
Seeded region growing (SRG) [111]	Segmentation by SRG algorithm, used in Spot software
Seeded region growing (SRG) [45]	Segmentation by SRG algorithm, used in SLOCO software
Watershed transform [89]	Segmentation by Watershed transform
Watershed transform [5]	Segmentation by Watershed transform
Watershed transform [88]	Morphological opening as denoising preprocessing
(a.3) Using edge detection	
Neighbourhood thresholding [35]	Pixel value comparison with four neighbours, thresholding the difference
Laplacian edge detection: Dapple [24]	Adaptive radius combined with Laplacian edge detection
Canny edge detection [105]	Strong/weak thresholding with Gaussian filter and Sobel edge detection
Hough Transform [19]	Merit to circle detection
Orientation Matching Transform (OMT) [6]	An extension of Hough Transform
Modulus Maxima [104]	Zero-crossing edge detection based on wavelet transform
(b) methods based on intensity information	
Mann-Whitney test [28]	Computing two sample mean with Mann–Whitney test
Modified Mann–Whitney test [42]	Reducing the number of iterations by choosing the search strategy
K-means clustering [18]	K-mean clustering of pixels
Hybrid K-means	K-mean clustering of pixels and removing outliers with mask matching
Partitioning around medoids (PAM) [94]	PAM clustering of pixels
Fuzzy C-means clustering [106]	Fuzzy clustering approach, an improvement of K-means by soft clustering
Markov random field (MRF)[50]	MRF modeling of pixels
Flexible Markov random field [38]	Allowing a wider range of spot shapes
Gaussian Mixture Model (GMM) [21]	Parametric or semiparametric modeling of pixels
Gaussian Mixture Model (GMM) [17]	Scalar GMM with two or three normal distributed components
Thresholding technique: TIGR-Spotfinder [96]	Global and local thresholding
(c) hybrid methods	
Clustering and morphology analysis [108]	K-means clustering combined with mathematical morphology
An ad-hoc pipeline [12]	K-means clustering combined with seeded region growing
Robust model-based segmentation [66]	Gaussian Mixture Model combined with spatial features
Matarray [103]	Iteratively updating of spot mask based on spatial and intensity information

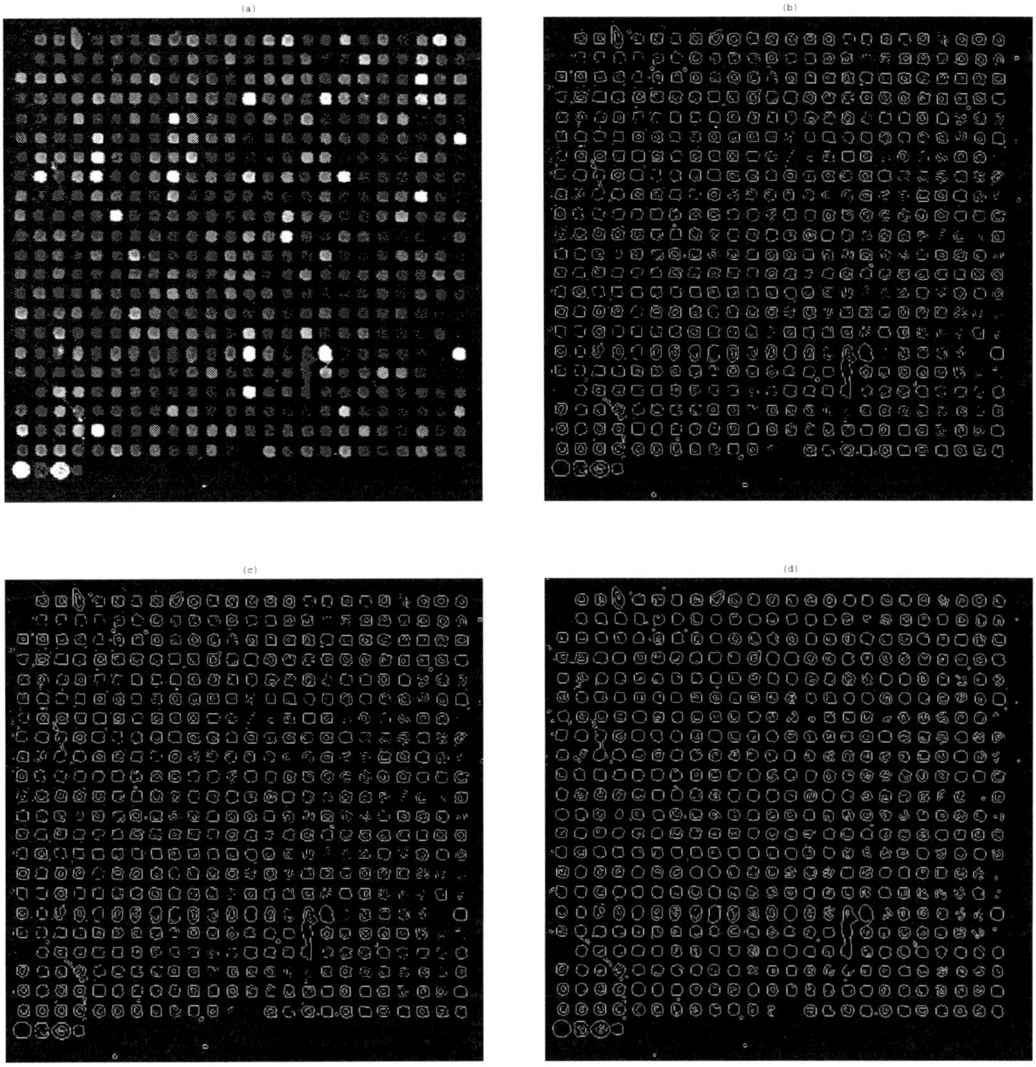

Figure 6. Edge detection of original microarray image (a), respectively, by zerocrossing algorithm (b), Laplacian of Gaussian algorithm (c) and Canny algorithm (d). Thresholds for (b) and (c) are identical, (d) with strong and weak two thresholds whereas the weak threshold= 0.4×the strong threshold.

include the mean μ_j and the variances σ_j of the jth Gaussian component $\mathcal{N}(x|\mu_j, \sigma_j)$. The scale coefficients ρ_j, which are also estimated, must satisfy the requirements $0 \leq \rho_j \leq 1$ and $\sum_{j=1}^{n} \rho_j = 1$ $(j = 1, \ldots, n)$. By denoting vectors $\Theta = \{\mu_1, \sigma_1, \ldots, \mu_n, \sigma_n\}$ and $\Pi = \{\rho_1, \ldots, \rho_n\}$, the density function within a spot mask is given by

$$p(x|\Theta, \Pi) = \sum_{j=1}^{n} \rho_j \mathcal{N}(x|\mu_j, \sigma_j). \tag{3}$$

In practical applications, the number of components is assumed $n = 2$ for background and foreground classes. If artifacts are considered as another class, then $n = 3$. Parameter estimation can be based on maximum likelihood (ML) and the Expectation-Maximization (EM) algorithm [40] is a popular tool.

2.4.3. Segmentation Based on Hybrid Methods

Hybrid methods use the spatial and intensity information. The Matarray toolbox [103] is a Matlab based package. Segmentation is performed by an iterative algorithm utilizing both spatial characteristics and intensity information. In Matarray, the spatial information updates the spot mask so that the spot is inside the circular mask, and the intensity information is used to define the foreground and the background. Another hybrid method [108] uses K-means ($K = 2$) clustering by employing morphological analysis in its initialization. The number of iterations is reduced comparing to the ordinary K-means clustering segmentation method. An improvement [84] is achieved when $K = 3$ is assumed. An improved GMM approach [66] combines spatial and density information to improve the spatial coherence of the estimated spots. It attempts to process irregular shape of spots such as donuts, scratches and blanks.

2.5. Quantification

2.5.1. Intensity and Expression Matrix

In the final stage, the intensity of foreground and background pixels need to be reliably extracted. Before we turn to the process of intensity extraction, let us clarify the relationship between the intensities and the gene expression. The important target of microarray image processing is to export a gene expression matrix, an example of which is shown in Table 2. When searching for a set of differentially expressed genes among biological samples, most investigators choose a threshold of difference in expression level. Let r and g denote the intensity measurement of red and green fluorescent channel at a particular location, respectively. They are obtained by $r = r_f - r_b$ and $g = g_f - g_b$ where the subscripts f and g represent values in the foreground and background, respectively. The gene expression is then calculated as the log ratio between the r and g values

$$L = \log_2 \frac{r}{g}. \tag{4}$$

When the value of the log ratio is 1, the expression level of a particular gene is 2, which means the intensity of r is as twice strong as that of g. In order to clearly display how a gene is expressed, we need to discuss the ways to extract intensities of spots in a mask generating the gene expression. When the green fluorescent channel data is zero ($g = 0$), the log ratio is not defined. To avoid this problem, the relative difference [99] is defined as follows

$$D = \frac{r - g}{\frac{1}{2}(r + g)}. \tag{5}$$

The percentage of relative difference has a straightforward interpretation. For example, a value of 15.8% means that there are 15.8 percent *more* red pixels in the average luminosity of the spot studied.

Table 2. An example of gene expression matrix, with rows corresponding to genes, columns corresponding to experiments. The values in the matrix are calculated as $L = \log_2 \frac{r}{g}$

log2 Ratio	Expt1	Expt2	Expt3	Expt4	Expt5	Expt6
Gene A	-2.32	-1.69	-0.87	-0.12	0.73	1.42
Gene B	2.71	2.09	1.24	0.71	0.08	0.25
Gene C	-0.55	-0.49	0.97	1.32	0.89	-0.38

2.5.2. Intensity Extraction

There are three spot types representing background area, foreground area and noisy area. After noise filtering, we can calculate the intensity of the foreground (r_f, g_f) and the background (r_b, g_b). A natural measure of the pixel intensity is by taking the median or average value of pixel intensities within a spot mask. The acquired data are finally collected as a matrix to export. Each column corresponds to an experiment and data in each row correspond to genes.

3. Missing Data Imputation

3.1. Introduction

Once the gene expression information is obtained, the next step in the microarray analysis work flow is to study the co-expression among genes by data mining methods. Commonly used data mining techniques such as hierarchical clustering, K-means clustering, self-organizing maps (SOM) and so on, are based on a complete expression matrix. However, an expression matrix with missing values could frequently happen in practice. Sources for missing values include: (1) the weak spots may be flagged missing expression, (2) failure in hybridization in the microarray fabrication process leads to low signal to noise ratio and (3) scratches and systematic errors on the slides may also be treated as missing values. It is reported [30] that there are between 0.8% and 10.6% missing values and genes with one or more missing values range from 3.8% to 94.7%. Therefore, estimating missing values is a key preprocessing step which is called imputation [70].

When choosing an imputation method for the estimation of missing values, we need to take the following into consideration: the experiment design, patterns of correlation in data, dimensionality and percentage of values missing. In order to evaluate the effectiveness of different methods, missing values are artificially introduced into an original microarray image (matrix) denoted $\mathbf{G}$. The result of an imputation process is another image of the same size, denoted $\mathbf{G}^*$. The quality of the imputation method is then measured by using the normalized root mean squared error (RMSE) defined as follows

$$\mathrm{RMSE} = \left(\frac{\sum_{k=1}^{M} (\mathbf{G}(k) - \mathbf{G}^*(k))^2}{\sum_{k=1}^{M} \mathbf{G}^2(k)} \right)^{\frac{1}{2}}, \tag{6}$$

where $\mathbf{G}(k)$ and $\mathbf{G}^*(k)$ are the kth elements of $\mathbf{G}$ and $\mathbf{G}^*$, respectively. There are other metrics such as false positive rate and false negative rate [62]. They are not discussed in this paper. In the rest of this section, we discuss different imputation methods and study the performance of five typical estimation methods.

3.2. Imputation Using Local Information

3.2.1. K-nearest Neighbour Imputation (KNNimpute) [98]

The KNN imputation method estimates the missing value from information from its K-nearest neighbours. KNNimpute is regarded as a typical method using local information. Because of its low computation cost, it is used in SAM [95], PAM [94] and MAANOVA [44]. The selection of K nearest genes is based on the Euclidean distance. The final estimate of a missing value is the weighted average of the selected K neighbours. It is pointed out in [98] that when $K = 10 \sim 20$, KNNimpute performs better than choosing other ranges of neighbours.

Key steps of the algorithm is listed as follows.

Step 1. Find a target containing a missing value in an experiment (Column 1 for example).

Step 2. Measure the distance between the target gene and all other genes having values present in the experiment.

Step 3. Choose K most similar genes with respect to the distance measurement.

Step 4. Compute weights for K nearest genes based on their similarity to the target gene.

Step 5. Use the weighted combination of K-nearest neighbours to estimate the missing value in the experiment.

Step 6. Estimate other missing values by the procedure from Step 1 to Step 5.

3.2.2. Local Least Squares Imputation (LLSimpute) [54]

To describe the LLSimpute method, we need to set up some terminologies. Let $\mathbf{g}_t$ be a $(N \times 1)$ vector[1] representing a target gene expression with missing values. Without losing generality, we assume the first n-element, denoted as vector $\mathbf{g}_e$, are missing. The rest of the vector $\mathbf{g}_t$ is denoted $\mathbf{g}_p$. Suppose we can find K other genes, denoted $\mathbf{g}_k$ $(k = 1, 2, \ldots, K)$, which are similar to $\mathbf{g}_t$. Each of these genes can also be partitioned into two parts, i.e., the first n-elements denoted $\mathbf{g}_{ek}$ and the rest elements denoted $\mathbf{g}_{pk}$. We collect these sub-vectors into two matrices $\mathbf{G}_p = \left[\mathbf{g}_{p1}\mathbf{g}_{p2}\cdots\mathbf{g}_{pK}\right]$ and $\mathbf{G}_e = \left[\mathbf{g}_{e1}\mathbf{g}_{e2}\cdots\mathbf{g}_{eK}\right]$. The idea of

[1]In a microarray a gene expression is usually represented as a row vector. Here we use a column vector representation to simplify notations.

LLSimpute is to determine an optimal linear combination of $\mathbf{g}_{pk}$ to approximate $\mathbf{g}_p$ in the least squares sense

$$\begin{aligned}\mathbf{w} &= \arg\min_{\mathbf{w}} \|\mathbf{g}_p - \mathbf{G}_p\mathbf{w}\|_2 \\ &= (\mathbf{G}_p^T\mathbf{G}_p)^{-1}\mathbf{G}_p^T\mathbf{g}_p. \end{aligned} \quad (7)$$

The missing values are then obtained by

$$\hat{\mathbf{g}}_e = \mathbf{G}_e\mathbf{w} \quad (8)$$

To determine the K similar genes, we can use the Euclidean distance and Pearson's correlation coefficient [61]. The Pearson's correlation coefficient between the target gene $\mathbf{g}_t$ and the other gene $\mathbf{g}_k$ is defined as follows

$$c = \frac{1}{N-n}\bar{\mathbf{g}}_p^T\bar{\mathbf{g}}_{pk}, \quad (9)$$

where $\bar{\mathbf{g}}_{pk} = (\mathbf{g}_{pk} - \mu_{pk})/\sigma_{pk}$ is the vector $\mathbf{g}_{pk}$ shifted by its mean μ_{pk} and scaled by its standard deviation σ_{pk}. The vector $\bar{\mathbf{g}}_p$ is defined in the same way. We note that in (9) we have taken the missing values into consideration and only use the available data in the target gene.

Key steps of the LLSimpute are listed as follows.

Step 1. Compute the K most similar genes based on the Euclidean distance or Pearson's correlation.

Step 2. Estimate the missing values by using (7) and (8).

Step 3. Repeat Step 1 to Step 2 to estimate all missing values.

Interested readers are referred to [54] for applications of more advanced regularized least squares methods in missing value imputation.

3.2.3. Gaussian Mixture Clustering Imputation (GMCimpute) [80]

In GMCimpute, the ith gene $\mathbf{g}_i$, represented as a row vector in the expression matrix $\mathbf{G}$, is modeled by a Gaussian mixture of L components. Each component is a multivariate Gaussian distribution with mean μ_k and covariance matrix Λ_k, i.e.,

$$p_k(\mathbf{g}_i|\mu_k, \Lambda_k) = \mathcal{N}(\mu_k, \Lambda_k). \quad (10)$$

Let ρ_k denote the mixing parameter of each component, where $0 \leq \rho_k \leq 1$ and $\sum_{k=1}^{L} \rho_k = 1$. It is assumed that each gene follows an i.i.d. Gaussian mixture distribution. As such, the likelihood function for the expression data is given by

$$\mathcal{L}(\mu_1, \Lambda_1, \ldots, \mu_k, \Lambda_k|\mathbf{G}) = \sum_{i=1}^{M}\sum_{k=1}^{L} \rho_k p_k(\mathbf{g}_i|\mu_k, \Lambda_k). \quad (11)$$

The complete set of parameters $\Theta = \{\rho_1, \ldots, \rho_L, \mu_1, \ldots, \mu_L, \Lambda_1, \ldots, \Lambda_L\}$ can be iteratively estimated by the Expectation-Maximization (EM) algorithm [40]. Each missing value has L estimates, then the final estimation is the average of them.

Key steps of the algorithm are listed as follows.

Step 1. Set the value for L (usually $L < 10$), the number of Gaussian components. The value can be determined by using the Bayesian information criterion (BIC) or the Bayes factor (BF) [107].

Step 2. Initialize the L means, $\mu_1, \ldots, \mu_L$ by using the technique proposed in [20]. Cluster the data by the Classification Expectation Maximization (CEM) algorithm [26].

Step 3. Set up a mixture distribution model from which the expression data are generated, i.e., $\sum_{k=1}^{L} \rho_k \mathcal{N}(\mu_k, \Sigma_k)$.

Step 4. Iteratively estimate the whole parameters

$$\Theta = \{\rho_1, \ldots, \rho_K, \mu_1, \ldots, \mu_K, \Sigma_1, \ldots, \Sigma_K\} \tag{12}$$

by using the EM algorithm.

Step 5. Calculate the missing value by averaging the L estimates.

3.3. Imputation Using Global Information

3.3.1. Singular Value Decomposition Imputation (SVDimpute) [98]

Singular value decomposition (SVD) can only be implemented on a complete matrix. In order to estimate missing values by SVDimpute, one needs to firstly replace the missing values with zeros or row averages. One then finds eigengenes by using SVD on the complete matrix. Finally one estimates the missing entry by the combination of eigengenes, where the combining coefficients are determined by regressing the target gene with the K eigengenes. SVDimpute ideally suits expressions which have low noise and are structured by a strong regulation [98]. When the matrix is not in this category, SVDimpute is less accurate and robustness comparing to KNNimpute. Recent applications of SVD-based genome data modeling and processing include [3], [112], [4], [2].

Key steps are listed as follows.

Step 1. Substitute all missing values with zeros or row averages.

Step 2. Calculate the SVD of the complete gene expression $\mathbf{G}$

$$\mathbf{G} = \mathbf{USV}^T. \tag{13}$$

Matrix $\mathbf{V}^T$ now contains eigengenes whose contribution to the expression matrix $\mathbf{G}$ is indicated by eigenvalues, i.e., the diagonal elements of matrix $\mathbf{S}$.

Step 3. Select K most significant eigengenes (eigen vectors in $\mathbf{V}^{\mathrm{T}}$) which correspond to the K-largest eigen values.

Step 4. Decide the combination parameters of the K significant eigengenes by regressing the target gene with the K eigengenes. Notice that the elements of eigengenes corresponding to the index of missing values are not used in regression.

Step 5. Estimate the missing value by a weighted combination of the K eigengenes.

Step 6. Repeat Step 2 to Step 5 until the total change in the matrix $\mathbf{G}$ is below a threshold.

3.3.2. Bayesian Principal Component Analysis (BPCA) [78]

The BPCA imputation method[2], which extends the SVDimpute by using a parametric probabilistic model, can determine the parameters and eigengenes in a Bayesian framework. BPCA imputation is based on the assumption that the estimation residual and the combining parameters are Gaussian random variables. An important element in the BPCA algorithm is the hierarchical prior which is called an automatic relevance determination (ARD) prior [16]. With this prior knowledge, we can depict the presence or absence of an eigengene by its hyperparameter. When the hyperparameter goes infinity, the corresponding principal vector will be peaked at its mean value which is set to zero, thus the corresponding element could be pruned from Eq. 14. This procedure is called *shrinkage*. All shrinked vectors are pruned from the model and regarded as redundant axes.

More specifically, we use a linear observation model $\mathbf{e} = \mathbf{g} - \mathbf{W}\mathbf{x}$ and setup the following statistical models: $p(\mathbf{e}) = \mathcal{N}(0, \frac{1}{\beta}\mathbf{I}_D)$, $p(\mathbf{x}) = \mathcal{N}(0, \mathbf{I}_k)$ and $p(\mathbf{w}_j|\alpha_j, \beta) = \mathcal{N}(0, \frac{1}{\alpha_j\beta}\mathbf{I}_m)$, where $\mathbf{w}_j$ is the jth column vector of $\mathbf{W}$ and $\mathbf{I}_D$ represents a $(D \times D)$ identity matrix. Similar interpretation applies to $\mathbf{I}_k$ and $\mathbf{I}_m$. BPCA uses an evidence based approach [75], [16] to infer $\mathbf{x}$, $\mathbf{W}$ and all model parameters. The idea is that we can first fixed $\mathbf{W}$ and model parameters and determine an MAP estimate of $\mathbf{x}$. We then fixed $\mathbf{x}$ and determine an ML or an MAP estimate of $\mathbf{W}$ and model parameters. This process is continued until stopping criteria are met. Key steps of PCA and BPCA imputation are listed as follows.

- PCA regression.

The target gene $\mathbf{g}_t$ is partitioned into two parts: $\mathbf{g}_e$ representing missing values and $\mathbf{g}_p$ representing observed data. Similarly we can partition eigengene matrix $\mathbf{W} = [\mathbf{W}_p, \mathbf{W}_e]$ to distinguish the observed data and missing data, where the eigengene matrix $\mathbf{W}$ is composed of K significant eigen vectors by SVD of covariance matrix of other observed gene expression data. The regression result for the missing values is given by

$$\mathbf{g}_e = \mathbf{W}_e\mathbf{x}, \tag{14}$$

where $\mathbf{x}$ is regressed from the observed data $\mathbf{g}_p$ and $\mathbf{W}_p$.

- Bayesian PCA construction.
  Step 1. Initialize the matrix $\mathbf{W}$ and all other parameters α and β
  Step 2. Determine an MAP estimate of the missing values as well as $\mathbf{x}$ by maximizing the log-posterior

$$\begin{aligned}\mathcal{L}(\mathbf{g}_t, \mathbf{x}) &= \log p(\mathbf{g}_t, \mathbf{x}|\mathbf{W}, \beta) \\ &= -\frac{\beta}{2}\|\mathbf{g}_t - \mathbf{W}\mathbf{x}\|^2 - \frac{1}{2}\|\mathbf{x}\|^2 + \frac{D}{2}\log\beta + constant. \end{aligned} \tag{15}$$

  Step 3. Estimate $\mathbf{W}$, α and β.

  Step 4. Repeat steps 2 and 3 until convergence.

[2] A Matlab toolbox is available on line: hawaii.aist-nara.ac.jp/~shige-o/tools/BPCAFill.html.

3.4. Combined Methods

Combining local and global information in predicting missing values appears to be prevalent in recent studies. In this subsection, we summarize four combined methods in imputation study. We begin with *robust least squares estimation with principal component* (RLSP) [113], then discuss *projection onto convex sets method* (POCS) [36], *meta-data based imputation* [46] and another interesting method named *LinCmb* [47] (most likely named after its property of a linear combination of various methods). From these methods, interested researchers would find novel imputation methods in a broader horizon, where new resources are considered, for example, meta-data from public database, biological knowledge of gene network and so on.

3.4.1. Robust Least Squares with Principal Components (RLSP) [113]

RLSP imputation may be considered as a variation to the LLSimpute method and hence an improvement of the least squares application in missing value estimation. There are three main steps in the algorithm, i.e., selection of K similar genes, principal component analysis based on the selected genes and robust regression using principal components. RLSP improves the LLSimpute by applying a principal component analysis to the K-nearest neighbors. Then it finds the weights of eigengenes under least squares optimization. From this viewpoint, it is a combination of LLSimpute and SVDimpute.

Key steps are listed as follows.

Step 1. Selection of K similar genes. (a) Measure the distance of a target gene to all complete genes. (b) Select K most similar neighbours. Here L_1 norm is used in distance measure, instead of L_2 norm or Pearson's correlation coefficients used in LLSimpute, which are sensitive to the outliers.

Step 2. Principal component analysis (PCA). Execute PCA on the K selected neighbours, in order to choose, say p, most significant genes among neighbourhood. When $p = K$, the regression result of RLSP is identical to LLSimpute.

Step 3. Regression. Decide the parameters to the p most significant neighbours in the same manner as in LLSimpute.

3.4.2. Projection onto Convex Sets for Missing Data Imputation (POCS) [36]

The POCS technique [91] has been used in various applications such as image processing and information recovery, but its first introduction to missing value estimation was due to Gan et al. [36]. We first briefly go through the fundamental concepts of POCS, then present its application in microarray data imputation.

A set C is defined *convex*, if $x = \xi x_1 + (1-\xi)x_2 \in C$ for all $x_1, x_2 \in C$ and $0 \leq \xi \leq 1$. In other words, the set C is convex if any point x on the segment between two points x_1 and x_2 also belongs to this set. Here all the points x, x_1 and x_2, and the set C are defined in a Hilbert space. The other vital concept in POCS theory is *projection*. It governs the projection of any x in Hilbert space (in or out of set C) to a point x^* in convex set C, with respect to the minimum distance in the Hilbert space. The central idea of POCS is to (1) define every piece of prior knowledge by a convex set in a Hilbert space, (2) successively project a matrix (a point in the Hilbert space) onto those sets by appropriate mathematical expression, and (3) obtain a final projection point on the intersection of all sets. One main advantage of the POCS method is its ability to embrace a diversity of prior knowledge in a recovery problem [91].

In order to search the correlation among co-expressed genes, three convex sets [36] are defined including the correlation in genes, the correlation in samples, as well as a biological phenomenon, namely *synchronization loss*. Thereafter, the imputation results lie in the intersection of three features. For a detailed illustration, interested readers are referred to [36].

Key steps of the algorithm are listed as follows.

Step 1. Capture the correlation among genes. First, execute the KNNimpute algorithm on genes, giving an estimate of all missing values denoted $\hat{K}(l)$ where $l \in I$ as a set of all missing values. Second, the convex set C_1 of this correlation is defined as a range

$$C_1 = \left\{Y : \hat{K}(l) - \varepsilon \leq Y(l) \leq \hat{K}(l) + \varepsilon, \, l \in I\right\}, \tag{16}$$

where ε is an estimation error related to the variance of residual error. Third, the projection onto the convex set is defined as

$$P_1 = \begin{cases} \hat{K}(l) - \varepsilon, & Y(l) < \hat{K}(l) - \varepsilon \\ \hat{K}(l) + \varepsilon, & Y(l) > \hat{K}(l) - \varepsilon \\ Y(l), & otherwise \end{cases}. \tag{17}$$

Step 2. Capture the correlation among samples. First, execute the SVDimpute algorithm on samples, giving an estimation of all missing values denoted $\hat{P}(l)$, $l \in I$. Second, define convex set C_2 as

$$C_2 = \left\{Y : \hat{P}(l)\varepsilon_1 \leq Y(l) \leq \hat{P}(l)\varepsilon_2, \, l \in I\right\}, \tag{18}$$

where ε_1 and ε_2 are again estimation errors. Third, the projection onto the convex set is defined as

$$P_2 = \begin{cases} \hat{P}(l)\varepsilon_1, & Y(l) < \hat{P}(l)\varepsilon_1 \\ \hat{P}(l)\varepsilon_2, & Y(l) > \hat{P}(l)\varepsilon_2 \\ Y(l), & otherwise \end{cases}. \tag{19}$$

Step 3. Capture the phenomenon of synchronization loss. In this case, it is assumed that the expression data belong to n different periods. Convex set C_3 and projection P_3 , which are related to biological phenomenon, are both defined on period-basis.

3.4.3. Meta-data Based Imputation [46]

The authors of this method argue that the expression matrix from current microarray experiment may not be most suitable for imputing by itself. Therefore, they search the correlated genes in public databases, rather than within a certain data set, which is so far used by most other methods. For example, in a yeast microarray analysis, researchers download vast quantity of data for yeast from public accessible database, such as Stanford Microarray Database (SMD). They can choose K nearest neighbours by the distance measure between a target gene and a *database matrix*. Then they regress the missing value by using GMCimpute. In practice, the iterations could be executed 10 times and $K = 40$ is recommended for a trade-off between imputation accuracy and computation cost. Meta-data based imputation is superior to imputation within one logical data set because of usage of information from external database matrix in addition to the data matrix of interest.

Key steps of the algorithm are listed as follows.

Step 1. Construct a database matrix obtained from a large amount of data in the same species.

Step 2. Calculate the distance based on the mean absolute Pearson's correlation coefficient.

Step 3. Choose K most similar genes, where K is pre-defined.

Step 4. Use GMCimpute to estimate the missing value.

Step 5. Repeat Step 3 to Step 4 to achieve better RMSE measure.

3.4.4. LinCmb [47]

Using LinCmb, missing values are estimated by a linear combination of various methods. In LinCmb, the missing values are regressed by model stacking. Let R, K, S, B and $\mathbf{G}_1 \cdots \mathbf{G}_5$ be the estimates by row average, KNNimpute, SVDimpute, BPCA and GCMimpute with one to five clusters, respectively. The missing matrix $\mathbf{G}^*$ is then given by

$$\mathbf{G}^* = r\mathbf{R} + k\mathbf{K} + s\mathbf{S} + b\mathbf{B} + \sum_{i=1}^{5} g_i \mathbf{G}_i \tag{20}$$

where the parameters are non-negative and satisfy the condition

$$r + k + s + b + \sum_{i=1}^{5} g_i = 1. \tag{21}$$

LinCmb begins with a KNNimpute to give a complete matrix, then it uses the least squares method to regress the parameters. The LinCmb procedure will be repeatedly performed in multi-loops (e.g. 30 loops). Eventually the missing values will be taken as the average of iterations. The LinCmb naturally adapts the structure of an expression matrix. That means it may be regarded as one of the local imputation methods (row average, KNNimpute and GMCimpute above) if the number of missing values is low, or one of the global imputation methods (SVDimpute and BPCA), otherwise. Therefore, it is a combination of the merits of various estimation methods.

3.5. Performance Evaluation

3.5.1. Data Sets

Three data sets are used for performance evaluation. Dataset A is "Yeast200" included in AMIADA software [109]. It is a sample of yeast data which consists of 200 genes and 17 experiments. Data set B is provided along with the BPCA imputation software [78]. The data comprises 758 genes and 50 samples. Data set C is liver cancer data which is provided in LinCmb [47] and consists of 6511 genes and 40 conditions. To evaluate the performance of different algorithms, all of three data sets were processed in this way. A proportion of artificial missing data was randomly generated in each data set, thereafter the prediction results would be compared with the original known data with respect to RMSE. The proportion of missing data was set to 2.5% for Data set A and Data set B, and was set to 5% for Data set C.

3.5.2. Performance Comparison

Using the three microarray data sets, we study the performance of five algorithms described in Sections 3.2. and 3.3.: KNNimpute, LLSimpute, GMCimpute ($K = 3$ clusters), SVDimpute and BPCA. The first three are local imputation methods, while the last two are global imputation methods. Results are shown in Figures 7–9. It can be clearly seen that local imputation methods are more suitable for Data set A, and that global imputation methods are more suitable for Data set B and Data set C. Therefore, the selection of an imputation method should be based on the structure of microarrays.

4. Data Classification

4.1. Introduction

Statistical learning methods are indispensable in microarray data analysis [63]. They can be generally divided into two categories: supervised learning and unsupervised learning [16]. The main difference between them is that the former classifies the objects using known reference labels, while the latter decides the classification without prior information, but merely depending on input data. Unsupervised classification is sometimes called "clustering". In Section 4.2., we introduce unsupervised data mining methods including K-means clustering, hierarchical clustering, self organizing maps (SOM), fuzzy K-nearest neighbors clustering and projection methods. Supervised methods, including support vector machines, relevance vector machines, and neural networks, are discussed in Section 4.3.. In Section 4.4., we discuss two representative Bayesian approaches, i.e., hierarchical Bayesian method and Empirical Bayesian method.

4.2. Unsupervised Classification

4.2.1. K-means Clustering

Key steps of K-means clustering are listed as follows.

Step 1. Choose a set of K vectors in a gene expression matrix as the initial guess of K-means.

Step 2. Assign every gene to one of K clusters, with respect to the Euclidean distance.

Step 3. Recalculate the mean vector in each cluster after assigning all genes.

Step 4. Repeat Step 2 and Step 3 until the change between two successive assignments is insignificant.

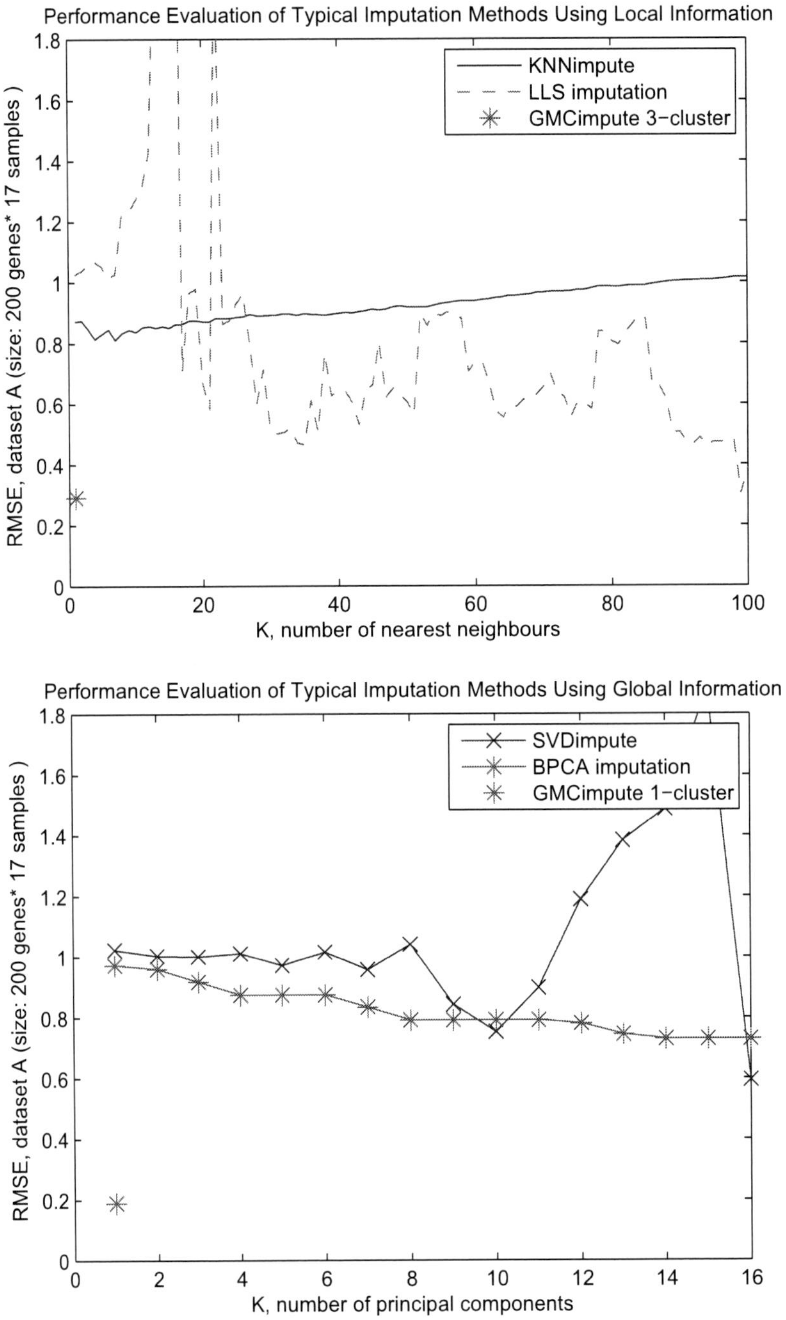

Figure 7. Performance comparison of five fundamental imputation methods applied on Data set A (200 genes, 17 samples). Upper panel is the performance of methods using local information, lower panel is the performance of methods using global information.

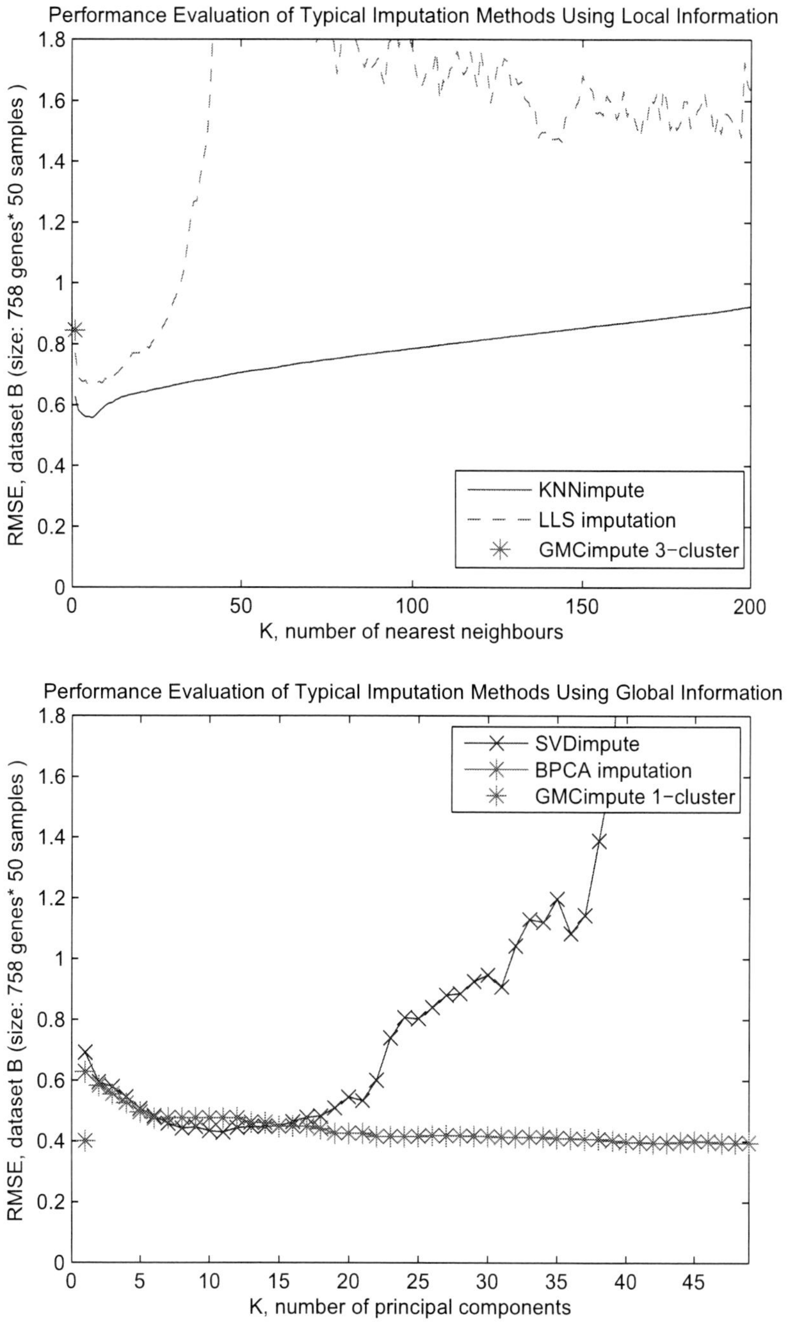

Figure 8. Performance comparison of five fundamental imputation methods applied on Data set B (758 genes, 50 samples). Upper panel is the performance of methods using local information, lower panel is the performance of methods using global information.

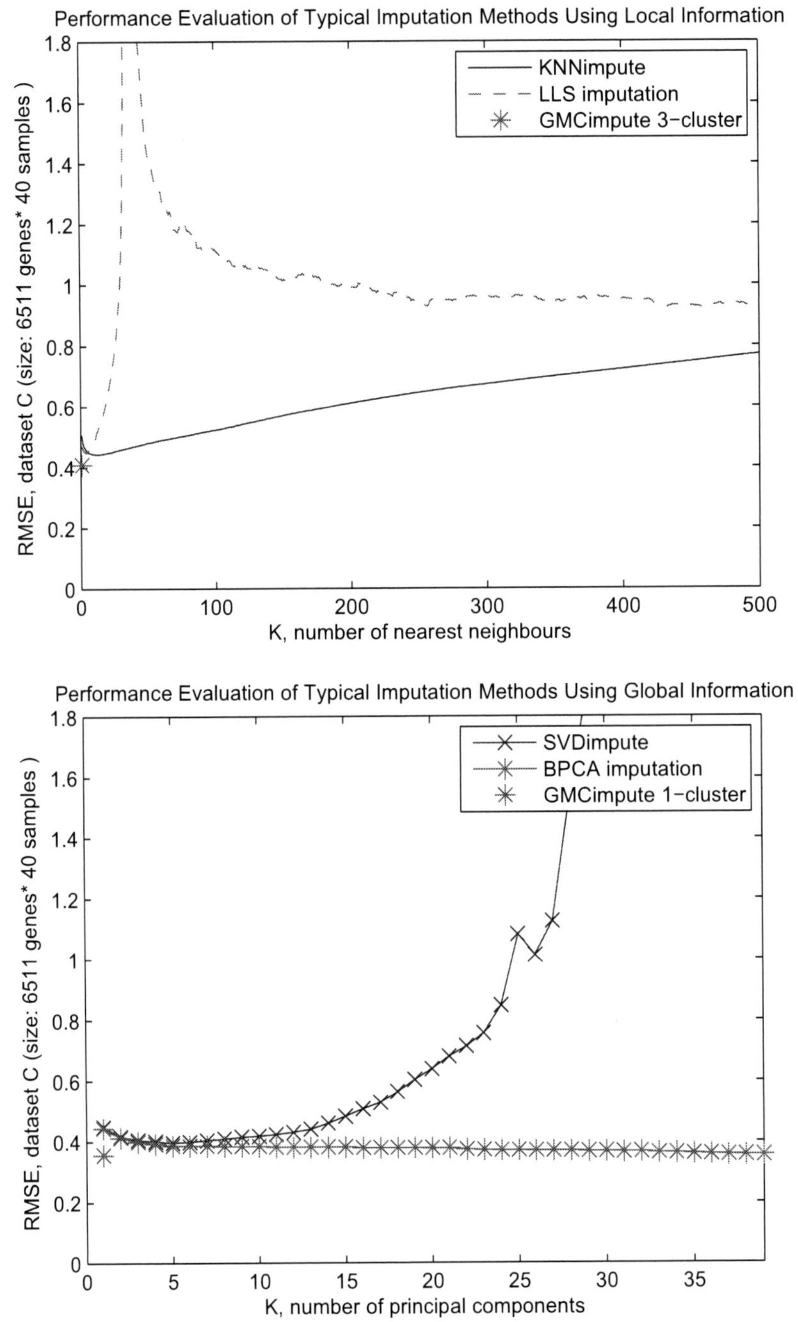

Figure 9. Performance comparison of five fundamental imputation methods applied on Data set C (6511 genes, 40 samples). Upper panel is the performance of methods using local information, lower panel is the performance of methods using global information.

4.2.2. Hierarchical Clustering

Hierarchical clustering can be performed in a bottom-up or top-down manner. Using the bottom-up method, each gene is initially regarded as a cluster by itself. In Data set A, for example, there are 200 clusters to start with. Then the two most similar genes are successively combined, until the number of clusters is reduced to one. In contrast, the top-down clustering method begins with one cluster including all genes. Then it successively splits the least similar gene from others, which forms a lower cluster. The result of hierarchical clustering is determined by the measure of similarity (linkage) of genes. It measures the distance between two different clusters by two points associated with each cluster. The common linkages include (1) single linkage, (2) complete linkage, (3) average linkage, etc. The single linkage (also called the *connectedness* or *minimum*) clustering method uses the nearest-neighbour distance to measure inter-cluster similarity so as to combine them in next step. The complete linkage (also called *diamater* or *maximum*) clustering method uses the furthest-neighbour distance to measure inter-cluster similarity. The average linkage clustering method considers the distance between one cluster to another as the average distance of all members in one cluster to all members in the other cluster. Because of its ability to reflect similarity in gene expression, the hierarchical clustering is currently used in many popular software packages such as GeneXpress, AMIADA, and Xcluster etc. However, one main drawback with hierarchical clustering is that it is not flexible to repair classification in previous steps [63].

4.2.3. Self Organizing Maps

Self organizing maps (SOM) [55] is able to train the high dimensional input to be represented in low dimensions, because it preserves a topology mapping from the high dimensional input to the low dimensional subspace output. A two-dimensional output is often used in microarray data mining. In SOM, the topology is described by prototype vectors $\mathbf{m}_j$ $(j = 1, \ldots, K)$. For example, there are 25 prototype vectors in a 5×5 two-dimensional grid topology. The error function of SOM [101] is given by

$$E = \sum_{i=1}^{n} \sum_{j=1}^{K} h_{bj} \|\mathbf{g}_i - \mathbf{m}_j\|^2, \tag{22}$$

where n is the number of training samples. The iterative training also updates the prototype units $\mathbf{m}_j$,

$$\mathbf{m}_j(t+1) = \mathbf{m}_j(t) + \alpha(t) \cdot h_{bj}(t)[\mathbf{g} - \mathbf{m}_j(t)], \tag{23}$$

where $\alpha(t)$ is called the learning rate, decreasing monotonically with time and $h_{bj}(t)$ is defined as

$$h_{bj}(t) = \exp\Big[- \|\mathbf{r}_b - \mathbf{r}_j\|^2 / (2\sigma^2(t)) \Big], \tag{24}$$

where $\mathbf{r}_b$ and $\mathbf{r}_j$ are positions of neurons b and j, respectively, on the SOM grid. An important observation of SOM is that if the neighbourhood kernel value $h_{bj}(t)$ is chosen as 1 for a neighbourhood prototype, or 0 otherwise, then the SOM is reduced to K-nearest neighbour clustering. Therefore, the SOM can be viewed as a robust version of KNN with high dimensions. Recent applications of SOM in microarray data analysis can be seen in [49], [56].

4.2.4. Fuzzy K-nearest Neighbors (KNN) Clustering

Fuzzy KNN clustering [51] improves traditional clustering methods in that it classifies an observation into multiple memberships, instead of a "1" or "0". In this sense, it models microarray data sets in a more accurate manner, because a single gene may perform multiple functions belonging to different clusters, not necessarily in only one class. According to [52], the membership of a gene, represented by the vector **g**, in class i is defined as

$$w_i(\mathbf{g}) = \frac{\sum_{j=1}^{K} w_{ij} \|\mathbf{g} - \mathbf{g}_j\|^{-\frac{2}{m-1}}}{\sum_{j=1}^{K} \|\mathbf{g} - \mathbf{g}_j\|^{-\frac{2}{m-1}}}, \tag{25}$$

for all i, where $\mathbf{g}_j$ $(j = 1, \ldots, K)$ denotes K neighbours of gene **g**, w_{ij} denotes the distance from **g** to $\mathbf{g}_j$, and m measures the effect of distance between **g** to $\mathbf{g}_j$. One difficulty is the choice of the scalar m [52]. When m is finite and $m \neq 1$, the smaller the value of m the greater influence the distance samples have. When m approaches $\pm\infty$, the results of fuzzy KNN approach those of KNN.

4.2.5. Projection Methods

There are two types of projection methods, principal component analysis (PCA) [15] and independent component analysis (ICA) [43]. Both methods project the data into a new subspace spanned by component genes, thus in this sense both of them are dimension-reduction methods for microarray data analysis. Another common property between PCA and ICA methods is that they represent each gene as a linear combination of multiple biological processes. This is actually a more meaningful biological model compared to clustering methods such as K-means clustering, hierarchical clustering and SOM discussed above. Major differences between the two methods are listed below.

- In PCA, a new component must be chosen orthogonal to its previous component so as to obtain the maximum information not present yet. To obtain the principal components, singular value decomposition (SVD) is performed onto normalized expression data $\mathbf{G}$ (assuming the rank is r), such that $\mathbf{G}$ is decomposed as $\mathbf{G} = \mathbf{U}\Sigma\mathbf{V}^T$, where $\mathbf{U}$ is an $M \times r$ orthogonal transformation matrix, $\mathbf{V}$ is an $N \times r$ orthogonal transformation matrix, and Σ is a $r \times r$ diagonal matrix whose elements are eigen values sorted in decreasing order. By SVD the high dimension of expression data can thus be reduced by r-dimensional eigengenes which contribute to the most significant variance in the matrix $\mathbf{G}$. There are recent studies of microarray data using PCA such as [31], [73].

- In ICA, its projection space is spanned by statistically independent components. We assume there are m independent components, $\mathbf{s}_1 \cdots \mathbf{s}_m$, governing the expression data, then the *j*th gene $\mathbf{g}_j$ can be described as a linear combination of the independent components, namely,

$$\mathbf{g}_j = \sum_{i=1}^{m} \alpha_{ij} \mathbf{s}_i \tag{26}$$

The central idea of ICA is modeling the observation data with a mixture of independent components, without prior knowledge either of parameters ($\alpha_{i1} \cdots \alpha_{im}$) or of independent components ($\mathbf{s}_1 \cdots \mathbf{s}_m$). The estimation of independent components and their parameters is based on the assumption that the gene expression data are non-Gaussian. Both linear ICA and nonlinear ICA have been applied to a microarray processing problem and have been compared with other unsupervised data analysis methods [64]. Additional information about ICA and its applications in microarray can be found on line: http://www.stanford.edu/~silee/ICA.

4.3. Supervised Classification

The aim of supervised learning is to construct the decision function $y = f(\mathbf{x}; \mathbf{w})$ based on the training data $\mathbf{D} = \{y_i, \mathbf{x}_i\}_{i=1:K}$, where y_i is the label of the gene $\mathbf{x}_i$. In a classification application, if a gene belongs to a functional group, then it is labeled "1", otherwise "0". For a new observation gene data $\mathbf{x}$ the function of this gene can be correctly classified by the decision function. There are two major considerations in supervised learning: the form of the decision function and the learning of the parameters, denoted $\mathbf{w}$. In the following, we only briefly describe decision function of three supervised learning methods.

4.3.1. Neural Networks

The simplest neural network [15] is single layer and consists of two basic components

$$y(\mathbf{x}; \mathbf{w}) = \sum_{i=1}^{K} w_i \phi_i(\mathbf{x}) + w_0, \tag{27}$$

and

$$u = f(y), \tag{28}$$

where $\phi_i(\mathbf{x})$ is the basis function, $\mathbf{w} = [w_0, w_1, \ldots, w_K]^T$ is the weight vector, w_0 is called the bias and the function $f(y)$ is a nonlinear mapping. A typical nonlinear function is a logistic sigmoid function taking the form $f(y) = (1 + e^{-y})^{-1}$. Multi-layer neural networks can be easily built from a single layer network. In microarray data analysis neural networks have been applied to disease classification. A simple two-layered linear neural network [53] can obtain accurate classification results when used for three types of cancers. More complex neural networks [79] have been successfully employed in determining patient prognosis. Neural networks have been extensively employed in gene expression study [13], [53], [85], [69], [71], [48].

4.3.2. Support Vector Machines

The support vector machine (SVM) is an attractive machine learning algorithm for two-category classification [100], [87], [14]. Given the training data $\mathbf{D} = \{y_i, \mathbf{x}_i\}_{i=1:K}$ (y_i is a binary label), the decision function can be written as

$$f(\mathbf{x}) = \text{sign}\left(\sum_{i=1}^{K} w_i y_i \mathcal{K}(\mathbf{x}, \mathbf{x}_i) - w_0 \right), \tag{29}$$

where $\text{sign}(x)$ is signum function, $\mathcal{K}(\mathbf{x},\mathbf{y})$ is called the kernel function and the $\mathbf{w} = [w_0,\ldots,w_K]$ is the weight vector. It is important to note that for a given kernel function there exists a mapping $\phi(\mathbf{x})$ such that $\mathcal{K}(\mathbf{x},\mathbf{y}) = \langle\phi(\mathbf{x}),\phi(\mathbf{y})\rangle$, where $\langle\mathbf{x},\mathbf{y}\rangle$ is the inner product. The function $\phi(\mathbf{x})$ maps $\mathbf{x}$ to a higher dimensional feature space. For example, a vector $\mathbf{x} = [x_1,x_2]^T$ can be mapped to $\phi(\mathbf{x}) = \left[x_1,x_2,x_1^2,x_2^2,x_1x_2\right]^T$. The use of a kernel function thus permits the support vector machine to operate efficiently in a nonlinear high-dimensional feature spaces without designing the mapping. A simple example of the kernel function is $\mathcal{K}(\mathbf{x},\mathbf{y}) = \exp\left(-\|\mathbf{x}-\mathbf{y}\|^2/2\sigma^2\right)$. The SVM decision function is a hyperplane in the feature space. The hyperplane corresponds to a nonlinear decision boundary in the input space. Learning of the parameters is through an optimization process that maximizes the hyperplane margin in the feature space [87], [14].

SVM has been exploited to identify unknown genes based on prior knowledge of gene functions [22]. As a supervised approach, SVM outperforms some unsupervised methods such as hierarchical clustering and SOM, since it takes the advantage of prior knowledge of the training set, whereas unsupervised methods may not cluster related genes if they stay far away with respect to the distance measurement. SVM has also be used in multiple classification cases [65]. The method was tested in a framework called one-versus-all (OVA). A significant improvement in accuracy has been achieved by using SVM in multiclass classification [65]. Recently, SVM has been actively exploited in the gene selection area. Gene selection is an advantageous preparation for classification. The final classification step can be performed on the result of gene selection [39], [22], [34], [115], [81], [93].

4.3.3. Relevance Vector Machines (RVM)

The RVM [97] can be described by using the following observation model

$$y(\mathbf{x};\mathbf{w}) = \sum_{i=1}^{K} w_i\phi_i(\mathbf{x}) + r, \tag{30}$$

where y is the label of $\mathbf{x}$, $\phi_i(\mathbf{x})$ is the kernel function and r is usually assumed i.i.d. zero mean Gaussian noise with variance β^{-1}. A key feature of the RVM is that a sparse weight vector $\mathbf{w}$ is sought from the training process. The sparseness constraint is enforced by using a heavy-tailed prior for w_i. The heavy-tailed distribution is expressed as $p(w_i|\alpha_i) = \mathcal{N}(0,\alpha_i^{-1})$ with a hyper-prior $p(\alpha_i|a,b) = \mathcal{G}(a,b)$, where $\mathcal{G}(a,b)$ is a Gamma distribution. We can easily prove that

$$p(w_i) = \int_0^\infty p(w_i|\alpha_i)p(\alpha_i|a,b)\,d\alpha_i \tag{31}$$

is a student's t distribution. In addition, a Gamma distribution is also assumed for the noise variance β^{-1}. Collecting the parameters α_i as a vector α, we can write the posterior of the parameters that we want to determine

$$p(\mathbf{w},\alpha,\beta|\mathbf{t}) = p(\mathbf{w}|\mathbf{t},\alpha,\beta)p(\alpha,\beta|\mathbf{t}) \tag{32}$$

A detailed description of the MAP estimate of the these parameters is beyond the scope of this paper. Interested readers should refer to [97], [16] for details.

RVM has already been applied in gene classification [59], [57], [60], [68], [58] and gene selection fields [67]. It was also combined with *recursive feature elimination* (RFE) framework [114], which repeatedly removes the "least useful" feature. In cancer gene selection, the RVM-RFE scheme is more accurate than linear RVM and leads to shorter running time than SVM-RFE [39].

4.4. Bayesian Methods

Bayesian methods are powerful statistical tools in microarray data analysis. They can be framed in either supervised or unsupervised learning based on the presence or absence of reference labels. For example, RVM (in Section 4.3.) is intrinsically a Bayesian supervised learning method. In the following, we briefly describe hierarchical Bayes and empirical Bayes approaches in microarray data analysis.

4.4.1. Hierarchical Bayes Clustering

Compared to conventional hierarchical clustering (Section 4.2.) which is a distance-based clustering method, hierarchical Bayes clustering is a model-based method and provides more flexibility. Using hierarchical Bayesian methods, parameters of prior distributions have their own distributions with hyper parameters. For example, a variable x, which has a prior $x \sim \mathcal{N}(u, S)$, is related to observations G . The mean μ is modeled as another Gaussian $u \sim \mathcal{N}(0, S_u)$ with a hyper parameter S_u. From the gene differential expression point of view, we can define, e.g., a dependent variable x [72], such that $x = 1$ if a gene is differentially expressed and $x = 0$ otherwise. Correspondingly, the posterior probability $p(x|G)$ stands for the probability of a gene being differentially expressed given intensities in red and green channels. The major computational task is the calculation of the posterior distribution of high dimensionality. This is usually implemented numerically by using Markov Chain Monte Carlo (MCMC) method [16].

A recent application of hierarchical Bayes models is called the weighted Chinese restaurant process [83] in which clustering is regarded as a sequential "seating arrangement" in a restaurant with an unlimited number of tables. This method is insensitive to missing data and is able to deal with time-shifted/-converted samples.

4.4.2. Empirical Bayes Classification

The empirical Bayes (EB) approach uses estimated hyper parameters based on the observation. Fundamentals of Bayes and EB methods for data analysis are available from [25]. Recent developments in EB applications to microarray data have appeared in [29], [92], [90], [72]. An intuitive method [102] combines EB analysis with thresholding technique. This is based on an observation that only a minority of genes are differentially expressed with non-zero log-ratio values. To identify and exploit the sparsity in the observed data, we can first threshold the data with a Bayesian estimate of the mean vector. After thresholding, the majority of the data is deemed zero and only the minority above the threshold value is recognized as differential expressions. An EB based thresholding procedure uses the software *EBayesThresh* which is available at
http:cran.r-project.org/src/contrib/Descriptions/EbayesThresh.html.

5. Conclusion

The microarray technology is broadly recognized as a latest breakthrough in experimental biology and creates a multilateral subject in bioinformatics. This paper mainly concentrates on the topics of image processing and data mining for microarray analysis. There is no doubt that, to a certain extent, existing methods have succeeded in disclosing correlations between genes. However, researchers are facing challenges in each procedure in the microarray analysis, including microarray image processing, missing data estimation and microarray data analysis. In microarray image processing, irregular shapes of spots may appear. Hence, an automatic and robust segmentation algorithm is worth studying. If one can not perform reliable analysis to donuts or scratches, these spots would be regarded as missing values in data analysis and be estimated in data imputation. Another challenge lies in choosing suitable estimation methods for missing values. Missing values frequently emerge in microarray experiments, due to factors such as hybridization failure or presence of dust, and they put difficulties in the subsequent data mining stage. In order to perform a reliable data imputation, one needs to carefully select an estimation method. As mentioned before, estimation methods using local or global information have distinguished properties to handle particular situations, whereas combined methods can take the advantages of local and global methods by putting different weights to the methods. When internal information is not adequate to predict the missing values, one could consider employing information from external data sets. In fact, combined methods in microarray data imputation have exploited robust models and performed heuristic attempts in imputation technique. Challenges in data mining include setting up empirical and robust models in order to achieve better clustering accuracy and integrating biological information into the data analysis process, because results are useful only when they are relevant in the biological context.

Acknowledgments

The authors are grateful to Dr. Ming Ouyang for providing the liver cancer data set and the Matlab code of the LinCmb imputation method. The authors thank Drs. Xuhua Xia and Shigeyuki Oba for providing the AMIADA and BPCA software as well as the data sets.

References

[1] K. Aas, Microarray data mining: a survey. Norway Computer Center, Oslo, Norway, *Tech. Rep.*, 2001.

[2] O. Alter, Discovery of principles of nature from mathematical modeling of DNA microarray data. *PNAS*, vol. 103, 2006.

[3] O. Alter, P. O. Brown, and D. Botstein, Singular value decomposition for genome-wide expression data processing and modeling. *PNAS*, vol. 97, pp. 10 101–10 106, 2000.

[4] O. Alter, P. O. Brown, and D. Botstein, Generalized singular value decomposition for comparative analysis of genome-scale expression data sets of two different organisms." *PNAS*, vol. 100, pp. 3351–3356, 2003.

[5] J. Angulo and J. Serra, Automatic analysis of DNA microarray images using mathematical morphology. *Bioinformatics*, vol. 19, no. 5, pp. 553–562, 2003.

[6] G. Antoniol and M. Ceccarelli, A Markov random field approach to microarray image gridding. in *Proc. IEEE 17th Int. Conf. Pattern Recognition*, vol. 3, Cambridge, UK, Aug. 2004, pp. 550–553.

[7] G. Antoniol, M. Ceccarelli, and A. Petrosino, Microarray image addressing based on the Radon transform. in *Proc. IEEE Int. Conf. Image Processing*, vol. 1, no. 11-14, Genoa, Italy, Sep. 2005, pp. I–13–16.

[8] *GenePix 4000A User's Guide*, Axon Instruments Inc., 2000. [Online]. Available: http://www.axon.com

[9] P. Bajcsy, Gridline: automatic grid alignment in DNA microarray scans. *IEEE Trans. Image Processing*, vol. 13, no. 1, pp. 15–25, 2004.

[10] P. Bajcsy, An overview of DNA microarray image requirements for automated processing. in *Proc. IEEE Conf. Computer Vision and Pattern recognition*, vol. 3, San Diego, CA, USA, Jun. 2005, pp. 147–152.

[11] P. Bajcsy, L. Liu, and M. Band, DNA microarray image processing. in *DNA array image analysis: Nuts and Bolts*, G. Kamberova, Ed. DNA Press LLC, 2008, in press.

[12] S. Battiato, G. D. Blasi, G. M. Farinella, G. Gallo, and G. C. Guarnera, Ad-hoc segmentation pipeline for microarray image analysis. in *Proc. IST/SPIE*, vol. 6064, San Jose, CA, USA, Jan. 2006, pp. 300–311.

[13] H. R. Birgit, Classifying cancers with DNA microarrays and artificial neural networks. Genome News Network, 2001. [Online]. Available: http://www.genomenewsnetwork.org/articles/06_01/Cancer_microarrays.shtml

[14] C. Bishop, *Pattern Recongition and Machine Learning*. Springer, 2006.

[15] C. M. Bishop, *Neural Networks for Pattern Recognition*. London: Oxford University Press, 1995.

[16] C. M. Bishop, *Pattern recognition and machine learning*. Singapore: Springer, 2006.

[17] K. Blekas, N. P. Galatsanos, A. Likas, and I. E. Lagaris, Mixture model analysis of DNA microarray images. *IEEE Trans. Med. Imag.*, vol. 24, no. 7, pp. 901–909, 2005.

[18] D. Bozinov and J. Rahnenfuhrer, Unsupervised technique for robust target seperation and analysis of DNA microarray spots through adaptive pixel clustering. *Bioinformatics*, vol. 18, no. 5, pp. 747–756, 2002.

[19] D. Bozinov and P. M. Seidel, Iterative gridding for automated microarray image analysis. in *Proc. 38th Asilomar Conf. Signals, Systems and Computers*, vol. 2, Pacific Grove, CA, USA, Nov. 2004, pp. 1635–1638.

[20] P. S. Bradley and U. M. Fayyad, Refining initial points for K-Means clustering. in *Proc. 15th Int. Conf. on Machine Learning*, J. Shavlik, Ed., San Francisco, USA, 1998, pp. 91–99.

[21] N. Brandle, H. Bischof, and H. Lapp, Robust DNA microarray image analysis. *Machine Vision and Applications*, vol. 15, no. 1, pp. 11–28, 2003.

[22] M. P. S. Brown and et. al., Knowledge-based analysis of microarray gene expression data by using support vector machine. *PNAS*, vol. 97, no. 1, pp. 262–267, 2000.

[23] M. J. Buckley, The Spot user's guide. 1999. [Online]. Available: http://www.cmis.csiro.au/IAP/Spot/spotmanual.htm

[24] J. Buhler, T. Ideker, and D. Haynor, Dapple: Improved techniques for finding spots on DNA microarrays. Univ. of Washington, WA, USA, Tech. Rep., 2000.

[25] B. P. Carlin and T. A. Louis, *Bayes and empirical Bayes methods for data analysis.* London, UK: Chapman and Hall, 1996.

[26] G. Celeux and G. Govaert, A classification EM algorithm for clustering and two stochastic versions. *Computational Statistics and Data Analysis*, vol. 14, no. 3, pp. 315–332, 1992.

[27] Y. Chen, E. R. Dougherty, and M. L. Bitter, Ratio-based decisions and the quantitative analysis of cDNA microarray images. *Biomedical Optics*, vol. 2, pp. 364–374, 1997.

[28] Y. Chen, E. R. Dougherty, and M. L. Bittner, Ratio-based decisions and the quantitative analysis of cDNA microarray images. *Biomedical Optics*, vol. 4, pp. 364–374, 1997.

[29] S. Datta and S. Datta, Empirical Bayes screening of many p-values with applications to microarry studies. *Bioinformatics*, vol. 2, no. 9, pp. 1987–1994, 2005.

[30] A. G. de Brevern, S. Hazout, and A. Malpertuy, Influence of microarrays experiments missing values on the stability of gene groups by hierarchical clustering. BMC Bioinformatics, 2004. [Online]. Available: http://www.biomedcentral.com/1471-2105/5/114

[31] J. R. de Haan, R. Wehrens, S. Bauerschmidt, E. Piek, R. C. van Schaik, and L. M. C. Buydens, Interpretation of ANOVA models for microarray data using PCA. *Bioinformatics*, vol. 23, no. 2, pp. 184–190, 2007.

[32] M. B. Eisen, Scanalyze software. 1999. [Online]. Available: http://rana.lbl.gov/EisenSoftware.htm

[33] R. Fabbri, L. da F. Costa, and J. Barrera, Towards non-parametric gridding of microarray images. in *Proc. 14th Int. Conf. Digital Signal Processing*, vol. 2, Santorini, Greece, Jul. 2002, pp. 623–626.

[34] T. S. Furey, N. Cristianini, N. Duffy, D. W. Bednarski, M. Schummer, and D. Haussler, Support vector machine classification and validation of cancer issue samples using microarray expression data. *Bioinformatics*, vol. 16, pp. 906–914, 2000.

[35] V. L. Galinsky, Automatic registration of microarray images. I. rectangular grid. *Bioinformatics*, vol. 19, no. 14, pp. 1824–1831, 2003.

[36] X. Gan, A. W. C. Liew, and H. Yan, Microarray missing data imputation based on a set theoretic framework and biological knowledge. *Nucleic Acids Research*, vol. 34, no. 5, pp. 1608–1619, 2006.

[37] N. Giannakeas and D. I. Fotiadis, An automated method for gridding in microarray images. in *Proc. 28th Annual Int. Conf. of IEEE Engineering in Medicine and Biology Society*, vol. Supplement, New York City, NY, USA, Aug. 2006, pp. 5876–5879.

[38] R. Gottardo, J. Besag, M. Stephens, and A. Murua, Probabilistic segmentation and intensity estimation for microarray images. *Biostatistics*, vol. 7, no. 1, pp. 85–99, 2006.

[39] I. Guyon, J. Weston, S. Barnhill, and V. Vapnik, Gene selection for cancer classification using support vector machines. *Machine Learning*, vol. 46, no. 1-3, pp. 389–422, 2002.

[40] T. Hastie, R. Tibshirani, and J. Friedman, *The element of statistical learning*. New York, NY, USA: Springer, 2001.

[41] R. J. Hirata, J. Barrera, R. F. Harshimoto, and D. O. Dantas, Microarray gridding by mathematical morphology. in *Proc. 14th Brazilian Symposium on Computer Graphichs and Image Processing*, Florianopolis, Brazil, Oct. 2001, pp. 112–119.

[42] J. Hua, Z. Xiong, Q. Wu, and K. Castleman, Fast segmentatioin and lossy-to-lossless compression of DNA microarray images. in *Proc. Workshop on Genomic Signal Processing and Statistics*, vol. 22, Raleigh, NC,USA, Oct. 2002.

[43] A. Hyvärinen, J. Karhunen, and E. Oja, *Independent component analysis*. John Wiley & Sons, Inc., 2001.

[44] *MAANOVA, Analysis of Variance for gene expression microarrays*, The Jackson Laboratory, 2000. [Online]. Available: http://www.jax.org/staff/churchill/labsite/software/anova/index.html

[45] R. Jornsten, Data compression and its statistical implications, with an application to the analysis of microarray images. Ph.D. dissertation, Univ. of California, Berkeley, USA, Dec. 2001.

[46] R. Jornsten, M. Ouyang, and H. Y. Wang, A meta-data based method for DNA microarray imputation. BMC Bioinformatics, 2007. [Online]. Available: http://www.biomedcentral.com/1471-2105/8/109

[47] R. Jornsten, H. Y. Wang, W. J. Welsh, and M. Ouyang, DNA microarray data imputation and significance analysis of differential expression. *Bioinformatics*, vol. 21, no. 22, pp. 4155–4161, 2005.

[48] N. Kasabov, L. Benuskova, and S. G. Wysoski, Computational neurogenetic modeling: integration of spiking neural networks, gene networks, and signal processing techniques. in *Artificial Neural Networks: Formal Models and Their Applications*. Berlin, Germany: Springer, 2005.

[49] T. Kato, K. Fujimura, and H. Tokutaka, Analysis of DNA microarray data by using self-organizing maps. *Genome Informatics*, vol. 14, pp. 328–329, 2003.

[50] M. Katzer, F. Kummert, and G. Sagerer, A Markov random field model of microarray gridding. *IEEE Trans. Pattern Anal. Machine Intell.*, vol. 25, no. 2, pp. 162–173, 2003.

[51] J. M. Keller, M. R. Gray, and J. A. Givens, A fuzzy k-nearest neighbour algorithm. *IEEE Trans. Syst., Man, Cybern.*, vol. 15, pp. 580–585, 1985.

[52] M. Kerwin, A fuzzy K-nearest neighbour algorithm, review and critical analysis. 2005. [Online]. Available: http://mattyk.nqitx.net/Honours/CP5090/FuzzyK-NN.pdf

[53] J. Khan *et al.*, Classification and diagnostic prediction of cancers using gene expression profiling and artificial neural network. *Natural Medicine*, vol. 7, no. 6, pp. 673–679, 2001.

[54] H. Kim, G. H. Golub, and H. Park, Missing value estimation for DNA microarray gene expression data: local least squares imputation. *Bioinformatics*, vol. 21, no. 2, pp. 187–198, 2005.

[55] T. Kohonen, *Self-Organizing Maps*. Berlin, Germany and NY, USA: Springer, 2001.

[56] M. Kotani, A. Sugiyama, and S. Ozawa, Analysis of DNA microarray data using self-organizing map and kernel based clustering. in *Proc. 9th Int. Conf. Neural Information Processing*, vol. 2, pp. 755–759, Singapore, Nov. 2002.

[57] B. Krishnapuram, L. Carin, and A. J. Hartemink, Joint classifier and kernel design. in *Proc. workshop on kernel methods in bioinformatics*, Berlin, Germany, Apr. 2003.

[58] B. Krishnapuram, L. Carin, and A. J. Hartemink, Gene expression analysis: joint feature selection and classifier design. in *Kernel Methods in Computational Biology*, B. Scholkopf, K. Tsuda, and J. P. Vert, Eds. MA, USA: MIT Press, 2004.

[59] B. Krishnapuram, A. J. Hartemink, and L. Carin, Logistic regression and RVM for cancer diagnosis from gene expression. in *Proc. Workshop Genomic Signal Processing and Statistics, 2002*, Raleigh, NC, USA, Oct. 2002.

[60] B. Krishnapuram, A. J. Hartemink, L. Carin, and M. A. T. Figueiredo, A Byesian approach to joint feature selection and classification design. *IEEE Trans. Pattern Anal. Machine Intell.*, vol. 26, no. 9, pp. 1105–1111, 2004.

[61] D. M. Lane, HyperStat online contents, Pearson's correlation. 1993. [Online]. Available: http://davidmlane.com/hyperstat/A34739.html

[62] D. M. Lane, HyperStat online contents, type I and II errors. 1993. [Online]. Available: http://davidmlane.com/hyperstat/A18652.html

[63] M. L. T. Lee, *Analysis of microarray gene expression data*. Boston, MA, USA: Kluwer Academic Publishers, 2004.

[64] S. I. Lee and S. Batzoglou, Application of independent component analysis to microarrays. Genome Biology, 2003. [Online]. Available: http://genomebiology.com/2003/4/11/R76

[65] Y. Lee and C. K. Lee, Classification of multiple cancer types by multicategory support vector machines using gene expression data. *Bioinformatics*, vol. 19, no. 9, pp. 1132–1139, 2003.

[66] Q. Li, C. Fraley, R. E. Bumgarner, K. Y. Yeung, and A. E. Raftery, Donuts, scratches and blanks: robust model-based segmentation of microarray images. *Bioinformatics*, vol. 21, no. 12, pp. 2875–2882, 2005.

[67] Y. Li, C. Campbell, and M. Tipping, Bayesian automatic relevance determination algorithms for classifying gene expression data. *Bioinformatics*, vol. 18, no. 10, pp. 1332–1339, 2002.

[68] Y. Li *et al.*, Establishing glucose- and ABA-regulated transcription networks in Arabidopsis by microarray analysis and promoter classification using a relevance vector machine. *Genome Research*, vol. 16, pp. 414–427, 2006.

[69] R. Linder *et al.*, The 'subsequent artificial neural network' (SANN) approach might bring more classificatory power to ANN-based DNA microarray analyses. *Bioinformatics*, vol. 20, pp. 3544–3552, 2004.

[70] R. J. A. Little and D. B. Rubin, *Statistical Analysis with Missing Data.* USA: John Wiley and Sons, 1987.

[71] B. Liu, Q. Cui, T. Jiang, and S. Ma, A combination feature selection and ensemble neural network method for classification of gene expression data. BMC Bioinformatics, 2004. [Online]. Available: http://www.biomedcentral.com/1471-2105/5/136

[72] I. Lonnstedt, Empirical Bayes methods for DNA microarray data. Ph.D. dissertation, Uppsala Univ., Sweden, Sep. 2005.

[73] Y. Lu, Q. Tian, M. Sanshez, and Y. Wang, Hybrid PCA and LDA analysis of microarray gene expression data. in *Proc. IEEE Symposium on Computational Intelligence in Bioinformatics and Computational Biology*, La Jolla, CA, USA, Nov. 2005, pp. 1–6.

[74] Y. Luo and S. Lonardi, Storage and transmission of microarray images. *Drug discovery today:Biosilico*, vol. 10, no. 23/24, pp. 1689–1695, 2005.

[75] D. J. C. Mackay, Bayesian interpolation. *Neural Computation*, vol. 4, no. 3, pp. 415–447, 1992.

[76] *Minimum Information About a Microarray Experiment-MIAME*, Microarray and Gene Expression DataMGED Society, 1999-2007. [Online]. Available: http://www.mged.org/Workgroups/MIAME/miame.html

[77] R. Nagarajan, Intensity based segmentation of microarray images. *IEEE Trans. Med. Imag.*, vol. 22, no. 7, pp. 882–889, 2003.

[78] S. Oba, M. A. Sato, I. Takemasa, M. Monden, K. I. Matsubara, and S. Ishii, A Bayesian missing value estimation method for gene expression profile data. *Bioinformatics*, vol. 19, no. 16, pp. 2088–2096, 2003.

[79] M. C. O'Neill and L. Song, Neural network analysis of lymphoma microarray data: prognosis and diagnosis near-perfect. BMC Bioinformatics, 2003. [Online]. Available: http://www.biomedcentral.com/1471-2105/4/13

[80] M. Ouyang, W. J. Welsh, and P. Georgopoulos, Gaussian mixture clustering and imputation of microarray data. *Bioinformatics*, vol. 20, no. 6, pp. 917–923, 2004.

[81] M. Pirooznia and Y. Deng, SVM classifier – a comprehensive java interface for support vector machine classification. BMC Bioinformatics, 2006. [Online]. Available: http://www.biomedcentral.com/1471-2105/7/S4/S25

[82] K. N. Plataniotis and A. N. Venetsanopoulos, *Color Image Processing and Applications*. NY,USA and Lodon, UK: Springer, 2000.

[83] Z. S. Qin, Clustering microarray gene expression data using weighted Chinese restaurant process. *Bioinformatics*, vol. 22, no. 16, pp. 1988–1997, 2006.

[84] L. Rueda and L. Qin, A new method for DNA microarray image segmentation. in *Proc. 2nd Int. Conf. Image Analysis and Recognition*, Toronto, Canada, Sep. 2005, pp. 886–893.

[85] T. Sawa and L. Ohno-Machado, A neural network-based similarity index for clustering DNA microarray data. *Comput. Biol. Med.*, vol. 33, no. 1, pp. 1–15, 2003.

[86] M. Schena, D. Shalon, R. W. Davis, and P. O. Brown, Quantitative monitoring of gene expression patterns with a complementary DNA microarray. *Science*, vol. 270, pp. 467–470, 1995.

[87] J. Shawe-Taylor and N. Cristianini, *Support vector machines and other kernel-based learning methods*. Cambridge University Press, 2000.

[88] C. L. Shih and H. W. Chiu, Automatic spot detection of cDNA microarray images using mathematical morphology methods. in *Proc. IEEE EMBS Asian-Pacific Conf. Biomedical Engineering*, Keihanna, Japan, Oct. 2003, pp. 70–71.

[89] K. I. Siddiqui, A. O. Hero, and M. M. Siddiqui, Mathematical morphology applied to spot segmentation and quantification of gene microarray images. in *Proc. 36th Asilomar Conf. Signal, Systems and Computers*, Pacific Grove, CA, USA, Nov. 2002, pp. 926–930.

[90] G. Smyth, Linear models and empirical Bayes methods for assessing differential expression in microarray experiments. Statistical Applications in Genetics and Molecular Biology, 2004. [Online]. Available: http://www.statsci.org/smyth/pubs/ebayes.pdf

[91] H. Stark and Y. Yang, *Vector space projections: a numerical approach to signal and image processing, neural nets, and optics*. Danvers, MA, USA: John Wiley and Sons, Inc., 1998.

[92] Y. C. Tai and T. P. Speed, A multivariate empirical Bayes statistics for replicated microarray time course data. *Annuals of statistics*, vol. 34, no. 5, pp. 2387–2412, 2006.

[93] E. K. Tang, P. N. Suganthan, and X. Yao, Gene selection algorithms for microarray data based on least squares support vector machine. BMC Bioinformatics, 2006. [Online]. Available: http://www.biomedcentral.com/1471-2105/7/95

[94] R. Tibshirani, PAM: Prediction Analysis for Microarrays. 2001. [Online]. Available: http://www-stat.stanford.edu/ tibs/PAM/

[95] R. Tibshirani, SAM: Significance Analysis of Microarrays. 2001. [Online]. Available: http://www-stat.stanford.edu/ tibs/SAM/

[96] TIGR, TIGR Spotfinder. 2000. [Online]. Available: http://www.tigr.org/software

[97] M. E. Tipping, Sparse Bayesian learning and the relevance vector machine. *J. of Machine Learning Res.*, vol. 1, pp. 211–244, 2001.

[98] O. Troyanskaya, M. Cantor, G. Sherlock, P. Brown, T. Hastie, R. Tibshirina, D. Botstein, and R. B. Altman, Missing value estimation methods for DNA microarrays. *Bioinformatics*, vol. 17, no. 6, pp. 520–525, 2001.

[99] A. Ultsch, Is log ratio a good value for identifying differential expressed genes in microarray experiments? 2003. [Online]. Available: http://www.mathematik.uni-marburg.de/forschung/publikationen/paper_info/bfi35.pdf

[100] V. Vapnik, *Statistical Learning Theory*. Wiley, 1998.

[101] J. Vestano and E. Alhoniemi, Cluster of the self-organizing map. *IEEE Trans. Neural Networks*, vol. 11, no. 3, pp. 586–600, 2000.

[102] R. E. Walls, S. Barber, J. T. Kent, and M. S. Gilthorpe, Analysis of microarray data by an adaptive empirical Bayesian thresholding procedure. in *Proc. Interdisciplinary Statistics and Bioinformatics 2006*, Leeds, UK, Jul. 2006, pp. 175–177.

[103] X. Wang, S. Ghosh, and S. W. Guo, Quantitative quality control in microarray image processing and data acquisition. Nucleic Acids Research, 2001. [Online]. Available: http://nar.oxfordjournals.org/cgi/reprint/29/15/e75.

[104] X. H. Wang, R. S. H. Istepanian, and Y. H. Song, Application of wavelet modulus maxima in microarray spots segmentation. *IEEE Trans. Nanobiosci.*, vol. 2, no. 4, pp. 190–192, 2003.

[105] Y. Wang, F. Y. Shih, and M. Q. Ma, Precise gridding of microarray images by detecting and correcting rotations in subarrays. in *Proc. 6th Int. Conf. Computer Vision, Pattern Recognition and Image Processing*, Salt Lake City, UT, USA, Jul. 2005.

[106] Y. P. Wang, M. R. Gunampally, and W. W. Cai, Automated segmentation of microarray spots using fuzzy clustering approaches. in *Proc. IEEE Workshop on Machine Learning for Signal Processing*, Mystic, CT, USA, Sep. 2005, pp. 387–391.

[107] D. L. Weakliem, A critique of the Bayesian information criterion for model selection. *Sociological Methods and Research*, vol. 27, no. 3, pp. 359–397, 1999.

[108] S. Wu and H. Yan, Microarray image processing based on clustering and morphological analysis. in *Proc. 1st Asia-Pacific Bioinformatics conf. on Bioinformatics*, vol. 19, Adelaide, Australia, Feb. 2003, pp. 111–118.

[109] X. Xia, AMIADA Analyzing Microarray Data. 2001. [Online]. Available: http://dambe.bio.utottawa.ca/amiada.asp

[110] Y. H. Yang, M. J. Buckley, S. Dudoit, and T. P. Speed, Comparison of methods for image analysis on cDNA microarray data. Department of Statistics, University of California, Berkeley, CA, USA, Tech. Rep., 2002.

[111] Y. H. Yang, M. J. Buckley, and T. P. Speed, Analysis of cDNA microarray images. *Briefings on Bioinformatics*, vol. 2, no. 4, pp. 341–349, 2001.

[112] M. K. S. Yeung, J. Tegnér, and J. J. Collins, Reverse engineering gene networks using singular value decomposition and robust regression. *PNAS*, vol. 99, no. 9, pp. 6163–6168, 2002.

[113] D. Yoon, E. K. Lee, and T. Park, Robust imputation method for missing values in microarray data. BMC Bioinformatics, 2007. [Online]. Available: http://www.biomedcentral.com/1471-2105/8/S2/S6

[114] W. Zhang and J. Liu, Gene selection for cancer classification using relevance vector machine. in *Proc. IEEE 1st Int. Conf. Bioinformatics and Biomedical Engineering*, Wuhan, China, Jul. 2007, pp. 184–187.

[115] X. Zhou and D. P. Tuck, MSVM-RFE: extensions of SVM-RFE for multiclass gene selection on DNA microarray data. *Bioinformatics*, vol. 23, pp. 1106–1114, 2007.

In: New Signal Processing Research
Editor: Takumi Maeda, pp. 127-158
ISBN 978-1-60456-479-2

Chapter 4

PROJECTION AND INTERPOLATION BASED TECHNIQUES FOR STRUCTURED AND IMPULSIVE NOISE FILTERING

Bishnu Lamichhane and Laura Rebollo-Neira
Aston University
Birmingham B4 7ET, United Kingdom

Abstract

In this chapter we present the relevant mathematical background to address two well defined signal and image processing problems. Namely, the problem of structured noise filtering and the problem of interpolation of missing data. The former is addressed by recourse to oblique projection based techniques whilst the latter, which can be considered equivalent to impulsive noise filtering, is tackled by appropriate interpolation methods.

1. Introduction

Structured noise filtering is a particular problem of signal separation, in which the subspaces hosting the signal components are assumed to be known and complementary. Thus, the filtering can be realized in a straightforward manner by recourse to an oblique projection onto the subspace where one of the signals belongs, and along the subspace hosting the other components. A number of signal processing applications in which this procedure is of assistance are discussed in [1].

Oblique projectors were introduced early [2, 3]. Recently there has been a renewed interest in their properties [4–6]. As a very small sample of the publications concerning signal processing application of oblique projections we could mention [7–10]. Unfortunately, given two complementary subspaces, it is not always possible to construct a numerically stable oblique projector onto one of the subspaces and along the other. If the angle between such subspaces is small, the numerical errors in the calculations are magnified and yield, thereby, an oblique projector of poor quality. In relation to the problem of structured noise filtering this may cause the failure to correctly filter the noise. Nevertheless, if the signal component one wishes to discriminate belongs to an 'unknown' subspace of the given one, and the construction of the oblique projector onto such smaller subspace is numerically stable, the problem of discriminating the signal component can be transformed into the one of finding the right subspace. This has motivated the adaptive Oblique Matching Pursuit approach for signal separation [11], extended in [12]. The approach is effectively implemented by recursive equations for adapting oblique projectors given in [13].

Impulsive noise is characterised by a linear combination of very sharp spikes. This type of noise can be regarded as a particular case of structured noise and in some situations its filtering could be accomplished by techniques used for structured noise. However, especially for images, impulsive noise can be effectively handled as a problem of interpolation of missing data or 'image inpainting' [14, 15]. This problem has been addressed from different points of view [14, 16, 17]. Here we focus on a method based on scattered data interpolation [18,19]. The interpolation is realised by using Delaunay triangulation [20,21]. Interpolation methods based on Delaunay triangulation have been previously applied to image processing in [22, 23].

The chapter is organised as follows: For the convenience of the readers all the elementary mathematical terms used throughout the chapter are defined in Section 2.. More advanced technical terms are defined in the sections where they are introduced. Section 3. discusses oblique projectors in the context of structured noise filtering. Section 4. provides the basic mathematical background relevant to polynomial and piecewise polynomial interpolation in one and two dimensions. Delaunay triangulation technique for scattered data interpolation is discussed in the same section. Such a technique is applied to the problem of filtering salt and pepper impulsive noise from an image.

List of Symbols

We shall use standard set-theoretic notation

$$\cup, \quad \cap, \quad \subseteq, \quad \subset, \quad \in$$

to denote 'union', 'intersection', 'subset of', 'proper subset of', 'belong(s) to', respectively. For the sets V_1 and V_2, the set $\{v \in V_1 \ : \ v \notin V_2\}$ is denoted by $V_1 \setminus V_2$.

The following standard notations and symbols will be used without defining them explicitly:

$\mathbb{N}$	:	set of all positive integers
$\mathbb{Z}$	:	set of all integers
$\mathbb{R}$	:	field of all real numbers
$\mathbb{C}$	:	field of all complex numbers
$\mathbb{F}$	:	field of real or complex numbers
$\Longrightarrow$	:	imply (implies)
$\Longleftrightarrow$	:	if and only if
$\rightarrow$	:	maps to

The Kronecker symbol is given by

$$\delta_{ij} = \begin{cases} 1 & \text{if } i = j \\ 0 & \text{otherwise.} \end{cases}$$

The characteristic function χ_S of a set S is defined as

$$\chi_S(x) = \begin{cases} 1 & \text{if } x \in S \\ 0 & \text{otherwise.} \end{cases}$$

The absolute value of a number $a \in \mathbb{F}$ is indicated as $|a|$.

If $a \in \mathbb{R}$

$$|a| = \begin{cases} a & \text{if } a \geq 0 \\ -a & \text{if } a < 0. \end{cases}$$

If $a \in \mathbb{C}$ its complex conjugate is denoted by $\overline{a}$ and $|a|^2 = a\overline{a}$.

Note: **Bold face** is used when a terminology is defined. *Italics* are used to emphasise a terminology or statement.

2. Elementary Definitions

A **vector space** over a field $\mathbb{F}$ is a set $\mathcal{V}$ together with two operations vector addition, denoted $v + w \in \mathcal{V}$ for $v, w \in \mathcal{V}$ and scalar multiplication, denoted $av \in \mathcal{V}$ for $a \in \mathbb{F}$ and $v \in \mathcal{V}$, such that the following axioms are satisfied:

1. $v + w = w + v, \ \ v, w \in \mathcal{V}$.
2. $u + (v + w) = (u + v) + w, \ \ u, v, w \in \mathcal{V}$.
3. There exists an element $0 \in \mathcal{V}$, called the zero vector, such that $v + 0 = v, \ \ v \in \mathcal{V}$.
4. There exists an element $\tilde{v} \in \mathcal{V}$, called the additive inverse of v, such that $v + \tilde{v} = 0, v \in \mathcal{V}$.
5. $a(bv) = (ab)v, \ \ a, b \in \mathbb{F}$ and $v \in \mathcal{V}$.
6. $a(v + w) = av + aw, \ \ a \in \mathbb{F}$ and $v, w \in \mathcal{V}$.
7. $(a + b)v = av + bv, \ \ a, b \in \mathbb{F}$ and $v \in \mathcal{V}$.
8. $1v = v, \ \ v \in \mathcal{V}$, where 1 denotes the multiplicative identity in $\mathbb{F}$.

The elements of a vector space are called **vectors**. A subset $\mathcal{S}$ of a vector space $\mathcal{V}$ is a **subspace** of $\mathcal{V}$ if it is a vector space with respect to the vector space operations on $\mathcal{V}$. A subspace which is a proper subset of the whole space is called a **proper subspace**. Two subspaces $\mathcal{V}_1$ and $\mathcal{V}_2$ are **complementary** or **disjoint** if $\mathcal{V}_1 \cap \mathcal{V}_2 = \{0\}$.

The sum of two subspaces $\mathcal{V}_1$ and $\mathcal{V}_2$ is the subspace $\mathcal{V} = \mathcal{V}_1 + \mathcal{V}_2$ of elements $v = v_1 + v_2, \ v_1 \in \mathcal{V}_1, \ v_2 \in \mathcal{V}_2$. If the subspaces $\mathcal{V}_1$ and $\mathcal{V}_2$ are complementary, $\mathcal{V} = \mathcal{V}_1 + \mathcal{V}_2$ is called **direct sum** and each element $v \in \mathcal{V}$ has a unique decomposition $v = v_1 + v_2$, $v_1 \in \mathcal{V}_1$, $v_2 \in \mathcal{V}_2$.

Let $\mathcal{V}_1$ and $\mathcal{V}_2$ be vectors spaces. A mapping $\hat{A} : \mathcal{V}_1 \to \mathcal{V}_2$ is a **linear operator** if

$$\hat{A}(v + w) = \hat{A}v + \hat{A}w, \quad \hat{A}(av) = a\hat{A}v,$$

for all $v, w \in \mathcal{V}_1$ and $a \in F$. $\mathcal{V}_1$ is called the **domain** of $\hat{A}$ and $\mathcal{V}_2$ its **codomain** or **image**. If the codomain of a linear operator is a scalar field, the operator is called a **linear functional** on $\mathcal{V}_1$. The set of all linear functionals on $\mathcal{V}_1$ is called the **dual space** of $\mathcal{V}_1$.

The **adjoint** of an operator $\hat{A} : \mathcal{V}_1 \to \mathcal{V}_2$ is the unique operator $\hat{A}^*$ satisfying that

$$\langle \hat{A}g_1, g_2 \rangle = \langle g_1, \hat{A}^* g_2 \rangle.$$

If $\hat{A}^* = \hat{A}$ the operator is **self-adjoint**.

An operator $\hat{A} : \mathcal{V}_1 \to \mathcal{V}_2$ has an **inverse** if there exists $\hat{A}^{-1} : \mathcal{V}_2 \to \mathcal{V}_1$ such that

$$\hat{A}^{-1}\hat{A} = \hat{I}_{\mathcal{V}_1} \quad \text{and} \quad \hat{A}\hat{A}^{-1} = \hat{I}_{\mathcal{V}_2},$$

where $\hat{I}_{\mathcal{V}_1}$ and $\hat{I}_{\mathcal{V}_2}$ denote the identity operators in $\mathcal{V}_1$ and $\mathcal{V}_2$, respectively. By a **generalised inverse** we shall mean an operator $\hat{A}^\dagger$ satisfying the following conditions

$$\begin{aligned} \hat{A}\hat{A}^\dagger\hat{A} &= \hat{A} \\ \hat{A}^\dagger\hat{A}\hat{A}^\dagger &= \hat{A}^\dagger. \end{aligned}$$

If $v_1, \ldots, v_n$ are some elements of a vector space $\mathcal{V}$, by a **linear combination** of $v_1, \ldots, v_n$ we mean an element in $\mathcal{V}$ of the form $a_1v_1 + \cdots + a_nv_n$, with $a_i \in \mathbb{F}, i = 1, \ldots, n$.

Let S be a subset of element of $\mathcal{V}$. The set of all *linear combinations* of elements of S is called the **span** of S and is denoted by span S.

A subset $S = \{v_i\}_{i=1}^n$ of $\mathcal{V}$ is said to be **linearly independent** if and only if

$$a_1v_1 + \cdots + a_nv_n = 0, \quad \Longrightarrow \quad a_i = 0, i = 1, \ldots, n.$$

A subset is said to be **linearly dependent** if it is not linearly independent.

$\mathcal{S}$ is said to be a **basis** of $\mathcal{V}$ if it is linearly independent and span $S = \mathcal{V}$. The **dimension** of a finite dimensional vector space $\mathcal{V}$ is the number of elements in a basis for $\mathcal{V}$. The number of elements in a set is termed the **cardinality** of the set.

Let $\{v_i\}_{i=1}^n$ be a basis for $\mathcal{V}$. For $v = a_1v_1 + \cdots, a_nv_n$ let $f_i(v) : \mathcal{V} \to \mathbb{F}$ be defined by

$$f_i(v) = a_i, \quad i = 1, \ldots, n.$$

Then f_i is a linear functional for each i. The linear functionals $f_1, \ldots, f_n$ are called **coordinate functionals** on $\mathcal{V}$ with respect to the basis $\{v_i\}_{i=1}^n$.

We denote the **space of polynomials** of degree $m \in \mathbb{N}$ on $\mathbb{R}$ by

$$\mathcal{P}_m(\mathbb{R}) = \Big\{p : p(x) = \sum_{i=0}^{m} a_i x^i, \ x \in \mathbb{R}\Big\}.$$

Such a space is a vector space.

Given a compact interval $[a, b]$ we define a **partition** of $[a, b]$ as the finite set of points

$$\Delta = \{x_i\}_{i=0}^n, \; n \in \mathbb{N}, \quad \text{such that} \quad a = x_0 < x_1 < \cdots < x_n = b.$$

An **inner product** on a vector space $\mathcal{V}$ is a map from $\mathcal{V}$ to $\mathbb{F}$ which satisfies the following axioms

1. $\langle v, v\rangle \geq 0, \; v \in \mathcal{V}$, and $\langle v, v\rangle = 0 \Longleftrightarrow v = 0$.
2. $\langle v + w, z\rangle = \langle v, z\rangle + \langle w, z\rangle, \; v, w, z \in \mathcal{V}$.
3. $\langle v, az\rangle = a\langle v, z\rangle, \; v, z \in \mathcal{V}$ and $a \in \mathbb{F}$.
4. $\langle v, w\rangle = \overline{\langle w, v\rangle}, \; v, w \in \mathcal{V}$.

A vector space $\mathcal{V}$ together with an inner product $\langle\cdot,\cdot\rangle$ is called an **inner product space**.

Two vectors v and w in an inner product space are said to be **orthogonal** if $\langle v, w\rangle = 0$.

Two subspaces $\mathcal{V}_1$ and $\mathcal{V}_2$ are orthogonal if $\langle v_1, v_2\rangle = 0$ for all $v_1 \in \mathcal{V}_1$ and $v_2 \in \mathcal{V}_1$. The sum of two orthogonal subspaces $\mathcal{V}_1$ and $\mathcal{V}_2$ is termed **orthogonal sum** and will be indicated as $\mathcal{V} = \mathcal{V}_1 \oplus \mathcal{V}_2$. The subspace $\mathcal{V}_2$ is called the **orthogonal complement** of $\mathcal{V}_1$ in $\mathcal{V}$. Equivalently, $\mathcal{V}_1$ is the orthogonal complement of $\mathcal{V}_2$ in $\mathcal{V}$.

A **norm** $\|\cdot\|$ on a vector space $\mathcal{V}$ is a function from $\mathcal{V}$ to $\mathbb{R}$ such that for every $v, w \in \mathcal{V}$ and $a \in \mathbb{F}$ the following three properties are fulfilled

1. $\|v\| \geq 0$, and $\|v\| = 0 \Longleftrightarrow v = 0$.
2. $\|av\| = |a|\,\|v\|$.
3. $\|v + w\| \leq \|v\| + \|w\|$.

A vector space $\mathcal{V}$ together with a norm is called a **normed vector space**.

Two vectors v and w are said to be **orthonormal** if they are *orthogonal* and $\|v\| = \|w\| = 1$.

The **Euclidean space** $\mathbb{R}^n$ is an inner product space with inner product defined by

$$\langle \mathbf{x}, \mathbf{y}\rangle = x_i\overline{y}_i + \ldots x_n\overline{y}_n,$$

with $\mathbf{x} = (x_1, \ldots, x_n)$ and $\mathbf{y} = (y_1, \ldots, y_n)$. The norm $\|\mathbf{x}\|$ is induced by the inner product

$$\|\mathbf{x}\| = \langle \mathbf{x}, \mathbf{x}\rangle^{\frac{1}{2}} = (x_i\overline{x}_i + \ldots x_n\overline{x}_n)^{\frac{1}{2}} = (|x_i|^2 + \ldots + |x_n|^2)^{\frac{1}{2}}.$$

The **space** $L^2[a, b]$ is an inner product space of functions on $[a, b]$ with inner product defined by

$$\langle x, y\rangle = \int_a^b x(t)\overline{y}(t)\,dt$$

and norm

$$\|x\| = \langle x, x\rangle^{\frac{1}{2}} = \left(\int_a^b |x(t)|^2\,dt\right)^{\frac{1}{2}}.$$

The **space** $C^k[a, b]$ is the space of functions on $[a, b]$ having continuous derivatives up to order $k \in \mathbb{N}$. The space of continuous functions on $[a, b]$ is denoted as $C^0[a, b]$.

3. Signal Representation, Reconstruction, and Projection

Throughout this chapter a **signal** is considered to be an element of an inner product space $\mathcal{H}$ with norm induced by the inner product, $\|\cdot\| = \langle \cdot, \cdot \rangle^{\frac{1}{2}}$. Moreover, for the problems to be addressed we assume that all the signals of interest belong to some finite dimensional subspace $\mathcal{V}$ of $\mathcal{H}$ spanned by the set $\{v_i \in \mathcal{H}\}_{i=1}^{M}$. Hence, a signal f can be expressed by a finite linear superposition

$$f = \sum_{i=1}^{M} c_i v_i,$$

where the coefficients c_i, $i = 1, \ldots, M$, are in $\mathbb{F}$.

We call **measurement** or **sampling** to the process of transforming a signal into a number. Hence a **measure** or **sample** is a *functional*. Because we restrict considerations to linear measures the associated functional is linear. Accordingly, making use of *Riesz theorem* [24] we can express a linear measure as

$$m = \langle w, f \rangle \quad \text{for some} \quad w \in \mathcal{H}.$$

We refer the vector w to as **measurement vector**.

Considering M measurements m_i, $i = 1, \ldots, M$, each of which is obtained by a measurement vector w_i, we have a numerical representation of f as given by

$$m_i = \langle w_i, f \rangle, \quad i = 1, \ldots, M.$$

Now we want to answer the question as to whether it is possible to reconstruct $f \in \mathcal{V}$ from these measurements. More precisely, we wish to find the requirements we need to impose upon the measurement vectors w_i, $i = 1, \ldots, M$, so as to use the concomitant measures $\langle w_i, f \rangle$, $i = 1, \ldots, M$, as coefficients for the signal reconstruction, i.e., we wish to have

$$f = \sum_{i=1}^{M} c_i v_i = \sum_{i=1}^{M} \langle w_i, f \rangle v_i. \tag{1}$$

By denoting

$$\hat{E} = \sum_{i=1}^{M} v_i \langle w_i, \cdot \rangle, \tag{2}$$

where the operation $\langle w_i, \cdot \rangle$ indicates that $\hat{E}$ acts by taking inner products, (1) is recast

$$f = \hat{E} f.$$

As will be discussed next, the above equation tells us that the measurement vectors w_i, $i = 1, \ldots, M$, should be such that the operator $\hat{E}$ is a projector onto $\mathcal{V}$.

An operator $\hat{E} : \mathcal{H} \to \mathcal{V}$ is a projector if it is *idempotent*, i.e.,

$$\hat{E}^2 = \hat{E}.$$

As a consequence, the projection is onto $\mathcal{R}(\hat{E})$, the range of the operator, and along $\mathcal{N}(\hat{E})$, the null space of the operator. Let us recall that

$$\mathcal{R}(\hat{E}) = \{f, \text{ such that } f = \hat{E}g, \; g \in \mathcal{H}\}.$$

Thus, if $f \in \mathcal{R}(\hat{E})$,

$$\hat{E}f = \hat{E}^2 g = \hat{E}g = f.$$

This implies that $\hat{E}$ behaves like the identity operator for all $f \in \mathcal{R}(\hat{E})$, regardless of $\mathcal{N}(E)$, which is the subspace of $\mathcal{H}$ defined as

$$\mathcal{N}(E) = \{g, \text{ such that } \hat{E}g = 0, \; g \in \mathcal{H}\}.$$

It is clear then that to reconstruct a signal $f \in \mathcal{V}$ by means of (1) the involved measurement vectors w_i, $i = 1, \dots, M$, should give rise to an operator of the form (2), which must be a projector onto $\mathcal{V}$. Notice that the required operator is not unique, because there exist many projectors onto $\mathcal{V}$ having different $\mathcal{N}(\hat{E})$. Thus, for reconstructing signals in the range of the projector its null space can be chosen arbitrarily. However, the null space becomes extremely important when the projector acts on signals outside its range. A popular projector, to be discussed below, is the orthogonal one. When $\mathcal{N}(\hat{E})$ happens to be equal to $\mathcal{R}(\hat{E})^\perp$, which indicates the orthogonal complement of $\mathcal{R}(\hat{E})$, the projector is called **orthogonal projector** onto $\mathcal{R}(\hat{E})$. This is the case if and only if the projector is *self adjoint*.

A projector which is not orthogonal is called an **oblique projector** and we need two subscripts to represent it. One subscript to indicate the range of the projector and another to represent the subspace along which the projection is performed. Hence the projector onto $\mathcal{V}$ along $\mathcal{W}^\perp$ is indicated as $\hat{E}_{\mathcal{V}\mathcal{W}^\perp}$.

The particular case $\hat{E}_{\mathcal{V}\mathcal{V}^\perp}$ corresponds to an *orthogonal projector* and we use the special notation $\hat{P}_{\mathcal{V}}$ to indicate such a projector. When a projector onto $\mathcal{V}$ is used for signal processing, $\mathcal{W}^\perp$ can be chosen according to the processing task. The examples below illustrate two different situations.

Example 1. Let us assume that the signal processing task is to approximate a signal $f \in \mathcal{H}$ by a signal $f_{\mathcal{V}} \in \mathcal{V}$. In this case, one normally would choose $f_{\mathcal{V}} = \hat{P}_{\mathcal{V}} f$ because this is the unique signal in $\mathcal{V}$ minimising the distance $||f - f_{\mathcal{V}}||$. Indeed, let us take another signal g in $\mathcal{V}$ and write it as $g = g + \hat{P}_{\mathcal{V}} f - \hat{P}_{\mathcal{V}} f$. Since $f - \hat{P}_{\mathcal{V}} f$ is orthogonal to every signal in $\mathcal{V}$ we have

$$||f - g||^2 = ||f - g + \hat{P}_{\mathcal{V}} f - \hat{P}_{\mathcal{V}} f||^2 = ||f - \hat{P}_{\mathcal{V}} f||^2 + ||\hat{P}_{\mathcal{V}} f - g||^2.$$

Hence $||f - g||$ is minimised if $g = \hat{P}_{\mathcal{V}} f$.

Example 2. Assume that the signal f to be analysed here is the superposition of two signals, $f = f_1 + f_2$, each component being produced by a different phenomenon we want to discriminate. Let us assume further that $f_1 \in \mathcal{V}$ and $f_2 \in \mathcal{W}^\perp$ with $\mathcal{V}$ and $\mathcal{W}^\perp$ disjoint subspaces. Thus, we can obtain, f_1 say, from f, by an oblique projector onto $\mathcal{V}$ and along $\mathcal{W}^\perp$. The projector will map to zero the component f_2 to produce

$$f_1 = \hat{E}_{\mathcal{V}\mathcal{W}^\perp} f.$$

3.1. Constructing Oblique Projectors for Structured Noise Filtering

Example 2 discusses the fact that a signal component can be discriminated from another components by an oblique projection. In this section we will focus on the actual construction of such a projector.

Given $\mathcal{V}$ and $\mathcal{W}^\perp$ disjoint, i.e., such that

$$\mathcal{V} \cap \mathcal{W}^\perp = \{0\},$$

in order to provide a prescription for constructing $\hat{E}_{\mathcal{V}\mathcal{W}^\perp}$ we proceed as follows. Firstly we define $\mathcal{S}$ as the direct sum of $\mathcal{V}$ and $\mathcal{W}^\perp$, which we express as

$$\mathcal{S} = \mathcal{V} + \mathcal{W}^\perp.$$

Let $\mathcal{W} = (\mathcal{W}^\perp)^\perp$ be the orthogonal complement of $\mathcal{W}^\perp$ in $\mathcal{S}$. Thus we have

$$\mathcal{S} = \mathcal{V} + \mathcal{W}^\perp = \mathcal{W} \oplus \mathcal{W}^\perp,$$

where the operation $\oplus$ indicates the orthogonal sum.

Considering that $\{v_i\}_{i=1}^M$ is a spanning set for $\mathcal{V}$ a spanning set for $\mathcal{W}$ is obtained as

$$u_i = v_i - \hat{P}_{\mathcal{W}^\perp} v_i = \hat{P}_{\mathcal{W}} v_i, \quad i = 1, \ldots, M.$$

Denoting as $\{\mathbf{e}_i\}_{i=1}^M$ the standard orthonormal basis in $\mathbb{F}^M$, i.e., the Euclidean inner product $\langle \mathbf{e}_i, \mathbf{e}_j \rangle = \delta_{ij}$, we define the operators $\hat{V} : \mathbb{F}^M \to \mathcal{V}$ and $\hat{U} : \mathbb{F}^M \to \mathcal{W}$ as

$$\hat{V} = \sum_{i=1}^M v_i \langle \mathbf{e}_i, \cdot \rangle, \qquad \hat{U} = \sum_{i=1}^M u_i \langle \mathbf{e}_i, \cdot \rangle.$$

Thus the adjoint operators $\hat{U}^*$ and $\hat{V}^*$ are

$$\hat{V}^* = \sum_{i=1}^M \mathbf{e}_i \langle v_i, \cdot \rangle, \qquad \hat{U}^* = \sum_{i=1}^M \mathbf{e}_i \langle u_i, \cdot \rangle.$$

It follows that $\hat{P}_{\mathcal{W}} \hat{V} = \hat{U}$ and $\hat{U}^* \hat{P}_{\mathcal{W}} = \hat{U}^*$ hence, $\hat{G} : \mathbb{C}^M \to \mathbb{C}^M$ defined as:

$$\hat{G} = \hat{U}^* \hat{V} = \hat{U}^* \hat{U}$$

is self-adjoint and its matrix representation, G, has elements

$$g_{i,j} = \langle u_i, v_j \rangle = \langle \hat{P}_{\mathcal{W}} u_i, v_j \rangle = \langle u_i, \hat{P}_{\mathcal{W}} u_j \rangle = \langle u_i, u_j \rangle, \quad i, j = 1, \ldots, M.$$

From now on we restrict our signal space to be $\mathcal{S}$, since we would like to build the oblique projector $\hat{E}_{\mathcal{V}\mathcal{W}^\perp}$ onto $\mathcal{V}$ and along $\mathcal{W}^\perp$ having the form

$$\hat{E}_{\mathcal{V}\mathcal{W}^\perp} = \sum_{i=1}^M v_i \langle w_i, \cdot \rangle. \tag{3}$$

Clearly for the operator to map to zero every vector in $\mathcal{W}^\perp$ vectors $\{w_i\}_{i=1}^M$ must span $\mathcal{W} = (\mathcal{W}^\perp)^\perp = \text{span}\{u_i\}_{i=1}^M$. This entails that for each w_i there exists a set of coefficients $\{b_{i,j}\}_{j=1}^M$ such that

$$w_i = \sum_{i=1}^{M} b_{i,j} u_i, \tag{4}$$

which guarantees that every w_i is orthogonal to all vectors in $\mathcal{W}^\perp$ and therefore $\mathcal{W}^\perp$ is included in the null space of $\hat{E}_{\mathcal{V}\mathcal{W}^\perp}$. Moreover, since every signal, g say, in $\mathcal{S}$ can be written as $g = g_\mathcal{V} + g_{\mathcal{W}^\perp}$ with $g_\mathcal{V} \in \mathcal{V}$ and $g_{\mathcal{W}^\perp} \in \mathcal{W}^\perp$, the fact that $\hat{E}_{\mathcal{V}\mathcal{W}^\perp} g = 0$ implies $g_\mathcal{V} = 0$. Hence, $g = g_{\mathcal{W}^\perp}$, which implies that the null space of $\hat{E}_{\mathcal{V}\mathcal{W}^\perp}$ restricted to $\mathcal{S}$ is $\mathcal{W}^\perp$. In order for $\hat{E}_{\mathcal{V}\mathcal{W}^\perp}$ to be a projector it is necessary that $\hat{E}^2_{\mathcal{V}\mathcal{W}^\perp} = \hat{E}_{\mathcal{V}\mathcal{W}^\perp}$. As will be shown in the next proposition, if the coefficients $b_{i,j}$ are the matrix elements of a *generalised inverse* of the matrix G defined above, this condition is fulfilled.

Proposition 1. *If the coefficients $b_{i,j}$ in (4) are the matrix elements of a generalised inverse of the matrix G, which has elements $g_{i,j} = \langle v_i, u_j \rangle$, $i, j = 1, \ldots, M$, the operator in (3) is a projector.*

Proof. For the measurement vectors in (4) to yield a projector of the form (3), the corresponding operator should be idempotent, i.e.,

$$\sum_{n=1}^{M}\sum_{m=1}^{M}\sum_{i=1}^{M}\sum_{j=1}^{M} v_i b_{i,j} \langle u_j, v_n \rangle \overline{b_{n,m}} \langle u_m, \cdot \rangle = \sum_{i=1}^{M}\sum_{j=1}^{M} v_i \overline{b_{i,j}} \langle u_j, \cdot \rangle. \tag{5}$$

Defining

$$\hat{B} = \sum_{i=1}^{M}\sum_{j=1}^{M} \mathbf{e}_i \overline{b_{i,j}} \langle \mathbf{e}_j, \cdot \rangle \tag{6}$$

and using the operators $\hat{V}$ and $\hat{U}^*$, as given above, the left hand side in (5) can be expressed as

$$\hat{V}\hat{B}^*\hat{U}^*\hat{V}\hat{B}^*\hat{U}^* \tag{7}$$

and the right hand side as

$$\hat{V}\hat{B}^*\hat{U}^*. \tag{8}$$

Assuming that $\hat{B}^*$ is a generalised inverse of $(\hat{U}^*\hat{V})$ indicated as $\hat{B}^* = (\hat{U}\hat{V})^\dagger$ it satisfies (c.f. Section 2.)

$$\hat{B}^*(\hat{U}^*\hat{V})\hat{B}^* = \hat{B}^*, \tag{9}$$

and therefore, from (7), the right hand side of (5) follows. Since $\hat{B}^* = (\hat{U}^*\hat{V})^\dagger = \hat{G}^\dagger$ and $\hat{G}^* = \hat{G}$, we have $\hat{B} = \hat{G}^\dagger$. Hence, if the elements $b_{i,j}$ determining $\hat{B}$ in (6) are the matrix elements of a generalised inverse on the matrix representation of $\hat{G}$, the corresponding vectors $\{w_i\}_{i=1}^n$ obtained by (4) yield an operator of the form (3), which is an oblique projector. □

Property 1. *Let $\hat{E}_{\mathcal{V}\mathcal{W}^\perp}$ be the oblique projector onto $\mathcal{V}$ and along $\mathcal{W}^\perp$ and $\hat{P}_\mathcal{W}$ the orthogonal projector onto $\mathcal{W} = (\mathcal{W}^\perp)^\perp$. Then the following relation holds*

$$\hat{P}_\mathcal{W}\hat{E}_{\mathcal{V}\mathcal{W}^\perp} = \hat{P}_\mathcal{W}.$$

Proof. $\hat{E}_{\mathcal{V}\mathcal{W}^\perp}$ given in (3) can be recast, in terms of operator $\hat{V}$ and $\hat{U}^*$, as:

$$\hat{E}_{\mathcal{V}\mathcal{W}^\perp} = \hat{V}(\hat{U}^*\hat{V})^\dagger\hat{U}^*.$$

Applying $\hat{P}_{\mathcal{W}}$ both sides of the equation we obtain:

$$\hat{P}_{\mathcal{W}}\hat{E}_{\mathcal{V}\mathcal{W}^\perp} = \hat{P}_{\mathcal{W}}\hat{V}(\hat{U}^*\hat{V})^\dagger\hat{U}^* = \hat{U}(\hat{U}^*\hat{V})^\dagger\hat{U}^* = \hat{U}(\hat{U}^*\hat{U})^\dagger\hat{U}^*,$$

which is a well known form for the orthogonal projector onto $\mathcal{R}(\hat{U}) = \mathcal{W}$. □

Notice that the operative steps for constructing an oblique projector are equivalent to those for constructing an orthogonal one. The difference being that in general the spaces span$\{v_i\}_{i=1}^M = \mathcal{V}$ and span$\{w_i\}_{i=1}^M = \mathcal{W}$ are different. For the special case $u_i = v_i$, $i = 1, \dots, M$, both sets of vectors span $\mathcal{V}$ and we have an orthogonal projector onto $\mathcal{V}$ along $\mathcal{V}^\perp$.

Example 3. Suppose that the chirp signal in the first graph of Figure 1 is corrupted by impulsive noise belonging to the subspace

$$\mathcal{W}^\perp = \text{span}\{y_j(t) = e^{-100000(t-0.05j)^2}, \quad t \in [0,1]\}_{j=1}^{200}.$$

The chirp after being corrupted by a realization of the noise consisting of 50 pulses taken randomly from elements of $\mathcal{W}^\perp$ is plotted in the second graph of Figure 1.

Consider that the signal subspace is well represented by $\mathcal{V}$ given by

$$\mathcal{V} = \text{span}\{v_{i+1}(t) = \cos \pi i t, \quad t \in [0,1]\}_{i=0}^{M=99}.$$

In order to eliminate the impulsive noise from the chirp we have to compute the measurement vectors $\{w_i\}_{i=1}^{100}$, here functions of $t \in [0,1]$, determining the appropriate projector. For this we first need a representation of $\hat{P}_{\mathcal{W}^\perp}$, which is obtained simply by transforming the set $\{y_j\}_{j=1}^{200}$ into an orthonormal set. Since the given set $\{y_j\}_{j=1}^{200}$ is linear independent its orthogonalisation can be achieved by a *Gram Schmidt*, or *Generalised Gram Schmidt*, procedure. For the extension to include the non-linear independent case, see [25]. The corresponding codes can be found at [26]. Using the orthonormal basis $\{o_j\}_{j=1}^{200}$ for $\mathcal{W}^\perp$ the orthogonal projector onto this subspace is the operator

$$\hat{P}_{\mathcal{W}^\perp} = \sum_{j=1}^{200} o_j \langle o_j, \cdot \rangle.$$

With this projector we construct vectors

$$u_{i+1}(t) = \cos \pi i t - \sum_{j=1}^{200} o_j(t)\langle o_j(t), \cos \pi i t\rangle, \quad i = 0, \dots, 99, \quad t \in [0,1]$$

using a quadrature formula [27] for computing the inner products involved in the above equations and in the elements, $g_{i,j}$, of matrix G

$$g_{i+1,j+1} = \int_0^1 u_{i+1}(t) \cos \pi j t \, dt, \quad i = 0, \dots, 99, \; j = 0, \dots, 99.$$

This matrix has an inverse, which is used to obtain functions $\{w_i(t), \; t \in [0,1]\}_{i=1}^{100}$ giving rise to the required oblique projector. The chirp filtered by such a projector is depicted in the last graph of Figure 1.

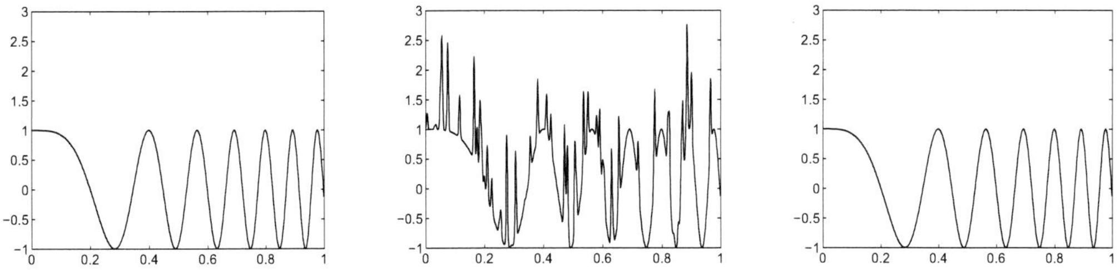

Figure 1. Chirp signal (first graph). Chirp corrupted by 50 randomly taken pulses (middle graph). Chirp denoised by oblique projection (last graph).

Example 4. Here we deal with the image of Lena shown in the first picture of Figure 4. This image is an array of 512×512 pixels that we process row by row. Each row is a vector $\mathbf{I}_i \in \mathbb{R}^{512}$, $i = 1, \ldots, 512$. The image is affected by structured noise produced when random noise passes through a channel characterised by a given matrix A having 100 columns and 512 rows. The model for each row $\mathbf{I}_i^o \in \mathbb{R}^{512}$ of the noisy image is

$$\mathbf{I}_i^o = \mathbf{I}_i + A\mathbf{h}_i, \quad i = 1, \ldots, 512,$$

where each $\mathbf{h}_i$ is a random vector in $\mathbb{R}^{100}$. The image plus noise is represented in the middle graph of Figure 4. In order to denoise the image we consider that every row $\mathbf{I}_i \in \mathbb{R}^{512}$ is well represented in a subspace $\mathcal{V}$ spanned by discrete cosines [28]. More precisely, we consider $\mathbf{I}_i \in \mathcal{V}$ for $i = 1, \ldots, 512$, where

$$\mathcal{V} = \text{span}\left\{\mathbf{x}_i = \cos\left(\frac{\pi(2j-1)(i-1)}{2L}\right), \quad j = 1, \ldots, 512\right\}_{i=1}^{300}.$$

The space of the noise is spanned by the 100 vectors in $\mathbb{R}^{512}$ corresponding to the columns of the given matrix A. The image, after being filtered row by row by the oblique projector onto $\mathcal{V}$ and along the space of the noise, is depicted in the last graph of Figure 4.

Figure 2. Image of Lena (first picture). Image plus structured noise (middle picture). The image obtained from the middle picture by an oblique projection (last picture).

Notice that the oblique projector onto $\mathcal{V}$ is independent of the selection of the spanning set for $\mathcal{W}$. Thus, one has a number of possibilities for expressing the oblique projector. Although all the forms are theoretically equivalent, they may not always be 'numerically' equivalent. This feature is illustrated in the next example.

Example 5. Consider that $\mathcal{W}^{\perp}$ is as in Example 3 and $\mathcal{V}$ is also as in that example but with $M = 250$ which, as illustrated by the experiment below, makes the construction of the corresponding oblique projector numerically unstable.

We consider four theoretically equivalent ways of computing vectors $\{w_i\}_{i=1}^{250}$.

i) $w_i = \sum_{j=1}^{250} \tilde{g}_{i,j} u_j$, where $\tilde{g}_{i,j}$ is the (i,j)-th element of the inverse of the matrix G having elements $g_{i,j} = \langle u_i, v_j \rangle$, $i,j = 1,\ldots,250$.

ii) Vectors $\{w_i\}_{i=1}^{250}$ are as in i) but the matrix elements of G are computed as $g_{i,j} = \langle u_i, u_j \rangle$, $i,j = 1,\ldots,250$.

iii) Orthonormalising $\{u_i\}_{i=1}^{250}$ to obtain $\{q_i\}_{i=1}^{250}$, vectors $\{w_i\}_{i=1}^{250}$ are then computed as

$$w_i = \sum_{j=1}^{250} \tilde{g}_{i,j} q_j,$$

with $g_{i,j} = \langle q_i, v_j \rangle$, $i,j = 1,\ldots,250$.

iv) Same as in iii) but $g_{i,j} = \langle q_i, u_j \rangle$, $i,j = 1,\ldots,250$.

We test the numerical quality of the corresponding projectors by comparing the following quantities

a) $\max_{g \in \mathcal{V}+\mathcal{W}^{\perp}} \frac{\|(\hat{E}_{\mathcal{V}\mathcal{W}^{\perp}} - \hat{E}^2_{\mathcal{V}\mathcal{W}^{\perp}})g\|}{\|g\|}$, $\|g\| \neq 0$,

b) $\max_{v_i \in \mathcal{V}} \frac{\|\hat{E}_{\mathcal{V}\mathcal{W}^{\perp}} v_i - v_i\|}{\|v_i\|}$, $\|v_i\| \neq 0$,

c) $\max_{s_i \in \mathcal{W}^{\perp}} \frac{\|\hat{E}_{\mathcal{V}\mathcal{W}^{\perp}} s_i\|}{\|s_i\|}$, $\|s_i\| \neq 0$.

The quality of the projector is indicated by the quantities a), b), and c), which should be small if the numerical representation of the projector is accurate. All the projectors produce equivalent results for quantity c). Nevertheless, there is a significant difference between case iii) and the others with respect to the quantities a) and b). Case iii) produces a value of a) and a value of b) which are 10^{-5} times smaller than the corresponding values produced by cases i), ii), and iv).

Unfortunately, even when the subspaces $\mathcal{V}$ and $\mathcal{W}^{\perp}$ are 'theoretically' complementary, in practice, due to the calculations being performed in finite precision arithmetics, the inaccuracy in the computation of the corresponding projector may cause the failure to correctly filter structured noise. This is discussed in [11, 12], where examples illustrating such situations are provided. As pointed out in those efforts, the problem can be overcome if the signal, $f_{\mathcal{V}}$, one is trying to discriminate from the noise, admits a sparse representation in a spanning set for $\mathcal{V}$. This implies that given $\{v_i\}_{i=1}^{M}$ there exists a subset of elements characterised by the set of indices J, of cardinality $K < M$, spanning the subspace $\mathcal{V}_K = \text{span}\{v_\ell\}_{\ell \in J}$ and such that $f_{\mathcal{V}} = \hat{E}_{\mathcal{V}_K \mathcal{W}^{\perp}} f$. Under the hypothesis that the numerical construction of $\hat{E}_{\mathcal{V}_K \mathcal{W}^{\perp}}$ is well posed, this projector will produce the correct signal splitting. The problem one has to address then is the one of finding the 'right' subspace $\mathcal{V}_K$. Assuming that $\{v_i\}_{i=1}^M$ is a basis for $\mathcal{V}$, out of it we have $\binom{M}{K}$ possible basis of dimension K. Hence the problem of finding that of the right subspace $\mathcal{V}_K$ is in general intractable. In order to reduce complexity one can make the search for the right subspace signal dependent. An adaptive approach is considered in the next section.

3.2. Adaptive Pursuit Strategy and Structured Noise Filtering

Given a signal f, and assuming that the subspaces $\mathcal{W}^\perp$ and $\mathcal{V}$, are known, the goal is to find a subset $\tilde{\mathcal{V}} \in \mathcal{V}$ such that $\hat{E}_{\tilde{\mathcal{V}}\mathcal{W}^\perp} f = \hat{E}_{\mathcal{V}\mathcal{W}^\perp} f$. Moreover, it is assumed that $\mathcal{V} = \text{span}\{v_i\}_{i=1}^M$ and $\tilde{\mathcal{V}} = \text{span}\{v_\ell\}_{\ell \in J}$, where J is a unknown subset of indices, out of the original set $i = 1, \ldots, M$. The cardinality of J is such that the construction of $\hat{E}_{\tilde{\mathcal{V}}\mathcal{W}^\perp}$ is well posed. This assumption restricts the class of signals that can be handled by the approach.

Under the above hypothesis, if the subspace $\tilde{\mathcal{V}}$ were known, one would have

$$\hat{E}_{\mathcal{V}\mathcal{W}} f = \hat{E}_{\tilde{\mathcal{V}}\mathcal{W}^\perp} f = \sum_{\ell \in J} v_\ell \langle w_\ell, f \rangle. \tag{10}$$

However, if the computation of $\hat{E}_{\mathcal{V}\mathcal{W}^\perp}$ is an ill posed problem, which is the situation we are considering, $\hat{E}_{\mathcal{V}\mathcal{W}^\perp} f$ is not available. In order to look for the subset of indices J yielding $\tilde{\mathcal{V}}$ one may proceed as follows: Applying $\hat{P}_{\mathcal{W}}$ on every term of (10) and using the properties $\hat{P}_{\mathcal{W}} \hat{E}_{\mathcal{V}\mathcal{W}^\perp} = \hat{P}_{\mathcal{W}}$ and $\hat{P}_{\mathcal{W}} \hat{E}_{\tilde{\mathcal{V}}\mathcal{W}^\perp} = \hat{P}_{\tilde{\mathcal{W}}}$, where $\tilde{\mathcal{W}} = \text{span}\{u_\ell\}_{\ell \in J}$, (10) becomes

$$\hat{P}_{\mathcal{W}} f = \hat{P}_{\tilde{\mathcal{W}}} f = \sum_{\ell \in J} u_\ell \langle w_\ell, f \rangle. \tag{11}$$

Since $\mathcal{W}^\perp$ is given and $\hat{P}_{\mathcal{W}} f = f - \hat{P}_{\mathcal{W}^\perp} f$, the left hand side of (10) is available. We are then in a position to look for the set $\{u_\ell\}_{\ell \in J}$, out of the set $\{u_i\}_{i=1}^M$, in a stepwise manner. This can be achieved by an adaptive pursuit approach, termed Optimised Orthogonal Matching Pursuit approach [29], which at each step, say the step $k+1$, selects the element $u_{\ell_{k+1}}$ minimising the norm of the residual error $\|\hat{P}_{\mathcal{W}} f - \hat{P}_{\mathcal{W}_{k+1}} \hat{P}_{\mathcal{W}} f\|^2$, where $\mathcal{W}_{k+1} = \text{span}\{u_{\ell_i}\}_{i=1}^k + u_{\ell_{k+1}}$ and $\{u_{\ell_i}\}_{i=1}^k$ is the set of elements that have been selected in the previous steps. Minimisation of $\|\hat{P}_{\mathcal{W}} f - \hat{P}_{\mathcal{W}_{k+1}} \hat{P}_{\mathcal{W}} f\|^2 = \|\hat{P}_{\mathcal{W}} f - \hat{P}_{\mathcal{W}_{k+1}} f\|^2$ is equivalent to choosing the index ℓ_{k+1} such that

$$\ell_{k+1} = \arg \max_{n \in J \setminus J_k} \frac{|\langle \gamma_n, \hat{P}_{\mathcal{W}} f \rangle|}{\|\gamma_n\|} = \arg \max_{n \in J \setminus J_k} \frac{|\langle \gamma_n, f \rangle|}{\|\gamma_n\|}, \quad \|\gamma_n\| \neq 0,$$

with $\gamma_n = u_n - \hat{P}_{\mathcal{W}_k} u_n$ and $J \setminus J_k$ the set of indices not selected in previous steps.

Details on the implementation of this approach and extended versions of it are presented in [30]. Further discussions in the context of structured noise filtering are given in [12] and relevant MATLAB codes are available at [31].

4. Signal Interpolation

As already mentioned, we wish to consider the problem of impulsive noise filtering, specially in the case of images, as a problem of 'missing data' to be addressed here by means of interpolation techniques. The interpolation methods we propose are based on particular piecewise polynomial interpolation. However, in order to assist the readers unfamiliar with approximation theory, we introduce in the next two sections some basic elements of polynomial and piecewise polynomial interpolation.

4.1. Polynomial Interpolation

Suppose that we are given data as a set of $n+1$ points $\{x_i\}_{i=0}^n$ and a function f having values $\{y_i = f(x_i)\}_{i=0}^n$. Suppose further that we want to approximate f by a simple function p with the properties

$$p(x_i) = y_i, \quad i = 0, \dots, n. \tag{12}$$

A function p satisfying the properties given in (12) is said to interpolate the function f at the points $\{x_i\}_{i=0}^n$ and is called an **interpolant**. In practice, the simple function p is *a polynomial, a piecewise polynomial or a rational function*. Different interpolation methods arise from the choice of the interpolating function. If the interpolating function is chosen to be a polynomial, the interpolation is called **polynomial interpolation**.

A polynomial interpolation problem can be stated as: Given a set of $n+1$ pairs of real numbers $\{(x_i, y_i)\}_{i=0}^n$, find a polynomial $p_m \in \mathcal{P}_m(\mathbb{R})$ such that $y_i = p_m(x_i)$, $i = 0, \dots, n$. If $n \neq m$, the problem is over or under-determined. The following theorem holds if $n = m$. We refer to [32, 33] for a proof.

Theorem 1. *If $\{x_i\}_{i=0}^n$ is a set of distinct points, for an arbitrary set $\{y_i\}_{i=0}^n$ of $n+1$ numbers there exists a unique polynomial $p_n \in \mathcal{P}_n(\mathbb{R})$ such that $p_n(x_i) = y_i$, $i = 0, \dots, n$.*

Assume that we are given a function f in $[a, b]$, and we have an interpolating polynomial p_n of degree n on a partition $\{x_i\}_{i=0}^n$ of $[a, b]$. Clearly, the function f and the polynomial p_n have exactly the same values at the interpolation points $\{x_i\}_{i=0}^n$. However, if we pick some arbitrary point $x \in [a, b]$ which is not an interpolating point, the function value $f(x)$ may be quite different from $p_n(x)$. Under the assumption that the function f is sufficiently smooth, the interpolation error is estimated in the following theorem, the proof of which can be found in many numerical analysis textbooks, e.g., [27, 33].

Theorem 2. *Suppose that f is an $(n+1)$-times continuously differentiable real-valued function on $[a, b]$. Suppose further that the interpolating points $\{x_i\}_{i=0}^n$ are distinct. Then, for $x \in [a, b]$,*

$$f(x) - p_n(x) = \frac{f^{(n+1)}(\xi)}{(n+1)!}\pi_{n+1}(x),$$

where $\xi \in [a, b]$ is a function of x, and $\pi_{n+1}(x) = (x - x_0)(x - x_1)\cdots(x - x_n)$. Denoting by M_{n+1} the maximum value of $|f^{(n+1)}(\xi)|$ in $[a, b]$, we can bound the above error as

$$|f(x) - p_n(x)| \leq \frac{M_{n+1}}{(n+1)!}|\pi_{n+1}(x)|. \tag{13}$$

Lagrange Form of the Interpolation Polynomial

The Lagrange form of the interpolation polynomial is obtained by using the Lagrange basis for the vector space $\mathcal{P}_n(\mathbb{R})$.

Definition 1. *A set of functions* $\{l_i\}_{i=0}^n$ *is said to be a* **Lagrange basis** *for the space of polynomials* $\mathcal{P}_n(\mathbb{R})$ *with respect to the set of distinct points* $\{x_i\}_{i=0}^n$ *if* $l_i(x_j) = \delta_{ij}$.

Explicitly, the Lagrange basis functions with respect to the set of distinct points $\{x_i\}_{i=0}^n$ is

$$\left\{ l_i(x) = \prod_{\substack{j=0 \\ j\neq i}}^{n} \frac{x - x_j}{x_i - x_j} \right\}_{i=0}^{n}.$$

In this case the interpolant p_n is given by

$$p_n(x) = \sum_{i=0}^{n} y_i l_i(x),$$

and is said to have the Lagrange form of polynomial interpolation.

Newton Form of the Interpolation Polynomial

Although the Lagrange form of the interpolation polynomial is suitable from the theoretical point of view, practically it is not the most convenient form. It is sometimes useful to start from an interpolation polynomial of lower degree and construct higher degree interpolation polynomials. In the case of the Lagrange polynomial there is no constructive relation between p_{n-1} and p_n. The Newton form of polynomial interpolation is designed to do that.

Let $\{x_i\}_{i=0}^n$ be a partition of $[a, b]$ and assume that we are given an interpolation polynomial p_{n-1} of degree $n-1$ for n pairs of numbers $\{(x_i, y_i)\}_{i=0}^{n-1}$ with $p_{n-1}(x_i) = y_i$, $i = 0, \dots, n-1$. We want to represent the interpolation polynomial p_n of degree n for a set of $n+1$ pairs of numbers $\{(x_i, y_i)\}_{i=0}^n$ as a sum of p_{n-1} and a polynomial of degree n with only one unknown coefficient depending on the set of points $\{x_i\}_{i=0}^n$. Thus, taking $q_n \in \mathcal{P}_n(\mathbb{R})$, we set

$$p_n(x) = p_{n-1}(x) + q_n(x).$$

Using the fact that $p_{n-1}(x_i) = y_i$, we have $q_n(x_i) = p_n(x_i) - p_{n-1}(x_i) = 0$, $i = 0, \dots, n-1$. Hence, q_n can be written as $q_n(x) = b_n \Pi_{i=0}^{n-1}(x - x_i)$, where b_n is to be determined. Assuming that $y_i = f(x_i)$, $i = 0, \dots, n$, the coefficient b_n can be found by setting $p_n(x_n) = f(x_n)$. Thereby

$$b_n = \frac{f(x_n) - p_{n-1}(x_n)}{\pi_n(x_n)}, \tag{14}$$

where $\pi_n(x) = \Pi_{i=0}^{n-1}(x - x_i)$. The coefficient b_n is called the n-th Newton divided difference and is denoted by

$$b_n = f[x_0, x_1, \dots, x_n], \quad n \geq 1. \tag{15}$$

The interpolation polynomial p_n can then be written as

$$p_n(x) = p_{n-1}(x) + \pi_n(x) f[x_0, \ldots, x_n].$$

Using recursion on n, we obtain the formula for the interpolation polynomial in Newton form

$$p_n(x) = \sum_{i=0}^{n} \pi_i(x) f[x_0, \ldots, x_i], \tag{16}$$

where $p_0(x) = f(x_0) = f[x_0] = y_0$ and $\pi_0 = 1$.

The Lagrange and Newton forms yield the same interpolation polynomial due to the uniqueness of the interpolation polynomial. The interpolation polynomial in the form (16) is called the **Newton divided difference** formula. There are many properties of the Newton divided differences which make them computationally efficient [27, 32].

It is interesting to compare the two forms of interpolation polynomial in terms of the basis of the polynomial space $\mathcal{P}_n(\mathbb{R})$. The interpolation polynomial of the Lagrange form is obtained by using the Lagrange basis $\{l_i\}_{i=0}^n$, whereas the Newton form is obtained by taking the basis $\{\pi_i\}_{i=0}^n$. In the case of Lagrange form, the coordinate functionals are simply the function values at the given set of points whereas the coordinate functionals in the Newton form are given by Newton divided differences.

Hermite Interpolation

If the polynomial interpolation consists of finding an interpolant which is required to take not only the function values but also the derivatives of the function at the prescribed points, it is called **Hermite interpolation**.

Assume that we require the polynomial of lowest degree which interpolates a function f and its derivatives at two distinct points, say, x_0 and x_1 with $x_0 \neq x_1$. Thus, we have to find a polynomial p which fulfils the properties:

$$p(x_i) = f(x_i), \quad p'(x_i) = f'(x_i), \quad i = 0, 1.$$

Since there are four conditions, we look for a solution in $\mathcal{P}_3(\mathbb{R})$. Let

$$l_0(x) = \frac{x - x_1}{x_0 - x_1}, \quad \text{and} \quad l_1(x) = \frac{x - x_0}{x_1 - x_0}$$

be the two Lagrange basis functions with respect to the partition $\{x_0, x_1\}$. We consider the following basis of $\mathcal{P}_3(\mathbb{R})$: $H_0(x) = (x - x_1)^2(1 + 2l_1(x))$, $H_1(x) = (x - x_0)^2(1 + 2l_0(x))$, $K_0(x) = (x - x_1)^2(x - x_0)$, $K_1(x) = (x - x_0)^2(x - x_1)$. It is immediate to verify that

$$H_i(x_j) = c\delta_{ij}, \quad H_i'(x_j) = 0, \quad K_i(x_j) = 0, \quad K_i'(x_j) = c\delta_{ij} \quad \text{with} \quad c = (x_1 - x_0)^2.$$

Hence, the required polynomial of lowest degree is simply

$$p(x) = \frac{1}{c}\left(\sum_{i=0}^{1} f(x_i) H_i(x) + f'(x_i) K_i(x)\right).$$

The general Hermite interpolation problem can be posed as follows: Given a partition $\Delta = \{x_i\}_{i=0}^n$ of $[a, b]$, find a polynomial p satisfying the conditions

$$\frac{d^j p}{dx^j}(x_i) = c_{ij}, \quad j = 0, \ldots, k_i - 1, \quad i = 0, \ldots, n, \tag{17}$$

where $\frac{d^j p}{dx^j} = p$ for $j = 0$. At node x_i, k_i interpolatory conditions are imposed, where k_i might change with respect to i. There are in total $m = \sum_{i=0}^n k_i$ conditions. The proof of the following theorem can be found in [32].

Theorem 3. *There exists a unique polynomial $p \in \mathcal{P}_{m-1}(\mathbb{R})$ fulfilling the Hermite conditions in* (17).

Example 6. We give a construction of Hermite interpolation with $k_i = 2$ for all i. Let $\Delta = \{x_i\}_{i=0}^n$ be a partition of $[a, b]$ with $n + 1$ points, and $\{y_i\}_{i=0}^n$ and $\{z_i\}_{i=0}^n$ two sets of real numbers. We wish to find a polynomial $p \in \mathcal{P}_{2n+1}(\mathbb{R})$ so that $p(x_i) = y_i$ and $p'(x_i) = z_i$ for $i = 0, \ldots, n$. Our construction is based on a polynomial basis

$$B = \{H_0, \ldots, H_n, K_0, \ldots, K_n\}$$

for $\mathcal{P}_{2n+1}(\mathbb{R})$ with the properties

$$H_j(x_i) = \delta_{ij}, \ H_j'(x_i) = 0, \ K_j(x_i) = 0, \ K_j'(x_i) = \delta_{ij}, \ i, j = 0, \ldots, n. \tag{18}$$

Assuming that $\{l_i\}_{i=0}^n$ is the Lagrange basis with respect to the partition Δ, if we define

$$H_i(x) = [l_i(x)]^2(1 - 2l_i'(x_i)(x - x_i)), \ K_i(x) = [l_i(x)]^2(x - x_i)$$

for $i = 0, \ldots, n$, the basis B satisfies the properties (18). Hence, the interpolant

$$p(x) = \sum_{i=0}^n a_i H_i(x) + b_i K_i(x)$$

fulfils the required conditions for $a_i = y_i$ and $b_i = z_i$, $i = 0, \ldots, n$.

Extension to Two Dimensions

The two-dimensional interpolation is concerned with interpolating a function of two variables. Assume that $\mathcal{G} = \{(x_i, y_i)\}_{i=0}^N$ is a set of points in $\mathbb{R}^2$, and a function f is given on $\mathcal{G}$ with $\{z_i = f(x_i, y_i)\}_{i=0}^N$. We are interested in an interpolation problem posed in a polygonal domain $\Omega \subset \mathbb{R}^2$.

Definition 2. *A* **domain or subdomain** *in two dimensions is an open bounded region. A* **polygonal domain or subdomain** *is an open and bounded region whose boundary consists of pieces of lines.*

In order to state the two-dimensional interpolation problem, we need definitions of a convex set and a convex hull.

Definition 3. *A set $S \subset \mathbb{R}^k$, $k \in \mathbb{N}$, is* **convex** *if for all $\mathbf{x}, \mathbf{y} \in S$ and all $t \in [0,1]$, the point $(1-t)\mathbf{x} + t\mathbf{y} \in S$.*

Definition 4. *The* **convex hull** *for a set of points $\mathcal{G}$ is the minimal open convex set containing $\mathcal{G}$.*

Definition 5. *The* **closure** *of a domain Ω is defined by $cl(\Omega) = \Omega \cup \partial\Omega$, where $\partial\Omega$ denotes the boundary of the domain Ω.*

A two-dimensional interpolation problem is then posed as follows: Given a set of points $\mathcal{G}$, the convex hull, Ω, of $\mathcal{G}$ and given a function f defined on $\mathcal{G}$, find a function $p : \mathrm{cl}(\Omega) \to \mathbb{R}$ with $\{z_i = p(x_i, y_i)\}_{i=0}^N$.

We only consider the situation for which p is a polynomial or a piecewise polynomial. If the set of points $\mathcal{G}$ has a tensor product structure, it is easy to extend the idea of one-dimensional construction to the multi-dimensional case. A tensor product partition is defined as follows:

Definition 6. *Assume that $\Delta_x = \{x_i\}_{i=0}^n$ is a partition of the closed interval $[a,b]$ and $\Delta_y = \{y_j\}_{j=0}^m$ is that of $[c,d]$. Then the set of points $\Delta_{xy} = \{(x_i, y_j)\}_{i=0,j=0}^{n,m}$ is called a* **tensor product partition** *of the rectangular region $[a,b] \times [c,d]$. In short, we write $\Delta_{xy} = \Delta_x \otimes \Delta_y$.*

Let $\Delta_{xy} = \Delta_x \otimes \Delta_y$ be a tensor product partition of $[a,b] \times [c,d]$. Assume that $\{l_i^{\Delta_x}\}_{i=0}^n$ is the Lagrange basis of $\mathcal{P}_n(\mathbb{R})$ with respect to the partition Δ_x, and $\{l_j^{\Delta_y}\}_{j=0}^m$ is that of $\mathcal{P}_m(\mathbb{R})$ with respect to the partition Δ_y. Then, given the values of the function $f(x,y)$ at the partition Δ_{xy}, the Lagrange interpolation polynomial of $f(x,y)$ with respect to the partition Δ_{xy} is of degree n in x and degree m in y, and is given by

$$\sum_{i=0}^{n} \sum_{j=0}^{m} f(x_i, y_j) l^{\Delta_x}(x) l^{\Delta_y}(y).$$

If $\mathcal{G}$ does not have a tensor product structure, we have to solve a global polynomial interpolation problem in a non-rectangular domain or non-tensor product partition, which is difficult and often ill-posed [27]. Furthermore, the following remark points out another limitation of the global polynomial interpolation.

Remark 1. *If the function f to be interpolated is not a polynomial, the quantity $M_{n+1}|\pi_{n+1}(x)|$ in* (13) *can be very large when n is large, leading to a severe limitation of the higher order polynomial interpolation. This problem is typically known as Runge's phenomenon and is explained by Runge's example* [27,32]. *If one has the freedom to choose the interpolating points, the expression $|\pi_{n+1}(x)|$ can be made small by choosing the set of interpolation points as the zeros or the maxima of a Chebyshev polynomial* [32,33]. *However, in many interpolation problems, the set of points is already given and one cannot use a different set of points.*

Piecewise polynomial interpolation to be discussed in the next section provides a flexible and efficient solution to the above discussed problems.

4.2. Piecewise Polynomial Interpolation

So far we have considered the approximation of a function by a global polynomial. This implies that if the definition of the function is to be modified at a point, the polynomial interpolant changes globally. On the contrary, the piecewise polynomial interpolant does not change globally if the definition of the function changes locally.

Assume that the interpolation problem is posed in a domain $\Omega \subset \mathbb{R}^k$ with $k = 1, 2$. The central idea of piecewise interpolation is to decompose the domain Ω into non-overlapping subdomains yielding its decomposition and define polynomial basis functions in each subdomain.

Definition 7. *Let $\Omega \subset \mathbb{R}^k$ be a domain with $k = 1, 2$. The collection of disjoint subdomains $\mathcal{T}$ with $cl(\Omega) = \cup_{T \in \mathcal{T}} cl(T)$ is called a* **decomposition** *of Ω.*

In the two-dimensional case, if the interpolant is to be at least continuous, the decomposition should be geometrically conforming.

Definition 8. *A decomposition $\mathcal{T}$ of $\Omega \subset \mathbb{R}^2$ is called* **geometrically conforming** *if the intersection between the boundaries of any two different subdomains $\partial T_l \cap \partial T_k$, $k \neq l$, $T_k, T_l \in \mathcal{T}$ is either empty, a vertex or a common edge.*

Four decompositions of the domain Ω are shown in Figure 3. The two on the left are geometrically conforming and the two on the right are geometrically non-conforming. In the one-dimensional case, the subdomains are always intervals. In the two-dimensional case, only quadrilaterals or triangles are allowed.

The polynomial space $\mathcal{P}_m(T)$ will denote three different spaces depending on T. If T is an interval,

$$\mathcal{P}_m(T) = \Big\{p : p(x) = \sum_{i=0}^{m} a_i x^i\Big\},$$

if T is a triangle,

$$\mathcal{P}_m(T) = \Big\{p : p(x, y) = \sum_{\substack{i,j=0 \\ i+j \leq m}}^{m} a_{ij} x^i y^j\Big\},$$

and finally, if T is a quadrilateral,

$$\mathcal{P}_m(T) = \Big\{p : p(x, y) = \sum_{i,j=0}^{m} a_{ij} x^i y^j\Big\}.$$

The smoothness of the interpolation polynomial is determined by the smoothness of the piecewise polynomial on the boundary of two subdomains.

Definition 9. *Let $m, k \in \mathbb{N}_0 = \mathbb{N} \cup \{0\}$. The* **piecewise polynomial space** *of degree m and smoothness k with respect to the decomposition $\mathcal{T}$ is defined as*

$$S_{m,k}(\mathcal{T}) = \big\{f \in C^k(\Omega) : \ f|_T \in \mathcal{P}_m(T), \ T \in \mathcal{T}\big\},$$

where $f|_T$ represents the restriction of the function f to the element T. The space of piecewise constant function with respect to the decomposition $\mathcal{T}$ is denoted by $S_0(\mathcal{T})$.

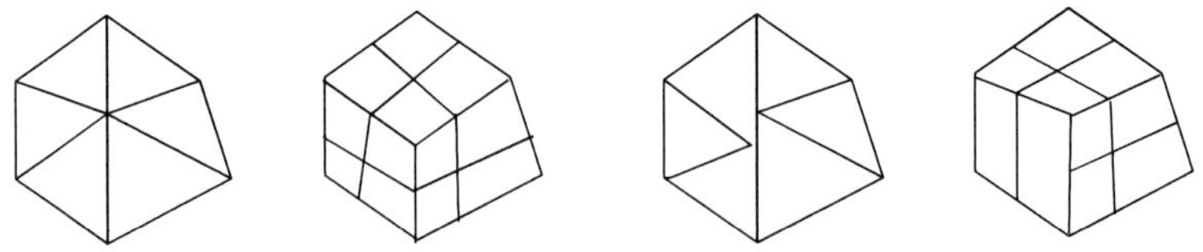

Figure 3. Four decompositions of the domain Ω: geometrically conforming (first two) and geometrically non-conforming (last two).

A generalisation of a Lagrange basis to the piecewise polynomial space is a nodal basis.

Definition 10. *Let $\mathcal{G} = \{(x_i, y_i)\}_{i=0}^{N}$ be a set of points in $\mathbb{R}^2$, and $\mathcal{T}$ be a decomposition of Ω = convex hull of $\mathcal{G}$. Then, a basis $\{\phi_i\}_{i=0}^{N}$ of $S_{m,k}(\mathcal{T})$ is called a* **nodal basis** *of $S_{m,k}(\mathcal{T})$ with respect to $\mathcal{G}$ if and only if $\phi_j(x_i, y_i) = \delta_{ij}$ for $i, j = 0, \ldots, N$.*

One-Dimensional Case

Before constructing some examples, we introduce a decomposition induced by a partition.

Definition 11. *Let $\Delta = \{x_i\}_{i=0}^{n}$ be a partition of the closed interval $[a, b]$, and $I_i = (x_i, x_{i+1})$ an interval. The decomposition $\mathcal{T} = \{I_i\}_{i=0}^{n-1}$ of the open interval (a, b) is called* **the decomposition** *of (a, b) induced by the partition Δ.*

Example 7. Assume that $\mathcal{T}$ is the decomposition of (a, b) induced by a partition $\Delta = \{x_i\}_{i=0}^{n}$ of $[a, b]$. Then, $S_{1,0}(\mathcal{T})$ is the space of linear splines on the decomposition $\mathcal{T}$. Let

$$\phi_0(x) = \begin{cases} \dfrac{x - x_1}{x_0 - x_1} & \text{if } x \in I_0 \\ 0 & \text{otherwise} \end{cases}, \quad \phi_n(x) = \begin{cases} \dfrac{x - x_{n-1}}{x_n - x_{n-1}} & \text{if } x \in I_{n-1} \\ 0 & \text{otherwise} \end{cases} \quad \text{and}$$

$$\phi_i(x) = \begin{cases} \dfrac{x - x_{i-1}}{x_i - x_{i-1}} & \text{if } x \in I_{i-1} \\ \dfrac{x - x_{i+1}}{x_i - x_{i+1}} & \text{if } x \in I_i \\ 0 & \text{otherwise} \end{cases}, \quad \text{for} \quad i = 1, \ldots, n-1.$$

The set $\{\phi_i\}_{i=0}^{n}$ forms a basis for the space $S_{1,0}(\mathcal{T})$. Thus, a function $s_l \in S_{1,0}(\mathcal{T})$ can be written as

$$s_l(x) = \sum_{i=0}^{n} c_i \phi_i(x),$$

where $c_0, \ldots, c_n$ are arbitrary constants. As $\phi_i(x_j) = \delta_{ij}$, the basis $\{\phi_i\}_{i=0}^{n}$ is a nodal basis of $S_{1,0}(\mathcal{T})$ with respect to the partition Δ. Therefore, the piecewise linear interpolation of a continuous function $f : [a, b] \to \mathbb{R}$ on the decomposition $\mathcal{T}$ is obtained by setting $c_i = f(x_i)$, $i = 0, \ldots, n$.

Example 8. *Nearest Neighbour Interpolation in $[a, b]$:* Assume that the values of a function f at the partition $\Delta = \{x_i\}_{i=0}^{n}$ are given. Associated with the partition Δ, we form a dual partition $\tilde{\Delta} = \{z_i\}_{i=0}^{n+1}$ with $z_0 = x_0$, $z_i = \frac{x_{i-1}+x_i}{2}$, $i = 1, \ldots, n$, and $z_{n+1} = x_n$. Let

χ_{I_i} be a characteristic function of the interval $I_i = [z_i, z_{i+1})$, $i = 0, \ldots, n$. If $\mathcal{T}$ is the decomposition induced by the partition $\tilde{\Delta}$, $S_0(\mathcal{T})$ is spanned by the basis $\{\chi_{I_i}\}_{i=0}^{n}$. Then, the nearest neighbour interpolation of the function f at the partition $\tilde{\Delta}$ is given by

$$N(x) = \sum_{i=0}^{n} f(x_i)\chi_{I_i}(x).$$

Example 9. *Piecewise Cubic Hermite Interpolation:* Given a partition $\Delta = \{x_i\}_{i=0}^{n}$ of the interval $[a, b]$ having $n+1$ number of points, we want to find a piecewise polynomial p with $p(x_i) = y_i$, and $p'(x_i) = z_i$ for $i = 0, \ldots, n$. Using the construction of Example 6, we define a piecewise polynomial p for $i = 0, \ldots, n-1$ as

$$p(x) = H_i(x)y_i + H_{i+1}(x)y_{i+1} + K_i(x)z_i + K_{i+1}(x)z_{i+1}, \quad x \in [x_i, x_{i+1})$$

with H_i, H_{i+1}, K_i, K_{i+1} as in Example 6, but l_i and l_{i+1} are now piecewise linear polynomials given by

$$l_i(x) = \frac{x - x_{i+1}}{x_i - x_{i+1}}, \quad l_{i+1}(x) = \frac{x - x_i}{x_{i+1} - x_i}.$$

Assume that no derivatives are provided at the partition, but only function values. A piecewise cubic polynomial can be constructed also in this case assigning some suitable derivatives of the function at the partition. The derivatives are assigned in such a way that the resulting piecewise curve is continuously differentiable. One such example can be found in [34] and is used in piecewise cubic interpolation of MATLAB. The derivatives z_i are assigned in such a way that the function is continuously differentiable and function values do not locally overshoot the data values. Let d_i be defined as

$$d_i = \frac{y_{i+1} - y_i}{x_{i+1} - x_i}, \quad i = 0, \ldots, n-1.$$

For an inner point x_i, if d_i and d_{i-1} are of opposite signs or if either of them is zero, x_i is the local extremum. Thus z_i is set to be zero. If d_i and d_{i-1} have the same sign, z_i is set to be a weighted harmonic mean of the two discrete slopes

$$\frac{1}{z_i} = \frac{1}{w_1 + w_2}\left(\frac{w_1}{d_{i-1}} + \frac{w_2}{d_i}\right)$$

with $w_1 = 2h_i + h_{i-1}$ and $w_2 = h_i + 2h_{i-1}$. For more details see [34].

Although in the one-dimensional case a decomposition can always be induced by a partition, this is not the case in two dimensions. Therefore, for simplicity, we divide the two-dimensional case into two parts depending on whether the decomposition has a tensor product structure or not.

Two-Dimensional Tensor Product Case

We start by defining a decomposition of a rectangular domain induced by a tensor product partition.

Definition 12. *Let $\Delta_{xy} = \Delta_x \otimes \Delta_y$ be a tensor product partition of the rectangular region $[a,b] \times [c,d]$, where $\Delta_x = \{x_i\}_{i=0}^n$ is a partition of $[a,b]$ and $\Delta_y = \{y_i\}_{i=0}^n$ is that of $[c,d]$. Let $I_{ij} = (x_i, x_{i+1}) \times (y_j, y_{j+1})$, $i = 0,\ldots,n-1$, $j = 0,\ldots,m-1$. Then $\mathcal{T} = \{I_{ij}\}_{i=0,j=0}^{n-1,m-1}$ is called* **the decomposition of the domain** *$(a,b) \times (c,d)$ induced by the partition Δ_{xy}.*

Example 10. Let $\mathcal{T}_x$ be the decomposition of (a,b) induced by a partition $\Delta_x = \{x_i\}_{i=0}^n$ and $\mathcal{T}_y$ that of (c,d) induced by a partition $\Delta_y = \{y_j\}_{j=0}^m$. Let $\{\phi_i\}_{i=0}^n$ and $\{\varphi_j\}_{j=0}^m$ be bases of $S_{1,0}(\mathcal{T}_x)$ and $S_{1,0}(\mathcal{T}_y)$, respectively, as constructed in Example 7. If $\mathcal{T}_{xy}$ is the decomposition of the domain $(a,b) \times (c,d)$ induced by the partition $\Delta_{xy} = \Delta_x \otimes \Delta_y$, then $\{\psi_{ij}\}_{i=0,j=0}^{n,m}$ with $\psi_{ij}(x,y) = \phi_i(x)\varphi_j(y)$ forms a nodal basis for $S_{1,0}(\mathcal{T}_{xy})$ with respect to the set of points Δ_{xy}. A piecewise linear interpolation of a continuous function $f : \mathbb{R}^2 \to \mathbb{R}$ can be obtained as

$$\sum_{i=0}^{n}\sum_{j=0}^{m} f(x_i, y_j)\psi_{i,j}(x,y).$$

Image Transformation through Interpolation

A digital image can be resized, rotated or distorted by using piecewise polynomial interpolation. Although the interpolation technique can be applied to any digital image, we only consider intensity images. An intensity image is an array of size $m \times n$, where each element of the array represents the intensity or gray scale of the pixel.

Associated with an image I of size $m \times n$ we define a tensor product partition Δ_{xy}^0 of the square $[0,1] \times [0,1]$ as

$$\Delta_{xy}^0 = \{(a_i, b_j)\}_{i=0,j=0}^{n-1,m-1}, \quad \text{where} \quad a_i = \frac{i}{n-1} \text{ and } b_j = \frac{j}{m-1}.$$

We also define an **image function** $I_f : \Delta_{xy}^0 \to \mathbb{R}$ in such a way that $I_f(a_i, b_j)$ gives the intensity of the pixel at the point (a_i, b_j).

Suppose that we are given an intensity image I of size $m \times n$, and we want to convert this image into another image $\tilde{I}$ of size $m_1 \times n_1$. The image function $\tilde{I}_f$ for the new image is to be defined on the new tensor product partition

$$\Delta_{xy}^1 = \{(c_i, d_j)\}_{i=0,j=0}^{n_1-1,m_1-1} \text{ with } c_i = \frac{i}{n_1-1} \text{ and } d_j = \frac{j}{m_1-1}.$$

The central idea of image resizing is to define a function on the square $[0,1] \times [0,1]$ and to sample it at the tensor product partition Δ_{xy}^1. This procedure leads to the image function $\tilde{I}_f : \Delta_{xy}^1 \to \mathbb{R}$.

In general, image resizing, rotation or distortion can be seen as a transformation of an image. Suppose that we are given an invertible transformation $\hat{T}$ to be applied to an intensity image of size $m \times n$. At the first step, we apply the transformation to the underlying tensor product partition Δ_{xy}^0 to obtain a new set of points Δ_{xy}^2, and find a suitable rectangular region $[a,b] \times [c,d]$ which contains all the points in Δ_{xy}^2.

Let

$$\Delta_{xy}^1 = \{(c_i, d_j)\}_{i=0,j=0}^{n_1-1,m_1-1}$$

Require: A digital image I and a transformation $\hat{T}$
Ensure: A transformed digital image $\tilde{I}$
1: Associate a tensor product partition Δ^0_{xy} and an image function $I_f : \Delta^0_{xy} \to \mathbb{R}$
2: Compute

$$\Delta^2_{xy} = \{\hat{T}\begin{bmatrix} a_i \\ b_j \end{bmatrix}, (a_i, b_j) \in \Delta^0_{xy}\}$$

3: Find a rectangle $[a, b] \times [c, d]$ with $\Delta^2_{xy} \subset [a, b] \times [c, d]$
4: Define a tensor product partition $\Delta^1_{xy} = \{(c_i, d_j)\}_{i=0,j=0}^{n_1-1,m_1-1}$ of $[a, b] \times [c, d]$, $n_1 \times m_1$ is the size of the new image
5: For each $(c_i, d_j) \in \Delta^1_{xy}$, find (x_i, y_j) with

$$\begin{bmatrix} x_i \\ y_j \end{bmatrix} = \hat{T}^{-1}\begin{bmatrix} c_i \\ d_j \end{bmatrix}$$

6: Compute an interpolation polynomial p of I_f on the unit square
7: Define an image function $\tilde{I}_f : \Delta^1_{xy} \to \mathbb{R}$ as

$$\tilde{I}_f(c_i, d_j) = \begin{cases} p(x_i, y_j) & \text{if } (x_i, y_j) \in \text{convex hull of } \Delta^0_{xy} \\ 0 & \text{otherwise} \end{cases}$$

8: The new digital image $\tilde{I}$ is then given by $\tilde{I}(i, j) = \tilde{I}_f(c_i, d_j)$

Algorithm 1: Image Transformation

be a tensor product partition of the region $[a, b] \times [c, d]$, where we want to define the new image function $\tilde{I}_f : \Delta^1_{xy} \to \mathbb{R}$. Notice that for image resizing $\hat{T}$ is an identity transformation and therefore Δ^0_{xy} and Δ^2_{xy} are the same. Since Δ^0_{xy} has a tensor product structure, it is easy to find a piecewise polynomial p on $[0, 1] \times [0, 1]$ which interpolates the image function I_f at the tensor product partition Δ^0_{xy} (see Example 10).

Let (c_i, d_j) be a point in the tensor product partition Δ^1_{xy} with

$$\begin{bmatrix} x_i \\ y_j \end{bmatrix} = \hat{T}^{-1}\begin{bmatrix} c_i \\ d_j \end{bmatrix}.$$

At the second step, the new image function $\tilde{I}_f$ is obtained by setting $\tilde{I}_f(c_i, d_j) = p(x_i, y_j)$ if (x_i, y_j) is inside the closure of the convex hull of the set of points Δ^0_{xy}, otherwise $\tilde{I}_f(c_i, d_j)$ is set to zero. The algorithm for image transformation is given in Algorithm 1. MATLAB codes are available at [31].

In Figure 4, we show three examples of image transformation by using linear interpolation of the test image. The top left picture of Figure 4 shows the original image of size 512×512. The top right one is the original image rotated 40° anti-clockwise. The pictures at the bottom correspond to the original image resized to 200×500 (left) and to 200×200 (right).

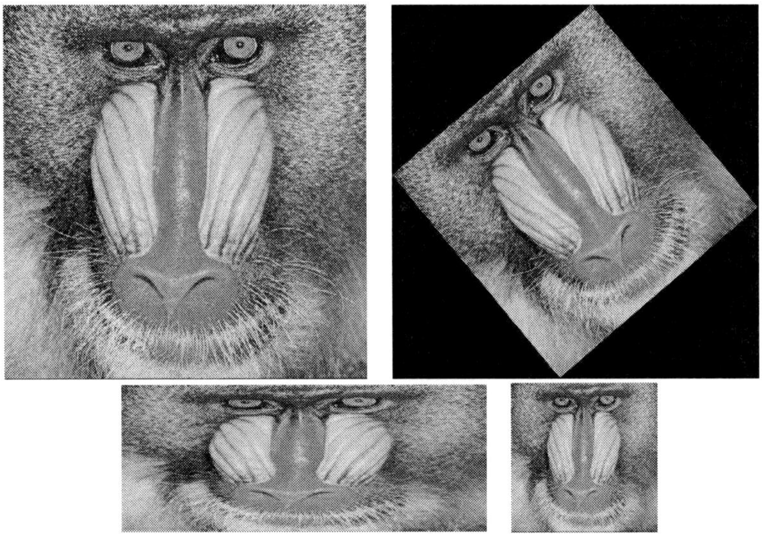

Figure 4. The original image of size 512×512 and image rotated $40°$(top left and right). Resized image of size 200×500 and of size 200×200 (bottom left and right).

Two-Dimensional Non-Tensor Product Case

The two-dimensional interpolation problem is more difficult if the interpolation points do not have a tensor product structure. In such a situation, the two-dimensional interpolation is called **scattered data interpolation**. There is a vast amount of literature devoted to scattered data interpolation. We refer to [18, 19] for extensive surveys on this subject. Here, we consider an approach based on decomposing the convex hull of $\mathcal{G}$ into triangles with vertices in $\mathcal{G}$ and piecewise polynomial interpolation. Therefore, in what follows we restrict ourselves to the case of piecewise interpolation on triangles. The most efficient and popular way of decomposing the polygonal domain Ω into triangles when Ω is the convex hull of the scattered points is the Delaunay triangulation.

Definition 13. *Given a set $\mathcal{G}$ of points in $\mathbb{R}^2$, a* **Delaunay triangulation** *for $\mathcal{G}$ is a decomposition $\mathcal{T}$ of convex hull of $\mathcal{G}$ into triangles such that no point in $\mathcal{G}$ is inside the circumcircle of any triangle in $\mathcal{T}$.*

A Delaunay triangulation of a finite set of points in the plane is a triangulation that minimises the standard deviations of the angles of the triangles. In this sense, the Delaunay triangulation is the most equi-angular triangulation. The dual graph of the Delaunay triangulation is a Voronoi diagram for the same set of points.

Definition 14. *For a set of points $\mathcal{G} \subset \mathbb{R}^2$, the* **Voronoi diagram** *is the decomposition of the plane into convex polygons such that each polygon contains exactly one generating point from $\mathcal{G}$ and every point in a given polygon is closer to its generating point than to any other point in $\mathcal{G}$. A convex polygon $V_{\mathbf{x}}$ associated with the generating point $\mathbf{x} \in \mathcal{G}$ is called the* **Voronoi cell** *for the point $\mathbf{x} \in \mathcal{G}$.*

In other words, the Voronoi cell $V_{\mathbf{x}}$ for the point $\mathbf{x} \in \mathcal{G}$ has the property that the distance from every $\mathbf{y} \in V_{\mathbf{x}}$ to $\mathbf{x}$ is less than or equal to the distance from $\mathbf{y}$ to any other point in $\mathcal{G}$. The circle circumscribed about a Delaunay triangle has its centre at the vertex of a Voronoi cell, see the right graph of Figure 5. The idea of Delaunay triangulation and Voronoi diagram is also extended to higher dimension. An efficient algorithm for computing Delaunay triangulations and Voronoi diagrams are presented in [35], see also [20, 36]. A lot of interesting materials can be found in websites [37, 38].

As an example of a Delaunay triangulation and Voronoi diagram, we define a set $\mathcal{G}_1 = \{(0.1, 0.4), (0.5, 0.1), (0.45, 0.5), (0.3, 0.6), (0.3, 0.3), (0.1, 0.4), (0.9, 0.8), (0.3, 0.9), (0.2, 0.1), (0.8, 0.9)\}$, and generate the Delaunay triangulation and the Voronoi diagram of $\mathcal{G}_1$. The Delaunay triangulation and the Voronoi diagram of $\mathcal{G}_1$ are shown in the left and middle graphs of Figure 5, respectively. The right graph depicts the circumcircle of a triangle with its centre at a vertex of the Voronoi diagram (filled circle).

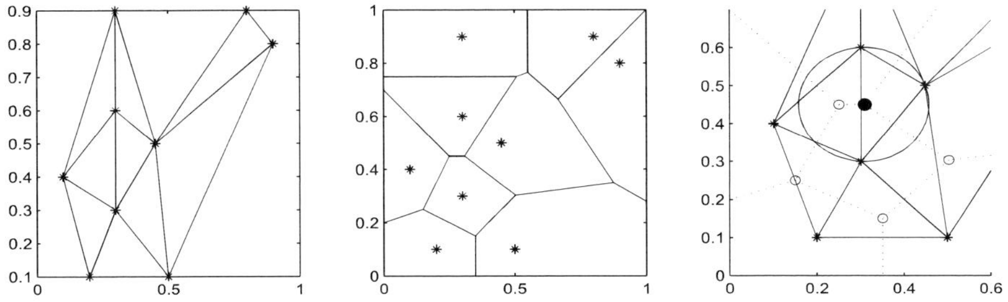

Figure 5. The Delaunay triangulation of the set $\mathcal{G}_1$ (left graph). The corresponding Voronoi diagram (middle graph). The circumcircle of a triangle with Delaunay triangulation and Voronoi diagram (last graph).

Example 11. Let $\mathcal{T}$ be a Delaunay triangulation of the set of points $\mathcal{G}$. We construct a nodal basis $\{\phi_i\}_{i=0}^{N}$ of $S_{1,0}(\mathcal{T})$ with each basis function ϕ_i corresponding to a point $(x_i, y_i) \in \mathcal{G}$ and satisfying $\phi_i(x_j, y_j) = \delta_{ij}$, $i, j = 0, \ldots, N$. A basis function ϕ_i is shown in the right graph of Figure 6. The basis function ϕ_i vanishes outside $\cup_{j=1}^{5}\text{cl}(T_j)$, where $T_j \in \mathcal{T}$, $j = 1, \ldots, 5$, are the triangles having the common vertex (x_i, y_i) as shown in the right graph of Figure 6. The key point for the construction is to define a reference triangle, where it is easy to compute polynomial basis functions, and map the triangle to the actual element by using an affine transformation. It is convenient to choose the right-angled triangle $\tilde{T} = \{(\tilde{x}, \tilde{y}) : \tilde{x} > 0,\ \tilde{y} > 0,\ \tilde{x} + \tilde{y} < 1\}$ as the reference triangle.

The local Lagrange basis functions associated with three vertices of $\tilde{T}$ for the linear interpolation are given by $l_1(\tilde{x}, \tilde{y}) = (1 - \tilde{x} - \tilde{y})$, $l_2(\tilde{x}, \tilde{y}) = \tilde{x}$, and $l_3(\tilde{x}, \tilde{y}) = \tilde{y}$. See the left graph of Figure 6. If $(\tilde{x}_j, \tilde{y}_j)$, $j = 1, \ldots, 3$ are the three vertices of the reference triangle, the basis functions satisfy $l_i(\tilde{x}_j, \tilde{y}_j) = \delta_{ij}$, $i, j = 1, \ldots, 3$. Let $T \in \mathcal{T}$ have three vertices (x_1, y_1), (x_2, y_2) and (x_3, y_3) as in the middle graph of Figure 6. The mapping $\hat{F}_T$ transforms a point in the reference triangle $\tilde{T}$ to a point in the actual triangle T as follows

$$\begin{bmatrix} x \\ y \end{bmatrix} = \hat{F}_T \begin{bmatrix} \tilde{x} \\ \tilde{y} \end{bmatrix} \quad \text{with} \quad \hat{F}_T \begin{bmatrix} \tilde{x} \\ \tilde{y} \end{bmatrix} = \begin{bmatrix} x_2 - x_1 & x_3 - x_1 \\ y_2 - y_1 & y_3 - y_1 \end{bmatrix} \begin{bmatrix} \tilde{x} \\ \tilde{y} \end{bmatrix} + \begin{bmatrix} x_1 \\ y_1 \end{bmatrix}.$$

The first two graphs of Figure 6 show a reference triangle $\tilde{T}$ and the triangle T.

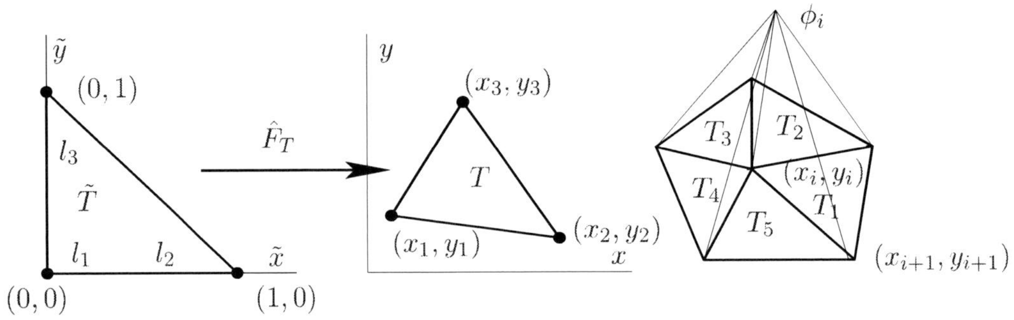

Figure 6. The reference triangle and an affine map $\hat{F}_T$ from the reference triangle to a generic triangle (first two graphs). A nodal basis function ϕ_i (last graph).

If the three vertices of the triangle T are not collinear, the determinant of the matrix

$$\begin{bmatrix} x_2 - x_1 & x_3 - x_1 \\ y_2 - y_1 & y_3 - y_1 \end{bmatrix}$$

does not vanish and hence $\hat{F}_T$ is invertible. The three global Lagrange basis functions on the triangle T are then given by $g_i(x, y) = l_i(\tilde{x}, \tilde{y})$, $i = 1, \ldots, 3$, with

$$\begin{bmatrix} \tilde{x} \\ \tilde{y} \end{bmatrix} = \hat{F}_T^{-1} \begin{bmatrix} x \\ y \end{bmatrix}.$$

By construction the global Lagrange basis functions g_i satisfy $g_i(x_j, y_j) = \delta_{ij}$, $i, j = 1, \ldots, 3$, and are linear in T.

Let us consider the construction of the basis function ϕ_i corresponding to the point (x_i, y_i) in more detail. The point (x_i, y_i) is a vertex of five triangles as shown in the right graph of Figure 6. The Lagrange basis function corresponding to the vertex (x_i, y_i) for each of these five triangles has value one at the point (x_i, y_i) and value zero on all edges of the triangle opposite to this vertex. Let $\tilde{\phi}_j$ be the Lagrange basis function for the triangle T_j, $j = 1, \ldots, 5$, corresponding to the point (x_i, y_i). The global basis ϕ_i corresponding to the point (x_i, y_i) is then defined by

$$\phi_i(x, y) = \begin{cases} \tilde{\phi}_j(x, y) & \text{if} \quad (x, y) \in T_j, \quad j = 1, \ldots, 5 \\ 0 & \text{otherwise.} \end{cases}$$

It is obvious that the basis function ϕ_i is piecewise linear and $\phi_i(x_j, y_j) = \delta_{ij}$. Furthermore, the following lemma shows that ϕ_i is continuous.

Lemma 1. *The basis function ϕ_i defined as above is continuous on the boundary of two triangles.*

Proof. Let us analyse the value of the function ϕ_i on the boundary of T_1 and T_5, which is a line joining two points (x_i, y_i) and (x_{i+1}, y_{i+1}), see the right graph of Figure 6. Here, both $\tilde{\phi}_1$ and $\tilde{\phi}_5$ take the values one at (x_i, y_i) and zero at (x_{i+1}, y_{i+1}). In between are linear. Since there is a unique linear polynomial having this property, they have exactly the same value on the boundary of T_1 and T_5. □

Each vertex in $\mathcal{G}$ has an associated basis function which has value one at this vertex and zero at all other vertices. This leads to a nodal basis $\{\phi_i\}_{i=0}^N$ of the piecewise polynomial space $S_{1,0}(\mathcal{T})$.

Example 12. *Nearest Neighbour Interpolation:* The nearest neighbour interpolation of a set of scattered data $\mathcal{G} = \{(x_i, y_i)\}_{i=0}^N$ can be realised by generating the Voronoi diagram of the set $\mathcal{G}$. Assume that we are given the values $\{z_i\}_{i=0}^N$ of a function at $\mathcal{G}$, and χ_{V_i} is the characteristic function of the Voronoi cell V_i corresponding to the point (x_i, y_i), $i = 0, \ldots, N$. Then, the nearest neighbour interpolant p of the given data is obtained as

$$p(x) = \sum_{i=0}^{N} z_i \chi_{V_i}(x).$$

Image Denoising through Interpolation

As an application of the scattered data interpolation, we propose a very simple method of denoising an image in those cases in which the noisy and pure pixels can be identified. For example, we can identify the noisy and pure pixels in an image corrupted with salt and pepper noise. The salt and pepper noise is of impulsive nature. In an intensity image the noisy pixels are randomly set either to white or black [15]. Here we apply the scattered data interpolation to filter this type of noise. An alternative approach for removing high density salt and pepper noise is proposed in [17]. However, the method of scattered data interpolation based on a Delaunay triangulation is easy and efficient.

Assume that a digital image I of size $m \times n$ is corrupted with salt and pepper noise. As in the previous case, we associate with the image a tensor product partition

$$\Delta_{xy}^0 = \{(a_i, b_j)\}_{i=0,j=0}^{n-1,m-1}, \quad \text{where} \quad a_i = \frac{i}{n-1} \text{ and } b_j = \frac{j}{m-1}$$

and the image function $I_f : \Delta_{xy}^0 \to \mathbb{R}$.

Let S^n be the set of points in Δ_{xy}^0 having corrupted pixels. Denoting by S^p the set of points where only the pure pixels of the image are located we split the image function I_f into two functions I_f^n and I_f^p, which are defined as $I_f^n = I_f|_{S^n}$ and $I_f^p = I_f|_{S^p}$, respectively. As the positions of the noisy pixels are random, the points in S^n have no structure.

The key idea for getting rid of noisy pixels is to find an interpolant which interpolates the scattered data in S^p. To such an end we generate a Delaunay triangulation of the scattered points in S^p and compute an interpolant p on the convex hull of S^p. Hence $p(x, y) = I_f^p(x, y)$ for all $(x, y) \in S^p$. Choosing a tensor product partition $\Delta_{xy}^1 = \{(c_i, d_j)\}_{i=0,j=0}^{n_1-1,m_1-1}$ we define a new image function $\tilde{I}_f : \Delta_{xy}^1 \to \mathbb{R}$ so that $\tilde{I}_f(x, y) = p(x, y)$ for all $(x, y) \in \Delta_{xy}^1$. If the point $(x, y) \in \Delta_{xy}^1$ does not belong to the closure of the convex hull of S^p, we need to extrapolate for that point. For this, we choose one point $(\tilde{x}, \tilde{y})$ in S^p, out of the nearest to (x, y), and set $\tilde{I}_f(x, y) = I_f^p(\tilde{x}, \tilde{y})$. We have canned the complete denoising procedure in Algorithm 2.

Assuming that we know the image before being corrupted with noise, we use peak signal-to-noise ratio (PSNR) to compare our results with the results obtained by using the

Require: A noisy digital image I

Ensure: A filtered digital image $\tilde{I}$

1: Associate a tensor product partition Δ^0_{xy} and an image function $I_f : \Delta^0_{xy} \to \mathbb{R}$ with the image I

2: Let S^p and S^n denote the set of points with pure and noisy pixels with $\Delta^0_{xy} = S^p \cup S^n$, respectively, and $I^p_f = I_f|_{S^p}$

3: Compute the Delaunay triangulation $\mathcal{T}$ of the set of points S^p

4: Find a piecewise interpolation polynomial p of I^p_f on the convex hull of S^p

5: Let $\Delta^1_{xy} = \{(c_i, d_j)\}_{i=0,j=0}^{n_1-1,m_1-1}$ be a tensor product partition of the square $[0,1] \times [0,1]$ where we want to define the new image function $\tilde{I}_f$ for the filtered image. Define the new image function $\tilde{I}_f : \Delta^1_{xy} \to \mathbb{R}$ as

$$\tilde{I}_f(c_i, d_j) = \begin{cases} p(c_i, d_j) & \text{if} \quad (c_i, d_j) \in \text{convex hull of } S^p \\ I^p_f(\tilde{x}_i, \tilde{y}_j) & \text{otherwise,} \end{cases}$$

where $(\tilde{x}_i, \tilde{y}_j) \in S^p$ and is nearest to (c_i, d_j)

6: The filtered image $\tilde{I}$ is then given by $\tilde{I}(i, j) = \tilde{I}_f(c_i, d_j)$

Algorithm 2: Impulsive Noise Removal

standard median filter [16]. The peak signal-to-noise ratio (PSNR) is defined by

$$PSNR = 10 \cdot \log_{10}\left(\frac{MAX_I^2}{MSE}\right) = 20 \cdot \log_{10}\left(\frac{MAX_I}{\sqrt{MSE}}\right),$$

where MAX_I is the maximum pixel value of the image, and MSE is the mean square error, i.e.,

$$MSE = \frac{1}{mn}\sum_{i=1}^{m}\sum_{j=1}^{n} \|IO(i,j) - IR(i,j)\|^2,$$

where IO is the original image before corruption and IR is the image recovered after removing the noise.

In Figure 7, we show an example of applying the linear interpolation to remove the salt and pepper noise. The first picture (from the left) of Figure 7 shows the original image. The second one shows the noisy image with noise density 50%. The third picture is the image reconstructed by using the linear interpolation, and the fourth one is the image denoised by the median filter. In this example, the nearest neighbour interpolation produces the image which is visually equivalent to that produced by the linear interpolation. We also observe that the image obtained by the median filter does not differ much from the previous ones. However, if the noise density is increased, the median filter does not recover even the main feature of the image. This is illustrated in Figure 8, where we deal with noise density 95%. The first picture of this figure shows the noisy image. The second picture shows the image reconstructed by using the linear interpolation. The third picture shows the image reconstructed by using the nearest neighbour interpolation, and the last one the reconstruction by using the median filter. We can see that here the median filter produces a meaningless image.

The peak signal-to-noise ratio (PSNR) for two test images are given in Table 1. The two images we have considered are Lena (first picture in Figure 7) and Baboon (top left picture of Figure 4). For both images, the linear interpolation produces the best PSNR values for all noise densities. The PSNR values from the nearest neighbour interpolation are not far from these values. The PSNR values from median filter are acceptable in the cases of low noise densities, whereas at high noise densities (above 50%) the PSNR values decrease considerably.

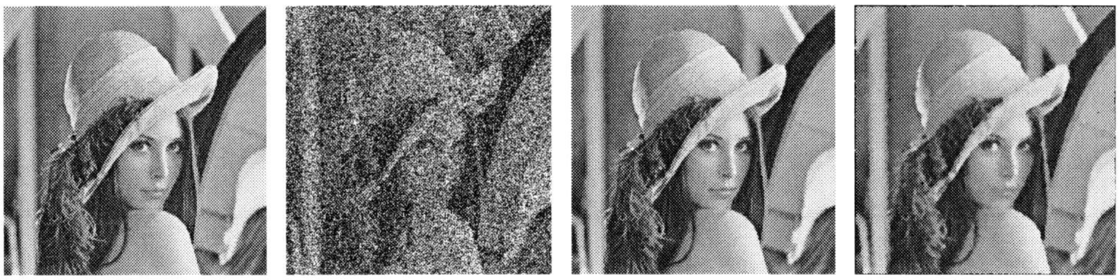

Figure 7. The original image. Noisy image (noise density 50%). Image denoised with linear interpolation and median filter.

Figure 8. Noisy image (noise density 95%). Image denoised with linear interpolation, nearest neighbour interpolation and median filter.

Table 1. PSNR of the test images with various noise densities

	PSNR for Lena's image				PSNR for Baboon's image			
	Noise Density				Noise Density			
Method of Denois.	30%	50%	70%	90%	30%	50%	70%	90%
Linear Interp.	38.57	35.20	32.12	27.73	26.60	23.91	21.76	19.34
Nearest Neighbour Interp.	34.43	31.53	29.18	25.61	23.79	21.74	20.10	17.89
Median Filter	26.11	23.68	19.34	9.47	19.01	18.14	17.23	9.36

As a final remark we would like to stress that the method of scattered data interpolation based on Delaunay triangulation provides a promising approach to remove high density impulsive noise. It is easy to implement and computationally efficient. Our numerical results show that linear and nearest-neighbour interpolation can outperform the standard median filter.

Acknowledgement

Support from EPSRC (EP/D062632/1) is gratefully acknowledged.

References

[1] R.T. Behrens and L.L. Scharf. Signal processing applications of oblique projection operators. *IEEE Transactions on Signal Processing*, **42**:1413–1424, 1994.

[2] S. N. Afriat. Orthogonal and oblique projectors and the characteristics of pairs of vectors spaces. *Philos. Soc.*, **53**:800–816, 1957.

[3] T. N. Greville. Solutions of the matrix equations $XAX = X$ and relations between oblique and orthogonal projectors. *SIAM Journal on Applied Mathematics*, **26**:828–832, 1974.

[4] G. Gorach, A. Maestripieri, and D Stojanoff. Oblique projection and abstract splines. *Journal of Approximation Theory*, **117**:189–206, 2002.

[5] A. Galantai. *Projectors and Projection Methods*. Kluwer Academic Publishers, London, 2004.

[6] G. Gorach, A. Maestripieri, and D Stojanoff. Projections in operators ranges. *Proc. Amer. Math. Soc*, **134**(3):765–788, 2005.

[7] M. Unser and A. Aldroubi. A general sampling theory for nonideal acquisition devices. *IEEE Transactions Signal Processing*, **42**:2915–2925, 1994.

[8] T. Blu and M. Unser. Quantitative fourier analysis of approximation techniques: part i–interpolations and projections. *IEEE Transactions Signal Processing*, **47**:2783–2795, 1999.

[9] Y.C. Eldar. Sampling with arbitrary sampling and reconstruction spaces and oblique dual frame vectors. *Journal of Fourier Analysis and Applications*, **9**:77–96, 2003.

[10] X. Yu and L. Tong. Joint channel and symbol estimation by oblique projections. *IEEE Transactions Signal Processing*, **49**:3074–3083, 2001.

[11] L. Rebollo-Neira. Oblique matching pursuit. *IEEE Signal Processing Letters*, **14**(10):703–706, 2007.

[12] B. Lamichhane and L. Rebollo-Neira. Sparse representations for structured noise filtering. *http://arxiv.org/abs/0709.2475v1*, 2007.

[13] L. Rebollo-Neira. Constructive updating/downdating of oblique projectors: a generalization of the GramSchmidt process. *Journal of Physics A: Mathematical and Theoritcal*, **40**:6381–6394, 2007.

[14] K.N. Plataniotis and A.N. Venetsanopoulos. *Color Image Processing and Applications*. Springer, first edition, 2000.

[15] T.F. Chan and J. Shen. *Image Processing And Analysis: Variational, PDE, Wavelet, And Stochastic Methods*. SIAM, Philadelphia, 2005.

[16] J.S. Lim. *Two-Dimensional Signal and Image Processing*. Englewood Cliffs, NJ, Prentice Hall, 1990.

[17] R.H. Chan, C. Ho, and M. Nikolova. Salt-and-pepper noise removal by median-type noise detectors and detail-preserving regularization. *IEEE Transactions on Image Processing*, **14**:1479–1485, 2005.

[18] R. Franke and G.M. Nielson. Scattered data interpolation and applications: A tutorial and survey. In H. Hagen and D. Roller, editors, *Geometric Modelling: Methods and Their Application*, pages 131–160. Springer-Verlag, 1991.

[19] I. Amidror. Scattered data interpolation methods for electronic imaging systems: a survey. *Journal of Electronic Imaging*, **11**:157–176, 2002.

[20] F. R. Preparata and M. I. Shamos. *Computational Geometry: An Introduction*. Springer-Verlag, 1985.

[21] N. Dyn, D. Levin, and S. Rippa. Data dependent triangulations for piecewise linear interpolation. *IMA Journal of Numerical Analysis*, **10**:137–154, 1990.

[22] E. Besdok. Impulsive noise suppression from images by using anfis interpolant and lillietest. *EURASIP Journal on Applied Signal Processing*, 2004:2423–2433, 2004.

[23] L. Demaret and A. Iske. Adaptive image approximation by linear splines over locally optimal Delaunay triangulations. *IEEE Signal Processing Letters*, **13**:281–284, 2006.

[24] M. Reed and B. Simon. *Functional Analysis*. Academic Press, London, 1980.

[25] L. Rebollo-Neira. Dictionary redundancy elimination. *IEE Proceedings – Vision, Image and Signal Processing*, **151**:31–34, 2004.

[26] Biorthogonal techniques for optimal signal representation. http://www.ncrg.aston.ac.uk/Projects/BiOrthog/.

[27] A. Quarteroni, R. Sacco, and F. Saleri. *Numerical Mathematics*. Springer, second edition, 2007.

[28] S. Mallat. *A Wavelet Tour of Signal Processing*. Academic Press, London, second edition, 1999.

[29] L. Rebollo-Neira and D. Lowe. Optimized orthogonal matching pursuit approach. *IEEE Signal Processing Letters*, **9**:137–140, 2002.

[30] M. Andrle and L. Rebollo-Neira. A swapping-based refinement of orthogonal matching pursuit strategies. *Signal Processing*, **86**:480–495, 2006.

[31] Highly nonlinear approximations for sparse signal representation. http://www.ncrg.aston.ac.uk/Projects/BiOrthog/.

[32] D. Kincaid and W. Cheney. *Numerical Analysis*. Brooks/Cole Publishing Company, second edition, 2002.

[33] E. Süli and D. Mayers. *An Introduction to Numerical Analysis*. Cambridge University Press, Cambridge, first edition, 2003.

[34] F.N. Fritsch and R.E. Carlson. Monotone piecewise cubic interpolation. *SIAM Journal on Numerical Analysis*, **17**:238–246, 1980.

[35] C. B. Barber, D.P. Dobkin, and H.T. Huhdanpaa. The quickhull algorithm for convex hulls. *ACM Transactions on Mathematical Software*, **22**:469–483, 1996.

[36] F. Aurenhammer and R. Klein. Voronoi diagrams. In J.-R. Sack and J. Urrutia, editors, *Handbook of Computational Geometry*, pages 201–290. North-Holland, Amsterdam, Netherlands, 2000.

[37] D. Eppstein. Nearest neighbors and voronoi diagrams. http://www.ics.uci.edu/ eppstein/junkyard/nn.html.

[38] Qhull computes the convex hull, Delaunay triangulation, Voronoi diagram and halfspace intersection about a point. http://www.qhull.org.

In: New Signal Processing Research
Editor: Takumi Maeda, pp. 159-196

ISBN 978-1-60456-479-2

Chapter 5

TIME-VARYING MULTIRATE SYSTEMS AND FILTER BANKS

Guangyu Wang
School of Communication Engineering,
Chongqing University of Posts and Communications, Chongqing/China

Abstract

In many applications of digital signal processing, such as audio and image processing it is found that the filter bank with the changed channel number and filter coefficients can provide better performance than the time-invariant filter bank. This observation motivated us to generally study the time-varying filter bank. We know that the theory of time-invariant filter banks is quite complete and has arrived a very high level. In comparison with time-invariant filter banks the research of time-varying filter banks is relative new. In our research on this topic we have established a series of theories to analyze the time-varying filter bank and multirate system [9-11]. In this Chapter we provide a summarization of the developed theories and introduce also some new research results on this topic.

1. Introduction

A general time-varying filter bank is defined as a filter bank, in which all parameters including number of channel and filter length change with time. Fig. 1 shows a general time-varying filter bank with maximal decimation. Note that we define that the filter bank varies with time index m instead of time index n, where m is the time index of the output in the analysis filter bank.

The reason to choose m as time index is that in the analysis side just the output at time m is sent to the synthesis side and used to reconstruct the input signal. The relationship between n and m is depicted in Figure 2. Mathematically we can indirectly describe the

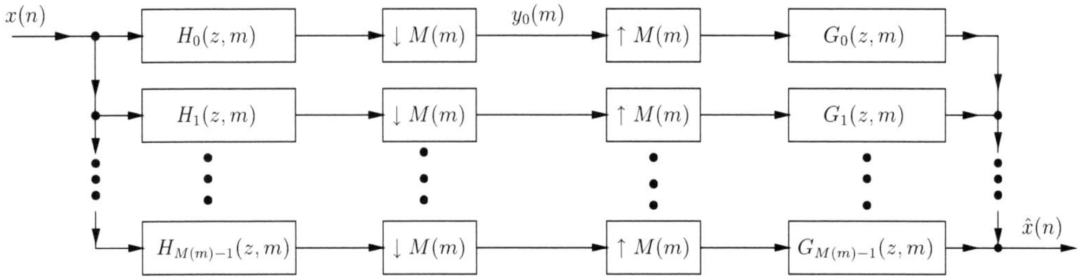

Figure 1. The most general time-varying filter bank with maximal decimation.

relationship between n and m as

$$d(m) = \begin{cases} \sum_{i=0}^{m-1} M(i), & m > 0 \\ 0, & m = 0 \\ -\sum_{i=m}^{-1} M(i), & m < 0 \end{cases}, \tag{1}$$

where $d(m)$ represents the distant in the n axis from zero point to the block that m corresponds to.

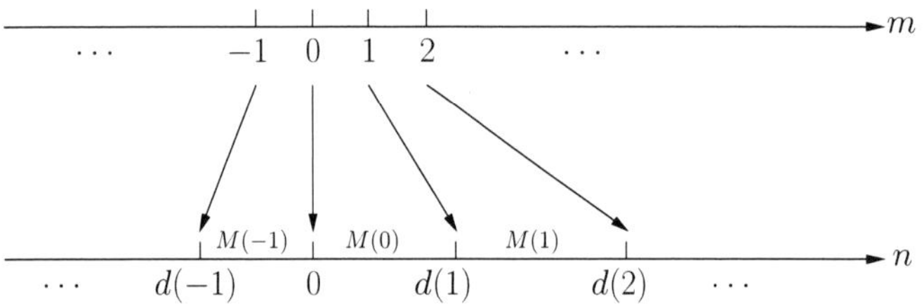

Figure 2. Graphic representation of the relationship between time index m and n.

Using commutators the time-varying filter bank in Figure 1 can be implemented like Figure 2. From Figure 2 we can see that a maximally decimated time-varying filter bank consists of infinite $M(m)$-channel time-invariant filter banks with $-\infty < m < \infty$. Each time-invariant analysis filter bank contributes just $M(m)$ values to the analysis output $\mathbf{y}$, where $\mathbf{y}$ is defined by

$$\mathbf{y} = [\cdots y_0(-1)y_1(-1)\cdots y_{M(-1)-1}(-1) \quad y_0(0)y_1(0)\cdots y_{M(0)-1}(0)\cdots]^T. \tag{2}$$

It is well-know that an M-channel time-invariant filter bank is a periodic time-varying system with period M, where the signal $y_i(m)$ is the output of i-subchannel filter with unchanged filter coefficients [1-2]. So far, all available theories and design methods for the time-invariant filter bank are based on the periodic time-varying system. However, the filter bank in Figure 1 is not periodic, and the elements in (2) are coming from the filters with changed coefficients. So, the available theories and design methods for the time-invariant filter bank cannot be directly used to analyze and design the time-varying filter bank. A new theory needs to be developed to analyze and design the time-varying filter bank. In our research we have established some new theories to analyze time-varying filter banks [9-11]. In this Chapter we first summarize the developed theories, then describe some new design approaches for the time-varying filter bank.

The Chapter begins with the analysis of basic building blocks of a time-varying multirate system, i.e. time-varying decimator and expander. Then, using the time-varying decimator and expander the time-varying filter will be analyzed in both time- and transform-domain. The general design methods, which are quite different from the classic design methods used in time-invariant filter banks, are discussed. At the end of Chapter we will give some research directions of time-varying multirate systems and filter banks.

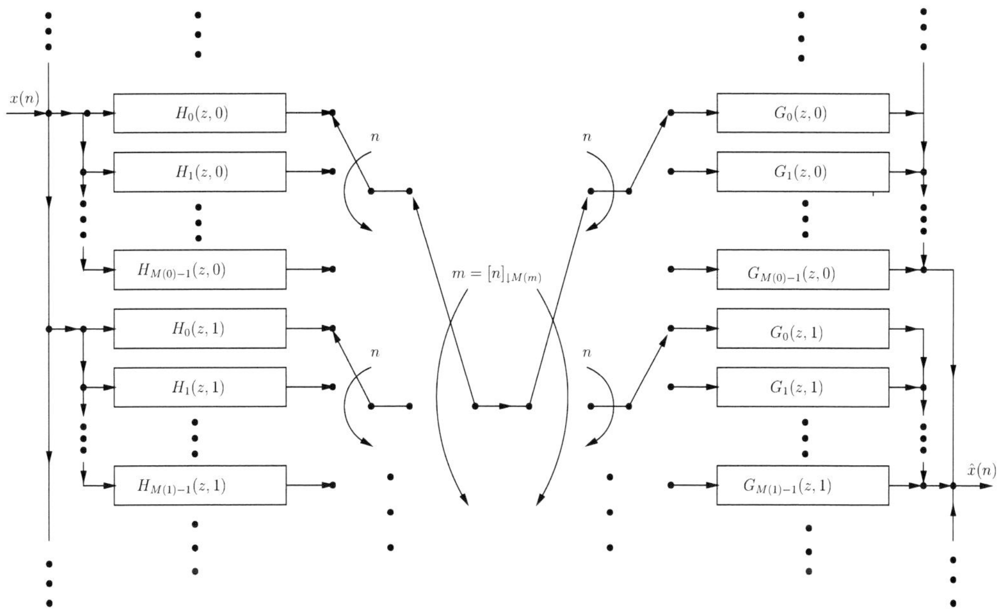

Figure 3. Implementation of the time-varying filter bank with commutators. The outer commutators have period $M(m)$ and represent LTI filter banks at time index m.

2. The Most General Time-Varying Building Blocks

Fig. 4 shows the most general time-varying building block consisting of an analysis-synthesis system, where the filters, the decimator and expander, all change with time index m.

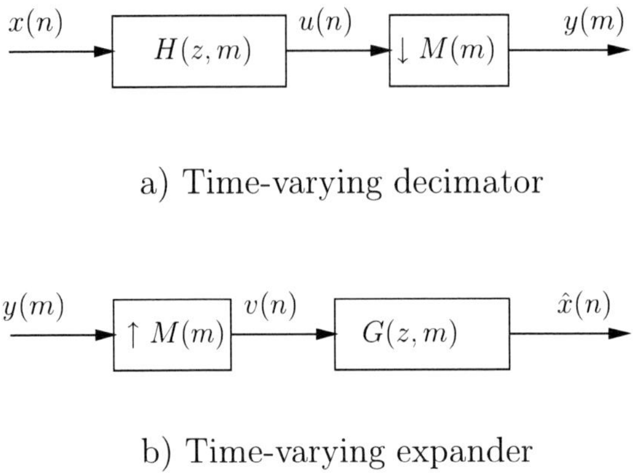

Figure 4. The most general time-varying building block.

In the analysis side, the output signal can be expressed in the time-domain using the matrix notations as

$$\underbrace{\begin{bmatrix} \vdots \\ y(m) \\ y(m+1) \\ \vdots \end{bmatrix}}_{\mathbf{y}} = \mathbf{H} \underbrace{\begin{bmatrix} \vdots \\ x[mM(m-1)] \\ x[mM(m-1)+1] \\ \vdots \end{bmatrix}}_{\mathbf{x}}, \tag{3}$$

where $\mathbf{H}$ is defined as

$$\mathbf{H} = \begin{bmatrix} & \vdots & \vdots & \vdots & \vdots & \\ \cdots & h(L(m)-1,m) & \cdots & h(L(m)-M(m),m) & h(L(m)-M(m)-1,m) & \cdots \\ \cdots & 0 & \cdots & 0 & h(L(m+1)-1,m+1) & \cdots \\ & \vdots & \vdots & \vdots & \vdots & \end{bmatrix}, \tag{4}$$

and $L(m)$ is the filter length at time m

In the transform-domain we cannot directly describe the time-varying system in Fig. 4 because the convolution rule in the transform-domain does not exist for the time-varying system. To make the description in the transform-domain for the time-varying system in Fig. 4 possible, we define following matrices

$$\mathbf{H}(m) = \begin{bmatrix} & \vdots & \vdots & \vdots & \vdots & \\ \cdots & h(L(m)-1,m) & \cdots & h(L(m)-M(m),m) & h(L(m)-M(m)-1,m) & \cdots \\ \cdots & 0 & \cdots & 0 & h(L(m)-1,m) & \cdots \\ & \vdots & \vdots & \vdots & \vdots & \end{bmatrix} \tag{5}$$

$$\mathbf{y}(l,m) = [\ \cdots\ y'(0,m)\ \ y'(1,m)\ \ y'(2,m)\ \ \cdots\ ]^T \tag{6}$$

$$\mathbf{w}(m) = diag\{\cdots 0\ 1\ 0\ \cdots\}, \tag{7}$$

where $\mathbf{w}(m)$ is a diagonal matrix and has just one non-zero element in the position corresponding to $y'(m,m) = y(m)$. With above definitions we can rewrite the output $\mathbf{y}$ in (3) as

$$\begin{aligned} \mathbf{y} &= \sum_{m=-\infty}^{\infty} \mathbf{w}(m)\, \mathbf{y}(l,m) \\ &= \sum_{m=-\infty}^{\infty} \mathbf{w}(m)\, \mathbf{H}(m)\, \mathbf{x}. \end{aligned} \tag{8}$$

In the equation (8), the operation $\mathbf{H}(m)\mathbf{x}$ describes a normal time-invariant filter operation plus a decimator. Actually, multiplying matrix $\mathbf{w}(m)$ is equal to multiplying $y'(l,m)$ with the Kronecker function $\delta(l-m)$ defined by

$$\delta(l-m) = \begin{cases} 1, & l = m \\ 0, & \text{otherwise} \end{cases}. \tag{9}$$

If we shift the multiplication from the output side of the decimator to the input side, the Kronecker function should be replaced with a window function defined as

$$w(n-d(m)) = \begin{cases} 1, & d(m) \le n \le d(m) + M(m) - 1 \\ 0, & \text{otherwise} \end{cases}, \tag{10}$$

where n and m are independent and $d(m)$ is defined in (1). With considering the effect of decimation the filter operation $\mathbf{y}(l,m) = \mathbf{H}(m)\mathbf{x}$ can be described in the transform-domain as

$$\begin{aligned} Y(z,m) &= \frac{1}{M(m)} \sum_{i=0}^{M(m)-1} U(W^i_{M(m)} z^{1/M(m)}, m) \\ &= \frac{1}{M(m)} \sum_{i=0}^{M(m)-1} H(W^i_{M(m)} z^{1/M(m)}, m) X(W^i_{M(m)} z^{1/M(m)}), \end{aligned} \tag{11}$$

where $W_{M(m)} = e^{-j2\pi/M(m)}$. Using the relationship

$$x(n)y(n) \Longleftrightarrow X(z) * Y(z), \tag{12}$$

where $*$ means the convolution in the transform-domain, we can express the output $y(m)$ in the transform-domain as

$$\begin{aligned} Y(z) &= \sum_{m=-\infty}^{\infty} Y(z,m) * (z^{-m}) \\ &= \sum_{m=-\infty}^{\infty} \frac{1}{M(m)} \sum_{i=0}^{M(m)-1} \Big\{ \Big[H(W^i_{M(m)} z^{1/M(m)}, m) X(W^i_{M(m)} z^{1/M(m)}) \Big] \\ &\quad * \Big[W(W^i_{M(m)} z^{1/M(m)}) (W^i_{M(m)} z^{1/M(m)})^{-d(m)} \Big] \Big\}, \end{aligned} \tag{13}$$

in which $W(z)$ represents the z-transform of the window function $w(n)$.

In the synthesis side, the filter operation can be described using matrices as

$$\begin{bmatrix} \vdots \\ \widehat{x}(mM(m-1)) \\ \widehat{x}(mM(m-1)+1) \\ \vdots \end{bmatrix} = \begin{bmatrix} & \vdots & \vdots & \\ \cdots & g(0,m) & 0 & \cdots \\ \cdots & \vdots & \vdots & \cdots \\ \cdots & g(M(m)-1,m) & 0 & \cdots \\ \cdots & g(M(m),m) & g(0,m+1) & \cdots \\ & \vdots & \vdots & \end{bmatrix} \begin{bmatrix} \vdots \\ y(m) \\ y(m+1) \\ \vdots \end{bmatrix}. \quad (14)$$

To make the structure in (14) more clear, we define L as the least common multiplier of $L(m)$

$$L = LCM\{L(m),\ m \in \mathbf{Z}\}. \quad (15)$$

All filter length at different m will be extended to L with adding zeros. Then, we rewrite (14) as

$$\underbrace{\begin{bmatrix} \vdots \\ \widehat{x}(mM(m-1)) \\ \widehat{x}(mM(m-1)+1) \\ \vdots \end{bmatrix}}_{\widehat{\mathbf{x}}} = \mathbf{G} \underbrace{\begin{bmatrix} \vdots \\ y(m) \\ y(m+1) \\ \vdots \end{bmatrix}}_{\mathbf{y}}, \quad (16)$$

where

$$\mathbf{G} = \begin{bmatrix} & \vdots & \vdots & \\ \cdots & g(0,m) & 0 & \cdots \\ \cdots & \vdots & \vdots & \cdots \\ \cdots & g(M(m),m) & g(0,m+1) & \cdots \\ & \vdots & \vdots & \\ \cdots & g(L-M(m-1),m) & g(L-M(m)-M(m-1),m+1) & \cdots \\ & \vdots & \vdots & \\ \cdots & g(L-1,m) & g(L-M(m)-1,m+1) & \cdots \\ \cdots & 0 & g(L-M(m),m+1) & \cdots \\ & \vdots & \vdots & \\ \cdots & 0 & g(L-1,m+1) & \cdots \\ & \vdots & \vdots & \end{bmatrix}. \quad (17)$$

Substituting (8) into (16), we get

$$\widehat{\mathbf{x}} = \sum_{m=-\infty}^{\infty} \mathbf{G}\mathbf{w}(m)\mathbf{y}(l,m). \quad (18)$$

The operation $\mathbf{Gw}(m)$ in (18) selects exactly the synthesis function at time m. The other columns in $\mathbf{G}$ beside $g(n,m)$ have no effect on the results of $\mathbf{Gw}(m)$. So, if we replace $\mathbf{G}$ in (18) with $\mathbf{G}(m)$ that contains just the synthesis filter coefficients at time m and is built similar with (5), the equation (18) does not change. After replacing $\mathbf{G}$ in (18) with $\mathbf{G}(m)$, we get

$$\widehat{\mathbf{x}} = \sum_{m=-\infty}^{\infty} \mathbf{G}(m)\mathbf{w}(m)\mathbf{y}(l,m). \tag{19}$$

With description in (19), the time-varying synthesis filter in Fig. 5 becomes time-invariant in respect to a fixed m. Based on the formulation in (8) and (19), we can now establish a model showed in Fig. 5 to analyze time-varying building blocks. Note that the indices n and m in Fig. 5 are independent.

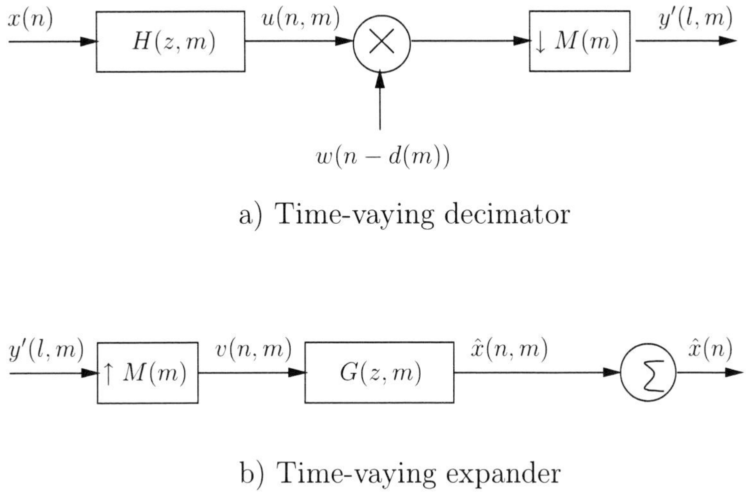

Figure 5. Model of the most general time-varying building block.

For a fixed m, the model is time-invariant, therefore, we can write the reconstructed signal $\widehat{x}(n)$ in the transform-domain as

$$\begin{aligned}
\widehat{X}(z) &= \sum_{m=-\infty}^{\infty} \widehat{X}(z,m)z^{-d(m-1)} \\
&= \sum_{m=-\infty}^{\infty} V(z,m)G(z,m)z^{-d(m-1)} \\
&= \sum_{m=-\infty}^{\infty} \frac{1}{M(m)} \sum_{i=0}^{M(m)-1} G(z,m)\Big\{\Big[H(W_{M(m)}^{i}z,m)X(W_{M(m)}^{i}z)\Big] \\
&\quad * \Big[W(W_{M(m)}^{i}z)(W_{M(m)}^{i}z)^{-d(m)}\Big]\Big\}z^{-d(m-1)},
\end{aligned} \tag{20}$$

where delay $z^{-d(m-1)}$ is introduced due to upsampling in the synthesis side.

Equation (20) is still complicate. From (20) it is not simple to see the relationship between $X(z)$ and $\widehat{X}(z)$. However, if we can remove the convolution operation in (20), we have possibility to simplify the formulation in (20). To do this, let us see again the definition of $\mathbf{G}$ in (17), where the last $M(m-1)$ elements from $g(L-M(m-1),m)$ to $g(L-1,m)$ in the column at time m must have at least one non-zero element for a maximally decimated time-varying filter bank. From left to right the non-zero elements of those rows are coming from column elements at time from m to $m+N(m)-1$, where $N(m)$ is decided by

$$\sum_{k=0}^{N(m)-2} M(m+k) < L \leq \sum_{k=0}^{N(m)-1} M(m+k). \tag{21}$$

Referencing to (3), (4) and (17) and using the following relationship

$$\widehat{\mathbf{x}} = \mathbf{G}\,\mathbf{H}\,\mathbf{x}, \tag{22}$$

we can see that the reconstructed signal from $\widehat{x}(d(m-1))$ to $\widehat{x}(d(m-1)+M(m-1)-1)$ are just determined by the analysis and synthesis filters at time from m to $m+N(m)-1$. Therefore, we can use a periodic time-varying system to calculate the reconstructed signal at time m. Such a periodic time-invariant system can be described based on (22) as

$$\widetilde{X}(z,m) = \frac{1}{L}\sum_{i=0}^{L-1}\sum_{k=0}^{N(m)-1}\left[G(z,m+k)z^{-e(k)}\right]\left[H(W_L^i z,m+k)(W_L^i z)^{-e(k)}\right]X(W_L^i z) \tag{23}$$

where $e(k)$ is defined by

$$e(k) = \begin{cases} 0, & k=0 \\ \sum_{i=0}^{k-1} M(m-i), & k>0 \end{cases}, \tag{24}$$

and describes the delay of filters at time $m+k$ with respect to $G(z,m)$ and $H(z,m)$ at time m.

If $M(m) = M, \forall m$, then $e(k) = kM$. Actually, equation (23) describes exactly an $N(m)$-channel time-invariant filter bank with decimation factor L and built at time m. To get the reconstructed signal at time m, we need just to simply multiply the output $\widetilde{X}(z,m)$ of the filter bank with a window function $w(n-d(m-1))$, i.e.

$$\widehat{x}(n,m) = \widetilde{x}(n,m)w(n-d(m-1)). \tag{25}$$

$\widehat{x}(n,m)$ is related with $\widehat{x}(n)$ by

$$\widehat{x}(n) = \sum_{m=-\infty}^{\infty} \widehat{x}(n,m). \tag{26}$$

Based on (23) we have another model for the most time-varying building block showed in Fig. 6.

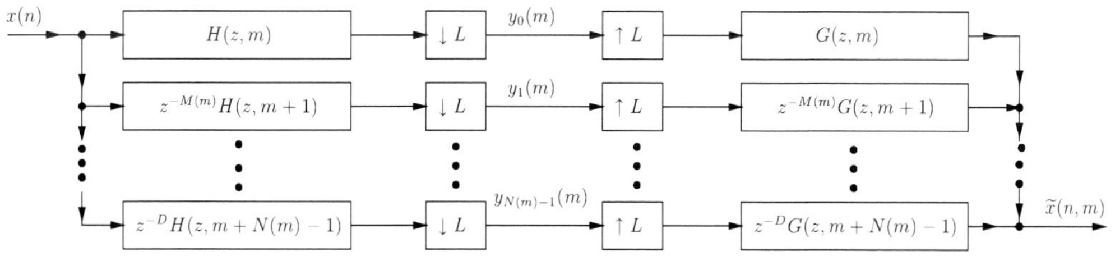

Figure 6. Filter bank implementation of the most general time-varying building block, where $D = e(N(m) - 1)$.

3. Analysis of the Most General Time-Varying Filter Bank

In this section we present analysis of the most general time-varying filter bank in the transform-domain based on the model in Fig. 6. Section 3.1 describes the periodic time-varying system, the model established for analysis of the time-varying filter bank. The polyphase and modulation analysis will be given in section 3.2 and section 3.3, respectively.

3.1. Description of Analysis Model

From equation (23) we know that the time-varying building block can be realized by a periodic time-varying system. Substituting Fig. 6 into Fig. 1, we get the model showed in Fig. 7 for the time-varying filter bank.

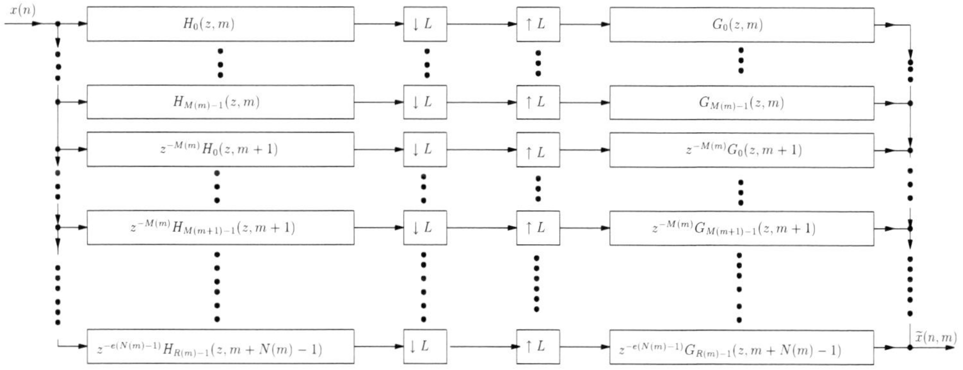

Figure 7. L-channel maximally decimated time-invariant filter bank: model for analyzing the most general time-varying filter bank.

Fig. 7 is actually a L-channel maximally decimated time-invariant filter bank consisting of the filters from time m to $m+N(m)-1$. But the filters at time $m+N(m)-1$ could not be completely included in the L-channel time-invariant filter bank, because of the relationship in (21). Just the first $R(m)$ channels of the filters at time $m + N(m) - 1$ could be used to build the L-channel filter bank, where $R(m)$ is determined by

$$R(m) = L - \sum_{k=0}^{N(m)-2} M(m+k) \leq M(m + N(m) - 1). \quad (27)$$

Referencing to (23) the periodic time-varying system in Fig. 7 can be described as

$$\begin{aligned}\widetilde{X}(z,m) &= \frac{1}{L}\sum_{i=0}^{L-1}\Bigg\{\sum_{k=0}^{N(m)-2}\sum_{l=0}^{M(m+k)-1}\left[G_l(z,m+k)z^{-e(k)}\right]\left[H_l(W_L^i z, m+k)(W_L^i z)^{-e(k)}\right] \\ &+ \sum_{l=0}^{R(m)-1}\left[G_l(z,m+N(m)-1)z^{-e(N(m)-1)}\right] \\ &\times \left[H_l(W_L^i z, m+N(m)-1)(W_L^i z)^{-e(N(m)-1)}\right]\Bigg\}X(W_L^i z). \end{aligned} \quad (28)$$

With definitions

$$H_l'(z,m+k) = H_l(z,m+k)z^{-e(k)} \quad (29)$$

$$G_l'(z,m+k) = G_l(z,m+k)z^{-e(k)}, \quad (30)$$

the equation (28) can be rewritten as

$$\begin{aligned}\widetilde{X}(z,m) = \frac{1}{L}\sum_{i=0}^{L-1}\Bigg\{\sum_{k=0}^{N(m)-2}\sum_{l=0}^{M(m+k)-1} G_l'(z,m+k)H_l'(W_L^i z, m+k) \\ + \sum_{l=0}^{R(m)-1} G_l'(z,m+N(m)-1)H_l'(W_L^i z, m+N(m)-1)\Bigg\}X(W_L^i z). \end{aligned} \quad (31)$$

To make the relationship between $\widetilde{x}(n,m)$ and $x(n)$ more clear, we can rewrite (31) through decomposing the input signal $x(n)$ as

$$\begin{aligned}X(z) &= \sum_{n=-\infty}^{\infty} x(n)z^{-n} \\ &= \sum_{r=0}^{N_1(m)-1}\left\{\sum_{i=0}^{M(m-1)-1}\left\{\sum_{l=-\infty}^{\infty} x(lL + rM(m-1) + i + d(m))z^{-d(m)}z^{-lL}\right\}z^{-i}\right\}z^{-rM(m-1)} \\ &= \sum_{r=0}^{N_1(m)-1}\left\{\sum_{i=0}^{M(m-1)-1} X_{ir}(z^L, m)z^{-i}\right\}z^{-rM(m-1)} \\ &= \sum_{r=0}^{N_1(m)-1} X_r(z,m)z^{-rM(m-1)}, \end{aligned} \quad (3$$

where $N_1(m) = L/M(m-1)$ and $X_r(z,m)$ is the r-th polyphase component decomposed at time m. Replacing (32) into (31) we have the new formulation for $\widetilde{X}(z,m)$

$$\widetilde{X}(z,m) = \frac{1}{L}\sum_{r=0}^{N_1(m)-1}\sum_{i=0}^{L-1}\left\{\sum_{k=0}^{N(m)-2}\sum_{l=0}^{M(m+k)-1} G_l'(z,m+k)H_l'(W_L^i z, m+k)\right.$$
$$\left. + \sum_{l=0}^{R(m)-1} G_l'(z,m+N(m)-1)H_l'(W_L^i z, m+N(m)-1)\right\}$$
$$X_r(W_L^i z, m)(W_L^i z)^{-rM(m-1)}. \tag{33}$$

Similar with (32), $\widetilde{X}(z,m)$ can be also decomposed like

$$\widetilde{X}(z,m) = \sum_{n=-\infty}^{\infty}\widetilde{x}(n,m)z^{-n}$$
$$= \sum_{r=0}^{N_1(m)-1}\left\{\sum_{i=0}^{M(m-1)-1}\left\{\sum_{l=-\infty}^{\infty}\widetilde{x}(lL+rM(m-1)+i+d(m),m)z^{-d(m)}z^{-lL}\right\}\right.$$
$$\left. z^{-i}\right\}z^{-rM(m-1)}$$
$$= \sum_{r=0}^{N_1(m)-1}\widetilde{X}_r(z,m)z^{-rM(m-1)}. \tag{34}$$

Comparing (34) with (25) we can see that the reconstructed signal $\widehat{x}(n,m)$ at time m is contained just in one of the polyphase components $\widetilde{X}_r(z,m)$. So, if the polyphase component $\widetilde{x}_r(z,m)$ is perfectly reconstructed, the time-varying filter bank is also perfectly reconstructed. In the case of perfect reconstruction we get

$$\widehat{x}(n,m) = x(n)w(n-d(m-1)). \tag{35}$$

Based on above analysis we have following theorem.

Theorem 1. *A general time-varying filter bank at time m can be modeled as a L-channel time-invariant filter bank that consists of L filters coming from the time-varying filter banks at time from m to $m+N(m)-1$, where L and $N(m)$ are determined by equations (15) and (21). The built L-channel filter banks are defined by (29) and (30). The reconstructed signal $\widehat{x}(n,m)$ at time m is just determined by the L-channel filter bank and contained in one of the polyphase components $\widetilde{x}_r(n,m)$. The PR condition of $\widehat{x}(n,m)$ is equivalent to PR condition of $\widetilde{x}_r(n,m)$.*

Theorem 1 implies that study of the most general time-varying filter bank is equal to study of the built L-channel time-invariant filter bank.

3.2. Polyphase-Domain Analysis

For describing the built L-channel filter bank in the polyphase-domain, we need to decompose the analysis and synthesis filter in Fig. 7. From (29) and (30), the analysis and synthesis filters can be written in the time-domain as

$$h'_l(n, m+k) = h_l(n - e(k), m+k) \tag{36}$$

$$g'_l(n, m+k) = g_l(n - e(k), m+k), \tag{37}$$

where $0 \leq l \leq M(m+k) - 1$ if $0 \leq k \leq N(m) - 2$, and $0 \leq l \leq R(m) - 1$ for $k = N(m) - 1$. Before decomposition of filter functions we need to shift the filter coefficients in order to make sure that the filters in Fig. 7 have the same delay $z^{-M(m-1)}$ of each other. For this purpose we define

$$D(k) = \sum_{i=1}^{k} M(m+i-1) - kM(m-1), \tag{38}$$

for $1 \leq k \leq N(m) - 1$. Based on (36), (37) and (38) we can decompose the L-channel filter bank as

$$H'_{(l+e(k))}(z, m+k) = \sum_{r=0}^{N_1(m)-1} \left\{ \sum_{i=0}^{M(m-1)-1} \left[H^{(r)}_{(l+e(k)),i}(z^L, m+k) z^{D(k)} \right] z^i \right\} z^{rM(m-1)} \tag{39}$$

$$G'_{(l+e(k))}(z, m+k) = \sum_{r=0}^{N_1(m)-1} \left\{ \sum_{i=0}^{M(m-1)-1} \left[G^{(r)}_{i,(l+e(k))}(z^L, m+k) z^{-D(k)} \right] z^{-i} \right\} z^{-rM(m-1)}, \tag{40}$$

where

$$H^{(r)}_{(l+e(k)),i}(z, m+k) = \sum_{j=-\infty}^{\infty} h_l[jL - rM(m-1) - i + e(k), m+k] z^{-j} \tag{41}$$

$$G^{(r)}_{i,(l+e(k))}(z, m+k) = \sum_{j=-\infty}^{\infty} g_l[jL - rM(m-1) + i - e(k), m+k] z^{-j}, \tag{42}$$

and analysis and synthesis filter banks are decomposed using type-3 and type-1 polyphase decomposition [2], respectively. With (41) and (42) we can define the elements of the polyphase matrices for the L-channel filter bank as

$$\left[H^{(r)}_p(z, m+k) \right]_{l,i} = H^{(r)}_{(l+e(k)),i}(z, m+k) \tag{43}$$

$$\left[G^{(r)}_p(z, m+k) \right]_{i,l} = G^{(r)}_{i,(l+e(k))}(z, m+k), \tag{44}$$

where $0 \leq r \leq N_1(m) - 1$.

Based on the polyphase decomposition for signal and system functions defined in (32), (43) and (44), the output of the L-channel time-invariant filter bank can be formulated by

$$\widetilde{X}(z, m) = [\, 1 \quad z^{-1} \quad \cdots \quad z^{-L+1} \,]\, \mathbf{G}_p(z^L, m) \mathbf{H}_p(z^L, m) \begin{bmatrix} \mathbf{x}_0(z^L, m) \\ \mathbf{x}_1(z^L, m) \\ \vdots \\ \mathbf{x}_{N_1(m)-1}(z^L, m) \end{bmatrix}, \tag{45}$$

where the elements of matrices $\mathbf{G}_p(z^L, m)$ and $\mathbf{H}_p(z^L, m)$ are

$$\mathbf{H}_p^{(r)}(z,m) = \begin{bmatrix} H_p^{(r)}(z,m) & H_p^{(r)}(z,m+1) & \cdots & H_p^{(r)}(z,m+N(m)-1) \end{bmatrix}^T \quad (46)$$

$$\mathbf{G}_p^{(r)}(z,m) = \begin{bmatrix} G_p^{(r)}(z,m) & G_p^{(r)}(z,m+1) & \cdots & G_p^{(r)}(z,m+N(m)-1) \end{bmatrix} \quad (47)$$

and input vector $\mathbf{x}_r(z, m)$ is defined as

$$\mathbf{x}_r(z,m) = \begin{bmatrix} X_{0,r}(z,m) & X_{1,r}(z,m) & \cdots & X_{M(m-1)-1,r}(z,m) \end{bmatrix}, \quad (48)$$

for $0 \le r \le N_1(m) - 1$.

In order to get the PR condition for the general time-varying filter bank we define

$$\begin{aligned} \mathbf{P}_{ir}(z,m) &= \mathbf{G}_p^{(i)}(z,m)\mathbf{H}_p^{(r)}(z,m) \\ &= \sum_{k=0}^{N(m)-1} G_p^{(i)}(z,m+k)H_p^{(r)}(z,m+k). \end{aligned} \quad (49)$$

Because according to theorem 1 just one of the polyphase components $\mathbf{x}_r(z, m)$ needs to be perfectly reconstructed at time m. So, the PR condition for the most general time-varying filter bank reads

$$\mathbf{P}_{N_1(m)-1,i}(z,m) = \mathbf{I}\,\delta(i-r),\ \forall m\ r \in \{0\ 1 \cdots N_1(m)-1\}. \quad (50)$$

If the PR condition (91) is satisfied we get the reconstructed signal

$$\begin{aligned} \widehat{X}(z) &= \sum_{m=-\infty}^{\infty} \left\{ \sum_{i=0}^{M(m-1)-1} \widetilde{x}(i+d(m-1),m)z^{-i} \right\} z^{-(L \pm rM(m-1))} z^{-d(m-1)} \\ &= \sum_{m=-\infty}^{\infty} \left\{ \sum_{i=0}^{M(m-1)-1} x(i+d(m-1),m)z^{-i} \right\} z^{-(L \pm rM(m-1))} z^{-d(m-1)} \end{aligned} \quad (51)$$

where delay $z^{-(L \pm rM(m-1))}$ is introduced.

The PR condition for the general time-varying filter bank described in (91) is quite different from it for time-invariant filter bank. The polynomial matrices $\mathbf{G}_p^{(N_1(m)-1)}(z, m)$ and $\mathbf{H}_p^{(r)}(z, m)$ are no more square matrices. Therefore, the design methods in the transform-domain used for time-invariant filter banks and based on polyphase matrices, such as factorization method, can not be directly used for design of time-varying filter banks.

3.3. Modulation-Domain Analysis

Referencing to equation (33), if we define

$$A_{ri}(z,m) = \sum_{k=0}^{N(m)-2} \sum_{l=0}^{M(m+k)-1} G_l'(z,m+k)H_l'(W_L^i z, m+k)$$

$$+\sum_{l=0}^{R(m)-1} G'_l(z, m+N(m)-1)H'_l(W_L^i z, m+N(m)-1), \qquad (52)$$

then $\widetilde{X}(z,m)$ can be reformulated with

$$\widetilde{X}(z,m) = \frac{1}{L}\sum_{r=0}^{N_1(m)-1}\sum_{i=0}^{L-1} A_{ri}(z,m)X_r(W_L^i z, m). \qquad (53)$$

In matrix notation, (53) can be rewritten as

$$\widetilde{X}(z,m) = \frac{1}{L}\left[\, \mathbf{A}_0(z,m) \quad \mathbf{A}_1(z,m)\cdots\mathbf{A}_{N_1(m)-1}(z,m)\,\right]\begin{bmatrix} \mathbf{x}_m^{(0)}(z,m) \\ \mathbf{x}_m^{(1)}(z,m) \\ \vdots \\ \mathbf{x}_m^{(N_1(m)-1)}(z,m) \end{bmatrix}, \qquad (54)$$

where $\mathbf{A}_r(z,m)$ $(0 \le r \le N_1(m)-1)$ is called distortion matrix of the time-varying filter bank and defined as

$$\mathbf{A}_r(z,m) = [\, A_{r,0}(z,m) \quad A_{r,1}(z,m)\cdots A_{r,L-1}(z,m)\,], \qquad (55)$$

and $\mathbf{x}_m^{(r)}(z,m)$ is the modulation input vector with

$$\mathbf{x}_m^{(r)}(z,m) = [\, X_r(z,m) \quad X_r(W_L z,m)\cdots X_r(W_L^{L-1}z,m)\,]^T. \qquad (56)$$

From (53) we can see that a distortion-free time-varying filter bank should satisfy

$$A_{ri}(z,m) = c\, z^{-l}\, \delta(i),\ \forall m,\ r \in \{0\ 1\cdots N_1(m)-1\}, \qquad (57)$$

where c and l are constants. If condition (57) is satisfied, we get

$$\widetilde{X}(z,m) = \frac{c}{L}\, X_r(z,m)\, z^{-l},\ \forall m, \qquad (58)$$

where $X_r(z,m)$ is the r-th polyphase component of $X(z)$ decomposed at time m and defined in (32).

4. Design of General Time-Varying Filter Banks

In the literature, the design of time-varying filter banks war addressed by several authors through researching the problem of switching between different PR LTI [3-8]. Basically, there are two classes of design methods, namely design method with boundary filters and without boundary filter banks. Usually, if we change from one LTI filter bank to another LTI filter bank, we need to add some boundary filters between two LTI filter banks to keep the PR condition. The disadvantage of using boundary filters is that the spectrum characteristics of the boundary filters are not good. But for time-varying cosine-modulated filter banks it is possible to achieve good frequency characteristics while keeping PR condition during the transition period[9]. In this case, a window switching method is introduced to design the prototype window[11].

The design method of directly switching the analysis filter banks was first proposed by Nayebi, Sodagar *et al.* [3,4], which allows directly switching analysis filter banks without inserting boundary filters in the analysis side. The errors will be eliminated through designing a transition synthesis filter bank in the synthesis side. But such method can only get almost perfect reconstruction. However, the synthesis errors could be minimized through properly choosing the optimization approach.In this section we introduce some design methods of the general time-varying filter bank based on out study results.

4.1. Design Formulations

Firstly, we give the general design formulation for the time-varying filter banks. Suppose we want to change the filter bank at time m from an $M(m)$-channel LTI filter bank with the filter length $L(m)$ to an $M(m+1)$-channel LTI filter bank with the filter length $L(m+1)$. According to (4) and (17), the analysis and synthesis matrix in the transition period can be written as

$$\mathbf{T}_{ma} = \begin{bmatrix} & \vdots & \vdots & \vdots & \vdots & \\ \cdots & \mathbf{0}_{M(m-1)} & [\mathbf{h}_0(m)]_{M(m)} & [\mathbf{h}_1(m)]_{M(m)} & [\mathbf{h}_2(m)]_{M(m)} & \cdots \\ \cdots & \cdots & \mathbf{0}_{M(m)} & [\mathbf{h}_0(m+1)]_{M(m+1)} & [\mathbf{h}_1(m+1)]_{M(m+1)} & \cdots \\ \cdots & \cdots & \cdots & \mathbf{0}_{M(m+1)} & [\mathbf{h}_0(m+2)]_{M(m+2)} & \cdots \\ & \vdots & \vdots & \vdots & \vdots & \end{bmatrix} \quad (59)$$

$$\mathbf{T}_{ms} = \begin{bmatrix} & \vdots & \vdots & \vdots & \\ \cdots & \mathbf{0}_{M(m-1)} & \cdots & \cdots & \cdots \\ \cdots & [\mathbf{g}_0(m)]_{M(m)} & \mathbf{0}_{M(m)} & \cdots & \cdots \\ \cdots & [\mathbf{g}_1(m)]_{M(m)} & [\mathbf{g}_0(m+1)]_{M(m+1)} & \mathbf{0}_{M(m+1)} & \cdots \\ \cdots & [\mathbf{g}_2(m)]_{M(m)} & [\mathbf{g}_1(m+1)]_{M(m+1)} & [\mathbf{g}_0(m+2)]_{M(m+2)} & \cdots \\ & \vdots & \vdots & \vdots & \end{bmatrix}, \quad (60)$$

where $\mathbf{h}_i(m)$ and $\mathbf{g}_i(m)$ have the forms

$$\mathbf{h}_i(m) =$$

$$\begin{bmatrix} h_0(L(m)-iM(m)-1,m) & \cdots & h_0(L(m)-iM(m)-M(m),m) \\ h_1(L(m)-iM(m)-1,m) & \cdots & h_1(L(m)-iM(m)-M(m),m) \\ \vdots & \ddots & \vdots \\ h_{M(m)-1}(L(m)-iM(m)-1,m) & \cdots & h_{M(m)-1}(L(m)-iM(m)-M(m),m) \end{bmatrix} \quad (61)$$

$$\mathbf{g}_i(m) = \begin{bmatrix} g_0(iM(m),m) & \cdots & g_{M(m)-1}(iM(m),m) \\ g_0(iM(m)+1,m) & \cdots & g_{M(m)-1}(iM(m)+1,m) \\ \vdots & \ddots & \vdots \\ g_0(iM(m)+M(m)-1,m) & \cdots & g_{M(m)-1}(iM(m)+M(m)-1,m) \end{bmatrix}, \quad (62)$$

and represent filter coefficient matrices of the time-varying analysis and synthesis filter bank, respectively. $M(m)$ and $L(m)$ are the time-varying decimator ratio and filter length. It is clear that a PR time-varying filter bank must satisfy

$$\mathbf{T}_{ms}\mathbf{T}_{ma} = \mathbf{I}, \tag{63}$$

which is also the design criterion. The design process is to choose the filter coefficients in the transition period to satisfy the condition in (63).

4.2. Determination of Filter Coefficients in Transition Period

The design of the time-varying filter banks is equivalent to determination of the time-varying analysis and synthesis filter coefficient $h_k(n,m)$ and $g_k(n,m)$ based on the perfect reconstruction condition in (63). From definitions of $\mathbf{T}_{ma}$ and $\mathbf{T}_{ms}$ in (59) and (60) we know that the matrix size $M(m)$ is different and varies with time m. To describe the matrix multiplication $\mathbf{T}_{ms}\mathbf{T}_{ma}$ mathematically in detail, we need to redefine the sub-matrices in $\mathbf{G}$ and $\mathbf{H}$ with the same size, just as done in (17). Here, we introduce a simple and practical way to determine the matrix elements in $\mathbf{T}_{ma}$ and $\mathbf{T}_{ms}$ which will be involved in calculation of analysis and synthesis filter coefficients $h_k(n,m)$ and $g_k(n,m)$ at time m. An example will be given in section 4.3 to declare how to use the introduced method.

Fig. 8 shows graphically the structure of the synthesis matrix $\mathbf{T}_{ms}$ around the time point m, where the square box represents schematically the matrix $\mathbf{g}_k(m)$ with size $M(m)$. $\mathbf{G}_0^T$, $\mathbf{G}_1^T$ and $\mathbf{G}_2^T$ are matrices that contain the $\mathbf{g}_0(m)$, $\mathbf{g}_1(m)$ and $\mathbf{g}_2(m)$, respectively. Before using the perfect reconstruction condition for calculating $\mathbf{g}_k(m)$ we need to determine the corresponding matrix elements in the analysis matrix $\mathbf{T}_{ma}$. With the aid of graphic matrix boxes the corresponding matrices in $\mathbf{T}_{ma}$that are required to calculate the synthesis coefficients $\mathbf{g}_0(m)$, $\mathbf{g}_1(m)$ and $\mathbf{g}_2(m)$ are depicted in Fig. 9, 10 and 11, respectively.

Based on the PR condition in (63), we have

$$\mathbf{G}_i^T\mathbf{P}_{i,0} = \mathbf{0}, \tag{64}$$

$$\mathbf{G}_i^T\mathbf{P}_{i,1} = \mathbf{I}_i, \tag{65}$$

$$\mathbf{G}_i^T\mathbf{P}_{i,2} = \mathbf{0}, \tag{66}$$

where $i = 0,1,2$, $\mathbf{I}_i$ is a unit matrix whose size is depending on the row size of $\mathbf{G}_i^T$. Combining (64), (65) and (66), the PR condition can be rewritten as

$$\underbrace{\begin{bmatrix} \mathbf{P}_{i,0}^T \\ \mathbf{P}_{i,1}^T \\ \mathbf{P}_{i,2}^T \end{bmatrix}}_{\mathbf{P}_i} \mathbf{G}_i = \underbrace{\begin{bmatrix} \mathbf{0} \\ \mathbf{I}_i \\ \mathbf{0} \end{bmatrix}}_{\mathbf{B}_i}. \tag{67}$$

Assuming that $\mathbf{a}$ and $\mathbf{b}$ are column vectors of matrices $\mathbf{G}_i$ and $\mathbf{B}_i$, we have

$$\mathbf{P}_i\,\mathbf{a} = \mathbf{b}. \tag{68}$$

Then, calculating synthesis filter coefficients becomes solving the linear equations described in (68). Based on the linear algebra theory, there are two approaches to solve the linear equations (68).

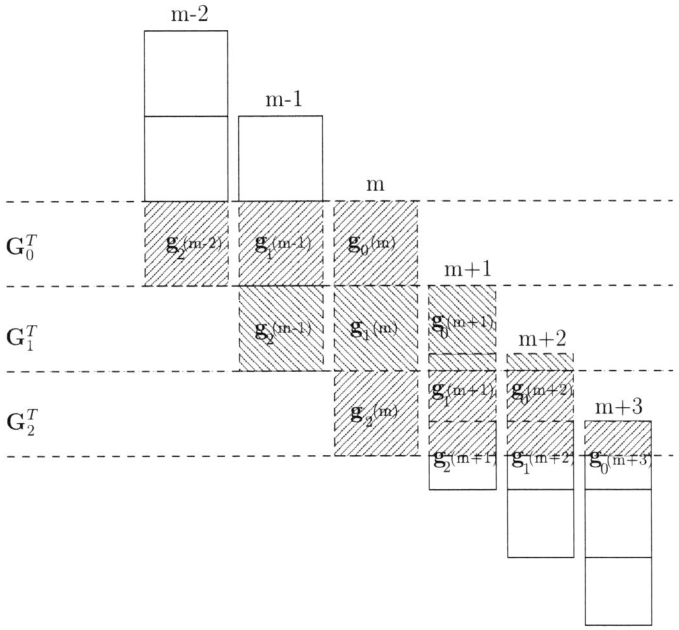

Figure 8. Graphic representation of the synthesis matrix $\mathbf{T}_{ms}$ of a maximally decimated time-varying filter bank around the time point m.

1) Gauss-Jordan Algorithm

Gauss-Jordan algorithm is a process which starts with a given matrix $\mathbf{A}$ and produce a matrix $\mathbf{B}$ in reduced row-echelon form and being row-equivalent to $\mathbf{A}$.

Suppose that $\mathbf{P}_i$ is $m \times n$ $(m \geq n)$, and $\mathbf{b}$ is $m \times 1$. The augmented matrix of the linear equation (68) is defined as

$$\mathbf{P} = [\,\mathbf{P}_i\ \mathbf{b}\,]. \tag{69}$$

Using Gauss-Jordan algorithm we can get a matrix $\mathbf{Q}$ of size $m \times (n+1)$ that is in reduced row-echelon form and row-equivalent to $\mathbf{P}$. Suppose that $\mathbf{Q}$ has r non-zero rows and that the leading entry 1 in the row i occurs in column number c_i, for $1 \leq i \leq r$. Then,

$$1 \leq c_1 < c_2 < \cdots < c_r \leq n+1. \tag{70}$$

Also assume that the remaining column number are $c_{r+1}, \cdots, c_n$, where

$$1 \leq c_{r+1} < c_{r+2} < \cdots < c_n \leq n+1. \tag{71}$$

Then, we have following solutions for the linear system in (68):

Case 1: $c_r = n+1$. The linear system is inconsistent and has no solution.

Case 2: $c_r \leq n$. The linear system has solution.

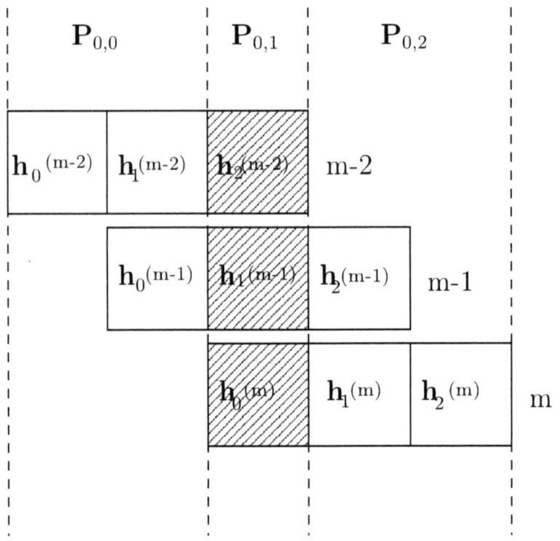

Figure 9. Matrix elements in the analysis matrix $\mathbf{T}_{ma}$ corresponding to $\mathbf{G}_0^T$.

If $r = n$, then $c_1 = 1, c_2 = 2, \cdots, c_n = n$, and

$$\mathbf{Q} = \begin{bmatrix} 1 & 0 & \cdots & 0 & d_1 \\ 0 & 1 & \cdots & 0 & d_2 \\ \vdots & \vdots & \vdots & \vdots & \vdots \\ 0 & 0 & \cdots & 1 & d_n \\ 0 & 0 & \cdots & 0 & 0 \\ \vdots & \vdots & \vdots & \vdots & \vdots \\ 0 & 0 & \cdots & 0 & 0 \end{bmatrix}. \tag{72}$$

There is a unique solution $a_1 = d_1, a_2 = d_2, \cdots, a_n = d_n$.

If $r < n$, there will be more than one solution. All solutions are obtained by taking the unknowns $a_{c_1}, a_{c_2}, \cdots, a_{c_r}$ as dependent unknowns and using the r equations corresponding to the non-zero rows of $\mathbf{Q}$ to express these unknowns in terms of the remaining independent unknowns $a_{c_{r+1}}, a_{c_{r+2}}, \cdots, a_{c_n}$, which can take an arbitrary values.

In general, we can say that the linear equations in (68) are solvable if and only if the rank of augmented matrix $\mathbf{P}$ is equal to the rank of $\mathbf{P}_i$, or the vector $\mathbf{b}$ is in the column space of $\mathbf{P}_i$.

2) Least Squares (LS) Solution

If the linear equations in (68) are consistent, we can always use the Gauss-Jordan algorithm to calculate the synthesis filters in the transition period which satisfy the biorthogonal condition in (63). But, linear equations in (68) are usually inconsistent and have no exact solutions. To make linear equations in (68) solvable the matrix $\mathbf{P}_i$ must be modified

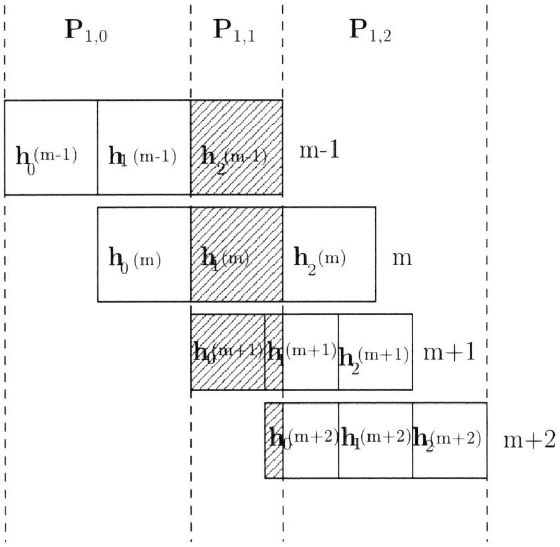

Figure 10. Matrix elements in the analysis matrix $\mathbf{T}_{ma}$ corresponding to $\mathbf{G}_1^T$.

by designing time-varying filter banks, which means that the analysis filter coefficients in $\mathbf{P}_i$ will be changed. Such modification of analysis filter coefficients destroys not only the frequency characteristics of analysis filter at time m, but also the PR properties at before and after time m. Generally, it is difficult to keep both PR and frequency characteristics by time-varying filter bank design. Keeping PR conditions will usually lose frequency properties of analysis filter banks, and vice versa. Under loss of PR condition there is a optimized solution for linear equations in (68) being called least squares (LS) solution and without changing the frequency properties of time-varying analysis filter banks. The least squares solution of equation (68) is the solution of the associated normal equation defined by

$$\mathbf{P}_i^T\ \mathbf{P}_i\ \mathbf{a} = \mathbf{P}_i^T\ \mathbf{b}. \tag{73}$$

It is well-known that the solution of equations in (73) is equal to

$$\mathbf{a} = \left[\mathbf{P}_i^T \mathbf{P}_i\right]^{-1}\ \mathbf{P}_i^T\ \mathbf{b}. \tag{74}$$

More generally, assuming that $\mathbf{P}_i$ has rank $k = n$ or $k = m$, we define

$$\begin{aligned} \mathbf{P}_i^+ &= \left[\mathbf{P}_i^T \mathbf{P}_i\right]^{-1}\ \mathbf{P}_i^T, \ \ k = n & (75)\\ \mathbf{P}_i^+ &= \mathbf{P}_i^T\ \left[\mathbf{P}_i \mathbf{P}_i^T\right]^{-1}, \ \ k = m & (76)\\ \mathbf{P}_i^+ &= \mathbf{P}_i^{-1}, \ \ k = n = m, & (77) \end{aligned}$$

where the matrix $\mathbf{P}_i^+$ is called the Moore-Penrose pseudoinverse.

With $\mathbf{P}_i^+$ the solution in (74) can be rewritten as

$$\mathbf{a} = \mathbf{P}_i^+\ \mathbf{b}. \tag{78}$$

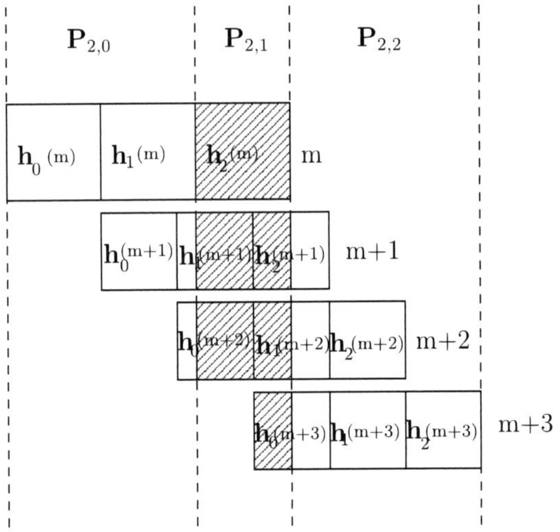

Figure 11. Matrix elements in the analysis matrix $\mathbf{T}_{ma}$ corresponding to $\mathbf{G}_2^T$.

Suppose that $\widehat{\mathbf{b}}$ represents the orthogonal project of $\mathbf{b}$ into the column space of $\mathbf{P}_i$. Then, we have

$$\widehat{\mathbf{b}} = \mathbf{P}_i \, \mathbf{P}_i^+ \, \mathbf{b}. \tag{79}$$

It is clear that

$$\| \, \mathbf{b} - \widehat{\mathbf{b}} \, \| > 0, \tag{80}$$

which means that synthesis filters obtained with LS solution do not satisfy the perfect reconstruction conditions. Therefore, the built basis functions based on LS solution are not real dual basis functions. However, because of no constraint of matrix $\mathbf{P}_i$, i.e. no limitation of time-varying analysis filter banks, and small reconstruction errors in comparison with direct filter bank switching, LS solution is very useful in the design of time-varying filter banks in the time-domain [3-4].

Finally we give the synthesis filter coefficient matrix with LS solution as

$$\mathbf{G}_i = \mathbf{P}_i^+ \, \mathbf{B}_i, \quad i = 0, 1, 2. \tag{81}$$

In the following we give an example to show how to calculate the synthesis filter coefficients in the transition period using the presented algorithm.

4.3. Example

We suppose that a time-varying filter bank uses a 3-channel perfectly reconstructed filter bank with filter length $L(m) = 6$ until time m. From time point m, the time-varying filter bank uses a 2-channel PR filter bank with filter length $L(m+1) = 8$. The analysis matrix around time m can be described as

$\mathbf{T}_{ma} =$

$$\begin{bmatrix} & \vdots & \vdots & \vdots & \vdots & \vdots & \vdots & \\ \cdots & h_0(5,m) & h_0(4,m) & h_0(3,m) & h_0(2,m) & h_0(1,m) & h_0(0,m) & \cdots \\ \cdots & h_1(5,m) & h_1(4,m) & h_1(3,m) & h_1(2,m) & h_1(1,m) & h_1(0,m) & \cdots \\ \cdots & h_2(5,m) & h_2(4,m) & h_2(3,m) & h_2(2,m) & h_2(1,m) & h_2(0,m) & \cdots \\ \cdots & 0 & 0 & 0 & h_0(3,m+1) & h_0(2,m+1) & h_0(1,m+1) & \cdots \\ \cdots & 0 & 0 & 0 & h_1(3,m+1) & h_1(2,m+1) & h_1(1,m+1) & \cdots \\ \cdots & 0 & 0 & 0 & 0 & 0 & h_0(3,m+2) & \cdots \\ \cdots & 0 & 0 & 0 & 0 & 0 & h_1(3,m+2) & \cdots \\ & \vdots & \vdots & \vdots & \vdots & \vdots & \vdots & \end{bmatrix}. \quad (82)$$

The corresponding synthesis matrix $\mathbf{T}_{ms}$ has the form

$\mathbf{T}_{ms} =$

$$\begin{bmatrix} & \vdots & \vdots & \vdots & \vdots & \vdots & \vdots & & \\ \cdots & g_0(0,m) & g_1(0,m) & g_2(0,m) & 0 & 0 & 0 & 0 & \cdots \\ \cdots & g_0(1,m) & g_1(1,m) & g_2(1,m) & 0 & 0 & 0 & 0 & \cdots \\ \cdots & g_0(2,m) & g_1(2,m) & g_2(2,m) & 0 & 0 & 0 & 0 & \cdots \\ \cdots & g_0(3,m) & g_1(3,m) & g_2(3,m) & g_0(0,m+1) & g_1(0,m+1) & 0 & 0 & \cdots \\ \cdots & g_0(4,m) & g_1(4,m) & g_2(4,m) & g_0(1,m+1) & g_1(1,m+1) & 0 & 0 & \cdots \\ \cdots & g_0(5,m) & g_1(5,m) & g_2(5,m) & g_0(2,m+1) & g_1(2,m+1) & g_0(0,m+2) & g_1(0,m+2) & \cdots \\ \cdots & 0 & 0 & 0 & g_0(3,m+1) & g_1(3,m+1) & g_0(1,m+2) & g_1(1,m+2) & \cdots \\ \cdots & \vdots & \vdots & \vdots & \vdots & \vdots & \vdots & \cdots & \\ \cdots & 0 & 0 & 0 & g_0(7,m+1) & g_1(7,m+1) & g_0(5,m+2) & g_1(5,m+2) & \cdots \\ \cdots & 0 & 0 & 0 & 0 & 0 & g_0(6,m+2) & g_1(6,m+2) & \cdots \\ \cdots & 0 & 0 & 0 & 0 & 0 & g_0(7,m+2) & g_1(7,m+2) & \cdots \\ & \vdots & \vdots & \vdots & \vdots & \vdots & \vdots & \vdots & \end{bmatrix}. \quad (83)$$

As an example we show how to calculate the synthesis coefficients at time m. Using the method of matrix classification depicted in Fig. 8, we have following matrices based on $\mathbf{T}_{ms}$

$$\mathbf{g}_0(m) = \begin{bmatrix} g_0(0,m) & g_1(0,m) & g_2(0,m) \\ g_0(1,m) & g_1(1,m) & g_2(1,m) \\ g_0(2,m) & g_1(2,m) & g_2(2,m) \end{bmatrix}, \quad (84)$$

$$\mathbf{g}_1(m) = \begin{bmatrix} g_0(3,m) & g_1(3,m) & g_2(3,m) \\ g_0(4,m) & g_1(4,m) & g_2(4,m) \\ g_0(6,m) & g_1(5,m) & g_2(5,m) \end{bmatrix}, \quad (85)$$

$\mathbf{G}_1^T =$

$$\begin{bmatrix} g_0(3,m) & g_1(3,m) & g_2(3,m) & g_0(0,m+1) & g_1(0,m+1) & 0 & 0 \\ g_0(4,m) & g_1(4,m) & g_2(4,m) & g_0(1,m+1) & g_1(1,m+1) & 0 & 0 \\ g_0(5,m) & g_1(5,m) & g_2(5,m) & g_0(2,m+1) & g_1(2,m+1) & g_0(0,m+2) & g_1(0,m+2) \end{bmatrix}. (86)$$

Because the 3-band filter bank is perfect reconstructed, the coefficient matrix $\mathbf{g}_0(m)$ does not need to be recalculated and can be transfered directly from 3-band filter bank. Just the coefficient matrix $\mathbf{g}_1(m)$ needs to be calculated. From Fig. 10 and $\mathbf{T}_{ma}$ in (82) we have

$$\mathbf{P}_{1,0} = \begin{bmatrix} h_0(5,m) & h_0(4,m) & h_0(3,m) \\ h_1(5,m) & h_1(4,m) & h_1(3,m) \\ h_2(5,m) & h_2(4,m) & h_2(3,m) \\ 0 & 0 & 0 \\ 0 & 0 & 0 \\ 0 & 0 & 0 \\ 0 & 0 & 0 \end{bmatrix}, \tag{87}$$

$$\mathbf{P}_{1,1} = \begin{bmatrix} h_0(2,m) & h_0(1,m) & h_0(0,m) \\ h_1(2,m) & h_1(1,m) & h_1(0,m) \\ h_2(2,m) & h_2(1,m) & h_2(0,m) \\ h_0(7,m+1) & h_0(6,m+1) & h_0(5,m+1) \\ h_1(7,m+1) & h_1(6,m+1) & h_1(5,m+1) \\ 0 & 0 & h_0(7,m+2) \\ 0 & 0 & h_1(7,m+2) \end{bmatrix}, \tag{88}$$

$$\mathbf{P}_{1,2} = \begin{bmatrix} 0 & 0 & 0 & 0 & 0 & 0 & 0 \\ 0 & 0 & 0 & 0 & 0 & 0 & 0 \\ 0 & 0 & 0 & 0 & 0 & 0 & 0 \\ h_0(4,m+1) & h_0(3,m+1) & h_0(2,m+1) & h_0(1,m+1) & h_0(0,m+1) & 0 & 0 \\ h_1(4,m+1) & h_1(3,m+1) & h_1(2,m+1) & h_1(1,m+1) & h_1(0,m+1) & 0 & 0 \\ h_0(6,m+2) & h_0(5,m+2) & h_0(4,m+2) & h_0(3,m+2) & h_0(2,m+2) & h_0(1,m+2) & h_0(0,m+2) \\ h_1(6,m+2) & h_1(5,m+2) & h_1(4,m+2) & h_1(3,m+2) & h_1(2,m+2) & h_1(1,m+2) & h_1(0,m+2) \end{bmatrix}. \tag{89}$$

According to (67) the linear equations are described by

$$\mathbf{P}_1\mathbf{G}_1 = \mathbf{B}_1, \tag{90}$$

where

$$\mathbf{P}_1^T = [\,\mathbf{P}_{1,0}\;\mathbf{P}_{1,1}\;\mathbf{P}_{1,2}\,], \tag{91}$$

$$\mathbf{B}_1 = [\,\mathbf{0}_{3\times3}\;\mathbf{I}_3\;\mathbf{0}_{3\times7}\,]^T, \tag{92}$$

and $\mathbf{0}_{3\times3}$ and $\mathbf{0}_{3\times7}$ represent zero matrices of the size 3×3 and 3×7, respectively, and $\mathbf{I}_3$ is a unit matrix with size 3. According to (81) the LS solution for this example is

$$\mathbf{G}_1 = \left[\mathbf{P}_1^T\mathbf{P}_1\right]^{-1}\mathbf{P}_1^T\mathbf{B}_1. \tag{93}$$

Table 1 and Table 2 list filter coefficients for the 3-band and 2-band filter banks that are used in the example, respectively. The calculated synthesis filter coefficients are given in Table 3. Because the linear equations (90) of this example are not exactly solvable, we have just LS solution.

Table 1. Coefficients of 3-band filter bank

$h_0(n,m)$	$h_1(n,m)$	$h_2(n,m)$	$g_0(n,m)$	$g_1(n,m)$	$g_2(n,m)$
-1.8301e-001	0.e+00	1.8301e-001	1.0566e-001	-2.1132e-001	1.0566e-001
-5.7735e-001	-5.7735e-001	-5.7735e-001	0.e+00	0.e+00	0.e+00
-6.8301e-001	9.4195e-016	6.8301e-001	-3.9434e-001	7.8868e-001	-3.9434e-001
-3.9434e-001	7.8868e-001	-3.9434e-001	-6.8301e-001	9.4195e-016	6.8301e-001
0.e+00	0.e+00	0.e+00	-5.7735e-001	-5.7735e-001	-5.7735e-001
1.0566e-001	-2.1132e-001	1.0566e-001	-1.8301e-001	0.e+00	1.8301e-001

Table 2. Coefficients of 2-band filter bank

$h_0(n,m)$	$h_1(n,m)$	$g_0(n,m)$	$g_1(n,m)$
1.0014e-001	4.1478e-002	4.1478e-002	-1.0014e-001
-1.4986e-001	-6.2076e-002	6.2076e-002	-1.4986e-001
-2.0852e-001	5.0342e-001	5.0342e-001	2.0852e-001
3.1208e-001	-7.5342e-001	7.5342e-001	3.1208e-001
7.5342e-001	3.1208e-001	3.1208e-001	-7.5342e-001
5.0342e-001	2.0852e-001	-2.0852e-001	5.0342e-001
6.2076e-002	-1.4986e-001	-1.4986e-001	-6.2076e-002
4.1478e-002	-1.0014e-001	1.0014e-001	4.1478e-002

Table 3. Synthesis filter coefficients in the transition period at time m with LS solution

$g_0(n,m)$	$g_1(n,m)$	$g_2(n,m)$
1.0566e-001	-2.1132e-001	1.0566e-001
0.e+00	0.e+00	0.e+00
-3.9434e-001	7.8868e-001	-3.9434e-001
-6.9653e-001	4.9872e-004	6.9753e-001
-5.7735e-001	-5.7735e-001	-5.7735e-001
-1.3256e-001	-1.8612e-003	1.2884e-001

4.4. Prefiltering and Postfiltering

Suppose that the transition synthesis filters have been designed. Before realization we should assign the designed filter coefficients. As the above example shows, the synthesis-filter coefficients on the synthesis side must be replaced by the correspondingly designed transition-filter coefficients. After doing so, we obtain the transition synthesis filters, whose coefficients come from both the original synthesis-filter bank and the designed transition filters. In the example we have three transition synthesis filters. Using commutators we have the implementation given in Fig. 12 for switching between two LTI filter banks.

In Fig. 12 the first analysis filter bank is directly switched to the second analysis filter bank at time m. On the synthesis side, the first synthesis filter bank is switched to the second synthesis filter bank after the transition synthesis filters.

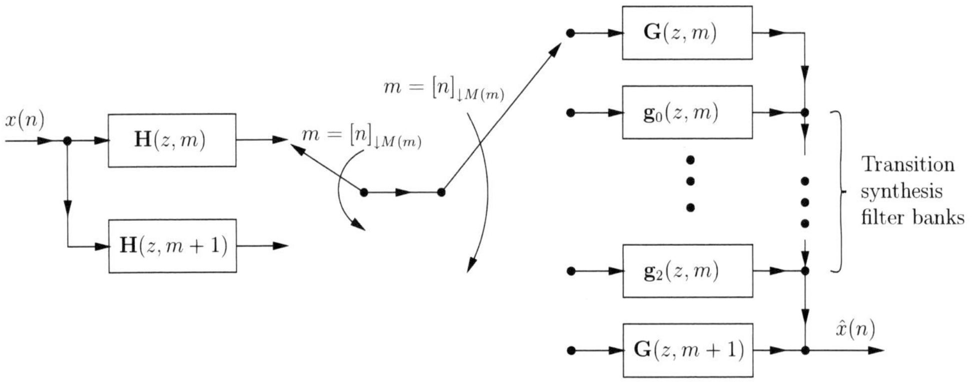

Figure 12. Implementation of the switch between two LTI filter banks based on commucators.

Besides the implementation in Fig. 12 we have another possibility to realize the transition synthesis filter banks, namely, prefiltering and postfiltering. The basic idea of the prefiltering is to move the transition synthesis filters at the front of the switch device. The postfiltering is used to process the reconstructed signal after the switching.

We know that the PR condition for the time-varying filter bank is $\mathbf{T}_{ms}\mathbf{T}_{ma} = \mathbf{I}$. Considering the switch between two PR LTI filter banks, the elements in $\mathbf{T}_{ms}$ and $\mathbf{T}_{ma}$ come from the first and the second PR filter banks besides the entries in the transition period. The usual method is to design transition synthesis filter banks to keep $\mathbf{T}_{ms}\mathbf{T}_{ma} = \mathbf{I}$ during the transition period, as we described above. Alternatively, we can insert a matrix $\mathbf{T}_R$ between $\mathbf{T}_{ms}$ and $\mathbf{T}_{ma}$, so that $\mathbf{T}_{ms}\mathbf{T}_R\mathbf{T}_{ma} = \mathbf{I}$, where $\mathbf{T}_R$ is a block-diagonal matrix in which all blocks are identity matrices besides the block in the transition period, which leads to

$$\mathbf{T}_R = \begin{bmatrix} \ddots & & & & \\ & \mathbf{I} & & & \\ & & \mathbf{I} & & \\ & & & \mathbf{R} & & \\ & & & & \mathbf{I} & \\ & & & & & \ddots \end{bmatrix} \tag{94}$$

where $\mathbf{I}$ is the identity matrix and $\mathbf{R}$ is the matrix in the transition period.

Then, we can write the PR condition in the transition period as

$$\mathbf{G}_s^T \mathbf{R} \mathbf{P}_s^T = \mathbf{B}^T \tag{95}$$

where matrices $\mathbf{G}_s^T$ and $\mathbf{P}_s^T$ contain the synthesis and analysis filter coefficients in the transition period. For the example in section 4.3, we have

$$\mathbf{G}_s^T = \mathbf{G}_1^T(m) \tag{96}$$

$$\mathbf{P}_s^T = [\, \mathbf{P}_{1,0}\, \mathbf{P}_{1,1}\, \mathbf{P}_{1,2} \,]. \tag{97}$$

$\mathbf{B}^T$ has similar definition in (67). $\mathbf{R}$ is called prefilter transform matrix, whose rows consist of prefilter coefficients. From (95), we obtain the prefilter as

$$\mathbf{R} = \mathbf{G}_s(\mathbf{G}_s^T\mathbf{G}_s)^{-1}\mathbf{B}^T(\mathbf{P}_s\mathbf{P}_s^T)^{-1}\mathbf{P}_s. \tag{98}$$

Using the prefilter, the transition synthesis filters in Fig. 12 can be removed. In this case, the first and second synthesis filter bank are allowed to be directly switched. Fig. 13 illustrates the position of the prefilter.

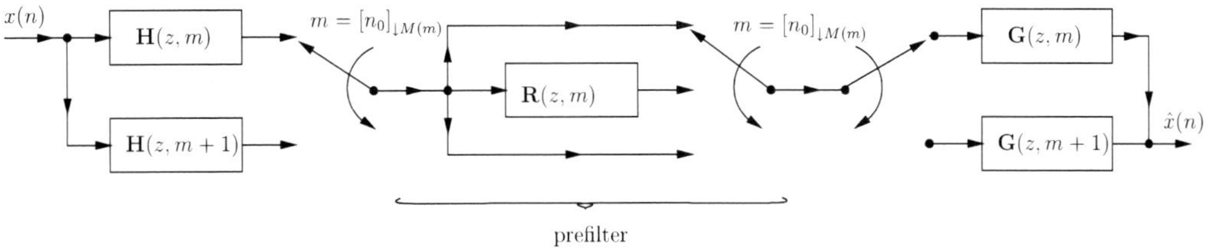

Figure 13. Implementation of switching between two LTI filter banks with the prefilter based on commucators.

Similarly, we have a postfilter implementation. To get the postfilter, we need to write the input-output relation in the transition period as. For this purpose, we define the input and output vectors during the transition as

$$\hat{\mathbf{x}}(m) = \mathbf{A}\mathbf{x}(m) \tag{99}$$

where $\mathbf{x}(m)$ and $\hat{\mathbf{x}}(m)$ are the input and output vectors in the transition period. For the example in section 4.3, $\mathbf{A}$ can be written by

$$\mathbf{A} = \begin{bmatrix} \mathbf{I} & \mathbf{0} & \mathbf{0} \\ \mathbf{A}_0 & \mathbf{A}_T & \mathbf{A}_1 \\ \mathbf{0} & \mathbf{0} & \mathbf{I} \end{bmatrix} \tag{100}$$

where

$$\mathbf{A}_0 = \mathbf{G}_1^T\mathbf{P}_{1,0} \tag{101}$$

$$\mathbf{A}_T = \mathbf{G}_1^T\mathbf{P}_{1,1} \tag{102}$$

$$\mathbf{A}_1 = \mathbf{G}_1^T\mathbf{P}_{1,2}. \tag{103}$$

In order to obtain the input $\mathbf{x}(m)$ from the reconstructed output $\hat{\mathbf{x}}(m)$, we invert (99), so that

$$\mathbf{x}(m) = \mathbf{A}^{-1}\hat{\mathbf{x}}(m) \tag{104}$$

where

$$\mathbf{A}^{-1} = \begin{bmatrix} \mathbf{I} & \mathbf{0} & \mathbf{0} \\ -\mathbf{A}_T^{-1}\mathbf{A}_0 & \mathbf{A}_T^{-1} & -\mathbf{A}_T^{-1}\mathbf{A}_1 \\ \mathbf{0} & \mathbf{0} & \mathbf{I} \end{bmatrix} \tag{105}$$

and $\mathbf{A}_T$ is a square matrix of the length $M(m)$. According to (105), we define the postfilter coefficient matrix as

$$\mathbf{F} = \left[\begin{array}{ccc} -\mathbf{A}_T^{-1}\mathbf{A}_0 & \mathbf{A}_T^{-1} & -\mathbf{A}_T^{-1}\mathbf{A}_1 \end{array} \right]. \tag{106}$$

Each row of $\mathbf{F}$ contains the coefficients from one postfilter.

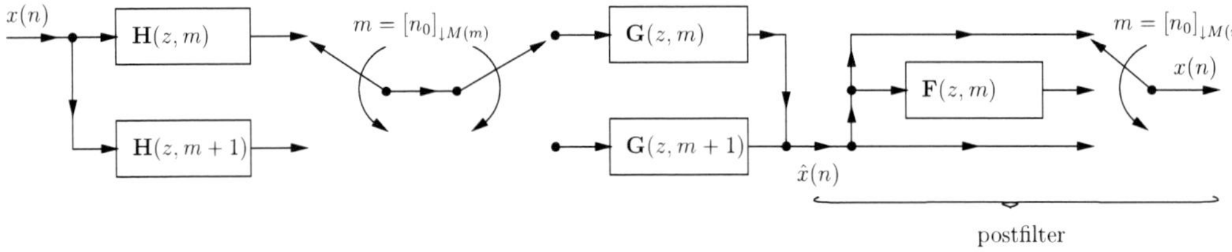

Figure 14. Switching between two PR LTI filter banks and postfilters based on commucators.

Now, we summarize the design procedures used for time-varying filter banks. For the direct realization scheme shown in Fig. 12, we first calculate the filter coefficients according to the design formulation given in (81); then, we replace the synthesis-filter coefficients in the transition period shown as shaded part in Fig. 8 with the designed filter coefficients. Together with the original synthesis-filter coefficients, we finally obtain the corresponding transition synthesis filter banks. For the pre- and postfilter implementation schemes, we only need to calculate the pre- and postfilter matrices $\mathbf{R}$ and $\mathbf{F}$ according to (98), (106).

4.5. Design of Time-Varying Cosine-Modulated Filter Banks

The cosine-modulated filter bank is an important class of filter banks. Due to its easy design and effective realization properties, cosine-modulated filter banks are widely used in practice [1,2]. In this section we consider only paraunitary time-varying cosine-modulated filter banks. Generally, the basis functions (the column vectors in the synthesis matrix $\mathbf{T}_{ms}$) of the cosine-modulated time-varying filter bank are defined by

$$p_k(n,m) = h(n,m)cos\left[\left(k+\frac{1}{2}\right)\left(n-\frac{L(m)-1}{2}\right)\frac{\pi}{M(m)} + \phi_k(m)\right] \tag{107}$$

with $k = 0, 1, \cdots, M-1$, and $n = 0, 1, \cdots, L(m)-1$, where m is the block index. In (107), there are four parameters which can vary with the block index m, namely, the prototype window $h(n,m)$, the filter length $L(m)$, the channel number $M(m)$, and the phase parameter $\phi_k(m)$ which controls the relative phases of the modulated cosines. By defining a time-varying cosine-modulated filter bank we have four parameters to be chosen in order to maintain the PR, whereas in time-invariant cosine-modulated filter banks the only design freedom is the choice of the prototype window $h(n)$. Therefore, the design of time-varying filter banks has more degrees of freedom than that of time-invariant filter banks. Generally speaking, giving up the constraint of shift-invariance for LTI system brings more design freedom.

In this section, we study only the behavior of $h(n,m)$, assuming that $M(m)$ and $\phi_k(m)$ do not change with m. Furthermore, $\phi_k(m)$ is chosen as

$$\phi_k(m) = \left(k + \frac{1}{2}\right)(N+1)\frac{\pi}{2} \tag{108}$$

where $N = 2K$ (K is an integer), and $L = NM = 2KM$. Substituting (108) into (107) and introducing a normalization factor $\sqrt{\frac{2}{M}}$, we have

$$p_k(n,m) = h(n,m)\sqrt{\frac{2}{M}}\, cos\left[\left(k+\frac{1}{2}\right)\left(n+\frac{M+1}{2}\right)\frac{\pi}{M}\right] \tag{109}$$

where the channel number M is a constant.

In the following we will show how to choose the prototype window $h(n,m)$ in order to satisfy the PR condition for the paraunitary time-varying filter bank.

The analysis process is similar to that one presented in [1] for the time-invariant cosine-modulated filter bank. For a convenient comparison with the time-invariant case, we use similar notations. Firstly, we define matrices

$$\begin{aligned} \mathbf{\Phi} &= \left[\ \mathbf{\Phi}_0^T \ \ \mathbf{\Phi}_1^T \ \ \cdots \ \ \mathbf{\Phi}_{N-1}^T\ \right]^T & (110)\\ \mathbf{h}(m) &= diag\{\ \mathbf{h}_0(m) \ \ \mathbf{h}_1(m) \ \ \cdots \ \ \mathbf{h}_{N-1}(m)\ \}, & (111)\end{aligned}$$

where

$$\begin{aligned} [\mathbf{\Phi}_i]_{nk} &= \sqrt{\frac{2}{M}}\, cos\left[\left(k+\frac{1}{2}\right)\left(n+iM+\frac{M+1}{2}\right)\frac{\pi}{M}\right] & (112)\\ \mathbf{h}_i(m) &= diag\{\ h(iM,m) \ \ h(iM+1,m) \ \ \cdots \ \ h(iM+M-1,m)\ \} & (113)\end{aligned}$$

with $i = 0, 1, \cdots, N-1$, $n = 0, 1, \cdots, M-1$, and $k = 0, 1, \cdots, M-1$.

Using (110) and (111), the synthesis matrix $\mathbf{P}^T(m)$ can be written as

$$\begin{aligned} \mathbf{P}^T(m) &= \mathbf{h}(m)\mathbf{\Phi} & (114)\\ \mathbf{P}_i^T(m) &= \mathbf{h}_i(m)\mathbf{\Phi}_i, \quad i = 0, 1, \cdots, N-1. & (115)\end{aligned}$$

For a paraunitary time-varying cosine-modulated filter banks the analysis and synthesis matrices in (59) and (60) are associated with $\mathbf{P}(m)$ by

$$\begin{aligned} \mathbf{h}_i(m) &= \mathbf{P}_i(m) & (116)\\ \mathbf{g}_i(m) &= \mathbf{P}_i^T(m), & (117)\end{aligned}$$

For $\mathbf{\Phi}_i$ there are the following properties [1]:

$$\begin{aligned} \mathbf{\Phi}_i\mathbf{\Phi}_{i+2s}^T &= (-1)^s[\,\mathbf{I} + (-1)^i\mathbf{J}\,] & (118)\\ \mathbf{\Phi}_i\mathbf{\Phi}_{i+2s+1}^T &= 0. & (119)\end{aligned}$$

Using equations (118) to (119), we have following expressions based on (63)

$$\sum_{i=0}^{N-1-l} \mathbf{P}_i^T(m+i)\mathbf{P}_{i+l}(m+i) = \sum_{i=0}^{2K-1-l} \mathbf{h}_i(m+i)\mathbf{\Phi}_i\mathbf{\Phi}_{i+l}^T\mathbf{h}_{i+l}(m+i)$$

$$= \begin{cases} \sum_{i=0}^{2(K-s)-1} (-1)^s \mathbf{h}_i(m+i)[\,\mathbf{I} + (-1)^i \mathbf{J}\,]\mathbf{h}_{i+2s}(m+i), & if\ l = 2s \\ 0, & otherwise \end{cases} \quad (120)$$

and

$$\sum_{i=0}^{N-1-l} \mathbf{P}_{i+l}^T(m+i+l)\mathbf{P}_i(m+i+l)$$
$$= \sum_{i=0}^{2K-1-l} \mathbf{h}_{i+l}(m+i+l)\mathbf{\Phi}_{i+l}\mathbf{\Phi}_i^T \mathbf{h}_i(m+i+l)$$
$$= \begin{cases} \sum_{i=0}^{2(K-s)-1} (-1)^s \mathbf{h}_{i+2s}(m+i+2s)[\,\mathbf{I} + (-1)^i \mathbf{J}\,]\mathbf{h}_i(m+i+2s), if\ l = 2s \\ 0, \quad otherwise \end{cases} \quad (121)$$

where $s = 0, 1, \cdots, K-1$ and $N = 2K$.

The key for a further simplification of equations (120) and (121) is the assumption of symmetrical window $h(n, m)$. In the time-invariant case, the symmetry of the prototype window is easy to understand, but in the time-varying case, the symmetry of $h(n, m)$ becomes complicated, because it contains different pieces of the $M \times 1$ window vector at the different blocks m. For $K = 1$, the symmetry of $h(n, m)$ can be simply written as

$$h(2M - n, m+1) = h(n, m), \quad 0 \leq n \leq M-1$$

which represents the symmetry of the second piece of the window in block $m-1$ with the first piece of the window in block m. The symmetry appears in the column direction of the analysis matrix for time-varying filter banks.

Under the assumption of the symmetry of $h(n, m)$, equations (120) and (121) are simplified as

$$\sum_{i=0}^{2(K-s)-1} \mathbf{h}_i(m+i)\mathbf{h}_{i+2s}(m+i) = \delta(s)\mathbf{I} \quad (122)$$

$$\sum_{i=0}^{2(K-s)-1} \mathbf{h}_{i+2s}(m+i+2s)\mathbf{h}_i(m+i+2s) = \delta(s)\mathbf{I} \quad (123)$$

where $s = 0, 1, \cdots, K-1$.

Equations (122) and (123) are the general PR conditions for M-channel time-varying modulated lapped transforms. If the block index m is removed in (122) and (123), the formulations reduce to the LTI MLTs case. So, we say that the results for time-invariant MLTs are a special case of time-varying MLTs.

According to (122) and (123), we formulate the following theorem:

Theorem 2. *A time-varying cosine-modulated filter bank, which has the constant filter length L and channel number M, but symmetric time-varying prototype window* $h(n, m)$, *has perfect reconstruction if and only if equations* (122) *and* (123) *are satisfied.*

This theorem suggests that it may be possible to maintain perfect reconstruction for the time-varying cosine-modulated filter bank by appropriately changing the prototype window. In particular, if we have a closed form prototype window, the design problem becomes that of a direct construction of the window instead of a numerical design.

An example is the modulated lapped transform (MLT). With $K = 1$, the PR condition is

$$h^2(n,m) + h^2(n+M, m+1) = 1, \quad 0 \leq n \leq M-1. \tag{124}$$

This is an important property for the time-varying MLT. Under the constraint (124), we can arbitrarily change the prototype window, while retaining the PR. For example, between two standard sine windows for the MLT given by Malvar [1], we can insert another window without losing the PR property. Fig. 15 schematically shows a possible window switching.

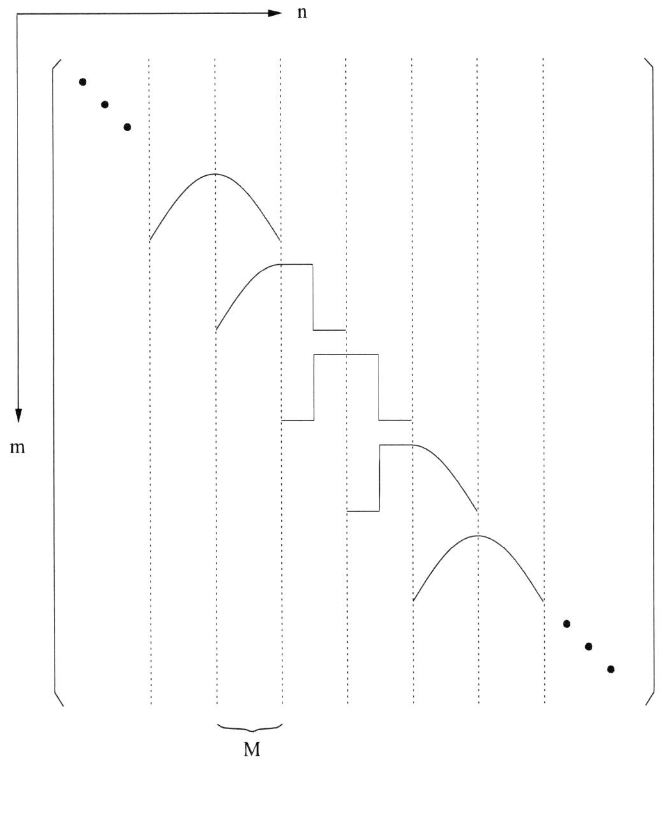

Figure 15. Window-matrix for the modulated lapped transform.

For the sake of an easy description we call the diagram showed in Fig. 15 as a *window-matrix*, though it does not represent any mathematic matrix. It is obtained by placing the corresponding window diagram at the matrix position in the analysis matrix $\mathbf{T}_{ma}$. For example, in order to build the window-matrix from the analysis matrix (59) in the last section, we need just to replace the mathematic matrix $\mathbf{P}_0$, $\mathbf{P}_1$ and $\mathbf{A}$ with their prototype windows, where $\mathbf{P}_0$ corresponds to the first piece of the sine prototype window, $\mathbf{P}_1$ corresponds to the second piece of the sine prototype window and $\mathbf{A}$ has a rectangular prototype window.

The above results are suitable for the cosine-modulated filter bank with constant values for L and M. However, in the case of varying L and M we can also switch the window such that PR is retained. For detailed discussion or window switching techniques please reference [9,11].

5. Time-Varying Discrete-Time Wavelet Transforms

From the point of digital signal processing of view, the time-varying discrete-time wavelet transform can be implemented by a time-varying octave-band filter bank with tree structure. Fig. 16 shows the most general time-varying discrete-time wavelet transform implemented with a time-varying octave-band filter bank, where the lowpass and highpass filter $H_l(z, m)$, $H_u(z, m)$, the stage number of the split-merge $J(m)$, all are varying with time index m. In other words, both the frequency characteristic and the time-frequency tiling of the discrete-time wavelet transform are varying with time. Fig. 17 shows the equivalent time-varying nonuniform filter bank implementation. The analysis of the time-varying discrete-time wavelet transform is equal to the analysis of the time-varying discrete-time octave-band filter bank.

Note that we define the time-varying discrete-time wavelet transform varying with index m which is equivalent to the output index at the last stage of octave-band filter banks. The time indices of the other output are related with m by

$$m_j = 2^{J(m)-i-j} \cdot m, \quad 0 \le j \le J(m) - 2. \tag{125}$$

In the literature there are some papers related with this topic by studying changes between two time-invariant filter banks [10,11,12]. In particular, in [10] the authors have discussed the transition behavior during the change between two time-invariant discrete-time wavelet transforms.Different from the existed publications, in this paper we analyze the general time-varying discrete-time wavelet transform in detail based on the octave-band filter bank and the nonuniform filter bank implementation.

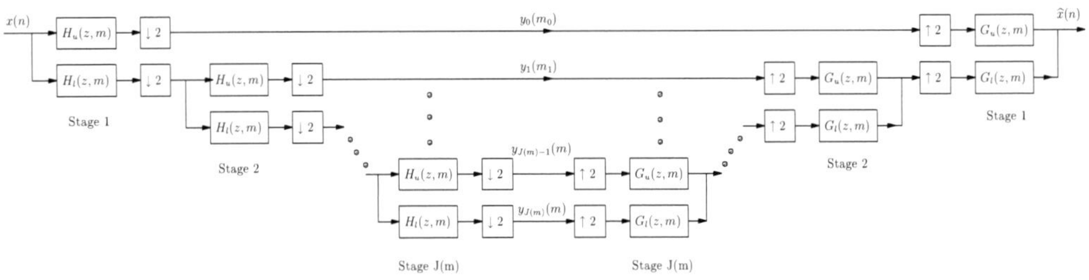

Figure 16. Time-varying discrete-time wavelet transform implemented with time-varying octave-band filter banks.

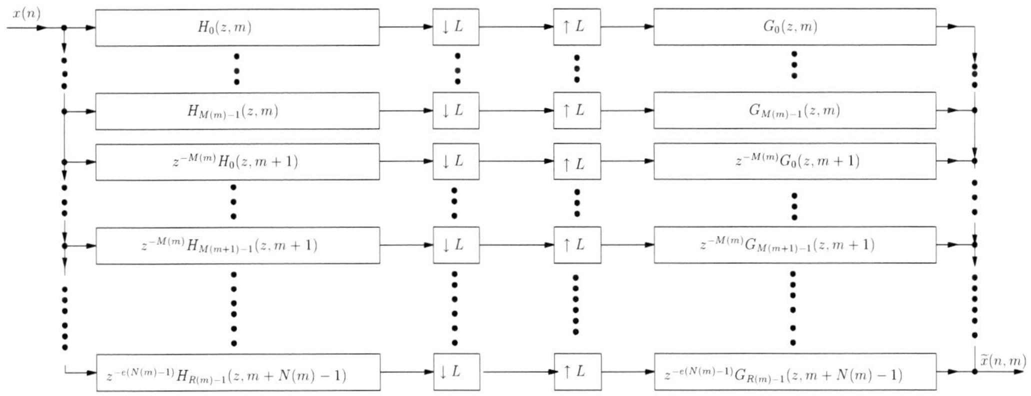

Figure 17. The equivalent time-varying nonuniform filter bank implementation, where $M(m) = 2^{J(m)}$.

5.1. Implementation with Time-Varying Octave-Band Filter Banks

To make the analysis simple, in the following analysis we suppose that the stage number J does change with time and is a constant. Then we get a J-stage time-varying octave-band filter bank. Just as depicted in Fig. 16, a J-stage octave-band time-varying filter bank consists of J stages of two-channel time-varying filter bank. In the analysis side, the input signal $x(n)$ is first split by the two-channel time-varying filter bank at the first stage, then the lowpass output is spilt again by the same two-band time-varying filter bank at the second stage. The process is ongoing until J-stage. In the synthesis side, the signal is merged to generate the reconstructed signal $\widehat{x}(n)$. From the theorem of time-invariant discrete-time wavelet transform [2], we know that if the individual two-channel filter bank,or each split-merge pair is perfectly reconstructed, the octave-band filter bank is as well. Such statement is also valid for the time-varying octave-band filter bank. Therefore, we have following theorem.

Theorem 3. *A time-varying discrete-time wavelet transform implemented with a time-varying octave-band filter bank is a biorthogonal time-varying transform if each two-channel time-varying filter bank is perfectly reconstructed.*

We cannot use the method used in the time-variant case to prove the above theorem because the system is time-varying. To prove theorem 1, we define analysis and synthesis matrices of the j-stage two-channel time-varying filter bank showed in Fig. 18 as

$$\mathbf{T}_{ma}^{(j)} =$$

$$\left[\begin{array}{cccccccc} & \vdots & \vdots & \vdots & \vdots & \vdots & \vdots & \\ \cdots & \mathbf{h}_0(m) & \mathbf{h}_1(m) & \cdots & \mathbf{h}_{N(j)-1}(m) & \mathbf{h}_{N(j)}(m) & \mathbf{h}_{N(j)+1}(m) & \cdots \\ \cdots & \mathbf{0} & \mathbf{h}_0(m) & \cdots & \mathbf{h}_{N(j)-2}(m) & \mathbf{h}_{N(j)-1}(m) & \mathbf{h}_{N(j)}(m) & \cdots \\ \cdots & \vdots & \vdots & \ddots & \vdots & \vdots & \vdots & \cdots \\ \cdots & \mathbf{0} & \mathbf{0} & \cdots & \mathbf{h}_0(m) & \mathbf{h}_1(m) & \mathbf{h}_2(m) & \cdots \\ \cdots & \mathbf{0} & \mathbf{0} & \cdots & \mathbf{0} & \mathbf{h}_0(m+1) & \mathbf{h}_1(m+1) & \cdots \\ & \vdots & \vdots & \vdots & \vdots & \vdots & \vdots & \end{array}\right] \tag{126}$$

$$\mathbf{T}_{ms}^{(j)} = \left[\begin{array}{ccccccc} & \vdots & \vdots & \vdots & \vdots & \vdots & \\ \cdots & \mathbf{g}_0(m) & \mathbf{0} & \cdots & \mathbf{0} & \mathbf{0} & \cdots \\ \cdots & \mathbf{g}_1(m) & \mathbf{g}_0(m) & \cdots & \mathbf{0} & \mathbf{0} & \cdots \\ \cdots & \vdots & \vdots & \ddots & \vdots & \vdots & \cdots \\ \cdots & \mathbf{g}_{N(j)-1}(m) & \mathbf{g}_{N(j)-2}(m) & \cdots & \mathbf{g}_0(m) & \mathbf{0} & \cdots \\ \cdots & \mathbf{g}_{N(j)}(m) & \mathbf{g}_{N(j)-1}(m) & \cdots & \mathbf{g}_1(m) & \mathbf{g}_0(m+1) & \cdots \\ \cdots & \mathbf{g}_{N(j)+1}(m) & \mathbf{g}_{N(j)}(m) & \cdots & \mathbf{g}_2(m) & \mathbf{g}_1(m+1) & \cdots \\ & \vdots & \vdots & \vdots & & & \end{array}\right], \tag{127}$$

where

$$N(j) = 2^{J-j}, \quad j = 1, 2, \dots, \tag{128}$$

and

$$\mathbf{h}_i(m) = \left[\begin{array}{cc} h_u(L-2i-1, m) & h_u(L-2i-2, m) \\ h_l(L-2i-1, m) & h_l(L-2i-2, m) \end{array}\right], \tag{129}$$

$$\mathbf{g}_i(m) = \left[\begin{array}{cc} g_u(2i, m) & g_l(2i, m) \\ g_u(2i11, m) & g_l(2i+1, m) \end{array}\right], \tag{130}$$

where L is the filter length.

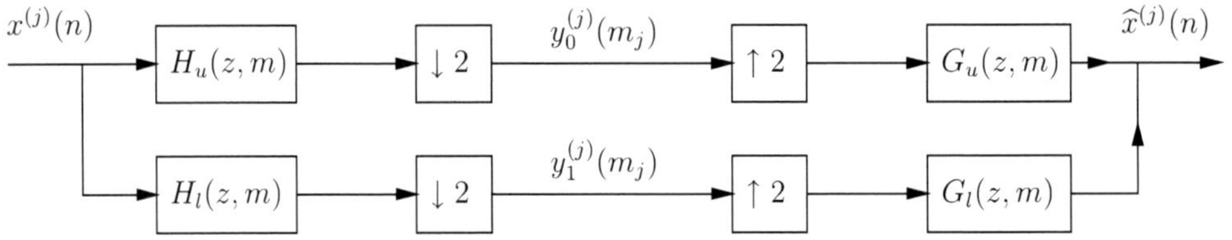

Figure 18. The j-th stage two-channel time-varying filter bank.

Furthermore, we define two special matrices $\mathbf{\Lambda}_0$ and $\mathbf{\Lambda}_1$

$$\mathbf{\Lambda}_0 = \left[\begin{array}{ccccccc} & \vdots & \vdots & \vdots & \vdots & \vdots & \\ \cdots & 1 & 0 & 0 & 0 & 0 & \cdots \\ \cdots & 0 & 0 & 1 & 0 & 0 & \cdots \\ & \vdots & \vdots & \vdots & \vdots & \vdots & \end{array}\right] \tag{131}$$

$$\mathbf{\Lambda}_1 = \begin{bmatrix} & \vdots & \vdots & \vdots & \vdots & \vdots & \\ \cdots & 0 & 1 & 0 & 0 & 0 & \cdots \\ \cdots & 0 & 0 & 0 & 1 & 0 & \cdots \\ & \vdots & \vdots & \vdots & \vdots & \vdots & \end{bmatrix}, \tag{132}$$

to extract the lowpass and highpass output like

$$\mathbf{y}_0^{(j)} = \mathbf{\Lambda}_0 \mathbf{y}^{(j)}, \tag{133}$$

$$\mathbf{y}_1^{(j)} = \mathbf{\Lambda}_1 \mathbf{y}^{(j)}, \tag{134}$$

where

$$\mathbf{y}^{(j)} = \left[\cdots y_0^{(j)}(-1)\, y_1^{(j)}(-1)\, y_0^{(j)}(0)\, y_1^{(j)}(0)\, y_0^{(j)}(1)\, y_1^{(j)}(1) \cdots\right]^T, \tag{135}$$

$$\mathbf{y}_0^{(j)} = \left[\cdots y_0^{(j)}(-1)\, y_0^{(j)}(0)\, y_0^{(j)}(1) \cdots\right]^T, \tag{136}$$

$$\mathbf{y}_1^{(j)} = \left[\cdots y_1^{(j)}(-1)\, y_1^{(j)}(0)\, y_1^{(j)}(1) \cdots\right]^T. \tag{137}$$

Based on the above matrix definitions we can describe the filter bank at the j-th stage showed in Fig. 18 as

$$\widehat{\mathbf{x}}^{(j)} = \mathbf{T}_{ms}^{(j)}\, \mathbf{T}_{ma}^{(j)}\, \mathbf{x}^{(j)}. \tag{138}$$

After adding the (j+1)-th stage with a biorthogonal time-varying two-channel filter bank showed in Fig. 19, we have

$$\begin{aligned} \widehat{\mathbf{x}}^{(j)} &= \mathbf{T}_{ms}^{(j)}\, \mathbf{y}^{(j)} \\ &= \mathbf{T}_{ms}^{(j)} \left\{ \mathbf{\Lambda}_0^T \mathbf{y}_0^{(j)} + \mathbf{\Lambda}_1^T \mathbf{y}_1^{(j)} \right\} \\ &= \mathbf{T}_{ms}^{(j)} \left\{ \mathbf{\Lambda}_0^T \mathbf{\Lambda}_0 \mathbf{T}_{ma}^{(j)} \mathbf{x}^{(j)} + \mathbf{\Lambda}_1^T\, \mathbf{T}_{ms}^{(j+1)}\, \mathbf{T}_{ma}^{(j+1)} \mathbf{\Lambda}_1 \mathbf{T}_{ma}^{(j)} \mathbf{x}^{(j)} \right\}. \end{aligned} \tag{139}$$

Because of

$$\mathbf{T}_{ms}^{(j+1)}\, \mathbf{T}_{ma}^{(j+1)} = \mathbf{I}, \tag{140}$$

$$\mathbf{\Lambda}_0^T \mathbf{\Lambda}_0 + \mathbf{\Lambda}_1^T \mathbf{\Lambda}_1 = \mathbf{I}, \tag{141}$$

we can rewrite (139) as

$$\begin{aligned} \widehat{\mathbf{x}}^{(j)} &= \mathbf{T}_{ms}^{(j)} \left\{ \mathbf{\Lambda}_0^T \mathbf{\Lambda}_0 + \mathbf{\Lambda}_1^T \mathbf{\Lambda}_1 \right\} \mathbf{T}_{ma}^{(j)}\, \mathbf{x}^{(j)} \\ &= \mathbf{T}_{ms}^{(j)}\, \mathbf{T}_{ma}^{(j)}\, \mathbf{x}^{(j)} \\ &= \mathbf{x}^{(j)} \end{aligned} \tag{142}$$

which means that the time-varying octave-band filter bank is still perfectly reconstructed after adding a stage of time-varying biorthogonal two-channel filter bank. In other words, theorem 3 is correct.

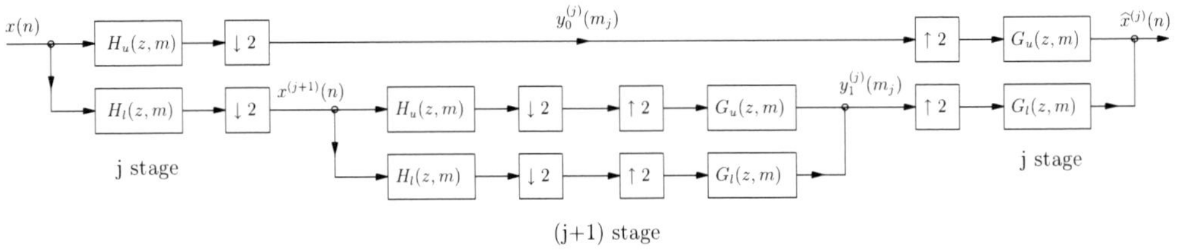

Figure 19. Adding the $(j+1)$-th stage.

5.2. Implementation with Time-Varying Nonuniform Filter Banks

Fig. 20 shows another implementation of a time-varying wavelet transform with $(J(m)+1)$-channel time-varying nonuniform filter bank. To make the analysis easy we suppose that $J(m)$ does not change with time and is equal to constant J. For analysis of the $(J+1)$-channel time-varying nonuniform filter bank we first reconstruct the nonuniform filter bank to a time-varying uniform filter bank through adding following filters between $H_i(z,m)$ and $H_{i+1}(z,m)$ $(0 \leq jJ-1)$

$$H_{i,k}(z,m) = z^{-k\cdot 2^{i+1}} H_i(z,m), \;\; 1 \leq k \leq 2^{J-i-1}-1. \tag{143}$$

After adding additional filters in the nonuniform filter bank in Fig. 20 the filter bank becomes M-channel time-varying uniform filter filter bank. The number of channel M is calculated by

$$\begin{aligned} M &= \sum_{i=0}^{J-2}(2^{J-i-1}-1)+(J+1) \\ &= 2^{J-1}\sum_{i=0}^{J-2}2^{-i}+2 \\ &= 2^J(1-2\cdot 2^{-J})+2 \\ &= 2^J. \end{aligned} \tag{144}$$

For the time-varying system in Fig. 20 we have following theorem.

Theorem 4 *A time-varying discrete-time wavelet transform implemented with a time-varying nonuniform filter bank is biorthogonal if each two-channel time-varying filter bank in its tree-structured implementation is perfectly reconstructed.*

To prove that theorem 4, we need to describe the filter $H_i(z,m)$ in Fig. 20 based on the tree structure in Fig. 19. In the time-invariant discrete-time wavelet transform the description of such filters can be simply got using the convolution role in the transform-domain. However, in the time-varying case, we cannot describe $H_i(z,m)$ as product of functions in the previous stages, like $H_0(z,m)H_1(z^2,m)$, because the system is time-varying and the

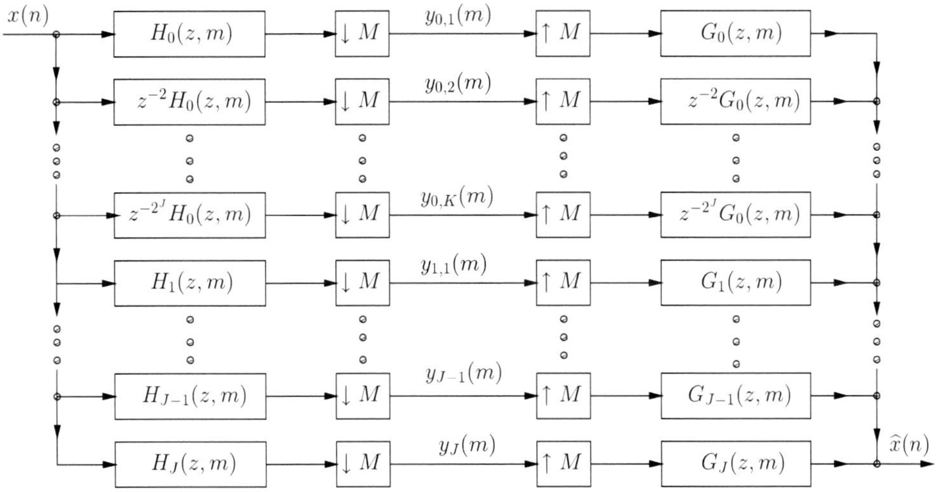

Figure 20. The equivalent M-channel time-varying uniform filter bank.

convolution role does not exist. Referencing to definitions of $\mathbf{T}_{ma}^{(j)}$ and $\mathbf{T}_{ms}^{(j)}$ in (126) and (127), we find that the analysis output $\mathbf{y}_0(m_0)$ can be expressed as

$$\mathbf{y}_0 = \mathbf{\Lambda}_0\, \mathbf{T}_{ma}^{(1)}\, \mathbf{x}, \tag{145}$$

where

$$\begin{aligned} \mathbf{y}_0 &= [\cdots \mathbf{y}_0(-1)\ \mathbf{y}_0(0)\ \mathbf{y}_0(1) \cdots]^T, & (146)\\ \mathbf{y}_0(m) &= [\, y_{0,1}(m)\ y_{0,2}(m)\ \cdots y_{0,K}(m)\,]^T, & (147) \end{aligned}$$

and $K = 2^{2^J-1} - 1$. In general, we have

$$\mathbf{y}_{j-1} = \underbrace{\mathbf{\Lambda}_0\, \mathbf{T}_{ma}^{(j)}\, \mathbf{\Lambda}_1\, \mathbf{T}_{ma}^{(j-1)}\, \cdots\, \mathbf{\Lambda}_1\, \mathbf{T}_{ma}^{(1)}}_{\mathbf{H}_{j-1}}\, \mathbf{x}, \tag{148}$$

where $1 \le j \le J-1$, and

$$\begin{aligned} \mathbf{y}_{J-1} &= \mathbf{\Lambda}_0\, \mathbf{T}_{ma}^{(J)}\, \mathbf{\Lambda}_1\, \mathbf{T}_{ma}^{(J-1)}\, \cdots\, \mathbf{\Lambda}_1\, \mathbf{T}_{ma}^{(1)}\, \mathbf{x} = \mathbf{H}_{J-1}\, \mathbf{x}, & (149)\\ \mathbf{y}_J &= \mathbf{\Lambda}_1\, \mathbf{T}_{ma}^{(J)}\, \mathbf{\Lambda}_1\, \mathbf{T}_{ma}^{(J-1)}\, \cdots\, \mathbf{\Lambda}_1\, \mathbf{T}_{ma}^{(1)}\, \mathbf{x} = \mathbf{H}_J\, \mathbf{x}, & (150) \end{aligned}$$

where

$$\begin{aligned} \mathbf{y}_{j-1} &= [\cdots \mathbf{y}_{j-1}(-1)\ \mathbf{y}_{j-1}(0)\ \mathbf{y}_{j-1}(1) \cdots]^T, & (151)\\ \mathbf{y}_{j-1}(m) &= [\, y_{j-1,1}(m)\ y_{j-1,2}(m)\ \cdots y_{j-1,K}(m)\,]^T, & (152) \end{aligned}$$

for $K = 2^{2^J-j} - 1$ and $1 \le j \le J-2$, and

$$\mathbf{y}_{J-1} = [\cdots\ y_{J-1}(-1)\ y_{J-1}(0)\ y_{J-1}(1)\cdots]^T, \quad (153)$$
$$\mathbf{y}_{J} = [\cdots\ y_{J}(-1)\ y_{J}(0)\ y_{J}(1)\cdots]^T. \quad (154)$$

At synthesis side, we have similar definitions as

$$\widehat{\mathbf{x}}_{j-1} = \underbrace{\mathbf{T}^{(1)}_{ms}\,\mathbf{\Lambda}_1^T\,\mathbf{T}^{(2)}_{ms}\ \cdots\ \mathbf{\Lambda}_1^T\,\mathbf{T}^{(j)}_{ms}\,\mathbf{\Lambda}_0^T}_{\mathbf{G}_{j-1}}\ \mathbf{y}_{j-1}, \quad (155)$$

$$\widehat{\mathbf{x}}_{J-1} = \mathbf{T}^{(1)}_{ms}\,\mathbf{\Lambda}_1^T\,\mathbf{T}^{(2)}_{ms}\ \cdots\ \mathbf{\Lambda}_1^T\,\mathbf{T}^{(J)}_{ms}\,\mathbf{\Lambda}_0^T\ \mathbf{y}_{J-1} = \mathbf{G}_{J-1}\,\mathbf{y}_{J-1}, \quad (156)$$
$$\widehat{\mathbf{x}}_{J} = \mathbf{T}^{(1)}_{ms}\,\mathbf{\Lambda}_1^T\,\mathbf{T}^{(2)}_{ms}\ \cdots\ \mathbf{\Lambda}_1^T\,\mathbf{T}^{(J)}_{ms}\,\mathbf{\Lambda}_1^T\ \mathbf{y}_{J} = \mathbf{G}_{J}\,\mathbf{y}_{J}, \quad (157)$$

Now, based on the definition in (147), we can build the analysis output vector for th time-varying filter bank in Fig. 20 as

$$\mathbf{y} = [\cdots\ \mathbf{y}_0(-1)\ \cdots\mathbf{y}_J(-1)\ \mathbf{y}_0(0)\ \cdots\mathbf{y}_J(0)\ \mathbf{y}_0(1)\ \cdots]^T. \quad (158)$$

Suppose that $\mathbf{T}_{ma}$ and $\mathbf{T}_{ms}$ are the analysis and synthesis matrices for the time-varying filter bank in Fig. 20. Referencing (158), $\mathbf{T}_{ma}$ is constructed by interleaving the rows from $\mathbf{T}^{(1)}_{ma}$ to $\mathbf{T}^{(J)}_{ma}$ with same time index m, $\mathbf{T}_{ms}$ is built with similar way, but interleaving the columns. Then, the production $\mathbf{T}^{(1)}_{ma}\mathbf{T}^{(J)}_{ma}$ can be expressed by

$$\mathbf{T}_{ms}\,\mathbf{T}_{ma} = \sum_{j=0}^{J} \mathbf{G}_j\,\mathbf{H}_j. \quad (159)$$

Substituting $\mathbf{H}_i$ and $\mathbf{G}_i$ defined in (148)-(150) and (155)-(106) into (159), and using properties in (140) and (141), we get

$$\mathbf{T}_{ms}\,\mathbf{T}_{ma} = \mathbf{I}, \quad (160)$$

which means that the time-varying nonuniform filter bank in Fig. 17 is perfectly reconstructed.

Finally, we give another property related with filter coefficients of the time-varying filter bank in Fig. 17. Suppose that $h_i(n,m)$ and $g_i(n,m)$ represent the analysis and synthesis filter coefficients in Fig. 17, then we have following equation

$$< g_i(n-kM, m+r),\ h_j(n-lM, m+s) >= \delta(k-l)\,\delta(i-j)\,\delta(r-s), \quad (161)$$

where $M = 2^J$. The proof of equation (161) can be simply got by using the PR condition in (160).

6. Conclusion

In this Chapter we have introduced our new research results in the area of time-varying multirate systems and filter banks, from theoretical analysis to design methods.From the point of the digital signal processing the time-varying multirate systems and filter banks are still a relative new research area. In particular, their applications need to be discovered. So far, a good application is found in audio coding [14]. In the image processing it is also found that time-varying cosine-modulated filter bank provides much better performance than the time-invariant filter bank [15]. So, we can estimate that the time-varying multirate systems and filter banks have good application perspective.

Regarding the theoretical research of time-varying multirate system, following topics can be further addressed: optimal design of time-varying filter banks, theory of multi-dimentional time-varying filter banks and time-varying wavelet transform.

References

[1] H. S. Malvar, *Signal Processing with Lapped Transforms.* Boston, MA: Artech House, 1992.

[2] P. P. Vaidyanathan, *Multirate Systems and Filter banks.* Englewood Cliffs, NJ: Prentice-Hall Signal Processing Series, 1993.

[3] I. Sodagar, K. Nayebi, and T. P. Barnwell, III, Time-varying filter banks and wavelets, *IEEE Trans. Signal Processing,* vol. 42, No. 11, pp. 2983–2996, Nov. 1994.

[4] I. Sodagar, K. Nayebi, T. Barnwell and M. Smith, Time-varying analysis-synthesis systems based on filter banks and post filtering, *IEEE Trans. Signal Processing,* vol. 43, No. 11, pp. 2512–2524, Nov. 1995.

[5] R. L. de Queiroz and K. R. Rao, Time-varying lapped transform and wavelet packets, *IEEE Trans. Signal Processing,* vol. 41, No. 12, pp. 3293–3305, Dec. 1993.

[6] C. Herley, Boundary filters for finite-length signals and time-varying filter banks, *IEEE Trans. Circuits and Systems-II: Analog and Digital Signal Processing,* vol. 42, No.2, pp. 102–114, Feb. 1995.

[7] S. Phoong and P. P. Vaidyanathan, Factorability of lossless time-varying filters and filter banks, *IEEE Trans. Signal Processing,* vol. 45, No. 8, pp. 1971–1986, Aug. 1997.

[8] S. Phoong and P. P. Vaidyanathan, Time-varying filters and filter banks: some basic principles, *IEEE Trans. Signal Processing,* vol. 44, No. 12, pp. 2971–2988, Dec. 1996.

[9] G. Wang, The most general time-varying filter bank and time-varying lapped transforms, *IEEE Trans. on Signal Proccesing*, vol. 54, No. 10, pp. 3775–3789, Oct. 2006.

[10] G. Wang and U. Heute, Time-varying MMSE modulated lapped transform and its applications to transform coding for speech and audio signals, *Signal Processing,* vol. 82, No. 9, pp. 1283–1304, Sept. 2002.

[11] G. Wang, Time-Varying Cosine-Modulated Filter Banks, *Digital Signal Processing,* vol. 15, No. 3, pp. 237–254, May 2005.

[12] A. Mertins, *Signal Analysis, Wavelets, Filter Banks, Time-Frequency Transforms and Application* England: John Wiley & Sons, 1999.

[13] X. Xia and B. Suter, Multirate filter banks with block sampling, *IEEE Trans. Signal Processing,* vol. 44, No. 3, pp. 484–496, March 1996.

[14] ISO/IEC JTCI/SC29, Information technology-coding of moving pictures and associated audio for digital storage media at up to about 1.5 Mbit/s-IS 11172-3 (audio), 1992.

[15] S. Imata, M. Ikehara and S. Takahashi, Cosine-Modulated Two-Dimensional Time-Varying Filter Banks, *Electronics and Communications in Japan,* vol. 80, No. 11, pp. 1977–1988, December 1997.

In: New Signal Processing Research
Editor: Takumi Maeda, pp. 197-231

ISBN: 978-1-60456-479-2

Chapter 6

FIR DIGITAL FILTERS DESIGN BASED ON HOPFIELD NEURAL NETWORKS

***Yue-Dar Jou*[1,*], *Fu-Kun Chen*[2] *and Chao-Ming Sun*[3]**

[1]Department of Electrical Engineering, R.O.C. Military Academy
Fengshan, Kaohsiung, 830, Taiwan
[2]Department of Computer Science and Information Engineering
Southern Taiwan University
Yung-Kang City, Tainan County, 710, Taiwan
[3]Department of Management Science, R.O.C. Military Academy
Fengshan, Kaohsiung, 830, Taiwan

Abstract

Filter is a simple system that can modify the components at certain frequencies, for example, the magnitude, phase, or group delay, for the input signals. Digital filters played an important role in the applications while using the digital signal processing. In the design of frequency-selective filters, the desired filter characteristics are specified in the frequency domain in terms of the desired magnitude and phase response of the filter. Many design algorithms have been proposed to meet the desired filter specifications. In general, linear algebra based methods have received considerable attention due to the powerful computation of digital computers. However, the typical linear algebra based methods require computationally intensive matrix inversion, and so these algorithms cannot meet the real-time requirement. Applying neural network to achieve the design of digital filter work is a new area of research [20]-[36]. In this chapter, the Hopfield neural network (HNN) based approaches are introduced to carry out the finite impulse response (FIR) filter design. The proposed HNN based approach uses an improved feedback neural network to implement the design task, and is comparatively more effective and practical than the previous neural-based methods presented by Bhattacharya. The proposed structure is unified, regular and applicable for the designs of differentiators, Hilbert transformers, all-pass phase equalizers, FIR Nyquist filters, and quadrature mirror filter, etc., with significant reduction of computational complexity and hardware cost. Simulation results indicate that the proposed technique can achieve better or comparable results as compared to existing methods. If the hardware implementation is

* E-mail address: ydjou@cc.cma.edu.tw

considered, for example, in a single IC chip, the proposed approach would provide a solution with real-time processing.

1. Introduction

Digital filters have been received much attention in different area of signal processing during the past several decades [1]-[4]. The main function of filters can be concluded into the two general purposes of signal separation and signal restoration [5]. For example, low-pass filters are used to reject noise frequencies higher than the cutoff frequencies of the signal. Digital differentiator can be used for detecting the edges of an image. In biomedical engineering, it is necessary to obtain the higher order derivatives of biomedical data at low frequency ranges [6]. Digital all-pass filter which possess unit-magnitude response and prescribed phase characteristics can be applied to communication system, multirate filtering, and group delay equalization [7][8]. Nyquist filters possess the property that their impulse response has uniformly spaced zero-crossings which results in a zero inter-symbol interference constraint. The design of Nyquist filters [9]-[10] has been received a number of applications in data transmission systems, perfect reconstruction filter banks, non-uniform sampling and interpolation filters over the past two decades. For many communication and speech signal processing systems, quadrature mirror filter (QMF) banks have been widely used to achieve the goals of subband coding and short-time spectral analysis [11]. Therefore, an efficient design method and optimal algorithm is needed to achieve a good performance for the above-mentioned applications.

Digital filters can be divided into two basic types: Finite Impulse Response (FIR) filters and Infinite Impulse Response (IIR) filters. To design a filter is just to find a set of coefficients that satisfies the desired specifications in time-domain or frequency-domain. Considering the stability thesis of systems, FIR filter is the best choice. It has been shown that an FIR filter with symmetric or anti-symmetric coefficients has a linear phase and therefore with a constant group delay [4]. These properties are very useful in the design of FIR filters and can be applied to such applications as audio signal processing and data transmission. However, a relatively higher order FIR filters is required to obtain the same characteristics as compared with an IIR filter. This leads to difficulties in realization for FIR filter. Therefore, how to select a suitable filter to design depends on the applications required. In this chapter, we will focus on the design of FIR filters.

The design focused on linear-phase FIR filters owing to its considerable importance during the past two decades. Much effort has been spent on designing filters based on minimum Chebyshev algorithm [12] or least-squares methods [13]-[15]. The minimum Chebyshev error approximation applies the optimization method such as Remez exchange algorithm or linear programming to fulfill the design specifications that is optimal in the minimax sense. The minimum Chebyshev error approximation is optimum in the minimax sense because it achieves both a given discrimination and a specified selectivity with a minimum length of the filter impulse response. Unfortunately, linear programming and exchange algorithm become computationally intensive as the filter length is increased.

FIR filters can be designed by minimizing a quadratic measure of the error [13][14] in the frequency domain. As a consequence, the filter coefficients are obtained by computing an eigenvector corresponding to the smallest eigenvalue of an appropriate matrix. In [15], the authors formulate the mentioned mean-square error as a quadratic function, thus resulting the filter coefficients are obtained by solving a system of linear equations involved a positive-definite matrix. The associated positive-definite matrix can be further decomposed as Toeplitz-plus-Hankel matrix form. Consequently, the designed filter coefficients are solved efficiently by using a recursive manner. Weighted least-squares (WLS) methods [16]-[19] have also attracted a myriad of research attention due to the flexible utilization for any type of filter designs. Moreover, the WLS method can obtain an optimal solution analytically with an appropriate weighting function, and the equiripple (i.e., optimal in the minimax sense) requirement is therefore completed.

Recently, several nonlinear techniques based on neural networks [20]-[36] have been developed for designing different kinds of digital filters including equiripple FIR filters, Hilbert transformers, digital differentiators, 2-D FIR filters, 2-D IIR filters, fractional delay filters, etc. Neural networks can use simple additions, multiplications, divisions and threshold operations to form a basic processing element. Based on the advantages of self-adjustment, high computational capability and ease of hardware implementation, neural networks can be implemented easily by using analog VLSI technology or optoelectronic devices. Moreover, combining an embedded DSP core with a circuit that performs the design of the filter coefficients can achieve the requirement of the time-varying filtering application, such as video and audio processing with system-on-chip (SOC) implementations. Bhattacharya *et al.* [22][25] focus on the implementation of FIR filters in real-time using feedback neural networks. These techniques construct the energy function of the network, which is equivalent to the error function of the filter optimization problem. When the network is initiated with proper inputs, it goes through an intermediate state and finally converges to the optimal solution. Nevertheless, the number of neurons required for the network is proportional to the sampling grid of frequency. Thus, the computational complexity and the hardware cost are considerable with finer sampling grid size. This motivates the researches for seeking a more efficient neural network based method.

2. Formulation of FIR Filter Design

The frequency response of an FIR digital filter with impulse response $h(n)$, $n = 0, 1, \cdots, N-1$, is generally characterized by

$$H(\omega) = \sum_{n=0}^{N-1} h(n) e^{-j\omega n} . \tag{1}$$

According to the properties of $h(n)$, i.e., symmetry or anti-symmetry, real-valued response or complex-valued response, different kinds of filters can be designed. For the least-squares design of FIR filter, the task is to find the impulse response $h(n)$ such that the weighted least-squares objective function

$$e = \int_{\Omega} W(\omega)\left|H(\omega) - H_d(\omega)\right|^2 d\omega \qquad (2)$$

is minimized, where $W(\omega)$ is a positive weighting function, $H_d(\omega)$ is the desired frequency response, Ω is the band of interest, and ω is the normalized frequency. In WLS method, the weighting function can be appropriately designed to achieve equiripple response which is almost optimal in the minimax response.

2.1. Linear-Phase Filter Design [32]-[34]

In order to design an FIR filter with linear phase response, the impulse response of the filter is required to be symmetrical or anti-symmetrical, i.e., $h(n) = \pm h(N-1-n)$, then the frequency response of the filter can be formulated as

$$H(\omega) = \sum_{n=0}^{N-1} h(n) e^{-j\omega n} = e^{-j\omega\frac{N-1}{2}} e^{js\frac{\pi}{2}} A(\omega) \qquad (3)$$

where $s = 0$ for $h(n)$ is symmetrical and $s = 1$ for $h(n)$ is anti-symmetrical. The frequency response of real-valued amplitude response $A(\omega)$ can be divided into four types according to length of the filter and symmetry of the impulse response. For odd N, $A(\omega)$ can be rewritten as

$$A(\omega) = \sum_{n=0}^{(N-1)/2} a_n \cos n\omega, \qquad (4)$$

where $a_0 = h[(N-1)/2]$ and $a_n = 2h[(N-1)/2 - n]$ for $1 \le n \le (N-1)/2$. The least-squares design of FIR filters is to find filter coefficients a_n such that the error response is minimized in the L_2 sense, i.e.,

$$E_2 = \sum_{l=0}^{L} \left|E(\omega_l)\right|^2 = \sum_{l=0}^{L} \left|A_d(\omega_l) - A(\omega_l)\right|^2 \qquad (5)$$

where $A_d(\omega_l)$ is the desired amplitude response, and L is the number of points sampled at frequency axis from 0 to π with. $\omega_l = l \cdot \pi / L$

2.2. Arbitrary Response Filter Design [30], [35]

The frequency response of an FIR digital filter with real impulse response $h(n)$, $n = 0, 1, \cdots, N-1$, can be characterized by

$$H(\omega) = \sum_{n=0}^{N-1} h(n) e^{-j\omega n} = \sum_{n=0}^{N-1} h(n)\cos(\omega n) - j\sum_{n=0}^{N-1} h(n)\sin(\omega n). \tag{6}$$

Therefore, the phase response of the filter is given by

$$\varphi(\omega) = -\tan^{-1}\left(\frac{\sum_{n=0}^{N-1} h(n)\sin(\omega n)}{\sum_{n=0}^{N-1} h(n)\cos(\omega n)}\right) \tag{7}$$

and the corresponding group delay is given by

$$\tau(\omega) = -\frac{d}{d\omega}\varphi(\omega). \tag{8}$$

The desired frequency response of an FIR all-pass phase equalizer $H_d(\omega)$ having an amplitude response $A_d(\omega)$ and a phase response $\rho(\omega)$ is characterized by

$$H_d(\omega) = \begin{cases} A_d(\omega) e^{j\rho(\omega)} & , \omega \in \Omega_p \\ 0 & , \omega \in \Omega_s \end{cases}, \tag{9}$$

where Ω_p and Ω_s are the passband and stopband regions, respectively. Consequently, the group delay is given by

$$\tau_\rho(\omega) = -\frac{d\,\rho(\omega)}{d\,\omega}. \tag{10}$$

The least-squares design of FIR filters is to find the impulse response $h(n)$ such that the error response is minimized in the L_2 sense, i.e.,

$$
\begin{aligned}
E_2 &= \int_{\Omega_p} (H_d(\omega) - H(\omega))^2 d\omega + \int_{\Omega_s} H^2(\omega)\, d\omega \\
&\approx \mathbf{h}^T \left\{ \sum_{l=1}^{L_p} \left[\mathbf{c}(\omega_l)\mathbf{c}^T(\omega_l) + \mathbf{s}(\omega_l)\mathbf{s}^T(\omega_l) \right] \right\} \mathbf{h} + \mathbf{h}^T \left\{ \sum_{l=1}^{L_s} \left[\mathbf{c}(\omega_l)\mathbf{c}^T(\omega_l) + \mathbf{s}(\omega_l)\mathbf{s}^T(\omega_l) \right] \right\} \mathbf{h} \\
&\quad - 2\mathbf{h}^T \sum_{l=1}^{L_p} A_d(\omega_l)\left[\cos\rho(\omega_l)\mathbf{c}(\omega_l) - \sin\rho(\omega_l)\mathbf{s}(\omega_l)\right] + \sum_{l=1}^{L_p} A_d^2(\omega_l)
\end{aligned}
\tag{11}
$$

where L_p and L_s are the number of points sampled at the passband and stopband regions, respectively, and $L = L_p + L_s$. Besides,

$$
\begin{aligned}
\mathbf{h} &= [h(0), h(1), \cdots, h(N-1)]^T \\
\mathbf{c}(\omega_l) &= [1, \cos(\omega_l), \cos(2\omega_l), \cdots, \cos((N-1)\omega_l)]^T \, . \\
\mathbf{s}(\omega_l) &= [0, \sin(\omega_l), \sin(2\omega_l), \cdots, \sin((N-1)\omega_l)]^T
\end{aligned}
\tag{12}
$$

2.3. Nyquist Filters Design

In order to design $H(\omega)$ as a Nyquist filter, the impulse response is required to be exactly zero-crossing at the Nyquist rate except for one point n_0

$$
h(n) = \begin{cases} c \neq 0, & n = n_0 = \dfrac{N-1}{2} \\ 0\ , & n = n_0 \pm kM, \quad k = 1, 2, \cdots \end{cases} ,
\tag{13}
$$

where n_0 and M are integers, M is the Nyquist interval, and n_0 corresponds to the desired group delay in the passband. By imposing the zero-crossing impulse and linear phase constraints, i.e., $h(n) = h(N-1-n)$, the frequency response can be written in the form

$$
H(\omega) = h(0) + \sum_{\substack{n=1 \\ n \notin Z}}^{\frac{N-1}{2}} 2h(n) \cos \omega n\, ,
\tag{14}
$$

where Z denotes the set of indices of $h(n)$'s which are constrained to be zero. The number of indices in the set Z is $\lfloor N/2M \rfloor$ where $\lfloor x \rfloor$ represents integer part of x, thus the number of unconstrained filter coefficient is

$$N_u = \frac{N-1}{2} - \left\lfloor \frac{N}{2M} \right\rfloor. \tag{15}$$

Therefore, (14) can be simplified as the general form

$$H(\omega) = \sum_{n=0}^{N_u-1} a_n \varphi_n(\omega), \tag{16}$$

where $\varphi_n(\omega) = \cos \omega n$, $n = 0,1,2,\cdots,N_u - 1$, $a_n = 2h(n_0 - n)$, $n = 1,2,\cdots,N_u - 1$ and $a_0 = h(n_0)$. The relationship between impulse response and unconstrained filter coefficients of the Nyquist filter can be expanded as

$$\begin{aligned}&\left[a_0, a_1, \cdots a_{M-1}, a_M, \cdots, a_{N_u-1}\right] \\ &= \left[h(n_0), 2h(n_0 - 1), \cdots 2h(n_0 - (M-1)), 2h(n_0 - (M+1)) \cdots, h(0)\right]\end{aligned} \tag{17}$$

Once the unconstrained filter coefficients have been obtained, we can insert the zeros that are required to obtain the impulse response of Nyquist filter. In general, an ideal Nyquist filter is required to be low-pass, and the desired frequency response is given by

$$H_d(\omega) = \begin{cases} 1, & 0 \le \omega \le \omega_p \\ 0, & \omega_s \le \omega \le \pi \end{cases}, \tag{18}$$

where the passband and stopband cut-off frequencies are $\omega_p = (1-\rho)\pi/M$ and $\omega_s = (1+\rho)\pi/M$, respectively, and ρ is a rolloff rate. It is evident that Nyquist filters satisfy the frequency edges condition $\omega_p + \omega_s = 2\pi/M$.

The least squares method for FIR Nyquist filter design requires finding filter coefficients a_n such that the error response, $E(\omega) = H_d(\omega) - H(\omega)$, is minimized in the L_2 sense. That is, the following function should be minimized,

$$\begin{aligned}E_2 &= \int_0^{\pi} \left|H_d(\omega) - H(\omega)\right|^2 d\omega \approx \sum_{l=1}^{L} \left|H_d(\omega_l) - H(\omega_l)\right|^2 \\ &= \sum_{i=0}^{N_u-1}\sum_{j=0}^{N_u-1}\left[\sum_{l=1}^{L} \varphi_i(\omega_l)\varphi_j(\omega_l)\right] a_i a_j - 2\sum_{i=0}^{N_u-1}\left[\sum_{l=1}^{L} H_d(\omega_l)\varphi_i(\omega_l)\right] a_i + \sum_{l=1}^{L} H_d^2(\omega_l)\end{aligned} \tag{19}$$

2.4. Two-Channel Quadrature Mirror Filter Design [37]-[39]

For the basic structure of a two-channel FIR filter banks (shown in Figure 1) with analysis low-pass filter H_0 and high-pass filter H_1, synthesis low-pass filter F_0 and high-pass filter F_1, the input-output relation of the system is given by

$$\hat{X}(z)=\frac{1}{2}[H_0(z)F_0(z)+H_1(z)F_1(z)]X(z)+\frac{1}{2}[H_0(-z)F_0(z)+H_1(-z)F_1(z)]X(-z) \tag{20}$$

where $H_i(z)$ and $F_i(z)$ are the transfer function of the various filters. As the mirror-image symmetry about the frequency $\omega=\pi/2$ exists between $H_0(z)$ and $H_1(z)$. We have the equality of $H_0(z)=H_1(-z)$. Therefore, (20) becomes

$$\hat{X}(z)=\frac{1}{2}[H_0(z)H_0(z)-H_0(-z)H_0(-z)]X(z). \tag{21}$$

Let the prototype analysis filter $H_0(z)$ be a linear-phase low-pass FIR filter with even length *N*, then the input-output frequency response of the QMF banks is obtained as

$$\hat{X}(\omega)=\frac{1}{2}e^{-j(N-1)\omega}\left[|H_0(\omega)|^2+|H_0(\omega+\pi)|^2\right]X(\omega). \tag{22}$$

In order to design QMF banks with perfect reconstruction constraint, the term, $D(\omega)=|H_0(\omega)|^2+|H_0(\omega+\pi)|^2$, must be equal to unity. In literature [37], the authors reformulate the nonlinear optimization problem of QMF banks into a quadratic form, thus yielding the optimal filter coefficients are obtained efficiently. The designed magnitude response $D(\omega)$ of the filter banks obtained at *k*-th iteration is approximately written as,

$$D'(\omega_l)=H_0^k(\omega_l)F_0(\omega_l)+H_0^k(\omega_l+\pi)F_0(\omega_l+\pi), \tag{23}$$

where $H_0^k(\omega_l)$ is the frequency response of the low-pass prototype filter calculated at the *k*-th iteration. Because the two frequency responses $H_0(\omega)$ and $F_0(\omega)$ are identical due to the constraint of perfect reconstruction [37]. The error function becomes a quadratic function of the low-pass synthesis filter coefficients $f_{0,n}$. Consequently, the design problem can be solved by minimizing the objective function

$$E_2 = \sum_{l=1}^{L} \left[D'(\omega_l) - 1\right]^2 + \alpha \sum_{l=s}^{L} \left|F_0(\omega_l)\right|^2 , \tag{24}$$

where α is the relative weight between the weighted reconstruction error and the stopband error, s is the discrete sampling grid corresponding to the stopband edge frequency ω_s, and L is the sampling grid.

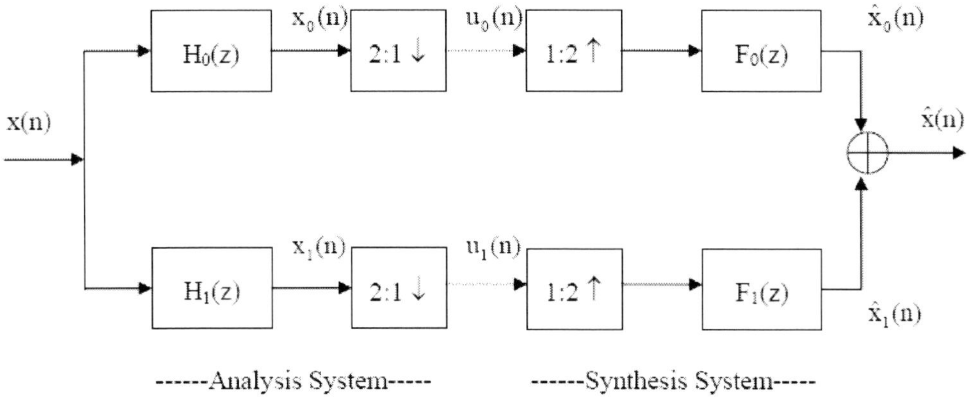

Figure 1. Standard structure of the two-channel QMF bank.

2.5. Complex Filter Design [31]

The frequency response of an FIR digital filter with complex impulse response $h(n)$, with the real part $h_r(n)$ and the imaginary part $h_i(n)$, is formulated as

$$H(\omega) = \sum_{n=0}^{N-1} h(n) e^{-j\omega n} = \mathbf{h}_c^T \mathbf{a}_1(\omega) + j\,\mathbf{h}_c^T \mathbf{a}_2(\omega) \tag{25}$$

where

$$\mathbf{h}_c = \left[h_r(0), h_r(1), \cdots, h_r(N-1), h_i(0), h_i(1), \cdots, h_i(N-1)\right]^T \tag{26}$$

$$\mathbf{a}_1(\omega) = \left[1, \cos\omega, \cdots, \cos(N-1)\omega, 0, \sin\omega, \cdots, \sin(N-1)\omega\right]^T \tag{27}$$

$$\mathbf{a}_2(\omega) = \left[0, -\sin\omega, \cdots, -\sin(N-1)\omega, 1, \cos\omega, \cdots, \cos(N-1)\omega \right]^T . \tag{28}$$

The phase and group delay responses of the filter are given respectively by

$$\varphi(\omega) = \tan^{-1}\left(\frac{\mathbf{h}_c^T \, \mathbf{a}_2(\omega)}{\mathbf{h}_c^T \, \mathbf{a}_1(\omega)}\right) \tag{29}$$

and

$$\tau(\omega) = -\frac{d}{d\omega}\varphi(\omega). \tag{30}$$

The desired complex-valued response $H_d(\omega)$, having an amplitude response $A_d(\omega)$ and a phase response $\rho(\omega)$, is given by $H_d(\omega) = A_d(\omega)e^{j\rho(\omega)}$. The least-squares design of a complex FIR filter requires finding the impulse response $h(n)$ such that the error response is minimized in the L_2 sense, i.e.,

$$\begin{aligned} E_2 &= \int_{2\pi} \left|H_d(\omega) - H(\omega)\right|^2 d\omega \;\; \cong \sum_{l=1}^{L} \left|H_d(\omega_l) - H(\omega_l)\right|^2 \\ &= \sum_{l=1}^{L} \mathbf{h}_c^T \left[\mathbf{a}_1(\omega_l)\mathbf{a}_1^T(\omega_l) + \mathbf{a}_2(\omega_l)\mathbf{a}_2^T(\omega_l)\right] \\ &= \sum_{l=1}^{L} \mathbf{h}_c^T \left[\mathbf{a}_1(\omega_l)\mathbf{a}_1^T(\omega_l) + \mathbf{a}_2(\omega_l)\mathbf{a}_2^T(\omega_l)\right] \mathbf{h}_c \\ &\quad - 2\sum_{l=1}^{L} \mathbf{h}_c^T \left[\mathbf{a}_1(\omega_l) A_d(\omega_l)\cos\rho(\omega_l) + \mathbf{a}_2(\omega_l) A_d(\omega_l)\sin\rho(\omega_l)\right] + \sum_{l=1}^{L} A_d^2(\omega_l) \end{aligned} \tag{31}$$

2.6. Equiripple Filter Design [32]

The basic WLS design of an FIR filter is started by initializing the weighting function defined in (2) with unity (i.e., $W_0(\omega_l) = 1, \quad 1 \le l \le L$). At each stage of the iteration, the weighting function is updated according to the previous error and then the optimization problem is solved again. The weight updating will affect the performance and convergent speed of the WLS approaches. However, there is still no analytical method for deriving the weighting function that will produce a minimax design [16]. At the (*k*+1)-th iteration, the following weights updating procedure has been used and applied to the design of equiripple FIR filters:

$$W_{k+1}(\omega_l) = W_k(\omega_l)\,\beta_k(\omega_l), \quad \text{for} \quad 1 \le l \le L, \tag{32}$$

where $\beta_k(\omega_l)$ is the error function calculated at the *k*-th iteration. Using this weight-updating function, we solve for the next vector of filter coefficients by using the HNN. The criterion for deriving the update of $\beta_k(\omega_l)$ is given as follows. First, the *q*-th extremal frequency $\omega_q\left(1 \le q < L\right)$ of the designed error $\left|E_k(\omega_l)\right| = \left|A_d\left(\omega_l\right) - A_k\left(\omega_l\right)\right|$ at the k-th iteration is computed to satisfy $\left|E_k(\omega_l)\right| > \left|E_k(\omega_{l-1})\right|$ and $\left|E_k(\omega_l)\right| > \left|E_k(\omega_{l+1})\right|$. Then the corresponding extremal value is stored as $Q_k(q) = \left|E_k(\omega_l)\right|$. The frequencies of $\omega = 0$ and $\omega = \pi$ are also considered as extremal frequencies. Second, the maximum and minimum values of $Q_k(q)$ are found and defined as $Peak\_\max = Max(Q_k(q))$ and $Peak\_\min = Min(Q_k(q))$, respectively. The updating formula of $\beta_k(\omega_l)$ is designed similarly to the methods of [16]-[18], which is given as

$$\beta_k(\omega_l) = \frac{1}{Peak\_\max} \frac{\left|E_k^{\alpha}(\omega_l)\right|}{\sum_{l=1}^{L}\left[W_k(\omega_l)\left|E_k^{\alpha}(\omega_l)\right|\right]} \tag{33}$$

The parameter α is an acceleration factor that affects the convergence and the convergence speed of the proposed algorithm. It can be found from several references in the literature [16]-[18] that the value of α in the range $1.0 < \alpha < 1.5$ has the best convergence performance.

To achieve an equiripple design as in WLS methods, the design process can be terminated when the following criterion is achieved,

$$\frac{Peak\_\max - Peak\_\min}{Peak\_\max} \le \varepsilon \tag{34}$$

where ε is a small positive constant (say 0.05). When the design process is stopped, the filter coefficients are designed, and the errors of response result in the equiripple situation.

Evidently, these objective functions to be minimized for different filters have the same quadratic form. Several methods have been proposed for solving the optimization problem. In this paper, we will develop a parallel method based on neural network for the design of FIR filters.

3. Neural Network Based Optimization

In order to solve the above-mentioned objective function efficiently, several methods have been developed for achieving the optimal solution. Nevertheless, these approaches are typically based on linear algebra methods, resulting computationally intensive matrix calculations, slow convergence, and can not meet real-time requirement, simultaneously. The Hopfield neural network (HNN) [40]-[42] is known as a powerful approach for solving the nonlinear optimization problem due to strong numerical approximation capabilities, fast convergence speed, massive parallelism and suitable for hardware implementation. Therefore, we take the advantages of HNN for solving the optimization of FIR filters satisfying the signal applications in real-time.

3.1. Hopfield Neural Network

The state dynamics of the HNN can be characterized by the following set of coupled non-linear differential equations [40]-[42],

$$\frac{du_i}{dt} = -\frac{u_i}{\tau_i} + \sum_{j=1}^{p} T_{ij} v_j + I_i \tag{35}$$

$$v_i = g(u_i) = \frac{1}{1+e^{-\frac{u_i}{\lambda}}} \tag{36}$$

where $\tau_i = C_i \rho_i$ is a time constant, ρ_i is the parallel combination of R_i and R_{ij} defined as $\frac{1}{\rho_i} = \frac{1}{R_i} + \sum_{j=1}^{p} R_{ij}$. The parameter R_{ij} is the resistive feedback between neuron *j* and *i* which must be positive, in general, $R_{ij} = \frac{1}{|T_{ij}|}$. u_i and v_i are the internal and output state of the *i*-th neuron, C_i and R_i are the input capacitance and resistance, *p* is the number of neurons required for the network, I_i is an external input to the neuron and T_{ij} is the interconnection strengths. The neuron activation function $g(u_i)$ is monotonously increasing with λ determining the slope of the function. The massive parallelism in physical topology makes this network especially suitable for analog VLSI implementation. The analog implementations have a network time constant and a convergence time, which are of the order of a few nanoseconds [41][42] that can meet the real-time requirement. The HNN would generate a good solution if one can map the objective function of any optimization problem with a Lyapunov energy function given as

$$E_L = -\frac{1}{2}\sum_{i=1}^{p}\sum_{j=1}^{p} T_{ij} v_i v_j - \sum_{i=1}^{p} I_i v_i + \sum_{i=1}^{p}\frac{1}{\tau_i}\int_0^{v_i} g^{-1}(v)dv \tag{37}$$

Using this simple observation, the problem of filter design is assessed in conjunction with the properties of HNN. For a simple HNN framework, the term u_i/τ_i in (35) can be ignored for improving the convergence speed which has bee shown in [41]. In such a case, the corresponding third term on the right side of (37) would accordingly vanish. If we can find the relation between the objective function and the filter optimization problem, such neural network will achieve an optimal solution with self-computed.

3.2. Link of Filter Design and Hopfield Neural Network

In this section, we will map the objective function for filter design to the Lyapunov energy function to obtain the required Hopfield related parameters.

3.2.1. Linear-Phase Filters

For designing FIR filter with linear phase, the Hopfield parameters of T_{ij} and I_i can be identified as

$$T_{ij} = -\sum_{l=0}^{L} \cos(i\omega_l)\cos(j\omega_l) \tag{38}$$

$$I_i = \sum_{l=0}^{L} A_d(\omega_l)\cos(i\omega_l) \tag{39}$$

where $0 \le i,\ j \le (N-1)/2$ and $p = (N+1)/2$. The relation between the filter coefficient a_i and the output state of Hopfield neural network is $v_{i+1} = a_i$, for $0 \le i \le (N-1)/2$.

3.2.2. Arbitrary Response Filters [30], [35]

For FIR filter with arbitrary amplitude and phase responses, the Hopfield parameters of T_{ij} and I_i can be identified as

$$T_{ij} = -\left[\sum_{l=1}^{L_p} \cos(i-j)\omega_l + \sum_{l=1}^{L_s} \cos(i-j)\omega_l\right] \tag{40}$$

$$I_i = \sum_{l=1}^{L_p} A_d(\omega_l)\cos[(i-1)\omega_l + \rho(\omega_l)] \tag{41}$$

where $v_i = h(i-1)$, $1 \le i,\ j \le N$ and $p = N$.

3.2.3. Nyquist Filters

The corresponding HNN parameters of T_{ij} and I_i for the design of FIR Nyquist filter can be identified as indicated as

$$T_{ij} = -\sum_{l=1}^{L} \varphi_{i-1}(\omega_l)\varphi_{j-1}(\omega_l) \tag{42}$$

$$I_i = \sum_{l=1}^{L} H_d(\omega_l)\varphi_{i-1}(\omega_l) \tag{43}$$

where $p = N_u$, $v_i = a_{i-1}, 1 \le i, j \le N_u$.

3.2.4. Two-Channel QMF Bank [37]-[39]

For the sake of simplicity, the Hopfield parameters of the designed filter banks can be written as a matrix-vector representation

$$\mathbf{T} = \left[T_{ij}\right]_{N/2\times N/2} = -(\mathbf{U}^T\,\mathbf{U} + \alpha\,\Phi_s^T\Phi_s) \tag{44}$$

$$\mathbf{I} = \left[I_i\right]_{N/2} = \mathbf{U}^T\,\mathbf{1} \tag{45}$$

where

$$\mathbf{U} = \mathbf{H}_0(\omega)\Phi_c(\omega) + \mathbf{H}_0(\omega+\pi)\Phi_c(\omega+\pi) \tag{46}$$

$$\mathbf{H}_0(\omega) = diag\left[H_0^k(\omega_1), H_0^k(\omega_2), \cdots, H_0^k(\omega_L)\right] \tag{47}$$

$$\Phi_c^{\,T}(\omega) = \left[\mathbf{c}(\omega_1), \cdots, \mathbf{c}(\omega_s), \cdots, \mathbf{c}(\omega_L)\right] \tag{48}$$

$$\Phi_s^{\,T}(\omega) = \left[\mathbf{c}(\omega_s), \mathbf{c}(\omega_{s+1}), \cdots, \mathbf{c}(\omega_L)\right] \tag{49}$$

$$\mathbf{c}(\omega) = 2\left[\cos((N-1)\omega/2),\ \cdots, \cos(\omega/2)\right]^T \tag{50}$$

and $\mathbf{1}$ is a $L\times 1$ vector whose elements are all equal to $1, v_{i+1} = f_{0,i}, 0 \le i \le N/2-1$ is the output state and $p = N/2$ is the required number of neurons.

3.2.5. Complex Filter

The Hopfield related parameters T_{ij} and I_i for implementing the least-squares design of complex FIR digital filters are obtained as followings:

$$\mathbf{T} = -\sum_{l=1}^{L}\left[\left|\mathbf{a}_1(\omega_l)\right|^2 + \left|\mathbf{a}_2(\omega_l)\right|^2\right] \tag{51}$$

$$\mathbf{I} = \sum_{l=1}^{L} A_d(\omega)\left[\mathbf{a}_1(\omega_l)\cos\rho(\omega) + \mathbf{a}_2(\omega_l)\sin\rho(\omega)\right] \tag{52}$$

where $\mathbf{v} = \mathbf{h}_c$ and $p = 2N$.

Table 1 shows the required neurons is proportional to the filter length. However, the previous neural based method depends heavily on the sampling grid in frequency. In general, the sampling grid is multiplies of the filter length. Therefore, the proposed architecture is simple and efficient.

Table 1. Comparisons of neurons required for neural based method

Filter Type	Bhattacharya [22]	Proposed
Linear-phase filter	$L+(N+1)/2$	$(N+1)/2$
Complex filter	$2(L+N)$	$2N$
Nyquist filter	$L+N_u$	N_u
All-pass filter	$2L+N$	N
Two-channel QMF bank	$L+N/2$	$N/2$

3.2.6. Equiripple Filters

In the case of equiripple filters design, the weighting function can be used to the adjustment of the Hopfield related parameters for different filters. Moreover, the Hopfiled parameters can be represented as the general form:

$$T_{ij} = -\sum_{l=1}^{L} W(\omega_l)\varphi_i(\omega_l)\varphi_j(\omega_l) \tag{53}$$

$$I_i = \sum_{l=1}^{L} W(\omega_l)A_d(\omega_l)\varphi_i(\omega_l) \tag{54}$$

where $\varphi(\omega)$ is a trigonometric function.

3.3. Close-Form Expression for Hopfield Related Parameters [33]

As seen from the Hopfield related parameters T_{ij} and I_i, they are function of sine and cosine. Therefore, some useful trigonometric properties and closed-form expressions can also be used to achieve a simplified structure and further to reduce the computational complexity of the Hopfield related parameters. Due to the fixed parameters used in the adaptation of design processing, all the iteration processing can be performed with the prepared parameters from a look-up table [33].

In the following, using the example of low-pass filter, the reduction and the closed-form for the parameters T_{ij} and I_i will be given. For the case of type I filter, we can use the trigonometric properties rewritten in the Appendix of [33] to obtain a closed-form expression for the elements of T_{ij}. The simplified expressions can be divided into four cases. The frequency sampling grid is assumed to $(L+1)$ sampled from 0 to π.

Case 1: $i = j = 0$.

$$T_{ij} = -\sum_{l=0}^{L} \cos^2\left(0 \cdot \frac{l}{L}\pi\right) = -\sum_{l=0}^{L} 1 = -(L+1) \tag{55}$$

Case 2: $i = j \neq 0$.

$$T_{ij} = -\sum_{l=0}^{L} \cos^2\left(i \cdot \frac{l}{L} \cdot \pi\right) = -\frac{L+1}{2} - \frac{1}{2}\sum_{l=0}^{L} \cos\left(2 \cdot i \cdot \frac{l}{L} \cdot \pi\right) = -\frac{L}{2} - 1. \tag{56}$$

Case 3: $i \neq j$ and $(i \pm j)$ is even.

$$T_{ij} = -\frac{1}{2}\sum_{l=0}^{L}\left[\cos\left((i+j)\frac{l}{L}\pi\right) + \cos\left((i-j)\frac{l}{L}\pi\right)\right] = -\left(\frac{1}{2} + \frac{1}{2}\right) = -1. \tag{57}$$

Case 4: $(i \pm j)$ is odd.

$$T_{ij} = -\frac{1}{2}\sum_{l=0}^{L}\left[\cos\left((i+j)\frac{l}{L}\pi\right) + \cos\left((i-j)\frac{l}{L}\pi\right)\right] = -(0+0) = 0. \tag{58}$$

As a result, the matrix $\mathbf{T} = [T_{ij}]$ can be written as a simplified form given by

$$\mathbf{T} = [T_{ij}] = -\begin{bmatrix} L+1 & 0 & 1 & 0 & \cdots & \\ 0 & \frac{L}{2}+1 & 0 & 1 & \cdots & \\ 1 & 0 & \frac{L}{2}+1 & 0 & \cdots & \\ \vdots & \vdots & \vdots & \vdots & \vdots & \frac{L}{2}+1 \end{bmatrix}. \tag{59}$$

It is noted that the parameter of I_i in (39) depends on the desired amplitude response, $A_d(\omega_l)$. Generally speaking, the symmetric filter can be used to design such filter as low-pass FIR filters. Assume the desired amplitude response with a constant gain is defined as

$$A_d(\omega_l) = \begin{cases} 1, & 0 \le \omega_l \le \omega_k = \dfrac{k}{L}\pi. \\ 0, & \omega_{k+1} < \omega_l \le \pi. \end{cases} \tag{60}$$

That is, there are k unit desired response points sampled at frequency axis. Then, the elements of I_i can also be divided as follows:

$$I_0 = \sum_{l=0}^{L} A_d(\omega_l) \cdot \cos\left(0 \cdot \frac{l}{L}\pi\right) = \sum_{l=0}^{k} 1 = k+1 \quad , \tag{61}$$

and for the case of $i \neq 0$

$$I_i = \sum_{l=0}^{L} A_d(\omega_l) \cdot \cos\left(i \cdot \frac{l}{L}\pi\right) = \sum_{l=0}^{k} \cos\left(\frac{il}{L}\pi\right) = \frac{\cos\left(\dfrac{ki\pi}{2L}\right) \cdot \sin\left(\dfrac{(k+1)i\pi}{2L}\right)}{\sin\left(\dfrac{i\pi}{2L}\right)}. \tag{62}$$

On the other hand, assume that the desired amplitude response is a linearly increasing function in the passband rather than a constant. This response can be given by

$$A_d(\omega_l) = \begin{cases} \dfrac{l}{L}, & 0 < \omega_l \le \omega_k = \dfrac{k}{L}\pi \\ 0, & \omega_{k+1} < \omega_l \le \pi. \end{cases} \tag{63}$$

Then, the elements of I_i are derived as

$$\begin{aligned} I_i &= \sum_{l=0}^{L} A_d(\omega_l) \cdot \cos\left(i \cdot \frac{l}{L}\pi\right) = \frac{1}{L}\sum_{l=1}^{k} l \cdot \cos\left(\frac{il}{L}\pi\right) = \frac{1}{L}\sum_{m=1}^{k}\sum_{l=m}^{k} \cos\left(\frac{il}{L}\pi\right) \\ &= \frac{1}{L}\left[\frac{k\sin\left(\dfrac{(2k+1)i\pi}{2L}\right)}{2\sin\left(\dfrac{i\pi}{2L}\right)} - \frac{\sin^2\left(\dfrac{ki\pi}{2L}\right)}{2\sin^2\left(\dfrac{i\pi}{2L}\right)}\right] \end{aligned} \quad . \tag{64}$$

Since the parameters of T_{ij} and I_i are all constants and can be obtained directly when the number of sampling points is determined. Therefore, the computations of these parameters can be achieved by using a look-up table. With the above discussions, the preprocessing of the Hopfield parameters can be simplified and concluded as closed-form expressions. Table 2 shows the closed-form expressions of the elements T_{ij} and I_i for the different four types of a low-pass FIR filter with constant gain in passband.

Table 2. Closed-form expressions of the related HNN parameters for the low-pass filter with constant gain [33]

Parameters / Type	T_{ij}		I_i	
I $0 \le i, j \le \frac{N-1}{2}$	(1) $i = j = 0$	$T_{ij} = -(L+1)$	(1) $i = 0$	$I_i = k+1$
	(2) $i = j \ne 0$	$T_{ij} = -\frac{L}{2} - 1$		
	(3) $i \ne j$ and $(i \pm j)$ is even	$T_{ij} = -1$	(2) $i \ne 0$	$I_i = \dfrac{\cos\left(\frac{ki\pi}{2L}\right) \cdot \sin\left(\frac{(k+1)i\pi}{2L}\right)}{\sin\left(\frac{i\pi}{2L}\right)}$
	(4) $(i \pm j)$ is odd	$T_{ij} = 0$		
II $1 \le i, j \le \frac{N}{2}$	(1) $i = j$	$T_{ij} = -\frac{(L-1)}{2}$	$I_i = \dfrac{\cos\left(\frac{(k+1)(i-1/2)\pi}{2L}\right) \cdot \sin\left(\frac{k(i-1/2)\pi}{2L}\right)}{\sin\left(\frac{(i-1/2)\pi}{2L}\right)}$	
	(2) $i \ne j$	$T_{ij} = \frac{1}{2}$		
III $1 \le i, j \le \frac{N-1}{2}$	(1) $i = j$	$T_{ij} = -\frac{L}{2}$	$I_i = \dfrac{\sin\left(\frac{ki\pi}{2L}\right) \cdot \sin\left(\frac{(k+1)i\pi}{2L}\right)}{\sin\left(\frac{i\pi}{2L}\right)}$	
	(2) $i \ne j$	$T_{ij} = 0$		
IV $1 \le i, j \le \frac{N}{2}$	(1) $i = j$	$T_{ij} = -\frac{(L+1)}{2}$	$I_i = \dfrac{\sin\left(\frac{k(i-1/2)\pi}{2L}\right) \cdot \sin\left(\frac{(k+1)(i-1/2)\pi}{2L}\right)}{\sin\left(\frac{(i-1/2)\pi}{2L}\right)}$	
	(2) $i \ne j$ and $(i \pm j)$ is even	$T_{ij} = -\frac{1}{2}$		
	(3) $(i \pm j)$ is odd	$T_{ij} = \frac{1}{2}$		

3.4. Practical Implementation

With a digital simulation of this HNN system implementing the filter design task, it has been shown by [43] that (35) can be approximated to

$$\Delta u_i = -\frac{u_i(t+\Delta t)-u_i(t)}{\tau_i} + \left(\sum_{j=1}^{p} T_{ij} v_j + I_i\right)\Delta t \approx \left(\sum_{j=1}^{p} T_{ij} v_j + I_i\right)\Delta t, \tag{65}$$

where Δt is a sufficiently small value, and $\left[u_i(t+\Delta t)-u_i(t)\right] << \tau_i$. In addition, the term u_i/τ_i was ignored for improving the convergence speed [41] [42]. Thus, the input state u_i can be iteratively updated at the $(t+1)$-th iteration by

$$u_i(t+1) = u_i(t) + \Delta u_i, \text{ for } 1 \le i \le p. \tag{66}$$

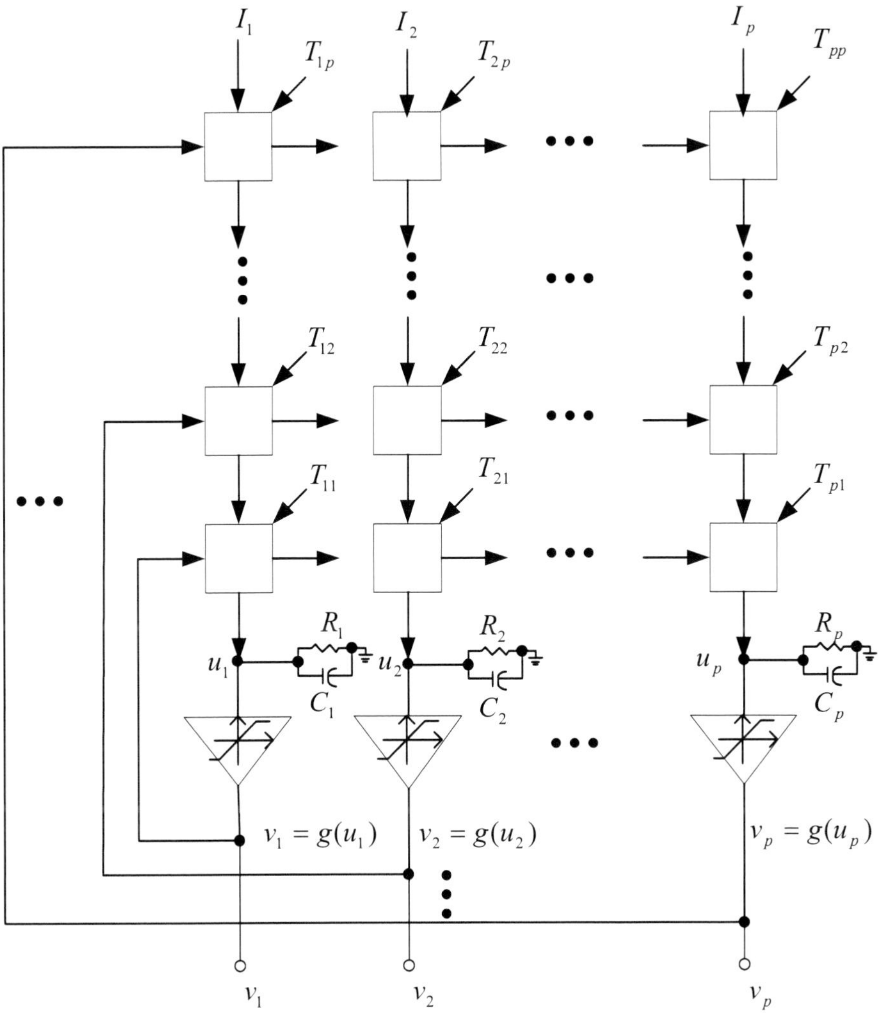

Figure 2. The proposed HNN for implementing the design of FIR digital filters.

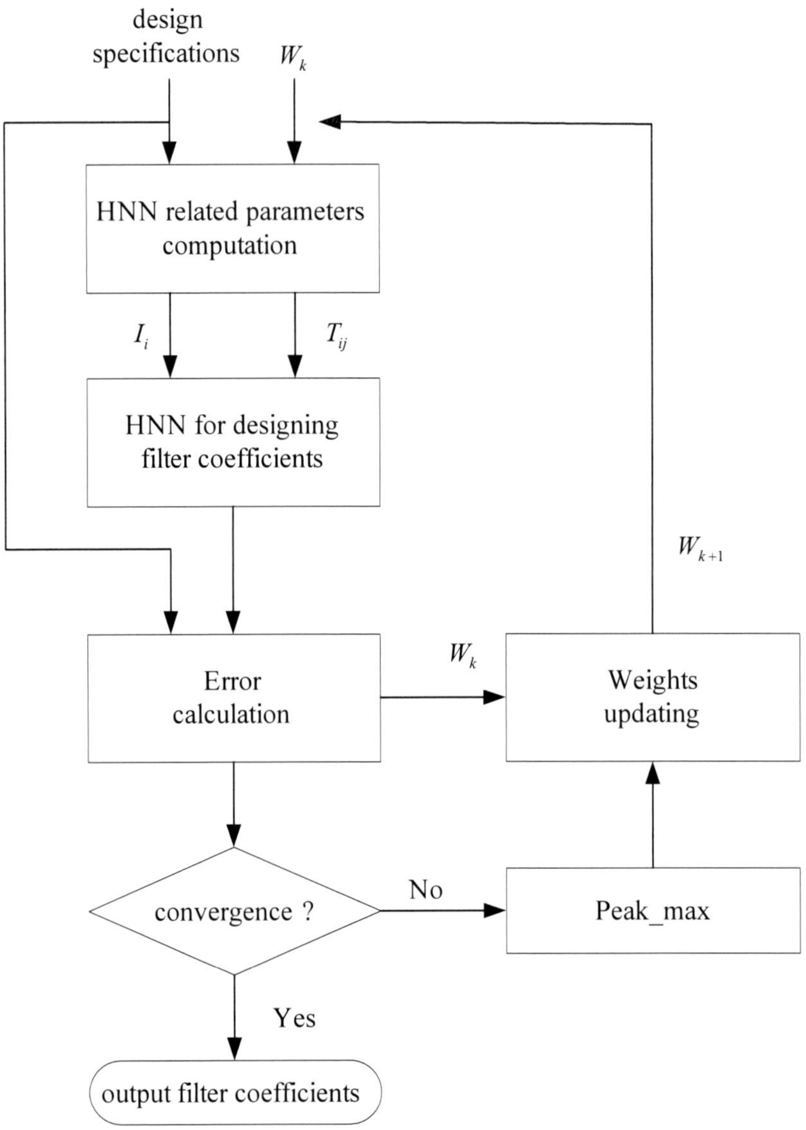

Figure 3. Flowchart for designing FIR filters using neural network.

Since the non-decreasing and nonlinear activation functions can guarantee the network to be gradient descent, any neuron with a non-negative slope can be selected such that the convergence to a local minimum is guaranteed only if $dv_i/du_i \geq 0$ for all i's. In the proposed approach, the output values are calculated using a soft limiter type neuron selected as

$$v_i = g(u_i) = \begin{cases} b, & u_i \geq b\lambda \\ \dfrac{u_i}{\lambda}, & |u_i| < b\lambda \\ -b, & u_i \leq -b\lambda \end{cases}, \tag{67}$$

where b and $1/\lambda$ are the dynamic range and slope of the soft limiter, respectively. The value of λ obviously affects the convergence speed and the value of b should be chosen larger than the maximum possible value of filter coefficients; otherwise, the HNN would be sometimes saturated. Figure 2 shows the structure of the filter design using Hopfield neural network. The flowchart for the WLS using neural network is also demonstrated in Figure 3.

4. Convergence Discussions

In this section, we will first prove the stability of the Hopfield neural network. Then the parameters that affect the convergence are investigated.

4.1. Proof of Convergence

In order to check the convergence of the network, we have to differentiate the Lyapunov energy function with respect to time, by the chain rule,

$$\frac{dE_L}{dt} = \sum_{i=1}^{p} \frac{\partial E_L}{\partial v_i}\frac{dv_i}{dt} = \sum_{i=1}^{p} \frac{\partial E_L}{\partial v_i}\frac{dv_i}{du_i}\frac{du_i}{dt}. \tag{68}$$

Based on (35), we obtain

$$\frac{\partial E_L}{\partial v_i} = -\sum_{j=1}^{p} T_{ij}v_j - I_i + \frac{u_i}{\tau_i} = -\frac{du_i}{dt}, \tag{69}$$

where $\frac{\partial}{\partial v_i}\left\{\sum_{i=1}^{p}\frac{1}{\tau_i}\int_0^{v_i} g^{-1}(v)dv\right\} = \frac{u_i}{\tau_i}$. Consequently, the result of (67) is identical to (35) except for the sign. Then, (66) can be rewritten as

$$\frac{dE_L}{dt} = -\sum_{i=1}^{p} \left(\frac{du_i}{dt}\right)^2 \frac{dv_i}{du_i} = -\sum_{i=1}^{p} \left(\frac{du_i}{dt}\right)^2 \frac{1}{\lambda} g(u_i)[1-g(u_i)]. \tag{70}$$

Since the sigmoid function described in (36) is bounded by $0 \le g(u_i) \le 1$, the term $g(u_i)(1-g(u_i)) \ge 0$. It is obvious that the sufficient condition for $\frac{dE_L}{dt} \le 0$ is

$$\frac{dv_i}{du_i} \ge 0 \qquad \text{for} \quad i = 1,2,\cdots,p \tag{71}$$

That is, the direction of the Lyapunov energy function always decreases as the system evolves. Consequently, the non-decreasing and nonlinear activation functions can always guarantee the neural network to be gradient descent except at the equilibrium points where the time derivative vanishes. Hence, the sigmoid function in (36) can be replaced with any monotonically increasing function without affecting the local minimum convergence property of the HNN when applied to any other applications. The equilibrium points of E_L occur at $\nabla E_L = -\frac{\partial E_L}{\partial v_i} = 0$. In order to determine the status of those equilibrium points, the Hessian matrix $\nabla^2 E_L$ or second partial derivative of E_L must be derived. It is shown in [44] that the equilibrium point is local minimum if the Hessian matrix of E_L is positive definite. Generally, the Hessian matrix is defined by

$$\nabla^2 E_L = \left[\frac{\partial^2 E_L}{\partial v_i\, \partial v_j}\right] = -\mathbf{T} = -[T_{ij}] \tag{72}$$

As a result, the interconnection strengths T_{ij} will determine the status of the equilibrium points. It has been shown in [45], for a symmetric connection matrix $\mathbf{T}$, (35) has a unique equilibrium point, and the equilibrium is locally stable if $\mathbf{T}$ is negative semidefinite. The simple properties can be stated as follows. Suppose $\mathbf{T}$ is negative semidefinite. Fix $g(\bullet)$ belongs to sigmoid function, and let $\mathbf{u}^{(1)} = \left[u_1^{(1)}, u_2^{(1)}, \cdots, u_p^{(1)}\right]^T$ and $\mathbf{u}^{(2)} = \left[u_1^{(2)}, u_2^{(2)}, \cdots, u_p^{(2)}\right]^T$ be two equilibrium points of (35). Then we obtain the result in a matrix-vector form

$$-\Gamma\mathbf{u}^{(1)} + \mathbf{T}\,g\left(\mathbf{u}^{(1)}\right) + \mathbf{I} = -\Gamma\mathbf{u}^{(2)} + \mathbf{T}\,g\left(\mathbf{u}^{(2)}\right) + \mathbf{I}, \tag{73}$$

where $\Gamma = diag\left[\frac{1}{\tau_1}, \frac{1}{\tau_2}, \cdots, \frac{1}{\tau_p}\right]$ is a diagonal matrix. Therefore, it implies that

$$\Gamma\left(\mathbf{u}^{(1)} - \mathbf{u}^{(2)}\right) = \mathbf{T}\left(g\left(\mathbf{u}^{(1)}\right) - g\left(\mathbf{u}^{(2)}\right)\right). \tag{74}$$

Making the inner product of both sides of (72) with vector $g\left(\mathbf{u}^{(1)}\right) - g\left(\mathbf{u}^{(2)}\right)$ yields

$$\left(g\left(\mathbf{u}^{(1)}\right) - g\left(\mathbf{u}^{(2)}\right)\right)^T \mathbf{T}\left(g\left(\mathbf{u}^{(1)}\right) - g\left(\mathbf{u}^{(2)}\right)\right) \leq 0\,. \tag{75}$$

Since $g(u)$ is a non-decreasing function and the fact that $\tau_i > 0$, one has $\frac{1}{\tau_i}\left(u_i^{(1)} - u_i^{(2)}\right)\left(g\left(u_i^{(1)}\right) - g\left(u_i^{(2)}\right)\right) \geq 0$ for all $i = 1, 2 \cdots, p$ with equality holds only for

$u_i^{(1)} = u_i^{(2)}$. Therefore, we must have $u_i^{(1)} = u_i^{(2)}$, for all $i = 1,2\cdots,p$ and consequently, $\mathbf{u}^{(1)} = \mathbf{u}^{(2)}$, i.e., the equilibrium point is unique.

Many literatures [45]-[51] have reported on the globally asymptotic stability of the HNN. In the following, we use the conclusions shown in [46]-[48] to illustrate that the unique equilibrium point is globally asymptotical stable. Let u_i^* be the unique equilibrium point of (35). Making the coordinate transformation $x_i = u_i - u_i^*$, (35) can be equivalently rewritten as the following system in term of x_i with a unique equilibrium point at $x_i = 0$,

$$\frac{dx_i}{dt} = \sum_{j=1}^{p} T_{ij} f_j(x_j) - \frac{x_i}{\tau_i}, i = 1,2,\cdots,p, \tag{76}$$

where $f_i(\rho) = g_i(\rho + u_i^*) - g_i(u_i^*)$ for $\rho \in R$. It has been shown in the literature [46]-[48] that $x_i = 0$ is global attractive, i.e.,

$$\lim_{t\to\infty} x_i(t;\ t_0, x_i^0) = 0, \tag{77}$$

where x_i^0 is the initial condition corresponding to $t = t_0$. Thus, the result $\lim_{t\to\infty} u_i(t;t_0,u_i^0) = u_i^*$ means that the unique equilibrium point is also global asymptotically stable.

4.2. Discussion of Convergence Parameters

From our simulation experiments, the sigmoid function plays an important role for the convergence. It indicates that a small value of λ results in faster convergence and smaller convergent error, but it will cause divergence occasionally. On the contrary, when λ is large, the algorithm converges eventually, but the convergence speed is slow. The property mentioned above was an inherent problem for HNN. In addition, to overcome this inherent problem of convergence, we can use the cooling process of simulated annealing [41][42], by starting from a low gain and gradually increasing the gain.

To speed the convergence, the following modification can be performed. Based on (35) and (65), it is found that for all resistance value (τ_i) and the slope of the soft limiter (λ), we have $-\frac{1}{\tau_i} + \frac{T_{ii}}{\lambda} < 0$ since $T_{ii} < 0$. In addition, in order to perform the non-decreasing nonlinear activation function, the term $\frac{1}{\tau_i}$ should be less than $\frac{|T_{ij}|}{\lambda}$ for all $i \neq j$, since T_{ij} are not always less than zero. Therefore, resistance value should satisfy $\tau_i \geq \frac{\lambda}{|T_{ij}|}$. In fact, the previous literatures always ignore the term $-\frac{u_i}{\tau_i}$, that is, $\tau_i = \infty$ (open circuit), to ensure the faster convergence of HNN. Therefore, the dynamic non-linear differential equations of the HNN can be expressed as [33],

$$C_i \frac{du_i}{dt} \approx \sum_{j=1}^{p} \frac{T_{ij}}{\lambda} u_j + I_i \,. \tag{78}$$

From (78), we can find that the value $\frac{T_{ij}}{\lambda}$ affects the convergence time of HNN. However, the values of T_{ij} only depend on the number of points, L, sampled at the frequency. Since the number of points , L, sampled at the frequency is fixed, it means the parameter λ controls the convergence speed and performance of HNN. The small values of λ will result in faster convergence and smaller convergence error. On the contrary, when λ is large, the proposed algorithm converges eventually, but the convergence speed is slow. To accelerate the convergence speed of network, the value of λ can be dynamically changed. For example, we can express a self-motivated function for λ, such as

$$\lambda(n) = \frac{0.1 + \varepsilon n}{1 + \varepsilon n} \mu L, \tag{79}$$

where the parameters ε and μ control the initial value and the finial values of λ. Thus values are $\lambda(n) \approx \frac{\mu L}{10}$ for the initial iterations and $\lambda(n) \approx \mu L$ for convergence achieved when ε is a small value. Figure 4 illustrates the convergence phenomenon with $\varepsilon = 0.001$, $\mu = 2$. It is clear that the self-motivated function converges fast initially.

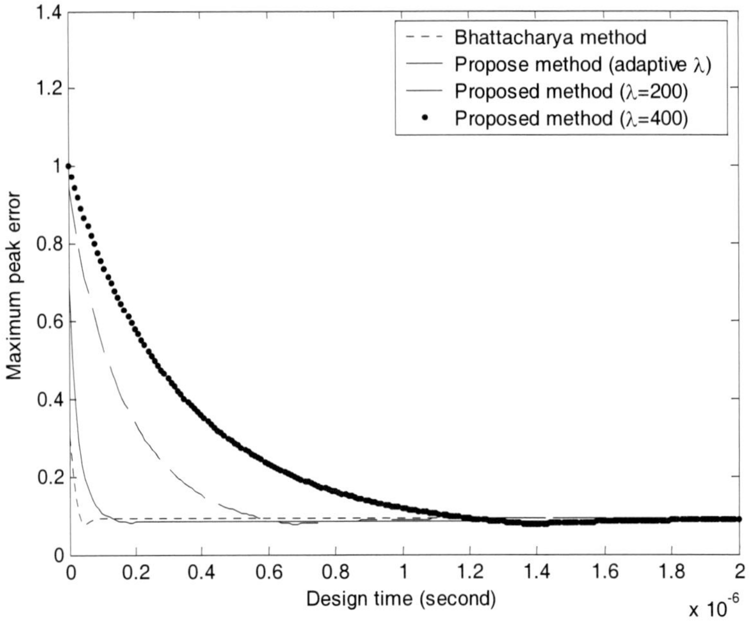

Figure 4. Plot of maximum peak error versus against evolution time.

As the system approaches to stable state, they have the same convergence performance. In addition, the proposed approach with such adaptive parameter in (79) outperforms the approach proposed by Bhattacharya *et al.*, as shown in Figure 4.

5. Design Examples

In this section, MATLAB programs are used to implement the least-squares design of FIR filters and QMF banks for evaluating the performance of the proposed neural-based approach. The Hopfield neural network is initiated with $u_i(0)=10^{-3}$, $v_i(0)=0$ for all *i*'s neuron. The software limiter is designed with $\lambda = 200$ and $b=3$. The input resistors and capacitors are chosen to be $C_i = 0.1\,\mu F$ and $R_i = 1\,\Omega$, respectively. The network is run for 5 μs, then the final outputs of the neurons are the optimal filter coefficients.

Example 1: Low-pass filter with linear phase response: The desired amplitude response $A_d(\omega)$ is a low-pass FIR filter with unity gain in the passband $\{0 \le \omega \le \omega_p = 0.30\pi\}$, zero gain in the stopband $\{\omega_s = 0.4\pi \le \omega \le \pi\}$. The filter length is chosen to be $N=31$ with the sampling frequency grid $(L+1)=180$. According to the proposed design procedure with the formulas in section 3.2.1, Figure 5 illustrates the amplitude response when the design process is completed.

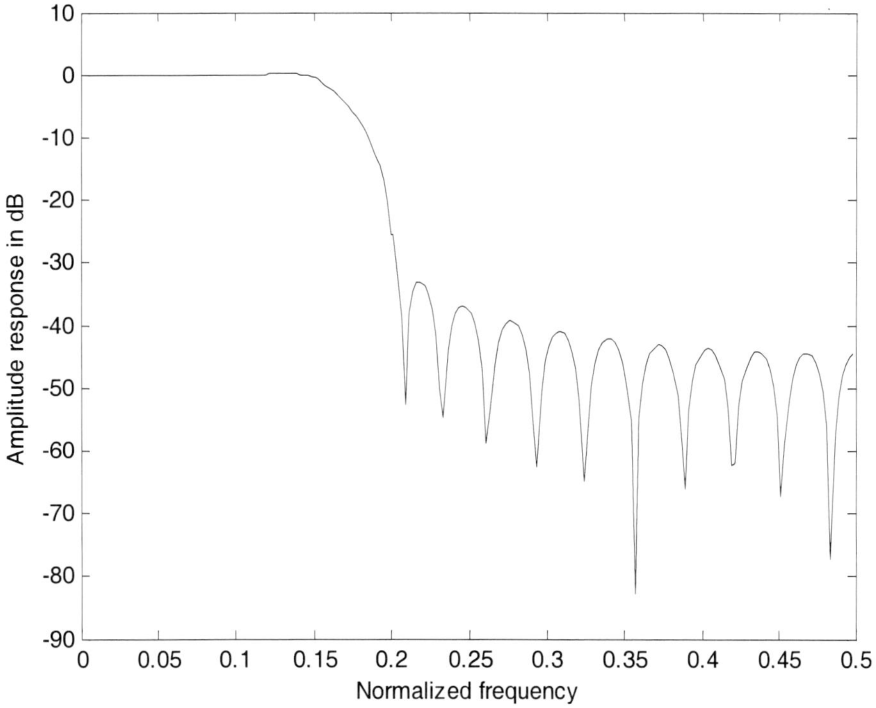

Figure 5. Amplitude response of an linear phase lowpass filter design with $N=31$.

Example 2: Chirp all-pass equalizer: For a FIR all-pass filter, the desired characteristics can be expressed by

$$H_d(\omega) = e^{j\rho(\omega)} = \cos(\rho(\omega)) + j\sin(\rho(\omega)), \qquad 0 \le \omega \le \pi \tag{80}$$

In this example, the desired phase specification of a digital chirp all-pass phase equalizer with $N = 61$ is given by

$$\rho(\omega) = -\left(\frac{N-1}{2}\right)\omega - \frac{8}{\pi}\left(\omega - \frac{\pi}{2}\right)^2 \tag{81}$$

Consequently, the desired group delay is obtained by

$$\tau_p(\omega) = -\frac{d}{d\omega}\rho(\omega) = \frac{N-1}{2} + \frac{16}{\pi}\left(\omega - \frac{\pi}{2}\right) \tag{82}$$

According to the proposed design procedure with the formulas in section 3.2.2, the amplitude and group delay responses are shown in Figure 6.

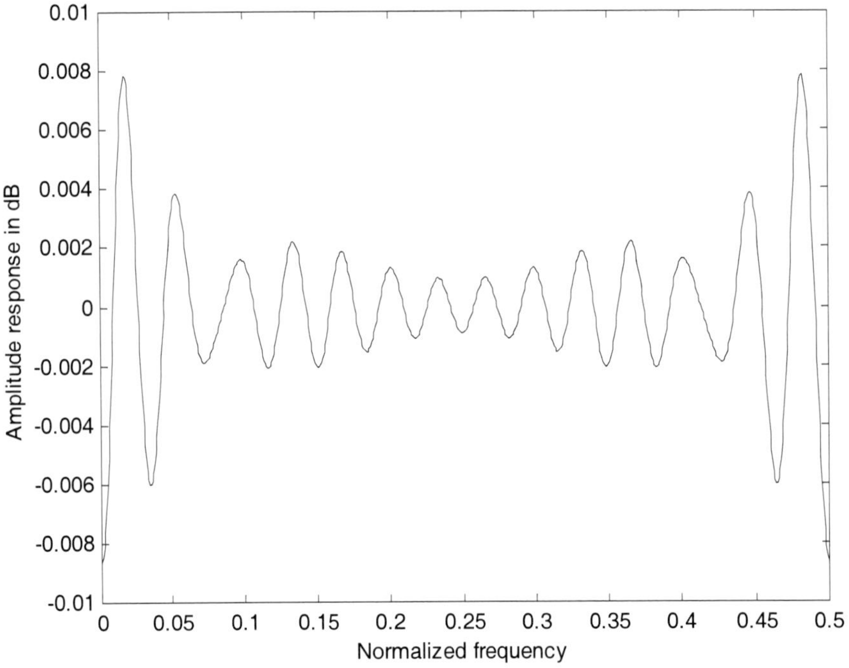

(a)

Figure 6. Continued on next page.

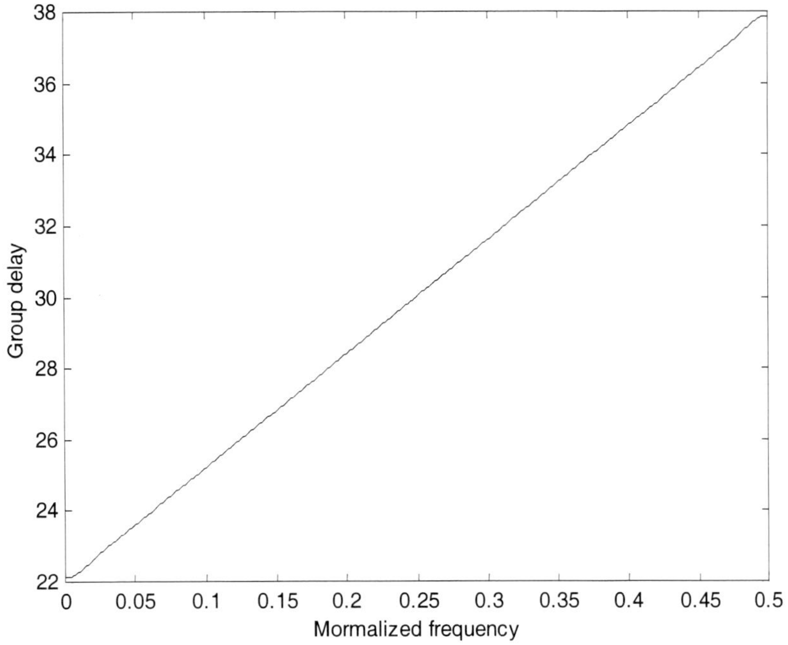

(b)

Figure 6. A digital chirp all-pass phase equalizer design with $N = 61$. (a) Amplitude response. (b) Group delay response.

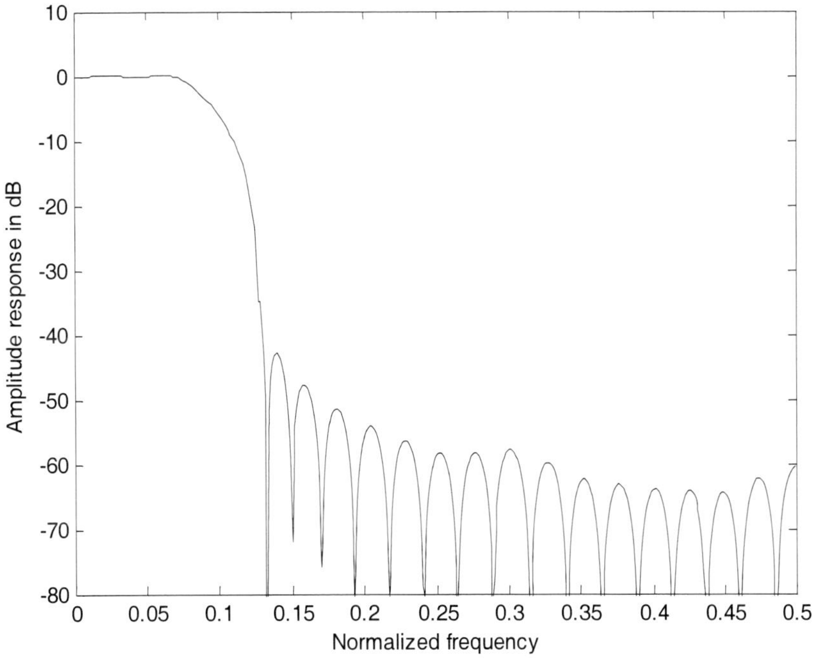

(a)

Figure 7. Continued on next page.

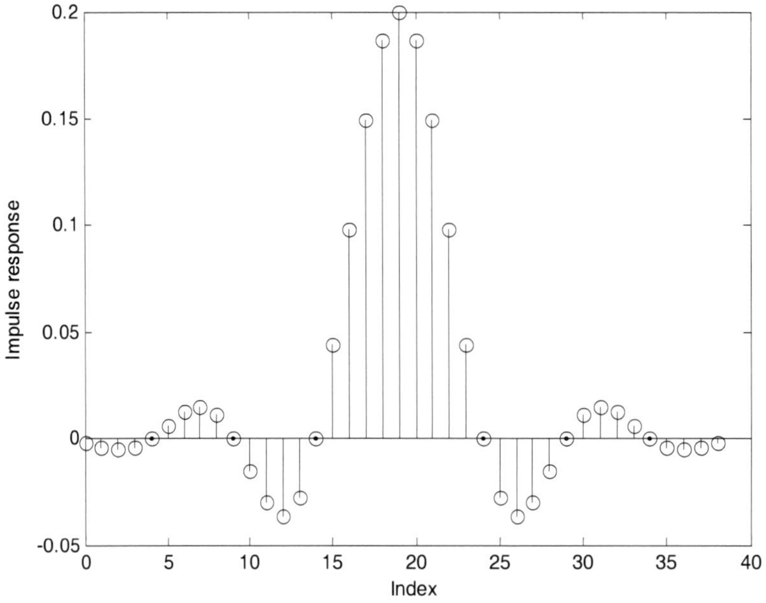

(b)

Figure 7. Design of the Nyquist filter with filter length $N = 39$, $\omega_p = 0.15\pi$, $\omega_s = 0.25\pi$, $M = 5$. (a) Magnitude response. (b) Impulse response.

Example 3: Nyquist filters: A Nyquist filter with filter length $N = 39$, $\omega_p = 0.15\pi$, $\omega_s = 0.25\pi$, $M = 5$ is designed using the neural-based approach. The frequency sampling grid is set $L = 10N$. According to the proposed design procedure with the formulas in section 3.2.3, Figure 7 illustrates the magnitude response and the corresponding impulse response of the FIR Nyquist filter.

Example 4: Two-channel QMF bank: The filter bank is designed with filter length $N = 80$ and sampling grid $L = 8N$. The relative weight of $\alpha = 1.0$ and the edge frequency of stopband is equal to $\omega_s = 0.55\pi$, respectively. The initial filter coefficients are set as $h_{0,n}^0 = 0.5$ for $n = N/2-1, N/2$ and $h_{0,n}^0 = 0$ else for starting the design process. According to the proposed design procedure with the formulas in section 3.2.4, the amplitude response of the analysis FIR filters and the overall magnitude response $D'(\omega)$ of the designed filter banks when the convergence is achieved are shown in Figure 8.

Example 5: Complex FIR filter: The desired single-passband complex FIR filter $D(\omega)$ has the following specifications:

$$D(\omega) = \begin{cases} e^{-j16\omega}, & \omega \in [-0.1\pi, 0.30\pi] \\ 0 \quad , & \omega \in [-\pi, \ -0.18\pi] \cup [0.38\pi, \pi] \end{cases} \tag{83}$$

The filter length is $N = 41$ with sampling grid of $L = 410$. According the proposed design procedure with the formulas in section 3.2.5, the amplitude and the group delay responses in the passband are shown in Figure 9. The maximum and minimum group delays in passband are 17.11 and 15.66, respectively.

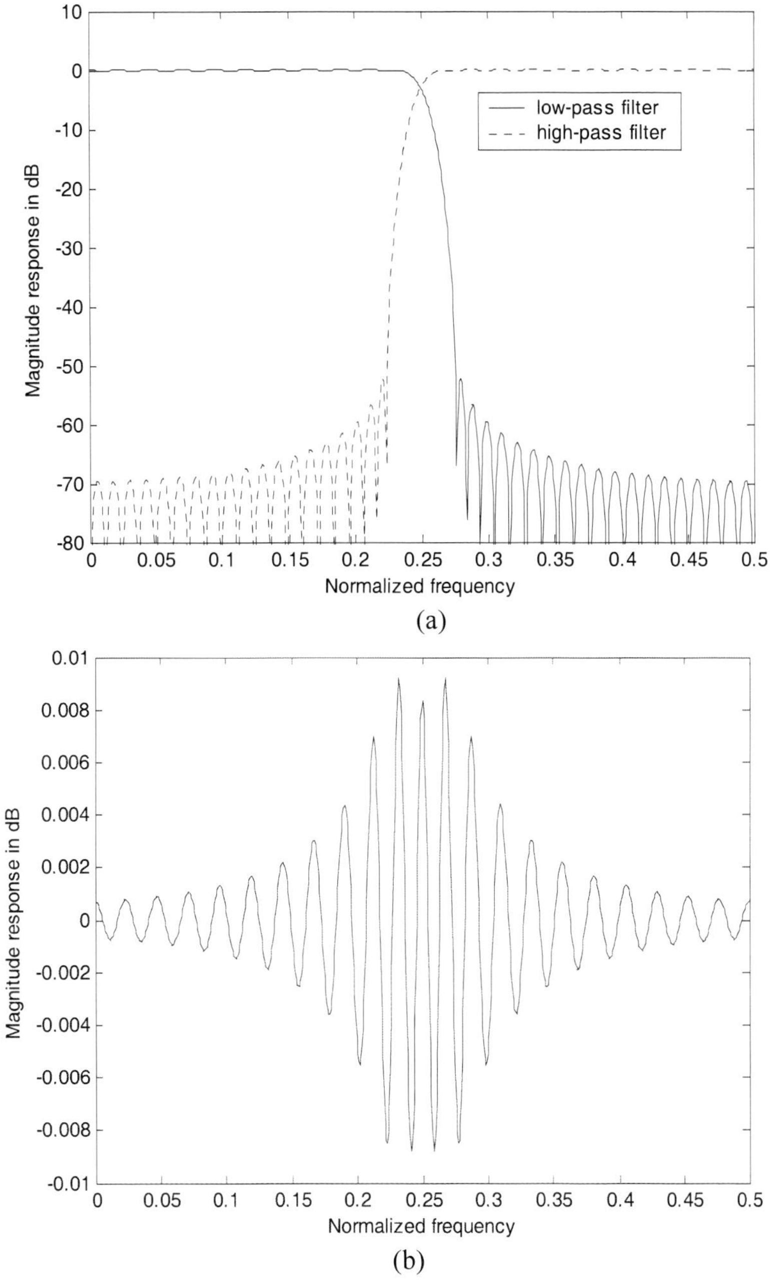

Figure 8. Design of the two-channel linear phase QMF bank with filter length $N = 80$ and edge frequency of stopband $\omega_s = 0.55\pi$. (a) Amplitude response of the analysis FIR filter. (b) Overall magnitude response of the designed filter bank.

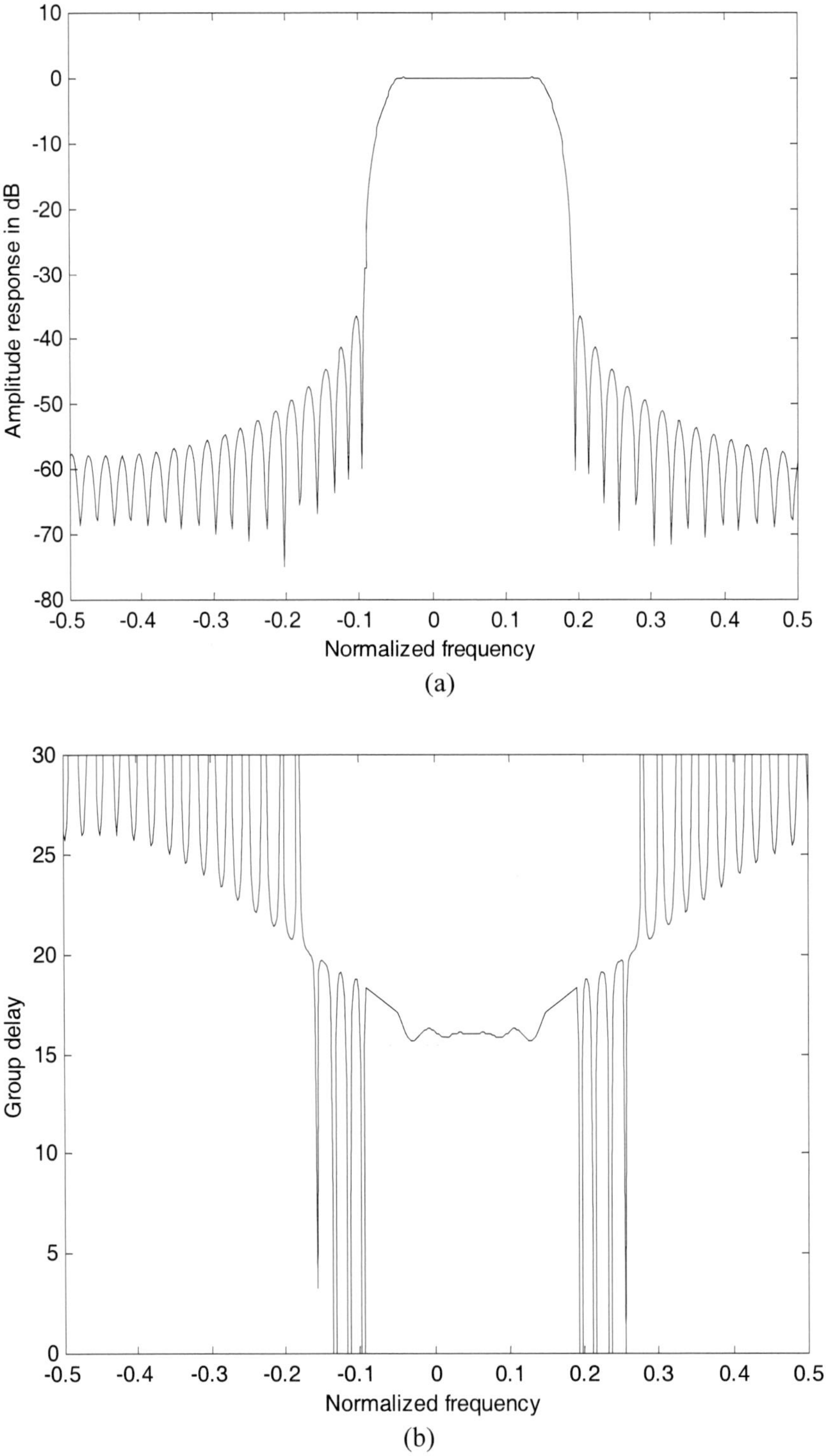

Figure 9. Design of the single-passband complex FIR filter with $N=41, \tau=16$. (a) Amplitude response in decibels (b) Group delay response.

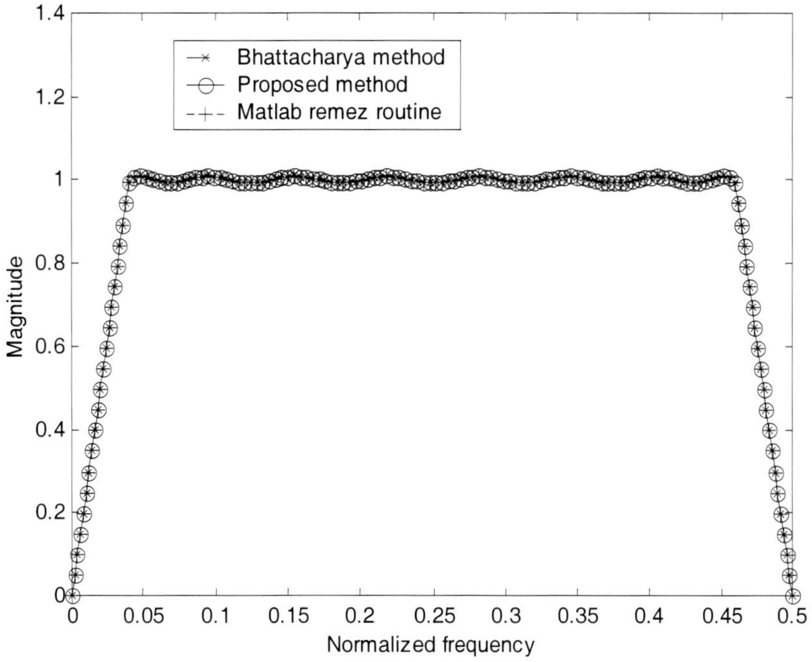

(a)

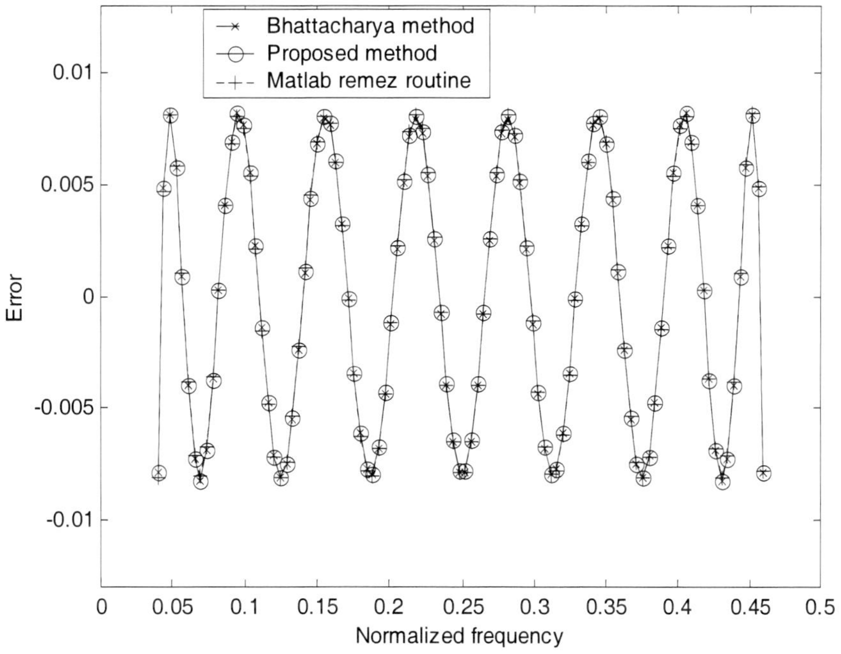

(b)

Figure 10. Equiripple design of the Hilbert transformer with $N = 31$, the lower and upper cutoff frequencies are chosen as $\omega_l = 0.08\pi$ and $\omega_h = 0.92\pi$, respectively. (a) Amplitude response. (b) Designed error response.

Example 6: Equiripple Design of Hilbert transformer: The desired amplitude response $A_d(\omega)$ is a Hilbert transformer with the gain given by

$$A_d(\omega) = \begin{cases} -1, & \omega_l \le \omega \le \omega_h \\ 1, & -\omega_h \le \omega \le \omega_l \end{cases} \tag{84}$$

where ω_l and ω_h represent the lower and upper cutoff frequencies, respectively. The filter length is chosen to be $N = 31$ with the lower and upper cutoff frequencies are chosen as $\omega_l = 0.08\pi$ and $\omega_h = 0.92\pi$, respectively. The sampling grid is taken to be 100, which is uniformly distributed in the band of interest. Nine iterations are required to reach the desired accuracy which approximates the equiripple response in this example. According to the proposed design procedure with the formulas in section 3.2.6, Figure 10 illustrates the amplitude response and the designed errors of the Hilbert transformer when the convergence is achieved. It can be seen that the proposed method can yield the same results as the previous neural-based method does.

Conclusion

A new and efficient method based on neural network structure has been developed to implement the least-squares design of different FIR filters and QMF banks. In the proposed approach, we first formulated the optimization problem of filter design to a Lyapunov energy function of the Hopfield neural network. Therefore, the filter coefficients can be obtained by dynamic nonlinear differential equations when the convergence is achieved. By using some useful trigonometric properties for the computation of Hopfield related parameters, the proposed approach not only significantly reduced the computational complexity but also outperformed another neural-based method. Furthermore, the proposed approach with inherent massive parallelism will provide a real-time processing structure for the implementation with a single IC chip.

References

[1] Oppenheim A. V., and Schafer, R. W. (1989). *Digital Signal Processing*. Englewood Cliffs, NJ: Prentice-Hall.

[2] Rabiner, L. R., and Gold, B. (1989). *Theory and Application of Digital Signal Processing*. Englewood Cliffs, NJ: Prentice-Hall.

[3] Mitra, S. K. (2001). *Digital Signal Processing: A Computer-Based Approach* (2nd Edition), McGraw-Hill.

[4] Shenoi, B. A. (2005). *Introduction to Digital Signal Processing and Filter Design*, John Wiley and Sons, INC.

[5] Smith, S. W. (1997). *The Scientist and Engineer's Guide to Digital Signal Processing*, California Technical Publishing.

[6] Tseng, C. C. (2000). Stable IIR digital differentiator design using iterative quadratic programming approach. *Signal Processing*, vol. 80, pp. 857-866.

[7] Regalia, P. A., Mitra, S. K., and Vaidyanathan, P. P. (1988). The digital all-pass filter: a versatile signal processing building block. *Proc. IEEE*, vol. 76, pp. 19-37.

[8] Tseng, C. C. (2003). Design of IIR digital all-pass filters using least pth phase error criterion. *IEEE Trans. Circuits and Systems–II: Analog and Digital Signal Processing*, vol. 50, no. 9, pp. 653-656.

[9] Samueli, H. (1988). On the design of optimal equiripple FIR digital filters for data transmission applications. *IEEE Trans.Circuits Syst.*, vol. 35, no. 12, pp. 1542-1546, 1988.

[10] Samueli, H. (1991), On the design of FIR digital data transmission filters with arbitrary magnitude specifications. *IEEE Trans. Circuits Syst.*, vol. 38, no. 12, pp. 1563-1567.

[11] Lee, J. H., and Yang, Y. H. (2003). Design of two-channel linear-phase QMF banks based on real IIR all-pass filters. *IEE Proc.-Vis. Image Signal Process.*, vol. 150, no. 5, pp. 331-338.

[12] McClellan, J. H., Parks J. W., and Rabiner, L. R. (1973). A computer program for designing optimum FIR linear phase digital filters. *IEEE Trans. Audio Electroacoust.*, vol. 21, pp. 506-526.

[13] Vaidyanathan, P. P., and Nquyen, T. Q. (1987). Eigenfilters: a new approach to least-squares FIR filter design and applications including Nyquist filters. *IEEE Trans. Circuits Syst.*, vol. 34, pp. 11-23.

[14] Pei, S, C., and Shyu, J. J. (1988). Design of FIR Hilbert transformers and differentiators by eigenfilter. *IEEE Trans. Circuits and Systems*, vol. 35, pp. 1457-1461.

[15] Sunder, S., and Ramachandran, R. P. (1994). Least-squares design of higher order nonrecursive differentiator. *IEEE Trans. Signal Processing*, vol. 42, pp. 956-961.

[16] Algazi, V. R., Suk, M., and Rim, C. S. (1986). Design of almost minimax FIR filters in one and two dimensions by WLS techniques. *IEEE Trans. Circuits and Systems*, vol. 33, pp. 590-596.

[17] Lim, Y. C., Lee, J. H., Chen, C. K., and Yang, R. H. (1992). A weighted least square algorithm for quasi-equiripple FIR and IIR digital filter design. *IEEE Trans. Signal Processing*, vol. 40, pp. 551-558.

[18] Chi, C. Y., and Chiou, S. L. (1992). A new WLS Chebyshev approximation method for the design of FIR digital filters with arbitrary complex frequency response. *Signal Processing*, vol. 29, pp. 335-347.

[19] Sunder, S. (1995). An efficient weighted least-squares design of linear-phase nonrecursive filters. *IEEE Trans. Circuits and Systems-II*, vol. 42, no. 5, pp. 359-361.

[20] Martinelli, G., and Perfetti, R. (1989). Neural network for real-time synthesis of FIR filters. *IEE Electronics Letters*, vol. 25, no. 17, pp. 1199-1200.

[21] Zhang X., and Iwakura, H. (1994). Design of FIR linear phase filters with discrete coefficients using Hopfield neural networks. *IEE Electronics Letters*, vol. 30, no. 13, pp. 1039-1040.

[22] Bhattacharya, D., and Antoniou, A. (1996). Real-time design of FIR filters by feedback neural networks. *IEEE Signal Processing Letters*, vol. 3, no. 5, pp. 158-161.

[23] Zhao, H., and Yu, J. (1997). A novel neural network-based approach for designing digital filters. *IEEE Int. Symposium on Circuits and Systems*, pp. 2272-2275.

[24] Zhao, H., and Yu, J. (1997). A novel neural network-based approach for designing 2-D FIR filters. *IEEE Trans. Circuits and Systems-I*, vol. 44, no. 11, pp. 1095-1099.
[25] Bhattacharya, D., and Antoniou, A. (1998). Design of equiripple FIR filters using a feedback neural network. *IEEE Trans. Circuits and Systems-II: Analog and Digital Signal Processing*, vol. 45, no. 4, pp. 527-531.
[26] Mladenov, V. M., and Mastorakis, N. E. (2001) Design of two-dimensional recursive filters by using neural networks. *IEEE Trans. Neural Networks*, vol. 12, no. 3, pp. 585-590.
[27] Li, X., Zhao, H., Li, M., and Yu, J. (2004). A design for variable fractional delay FIR filters. *Proc. of Int. Conf. on Communications, Circuits and Systems*, vol. 2, pp. 1131-1135.
[28] Jou, Y. D. (2005). Neural least-squares design of higher order digital differentiators. *IEEE Signal Processing Letters*, vol. 12, no. 1, pp. 9-12.
[29] Jou, Y. D. (2005). Least-squares design of digital differentiators using neural networks with closed-form derivations. *IEEE Signal Processing Letters*, vol. 12, no. 11, pp. 760-763.
[30] Jou, Y. D. (2006). Design of real FIR filters with arbitrary magnitude and phase specifications using a neural-based approach. *IEEE Trans. Circuit and Systems-II*, vol. 53, no. 10, pp. 1068-1072.
[31] Jou, Y. D., and Chen, F. K. (2006). On the use of Lyapunov functions for the design of complex FIR digital filters. 2006 *IEEE Asia-Pacific Conference on Circuits and Systems* (APCCAS 2006), pp. 741-744.
[32] Jou, Y. D., and Chen, F. K. (2007). Design of FIR digital filters using Hopfield neural network. *IEICE Trans. on Fundamentals of Electronics, Communications and Computer Sciences*, vol. E90-A, no.2, pp. 439-447.
[33] Jou, Y. D., and Chen, F. K. (2007). Least-squares design of FIR filters based on a compacted feedback neural network. *IEEE Trans. Circuit and Systems-II*, vol. 54, no.5, pp. 427-431.
[34] Jou, Y. D., Chen, F. K., Su, L. C., and Wang, S. M. (2006). Design of FIR digital filters and filter banks by neural networks. *IEEE TENCON* 2007, 30 Oct.-2 Nov. Taiwan.
[35] Jou, Y. D., Chen, F. K., and Su, L. C. (2007). An improved feedback neural network for the design of all-pass phase equalizers. 2007 *IEEE Int. Conference on Information, Communications and Signal Processing* (ICICS2007), 10-13 Dec. Singapore.
[36] Jou, Y. D. (2007). Design of two-channel linear-phase quadrature mirror filter banks based on neural networks. *Signal Processing*, vol. 87, no. 5, pp. 1031-1044.
[37] Chen, C. K., and Lee, J. H. (1992). Design of quadrature mirror filters with linear phase in the frequency domain. *IEEE Trans. Circuits and Systems-II: Analog and Digital Signal Processing*, vol. 39, pp. 593-605.
[38] Lim, Y. C., Yang, R. H., and Koh, S. N. (1993). The design of weighted minimax quadrature mirror filters, *IEEE Trans. Signal Processing*, vol. 41, no. 5, pp.1780-1789.
[39] Yang, R. H., and Lim, Y. C. (1994). Novel efficient approach for the design of equiripple quadrature mirror filters. *IEE Proc. Vision, Image and Signal Processing*, vol. 141, no. 2, pp. 95-100.
[40] Tank, D. W., and Hopfield, J. J. (1986). Simple neural optimization networks: An A/D converter, signal decision circuit, and a linear programming circuit. *IEEE Trans. Circuits and Systems*, vol. CAS-33, pp. 533-541.

[41] Hopfield, J. J. (1984). Neurons with graded response have collective computational properties like those of two-state neurons. *Proc. Nat. Sci.*, vol. 81, pp. 3088-3092.
[42] Hopfield, J. J., and Tank, D. W. (1985). Neural computation of decisions in optimization problem. *Biol. Cybern.*, vol. 52, pp. 141-152.
[43] Freeman, J. A., and Skapura, D. M. (1991). *Neural Network-Algorithms, Applications and Programming Techniques*, Addison-Wesley Publishing Company, ch. 4, pp. 144-156.
[44] Luenberger, D. G. (1984). *Linear and Nonlinear Programming* (2nd Edition) Reading, MA: Addison-Wesley.
[45] Liang, X. B., and Wu, L. D. (1998). A simple proof of a necessary and sufficient condition for absolute stability of symmetric neural networks. *IEEE Trans. Circuits and Systems-I*, vol. 45, no. 9, pp. 1010-1011.
[46] Forti, M., Manetti, S., and Marin, M. (1992). A condition for global convergence of a class of symmetric neural circuits. *IEEE Trans. Circuits and Systems-I*, vol. 39, no. 6, pp. 480-483.
[47] Forti, M., Manetti, S., and Marini, M. (1994). Necessary and sufficient condition for absolute stability of neural networks. *IEEE Trans. Circuits and Systems-I*, vol. 41, no. 7, pp. 491-494.
[48] Liang, X. B., and Wu, L. D. (1999). Global exponential stability of a class of neural circuits. *IEEE Trans. Circuits and Systems-I*, vol. 46, no. 6, pp. 748-751, 1999.
[49] Chen, L., and Aihara K. (1997). Chaos and asymptotically stability in discrete-time neural networks. *Physica D*, vol. 104, no. 3/4, pp. 286-325.
[50] Liu X., and Chen, T. (2002). A new result on the global convergence of Hopfield neural networks. *IEEE Trans. Circuits and Systems-I*, vol. 49, no. 10, pp. 1514-1516.
[51] Zhao, W., Lin, W., Liu, R., and Ruan, J. (2002). Asymptotical stability in discrete-time neural networks. *IEEE Trans. Circuits and Systems-I*, vol. 49, no. 10, pp. 1516-1520.

In: New Signal Processing Research
Editor: Takumi Maeda, pp. 233-248
ISBN: 978-1-60456-479-2

Chapter 7

NON-GAUSSIAN ADAPTIVE FILTERING

Emilio Soria-Olivas*[*]*, José David Martín-Guerrero*[+]*,
Marcelino Martínez, Antonio José Serrano López
and Rafael Magdalena
Department of Electronic Engineering,
Escola Tècnica Superior d'Enginyeria,
University of Valencia, Spain

Abstract

Adaptive filtering has become a more and more intense field of research for the last forty years. One of the starting points in almost all of this activity has been based on assuming Gaussian random signals that are then processed by the adaptive systems. Nonetheless, during the last few years, a huge number of theoretical advances and practical applications have been researched and developed without assuming the signals to be Gaussian. Some examples are underwater acoustic noise, low frequency atmospheric noise and many types of man-made noise. In addition, the presence of outliers may cause the adaptive algorithm to become unstable if the used algorithms are based on the Gaussian assumption.

The adaptive processing field has also experienced a change in its usual approaches the last few years since many new algorithms have been proposed in a non-Gaussian framework. One of the most relevant characteristics of these new algorithms is their robustness to impulsive (non-Gaussian) noise; this chapter reviews these algorithms.

Introduction

The field of signal processing has experienced a huge increase in both the number of algorithms and applications the last few years (Proakis and Manolakis, 2006), (Stein, 2000), partly due to the needs of multimedia information processing. It is remarkable that most of the

[*] E-mail address: soriae@uv.es; Telephone: +34-963543341- Fax: +34-963544353; C/ Dr. Moliner, 50. 46100 - Burjassot (València). Spain
[+] E-mail address: jdmg@uv.es; Telephone: +34-963544022- Fax: +34-963544353; C/ Dr. Moliner, 50. 46100 - Burjassot (València). Spain.

algorithms assume, as a starting point, that some of the random variables follow a Gaussian distribution (Figure 1).

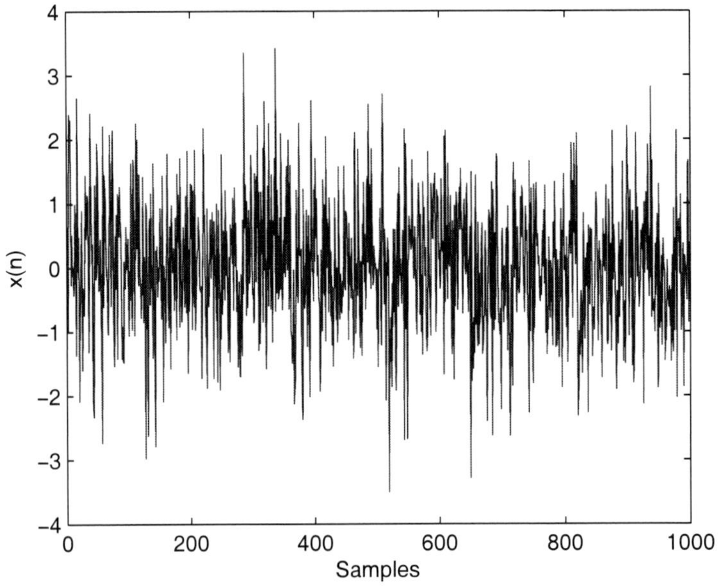

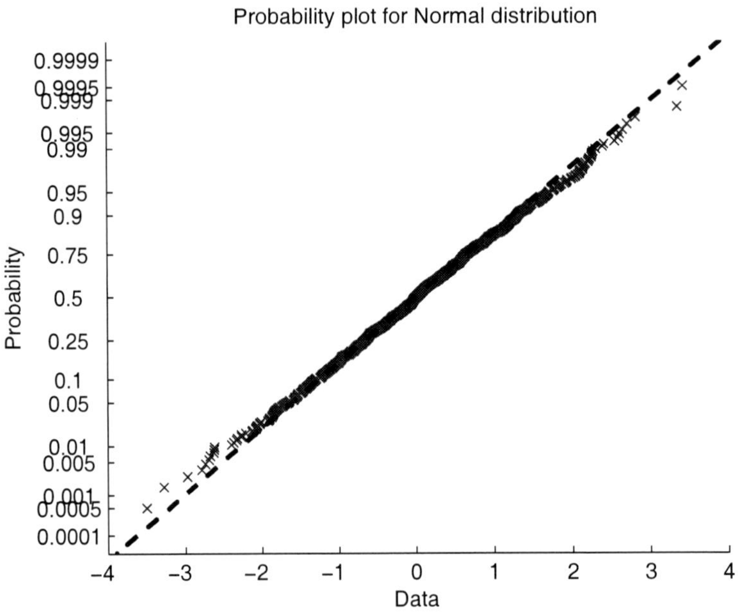

Figure 1. Continued on next page.

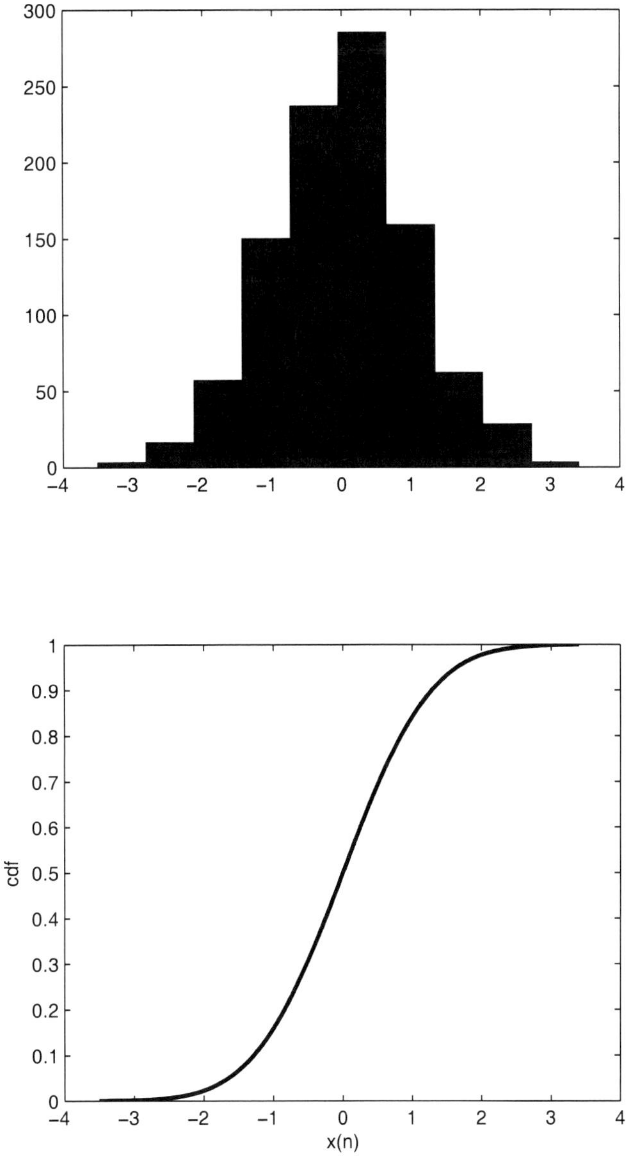

Figure 1. Different plots for a Gaussian signal: temporal samples, histogram, probability distribution function, and cumulative distribution.

The Gaussian assumption is mainly based on two reasons:

1. *Operational.* The Gaussian assumption together with the hypothesis of independent and identically distributed (*i.i.d.*) samples for the variables involved in the discrete system, guarantees reaching linear expressions in the parameters. Therefore, fast and robust algorithms can be used to solve these linear equations (some of them can solve the equation in only one iteration). However, if other distributions are taken into account, the final expressions may be non-linear, and therefore, this kind of algorithms cannot be used (Ljung, 1999).

2. *Central Limit Theorem.* This theorem has been the cornerstone of digital signal processing since its very beginning. It can be stated as follows; given a set of random variables $X_1, X_2, \ldots, X_N$ that are independent and identically distributed, *i.i.d.*, with mean value equal to zero and variance σ^2, it is shown that:

$$S = \lim_{N\to\infty} \frac{1}{\sqrt{N}} \sum_i X_i \quad (1)$$

Converges to a Gaussian variable with mean value equal to zero and variance σ^2. In other words, this theorem is a justification for the use of the Gaussian distribution by considering that any system may be affected by different interferences, so that, the final cumulative signal follows a Gaussian distribution.

The first reason is relevant when there is not availability of computational systems that are able to solve the obtained expressions after assuming the non-Gaussian hypothesis. However, the current evolution of computer systems allows solving non-Gaussian systems very quickly.

The following arguments can be posed against the second reason (Arce, 2004), (Erdogmus and Principe, 2006);

- A Gaussian distribution needs that the variance of all the variables that appear in (1) is finite, and that is not always true. If the variance is not finite, (1) gives as a result non-Gaussian impulsive variables (Arikan, Erzin and Cetin, 1994).
- It is necessary an infinite sum of variables, which is not a real situation.

It is important to emphasize the reasons for considering a non-Gaussian signal processing (Arce, 2004):

- A Gaussian model cannot model processes that show huge amplitude random values, since very low probability values are assigned to these instances.
- The values of the probability density functions may decrease quicker than the description provided by the Gaussian model.
- There are processes, such as atmospheric noise, underwater acoustic noise, low frequency electromagnetic disturbances on telephone lines, noise on old audio recordings, and urban vehicle noise that show impulsive distributions, very different from Gaussian ones (Nikias and Shao, 1995).
- Gaussian distributions cannot model asymmetric phenomena, since Gaussian distributions are unbiased.

The last few years, digital signal processing has been focused on non-Gaussian signals partly due to the reasons quoted above. In particular, it is remarkable the advances within the field of Alpha-Stables processes (Nikias and Shao, 1995), High-Order Stastistics (Mendel, 1991), Independent Component Analysis (Cichocki and Amari, 2002), (Hyvärinen, A., Karhunen, and Oja, 2001) and the use of Machine Learning techniques in digital signal processing, such as Artificial Neural Networks, Support Vector Machines and Fuzzy Systems (Kosko, 1991), (Hu and Hwang, 2001). All these advances have been partly based on the non-Gaussian assumption in some of the analyzed variables.

Adaptive Filtering

Adaptive filtering is a field within Digital Signal Processing that deals with systems that are able to adapt to the environment in order to carry out an optimum filtering. Adaptive filtering has many real applications (Haykin, 2001), (Sayed, 2003). A typical adaptive system is formed by the elements shown in Figure 2:

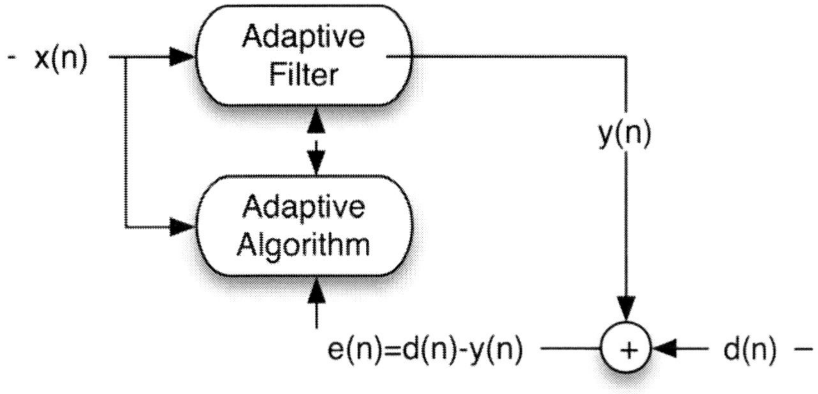

Figure 2. Schematic of an adaptive system.

- The signals that are processed by the filter: the input signal, x(n); the output signal, y(n); the desired signal, d(n); and the error signal, e(n), which is the difference between the desired and the output signals, and 'n' denotes the discrete time index.
- The structure that defines how the output signal of the filter is computed from its input signal.
- The parameters within this structure, which are the coefficients that define the transfer function of the adaptive filter. From now on, we will assume the system to be a Finite Impulse Response (FIR) system, which are the most used systems within adaptive filtering.
- The adaptive algorithm that describes how the parameters are controlled at each time instant. This algorithm is based on the minimization of a certain cost function.

The goal of an adaptive system is to achieve an error equal to zero. Its usual way of processing is the following (Sayed, 2003), (Haykin, 2001):

1. Random initialization of the parameters of the adaptive system. As quoted above, a FIR system will be used, with the following parameters (impulse response of the system) $W_n = [w_n(0) \quad w_n(1) \quad \ldots \quad w_n(L-1)]^t$ where the subscript stands for the time instant, t means transposed and L is the length of the filter. In addition, a value is set for the particular constants of the used algorithm.

2. Given a certain input $X_n = [x(n) \quad x(n-1) \quad \ldots \quad x(n-L+1)]^t$, compute its corresponding output after the adaptive filtering that will be given by $y(n) = W_n^t \cdot X_n$
3. Computation of the error committed by the adaptive filtering: $e(n) = d(n) - y(n)$.
4. Update of the coefficients of the adaptive filtering. One of the most usual updates is given by the general expression $W_{n+1} = W_n - \alpha \cdot \nabla J$, known as Delta Rule (Haykin, 2001), being ∇J the gradient of the cost function J and α a positive constant known as learning rate. The cost function is the error function that is minimized by the adaptive algorithm.
5. If the stopping criterion (error lower than a certain threshold, a maximum number of iterations, etc.) is not fulfilled, go to step 2, replacing n by $n+1$.

Most of the classical algorithms tend to take the error signal as a Normal distribution. This assumption leads to a quadratic cost function, and this, in turn, leads to a Least Minimum Square (LMS) problems in order to obtain the optimum coefficients of the adaptive system (Sayed, 2003). The Gaussian assumption cannot be accepted in some applications being this problem present in two parts of the system:

- Since the signals that are the inputs to the adaptive filter might be non-Gaussian, there are many approaches that cannot be used (Haykin, 2001).
- The error signal is the responsible of updating the weights of the adaptive filter and, hence, the characteristics of the adaptive system depend on this signal. If the error signal follows a non-Gaussian distribution, it is not suitable to propose a solution based on Gaussianity. For instance, if the error signal is affected by impulsive noise (a clear example of a non-Gaussian signal), the classical LMS algorithm is not suitable since it is based on the Gaussian assumption (Haykin, 2001), (Sayed, 2003).

Next sections will be devoted to reviewing the main approaches in the field of non-Gaussian adaptive processing.

Non-Gaussian Adaptive Processing

Gaussian adaptive processing is based on assuming that the signal error follows a Normal distribution. If the errors are *i.i.d.* with a Normal distribution of mean value equal to zero and variance σ^2, then the probability of obtaining the desired signal conditioned to the given inputs, equals (Rumelhart et alt, 1995):

$$P(d \mid x) = \frac{1}{[2 \cdot \pi]^{N/2} \cdot \sigma^N} \prod_{k=1}^{N} e^{-\frac{(d(k)-y(k))^2}{2 \cdot \sigma^2}} \qquad (2).$$

where N is the number of instances. Taking logarithms:

$$\log[P(d \mid x)] = K - \sum_{k=1}^{N} \frac{(d(k) - y(k))^2}{2 \cdot \sigma^2} \tag{3}$$

Being K a constant with the following value:

$$K = -N \cdot \left[\frac{1}{2} \cdot \log(2 \cdot \pi) + \log(\sigma) \right] \tag{4}$$

The goal is the maximization of the probability defined in (2). Therefore, the goal is the minimization of the following cost function:

$$J = \sum_{k=1}^{N} (d(k) - y(k))^2 \tag{5}$$

Those cost functions that are different from (5), are implicitly assuming that the error distribution is not Normal. It is also important to emphasize that when only the last value of the error is used to update the coefficients of the adaptive filter, the used cost function is the instant version of (5):

$$J(n) = (d(n) - y(n))^2 \tag{6}$$

This cost function together with the Delta rule is the basis of the so-called LMS algorithm (Haykin, 2001), (Sayed, 2003). This algorithm updates the coefficients of the adaptive filter, as follows:

$$W_{n+1} = W_n + 2 \cdot \alpha \cdot e(n) \cdot X_n \tag{7}$$

There are two main approaches to the non-Gaussian adaptive processing that summarize the research works carried out the last years:

Probabilistic approaches. The committed errors are supposed to be non-Normal, *i.i.d.* with mean value equal to zero and a certain variance. Two main proposals have been especially relevant:

- *Stable processes.* They are especially indicated when the signals are impulsive (Nikias and Shao, 1995).
- *Information Theory.* This proposal is based on using magnitudes from Information Theory, and in particular, Renyi entropy, that is a generalization of Shannon's entropy. This approach is usually called *Information Theoretic Learning* (ITL) (Principe, Xu and Fisher III, 2001), (Erdgomus and Principe, 2006).

3. *Computational approaches*. These approaches take the cost function as an error penalization function, and no probability density function of the error committed by the adaptive system is assumed. Within these approaches, it is remarkable those cost functions that consist of a combination of elementary cost functions (Cichocki and Amari, 2002).

Probabilistic Approach

One of the first adaptive processing algorithms proposed a variant to the quadratic cost function, the so-called sign-error LMS variant (Haykin, 2001), (Sayed, 2003). The used cost function, in its instant version is given by:

$$J(n) = |e(n)| \tag{8}$$

It is trivial to reproduce the steps from (5) to (2) to draw the conclusion that, the errors follow a Laplace distribution:

$$P(\alpha) = \frac{1}{2 \cdot \lambda} \cdot e^{-\frac{|\alpha|}{\lambda}} \tag{9}$$

This distribution is more sharply peaked than a Normal distribution. Eq. (8) shows that the contribution of high values of the errors is lower. Therefore, this distribution is immune to impulsive noise (Sayed, 2003). The cost function (8) leads to the following update of the coefficients of the adaptive filter, known as sign-variant LMS:

$$W_{n+1} = W_n + \alpha \cdot sign[e(n)] \cdot X_n \tag{10}$$

A generalization of the cost function shown in Eq. (8) will be given by the following expression:

$$J(n) = |e(n)|^p \tag{11}$$

This cost function appears when stable distributions are considered instead of Gaussian distributions (Shao and Nikias, 1993). A stable random variable is defined as follows (Nolan, 2007):

A random variable X is stable, or wide-sense stable if, for two independent copies of X, namely X_1 and X_2, for all constants a and b, for all positive constant c, and $e \in \Re$:

$$a \cdot X_1 + b \cdot X_2 \overset{d}{=} c \cdot X + e \tag{12}.$$

where $\overset{d}{=}$ stands for the distribution equality (Nolan, 2007). Some particular examples of stable distributions are the Normal distribution, Cauchy distribution and Levy distribution (Nolan, 2007). Eq. (13) shows the probability density function (*pdf*) of the Cauchy distribution, whereas Eq. (14) shows the *pdf* of the Levy distribution. These distributions are graphically represented in Figure 3.

$$pdf(x) = \frac{1}{\pi} \cdot \frac{\gamma}{\gamma^2 + (x-\delta)^2} \quad -\infty < x < \infty \tag{13}$$

$$pdf(x) = \sqrt{\frac{\gamma}{\pi}} \frac{1}{(x-\delta)^{3/2}} \cdot e^{-\left[\frac{\gamma}{2\cdot(x-\delta)}\right]} \quad \delta < x < \infty \tag{14}$$

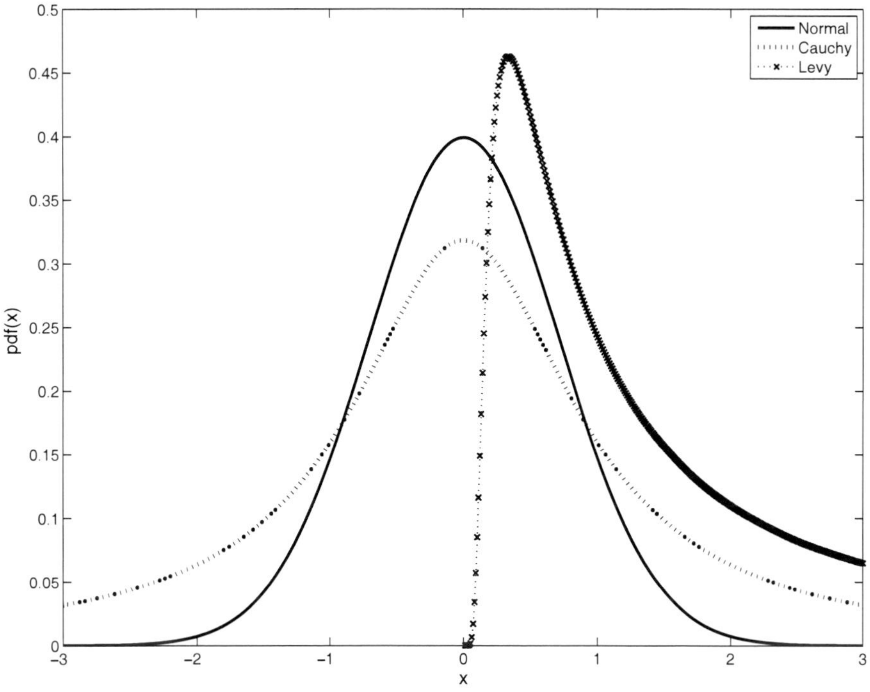

Figure 3. Probability density functions of a Normal distribution, with mean value equal to zero and unity variance, a Cauchy distribution (γ=1 , δ=0) and a Levy distribution (γ=1 , δ=0).

Stable random variables are characterized by four parameters (Shao and Nikias, 1993), (Arce, 2004):

3. Stability index, which measures the tail thickness $\alpha \in (0; 2]$
4. Scale parameter $\gamma > 0$
5. Skewness parameter $\delta \in [-1; 1]$
6. Location parameter $\beta \in R$.

It should be emphasized that there is not any expression for the *pdf* of a random variable except for particular cases, such as $\alpha=1$ for the Cauchy distribution, $\alpha=0.5$ and $\delta=1$ for the Levy distribution, and $\alpha=2$ for the Normal distribution.

In addition to the capability of modeling impulsive signals, stable random signals have many applications in signal processing due to the *Generalized Central Limit Theorem.* The main difference with the Central Limit Theorem is that it is not necessary that the i.i.d. sum gives a finite variance, as a result having a stable distribution. Therefore, stable random variables appear as the sum of other random variables (Shao and Nikias, 1993). In order to set the boundaries of the parameter p in Eq. (11), it can be shown that for a stable random variable, convergence is only achieved for the moments of *p*-order, with $0<p<\alpha\leq 2$ (Shao and Nikias, 1993), (Arikan, Cetin and Erzin 1994). This condition has to be satisfied in Eq. (11).

Update of the coefficients of the adaptive filter is carried out by the Delta rule:

$$W_{n+1} = W_n + p\cdot\alpha\cdot\left|e(n)\right|^{p-1}\cdot sign\left[e(n)\right]\cdot X_n \tag{15}$$

Eq. (15) is particularized to the sign-variant LMS shown in Eq. (8), as a particular case for $p=1$.

Eq. (11) can also be obtained from another distribution that is a generalization of the Normal distribution:

$$pdf(x) = \frac{p}{2\cdot\sigma\cdot\Gamma\left(\frac{1}{p}\right)}\cdot e^{-\left[\frac{|x-\beta|}{\sigma}\right]^{p}} \tag{16}$$

where Γ is the Gamma function (Wolfram, 1999). Eq. (16) has as particular cases, the Normal distribution if $p=2$ and the Laplace distribution if $p=1$. Figure 4 shows the probability density function of Eq. (16) for different values of p.

If $p=4$, the resulting algorithm is the so-called Least Mean Fourth (LMF) (Walach and Widrow, 1984). LMF algorithm shows problems to set the value of the learning rate since the initial weights of the coefficients of the learning rate have a strong dependence on the stability of the algorithm (Hübscher, Bermudez and Nascimento, 2007). In the LMF algorithm, the weight update is given by:

$$W_{n+1} = W_n + \alpha\cdot\left|e(n)\right|^{3}\cdot sign\left[e(n)\right]\cdot X_n \tag{17}$$

In Eq. (17), it is assumed that the error has a continuous range of values and it is not quantified. However, if the adaptive filter is applied to a classification problem (Proakis, 2000), (Haykin, 2000), then the error committed by the adaptive filter will follow a Binomial distribution:

$$P(d\mid x) = \prod_{k=1}^{N}\left[\frac{1+o(k)}{2}\right]^{\left[\frac{1+d(k)}{2}\right]}\cdot\left[\frac{1-o(k)}{2}\right]^{\left[\frac{1-d(k)}{2}\right]} \tag{18}$$

where o(k) is the output of the adaptive filter after using a non-linear function, such as the hyperbolic tangent in order to guarantee that the values of the output will be within ±1. (according to a classification problem). This approach was used to obtain recursive algorithms in classification problems (Soria-Olivas, Camps et alt, 2006a).

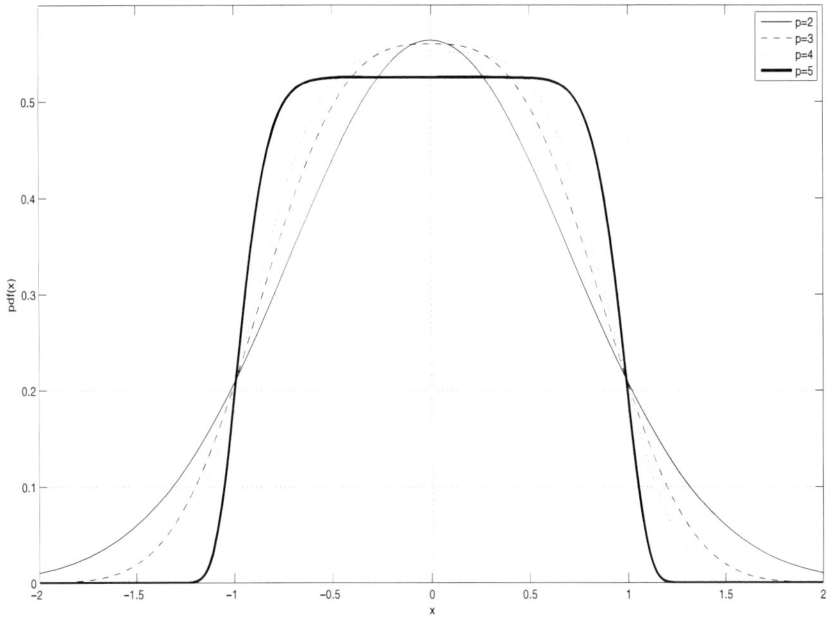

Figure 4. Gaussian density function generalized for different values of the p parameter.

To end up this section, the approach known as ITL *(Information Theoretic Learning)* will be mentioned. In this case, the cost function is replaced by other magnitudes related to Information Theory (entropy) (Principe, Xu and Fisher, III, 2001), (Erdgomus and Principe, 2006). Renyi entropy is used instead of the classical Shannon entropy because Renyi entropy can be estimated using a non-parametric estimator, such as the Parzen window (Webb, 2002). The computational burden is linear with the number of parameters while it is quadratic ig the classical entropy is used. This estimator plays a key role in the ITL theory since:

- It is not necessary to know the probability density of the analyzed process.
- The Parzen window is a consistent and efficient method and it is a "soft" estimator, too.
- It can be integrated within the gradient-descent formalism.
- It gives a gateway to more advanced techniques, e.g., Kernel Learning and Support Vector Machines, SVMs (Hu and Hwang, 2001).

ITL is a methodology that replaces the use of statistical moments of the error by a geometric interpretation of data in a probability space. Therefore, some magnitudes are changed: variance is replaced by entropy, mean square error by the minimum value of the error entropy and distances in a data space is replaced by distances in a probability space. It should be pointed out that ITL approach has a huge number of applications, such as density modeling, system identification, blind source separation, dimensionality reduction, image registration, and data clustering.

Computational Approach

In this approach, neither the distribution nor the possible dependence on the committed errors is taken into account. Instead, the cost function is analyzed from a computational point of view, i.e., it is analyzed how the cost function penalizes the committed errors.

The first cost function to be analyzed will be defined in Eq. (11). From an operational point of view, the adaptation of the coefficient p is studied in order to improve the results of the adaptive system (Rusu and Cowan, 2000). In this work, it was proved that the adaptation of the parameter p leads to an increase in the convergence speed and a decrease in the error. In (Sha and Cowan, 1995), it is shown that an adaptive algorithm with $2<p<3$ improves the stability of the system.

Another approach that comes from Eq. (11) is the combination of cost functions, especially remarkable those focused on decreasing the interference of impulsive noise:

$$J(n) = \lambda(n) \cdot |e(n)|^2 + [1 - \lambda(n)] \cdot |e(n)| \qquad (19).$$

where $\lambda(n)$ is a parameter that varies within the range (0,1). There are a number of algorithms that propose different variations for this parameter (Chambers and Avlonitis, 1997), (Papoulis and Stathaki, 2004) . Also related to this combination, there is an approach in which a convex combination of two outputs of different adaptive filters is proposed (Arenas-García, Figueiras-Vidal, and Sayed, 2006). Another work related is (Zang and Chambers, 2006) whose goal is the optimization of the length of the adaptive filter.

The first goal that led to Eq. (19) was the development of algorithms immune to impulsive noise. A cost function that is robust to *outliers*, and therefore, has been widely used in the adaptive signal processing of impulsive noise signals is given by (Cichocki andAmari, 2002):

$$J(n) = \log[\cosh(e(n))] \qquad (20)$$

The derivative of the cost function, that is involved in the update of the coefficients of the adaptive filter, is the hyperbolic tangent. In (Soria-Olivas et alt. 2006b), a fuzzy modeling of this function is proposed that leads to the following update of the coefficients of the adaptive filter:

$$W_{n+1} = \begin{cases} W_n + \alpha \cdot sign[e(n)] \cdot X_n & if\ |e(n)| > 1/\beta \\ W_n + \alpha \cdot [2 \cdot \beta - \beta^2 \cdot |e(n)|] \cdot e(n) \cdot X_n & f\ |e(n)| \le 1/\beta \end{cases} \qquad (21)$$

Parameter β is related to the error standard deviation. Therefore, this adaptive algorithm sets a threshold for the update of the coefficients, thus avoiding the influence of outliers. There are other cost functions that also propose different updates of the coefficients depending on a parameter β, e.g., Talvar and Hampel functions (Cichocki andAmari, 2002). All these three functions have been widely applied in adaptive signal processing, and in recent years, also in learning algorithms for SVMs (Rojo-Alvarez, et alt, 2005).

$$J = \begin{cases} e^2/2 & \text{for } |e| < \beta \\ \beta \cdot |e| - \dfrac{\beta^2}{2} & otherwise \end{cases} \qquad \text{Huber}$$

$$J = \begin{cases} e^2/2 & \text{for } |e| < \beta \\ \dfrac{\beta^2}{2} & otherwise \end{cases} \qquad \text{Talvar} \qquad (22)$$

$$J = \begin{cases} \dfrac{\beta^2}{\pi} \cdot \left[1 - \cos\left(\dfrac{\pi \cdot e}{\beta} \right) \right] & \text{for } |e| < \beta \\ \dfrac{2 \cdot \beta^2}{\pi} & otherwise \end{cases} \qquad \text{Hampel}$$

These cost functions lead to the following update of the synaptic weights:

$$W_{n+1} = \begin{cases} W_n + \alpha \cdot e(n) \cdot X_n & if\ |\mathrm{e(n)}| < \beta \\ W_n + \alpha \cdot \beta \cdot sign[e(n)] \cdot X_n & if\ |\mathrm{e(n)}| \geq \beta \end{cases} \qquad \text{Huber}$$

$$W_{n+1} = \begin{cases} W_n + \alpha \cdot e(n) \cdot X_n & if\ |\mathrm{e(n)}| < \beta \\ W_n & if\ |\mathrm{e(n)}| \geq \beta \end{cases} \qquad \text{Talvar} \qquad (23)$$

$$W_{n+1} = \begin{cases} W_n + \alpha \cdot \beta \cdot \sin\left[\dfrac{\pi \cdot e(n)}{\beta} \right] \cdot X_n & if\ |\mathrm{e(n)}| < \beta \\ W_n & if\ |\mathrm{e(n)}| \geq \beta \end{cases} \qquad \text{Hampel}$$

The main problem of these functions is the definition of the threshold. Moreover, it should be emphasized that in the three last algorithms, a "crisp" commutation criterion is applied, i.e., either one or other function is applied depending on the value of the error. A simple variation consists on the application of a fuzzy threshold for this commutation (Soria-Olivas, 2007c).

In (Kardec, Príncipe, Takeuchi and Ohnishi, 2006), the approach of Eq. (19) is carried out based on a linear combination of even powers of the error, with no limit for the number of functions taking part in the linear combination:

$$J = \sum_{k=1}^{N} a_k \cdot e^{2 \cdot k} \qquad (24)$$

Being a_k constant parameters. This cost function leads to the following update:

$$W_{n+1} = W_n + \alpha \cdot \left[\sum_{k=1}^{N} a_k \cdot e^{2 \cdot k - 1} \right] \cdot X_n \qquad (25)$$

Another research line within the computation approach to non-Gaussian adaptive filtering is based on considering the learning rate as an error function, in which case, the result will not be a quadratic function. There are many works related to this topic, basically based on increasing the convergence speed while decreasing the committed error in steady state according to the following principle: "Increase the learning rate far away from the minimum, and decrease it close to the minimum".

The variation of the error function depending on its evolution can also be applied to reduce the influence of impulsive noise in adaptive filters. In (Soria et alt, 2007b), the last errors committed by the adaptive filter are used to avoid the influence of outliers. The idea is based on arranging the last errors committed by the adaptive filter, and according to this arrangement, update the coefficients, as follows:

$$W_{n+1} = W_n + \alpha \cdot \left[\sum_{k=0}^{N-1} h_k \cdot e(n-k) \right] \cdot X_n \qquad (26)$$

where factor h_k depends on the location of the error in the arrangement quoted above. A related algorithm is the so-called Median LMS (MLMS), whose update is given by:

$$W_{n+1} = W_n + \alpha \cdot median\left[e(n) \cdot X_n, e(n-1) \cdot X_{n-1}, \ldots e(n-L) \cdot X_{n-L} \right]$$

In this algorithm, the median of the last gradients of the cost function is computed in order to use it in the update of the coefficients of the adaptive filter. This median algorithm belongs to the so-called Order Statistic LMS algorithms (Haweel and Clarkson, 1992), whose main characteristic is the filtering of the gradient of the error function. Therefore, the update is different from the classical LMS update, and hence, it can also be considered as a non-Gaussian adaptive algorithm.

Conclusions

This chapter has provided a general review on non-Gaussian adaptive processing. This field of research has been very active and prolific recently. This is partly due to the fact that in many modern applications, the Central Limit Theorem requirements are not fulfilled. For instance, the variance is not finite in impulsive interferences, what leads to the Generalized Central Limit Theorem, and this in turn, to Alpha-Stable distributions.

Another approach mentioned in the chapter is the use of Information Theory for the update of the coefficients of the adaptive system. In particular, no assumption with regard to the error distribution is taken into account, and Renyi entropy is used to update the coefficients of the adaptive filter.

The computational approach of the cost function to update the coefficients of the adaptive filter has also been taken into account. The cost function under the influence of different kinds of errors has been analyzed, and several improvements have been proposed without assuming any error distribution.

To end up this chapter, the authors would like to stress the wide range of algorithms that can be applied basically due to fact that different error distributions may need different solutions.

References

Arce, G (2004). *Nonlinear Signal Processing: A Statistical Approach*, Prentice, Wiley.

Arenas-García, J., Figueiras-Vidal, A., Sayed, A (2006). Mean Square Performance of a Convex Combination of Two Adaptive Filters. *IEEE Transactions on Signal Processing*, **54** (3), 1078-1090.

Arikan, O., Cetin, E., Erzin, E. (1994). Adaptive Filtering for Non-Gaussian Stable Processes. *IEEE Signal Processing Letters*, **1** (11),163-165.

Chambers, J., Avlonitis, A. (1997). A Robust Mixed-Norm Adaptive Filter Algorithm, *Signal Processing Letters,* **4** (2), 46-48.

Cichocki, A., Amari, S. (2002). *Adaptive Blind Signal and Image Processing*, Wiley.

Erdogmus, D., Principe, J.C. (2006). From Linear Adaptive Filtering to Nonlinear Information Processing. *IEEE Signal Processing Magazine*, **33** (6) 14-32.

Haykin, S. (2000). *Communication Systems*. Wiley, 2000.

Haykin, S. (2001). *Adaptive Filter Theory*, (4th Edition), Prentice Hall.

Haweel, T.I., Clarkson, P.M. (1992). A Class of Order Statistic LMS Algorithms. *IEEE Transactions on Signal Processing*, **40** (1), pp 44-53.

Hu, Y.H., Hwang, J.N. (Editors) (2001) *Handbook of Neural Network Signal Processing*, CRC Press, 2001

Hübscher, P.I., Bermudez, J.C.M., Nascimento, V.H. (2007). A Mean-Square Stability Analysis of the Least Mean Fourth Adaptive Algorithm. *IEEE TRansactions on Signal Processing,* **55** (8), 4018-4028.

Hyvärinen, A., Karhunen, J., Oja, E. (2001) *Independent Component Analysis*, Wiley.

Kardec, A., Príncipe, J.C., Takeuchi, Y., Ohnishi, N (2006). Using Non-Linear Even Functions for Error Minimization in Adaptive Filters. *Neurocomputing*, **70** (1-3), 9-13.

Kosko, B. (1991). *Neural Networks for Signal Processing*, Prentice Hall, 1991.

Ljung, L. (1999). *System Identification. Theory for the user*. Upper Saddle River, N.J.: Prentice Hall.

Mendel, J.M. (1991) (Tutorial on Higher-Order Statistics (Spectra) in Signal Processing and System Theory:Theoretical Results and Some Applications, *Proceedings of IEEE*, **19** (3), 278-305.

Nikias, C.L., Shao, M. (1995). *Signal Processing with Alpha-Stable Distributions and Applications (Hardcover),* Wiley.

Nolan, J. (2007). Stable Distributions: Models for Heavy-Tailed Data, Birkhauser

Oppenheim, A., V. Schafer, R.W., Back, J.R (1999). *Discrete-Time Signal Processing* (2nd Edition), Prentice Hall.

Papoulis, E., Stathaki, T (2004). A Normalized Robust Mixed-Norm Adaptive Algorithm for System Identification. *IEEE Signal Processing Letters*, **11** (1), 56-59.

Principe, J.C., Dongxin, Xu, Fisher III, J.W. (2001). *Information Theoretic Learning in Unsupervised Adaptive Filterering* edited by Haykin, S., Wiley.

Proakis J.G., Manolakis, D.K (2006). Digital Signal Processing (4th Edition), Prentice Hall.

Proakis, J. (2000) *Digital Communications*, McGraw-Hill.

Rojo-Alvarez, J.L., Camps-Valls, G., Martínez-Ramón, M., Soria-Olivas, E., Navia-Vazquez, A., Figueiras-Vidal, A (2005). *Support Vector Machines Framework for Linear Signal Processing*, (85), **12**, 2316-2326.

Rumelhart, D., Durbin, R., Golden, R., Chauvin Y. (1995) *Backpropagation: Theory, Architectures and Applications*. Lawrence Erlbaum.

Rusu, C., Cowan, C.F.N, (2000). Adaptive Data Echo Cancellation Using Cost Function Adaptation, *Signal Processing*, **80** (11), 2457-2473.

Sayed, A. (2003). *Fundamentals of Adaptive Filtering*, Wiley, 2003.

Shah, S.A., Cowan, C.F.N. (1995) Modified Stochastic Gradient Algorithm using Nonquadratic Cost Functions for Data Echo Cancellation, *IEE Proceedings in Vision, Image and Signal Process*, **142** (3), 187-191.

Shao, M., Nikias, C.L. (1993). Signal Processing with Fractional Lower Order Moments: Stable processes and Their Applications. *Proceedings of the IEEE*, **81** (7), 986-1010.

Soria-Olivas, E., Martínez, M., Calpe, J., Guerrero, J., A.J. Serrano. (1999). Nuevo Algoritmo Recursivo para la Obtención del ECG FEal, *Revista Brasileira de Engenharia Biomédica*, **15** (3) 135-140.

Soria-Olivas, E., Camps, G., Martín-Guerrero, J.D., Calpe, J., Vila, J., Serrano, A.J. (2006a). Non-linear RLS-based Algorithm for Pattern Classification. *Signal Processing*, **86** (5), 1104-1108.

Soria-Olivas, E., Martín-Guerrero, J.D., Calpe, Serrano, A.J., Chambers, J (2006b). Robust Adaptive Algorithm with Low Computational Cost, *Electronic Letters*, **42** (1), 60-62.

Soria-Olivas, E., Martín-Guerrero, J.D., Serrano, A.J, Calpe, J., Chambers, J (2007a). Steady-State and Tracking Analysis of a Robust Adaptive Filter with Low Computational Cost. *Signal Processing*, **87** (1), 210-215.

Soria-Olivas, E., Martín-Guerrero, J.D., Martínez, M., Rosado, A., Calpe, J (2007b). *A Family of Adaptive Algorithms Robust to Impulsive Noise, 15 th International Conference on Digital Signal Processing* 2007, Cardiff, UK.

Soria-Olivas, E., Martín-Guerrero, J.D., Serrano, A.J., Martínez, M., Magdalena, R (2007c). A Fuzzy Logic Approach to Robust Cost Functions in Adaptive Systems. *WSEAS*, World Scientific and Engineering Academy and Society, Venice.

Stein, J (2000). *Digital Signal Processing: A Computer Science Perspective*, Wiley.

Webb, A.R. (2002) *Statistical Pattern Recognition. Second Edition*, Wiley.

Walach, E., Widrow, B. (1984). The Least Mean Fourth (LMF) Adaptive Algorithm and Its Family. *IEEE Transactions on Information Theory*, **30** (2), 275-283.

Wolfram, S (1999). The Mathematica Book, Cambridge University Press.

Zang, Y., Chambers, J.A (2006). Convex Combination of Adaptive Filters for Variable Tap-Length LMS Algorithm. *IEEE Signal Processing Letters*, **13** (10), 628-631.

In: New Signal Processing Research
Editor: Takumi Maeda, pp. 249-278
ISBN 978-1-60456-479-2

Chapter 8

EXPLORING THE LINK BETWEEN BOLSTERED CLASSIFICATION ERROR AND DATASET COMPLEXITY FOR GENE EXPRESSION-BASED CANCER CLASSIFICATION

***Oleg Okun*[1]*, Giorgio Valentini*[2] *and Helen Priisalu*[3]**
[1]University of Oulu, Finland
[2]University of Milan, Italy
[3]Teradata, Finland

Abstract

Gene expression profiles were shown to be useful in genomic signal processing when discriminating between cancer and normal (healthy) examples and/or between different types of cancer. K-nearest neighbors (k-NN) is one of the classification algorithms that demonstrated good performance for gene expression based cancer classification. Given that distance metric is fixed, the conventional k-NN has a single parameter (k - the number of nearest neighbors for each example) to set, which makes k-NN a very attractive choice in addition to the fact that it does not need training.

Classification performance of any classifier, including a k-NN, is typically characterized by classification error achieved on independent examples, which are often unavailable for the considered task. Thus, unbiased and low-variance error estimation is of ultimate importance in this case. We found that bolstered error satisfies these requirements and it was therefore chosen for our study. Bolstered error estimation is built on random sampling in the neighborhood of each example (with example-dependent neighborhood radius) and computing the number of errors made on such artificially created data. Because of random sampling, all examples can be employed in assessing the error, unlike cross-validation or bootstrap procedures.

In this work, we investigate the link between k-NN bolstered error and dataset complexity characterizing how difficult to classify a certain dataset. Our measure for the dataset complexity is the normalized Wilcoxon rank sum statistic. Through extensive simulation coupled with the copula method for analysis of association in bivariate data, we show that dataset complexity and bolstered error are related in terms of several dependence types such as positive quadrant dependence, tail monotonicity, and stochastic monotonicity.

As a result, we propose a new scheme for generating ensembles of k-NN classifiers, which is based on the selection of low complexity feature subsets for k-NNs in the ensemble, which constitutes to choosing accurate k-NNs according to the found dependence relation. The candidate subsets are randomly sampled from the whole set of the original features in order to make predictions of individual k-NNs diverse.

Experiments carried out on eight gene expression datasets containing different types of cancer demonstrate that our ensemble generating scheme is superior (in terms of bolstered resubstitution error) to a single best classifier in the ensemble and to the traditional ensemble construction scheme that is ignorant of dataset complexity. It also outperforms the redundancy-based filter, especially designed to remove irrelevant genes.

1. Introduction

According to [1], genomic signal processing (GSP) is the engineering discipline that studies the processing of genomic signals, i.e. the measurable events carried out by the genome. Gene expression is a two-stage process including the transcription of deoxyribonucleic acid (DNA) into messenger ribonucleic acid (mRNA) which is then translated into protein by the ribosome. When a protein is produced, a gene is said to be expressed. Proteins are large compounds of amino acids joined together in a chain and they are essential parts of organisms and participate in every process within cells. GSP deals with extracting information from gene expression measurements, which is then processed, analyzed and used for gaining biological and medical knowledge. Recent advances in microarray technology facilitate measurement of gene expression levels for thousands of genes at once.

Cancer classification based on gene expression levels, which is the subject of our study, is one of the topics of intensive research in GSP, since it was shown in numerous works [2, 3, 4] that expression levels provide valuable information for discrimination between normal and cancer examples.

However, the classification task is not easy since there are typically thousands of expression levels versus few dozens of examples. In addition, expression levels are noisy due to the complex procedures and technologies involved in the measurements of gene expression levels, thus causing ambiguity in classification. Hence, the original set of genes must be reduced to those genes that are relevant to discrimination between different classes. This operation is called feature or gene selection[1]. Genes preserved or selected as a result of feature selection are then used to classify data.

Basically, general feature selection methods widely applied to the datasets with many more examples than features can be (and they are) utilized for gene selection, too. These methods can be approximately divided into filters and wrappers. Wrappers base their decisions on which feature(s) to select by employing a classifier. For small sample size gene expression datasets, they can easily introduce the induction bias when some genes are preferred over the others. Since those preferred genes might not be always relevant to cancer classification, we turn our attention to the filters that do gene selection independently of any classifier and solely based on data characteristics. Hence, they are less prone to the

[1]Further, the words 'feature' and 'gene' will be used interchangeably, since they have the same meaning in this work.

induction bias. It should be however noted that even filters might not be able to completely avoid this bias since class labels of examples often guide gene selection in the filter model.

This brief analysis prompted us to concentrate on random gene selection where genes to be used with a classifier, are randomly sampled from the original set of genes, irrespectively of class information and a classifier. The additional fact that caused us to make such a decision was the work [5], where it was concluded that *differences in classification performance among feature selection algorithms are less significant than performance differences among the error estimators used to implement these algorithms* . In other words, the way of how error is computed has a larger influence on classification accuracy than the choice of a feature selection algorithm. Among several error estimators we opted for the bolstered resubstitution error because it provides a low-bias, low-variance estimate of classification error, which is what is needed for high dimensional gene expression data [6].

However, a single random sample cannot guarantee that sampled genes will lead to good classification results. Hence, we need to sample genes several times to be more certain about the outcome, which, in turn, implies several classifications have to be done. Thus, it is natural to combine predictions of several classifiers into a single prediction. Such a scheme is termed an ensemble of classifiers in the literature [7]. It is well known that under certain conditions an ensemble can outperform its most accurate member. In the context of dimensionality reduction, an ensemble composed of a small number of classifiers, each working with a small subset of genes, results in the desired effect. For instance, if the original set comprises 1000 genes, five classifiers, each employing 20 genes [2], lead to a 10-fold dimensionality reduction. Thus, using an ensemble instead of a single classifier can be beneficial for both dimensionality reduction and classification performance.

As a base classifier in the ensemble, a k-nearest neighbor (k-NN) is used because it performed well for cancer classification, compared to more sophisticated classifiers [8]. Besides, it is a simple method that has a single parameter (the number of nearest neighbors) to be pre-defined, given that the distance metric is Euclidean.

Eight gene expression datasets containing different types of cancer are utilized in our work. We begin with the recently proposed redundancy-based filter [9], especially designed to filter out irrelevant genes. As this filter turned to be quite aggressive in removing genes, we proceed to experiment with k-NN ensembles. In particular, we compare two ensemble schemes: one relying on the concept called dataset complexity when choosing which subsets of features to include into an ensemble and another ignoring dataset complexity. Based on the copula method [10, 11, 12], which is useful in exploring association (dependence or concordance) relations in multivariate data, we hypothesize that there is positive dependence between dataset complexity measured by the Wilcoxon rank sum statistic [13] and the bolstered resubstitution error [6], with low (high) complexity associated with small (large) error. As a result, selecting a low-complexity subset of genes implies an accurate k-NN, which, in turn, implies an accurate k-NN ensemble. Experimental results clearly favor the complexity-based scheme of k-NN ensemble generation over 1) the complexity-ignorant scheme, 2) a single best k-NN in the ensemble, and 3) the redundancy-based filter, thus confirming our hypothesis.

The chapter has the following structure. Section 2. describes gene expression datasets

[2]Let us assume that there is no overlap between different subsets of genes.

employed in experiments. The redundancy-based filter is briefly introduced in Section 3.. Section 4. defines the dataset complexity characteristic while Section 5. defines bolstered resubstitution error. The link between the two is explored and analyzed in Section 6.. Uncertainty of single classification is discussed in Section 7.. Two ensemble generating schemes based on random feature selection are presented in Section 8. and experimental results obtained with them on eight datasets are given in Section 9.. Finally, Section 10. concludes the paper. Programming code for random number generation and mathematical derivations related to copulas are placed in two appendices.

2. Gene Expression Datasets

The following eight datasets were chosen for experiments.

2.1. SAGE Dataset 1

SAGE stands for Serial Analysis of Gene Expression [14, 15]. This is technology alternative to microarrays (complementary DNA and oligonucleotides). Though SAGE was originally conceived for use in cancer studies, there is not much research using SAGE datasets regarding ensembles of classifiers. SAGE provides a statistical description of the mRNA population present in a cell without prior selection of the genes to be studied [16]. This is the main distinction of SAGE over microarray approaches (cDNA and oligonucleotide) that are limited to the genes represented in the chip. SAGE "counts" the number of transcripts or tags for each gene, where the tags substitute the expression levels. As a result, counting sequence tags yields positive integer numbers in contrast to microarray measurements.

In the chosen dataset [17], there are expressions of 822 genes in 74 cases (24 cases are normal while 50 cases are cancerous) [18]. The dataset contains 9 different types of cancer. We decided to ignore the difference between cancer types and to treat all cancerous cases as belonging to a single class. No preprocessing was done.

2.2. Colon Dataset

This microarray (oligonucleotide) dataset [19], introduced in [2], contains expressions of 2000 genes for 62 cases (22 normal and 40 colon tumor cases). Preprocessing includes the logarithmic transformation to base 10, followed by normalization to zero mean and unit variance as usually done with this dataset.

2.3. Brain Dataset 1

This microarray (oligonucleotide) dataset [20] introduced in [3] contains two classes of brain tumor. The dataset (also known as Dataset B) contains 34 medulloblastoma cases, 9 of which are desmoplastic and 25 are classic. Preprocessing consists of thresholding of gene expressions with a floor of 20 and ceiling of 16000; filtering with exclusion of genes with $\max / \min \leq 3$ or $max - min < 100$, where max and min refer to the maximum and minimum expressions of a certain gene across the 34 cases, respectively; base 10 logarithmic transformation; normalization across genes to zero mean and unit variance. As a result, 5893 out of 7129 original genes are only retained.

2.4. SAGE Dataset 2

This is a larger SAGE dataset [17], containing 31 normal and 59 cancer (10 types of cancer) cases with 27679 expressed genes. As with the smaller dataset, no preprocessing was done and all cancer types were assigned to a single class.

2.5. Prostate Dataset 1

This microarray (oligonucleotide) dataset [21] introduced in [4] includes the expressions of 12600 genes in 52 prostate and 50 normal cases. No preprocessing of the data was done.

2.6. Prostate Dataset 2

This dataset [21] was obtained independently of the one described in the previous section. It has 25 prostate and 9 normal cases with 12600 expressed genes. No preprocessing was done.

2.7. Brain Dataset 2

This (oligonucleotide) dataset [20] known as Dataset C in [3] contains 60 medulloblastoma cases, corresponding to 39 survivors and 21 nonsurvivors according to the patient status. Preprocessing includes thresholding of gene expressions with a floor of 100 and ceiling of 16000; filtering with exclusion of genes with $\max / \min \leq 5$ or $max - min < 500$, where $\max$ and $\min$ refer to the maximum and minimum expressions of a certain gene across the 60 cases, respectively; base 10 logarithmic transformation; normalization across genes to zero mean and unit variance. As a result, 4459 out of 7129 original genes are only retained.

2.8. Diffuse Large B-Cell Lymphoma (DLBCL) Dataset

This (oligonucleotide) dataset [22] described in [23] contains 6149 gene expression levels characterizing 58 patients with diffuse large B-cell lymphoma according to their health status: 32 cured patients or those who died from other than lymphoma causes ('cured' class) and 26 patients who died of lymphoma or whose disease is either progressive or recurrent refractory ('fatal/refractory' class). Preprocessing includes thresholding of gene expressions with a floor of 20 and ceiling of 16000; filtering with exclusion of genes with $\max / \min < 3$ or $max - min < 100$, where $\max$ and $\min$ refer to the maximum and minimum expressions of a certain gene across the 60 cases, respectively. As a result, 6149 out of 7129 original genes are only retained.

2.9. Dataset Summary

Table 1 provides a summary for all datasets

Table 1. Summary of eight gene expression datasets.

Dataset no.	Cancer type(s)	Ref.	# expression levels	# cases
1	Multiple	[18]	822	74
2	Colon	[2]	2000	62
3	Brain	[3]	5893	34
4	Multiple	[18]	27679	90
5	Prostate	[4]	12600	102
6	Prostate	[4]	12600	34
7	Brain	[3]	4459	60
8	Lymphoma	[23]	6149	58

3. Redundancy-Based Filter

The redundancy-based filter (RBF) [9] is based on the concept of an approximate Markov blanket (AMB). Finding the complete Markov blanket is computationally prohibitive for high dimensional gene expression data, thus the approximation is used instead. The goal is to find for each gene F_i an AMB M_i that subsumes the information content of F_i. In other words, if M_i is a true Markov blanket for F_i, the class C is conditionally independent of F_i given M_i, i.e. $p(C|F_i, M_i) = p(C|M_i)$.

To efficiently find an AMB two types of correlations are employed: 1) individual C-correlation between a gene F_i and the class C and 2) combined C-correlation between a pair of genes F_i and F_j ($i \neq j$) and the class C. Both correlations are defined through symmetrical uncertainty $SU(X, C)$, where X is either F_i (individual C-correlation) or $F_{i,j}$ (combined C-correlation), with $SU(X, C)$ defined as

$$SU(X,C) = 2\left[\frac{IG(X|C)}{H(X)+H(C)}\right],$$

where $H(\cdot)$ is entropy, $IG(X|C) = H(X) - H(X|C)$ is information gain from knowing the class information. SU is a normalized characteristic whose values lie between 0 and 1, where 0 indicates that X and C are independent.

To reduce the variance and noise of the original data, continuous expression levels were converted to nominal values -1, 0, and +1, representing the under-expression, baseline, and over-expression of genes, which correspond to $(-\infty, \mu - \sigma/2)$, $[\mu - \sigma/2, \mu + \sigma/2]$, and $(\mu + \sigma/2, +\infty)$, respectively, with μ and σ being the mean and the standard deviation of all expression levels for a given gene. For nominal variables, the entropies needed in the formulas above are computed as follows:

$$H(X) = -\sum_i P(x_i)\log_2(P(x_i)),$$

$$H(X|C) = -\sum_k P(c_k)\sum_i P(x_i|c_k)\log_2(P(x_i|c_k)),$$

where $P(x_i)$ is the probability that $X = x_i$ and $P(x_i|c_k)$ is the probability that $X = x_i$ given $C = c_k$. For combined C-correlation, x_i in these formulas should be replaced with the pair (x_i, x_j) so that

$$H(X) = -\sum_{i,j} P(x_i, x_j) \log_2(P(x_i, x_j)),$$

$$H(X|C) = -\sum_{k} P(c_k) \sum_{i,j} P(x_i, x_j|c_k) \log_2(P(x_i, x_j|c_k)).$$

Since x_i (x_j) can take only three values: -1, 0, +1, there are nine pairs ((-1,-1), (-1,0), (-1,+1), (0,-1), (0,0), (0,+1), (+1,-1), (+1,0), (+1,+1)) for which probabilities (and hence, entropies) need to be evaluated.

RBF starts from computing individual C-correlation for each gene and sorting all correlations in descending order. The gene with the largest correlation is considered as predominant (no AMB exists for it) and hence it is put to the list S of the selected genes and used to filter out other genes. After that, the iteration begins with picking the first gene F_i from S and proceeds as follows. For all remaining genes, if F_i forms an AMB for F_j, the latter is removed from further analysis. The following conditions must be satisfied for this to happen: 1) individual C-correlation for F_i must be larger than or equal to individual C-correlation for F_j, which means that a gene with a larger individual correlation provides more information about the class than a gene with a smaller individual correlation, and 2) individual C-correlation for F_i must be larger than or equal to combined C-correlation for F_i and F_j, which means that if combining F_i and F_j does not provide more discriminative power than F_i alone, F_j is decided to be redundant. After one round of filtering, RBF takes the next (according to the magnitude of individual C-correlation) still unfiltered gene and the filtering process is repeated again. Since a lot of genes are typically removed at each round (gene expression data contain a lot of redundancy) and removed genes do not participate in the next rounds, the RBF is much faster than the typical hill climbing (greedy forward or backward search).

4. Dataset Complexity

It is known that the performance of classifiers is strongly data-dependent. To gain insight into a supervised classification problem[3], one can adopt dataset complexity characteristics. The goal of such characteristics is to provide a score reflecting how well classes of the data are separated. Given a set of features, the data of each class are projected onto the diagonal linear discriminant axis by using only these features (for details, see [24]). Projection coordinates then serve as input for the Wilcoxon rank sum test for equal medians [13] (the null hypothesis of this test is two medians are equal at the 5% significance level). Given a sample divided into two groups according to class membership, all the observations are ranked as if they were from a single sample and the rank sum statistic W is computed as the sum of the ranks in the smaller group. The value of the rank sum statistic is employed as a score characterizing separability power of a given set of features. The higher this score, the larger the overlap in projections of two classes, i.e. the worse separation between classes. To compare W coming from different datasets, each W can be normalized by the sum of all

[3]Two-class problems are assumed.

ranks, i.e. if N is the sample size, then the sum of all ranks will be $\sum_{i=1}^{N} i$. The normalized W lies between 0 and 1.

Our complexity characteristic is classifier-independent, i.e. it does not depend on a certain classifier. Employing a classifier-dependent characteristic would not provide an absolute scale for comparison. For example, it is well known that a nearest neighbor classifier can sometimes easily classify nonlinearly separable data.

Our choice for such a dataset complexity characteristic was not accidental. Since gene expression data are very high dimensional, it is not surprising that two classes could be linearly separable in high dimensional space (however, this does not make the classification task easy as interclass distances could be still smaller than intraclass distances, thus giving rise to classification errors). That is, the k-NN decision boundary can be assumed to be a hyperplane. On the other hand, the complexity characteristic we employ belongs to the class of linear discriminants estimating how well a line separates two classes. As a result, we have a good match between the behavior of a classifier and the model of class separability encoded in the complexity characteristic. To confirm our hypothesis, we computed the index of linear separability $L1$ [25], which is the objective function value for the linear programme, by using all features for each dataset. The closer $L1$ to zero, the more a given dataset linearly separable. It can be seen in Table 2 that all datasets can be considered to be linear separable. However, SAGE 1 seems to be much less linearly separable than Prostate 2.

Table 2. Dataset ranking based on the linear separability index $L1$ multiplied by 10^{-15}. The lower the rank is (1 - lowest rank, 8 - highest rank), the less linearly separable two classes are.

Dataset no.	1	2	3	4	5	6	7	8
$L1$	2.2	0.77	0.027	0.15	0.19	1.2e-019	0.054	4.8e-010
rank	1	2	6	4	3	8	5	7

5. Bolstered Resubstitution Error

This is a low-variance and low-bias classification error estimation method proposed in [6]. Unlike the cross-validation techniques reserving a part of the original data for testing, it permits to use the whole dataset. Since sample size of gene expression datasets is very small compared to the data dimensionality, using all available data is an important positive factor. However, one should be aware of the effect of overfitting in this case when a classifier demonstrates excellent performance on the training data but fails on independent unseen data. Braga-Neto and Dougherty [6] avoided this pitfall by randomly generating a number of artificial points (examples) in the neighborhood of each training point. These artificial examples then act as a test set and classification error on this set is called bolstered. In this paper, we utilize the bolstered variant of the conventional resubstitution error known as *bolstered resubstitution error*.

Briefly, bolstered resubstitution error is estimated as follows [6]. Let A_0 and A_1 be two decision regions corresponding to the classification generated by a given algorithm, N be the number of training points, and M_{MC} be the number of random samples drawn from the D-variate normal distribution per training point ($M_{MC} = 10$ as advocated in [6]). The bolstered resubstitution error is then defined as

$$\epsilon_{bresub} \approx \frac{1}{NM_{MC}} \sum_{i=1}^{N} \left(\sum_{j=1}^{M_{MC}} I_{x_{ij} \in A_1} I_{y_i=0} + \sum_{j=1}^{M_{MC}} I_{x_{ij} \in A_0} I_{y_i=1} \right), \tag{1}$$

where $\{x_{ij}\}_{j=1,\ldots,M_{MC}}$ are samples drawn from $1/((2\pi)^{D/2}\sigma_i^D)e^{-\|x\|^2/(2\sigma_i^2)}$. The bolstered resubstitution error is thus equal to the sum of all error contributions divided by the number of points. Samples are drawn based on the Marsaglia polar normal random number generator (see Appendix B.).

In a 2-D space, samples come from a circle centered at a particular training point. In a D-dimensional case, they are drawn from a hypersphere. Hence, the radius of this hypersphere, determined by σ_i, is of importance since its selection amounts to choosing the degree of bolstering. Typically, σ_i should vary from point to point in order to be robust to the data. In [6] $\sigma_i = \hat{d}(y_i)/cp$ for $i = 1, \ldots, N$, where $\hat{d}(y_i)$ is the mean minimum distance between points belonging to class of y_i (y_i can be either 0 or 1)[4], and cp is the constant called the correction factor defined as the inverse of the chi-square cdf (cumulative distribution function) with parameters 0.5 and D, because interpoint distances in the Gaussian case are distributed as a chi random variable with D degrees of freedom. Thus, cp is the function of the data dimensionality. The parameter 0.5 is chosen so that points inside a hypersphere will be evenly sampled.

6. Link Between Dataset Complexity and Classification Error

Our main idea to build ensembles of k-NNs is based on the hypothesis that *the dataset complexity and bolstered resubstitution error are related*. To verify our hypothesis, 10000 feature subsets were randomly sampled for each dataset (subset size ranged from 1 to 50) and both complexity and bolstered resubstitution error for 3-NN were computed. Anomalous complexity values lying three standard deviations from the average complexity were treated as outliers and therefore removed from further analysis. The result of such simulation is shown in Figs. 1-8 together with marginal histograms for each variable. It can be observed that univariate distributions vary from dataset to dataset and often they are non-Gaussian. Besides, the complexity and error distributions for a certain dataset belong to different types, e.g. one is normal while another is exponential. However, the dependence between complexity and error is clearly detectable when looking at Figs. 1-8. Besides, one characteristic important for successful ensemble generation is present: diversity among predictions since one complexity value corresponds to several different error values.

To quantify this dependence, the rank correlation coefficients Spearman's ρ and Kendall's τ were computed (see Table 3) and the test on positive correlation at the significance level 0.05 was done which confirmed the existence of such correlation (all p-values were equal to zero). The rank correlations measure the degree to which large (small) values

[4] $\hat{d}(y_i)$ is determined by first computing the minimum distance from each point x_i to all other points x_j ($j \neq i$) of the same class as that of x_i and then by averaging thus obtained minimum distances.

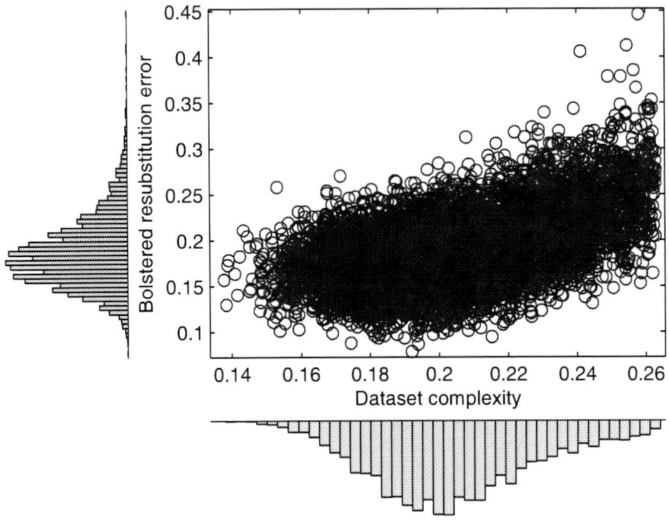

Figure 1. (SAGE 1) Bivariate distribution of normalized complexity and bolstered resubstitution error and univariate marginal histograms.

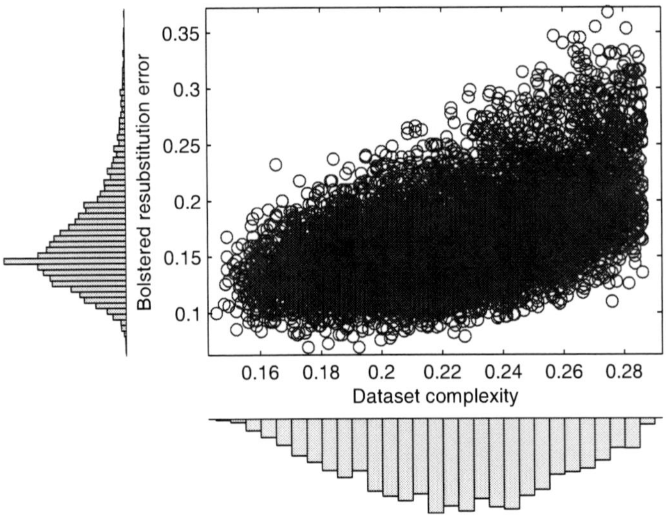

Figure 2. (Colon) Bivariate distribution of normalized complexity and bolstered resubstitution error and univariate marginal histograms.

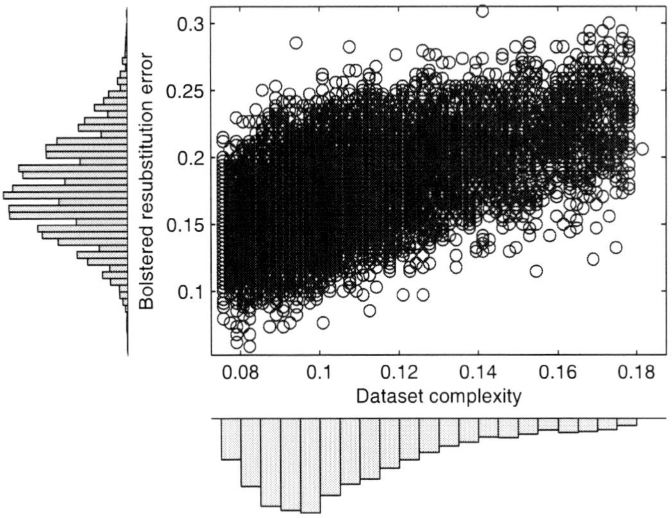

Figure 3. (Brain 1) Bivariate distribution of normalized complexity and bolstered resubstitution error and univariate marginal histograms.

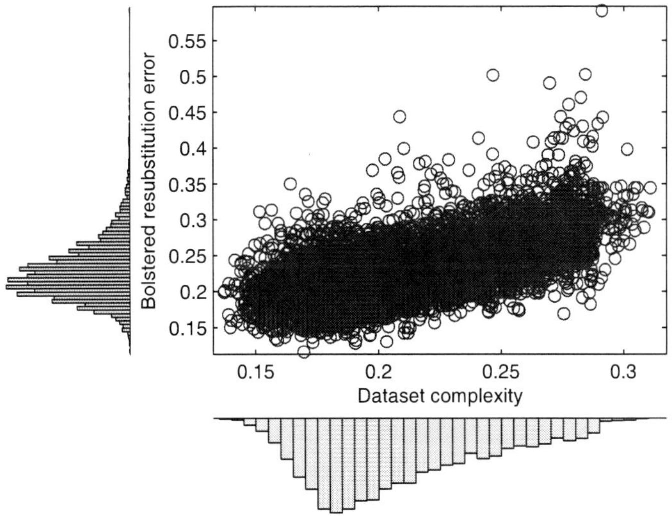

Figure 4. (SAGE 2) Bivariate distribution of normalized complexity and bolstered resubstitution error and univariate marginal histograms.

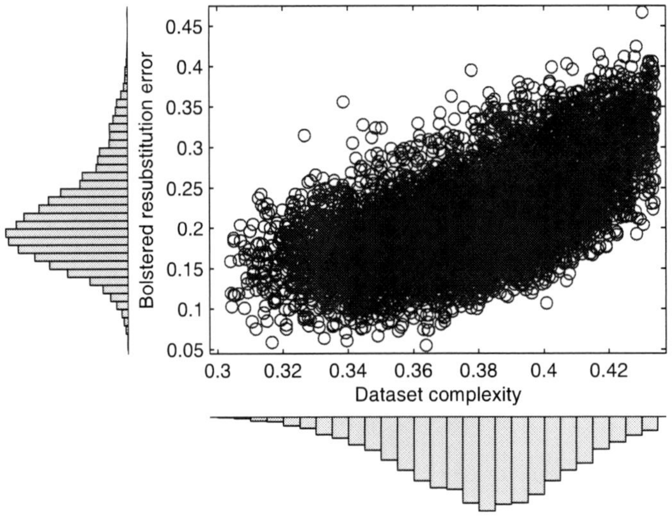

Figure 5. (Prostate 1) Bivariate distribution of normalized complexity and bolstered resubstitution error and univariate marginal histograms.

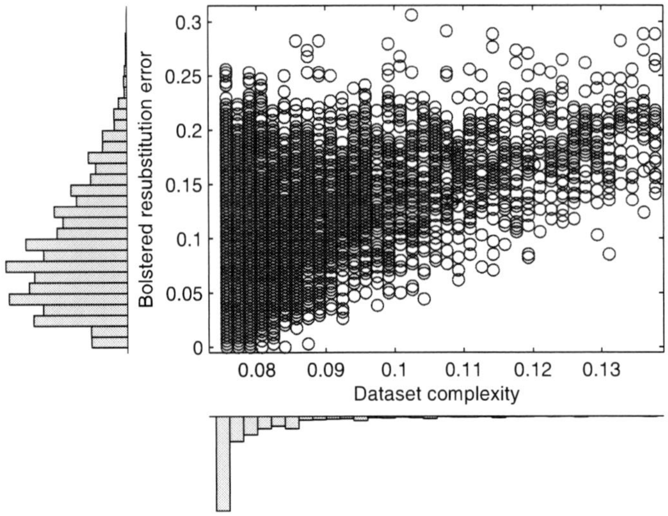

Figure 6. (Prostate 2) Bivariate distribution of normalized complexity and bolstered resubstitution error and univariate marginal histograms.

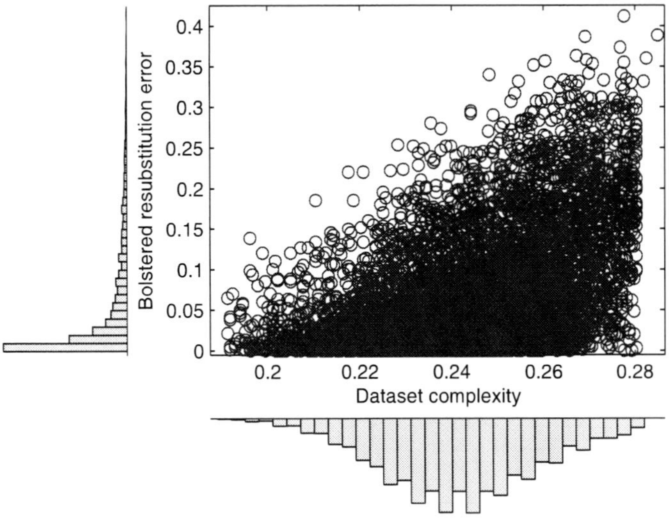

Figure 7. (Brain 2) Bivariate distribution of normalized complexity and bolstered resubstitution error and univariate marginal histograms.

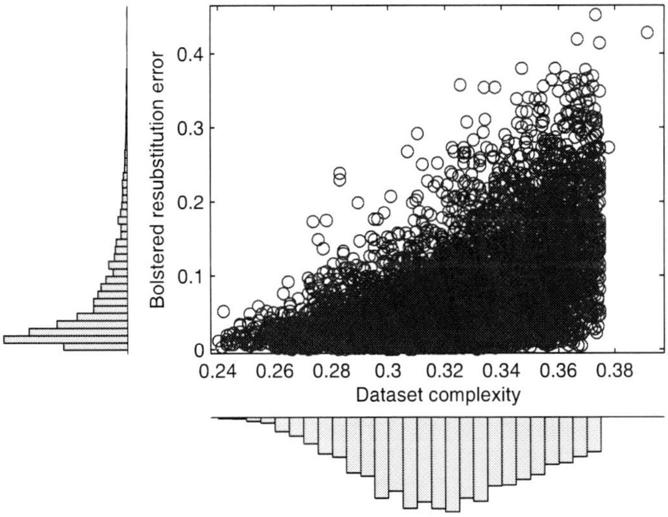

Figure 8. (Lymphoma) Bivariate distribution of normalized complexity and bolstered resubstitution error and univariate marginal histograms.

of one random variable correspond to large (small) values of another variable (concordance relations[5] among variables). They are useful descriptors in our case since high (low) complexity implies that the data are difficult (easy) to accurately classify, which, in turn, means high (low) classification error. Unlike the linear correlation coefficient, ρ and τ are preserved under any monotonic (strictly increasing) transformation of the underlying random variables.

Table 3. Spearman's ρ and Kendall's τ estimated for all datasets

Dataset no.	τ	ρ
1	0.3100	0.4468
2	0.3446	0.4964
3	0.3991	0.5581
4	0.4173	0.5864
5	0.4288	0.6006
6	0.3887	0.5107
7	0.3117	0.4486
8	0.3993	0.5667

To deeply explore dependence relations, we employed the copula method [10, 11, 12]. The word copula is a Latin noun which means 'a link, tie or bond' was first introduced by Abe Sklar in [10]. Copulas are functions that describe dependencies among variables and allow to model correlated multivariate data by combining univariate distributions. Using copulas is an appropriate solution since the assumption that the joint distribution of random variables is normal often does not hold for multivariate data in practice even if the marginal distributions are normal.

A copula is a multivariate probability distribution, where each random variable has a uniform marginal distribution on the interval [0,1]. The dependence between random variables is completely separated from the marginal distributions in the sense that random variables can follow any marginal distributions, and still have the same rank correlation. This is one of the main appeals of copulas: they allow separation of dependence and marginal distribution. Though there are multivariate copulas, we will only talk about bivariate ones since our dependence relation includes two variables.

Sklar's theorem, which is the foundation theorem for copulas, states that for a given joint multivariate distribution function $H(x, y) = P(X \leq x, Y \leq y)$ of a pair of random variables X and Y and the relevant marginal distributions $F(x) = P(X \leq x)$ and $G(y) = P(Y \leq y)$, there exists a copula function C relating them, i.e. $H(x, y) = C(F(x), G(y))$. If F and G are continuous, C is unique. Otherwise, C is uniquely determined on Ran$X \times$Ran$Y$, where 'RanX' ('RanY') stands for the range of X (Y). In other words, for each pair of real numbers (x, y) there are three numbers $F(x)$, $G(y)$, and $H(x, y)$ lying in the interval [0,1]. Alternatively, each pair (x, y) is matched by a point $(F(x), G(y))$ in the unit square $\mathbf{I}^2 : [0, 1] \times [0, 1]$, and this ordered pair in turn is associated with a number $H(x, y)$ in [0,1]. The correspondence assigning the value of the joint

[5]Since the definitions of these relations by ρ and τ are different, there is a difference in absolute values in Table 3.

distribution function to each ordered pair of values of the individual distribution functions is a copula function C [11].

Thus, a copula is a function C from $\mathbf{I}^2$ to $\mathbf{I}$ with the following properties [11]:

1. For every u, v in $\mathbf{I}$,
$$C(u,0) = 0 = C(0,v)$$
and
$$C(u,1) = u, C(1,v) = v.$$

2. for every u_1, u_2, v_1, v_2 in $\mathbf{I}$ such that $u_1 \leq u_2$ and $v_1 \leq v_2$,
$$C(u_2,v_2) - C(u_2,v_1) - C(u_1,v_2) + C(u_1,v_1) \geq 0.$$

If F and G are continuous, the following formula is used to construct copulas from the joint distribution functions: $C(u,v) = H(F^{-1}(u), G^{-1}(v))$ [11], where F^{-1} means a quasi-inverse of F, G^{-1} means a quasi-inverse of G, and U and V are uniform random variables distributed between 0 and 1. That is, the typical copula-based analysis of multivariate (or bivariate) data starts with the transformation from the (X,Y) domain to the U, V domain, and all manipulations with data are then done in the latter. Such a transformation to the copula scale (unit square $\mathbf{I}^2$) can be achieved through a kernel estimator of the cumulative distribution function (cdf) (we used the MATLAB function *ksdensity*). After that the copula function $C(u,v)$ is generated according to the appropriate definition for a certain copula family (see, e.g. Eq. 2 below).

In [26] it was shown that Spearman's ρ and Kendall's τ can be expressed solely in terms of the copula function as follows:

$$\rho = 12\int\int (C(u,v) - uv)dudv = 12\int\int C(u,v)dudv - 3,$$
$$\tau = 4\int\int C(u,v)dC(u,v) - 1,$$

where integration is over $\mathbf{I}^2$.

The integrals in these formulas can be interpreted as the expected value of the function $C(u,v)$ of uniform [0,1] random variables U and V whose joint distribution function is C, i.e.

$$\rho = 12E(UV) - 3, \quad \tau = 4E(C(u,v)) - 1.$$

As a consequence, ρ for a pair of continuous random variable X and Y is identical to Pearson's linear correlation coefficient for random variables $U = F(X)$ and $V = G(Y)$ [11].

In general, the choice of a particular copula may be based on the observed data. Among numerous copula families, we preferred the Frank copula belonging to the Archimedean family based on the visual look of plots in Figs. 1-8 and for dependence in the tail. Besides, this copula type permits negative as well as positive dependence. We are particularly concerned with lower tail dependence when low complexity is associated with small classification error as this forms the basis for ensemble construction in our approach. The Frank copula is a one-parameter (θ is a parameter, $\theta \in]-\infty, +\infty[\setminus 0)$ copula defined for uniform variables U and V (both are defined over the unit interval) as

$$C_\theta(u, v) = -\frac{1}{\theta} \ln \left(1 + \frac{(e^{-\theta u} - 1)(e^{-\theta v} - 1)}{e^{-\theta} - 1}\right), \quad (2)$$

with θ determining the degree of dependence between the marginals (we set θ to Pearson's correlation coefficient between U and V so that as θ increases, the positive dependence also increases). Fig. 9 shows 500 random points generated from the Frank copula when $\theta = 8$.

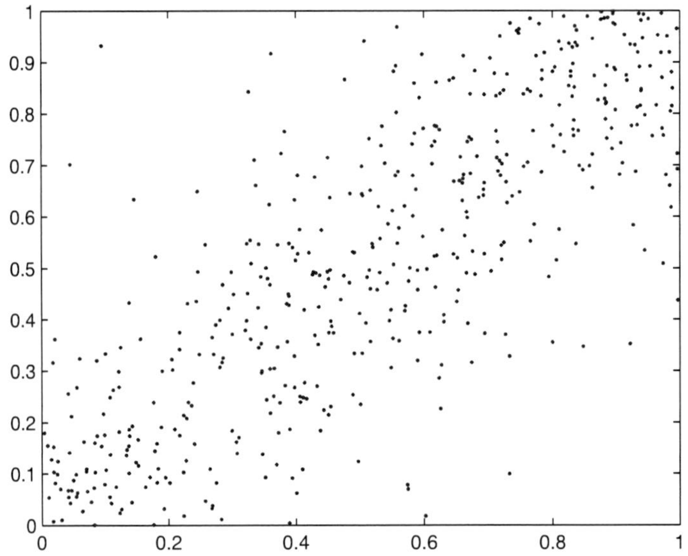

Figure 9. A random sample (500 points) generated from the Frank copula with $\theta = 8$.

Correlation coefficients measure the overall strength of the association, but give no information about how that varies across the distribution. The magnitude of τ or ρ is not an absolute indicator of such strength since for some distributions the attainable interval can be very small, say between -0.1 and +0.2 so that finding correlation of 0.2 and concluding that there is only weak dependence between variables would be a mistake since these variables are actually perfectly related. Hence, additional characteristics of dependence structure are necessary. They are quadrant dependence, tail monotonicity, stochastic monotonicity, with quadrant dependence being the weakest form of association among all the three.

6.1. Quadrant Dependence

Random variables X and Y are positively quadrant dependent (PQD) if $\forall(x, y)$ in $\mathbf{R}^2$, either inequality holds [11]:

$$P(X \leq x, Y \leq y) \geq P(X \leq x)P(Y \leq y),$$
$$P(X > x, Y > y) \geq P(X > x)P(Y > y).$$

X and Y are PQD if the probability that they are simultaneously small (or simultaneously large) is at least as it would be were they independent. In terms of C, the PQD conditions can be written as $C(u,v) \geq uv$ for all (u,v) in $\mathbf{I}^2$. By checking the last inequality, we found that complexity and bolstered resubstitution error are PQD for all datasets. Spearman's ρ (or, to be precise, $\rho/12$) can be interpreted as a measure of "average" quadrant dependence (both positive and negative) for random variables whose copula is C [11].

It is interesting to ask when one continuous bivariate distribution H_1 is more PQD (more concordant) than another H_2. The answer is readily provided by comparing ρ or τ [12]: if $\rho(H_1) \leq \rho(H_2)$ or $\tau(H_1) \leq \tau(H_2)$, then H_2 is more PQD (more concordant) than H_1. From Table 3 it can be seen that Prostate 1 is more PQD than other datasets, i.e. concordance relations between complexity and bolstered resubstitution error are much stronger for this data than those for other datasets.

6.2. Tail Monotonicity

As we mentioned above, we are interested in tail dependence when low (high) complexity associates small (large) classification error. Tail monotonicity reflects this type of association and it is a stronger condition for dependence than PQD. Let X and Y be random variables. Then four types of tail monotonicity can be defined as follows [11]:

- Y is left tail decreasing in X (LTD(Y|X)) if $P(Y \leq y|X \leq x)$ is a nonincreasing function of x for all y.
- X is left tail decreasing in Y (LTD(X|Y)) if $P(X \leq x|Y \leq y)$ is a nonincreasing function of y for all x.
- Y is right tail increasing in X (RTI(Y|X)) if $P(Y > y|X > x)$ is a nondecreasing function of x for all y.
- X is right tail increasing in Y (RTI(X|Y)) if $P(X > x|Y > y)$ is a nondecreasing function of y for all x.

In terms of a copula and its first-order partial derivatives these conditions are equivalent to

- LTD(Y|X) iff for any v in $\mathbf{I}$, $\partial C(u,v)/\partial u \leq C(u,v)/u$ for almost all u.
- LTD(X|Y) iff for any u in $\mathbf{I}$, $\partial C(u,v)/\partial v \leq C(u,v)/v$ for almost all v.
- RTI(Y|X) iff for any v in $\mathbf{I}$, $\partial C(u,v)/\partial u \leq (v - C(u,v))/(1-u)$ for almost all u.
- RTI(X|Y) iff for any u in $\mathbf{I}$, $\partial C(u,v)/\partial v \leq (u - C(u,v))/(1-v)$ for almost all v.

For the Frank copula, the first-order partial derivatives are (see Appendix A.)

$$\frac{\partial C_\theta(u,v)}{\partial u} = \frac{e^{-\theta u}(e^{-\theta v}-1)}{e^{-\theta}-1+(e^{-\theta u}-1)(e^{-\theta v}-1)}, \tag{3}$$

$$\frac{\partial C_\theta(u,v)}{\partial v} = \frac{e^{-\theta v}(e^{-\theta u}-1)}{e^{-\theta}-1+(e^{-\theta u}-1)(e^{-\theta v}-1)}. \tag{4}$$

Tail monotonicity is also guaranteed if $\rho \geq \tau \geq 0$ is met [11].

We verified that for all datasets bolstered resubstitution error is left tail decreasing in complexity, complexity is left tail decreasing in bolstered resubstitution error, bolstered resubstitution error is right tail increasing in complexity, and complexity is right tail increasing in bolstered resubstitution error. Thus, dependence in the tail between these two variables exists.

6.3. Stochastic Monotonicity

Stochastic monotonicity is stronger than tail monotonicity. According to [11],

- Y is stochastically increasing in X (SI(Y|X)) if $P(Y > y|X = x)$ is a nondecreasing function of x for all y.
- X is stochastically increasing in Y (SI(X|Y)) if $P(X > x|Y = y)$ is a nondecreasing function of y for all x.

Alternatively, stochastic monotonicity can be expressed as

- SI(Y|X) iff for any v in **I**, $C(u, v)$ is a concave function of u.
- SI(X|Y) iff for any u in **I**, $C(u, v)$ is a concave function of v.

A concave function implies that the second-order derivatives must be less than or equal to zero. For the Frank copula, these derivatives are (see Appendix A.)

$$\frac{\partial^2 C_\theta(u,v)}{\partial u^2} = \frac{\theta e^{-\theta u}(e^{-\theta v}-1)(e^{-\theta v}-e^{-\theta})}{\left[e^{-\theta}-1+(e^{-\theta u}-1)(e^{-\theta v}-1)\right]^2}, \tag{5}$$

$$\frac{\partial^2 C_\theta(u,v)}{\partial v^2} = \frac{\theta e^{-\theta v}(e^{-\theta u}-1)(e^{-\theta u}-e^{-\theta})}{\left[e^{-\theta}-1+(e^{-\theta u}-1)(e^{-\theta v}-1)\right]^2}. \tag{6}$$

Since $\theta > 0$ in our case (positive dependence as expressed by the rank correlation coefficients), it is easy to verify that $\frac{\partial^2 C_\theta(u,v)}{\partial u^2} \leq 0$ and $\frac{\partial^2 C_\theta(u,v)}{\partial v^2} \leq 0$, which, in turn, implies that $C_\theta(u, v)$ is concave (see also Appendix A.). Thus, for all datasets in our study, bolstered resubstitution error is stochastically increasing in complexity and complexity is stochastically increasing in bolstered resubstitution error.

7. Uncertainty of Single Classification

There exist many classifiers (k-NN, linear and quadratic discriminant analysis, support vector machine) showing good performance on gene expression data [8]. Each of them has its strong and weak points and that is why none of them is superior to others. In addition, when feature selection precedes classification, there can be multiple subsets of genes resulting in the same error rate. The fact that different subsets of genes can be equally relevant when predicting cancer has been already highlighted in several works [27, 28, 29]. It was argued that one of the possible explanations for such multiplicity and non-uniqueness is a strong influence of the training set on gene selection. In other words, different groups of patients can lead to different gene importance rankings due to genuine differences between patients (cancer grade, stage, etc.).

To mitigate this problem (complete alleviation seems to be currently impossible due to small sample size of gene expression datasets), we propose to employ an ensemble of classifiers instead of a single classifier, where each classifier in the ensemble works with its own feature subset. Potential gains in doing so are twofold: error rate can be significantly reduced and fewer biologically relevant genes can be missed when all subsets of genes are combined together for further analysis.

8. Ensemble of Classifiers

An ensemble of classifiers consists of several base classifiers (members) that make predictions independently of each other. After that, these predictions are combined together to produce the final prediction. Though ensemble members can belong to different types of algorithms, because of our interest in k-NN classifiers we utilize only this algorithm. Moreover, the value of k is fixed to 3 for all ensemble members[6]. As a combination technique, the conventional majority vote was selected in order to demonstrate that ensembles built with our approach demonstrate good performance even when employing simple non-trainable combiners.

It is well known that an ensemble is able to outperform its best performing member if ensemble members make mistakes of different samples so that their predictions are uncorrelated and diverse as much as possible. On the other hand, an ensemble must include a sufficient number of accurate classifiers since if there are only few good votes, they can be easily drowned out among many bad votes. As a result, an ensemble can predict wrongly most of the time.

So far many definitions of diversity were proposed [7, 30], but unfortunately the precise definition is still largely illusive and as commented in [31], the link between diversity of the ensemble members and prediction accuracy of an ensemble is not straightforward. Because of this fact, we decided not to follow any *explicit* definition of diversity, but introduce diversity implicitly instead. Since we fixed the base classifier and its parameter, one of the solutions is to let each ensemble member to work with its own feature subset.

Feature subset selection can be done in two ways: either applying a certain feature selection algorithm or a group of such algorithms, or randomly sampling features from the original feature set. As concluded in [5], differences in classification performance among feature selection algorithms are less significant than performance differences among the error estimators used to implement these algorithms. In other words, the way of how error is computed has a larger influence on classification accuracy than the choice of a feature selection algorithm. Since bolstered resubstitution error is a low-biased and low-variance estimate of classification error, which is what is needed for high dimensional gene expression data, we opt for random feature selection. Figs. 1-8 show that random feature selection leads to diversity since one complexity value corresponds to several different errors. Given that it is difficult to carry out biological analysis of many genes, we restricted the number of genes to be sampled to 50, i.e. each ensemble member works with 1 to 50 randomly

[6]In our opinion, $k = 1$ tends to lead to optimistic estimation of bolstered resubstitution error.

selected (sampled with replacement) genes. This will ensure that the combined list of all genes is not too long.

Based on the abovementioned, two approaches to form ensembles consisting of L classifiers are explored:

1. Randomly select L feature subsets, one subset per classifier, as described above. Classify the data with each classifier and combine votes.

2. Randomly select $M > L$ (e.g. $M = 100$) feature subsets and compute the dataset complexity for each of them. Rank subsets according to their complexity and select L least complex subsets while ignoring the others. Classify the data with each classifier and combine votes.

We will call the first approach conventional to distinguish it from ours, which is the second approach. The typical (and perhaps the earliest) example of the former is [32]. As one can see, the main difference between two approaches lies in the way of choosing feature subsets: in the conventional approach, subsets are chosen regardless of their classification power. As a result, one may equally expect both very good and very bad ensemble predictions. In contrast, in our approach, subsets are chosen based on the measure *directly* related to classification performance. As lower complexity is associated with smaller bolstered resubstitution error as shown in Section 6., selection of the subsets of smaller complexity implies more accurate classifiers included into an ensemble. Thus, with our approach, both diversity and accuracy requirements for ensembles are satisfied. Hence, we can expect better *average* classification performance with our approach compared to the conventional approach.

9. Experimental Results

In ensemble applications to bioinformatics problems, a small and accurate ensemble is of importance, since too many ensemble members would complicate biological understanding of relations among genes. Bearing this in mind, we set the number of 3-NNs (L) in the ensemble to be equal 3, 5, 7, 9, and 11.

Table 4 represents the dataset complexity as estimated by the normalized rank sum statistic W (see Section 4.) for different values of L when ensembles were built with our approach. For each dataset, two values are given: average minimum and average maximum complexity (averaging over 100 runs) of the selected feature subsets. It can be observed that complexity for each dataset is rather stable as L grows. Prostate 1 appears to be far more complex than the other datasets while Prostate 2 seems to be the least complex. For the latter, the minimum and maximum complexity stays the same, which implies that the complexity reached saturation during ensemble generation. The fact that saturation happened at L as low as 3 implicitly points to low complexity of Prostate 2. For the conventional ensemble approach 'avr.max' often went to a very big value, meaning poor class separation according to the Wilcoxon rank sum test. For comparison, Table 5 lists dataset complexity when all features are considered in computing W. Again Prostate 1 looks the most complex while Prostate 2 and Brain 1 are among least complex.

Table 4. Average minimum and maximum normalized W for feature subsets selected with our ensemble generating approach for various values of L

Dataset no.		$L=3$	$L=5$	$L=7$	$L=9$	$L=11$
1	avr.min	0.1544	0.1517	0.1551	0.1546	0.1547
	avr.max	0.1638	0.1681	0.1725	0.1749	0.1776
2	avr.min	0.1587	0.1584	0.1581	0.1587	0.1584
	avr.max	0.1676	0.1723	0.1769	0.1800	0.1845
3	avr.min	0.0761	0.0760	0.0761	0.0760	0.0759
	avr.max	0.0781	0.0803	0.0818	0.0831	0.0841
4	avr.min	0.1501	0.1494	0.1493	0.1503	0.1498
	avr.max	0.1584	0.1619	0.1655	0.1683	0.1707
5	avr.min	0.3105	0.3117	0.3121	0.3105	0.3104
	avr.max	0.3274	0.3349	0.3417	0.3447	0.3480
6	avr.min	0.0756	0.0756	0.0756	0.0756	0.0756
	avr.max	0.0756	0.0756	0.0756	0.0756	0.0756
7	avr.min	0.1987	0.1979	0.1990	0.1978	0.1995
	avr.max	0.2086	0.2140	0.2174	0.2190	0.2217
8	avr.min	0.2566	0.2565	0.2554	0.2553	0.2563
	avr.max	0.2692	0.2756	0.2798	0.2837	0.2867

Table 5. Unnormalized and normalized rank sum statistic W when all features are used

Dataset no.	1	2	3	4	5	6	7	8
W	465	409	48	496	1959	45	410	439
N	74	62	34	90	102	34	60	58
normalized W	0.17	0.21	0.08	0.12	0.37	0.009	0.22	0.26

Table 6 summarizes the average bolstered resubstitution error (over 100 runs) and its standard deviation achieved with two ensemble schemes. 'C' and 'O' stand for the conventional and our approaches to ensemble construction, respectively. It is clearly noticable that both the average error and its standard deviation are smaller for our approach, regardless of the number of k-NNs in the ensemble. It should be noted that one should not seek dependence between ensemble error in Table 6 and feature subset complexity in Table 4, since our hypothesis is only applied to the error of the individual classifiers.

For comparison, we also included experiments with RBF [9], followed by 3-NN classification using selected genes. Table 7 lists the average bolstered resubstitution error and its standard deviation computed over 100 runs when RBF was applied to each dataset prior to 3-NN classification. The third column contains the number of genes retained after filtering. Results of 3-NN classification without prior gene selection are given in the last column.

It can be observed that our ensemble scheme almost always outperforms RBF+3-NN,

Table 6. Average bolstered resubstitution error and its standard deviation for two ensemble schemes for different values of L

		$L=3$	$L=5$	$L=7$	$L=9$	$L=11$
1	C	0.141±0.025	0.125±0.024	0.119±0.021	0.110±0.018	0.111±0.018
	O	0.119±0.016	0.105±0.017	0.098±0.014	0.094±0.014	0.088±0.015
2	C	0.110±0.025	0.091±0.019	0.080±0.013	0.074±0.015	0.068±0.013
	O	0.092±0.014	0.077±0.012	0.071±0.014	0.066±0.010	0.064±0.010
3	C	0.129±0.035	0.117±0.032	0.101±0.031	0.092±0.027	0.088±0.028
	O	0.081±0.022	0.062±0.019	0.054±0.017	0.047±0.016	0.045±0.015
4	C	0.177±0.034	0.160±0.040	0.152±0.040	0.143±0.039	0.151±0.043
	O	0.130±0.023	0.113±0.022	0.098±0.019	0.093±0.017	0.089±0.016
5	C	0.141±0.034	0.111±0.026	0.096±0.020	0.084±0.016	0.076±0.014
	O	0.101±0.022	0.078±0.016	0.071±0.012	0.066±0.011	0.063±0.010
6	C	0.046±0.039	0.027±0.026	0.015±0.014	0.013±0.014	0.013±0.013
	O	0.023±0.018	0.011±0.012	0.007±0.008	0.004±0.005	0.004±0.005
7	C	0.172±0.024	0.147±0.021	0.133±0.020	0.128±0.019	0.120±0.016
	O	0.145±0.017	0.119±0.015	0.104±0.014	0.098±0.014	0.092±0.013
8	C	0.188±0.036	0.150±0.029	0.122±0.025	0.103±0.023	0.087±0.019
	O	0.164±0.027	0.130±0.022	0.099±0.020	0.087±0.021	0.075±0.020

Table 7. Average bolstered resubstitution error and its standard deviation 1) when RBF was applied before 3-NN classification (RBF+3-NN) and 2) with 3-NN classification without gene selection

Dataset no.	RBF+3-NN	#genes	3-NN
1	0.199±0.011	12	0.160±0.005
2	0.107±0.010	3	0.098±0.006
3	0.055±0.010	6	0.074±0.008
4	0.145±0.005	152	0.132±0.001
5	0.117±0.008	2	0.099±0.002
6	0.003±0.003	1	0.029±0.000
7	0.173±0.013	15	0.216±0.009
8	0.217±0.014	37	0.174±0.005

except for Brain 1 data[7], which were easy to classify according to dataset complexity. In contrast, the conventional scheme was inferior to RBF+3-NN on many more occasions, which again confirms the superiority of our approach to ensemble construction.

We also provide a comparison of both conventional and our approaches to ensemble construction versus a single best classifier (SBC) in each case. Let e_{SBC} and e_{ENS} be bolstered resubstitution error achieved with a SBC and an ensemble, respectively. To meet our goal, the following statistics widely used in machine learning and data mining were

[7] $L = 3, 5$.

computed over 100 ensemble generations:

- win-tie-loss count, where 'win'/'tie'/'loss' means the number of times when an ensemble was superior/equal/inferior in terms of bolstered resubstitution error to a SBC in the ensemble (in other words, the number of times when $e_{ENS} < e_{SBC}$, $e_{ENS} = e_{SBC}$, $e_{ENS} > e_{SBC}$, respectively).

- 'min. win', 'max. win', 'avr. win' (minimum, maximum, and average differences $e_{SBC} - e_{ENS}$ when an ensemble outperforms its SBC,

- 'min. loss', 'max. loss', 'avr. loss' (minimum, maximum, and average differences $e_{ENS} - e_{SBC}$ when a SBC outperforms an ensemble.

Tables 8-12 contain values of these statistics. If there were no losses, this fact is marked as 'no'. As one can see, both ensemble schemes were superior to a SBC on all eight datasets for the most part. The degree of success, however, varied, depending on dataset complexity. For example, Prostate 2 was much easier to classify compared to other datasets and therefore a SBC often reached the top performance so that an ensemble had nothing to improve on. When analyzing the performance of two ensemble schemes, it was observed that on average, our approach yields better results in the sense that its win (loss) count is typically higher (lower) and the absolute losses to a SBC are lower, too. In contrast, the conventional ensemble generating approach sometimes shows spectacular results (e.g. the high max.win count), but it also suffers many defeats from a SBC. That is, its results are less predictable since there is no control over complexity of the selected feature subsets and hence, if such 'complex' subsets are selected, a SBC can render ensemble efforts to further lower error fruitless. With the explicit selection of least complex subsets, our approach is able to succeed where the comparative approach failed.

10. Conclusion

We proposed a new ensemble generating scheme using a k-NN as a base classifier. Our approach leads to lower bolstered resubstitution error compared to the conventional ensemble approach, purely based on random selection of features, and to a single best classifier in the ensemble. In addition, our scheme outperforms a 3-NN preceded by the RBF algorithm [9], especially proposed to deal with redundancy among genes.

Our approach originates from the link between dataset complexity and bolstered resubstitution error established through the copula method. We found that there is positive dependence between complexity and error, where low (high) complexity corresponds to small (large) error. Hence, the dataset complexity serves as a reliable indicator of the expected classification performance. As a result, selection of least complex subsets of features implies more accurate ensemble members and therefore it ensures better ensemble performance. Extensive experiments with eight gene expression datasets containing different types of cancer show feasibility of our approach. Its extra attractiveness comes from the fact that good ensemble performance is achieved with a few 3-NNs (3 to 11), which limits the number of genes to further analyze.

Table 8. Comparison of a SBC and two ensembles when $L = 3$

		win-tie-loss	min.win	max.win	avr.win	min.loss	max.loss	avr.loss
1	C	84/1/15	0.0014	0.0608	0.0234	0.0014	0.0446	0.0137
	O	90/1/9	0.0014	0.0649	0.0277	$< 10^{-4}$	0.0135	0.0066
2	C	84/3/13	0.0032	0.0629	0.0275	$< 10^{-4}$	0.0548	0.0166
	O	96/2/2	0.0016	0.0532	0.0232	0.0065	0.0145	0.0105
3	C	79/2/19	0.0029	0.0824	0.0324	0.0029	0.0324	0.0135
	O	98/0/2	0.0029	0.0794	0.0428	0.0059	0.0059	0.0059
4	C	81/1/18	0.0011	0.0700	0.0355	0.0011	0.0578	0.0252
	O	99/0/1	0.0022	0.0856	0.0434	0.0056	0.0056	0.0056
5	C	86/0/14	0.0039	0.0824	0.0360	0.0020	0.0735	0.0212
	O	92/1/7	0.0020	0.0657	0.0311	0.0020	0.0265	0.0116
6	C	55/7/38	$< 10^{-4}$	0.0618	0.0224	0.0029	0.1412	0.0292
	O	68/4/28	$< 10^{-4}$	0.0559	0.0195	0.0029	0.0382	0.0123
7	C	91/1/8	0.0067	0.0750	0.0358	0.0017	0.0133	0.0081
	O	100/0/0	0.0100	0.0750	0.0426	no	no	no
8	C	93/1/6	0.0052	0.0931	0.0429	0.0172	0.0569	0.0279
	O	99/0/1	0.0017	0.1000	0.0499	0.0103	0.0103	0.0103

Table 9. Comparison of a SBC and two ensembles when $L = 5$

		win-tie-loss	min.win	max.win	avr.win	min.loss	max.loss	avr.loss
1	C	89/0/11	0.0027	0.0716	0.0310	0.0014	0.0581	0.0219
	O	96/0/4	0.0027	0.0703	0.0361	0.0068	0.0405	0.0196
2	C	93/1/6	0.0016	0.0694	0.0359	0.0032	0.0274	0.0137
	O	100/0/0	0.0016	0.0597	0.0340	no	no	no
3	C	76/2/22	0.0029	0.0853	0.0352	0.0029	0.0735	0.0217
	O	99/1/0	0.0059	0.1000	0.0536	no	no	no
4	C	84/0/16	0.0033	0.1022	0.0449	0.0022	0.0800	0.0296
	O	99/0/1	0.0122	0.1089	0.0578	0.0022	0.0022	0.0022
5	C	96/0/4	0.0029	0.0902	0.0455	0.0078	0.0333	0.0159
	O	92/0/8	0.0010	0.0784	0.0386	$< 10^{-4}$	0.0167	0.0056
6	C	60/11/29	$< 10^{-4}$	0.0765	0.0232	0.0029	0.0941	0.0221
	O	80/4/16	$< 10^{-4}$	0.0588	0.0201	0.0029	0.0176	0.0083
7	C	100/0/0	0.0033	0.0883	0.0484	no	no	no
	O	100/0/0	0.0200	0.0933	0.0602	no	no	no
8	C	99/0/1	0.0121	0.1121	0.0682	0.0121	0.0121	0.0121
	O	100/0/0	0.0328	0.1224	0.0763	no	no	no

Table 10. Comparison of a SBC and two ensembles when $L = 7$

		win-tie-loss	min.win	max.win	avr.win	min.loss	max.loss	avr.loss
1	C	90/2/8	0.0014	0.0649	0.0308	0.0014	0.0297	0.0084
	O	96/1/3	0.0054	0.0662	0.0353	0.0081	0.0270	0.0158
2	C	99/0/1	0.0032	0.0726	0.0400	0.0113	0.0113	0.0113
	O	99/1/0	0.0016	0.0694	0.0353	no	no	no
3	C	88/2/10	0.0029	0.0971	0.0394	0.0088	0.0412	0.0235
	O	99/0/1	0.0088	0.1088	0.0566	0.0029	0.0029	0.0029
4	C	80/0/20	0.0022	0.1256	0.0474	0.0044	0.0700	0.0267
	O	100/0/0	0.0178	0.1033	0.0652	no	no	no
5	C	96/0/4	0.0039	0.1078	0.0502	0.0069	0.0363	0.0223
	O	96/0/4	0.0039	0.0843	0.0391	0.0029	0.0147	0.0078
6	C	73/3/24	$< 10^{-4}$	0.0529	0.0194	0.0029	0.0618	0.0105
	O	79/11/10	0.0029	0.0500	0.0176	0.0029	0.0176	0.0068
7	C	100/0/0	0.0183	0.0983	0.0595	no	no	no
	O	100/0/0	0.0267	0.1067	0.0701	no	no	no
8	C	99/0/1	0.0241	0.1379	0.0874	0.0483	0.0483	0.0483
	O	100/0/0	0.0414	0.1362	0.0956	no	no	no

Table 11. Comparison of a SBC and two ensembles when $L = 9$

		win-tie-loss	min.win	max.win	avr.win	min.loss	max.loss	avr.loss
1	C	92/1/7	0.0014	0.0811	0.0329	0.0027	0.0270	0.0097
	O	97/0/3	0.0027	0.0824	0.0375	0.0014	0.0203	0.0086
2	C	99/0/1	0.0016	0.0903	0.0424	0.0226	0.0226	0.0226
	O	100/0/0	0.0032	0.0677	0.0363	no	no	no
3	C	91/1/8	0.0059	0.1206	0.0405	0.0029	0.0382	0.0158
	O	100/0/0	0.0176	0.1029	0.0598	no	no	no
4	C	83/1/16	0.0011	0.1011	0.0491	0.0011	0.1022	0.0326
	O	100/0/0	0.0322	0.1056	0.0709	no	no	no
5	C	98/0/2	0.0020	0.0990	0.0541	0.0020	0.0039	0.0029
	O	100/0/0	0.0029	0.0765	0.0420	no	no	no
6	C	72/5/23	$< 10^{-4}$	0.0618	0.0189	0.0029	0.0324	0.0106
	O	76/8/16	0.0029	0.0500	0.0173	0.0029	0.0118	0.0051
7	C	100/0/0	0.0167	0.1017	0.0611	no	no	no
	O	100/0/0	0.0400	0.1183	0.0767	no	no	no
8	C	100/0/0	0.0310	0.1655	0.1036	no	no	no
	O	100/0/0	0.0207	0.1603	0.1038	no	no	no

Table 12. Comparison of a SBC and two ensembles when $L = 11$

		win-tie-loss	min.win	max.win	avr.win	min.loss	max.loss	avr.loss
1	C	91/0/9	0.0014	0.0635	0.0304	0.0014	0.0284	0.0135
	O	97/2/1	0.0054	0.0811	0.0401	0.0014	0.0014	0.0014
2	C	100/0/0	0.0016	0.0806	0.0469	no	no	no
	O	100/0/0	0.0081	0.0726	0.0376	no	no	no
3	C	83/6/11	0.0029	0.0853	0.0422	0.0059	0.0676	0.0193
	O	98/0/2	0.0147	0.0941	0.0562	0.0029	0.0029	0.0029
4	C	82/0/18	0.0022	0.0967	0.0422	0.0056	0.0911	0.0380
	O	100/0/0	0.0256	0.1133	0.0741	no	no	no
5	C	100/0/0	$< 10^{-4}$	0.0990	0.0566	no	no	no
	O	98/0/2	0.0059	0.0794	0.0412	0.0088	0.0108	0.0098
6	C	76/6/18	0.0029	0.0353	0.0168	0.0029	0.0412	0.0108
	O	67/21/12	0.0029	0.0441	0.0158	0.0029	0.0265	0.0074
7	C	100/0/0	0.0200	0.1083	0.0653	no	no	no
	O	100/0/0	0.0383	0.1167	0.0766	no	no	no
8	C	100/0/0	0.0448	0.1603	0.1126	no	no	no
	O	100/0/0	0.0603	0.1603	0.1112	no	no	no

A. Derivatives for the Frank Copula

The definition of the Frank copula is

$$C_\theta(u,v) = -\frac{1}{\theta}\ln\left(1 + \frac{(e^{-\theta u}-1)(e^{-\theta v}-1)}{e^{-\theta}-1}\right). \tag{7}$$

The first-order derivative of $C_\theta(u,v)$ wrt u is

$$\frac{\partial C_\theta(u,v)}{\partial u} = -\frac{1}{\theta}\frac{1}{1+\frac{(e^{-\theta u}-1)(e^{-\theta v}-1)}{e^{-\theta}-1}}\frac{e^{-\theta v}-1}{e^{-\theta}-1}(-\theta)e^{-\theta u}.$$

After simplifications, we obtain that

$$\frac{\partial C_\theta(u,v)}{\partial u} = \frac{e^{-\theta u}(e^{-\theta v}-1)}{e^{-\theta}-1+(e^{-\theta u}-1)(e^{-\theta v}-1)}. \tag{8}$$

By analogy, the first-order partial derivative $C_\theta(u,v)$ wrt v is

$$\frac{\partial C_\theta(u,v)}{\partial v} = \frac{e^{-\theta v}(e^{-\theta u}-1)}{e^{-\theta}-1+(e^{-\theta u}-1)(e^{-\theta v}-1)}. \tag{9}$$

Then, the second-order partial derivative $C_\theta(u,v)$ wrt u is

$$\begin{aligned}\frac{\partial^2 C_\theta(u,v)}{\partial u^2} &= \frac{(e^{-\theta v}-1)}{\left[e^{-\theta}-1+(e^{-\theta u}-1)(e^{-\theta v}-1)\right]^2}\Big[\theta e^{-2\theta u}(e^{-\theta v}-1)\\ &- \theta e^{-\theta u}\left(e^{-\theta}-1+(e^{-\theta u}-1)(e^{-\theta v}-1)\right)\Big]\\ &= \frac{\theta e^{-\theta u}(e^{-\theta v}-1)}{\left[e^{-\theta}-1+(e^{-\theta u}-1)(e^{-\theta v}-1)\right]^2}\Big[e^{-\theta u}(e^{-\theta v}-1)\\ &- e^{-\theta}+1-(e^{-\theta u}-1)(e^{-\theta v}-1)\Big]\\ &= \frac{\theta e^{-\theta u}(e^{-\theta v}-1)}{\left[e^{-\theta}-1+(e^{-\theta u}-1)(e^{-\theta v}-1)\right]^2}\Big[e^{-\theta(u+v)}-e^{-\theta u}\\ &- e^{-\theta}+1-e^{-\theta(u+v)}+e^{-\theta v}+e^{-\theta u}-1\Big].\end{aligned}$$

After some terms cancel out each other, we obtain that

$$\frac{\partial^2 C_\theta(u,v)}{\partial u^2} = \frac{\theta e^{-\theta u}(e^{-\theta v}-1)(e^{-\theta v}-e^{-\theta})}{\left[e^{-\theta}-1+(e^{-\theta u}-1)(e^{-\theta v}-1)\right]^2}, \tag{10}$$

and by analogy, the second-order partial derivative $C_\theta(u,v)$ wrt v is

$$\frac{\partial^2 C_\theta(u,v)}{\partial v^2} = \frac{\theta e^{-\theta v}(e^{-\theta u}-1)(e^{-\theta u}-e^{-\theta})}{\left[e^{-\theta}-1+(e^{-\theta u}-1)(e^{-\theta v}-1)\right]^2}. \tag{11}$$

Given that $\theta > 0$ (positive dependence observed between dataset complexity and bolstered resubstitution error), the following pairs of inequalities hold $\forall u, v \in [0,1]$:

$$\begin{aligned} e^{-\theta v}-1 \le 0, \quad e^{-\theta v}-e^{-\theta} \ge 0,\\ e^{-\theta u}-1 \le 0, \quad e^{-\theta u}-e^{-\theta} \ge 0.\end{aligned}$$

Hence, the product of the inequalities in each row above is less than or equal to zero. Given that other terms in Eqs. 10-11 are positive, it means that $\frac{\partial^2 C_\theta(u,v)}{\partial u^2} \le 0$ and $\frac{\partial^2 C_\theta(u,v)}{\partial v^2} \le 0$, which, in turn, implies that $C_\theta(u,v)$ is concave.

B. Marsaglia Polar Method

This is the polar form of the Box-Müller transformation [33] intended to generate Gaussian pseudo-random numbers from the uniform pseudo-random numbers. Its C-like pseudo-code is given below, where *rand()* is the function for uniform [0,1] random number generation, *p* is the data dimensionality, *n* is equal to M_{MC} (see Section 5.), *log()* is the natural logarithm, *m* and *s* are the mean and the standard deviation, respectively. After each iteration over *i* two samples are generated and stored in *X* so that after $n/2$ iterations, we have $2M_{MC}/2 = M_{MC}$ samples.

```
for (i = 0; i < n/2; i++)
    {
        /* Generate normal random numbers */
        for (j = 0; j < p; j++)
        {
            do
            {
                u1 = 2.0*rand()/RAND_MAX - 1;
                u2 = 2.0*rand()/RAND_MAX - 1;
                r = u1*u1 + u2*u2;
            } while(r == 0 || r >= 1);
            r = sqrt(-2*log(r)/r);
            X[i*p+j)] = m[j] + s[j]*r*u1;
            X[(i+n/2)*p+j] = m[j] - s[j]*r*u2;
        }
    }
```

References

[1] Dougherty, E. R., Shmulevich, I., Chen, J. & Wang, Z. J. (2005) *Genomic Signal Processing and Statistics*. New York, NY: Hindawi Publishing Corporation.

[2] Alon, U., Barkai, N., Notterman, D. A., Gish, K., Ybarra, S., Mack, D. & Levine, A. J. (1999) Broad patterns of gene expression revealed by clustering analysis of tumor and normal colon tissues probed by oligonucleotide arrays, *Proceedings of the National Academy of Sciences* **96** 6745–6750.

[3] Pomeroy, S. L., Tamayo, P., Gaasenbeek, M., Sturla, L. M., Angelo, M., McLaughlin, M. E., Kim, J. Y. H., Goumnerova, L. C., Black, P. M., Lau, C., Allen, J. C., Zagzag, D., Olson, J. M., Curran, T., Wetmore, C., Biegel, J. A., Poggio, T., Mukherjee, S., Rifkin, R., Califano, A., Stolovitzky, G., Louis, D. N., Mesirov, J. P., Lander, E. S., & Golub, T. R. (2002) Prediction of central nervous system embryonal tumour outcome based on gene expression, *Nature* **415** 436–442.

[4] Singh, D., Febbo, P. G., Ross, K., Jackson, D. G., Manola, J., Ladd, C., Tamayo, P., Renshaw, A. A., D'Amico, A. V., Richie, J. P., Lander, E. S., Loda, M., Kantoff, P. W., Golub, T. R., & Sellers, W. R. (2002) Gene expression correlates of clinical prostate cancer behavior, *Cancer Cell* **1** 203–209.

[5] Sima, C., Attoor, S., Braga-Neto, U., Lowey, J., Suh, E., & Dougherty, E. R. (2005) Error estimation confounds feature selection in expression-based classification, *Proceedings of the IEEE International Workshop on Genomic Signal Processing and Statistics* (Newport, Rhode Island).

[6] Braga-Neto, U. & Dougherty, E. R. (2004) Bolstered error estimation, *Pattern Recognition* **37** 1267–1281.

[7] Kuncheva, L. (2004) *Combining Pattern Classifiers: Methods and Algorithms* . Hoboken, NJ: John Wiley & Sons.

[8] Dudoit, S. & Fridlyand, J. (2003) Classification in microarray experiments, in: T. Speed, ed., *Statistical Analysis of Gene Expression Microarray Data* . Boca Raton, FL: Chapman & Hall\CRC Press, 93–158.

[9] Yu, L. (2008) Feature selection for genomic data analysis, in H. Liu & H. Motoda, eds., *Computational Methods of Feature Selection*. Boca Raton, FL: Chapman & Hall\CRC, 337–354.

[10] Sklar, A. (1959) Fonctions de répartition à n dimensions et leurs marges, *Publications of the Institute of Statistics, University of Paris* 229–231.

[11] Nelsen, R. B. (2006) *An Inroduction to Copulas*. New York, NY: Springer Science+Business Media.

[12] Joe, H. (1997) *Multivariate Models and Dependence Concepts*. Boca Raton, FL: Chapman & Hall\CRC Press.

[13] Zar, J. H. (1999) *Biostatistical Analysis* . Upper Saddle River, NJ: Prentice Hall.

[14] http://www.sagenet.org

[15] Velculescu, V. E., Zhang, L., Vogelstein, B., & Kinzler, K. W. (1995) Serial analysis of gene expression, *Science* **270** 484–487.

[16] Aldaz, M. C. (2003) Serial analysis of gene expression (SAGE) in cancer research, in: M. Ladanyi and W.L. Gerald, eds., *Expression Profiling of Human Tumors: Diagnostic and Research Applications* . Totowa, NJ: Humana Press, 47–60.

[17] http://lisp.vse.cz/challenge/ecmlpkdd2004

[18] Gandrillon, O. (2004) Guide to the gene expression data, in: P. Berka and B. Crémilleux, eds., *Proceedings of the ECML/PKDD Discovery Challenge Workshop* (Pisa, Italy, 2004) 116–120.

[19] http://microarray.princeton.edu/oncology/affydata/index.html

[20] http://www.broad.mit.edu/mpr/CNS/

[21] http://www.broad.mit.edu/cgi-bin/cancer/publications/pub_paper.cgi?mode/ =view&paper_id=75

[22] http://www.broad.mit.edu/mpr/lymphoma/

[23] Shipp, M. A., Ross, K. N., Tamayo, P., Weng, A. P., Kutok, J. L., Aguiar, R. C. T., Gaasenbeek, M., Angelo, M., Reich, M., Pinkus, G. S., Ray, T. S., Koval, M. A., Last, K. W., Norton, A., Lister, T. A., Mesirov, J., Neuberg, D. S., Lander, E. S., Aster, J. C., & Golub, T. R. (2002) Diffuse large B-cell lymphoma outcome prediction by gene expression profiling and supervised machine learning, *Nature Medicine* **8** 68–74.

[24] Bø, T. H. & Jonassen, I. (2002) New feature subset selection procedures for classification of expression profiles, *Genome Biology* **3** 0017.1–0017.11.

[25] Ho, T. K. & Basu, M. (2002) Complexity measures of supervised classification problems, *IEEE Transactions on Pattern Analysis and Machine Intelligence* **24** 289–300.

[26] Schweizer, B. & Wolff, E. F. (1981) On nonparametric measures of dependence for random variables, *The Annals of Statistics* **9** 879–885.

[27] Ein-Dor, L., Kela, I., Getz, G., Givol, D., & Domany, E. (2005) Outcome signature genes in breast cancer: is there a unique set?, *Bioinformatics* **21** 171–178.

[28] Michiels, S., Koscielny, S., & Hill, C. (2005) Prediction of cancer outcome with microarrays: a multiple random validation strategy. *Lancet* **365** 488–492.

[29] Díaz-Uriarte R. & Alvarez de Andrés, S. (2006) Gene selection and classification of microarray data using random forest, *BMC Bioinformatics* **7**.

[30] Kuncheva L. & Whitaker, C. J. (2003) Measures of diversity in classifier ensembles, *Machine Learning* **51** 181–207.

[31] Kuncheva L. & Rodríguez, J. J. (2007) Classifier ensembles with a random linear oracle, *IEEE Transactions on Knowledge and Data Engineering* **19** 500–508.

[32] Bay, S. (1999) Nearest neighbor classification from multiple feature sets, *Intelligent Data Analysis* **3** 191–209.

[33] Box G. E. P. & Müller, M. E. (1958) A note on the generation of random normal deviates, *The Annals of Mathematical Statistics* **29** 610–611.

In: New Signal Processing Research
Editor: Takumi Maeda, pp. 279-343
ISBN: 978-1-60456-479-2

Chapter 9

SIGNAL PROCESSING IN MULTIPLE-SENSOR DISTRIBUTED DETECTION SYSTEMS WITH DATA FUSION

*Ashraf M. Aziz**
Elec. Eng. Dept., Military Technical College, Cairo, Egypt

Abstract

In recent years, signal processing with distributed multiple-sensor systems has attracted the attention of several investigators and has been gaining importance. This interest has been sparked by the requirement of military surveillance systems to be more reliable and immune to electronic attack than single sensor systems. In this chapter, signal processing with distributed multiple-sensor detection systems with data fusion is considered. The distributed sensors monitor a common volume and send their local decisions to a fusion center, which further combines them into a final global decision. There are two main approaches for signal processing in distributed multiple-sensor detection systems. The first is centralized signal processing, where all sensor observations are transmitted to a central processor that performs optimal detection of targets based on sensors observations. The second approach is decentralized signal processing, where the signal processing is distributed among the sensors and a fusion center. The local sensors monitor the same object scene and pass their local decisions about the same hypothesis to the fusion center that combines them into a final global decision. The contribution of this chapter is divided into two main parts. In the first part, the problem of obtaining and comparing the global performance improvement in centralized and decentralized approaches is considered and analyzed. The needed number of sensors in the decentralized approach, to achieve a performance non inferior to the centralized approach, for the same signal to noise ratio, is determined. The dependence of global performance improvement on the desired false alarm probability, the number of sensors, and the signal to noise ratio at each individual sensor, is also addressed. In the second part of the contribution, we propose a simple algorithm for optimizing multiple sensor detection system with data fusion, in terms of both the sensors and the fusion center, according to Neyman-Pearson strategy. The proposed algorithm determines, for a given false alarm probability, the corresponding optimum setting of the thresholds both at the fusion center and at the sensors. It scans all possible solutions and picks the solution that gives the largest detection probability.

* E-mail address: amaziz64@yahoo.com

The proposed algorithm is found to be simple, accurate and fast. The performance analysis is provided using Monte Carlo simulation in case of Gaussian and Rayleigh distributed observations.

I. Introduction

Multiple-sensor distributed sensor systems were originally motivated by their military applications with respect to command, control, and communications but are now being employed in a wide variety of military and civilian applications. Interest in signal processing with multiple-sensor systems has surfaced with anticipated applications in target detection using geographically distributed sensors [1-6]. This problem is called signal processing in geographically distributed sensor detection systems. Such systems are expected to increase the reliability of the detection and be immune to noise interference and failure [7, 8]. The distributed detection system employs several geographically separated local sensors and a central processor. The local sensors monitor the same object scene and pass their information about the same hypothesis to the central processor. This is considered as a binary hypothesis testing problem with two hypotheses; H_0 designating one hypothesis (signal absent) and H_1 the alternative (signal present). The central processor combines all the received information into a final global decision.

There are two main approaches for signal processing in multiple-sensor distributed detection systems (DDSs) [1, 3]. The first is centralized signal processing, where all sensor observations are transmitted to a central processor that performs optimal detection of targets based on conventional statistical techniques. In this approach, the central processor is the only place for making decisions and no local decisions are obtained at the sensors level. In this case, all sensor observations are transmitted to a central processor to derive a global decision. This requires transmission of sensor information without delay which requires a large communication bandwidth.

The second approach is decentralized signal processing, where the signal processing is distributed among the sensors and a fusion center. In this case, some preliminary processing of data is carried out at each sensor and the results are available locally. The local sensors monitor the same object scene and pass their local decisions about the same hypothesis to the fusion center. The local decisions are then sent to a fusion center that combines them into a final global decision. This situation arises when there are constraints on the amount of information that can be sent to the data fusion center. However, in the absence of this limitation, the optimal strategy for each sensor is simply to transmit its observation for centralized decision making. Unlike the central processor in centralized systems, the fusion center in decentralized systems has only partial information as communicated by the sensors. The result is a loss of performance in decentralized systems as compared to centralized systems. Despite the performance loss, which could result from the decentralized systems, some of their advantages are reduced communication bandwidth requirement, increased reliability, and reduced cost. Because of such considerations, or even the problem of flooding the central processor with more information than it can process, totally centralized processing of multiple sensor is abandoned and never implemented in practice [3, 6, 9].

The decision fusion in distributed sensor detection systems has been considered extensively in the literature. Tenney and Sandell [1] have extended the Bayesian hypothesis testing formulation and have solved the binary distributed decision problem for the case of two sensors. They assumed that the probability distributions under both hypotheses are known and the sensor observations are statistically independent. They have treated the distributed detection problem without considering the design of data fusion algorithms. The cost assignment in [1] may include the effect of fusion, but the design of a fusion rule itself was not considered. Sadjadi [10] has extended the results of [1] to include multiple hypotheses and more than two sensors. Chair and Varshney [11] have derived an optimum fusion rule under the assumptions that the a-priori probabilities are known and each sensor uses the likelihood ratio (L-R) test to arrive at its own decision. Thomopoulos et al. [12] have derived the optimum fusion rule for unknown a-priori probabilities in terms of the Neyman-Pearson (N-P) test at the local sensors as well as at the fusion center. According to Neyman-Pearson strategy, the detection probability is maximized for a desired false alarm probability. Srinivasan [13] has shown that the globally optimal solution to the fusion problem that maximizes the probability of detection for fixed probability of false alarm when sensors transmit independent, binary decisions to the fusion center, consists of L-R tests at all sensors and a N-P test (or a randomized N-P test) at the fusion center. The proof of the optimality of the N-P/L-R test in [13] is based on the first-order Lagrange multipliers methods. Unfortunately, this method does not always yield the optimal solution. The reason for this is explained in [2, 14]. Zhang et al. [15] have addressed the concept of quasi-convexity and have showed that the use of the Lagrange multiplier in case of independent identical sensors yields the optimal solution in the Neyman-Pearson sense. In this case, the optimum fusion rule is a k-out-of-n fusion rule (majority-rule voting). This means that the fusion center adopts the presence of a target when at least k sensors favor that hypothesis. For every desired value of global false alarm probability (at the fusion center), there is an optimum integer k that maximizes the global detection probability. A case of dependent sensor observations is presented in [16]. Specific solutions have been obtained for a random signal in additive noise detection problem with two sensors. In [17], it has been shown that for cases with dependent observations from sensor to sensor, the optimum sensor test statistics are generally not the likelihood ratios of the sensor observations. A data fusion model consisting of several levels of parallel decision fusion has been proposed in [16]. It has been shown that the parallel data fusion model of several levels has better performance than the parallel and serial fusion models. A Bayesian model has been presented in [17] for fusing detection probabilities obtained from a distributed detection system. In this model, each local sensor generates a probability that represents its confidence on the signal present hypothesis. The fusion center combines all the reported probabilities and makes a global decision. This model is equivalent to multilevel decision. In general, the optimal solution to the distributed detection problem is very complicated and computationally complex especially with multilevel quantization [2, 18, 19]. A simple and efficient suboptimal multilevel quantization approach in geographically distributed sensor systems is proposed in [20] to reduce the performance loss between the centralized and the decentralized approaches. Chamberland and Veeravalli [21] have considered the case of identical sensors decision fusion when the distributed detection system is constrained by the capacity of the channel over which the sensors are transmitting. A study of composite hypothesis testing has been addressed in [22, 23]. In composite hypothesis testing, unlike in simple hypothesis testing, the problem is to design a decision rule, for deciding in favor of one out of several hypotheses, under some uncertainty in the parameters of probability distribution functions associated with these hypotheses.

Many previous approaches to the Neyman-Pearson fusion of statistically independent sensor decisions have concentrated on optimizing system performance at the sensors or at the fusion center but not both [2, 10-13, 24-28]. The global optimum solution of a distributed detection system consists of a set of strongly coupled conditions. Thus explicit solutions of sensor thresholds and fusion rule are not obtained. Instead, the solution of a distributed sensor system is usually obtained by adopting various Boolean algebraic combinations as fusion strategies [13, 14, 26]. Since the Boolean algebraic fusion rules are optimum only in case of identical sensors, the use of such fusion rules does not obtain the optimum solutions in case of non identical sensors (practical case).

In this chapter, the problem of obtaining and comparing the global performance improvement in centralized and decentralized approaches is considered. This open problem is suggested to be investigated by [2] and is not previously considered. We consider more complicated examples by considering large number of sensors (not two or three sensors as considered in most previous publications) to characterize the performance trade-off between the centralized and decentralized approaches. Thus the communication requirements can be compared to choose a preferred architecture. It is found that the increments in the global performance improvement are depending on the desired global false alarm probability, the number of sensors, and the signal to noise ratio at each individual sensor. It is found that for very small values of global false alarm probabilities it is impractical to use a large number of sensors to avoid additional hardware costs without significant performance improvement. It is also found that the performance loss between the centralized and decentralized approaches increases as the number of sensors increases. The needed number of sensors in the decentralized approach, to achieve a performance non inferior to the centralized approach, with n sensors, for the same signal to noise ratio, is found to be $2n$ sensors.

Also, a simple algorithm for optimizing multiple sensor detection system with data fusion, in terms of both the sensors and the fusion center, according to Neyman-Pearson strategy, is proposed. The proposed algorithm determines, for a given false alarm probability, the corresponding optimum setting of the thresholds both at the fusion center and at the sensors. It scans all possible solutions and picks the solution that gives the largest detection probability. The proposed algorithm is found to be simple, accurate and fast. It is worth noting that the execution time of the proposed algorithm is mainly dependent on the number of considered sensors no matter they are identical or not, although from an analytic point of view finding optimum operating point of n sensors (2n unknowns) is far more difficult than finding a common optimum operating point (2 unknowns) in case of identical sensors. It is also shown that the proposed algorithm can be used to optimize distributed detection systems according to criteria other than the Neyman-Pearson one, such as the Ideal Observer criterion for example. The performance analysis is provided using Monte Carlo simulation in case of Gaussian and Rayleigh distributed observations.

The remainder of this chapter is organized as follow. Section II is devoted to a brief review of the distributed detection systems. In this section, we describe the structure of the distributed detection systems including the centralized and decentralized signal processing and define their mathematical models. The performance of centralized and decentralized detection systems is presented in section III. The performance analysis is provided in case of Gaussian and Rayleigh distributed observations. A simple and efficient solution to distributed detection systems with data fusion is proposed in section IV. It determines, for a given global false alarm probability, the corresponding optimum setting of the thresholds both at the fusion center and at the sensors. It scans all possible solutions and picks the solution that gives the largest global detection probability. In section IV we show that the proposed solution can be used to optimize distributed detection systems according to criteria other than the Neyman-Pearson one, such as the Ideal Observer criterion for example. Finally, concluding remarks are given in Section V.

II. Centralized and Decentralized Signal Processing

II.1. Centralized Signal Processing

The centralized distributed detection system is shown in Figure 1. In this system, geographically distributed sensors independently observe the same environment. Each sensor $i,\ i = 1, 2, \ldots\ldots, n$, sends its own observation vector $\mathbf{y}_i$, to a central processor. The central processor has access to the exact observations taken by the distributed sensors and no local decisions are made by the sensors. We assume that the received $\mathbf{y}_i$'s at the n sensors are statistically independent, conditioned on each hypothesis. It is assumed that the observation at the ith sensor is a scalar y_i. If a random sample of size p is available at a sensor, it is assumed that by a suitable processing of the same, a one-dimensional statistic y_i will be obtained. Existence of one-dimensional sufficient statistic implies no loss of information [29] (this depends on the distribution of the random samples). In this approach, all sensor observations are transmitted to a central processor in order to derive a global decision and no local decisions are made by the sensors. Thus the central processor derives a global decision u_0, based on the received sensor observations $y_1, y_2, \ldots\ldots, y_n$. Under each hypothesis the sensor observations have known joint probability densities $P(y_1, y_2, \ldots\ldots y_n \mid H_0)$ and $P(y_1, y_2, \ldots\ldots y_n \mid H_1)$ as functions of the observations $y_i, i = 1, 2, \ldots\ldots, n,$ where y_i's are random vectors representing the sensor observations. The crux of the centralized hypothesis testing problem is to derive a decision strategy of the form

$$u_0 = \begin{cases} 0: \text{decide signal absent, i.e. } \mathrm{H}_0 \\ 1: \text{decide signal present, i.e. } \mathrm{H}_1 \end{cases} \quad . \tag{1}$$

where the global decision u_0 depends on the observations $y_1, y_2, \ldots\ldots, y_n$. According to Neyman Pearson criterion, it is required to find a decision strategy expressed as a density function $P(u_0 \mid y_1, y_2, \ldots\ldots y_n)$ which maximizes the global detection probability (GDP) for a desired global false alarm probability (GFAP), where

$$GFAP = \Pr\{u_0 = 1 | H_0\} = 1 - \Pr\{u_0 = 0 | H_0\}, \tag{2}$$

$$GDP = \Pr\{u_0 = 1 | H_1\} = 1 - \Pr\{u_0 = 0 | H_1\}. \tag{3}$$

The solution of the centralized problem is [1, 3]:

(a) Deterministic, so that the decision rule is a function

$$\gamma(y_1, y_2, ...y_n) \rightarrow \{0,1\}, \tag{4}$$

where $u_0 = i$ is interpreted as choosing H_i, and

(b) Given by a likelihood ratio test (LHR)

$$\gamma(y_1, y_2, ...y_n) = \begin{cases} 0 & \text{if } LHR(y_1, y_2, ...y_n) < \lambda_0 \\ 1 & \text{if } LHR(y_1, y_2, ...y_n) \geq \lambda_0 \end{cases}, \tag{5}$$

where

$$LHR(y_1, y_2, ...y_n) = \frac{P(y_1, y_2, ...y_n | H_1)}{P(y_1, y_2, ...y_n | H_0)}. \tag{6}$$

(c) The threshold of the fusion center, λ_0, is determined according to the desired GFAP.

II.2. Decentralized Signal Processing

The decentralized detection system is shown in Figure 2. In this approach, a number of sensors n receive and process the observations {y_i's} to generate the sensor decisions {u_i's}. Thus each sensor derives a local decision u_i based on its own observation $y_i, i = 1, 2, ..., n$. Each local decision may take the value 0 or 1, depending on whether the sensor i decide H_0 or H_1, i.e.

$$u_i = \begin{cases} 0: \text{decide signal absent, i.e. } H_0 \\ 1: \text{decide signal present, i.e. } H_1 \end{cases}. \tag{7}$$

The hard-decision information {u_i's} are allowed to be sent to a data fusion center considering the limitations of communication bandwidth. The fusion center derives a global decision u_0, based on the received sensor decisions $u_1, u_2, ..., u_n$. We assume that the received y_i's at the n sensors are statistically independent, conditioned on each hypothesis. This implies that the received decisions at the fusion center are independent conditioned on each hypothesis. This approach greatly reduces channel capacity for two reasons. First, a report of a decision is a simpler message than a sensor observation, and second, most observations need not be reported at all since they don't correspond to a detection.

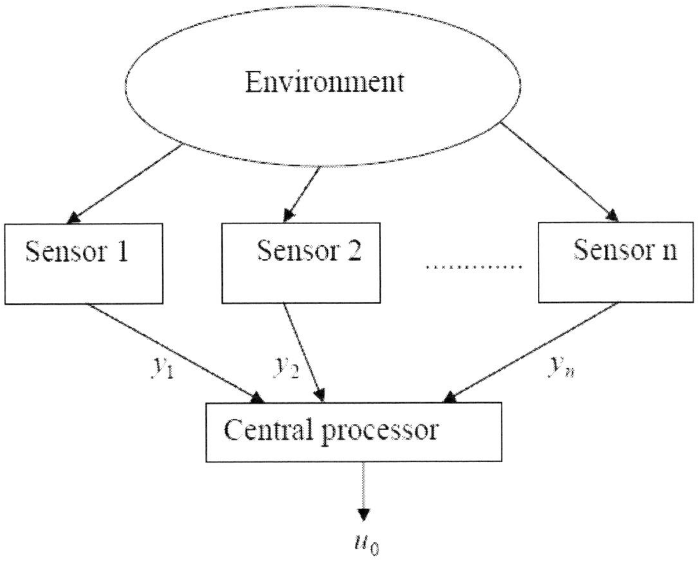

Figure 1. Centralized distributed detection system.

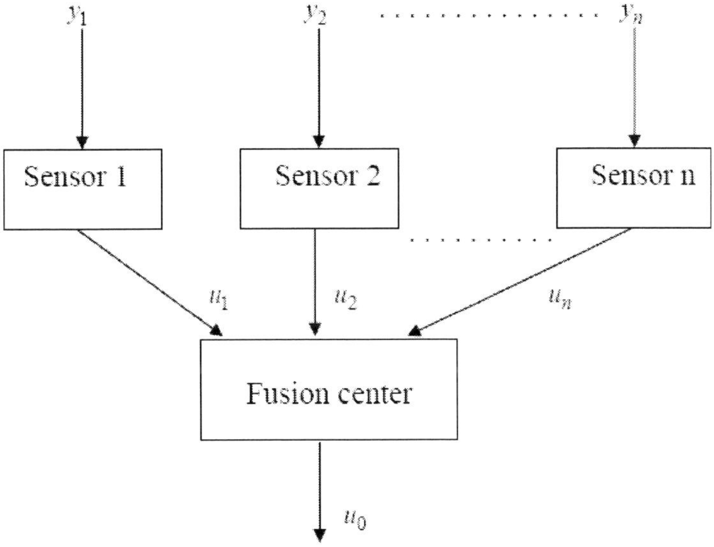

Figure 2. Decentralized distributed detection system.

Optimality criteria for the decentralized DDS are known from previous studies [11]-[13], [30]. According to the Neyman-Pearson criterion: maximizing the overall detection probability while keeping the overall false alarm probability below a given value, quantization at the sensors should be performed on the likelihood ratio of the observations. The one-bit quantization (hard-decision) rule at each sensor can be described as

$$u_i = \begin{cases} 1, & \text{if } L_i(y_i) \geq t_i \\ 0, & \text{otherwise} \end{cases}, \tag{8}$$

where $L_i(y_i)$ is the likelihood ratio test at the i th sensor and is given in terms of the ratio of the probability density functions of the observations under both hypotheses as

$$L_i(y_i) = \frac{P(y_i \mid H_1)}{P(y_i \mid H_0)}, \tag{9}$$

and the sensor's threshold, t_i, is determined by the probability of false alarm of the i th local sensor. At each sensor, the false alarm probability is

$$pf_i = \Pr(u_i = 1 | H_0) = \int_{t_i}^{\infty} P(y_i \mid H_0)\, dy, \tag{10}$$

and the detection probability is

$$pd_i = \Pr(u_i = 1 | H_1) = \int_{t_i}^{\infty} P(y_i \mid H_1)\, dy. \tag{11}$$

The performance of each sensor is characterized by the plot of pd_i versus pf_i, known as the receiver operating characteristic curve (ROC). The global performance of the fusion center is characterized by the plot of GDP versus $GFAP$.

The optimal Neyman-Pearson test at the fusion center must almost always be randomized to exactly meet the desired global false alarm probability of the fusion center (GFAP). The fusion center implements the N-P test using all the sensor decisions ($\mathbf{u} = \{u_1, u_2, \ldots\ldots\ldots\ldots, u_n\}$) that the individual sensors have communicated, i.e., it formulates the LR test [2, 11]

$$L_0(\mathbf{u}) = \frac{P(\mathbf{u} | H_1)}{P(\mathbf{u} | H_0)} = \frac{P(u_1, u_2, \ldots\ldots\ldots\ldots, u_n \mid H_1)}{P(u_1, u_2, \ldots\ldots\ldots\ldots, u_n \mid H_0)} \overset{H_1}{\underset{H_0}{\gtreqless}} t (\Rightarrow H_1 \text{ with probability } \beta), \tag{12}$$

where β is a randomization parameter and is determined to meet a given GFAP. The threshold t (of the fusion center) is determined by the desirable GFAP as

$$\sum_{L_0(\mathbf{u})>t^*} P(L_0(\mathbf{u})|H_0) = GFAP, \tag{13}$$

and the corresponding global detection probability (GDP) of the fusion center is

$$GDP = \sum_{L_0(\mathbf{u})>t^*} P(L_0(\mathbf{u})|H_1) \, . \tag{14}$$

Due to the independent assumption, (12) can be rewritten as [32]

$$\sum_{i=1}^{n} w_i \, u_i \underset{H_0}{\overset{H_1}{\gtrless}} t \, , \tag{15}$$

where

$$w_i = \begin{cases} \log\left(\dfrac{pd_i}{pf_i}\right), & \text{if } u_i = 1, \, i = 1,2,\ldots\ldots\ldots, n \\ \log\left(\dfrac{1-pd_i}{1-pf_i}\right), & \text{if } u_i = 0, \, i = 1,2,\ldots\ldots\ldots, n \, , \end{cases} \tag{16}$$

i.e., the individual decisions are weighted according to the detection and false alarm probabilities of each sensor.

When all sensors are identical and operate at the same level of detection probability and false alarm probability, i.e., $pd_i = pd_j = pd$ and $pf_i = pf_j = pf$ for all i and j, (15) reduces to k-out-of-n fusion rule. In this case, the GFAP and the GDP for an n-sensor DDS is given by [3, 28, 31]

$$GFAP = \sum_{i=k}^{n} c_i^n \, pf^i \, (1-pf)^{n-i} \, , \tag{17}$$

$$GDP = \sum_{i=k}^{n} c_i^n \, pd^i \, (1-pd)^{n-i} \, , \tag{18}$$

where c_i^n is the binomial coefficient,

$$c_i^n = \frac{n!}{i!(n-i)!} \, . \tag{19}$$

For $k=1$, the optimum fusion rule reduces to an OR fusion rule, while for $k=n$ it becomes an AND fusion rule. For a specified value of GFAP, there is an optimum integer k that maximizes the GDP. This means that if k or more sensors decide hypothesis H_1, then the global decision will be H_1, i.e.

$$u_0 = \begin{cases} 1, & \text{if } \sum_{i=1}^{n} u_i \geq k \\ 0, & \text{otherwise} \end{cases} \quad . \tag{20}$$

In [12], it was proved that if the N-P test is used at the fusion center in case of identical sensors, the probability of detection at the fusion center (GDP) exceeds that of the individual sensor for the same probability of false alarm. If the sensors are not identical, no general assessment can be made about the global performance of the fusion center since the global performance depends on how far the operating points of the sensors are from each other. The asymptotic global performances of k-out-of-n fusion rules have been addressed in [28].

III. Performance of Centralized and Decentralized Systems

III.1. Gaussian Distributed Observations

We assume the case of n sensors with Gaussian distributed observations; i.e.

$$\begin{aligned} P(y_i \mid H_0) &= \frac{1}{\sqrt{2\pi}} e^{-y_i^2/2}, \\ P(y_i \mid H_1) &= \frac{1}{\sqrt{2\pi}} e^{-(y_i - s_i)^2/2}, \; s_i > 0, \; i = 1, 2, \ldots\ldots, n, \end{aligned} \tag{21}$$

where the mean value of the received signal is zero under hypothesis H_0 and is s_i under hypothesis H_1. The Neyman-Pearson test, utilizes all of the received observations's (raw data, not decisions) in case of centralized distributed detection systems, with identical sensors ($s_1 = s_2 = \ldots.. = s_n = s$), will have the form [26, 27]

$$u_0 = \begin{cases} 0 & \text{if} \sum_{i=1}^{n} y_i < t_0 \\ 1 & \text{if} \sum_{i=1}^{n} y_i \geq t_0 \end{cases} , \tag{22}$$

where t_0 is the threshold of the fusion center. To achieve a desired global false alarm probability $(GFAP)$ at the fusion center, a threshold of

$$t_0 = \sqrt{n}\phi^{-1}(GFAP), \tag{23}$$

is needed at the fusion center, where the function is defined as

$$\phi(x) = \int_x^{\infty} \frac{1}{\sqrt{2\pi}} e^{-z^2/2} dz. \tag{24}$$

The corresponding global detection probability (GDP) is given by

$$GDP = \phi\left(\frac{t_0 - ns}{\sqrt{n}}\right). \tag{25}$$

The decision rules of the sensors in case of decentralized distributed detection systems are given by

$$u_i = \begin{cases} 0 & \text{if } LHR(y_i) = \dfrac{p(y_i \mid H_1)}{p(y_i \mid H_0)} < t_i \\ 1 & \text{if } LHR(y_i) = \dfrac{p(y_i \mid H_1)}{p(y_i \mid H_0)} \geq t_i. \end{cases}, \tag{26}$$

where t_i is the threshold of sensor i and is determined according to the sensor false alarm probability. The corresponding false alarm and detection probabilities are

$$pf_i = \phi(t_i), \tag{27}$$

$$pd_i = \phi(t_i - s_i), \tag{28}$$

where pf_i and pd_i are the ith sensor false alarm and detection probabilities, respectively. The fusion center implements the Neyman Pearson test using all the sensor hard decisions ($\mathbf{u} = \{u_1, u_2, \ldots\ldots\ldots\ldots, u_n\}$) that the individual sensors have communicated, i.e., the fusion rule has the form [1, 3, 12]

$$u_0 = \begin{cases} 1, & \text{if } \dfrac{P(y_1, y_2, \ldots\ldots, y_n | H_1)}{P(y_1, y_2, \ldots\ldots, y_n | H_0)} \geq t_0 \\ 0, & \text{otherwise} \end{cases}, \tag{29}$$

where t_0 is the threshold of the fusion center and is determined to meet a given (desired) global false alarm probability (GFAP). Equation (29) yields to equations (15) and (16). The optimum fusion rule in case of identical sensors can be obtained as a k-out-of-n fusion rule as given in [3, 13, 28, 30] and equation (20).

III.2. Rayleigh Distributed Observations

In case of Rayleigh distributed observations, the probability distributions are given by

$$P(y_i \mid H_0) = \exp(-y_i)\, U(y_i), \tag{30}$$

$$P(y_i \mid H_1) = d_i \exp(-d_i\, y_i)\, U(y_i), \tag{31}$$

$$d_i = \frac{1}{1+r_i}, \tag{32}$$

where r_i is the signal to noise ratio at the ith sensor, and $U(.)$ is the unit step function,

$$U(x) = \begin{cases} 1, & x \ge 0 \\ 0, & x < 0 \end{cases}. \tag{33}$$

The corresponding detection and false alarm probabilities are related as [26, 34]

$$pd_i = pf_i^{\,1/(1+r_i)}. \tag{34}$$

In case of centralized distributed detection systems with Rayleigh distributed observations, the optimum global decision is based on integration of n local observations (we assume Swerling II model and identical sensors). For $n >> 1$ ($n \ge 10$), the threshold at the centralized fusion center and the GDP, for a given $GFAP$, are approximated by [27]

$$t_0 \cong \sqrt{n}\, \phi^{-1}(GFAP) + n, \tag{35}$$

$$GDP \cong \phi\left[\frac{t_0 - n(1+r)}{\sqrt{n}\,(1+r)}\right]. \tag{36}$$

The optimum fusion rule in case of decentralized distributed detection systems is given by (15) and (16).

III.3. Simulation Results and Analysis

Figures 3-6 show the improvement in the global receiver operating characteristic (GDP versus $GFAP$) in case of the centralized detection systems, in case of Gaussian distributed observations, as the number of sensors n increases, assuming -5, -3, 0 and 3 dB SNR per sensor observation, respectively. Figures 7-10 show the same receiver operating characteristics (ROC's) in case of the decentralized detection systems. Clearly, increasing SNR leads to improvement in both cases, but it is worth noting that the global performance advantage of the centralized approach over the decentralized one increases with SNR. This is a reasonable result of the centralized decision rule because it makes more efficient use of the increased amount of information. It is clear that the increments in the (ROC's) decrease as n increases. For very small desired values of $GFAP$, it is impractical to use a large number of sensors to avoid additional hardware costs without significant performance improvement.

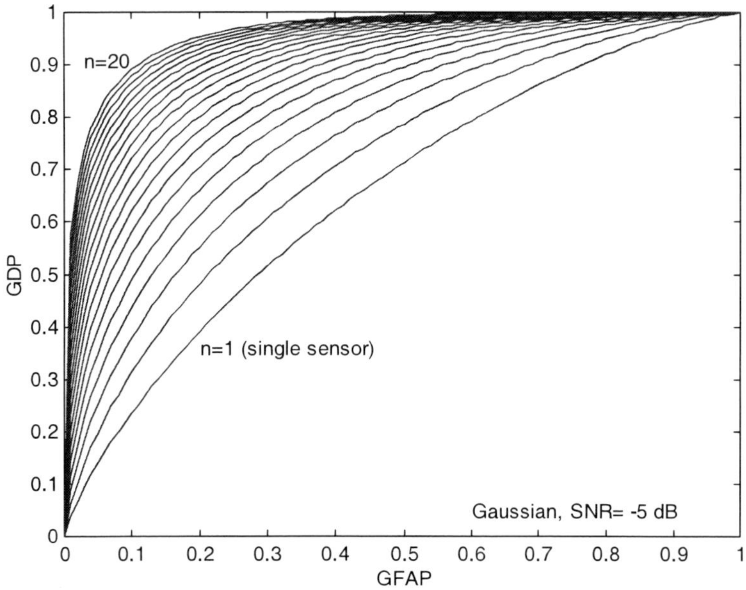

Figure 3. Performance of centralized detection system for different n.

Figures 11-24 compare the global performance improvement in the centralized and decentralized detection systems for different SNR in case of Gaussian distributed observations. These figures compare the performance of the centralized and decentralized systems as well as the ROC of a single sensor system. Figures 11-15 show the performance in case of 5, 8, 12, 15 and 20 sensors, respectively, assuming -5 dB SNR per sensor observation. Figures 16-20 show the performance in case of 5, 8, 12, 15 and 20 sensors, respectively, assuming 0 dB SNR per sensor observation. Figures 21-24 show the performance in case of 5, 8, 12 and 15 sensors, respectively, assuming 3 dB SNR per sensor observation. In all these figures (Figures 11-24), the performance loss due to decentralized systems is obvious. Clearly the degradation in the global performance of the decentralized detection systems, compared to the centralized detection systems, is caused by transmitting highly condensed information from the sensors to the central processor. The performance loss increases as n increases.

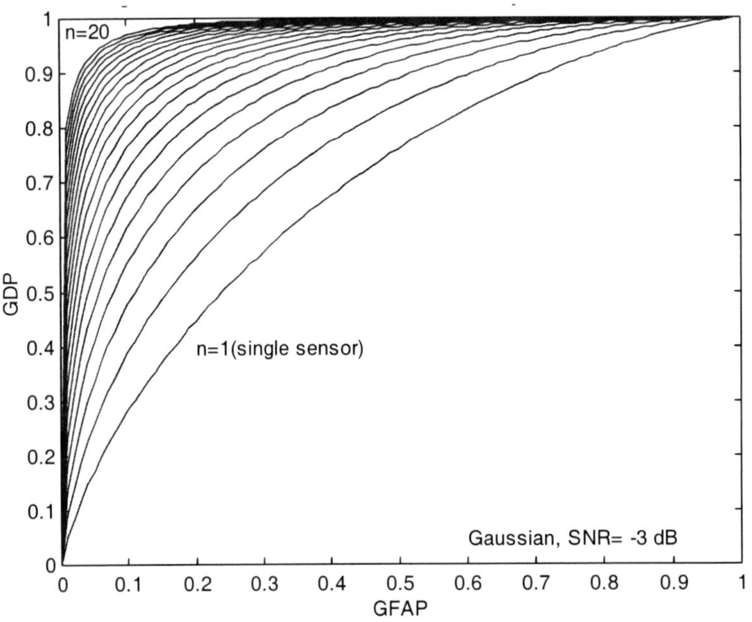

Figure 4. Performance of centralized detection system for different n.

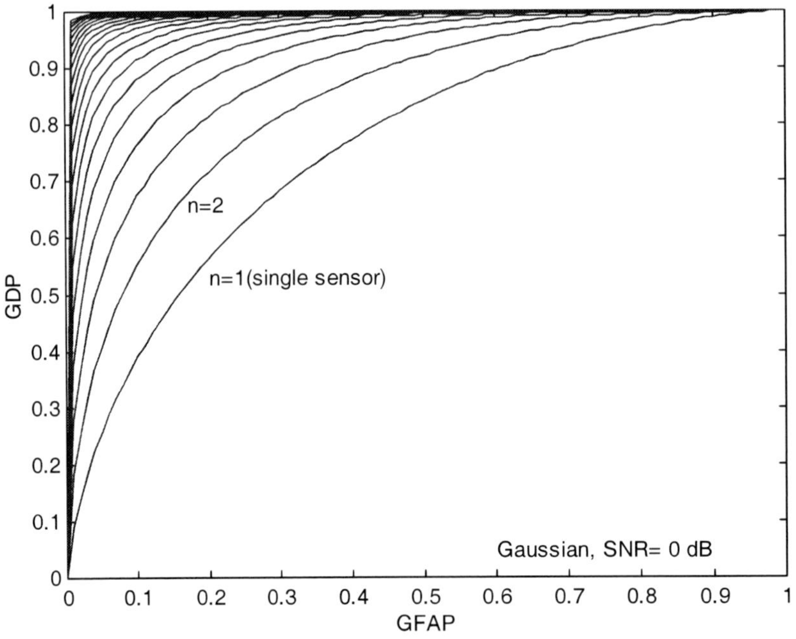

Figure 5. Performance of centralized detection system for different n.

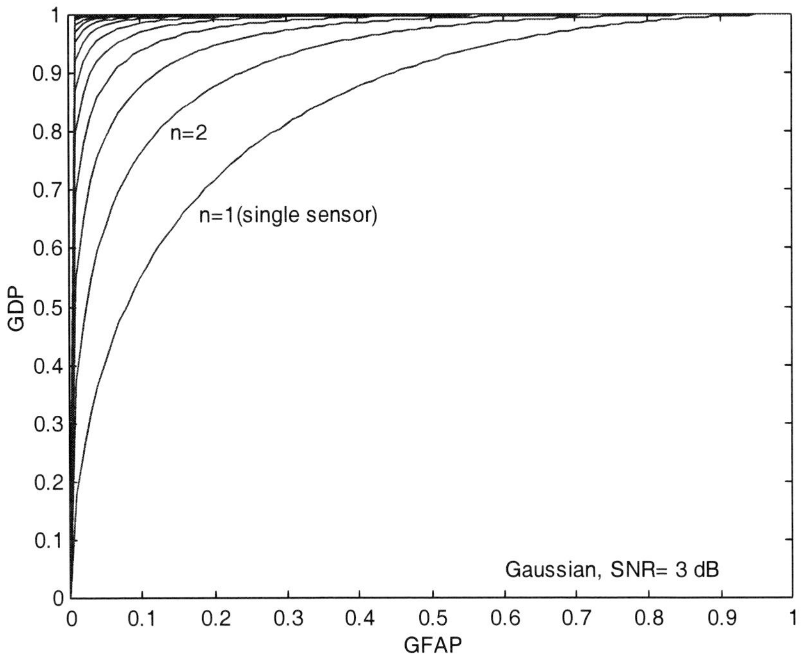

Figure 6. Performance of centralized detection system for different n.

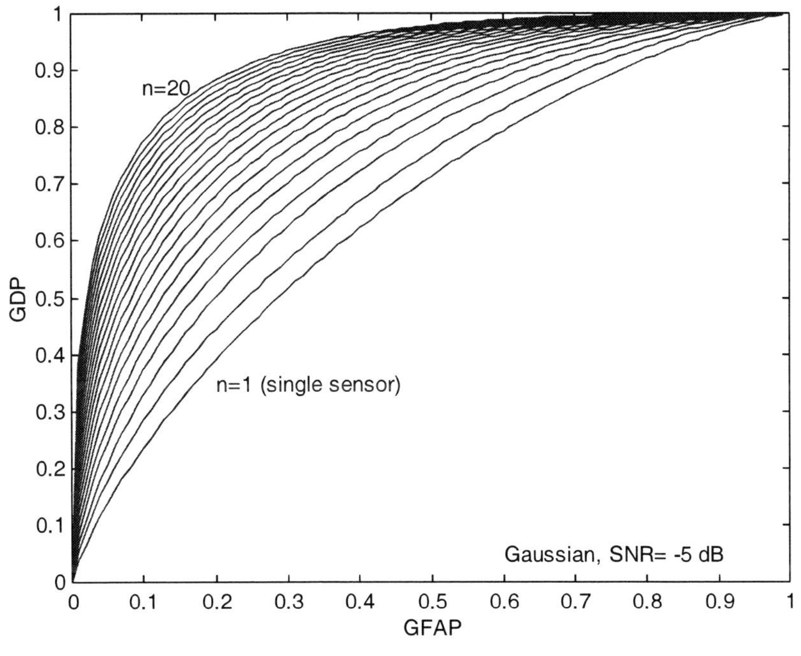

Figure 7. Performance of decentralized detection system for different n.

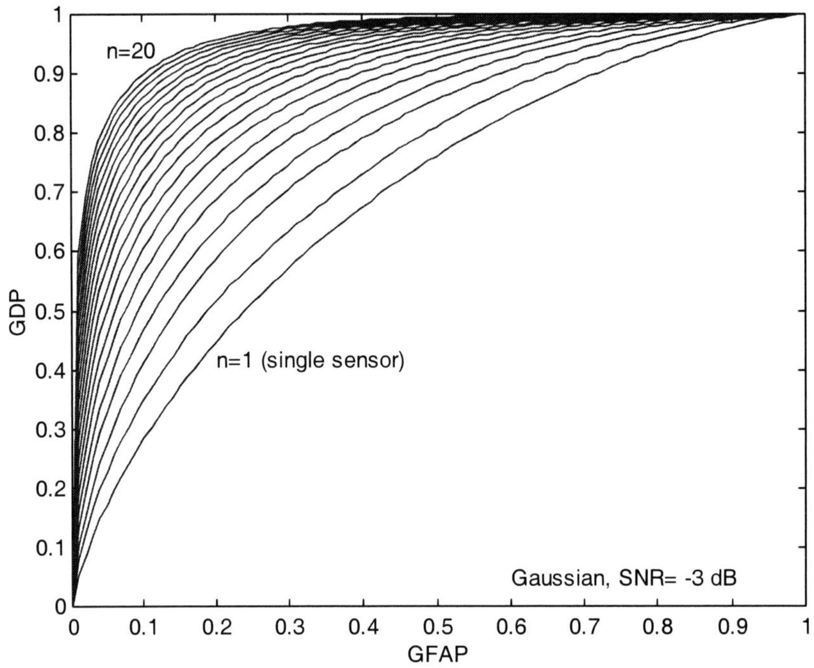

Figure 8. Performance of decentralized detection system for different n.

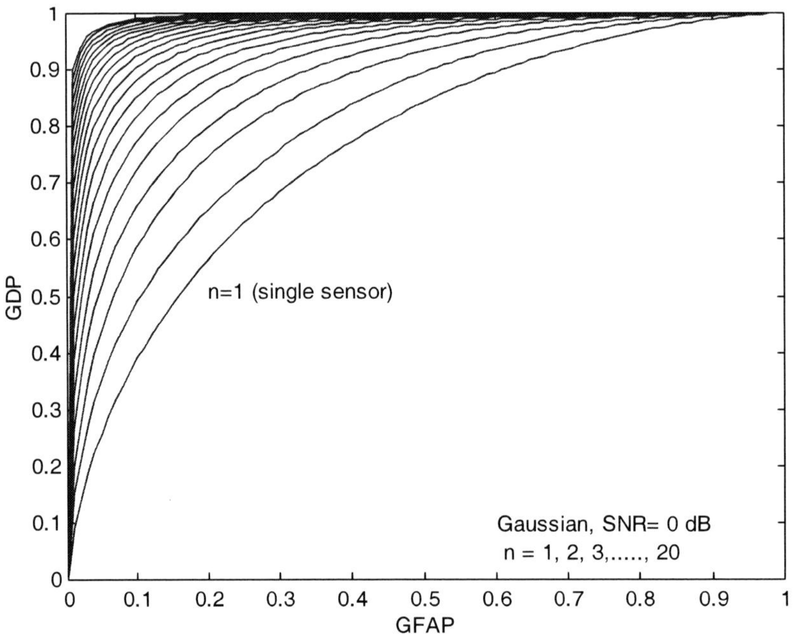

Figure 9. Performance of decentralized detection system for different n.

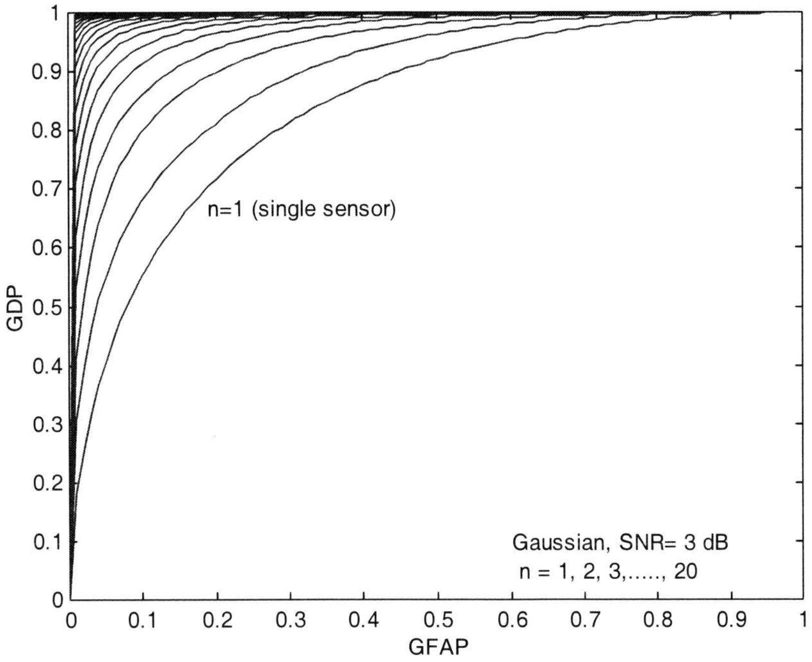

Figure 10. Performance of decentralized detection system for different n.

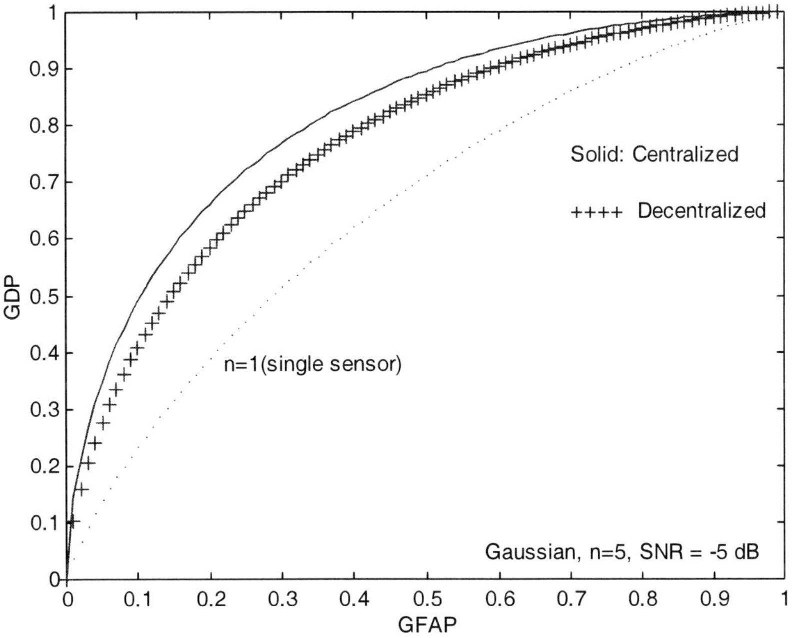

Figure 11. Comparison of centralized and decentralized systems.

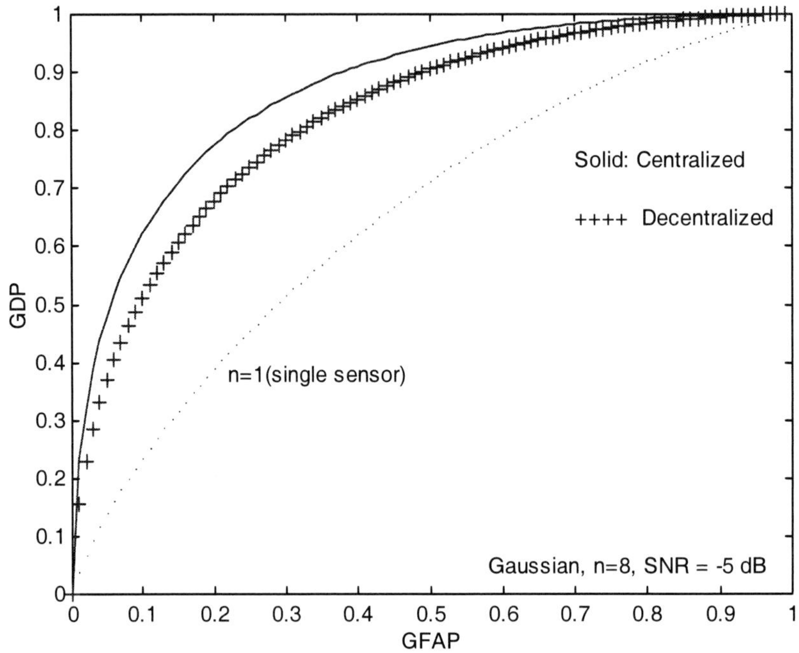

Figure 12. Comparison of centralized and decentralized systems.

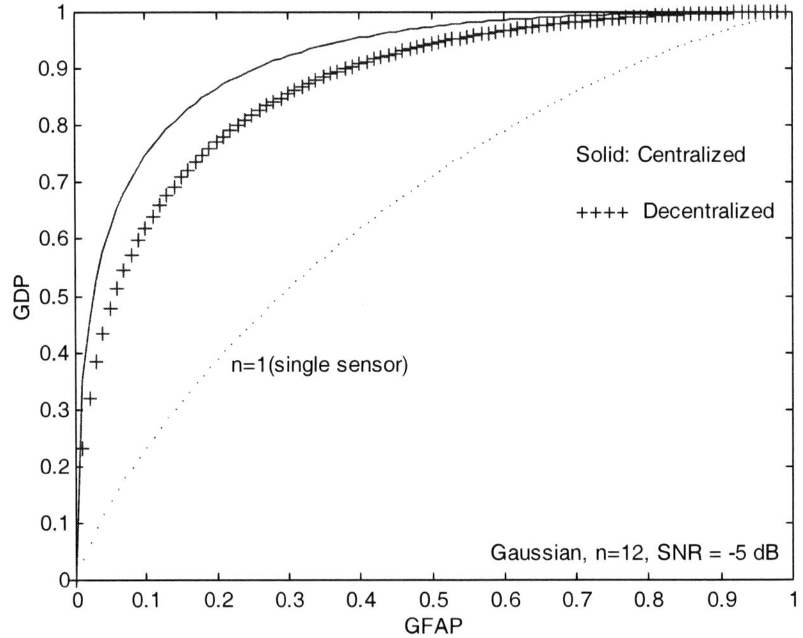

Figure 13. Comparison of centralized and decentralized systems.

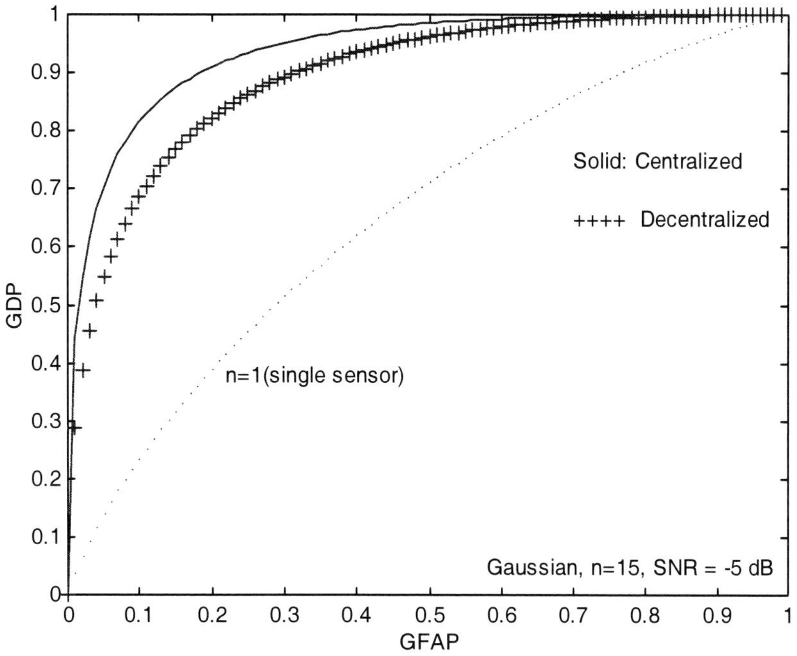

Figure 14. Comparison of centralized and decentralized systems.

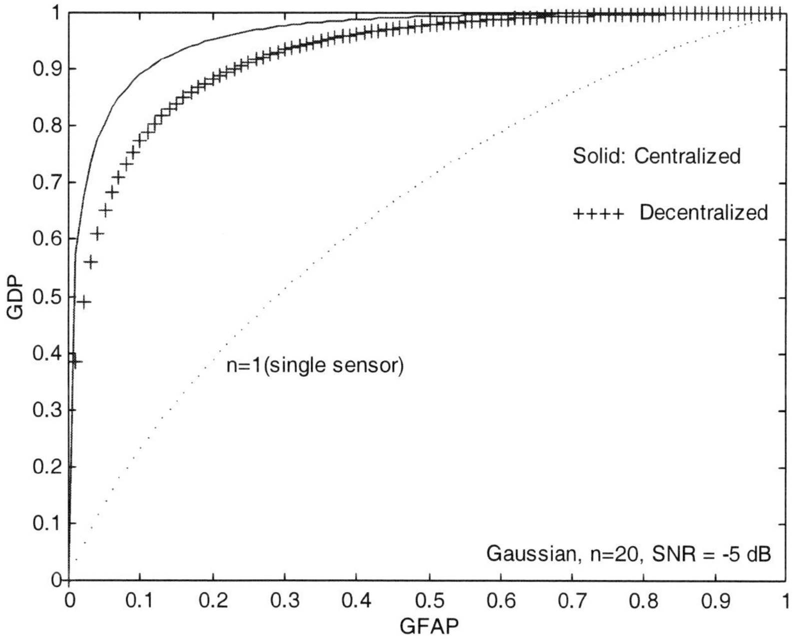

Figure 15. Comparison of centralized and decentralized systems.

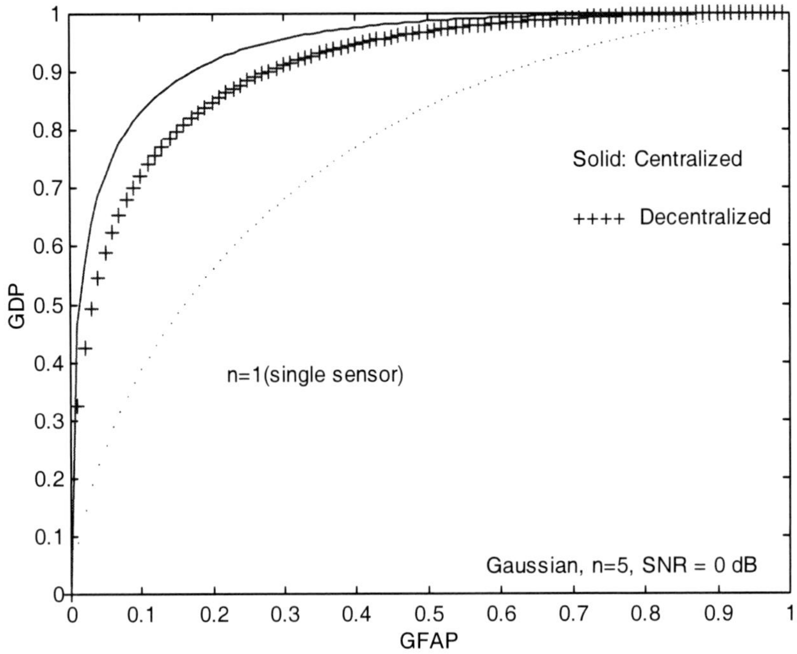

Figure 16. Comparison of centralized and decentralized systems.

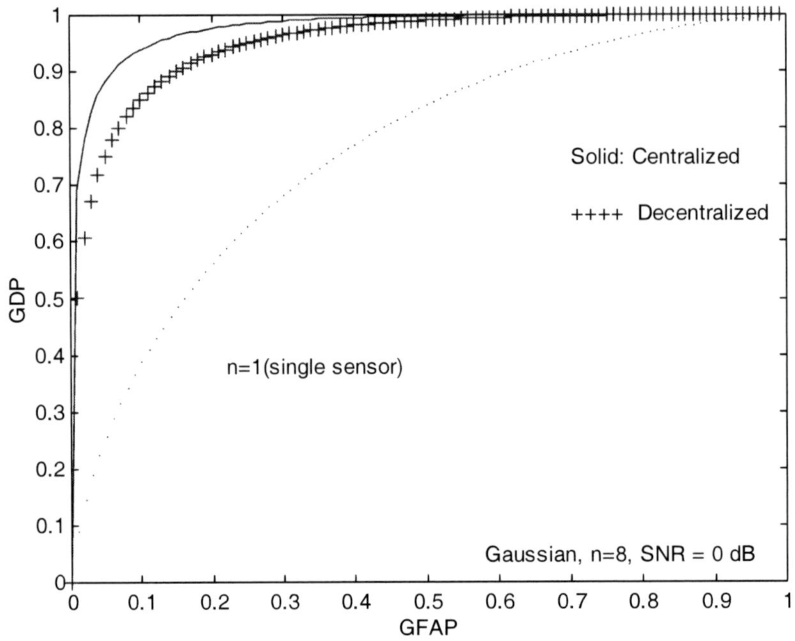

Figure 17. Comparison of centralized and decentralized systems.

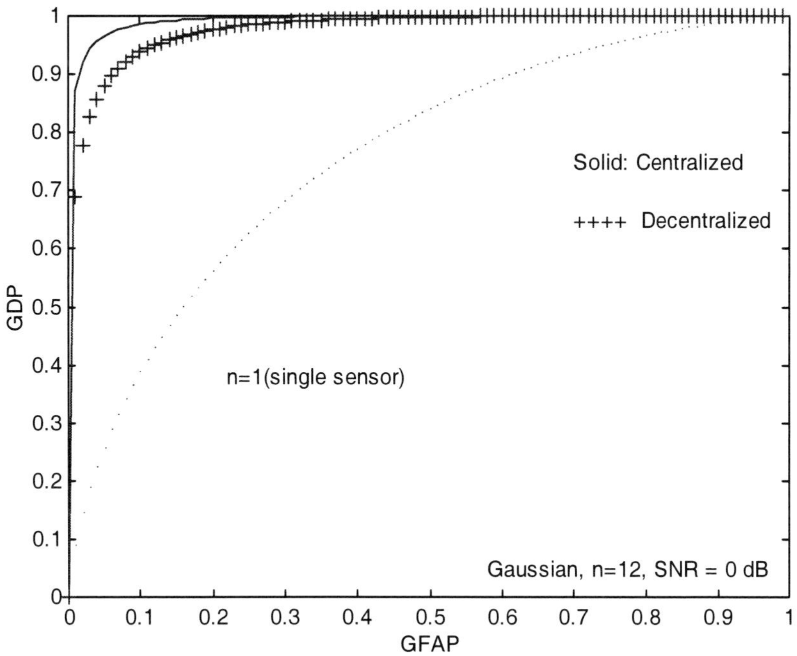

Figure 18. Comparison of centralized and decentralized systems.

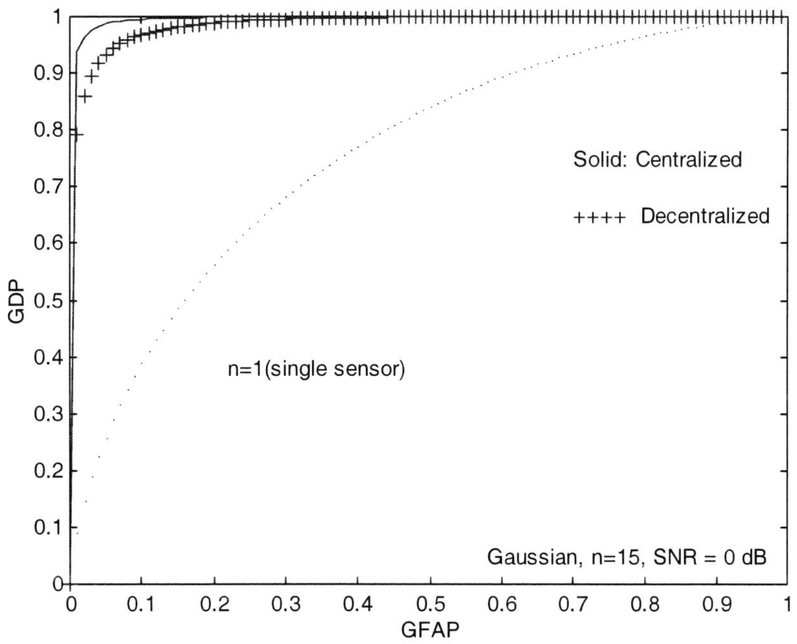

Figure 19. Comparison of centralized and decentralized systems.

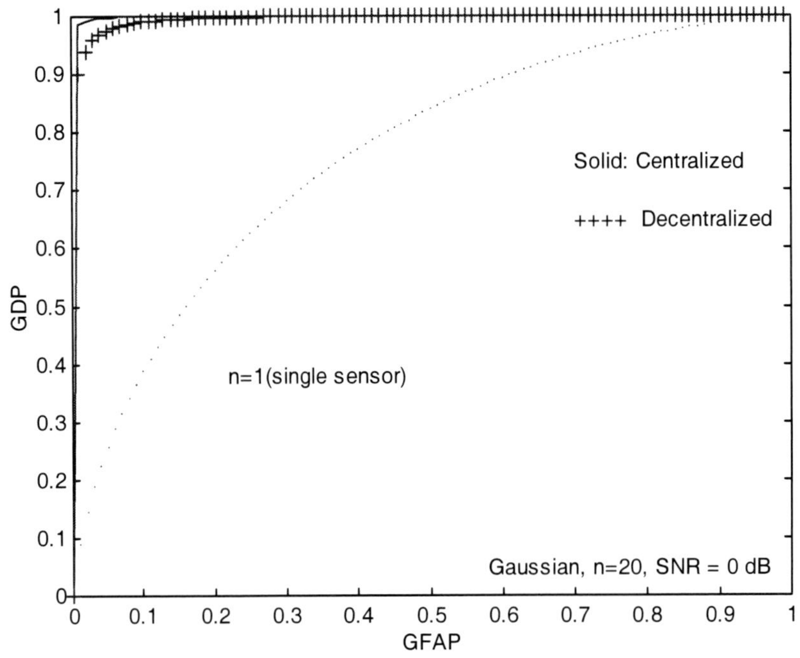

Figure 20. Comparison of centralized and decentralized systems.

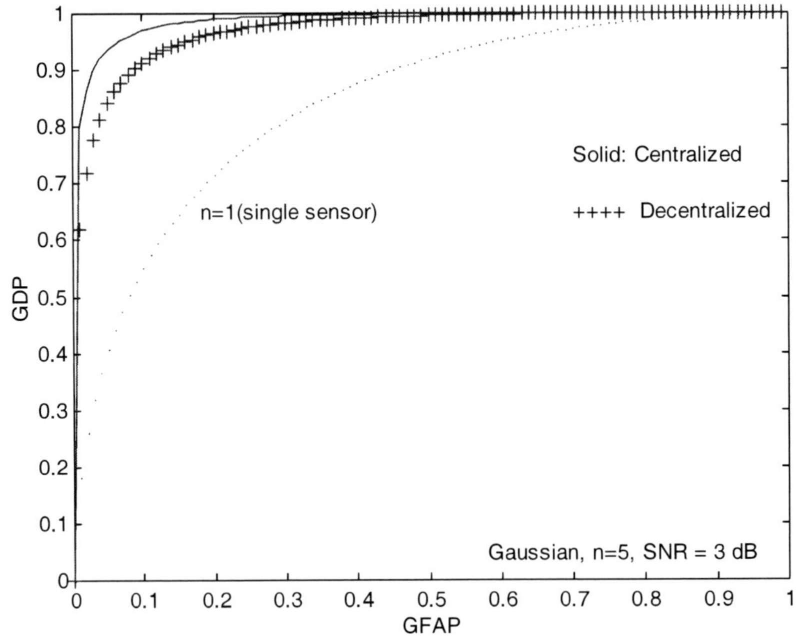

Figure 21. Comparison of centralized and decentralized systems.

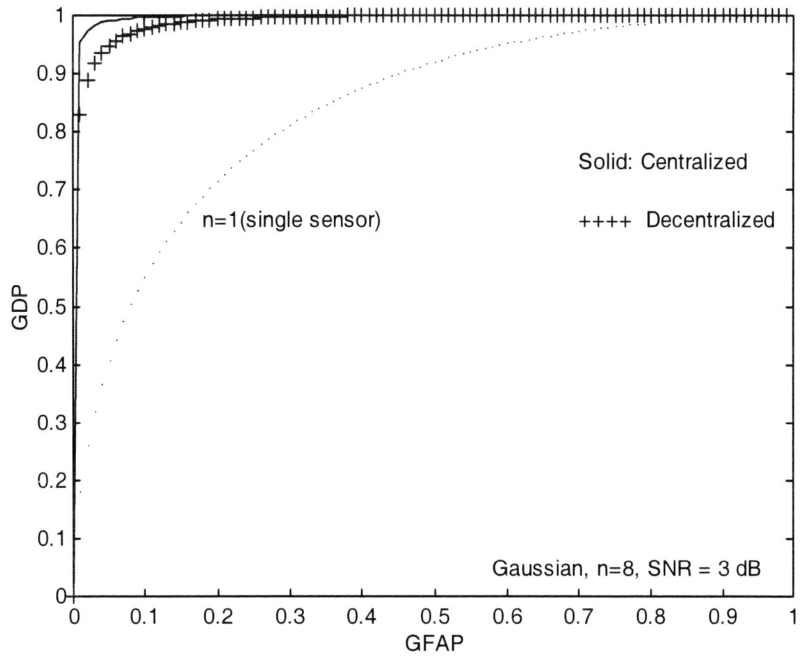

Figure 22. Comparison of centralized and decentralized systems.

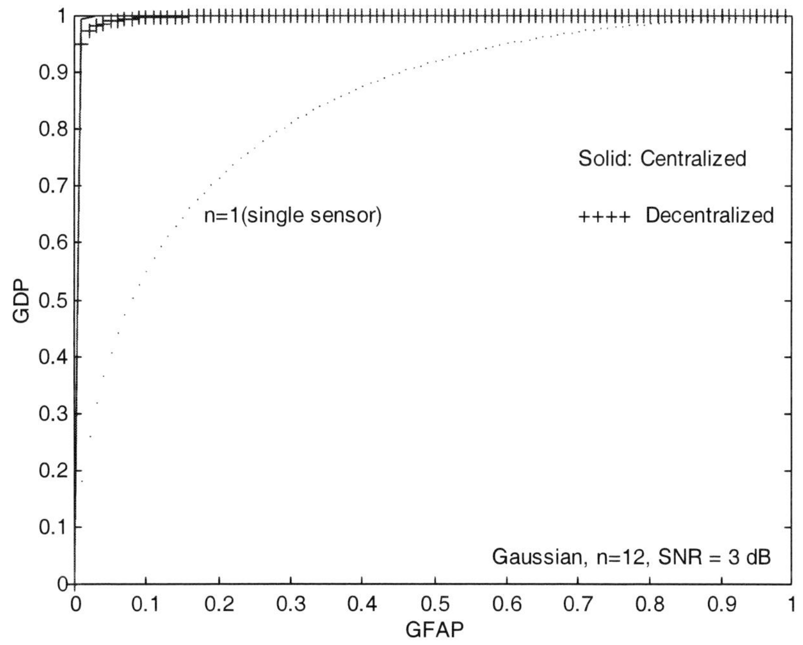

Figure 23. Comparison of centralized and decentralized systems.

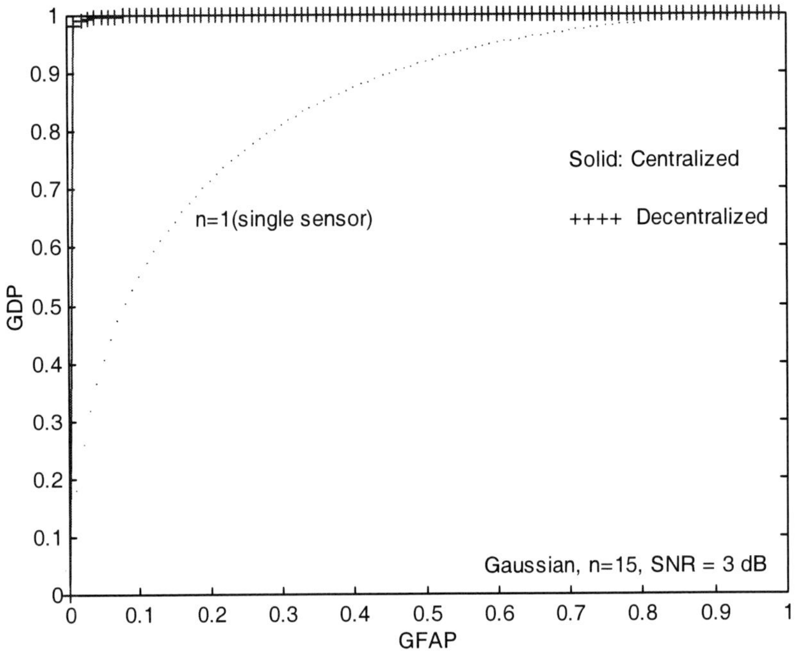

Figure 24. Comparison of centralized and decentralized systems.

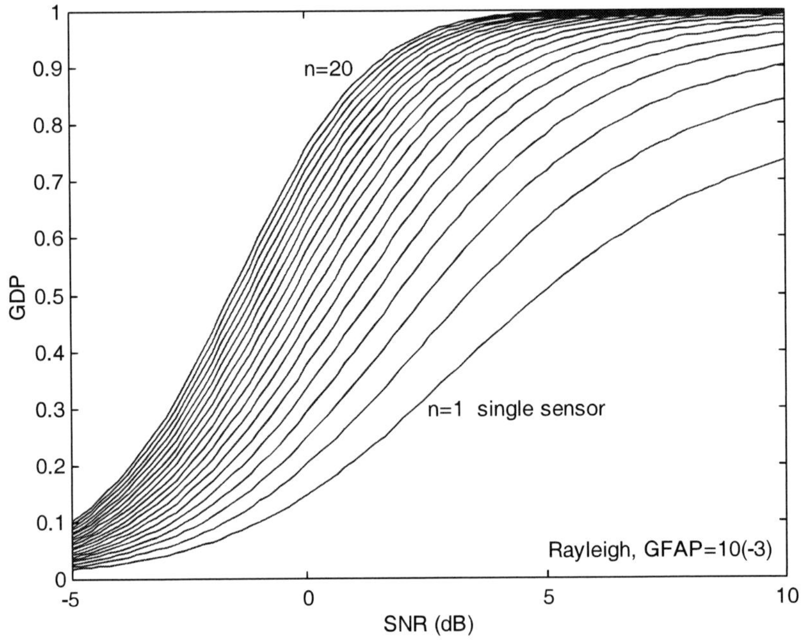

Figure 25. Performance of centralized detection system for different n.

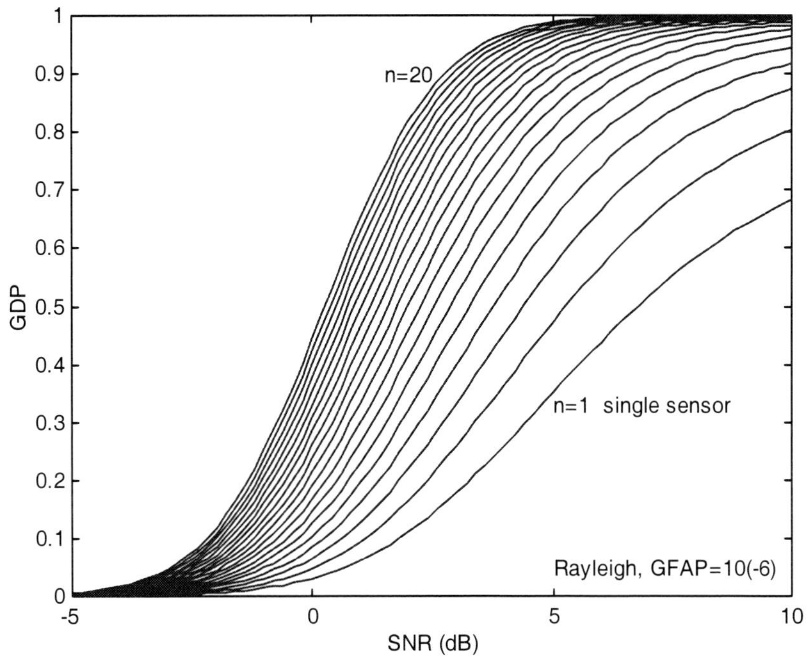

Figure 26. Performance of centralized detection system for different n.

Figures 25-40 show the results in case of Rayleigh distributed observations. The performances are evaluated in terms of the global detection probability (GDP) versus the SNR per sensor observation for a specified global false alarm probability $(GFAP)$. Figures 25-28 show GDP versus SNR in case of centralized detection systems with different n assuming $GFAP = 10^{-3}, 10^{-6}, 10^{-9}$ and 10^{-12}, respectively. Figures 29-32 show the same plots in case of decentralized detection systems. Figures 25-32 show the global performance improvement due to centralized and decentralized detection systems, as the number of sensors increases. Clearly, increasing the number of sensors leads to improvement in both cases, but the global performance improvement of the centralized approach over the decentralized one increases with the number of sensors. Figures 33-40 compare GDP versus SNR of the centralized and the decentralized detection systems as well as the performance of a single sensor system, for a specified $GFAP$. Figures 33-36 show these plots in case of $GFAP = 10^{-6}$ when $n = 10, 12, 15$ and 20 respectively. Figures 37-40 show these plots in case of $GFAP = 10^{-9}$ when $n = 10, 12, 15$ and 20 respectively. Figures 33-40 show the global performance improvement of the centralized and the decentralized detection systems over the single sensor systems. They also show the degradation in the global performance of the decentralized detection systems, compared to the centralized detection systems.

A large number of computer simulations was carried out to obtain the needed number of sensors in case of decentralized approaches to achieve a performance non inferior to centralized approaches, with n sensors, for the same SNR per sensor observations. Some of these results are shown in Figures 41-66. Figures 41-63 show some of the results in case of Gaussian distributed observations. Figures 41-44 compare the global performance of the centralized approach of four sensors $(n_{\text{cent}} = 4)$, with the decentralized approach of five, six, seven, and eight sensors $(n_{\text{decent}} = 5, 6, 7, \text{ and } 8)$, assuming -5, -3, 0 and 3 dB *SNR* per sensor observation, respectively. Letting the number of sensors be greater than or equal to 6 $(2n_{\text{cent}} - 2)$ reliably ensures a performance of the optimum decentralized schemes non inferior to that of the optimum centralized schemes, with 4 sensors, for the same *SNR* per sensor observations. This result is independent of the *SNR* per sensor observations and of the desired *GFAP*. Figures 45-48 compare the global performance of the centralized $(n_{\text{cent}} = 8)$ and the decentralized $(n_{\text{decent}} = 9, 10, \ldots\ldots, 16)$ approaches at very low *GFAP*s assuming -3, 0, 3 and 5 dB *SNR* per sensor observation, respectively. Letting the number of sensors be greater than or equal to 12 $(2n_{\text{cent}} - 4)$ reliably ensures a performance of the optimum decentralized schemes non inferior to that of the optimum centralized schemes, with 8 sensors, for the same *SNR* per sensor observations. Again, this result is independent of the *SNR* per sensor observations and of the desired *GFAP*.

Figures 49-51 compare the global performance of the centralized $(n_{\text{cent}} = 10)$ and the decentralized $(n_{\text{decent}} = 11, 12, \ldots\ldots, 16)$ approaches at very low *GFAP*s assuming 3, 6 and 9 dB *SNR* per sensor observation, respectively. Letting the number of sensors be greater than or equal to 16 $(2n_{\text{cent}} - 4)$ reliably ensures a performance of the optimum decentralized schemes non inferior to that of the optimum centralized schemes, with 10 sensors, for the same *SNR* per sensor observations.

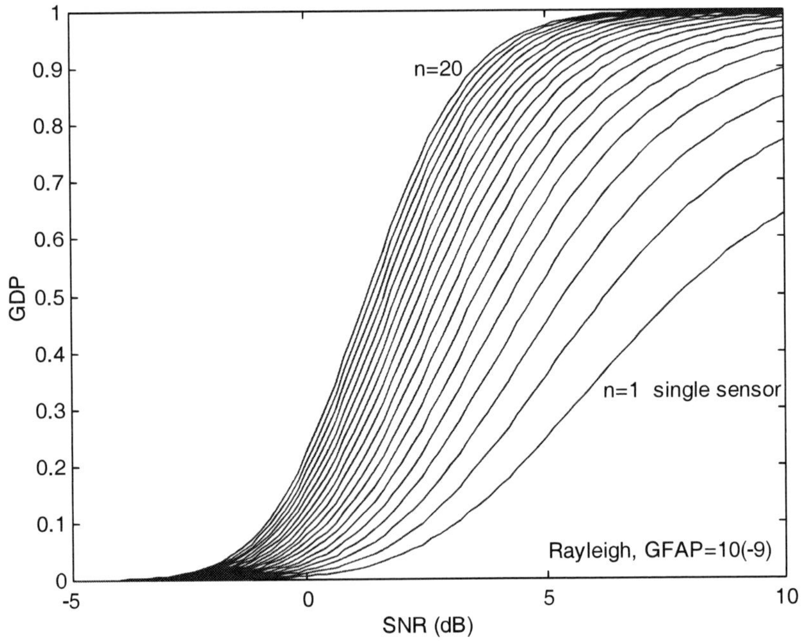

Figure 27. Performance of centralized detection system for different n.

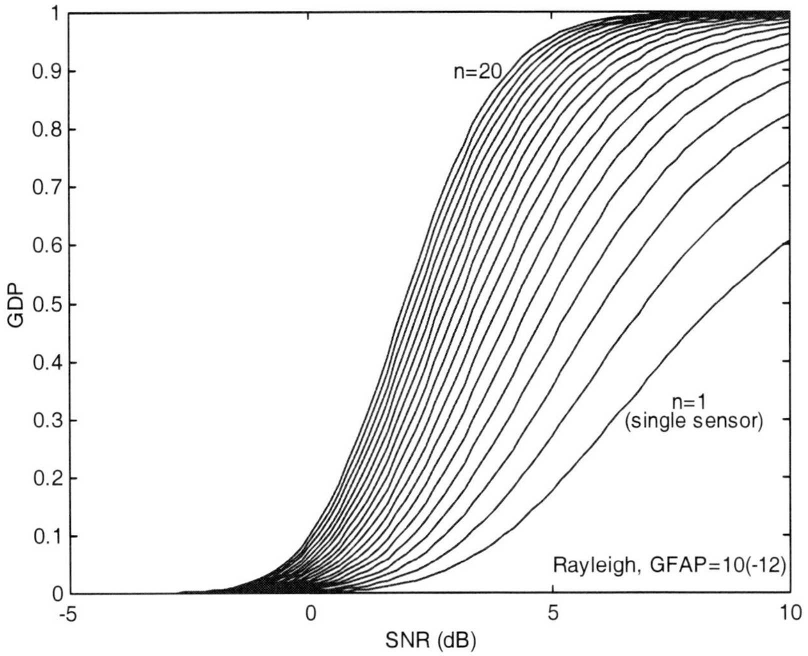

Figure 28. Performance of centralized detection system for different n.

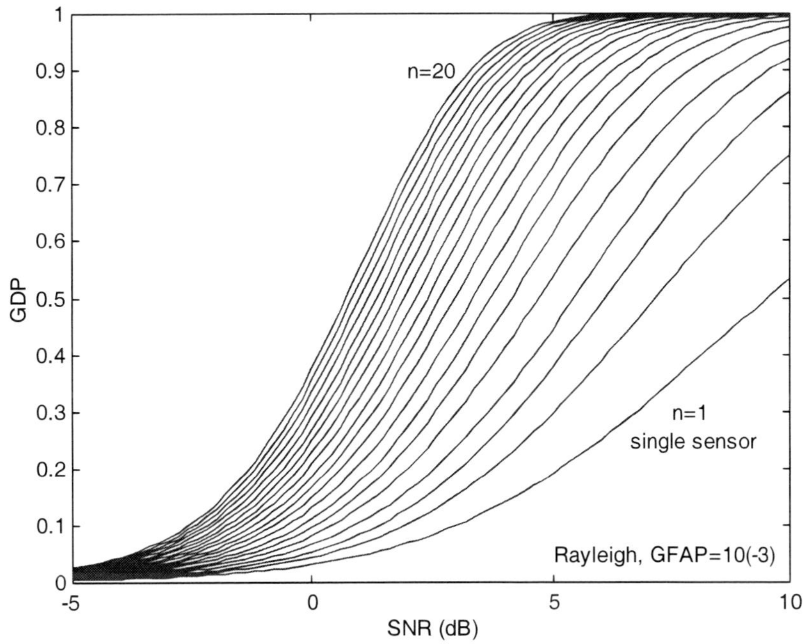

Figure 29. Performance of decentralized detection system for different n.

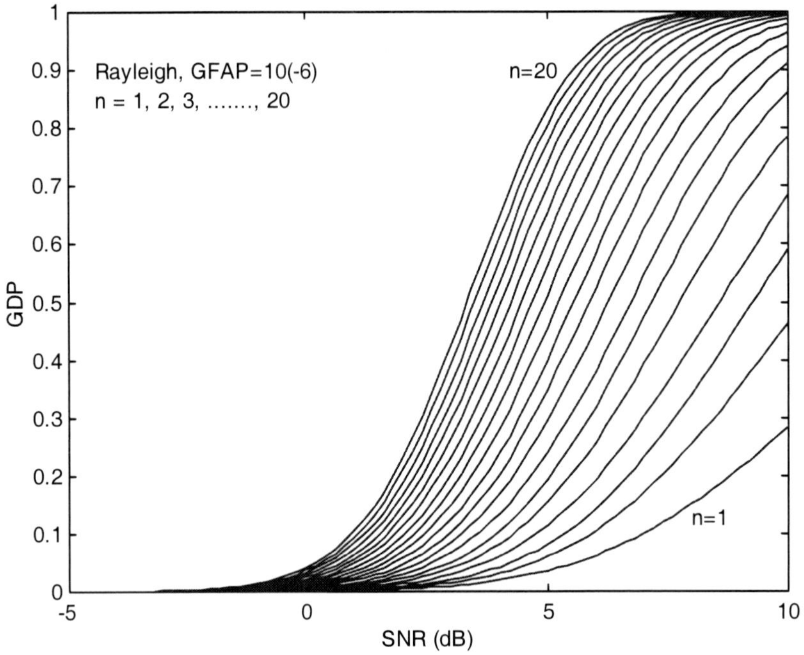

Figure 30. Performance of decentralized detection system for different n.

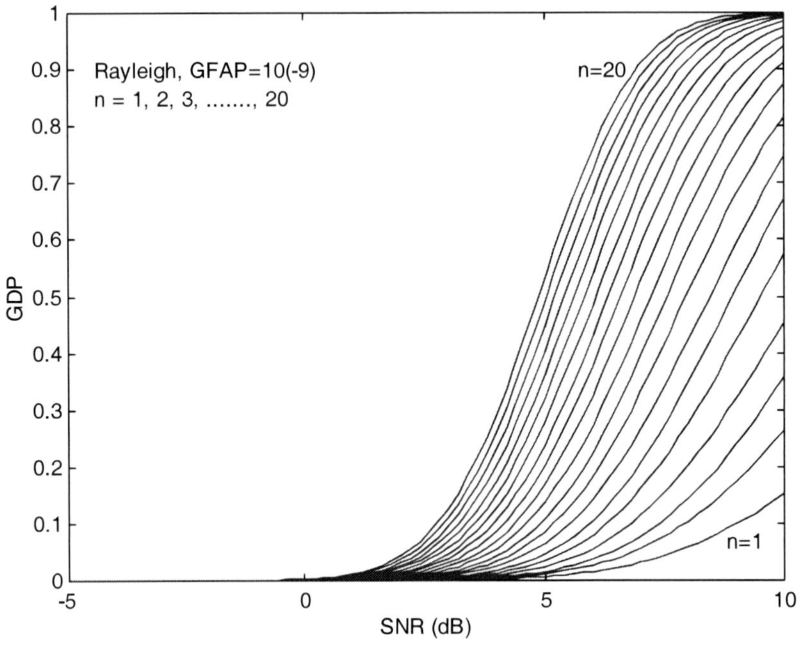

Figure 31. Performance of decentralized detection system for different n.

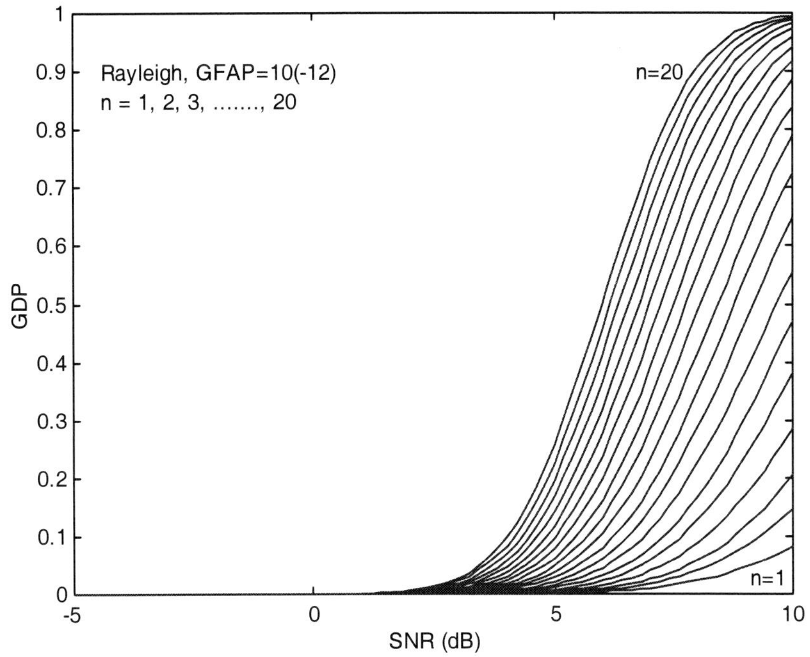

Figure 32. Performance of decentralized detection system for different n.

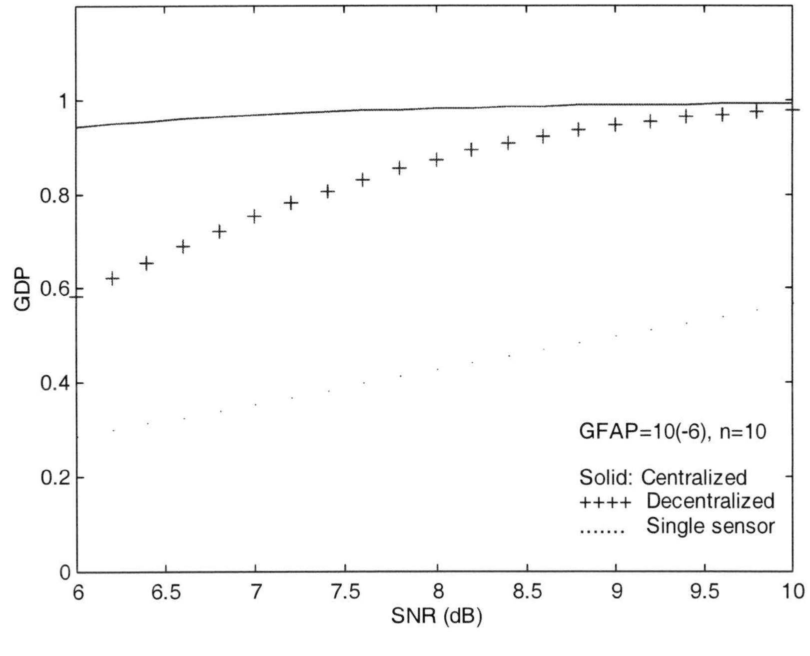

Figure 33. Global performance in Rayleigh distributed observations.

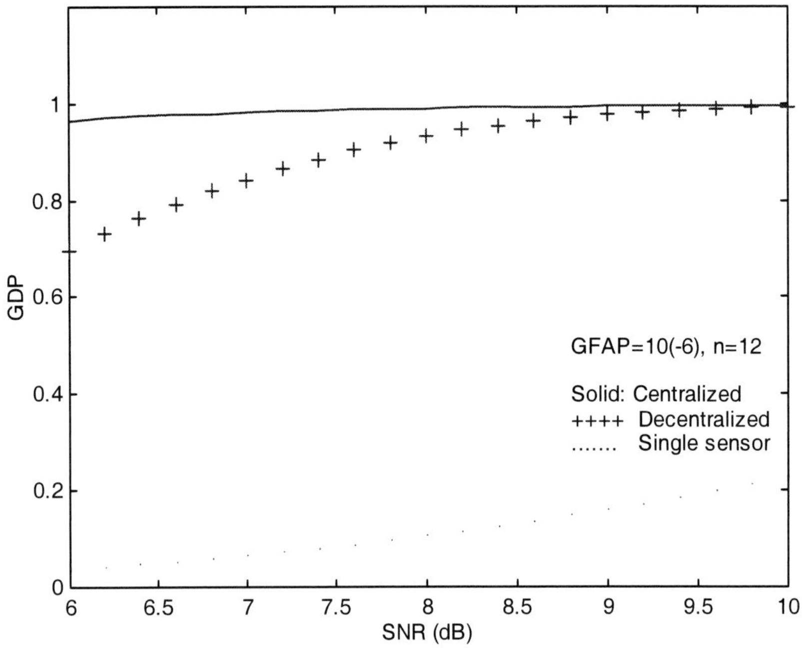

Figure 34. Global performance in Rayleigh distributed observations.

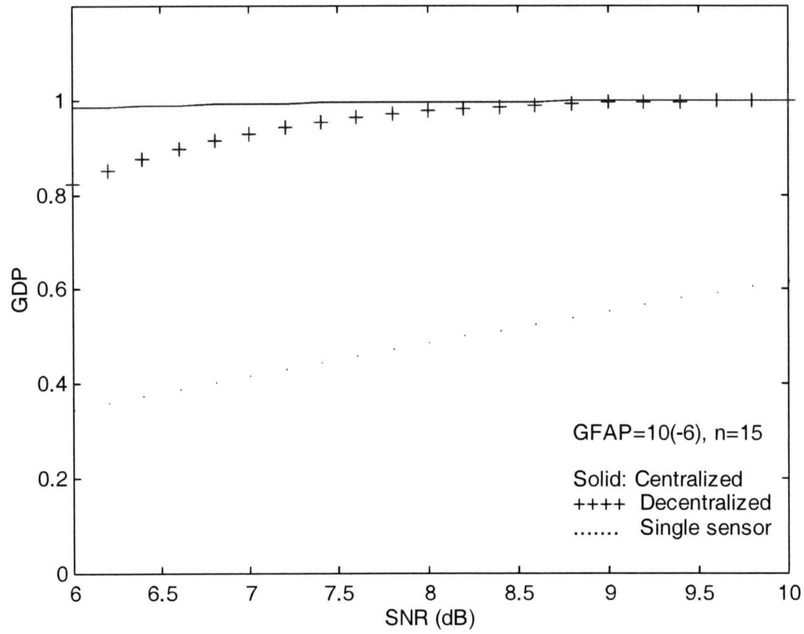

Figure 35. Global performance in Rayleigh distributed observations.

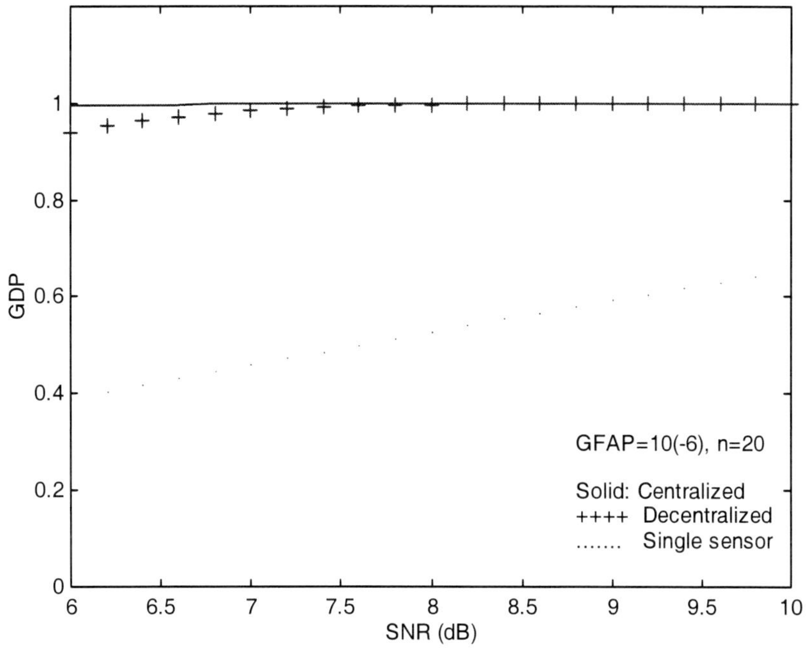

Figure 36. Global performance in Rayleigh distributed observations.

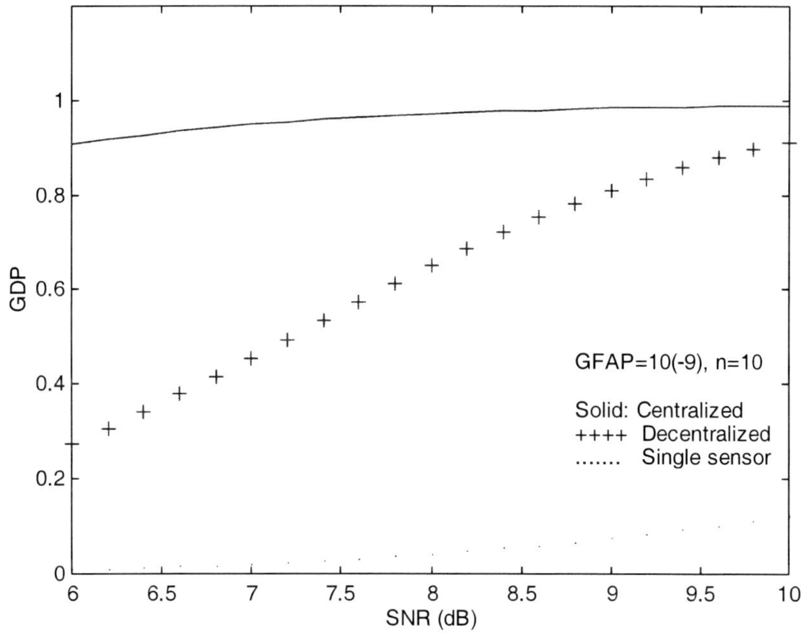

Figure 37. Global performance in Rayleigh distributed observations.

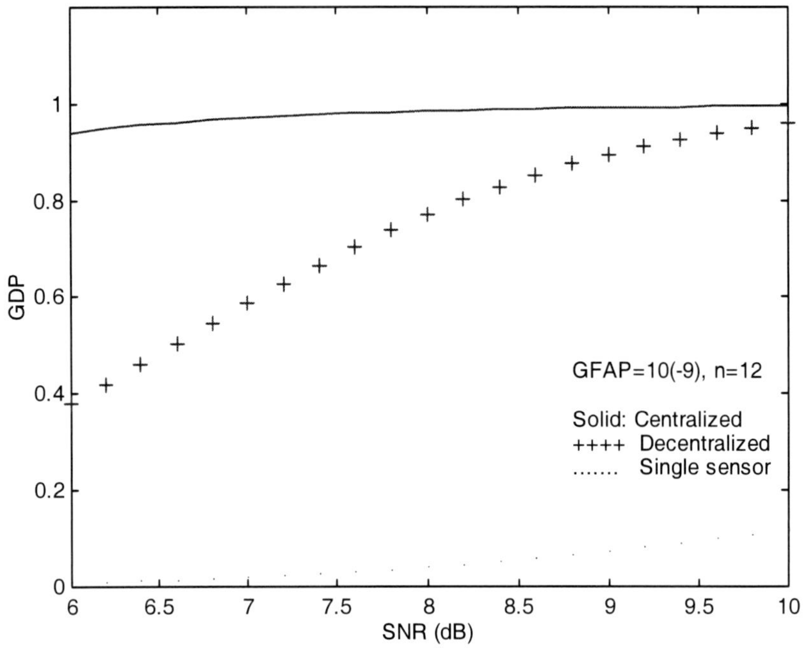

Figure 38. Global performance in Rayleigh distributed observations.

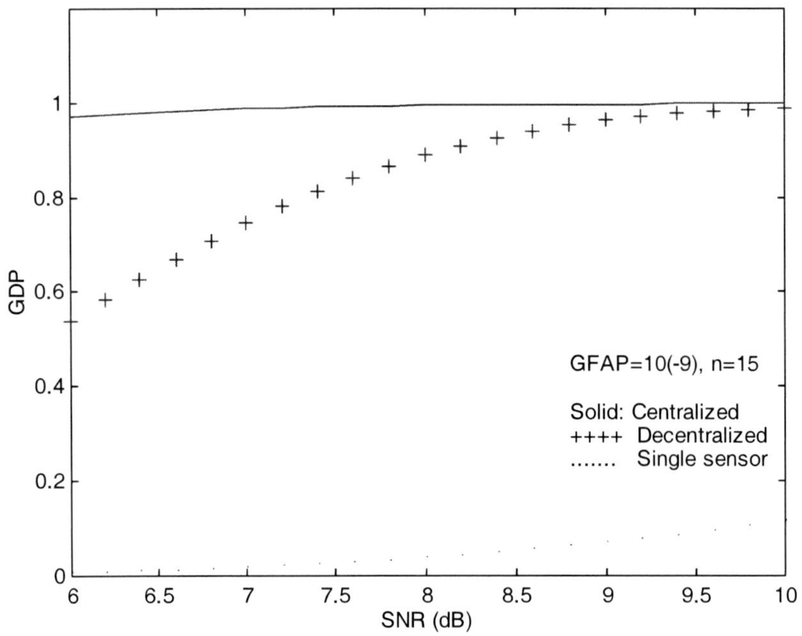

Figure 39. Global performance in Rayleigh distributed observations.

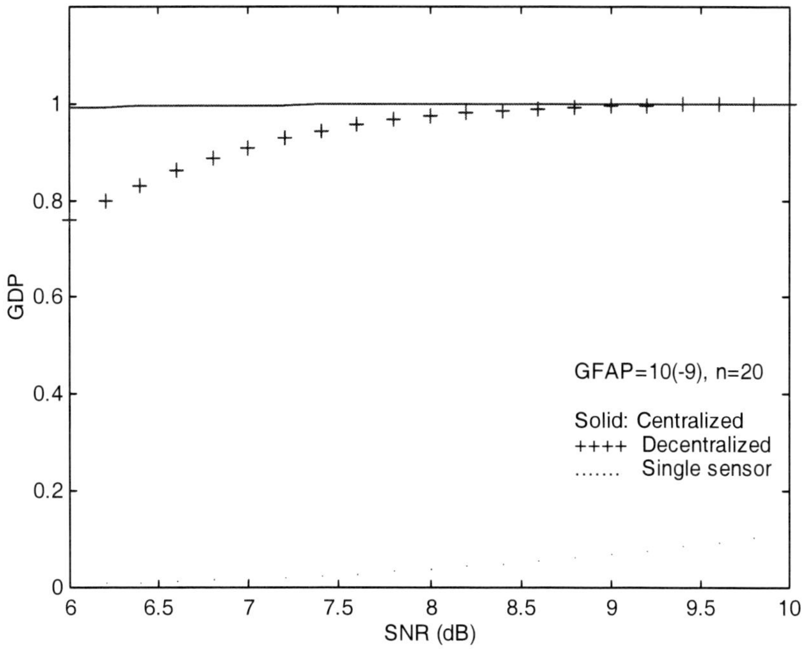

Figure 40. Global performance in Rayleigh distributed observations.

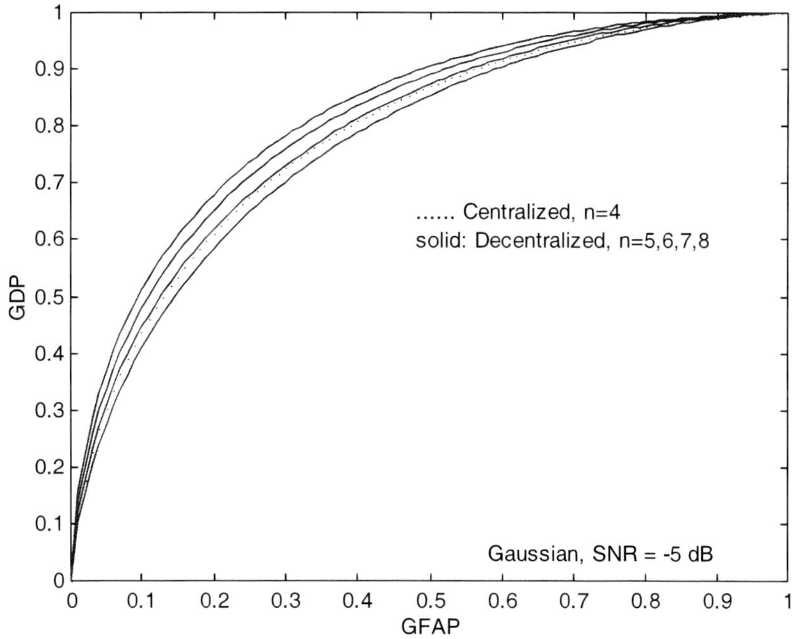

Figure 41. Comparison of centralized and decentralized systems.

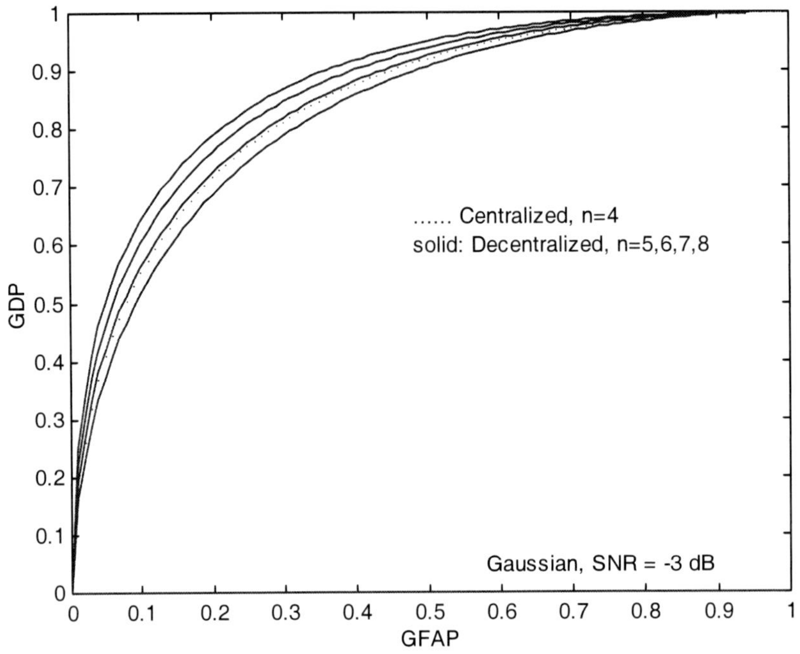

Figure 42. Comparison of centralized and decentralized systems.

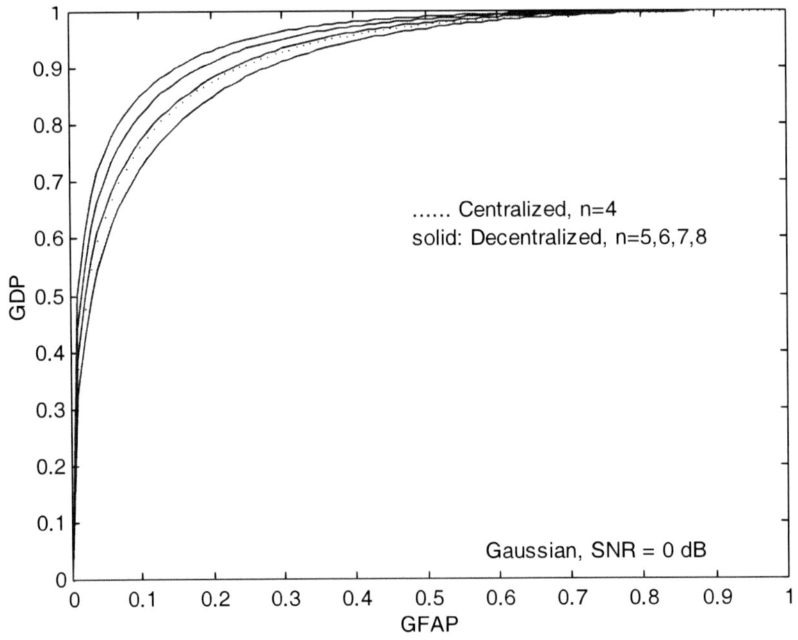

Figure 43. Comparison of centralized and decentralized systems.

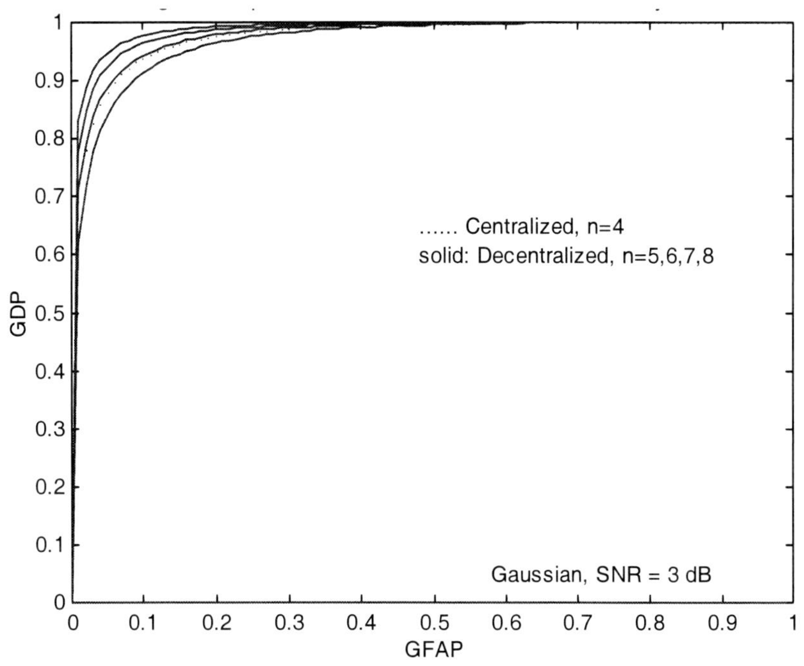

Figure 44. Comparison of centralized and decentralized systems.

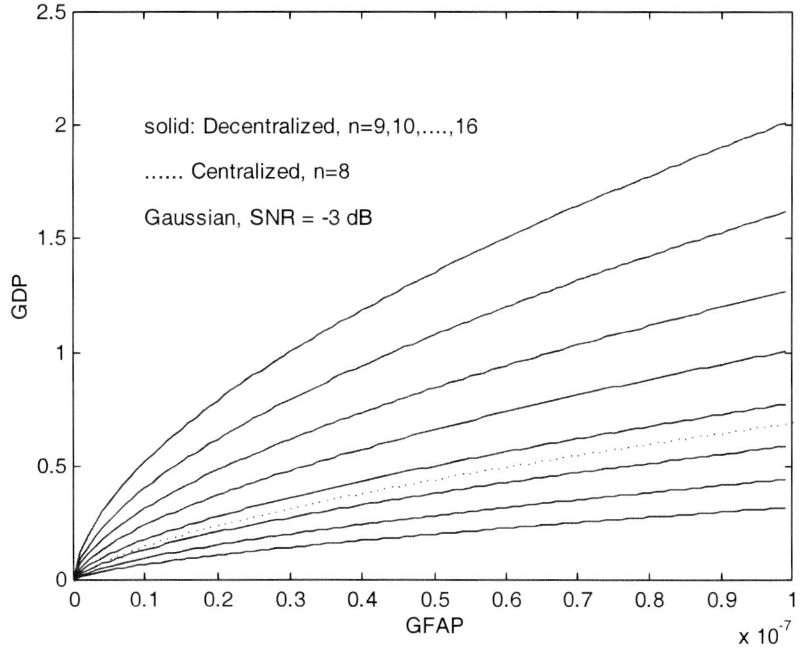

Figure 45. Comparison of centralized and decentralized systems.

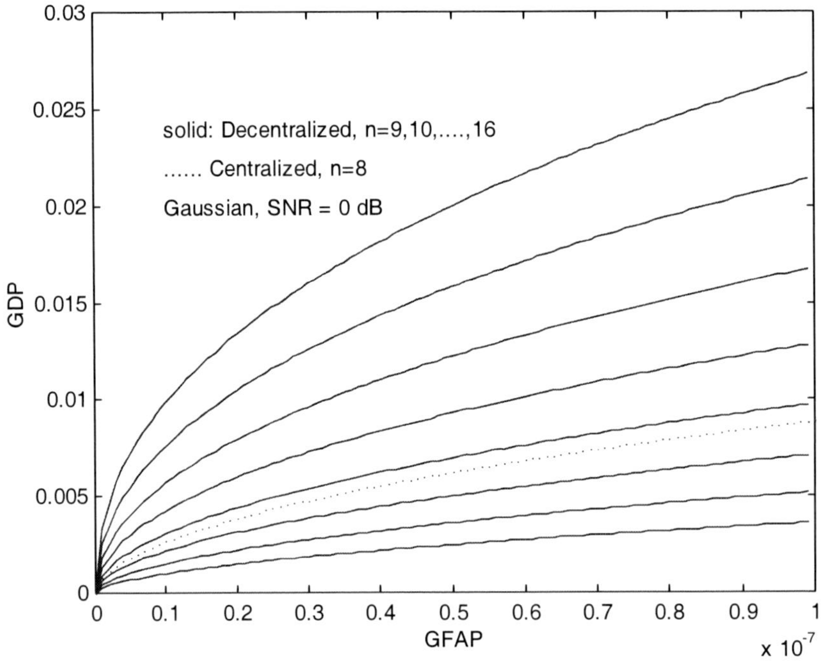

Figure 46. Comparison of centralized and decentralized systems.

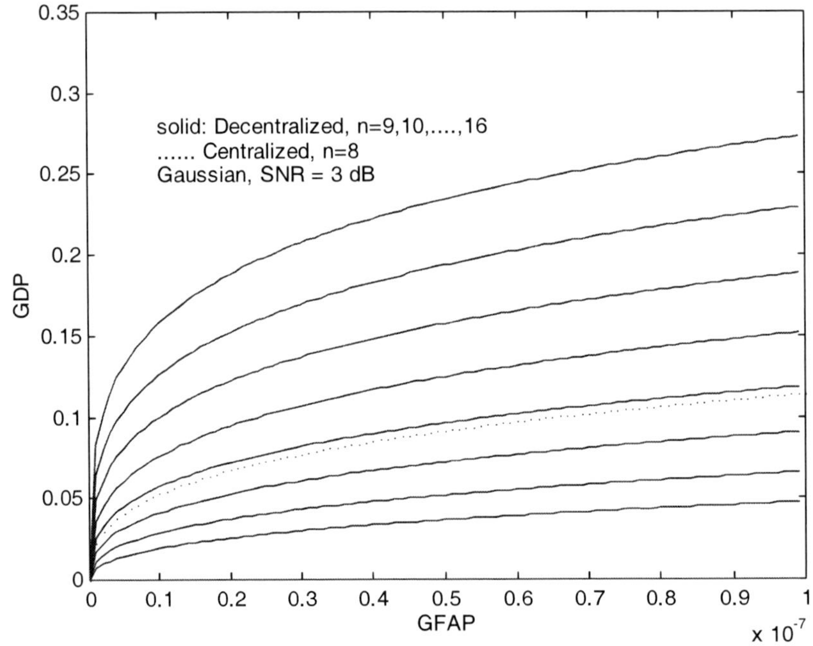

Figure 47. Comparison of centralized and decentralized systems.

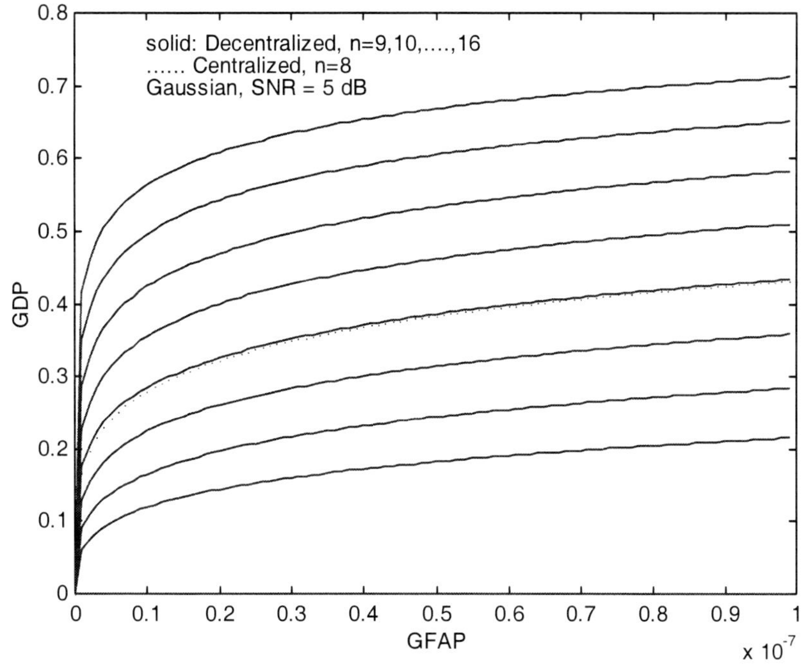

Figure 48. Comparison of centralized and decentralized systems.

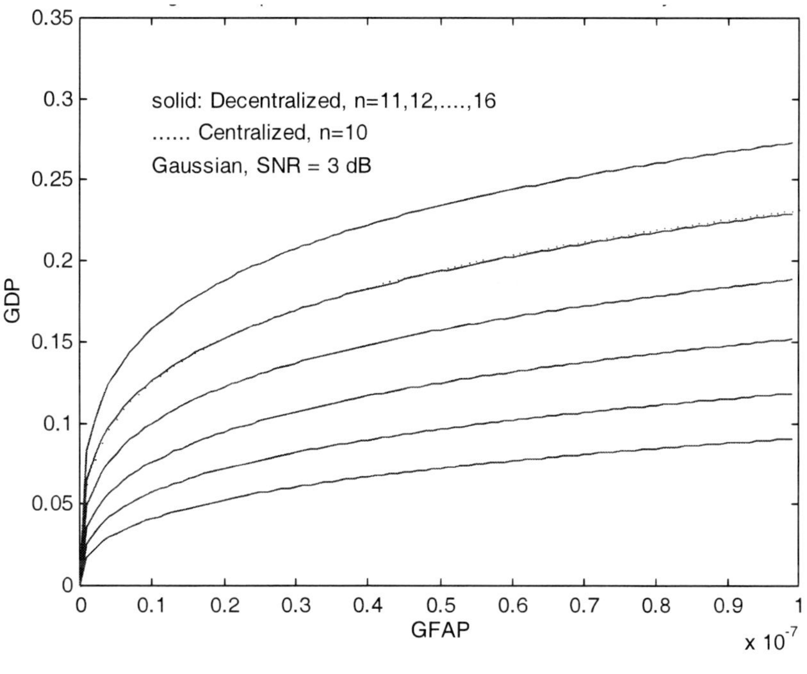

Figure 49. Comparison of centralized and decentralized systems.

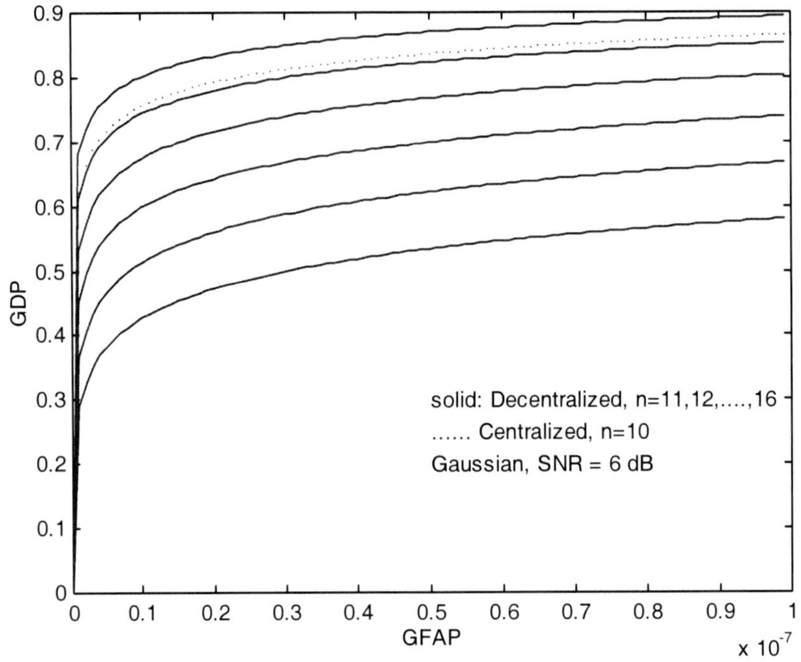

Figure 50. Comparison of centralized and decentralized systems.

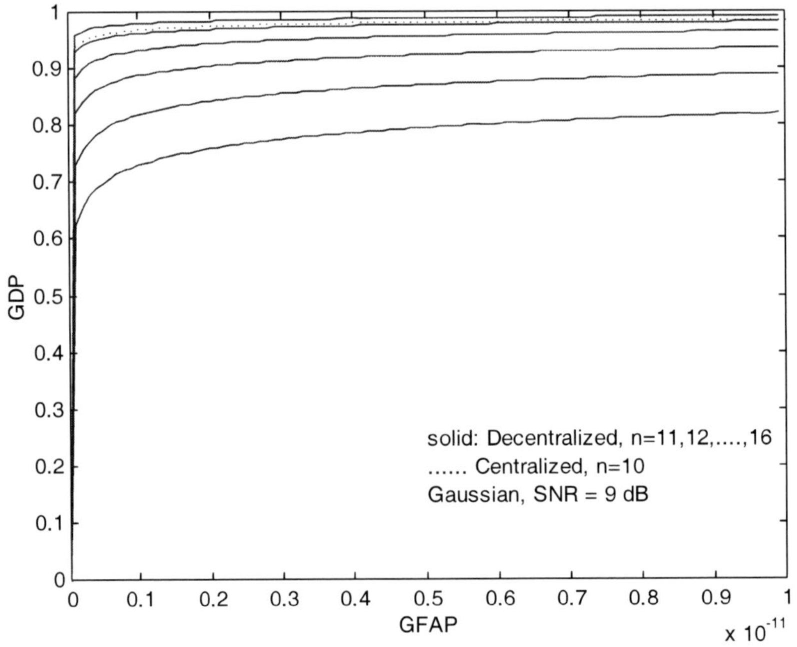

Figure 51. Comparison of centralized and decentralized systems.

Figures 52-54 show the same plot when $n_{\text{cent}} = 15$ and $n_{\text{decent}} = 16,17,\ldots\ldots, 25$ assuming 3, 6 and 9 dB *SNR* per sensor observation, respectively. In this case, letting the number of sensors be greater than or equal to 24 $(2n_{\text{cent}} - 6)$ reliably ensures a performance of the optimum decentralized schemes non inferior to that of the optimum centralized schemes, with 15 sensors, for the same SNR per sensor observations. Figures 55-60 show similar plots with different values of n_{cent}, n_{decent}, SNR and $GFAP$s. In this case, letting the number of sensors be greater than or equal to $(2n_{\text{cent}} - 1)$ reliably ensure a performance of the optimum decentralized schemes non inferior to that of the optimum centralized schemes. Figures 61-63 compare the performance of the centralized and the decentralized detection systems for different values of the same number of sensors. It is clear that the performance loss increases as the number of sensors increases. Figures 64-66 show some of the results in case of Rayleigh distributed observations. Figure 64 compare GDP versus SNR of the centralized approach with $n_{\text{cent}} = 10$ and the decentralized approach with $n_{\text{decent}} = 11,12,......., 21$ when $GFAP = 10^{-6}$. Figure 65 compare GDP versus SNR of the centralized approach with $n_{\text{cent}} = 11$ and the decentralized approach with $n_{\text{decent}} = 12,13,......., 23$ when $GFAP = 10^{-5}$. From Figures 64 and 65, letting the number of sensors be greater than or equal to $(2n_{\text{cent}})$ reliably ensure a performance of the optimum decentralized schemes non inferior to that of the optimum centralized schemes.

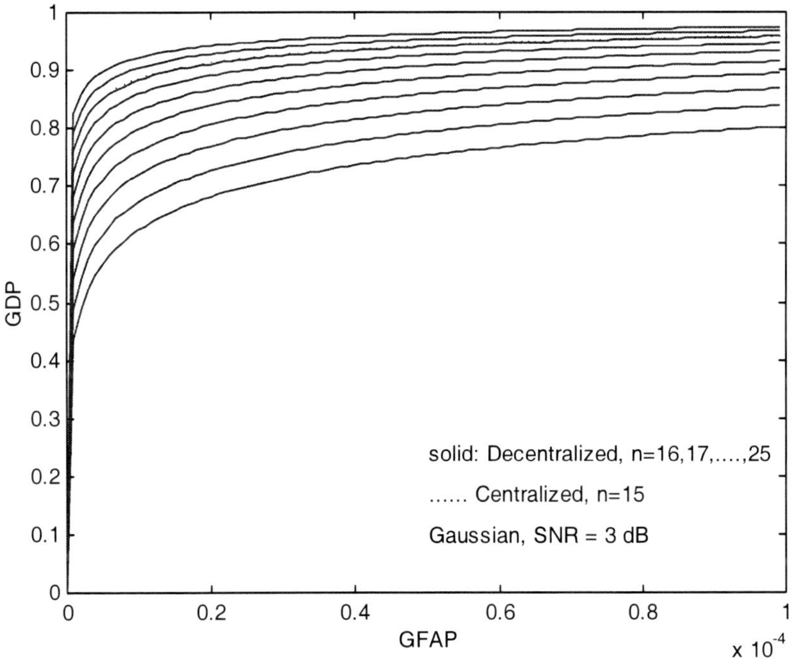

Figure 52. Comparison of centralized and decentralized systems.

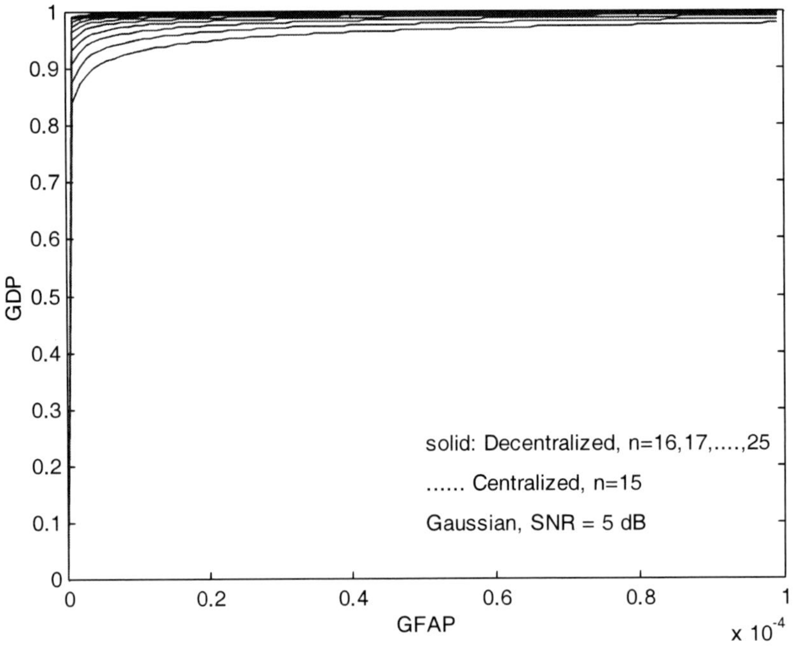

Figure 53. Comparison of centralized and decentralized systems.

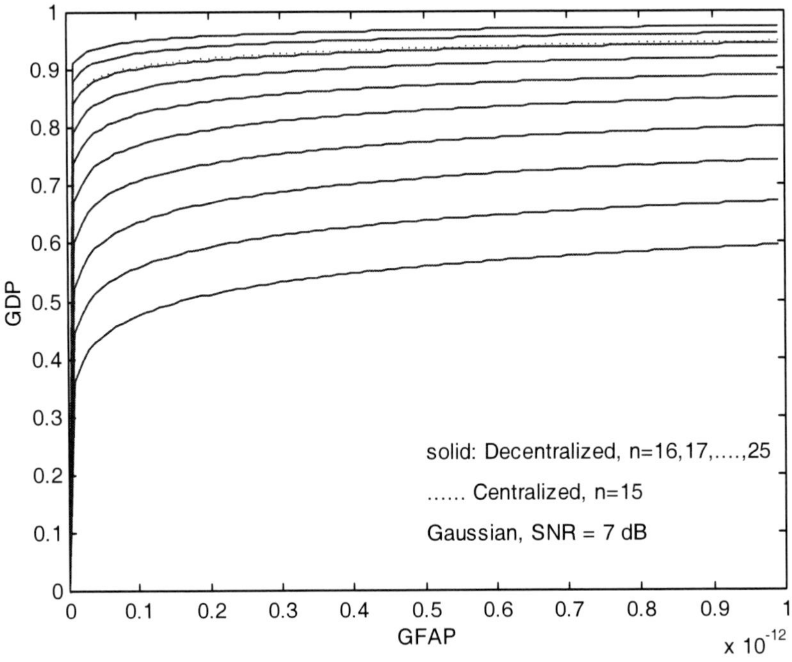

Figure 54. Comparison of centralized and decentralized systems.

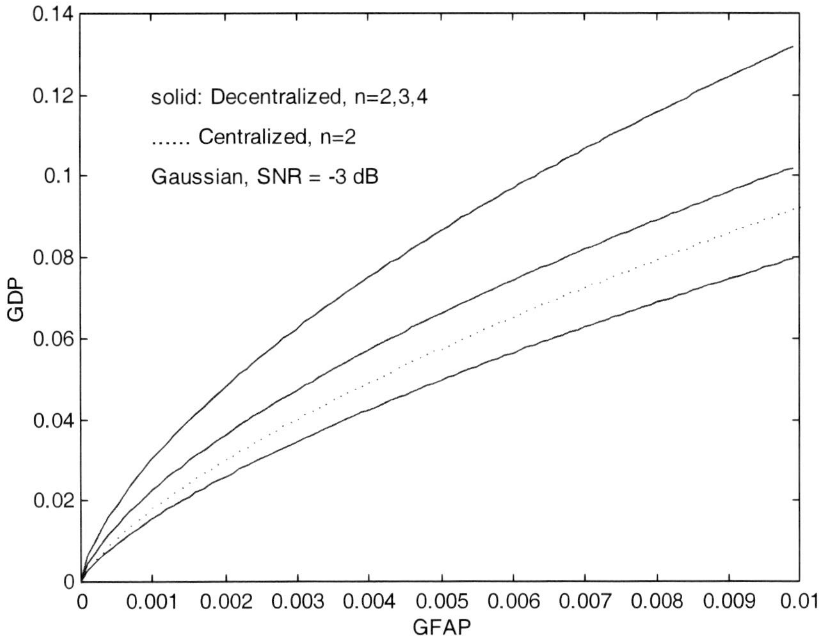

Figure 55. Comparison of centralized and decentralized systems.

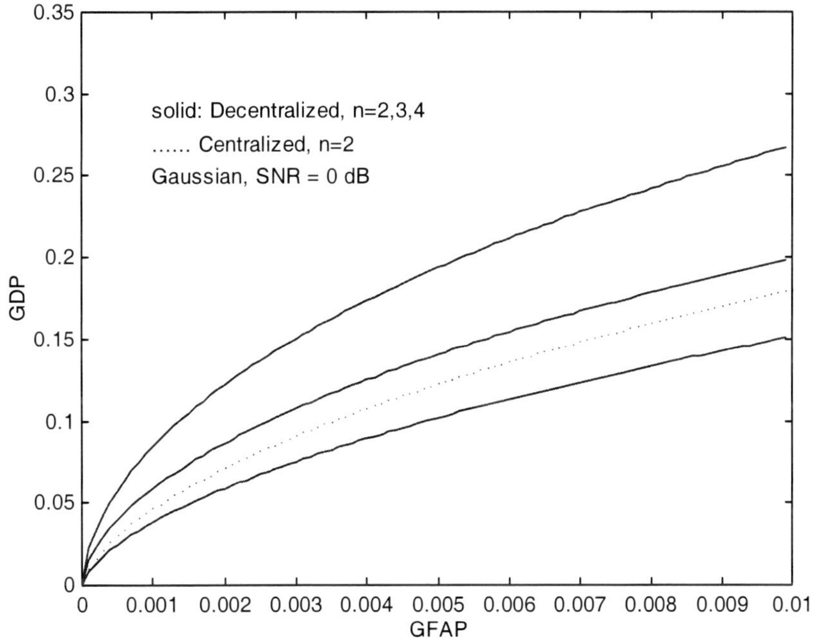

Figure 56. Comparison of centralized and decentralized systems.

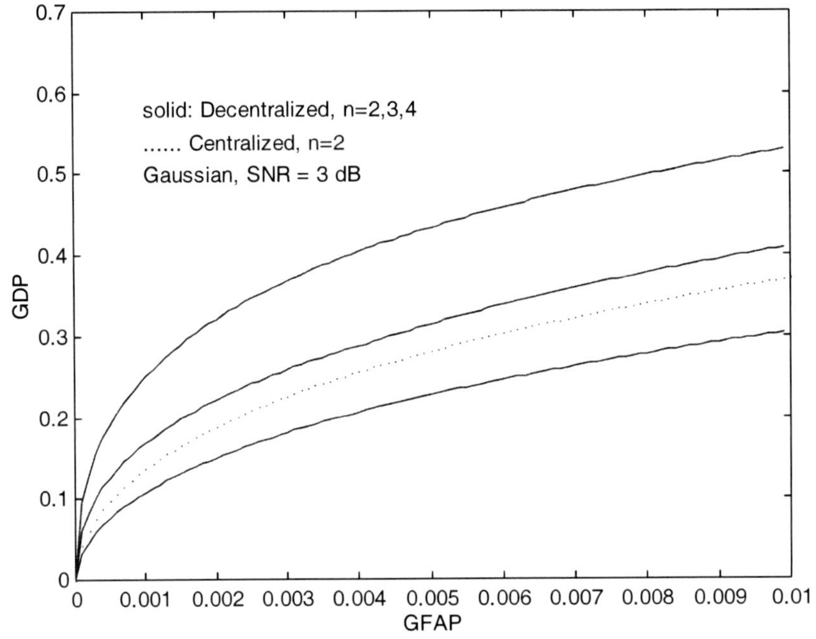

Figure 57. Comparison of centralized and decentralized systems.

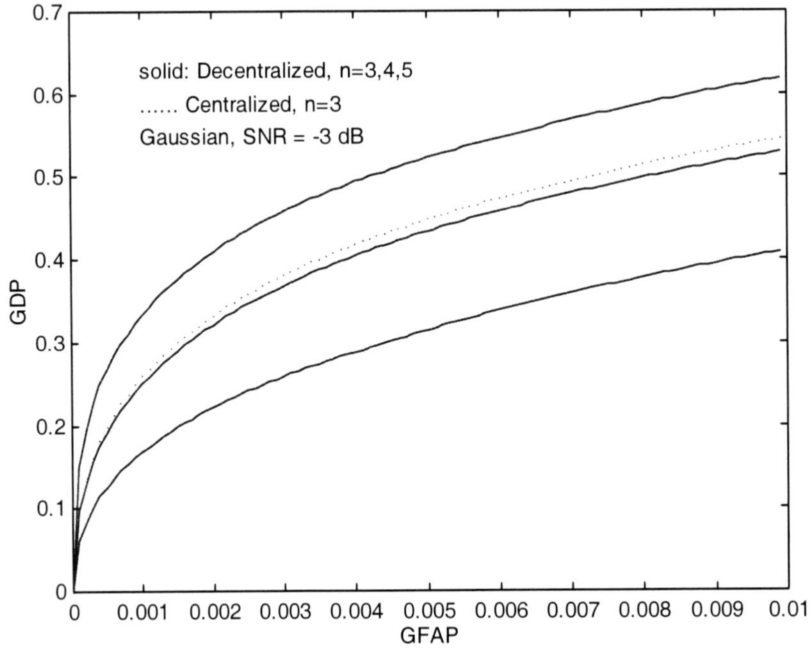

Figure 58. Comparison of centralized and decentralized systems.

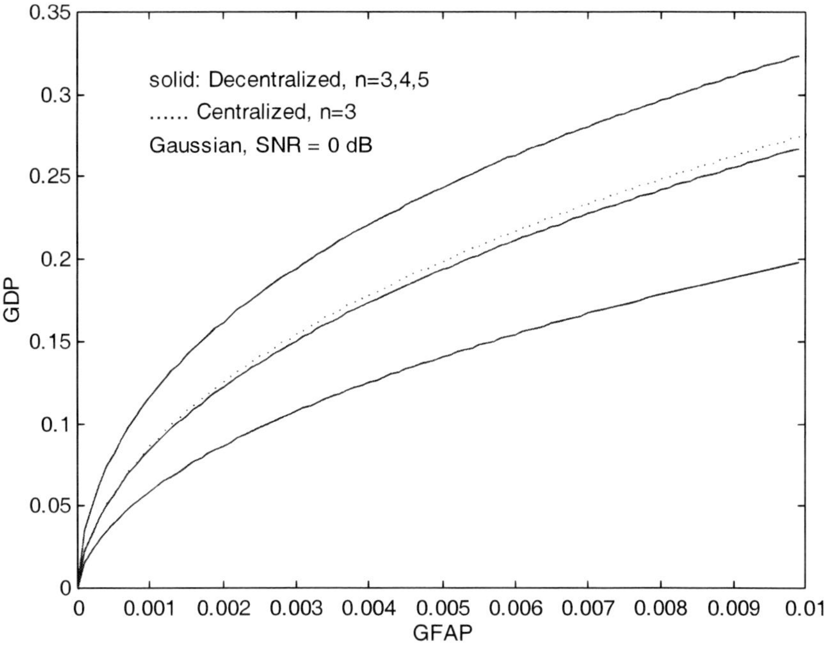

Figure 59. Comparison of centralized and decentralized systems.

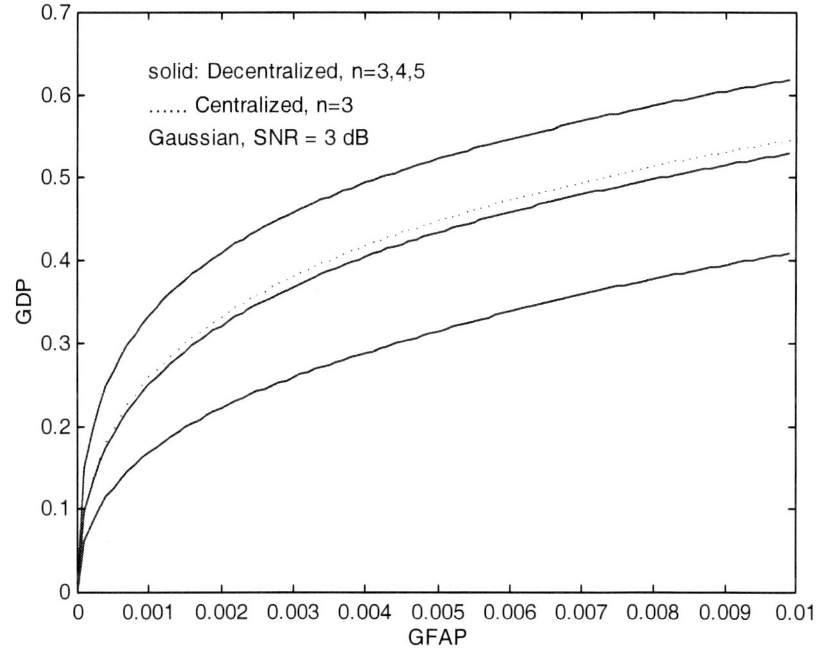

Figure 60. Comparison of centralized and decentralized systems.

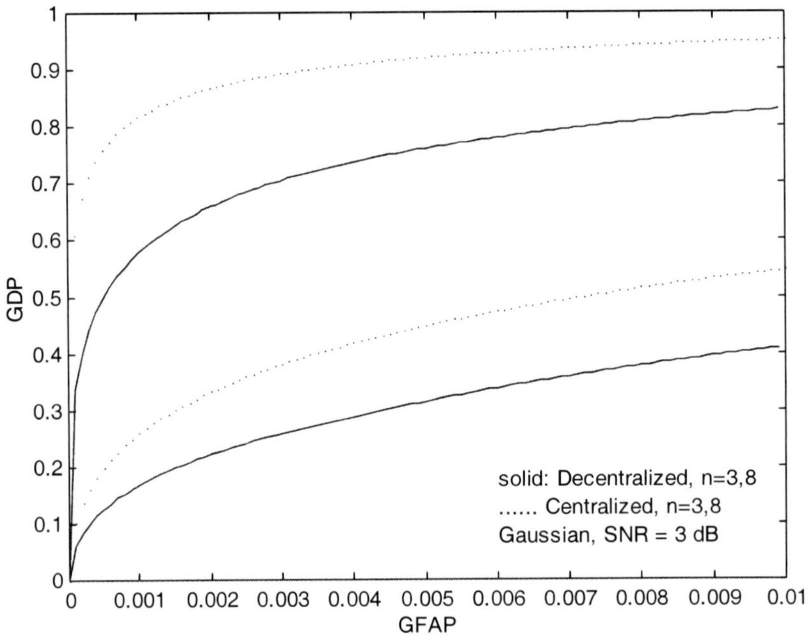

Figure 61. Comparison of centralized and decentralized systems.

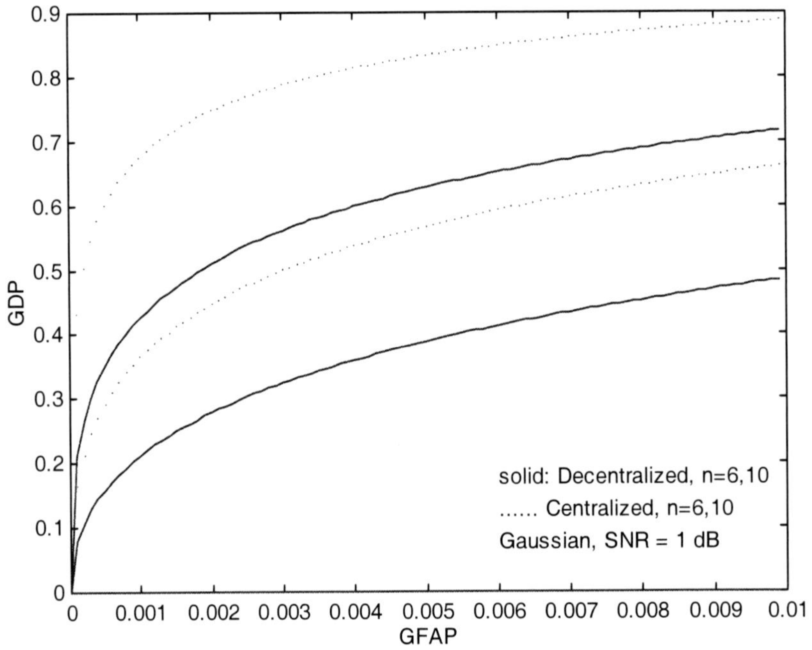

Figure 62. Comparison of centralized and decentralized systems.

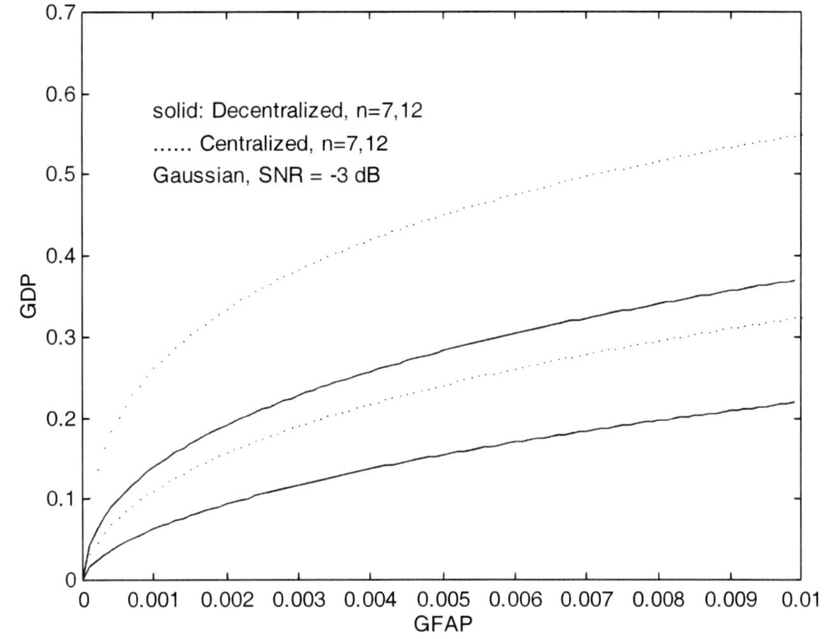

Figure 63. Comparison of centralized and decentralized systems.

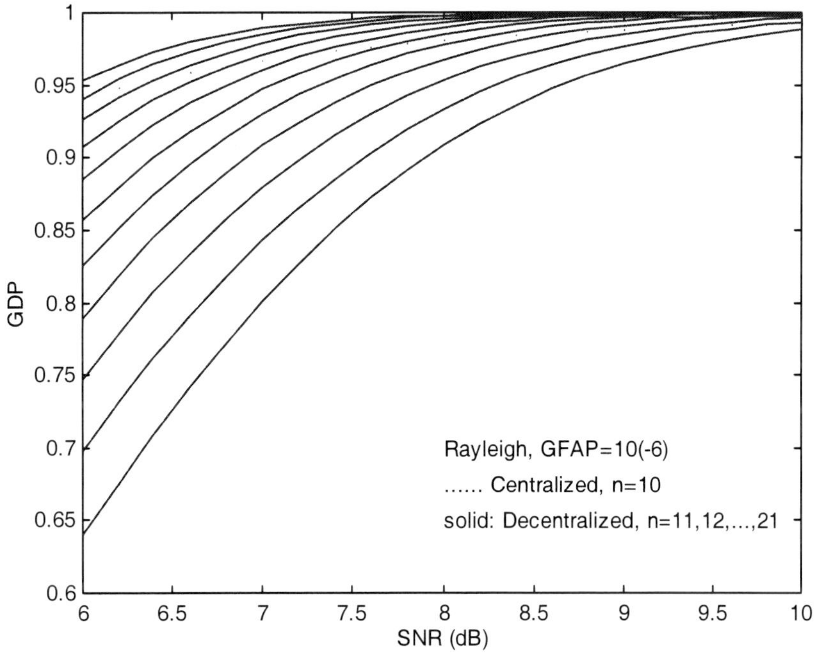

Figure 64. Comparison of centralized and decentralized systems.

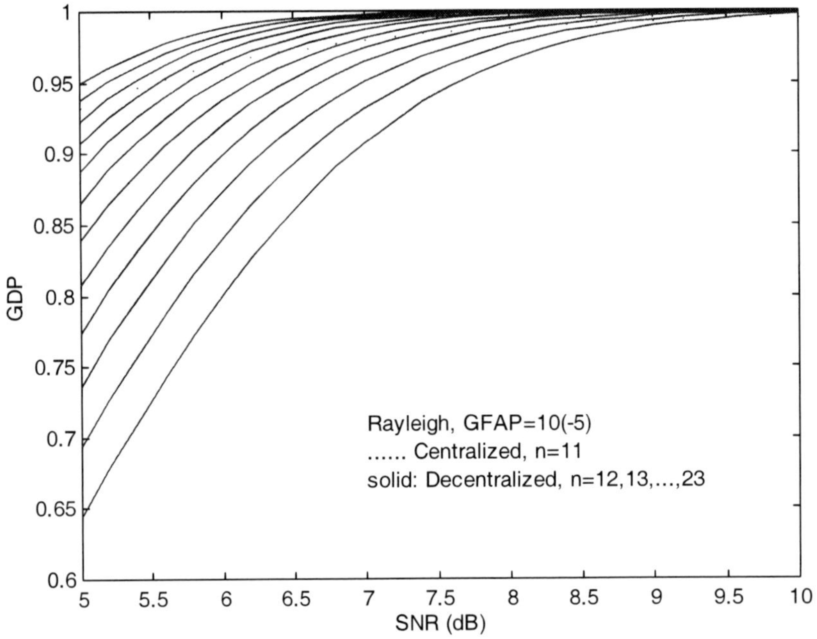

Figure 65. Comparison of centralized and decentralized systems.

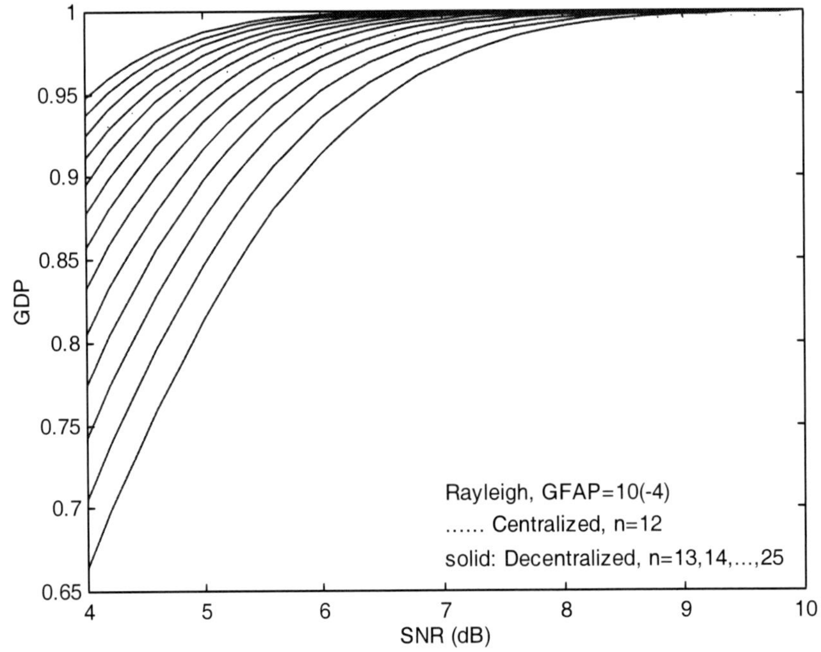

Figure 66. Comparison of centralized and decentralized systems.

Figure 66 compare GDP versus SNR of the centralized approach with $n_{\text{cent}} = 12$ and the decentralized approach with $n_{\text{decent}} = 13, 14, \ldots\ldots, 25$ when $GFAP = 10^{-4}$. In this case, letting the number of sensors be greater than or equal to $(2n_{cent} - 1)$ reliably ensure a performance of the optimum decentralized schemes non inferior to that of the optimum centralized schemes.

Based on the previous results (and additional large number of simulations), letting the number of sensors be greater than or equal to $2n$ reliably ensures a performance of the optimum decentralized schemes non inferior to that of the optimum centralized schemes, with n sensors, for any value of n, SNR per sensor observations and $GFAP$. It means that this result is independent of the distributed observation, the specified global false alarm probability, the signal to noise ratio and the number of sensors of the decentralized approach. Note that the performance of decentralized schemes with $2n$ sensors is the cost afforded in order to compensate for the use of very low bandwidth data links. It is worth noting that data transmission over small communication bandwidth provides desirable system engineering features such as low cost, immunity to noise and longer communication range.

IV. A Simple Solution to Distributed Detection Systems with Data Fusion

IV.1. Problem Formulation

Given n-sensor distributed detection system with data fusion, where the sensors local decisions $\{u_i\text{'s}\}$ are combined in a fusion center to derive a global decision u_0, it is required to optimize the structures of both the sensors and the fusion center. This is a distributed hypothesis testing problem with two hypotheses H_0 and H_1 [13, 35-37] . The Neyman Pearson (NP) optimization strategy maximizes the GDP for a desired GFAP. Srinivasan [13] derived a set of strongly coupled conditions defining solutions to a Lagrange multiplier optimization problem which is sometimes equivalent to the distributed NP hypothesis testing problem. He could not obtain explicit solutions to the derived set of conditions. He obtained, instead, generic expressions by adopting various Boolean algebraic combinations as fusion strategies. The combinations chosen in [13] are intuitive and require theoretical justification to be considered as candidates for optimality. Srinivasan's conditions have not been widely used because they are difficult to solve. Moreover, several authors [17, 36, 38] mentioned that there are cases which the solution of the NP distributed hypothesis testing problem, with maximum GDP, does not belong to the set of solutions provided by Srinivasan's conditions [13]. Certain assumptions need to be valid to avoid these cases. Fortunately, these assumptions are valid in some radar and communication applications of interest. Helstrom [31] presented a gradient algorithm for determining the quantization levels in a DDS satisfying the NP criterion. For any preassigned GFAP and starting by a variety of initial trial values, Helstrom's algorithm converges to solutions yielding local maxima of the GDP.

This section presents an algorithm for solution of Srinivasan's conditions. We suppose that the assumptions outlined in [17, 36], which are necessary for Srinivasan's conditions to yield an optimum solution to the NP distributed hypothesis testing problem, are valid. Further, the algorithm can easily be modified to implement the necessary conditions in [17] which are optimum in the more general case, since these conditions are not much different from Srinivasan's conditions. The convergence behavior of the algorithm is studied through elaboration of a number of examples. The results of global optimization of several DDSs are presented. The proposed algorithm proved to be simple, accurate and fast. However, an analytic proof of convergence has not yet been found.

IV.2. Review of Optimization Conditions

The prior knowledge needed for applying the NP strategy comprises the desired GFAP, and the probability density functions of the sensor observations conditioned to hypotheses H_0 and H_1; i.e. $P(y_k \mid H_0)$ and $P(y_k \mid H_1)$, $k = 1,2,....,n$. In [13], the sensor observations are assumed to have continuous densities and to be statistically independent when conditioned to the hypotheses. Each sensor decision is conditionally, given the observation, independent of the hypotheses and of the other sensor observations. Similarly, the decision of the fusion center is conditionally, given the sensor decisions, independent of the hypotheses.

Let $\mathbf{u}$ be the vector formed of the decisions $(u_1, u_2, ..., u_n)$. Thus $\mathbf{u}$ belongs to the n- dimensional binary space U^n. The global missed-detection and false alarm probabilities are respectively the probabilities $\Pr(u_0 = 0 | H_1)$ and $\Pr(u_0 = 1 | H_0)$. According to the NP criterion, it is desired to determine the fusion strategy $\Pr(u_0 = 1 \mid \mathbf{u})$ and the sensor decision strategies $\Pr(u_k = 1 \mid y_k), 1 \le k \le n,$ which minimizes $\Pr(u_0 = 0 | H_1)$ under the constraint

$$\Pr(u_0 = 1 | H_0) \le GFAP \tag{37}$$

Let us introduce the following notations:

$$\mathbf{S}_k = \{\mathbf{u} : \ u_k = 1\}, \tag{38}$$

$$\overline{\mathbf{S}}_k = U^n - \mathbf{S}_k, \tag{39}$$

$$\mathbf{u}_{-k} = \{u_1, u_2, ..., u_{k-1}, u_k, u_{k+1}, ..., u_n\}, \tag{40}$$

$$U_{-k}^{n-1} = \{\mathbf{u}_{-k}\}, \tag{41}$$

$$\Delta_{k1} = \mathrm{Proj}(\mathbf{D} \cap \mathbf{S}_k) \,\mathrm{onto}\, U_{-k}^{n-1}, \tag{42}$$

$$\Delta_{k0} = \mathrm{Proj}(\mathbf{D} \cap \overline{\mathbf{S}}_k) \,\mathrm{onto}\, U_{-k}^{n-1}, \tag{43}$$

$$C_{ki} = \sum_{\Delta_{k1}} \mathrm{pr}(\mathbf{u}_{-k} \mid H_i) - \sum_{\Delta_{k0}} \mathrm{pr}(\mathbf{u}_{-k} \mid H_i), \quad i = 0,1. \tag{44}$$

The optimal fusion strategy is given by [13]

$$\mathrm{pr}(u_0 = 1 \mid \mathbf{u}) = \begin{cases} 1 & \text{if } \mathrm{pr}(\mathbf{u} \mid H_1) - \lambda\, \mathrm{pr}(\mathbf{u} \mid H_0) > 0 \\ 0 & \text{if } \mathrm{pr}(\mathbf{u} \mid H_1) - \lambda\, \mathrm{pr}(\mathbf{u} \mid H_0) \le 0\ , \end{cases} \tag{45}$$

where λ is determined from the constraint (37). It is clear from(45) that the optimal fusion strategy is deterministic and it partitions the space $\bigcup^n$ into two disjoint subspaces $\mathbf{D}$ and $\overline{\mathbf{D}}$ such that

$$L(\mathbf{u}) = \frac{\mathrm{pr}(\mathbf{u} \mid H_1)}{\mathrm{pr}(\mathbf{u} \mid H_0)}, \tag{46}$$

$$\mathbf{D} = \{\mathbf{u} : L(\mathbf{u}) > \lambda^*\}, \tag{47}$$

where λ^* is the solution of

$$\sum_{\mathbf{D}} \mathrm{pr}(\mathbf{u} \mid H_0) \le GFAP. \tag{48}$$

The optimal decision strategy for the kth sensor is [13]

$$\mathrm{pr}(u_k = 1 \mid y_k) = \begin{cases} 1 & \text{if } \mathrm{P}(y_k \mid H_1) C_{k1} - \lambda^* \mathrm{P}(y_k \mid H_0) C_{k0} > 0 \\ 0 & \text{if } \mathrm{P}(y_k \mid H_1) C_{k1} - \lambda^* \mathrm{P}(y_k \mid H_0) C_{k0} \le 0\ , \end{cases} \tag{49}$$

or equivalently,

$$L(y_k) C_{k1} \underset{H_0}{\overset{H_1}{\gtrless}} \lambda^* C_{k0}, \quad 1 \le k \le n, \tag{50}$$

where

$$L(y_k) = \frac{P(y_k \mid H_1)}{P(y_k \mid H_0)}, \quad 1 \le k \le n. \tag{51}$$

It is clear from (50) and 951) that the kth sensor decision strategy, $1 \le k \le n$, is deterministic and it partitions the observation space $\mathbf{Y}$ into two disjoint subspaces, $\mathbf{Y}_k$ and $\hat{\mathbf{Y}}_k$ such that

$$\mathbf{Y}_k = \{y_k : \; L(y_k) C_{k1} > \lambda^* C_{k0}\}, \quad 1 \le k \le n. \tag{52}$$

Thus we can write

$$\mathrm{pr}(u_k = 1 \mid H_i) = (pd_k)^i \, (pf_k)^{1-i} = \int_{\mathbf{Y}_k} P(y_k \mid H_i) \, dy_k, \quad i = 0,1, \tag{53}$$

where pf_k and pd_k are respectively the false alarm and detection probabilities of the kth sensor.

Since the sensor decisions are independent, it follows that

$$\mathrm{pr}(\mathbf{u} \mid H_i) = \prod_{k=1}^{n} P(u_k \mid H_i), \quad i = 0,1. \tag{54}$$

In [13], it is shown that the fusion strategy (45) is equivalent to

$$V(\mathbf{u}) = \sum_{k=1}^{n} a_k \, u_k \underset{u_0=0}{\overset{u_0=1}{\gtrless}} t_0 \, , \tag{55}$$

where

$$a_k = \ln\left[\frac{pd_k \, (1 - pf_k)}{pf_k \, (1 - pd_k)}\right], \qquad 1 \le k \le n, \tag{56}$$

$$t_0 = \ln \lambda^* - \sum_{k=1}^{n} \ln\left(\frac{1 - pd_k}{1 - pf_k}\right). \tag{57}$$

Clearly, the subspace $\mathbf{D}$ can alternatively be defined as

$$\mathbf{D} = \{\mathbf{u} : \; V(\mathbf{u}) > t_0\}. \tag{58}$$

IV.3. Proposed Solution for Global Optimization

The solution of global optimization conditions is not straightforward because they have implicit analytic forms and are strongly coupled. Note that the direct solution of the set of inequalities and equations defining $\mathrm{pr}(u_0 = 1 \mid \mathbf{u})$, λ^*, and $C_{ki}, i = 0,1, k = 1,2,......,n$ is extremely difficult because of the strong interdependence among them. From (44) we note that evaluation of $C_{ki}, i = 0,1, k = 1,2,......,n$, requires the knowledge of the fusion center fusion rule through the rule $\mathrm{pr}(u_0 = 1 \mid \mathbf{u})$. The optimum fusion rule (combining strategy) as specified by (45) depends on the sensors thresholds. It therefore appeared to most publications that it is not possible to find explicit solutions for optimum thresholds and strategies without specification of the combining strategy environment. For example, [13] derived generic expressions by assigning various combining strategies to the fusion center. The combining strategies chosen in [13] (and in most publications) are mainly intuitive and require theoretical justification to be considered as candidates for optimality.

In this subsection, an algorithm is devised to determine for a given $GFAP$, the corresponding optimum setting of the fusion center threshold (t_0) and of the operating points of the sensors (pf_k, pd_k), $1 \leq k \leq n$, as well as the achieved GDP. Before presenting the algorithm, we note that if the distributions of the local sensor likelihood ratios have point masses of probabilities, it may not be possible to make the desired adjustment of the local sensor thresholds as implied by the global optimization of the DDS at a given desired $GFAP$ [39]. In such a case, the global optimization conditions (subsection IV.2) may be inconsistent; i.e. have no common solution, and it may be necessary to consider randomization both at the fusion center and at the sensors [36]. Conversely, if the distributions of the local sensor likelihood ratios have no point masses of probabilities, randomization is never optimal [39]. This is the case considered here and considered by [13] who was interested in distributed radar detection theory where the sensor observations are almost always continuous.

The key to the suggested algorithm is the remark that the values of the sufficient statistic $V(\mathbf{u})$ on the left hand side of (55) are discrete. Thus there is a continuous range of values of the threshold t_0 satisfying the constraint $\mathrm{pr}[V(\mathbf{u}) > t_0 \mid H_0] \leq GFAP$. Any value belonging to this range can be expressed as a convex linear combination of the boundary values [Equation (63)] and through the use of (57), (44), (52) and (54) it may or may not correspond to a possible solution of the considered DDS. In the following a parameter w is introduced to scan the permissible values of t_0 and to identify those for which the set of inequalities and equations (55), (57), (48), (44) and (51) are consistent. Thus w varies from δ to 1 in steps of δ, where ($1/\delta$) is a large positive integer. Note that the parameter w does not imply any randomization fusion. It only helps us scan all possible solutions of global optimization conditions [13] and pick the optimal one.

The proposed algorithm has the following steps

1. For every sensor determine the ROC corresponding to a likelihood ratio decision rule either explicitly, $pd_k = \phi_k(pf_k)$, or parametrically $pd_k = \beta_k(\lambda_k)$, $pf_k = \alpha_k(\lambda_k)$, with the threshold λ_k as a parameter, $1 \le k \le n$.
2. Assign w its smallest possible value δ.
3. Select arbitrary operating points (pf_k, pd_k) on the sensors ROCs such that
4. $$\prod_{k=1}^{n} pf_k \le GFAP \cdot \qquad (59)$$
5. Associate with every vector $\mathbf{u}$ of $\bigcup^n$ the corresponding value of $V(\mathbf{u})$ in (55) and the corresponding probabilities $\mathrm{pr}(\mathbf{u} \mid H_0)$ and $\mathrm{pr}(\mathbf{u} \mid H_1)$ given by (54).
6. Rearrange the values of $V(\mathbf{u})$ in a descending order and rearrange the probabilities $\mathrm{pr}(\mathbf{u} \mid H_0)$ and $\mathrm{pr}(\mathbf{u} \mid H_1)$ accordingly. Denote the reordered values of $V(\mathbf{u})$ by v_i's $(v_1 > v_2 > v_3 > \ldots\ldots\ldots)$.
7. Find the minimum value v_ℓ such that
8. $$\mathrm{pr}[V(\mathbf{u}) > v_\ell \mid H_0] \le GFAP, \qquad (60)$$
9. and define the subspace $\mathbf{D}$ as
10. $$\mathbf{D} = \{\mathbf{u} : V(\mathbf{u}) \ge v_\ell\}. \qquad (61)$$
11. Evaluate the probability $G\hat{D}P$ as
12. $$G\hat{D}P = \sum_{\mathbf{u} \in \mathbf{D}} \mathrm{pr}(\mathbf{u} \mid H_1). \qquad (62)$$
13. Calculate the threshold t_0 as
14. $$t_0 = w v_\ell + (1 - w) v_{\ell-1}, \qquad 0 < w \le 1. \qquad (63)$$
15. Substitute the calculated value of t_0 into (57) and solve for λ^*.
16. For every sensor calculate the coefficients $C_{k,i}, i = 0,1,\ k = 1,2,\ldots\ldots,n$, from (44). Update the operating points (pf_k, pd_k) according to the new thresholds in the set of decision rules (50). If the new values of the coefficients $C_{k,i}$ of the kth sensor happen to be zero, leave the operating point of this particular sensor unchanged. This is because the coefficients $C_{k,i}$ are equal to zero when the fusion center ignores the decisions of the kth sensor and hence the global decision is not influenced by them.
17. For the assigned value of w repeat the steps from (4) to (10) often enough to reach one of the following states: (a) it is impossible to find a value v_ℓ satisfying (60), (b) the difference between the estimated values of GDP in successive iterations alternates cyclically between two or more values, and (c) the difference between the estimated values of GDP in successive iterations becomes zero. Occurrence of the

first state indicates divergence of the algorithm for the considered w. The second state occurs when the algorithm does not converge to a common solution of the set of inequalities and equations (55), (57), (48), (44) and (51). The third state happens only when there is a convergence to a common solution. Skip the value of w if either the first or the second state occurs. Otherwise, record the value of w together with the reached steady state estimate $\hat{GDP}$.

18. Increment w by δ and repeat the steps from (3) to (12) till all the possible values of w are exhausted.
19. Scan the recorded values of w and their associated steady state estimates $\hat{GDP}$. The value of w which results in the largest value of $\hat{GDP}$ is the optimum choice ($w_{\text{opt}}, \hat{GDP}_{\max}$). Hence for this choice, print out the reached steady state operating points of the sensors (pf_k, pd_k), the fusion coefficients a_k, defined by (56), and the threshold of the fusion center t_0 given by (63).

Through elaboration of a very large number of examples it was possible to determine the necessary conditions to ensure the following desired features of the proposed algorithm:

1. If for a given choice of w the algorithm converges to a common solution, then the reached steady state estimate $\hat{GDP}$ should be independent on the initially selected sensors operating points (pf_k, pd_k) satisfying (59).
2. The dependence of the desired precision of determination of $\hat{GDP}_{\max}$ on the precision with which w_{opt} is determined should be known.

To ensure the first feature, it was found necessary to insert between steps (6) and (7) the condition that if $\text{pr}[V(\mathbf{u}) = v_{\ell+1} \mid H_0] > GFAP$, then the value of the subscript ℓ, determined in step (6), should be decreased by one. With this condition, the authors verified for every considered example that the algorithm possessed the first desired feature. However, they found that the number of iterations, needed in step (11) of the algorithm to reliably decide which state prevailed, depended on the initially selected operating points of the sensors. In the examples considered by the authors, this number varied from 40 to 80 iterations. Regarding the second desired feature mentioned above, it was found that letting $\delta = 10^{-m}$ where m is a positive integer larger than one, ensures the precise determination of $\hat{GDP}_{\max}$ up to $(m-1)$-decimal digits. Thus for example, in order to determine $\hat{GDP}_{\max}$ precisely up to five decimal digits, it is necessary to determine the optimum value of w (w_{opt}) up to six decimal digits. From the elaborated examples, it was also found that in order to speed up the algorithm; it is possible to start with $m = 2$ and scans the values of w between 0.01 and 1 in steps of 0.01. Let the optimum 2-digit value of w be $w_{\text{opt}}^{(2)}$. Increment

m by 1 and scan the values of w between $w_{\text{opt}}^{(2)} - 0.009$ and $w_{\text{opt}}^{(2)}$ in steps of 0.001 to determine the optimum 3-digit value of w and repeat this procedure till you reach the desired precision. It is important to note that the execution time of the proposed algorithm is mainly dependent on the number of considered sensors no matter the sensors are identical or not, although from analytical point of view finding optimum operating points of n sensors ($2n$ unknowns) is far more difficult than finding a common optimum operating point (2 unknowns) in case of identical sensors.

IV.4. Optimization Examples

We consider the case of Rayleigh distributed observations. The probability distributions under both hypotheses are given by [3, 26, 34].

$$P(y_k \mid H_0) = \exp(-y_k), \tag{64}$$

$$P(y_k \mid H_1) = d_k \exp(-d_k\, y_k), \quad d_k > 0, k = 1,2,\ldots\ldots\ldots, n, \text{ for } y_k \geq 0, \tag{65}$$

$$P(y_k \mid H_i) = 0, \;\; k = 1,2,\ldots\ldots\ldots, n, \; i = 0,1, \;\text{ elsewhere.} \tag{66}$$

The coefficient $d_k, k = 1,2,\ldots\ldots\ldots, n,$ is related to the signal to noise ratio at each individual sensor (r_k) as

$$d_k = \frac{1}{1 + r_k}, \; k = 1,2,\ldots\ldots\ldots\ldots\ldots, n. \tag{67}$$

The detection and false alarm probabilities are related, in terms of the coefficient d as:

$$pd_k = pf_k^{\,1/(1+r_k)} = pf_k^{\,d_k}, \quad k = 1,2,\ldots\ldots\ldots\ldots, n. \tag{68}$$

First, two special cases of DDSs are optimized using other procedures in order to compare the exact true values of GDP with $G\hat{D}P\max$. The first case considers a DDS with two non-identical sensors. In this case, the global optimization of the DDS is obtained as follows [3, 40].

Consider any two sensors with receiver operating characteristics, $pd_1 = g(pf_1)$ and $pd_2 = h(pf_2)$. To derive the optimum threshold setting for each sensor, we assume that the sensor's decisions are conditionally independent under hypothesis H_0 and H_1. In the case of AND fusion rule, the global false alarm and detection probabilities are given by:

$$GFAP = pf_1\, pf_2\,, \tag{69}$$

$$GDP = pd_1\, pd_2\,. \tag{70}$$

Equation (70) can be rewritten as:

$$GDP = g(pf_1)\; h(pf_2). \tag{71}$$

Substituting for pf_2 from Equation (69), we obtain

$$GDP(GFAP) = g(pf_1)\; h\left(\frac{GFAP}{pf_1}\right), \tag{72}$$

where the dependence $GDP(GFAP)$ is the global receiver operating characteristic of the distributed detection system. Clearly, from Equation (69), $pf_1 > GFAP$, Equation (71) will be

$$GDP(GFAP) = \operatorname*{Sup}_{GFAP < pf_1 < 1}\left(g(pf_1)\; h\left(\frac{GFAP}{pf_1}\right)\right). \tag{73}$$

In the case of OR fusion rule, we have

$$GFAP = pf_1 + pf_2(1 - pf_1), \tag{74}$$

which yields

$$pf_2 = \frac{(GFAP - pf_1)}{(1 - pf_1)}. \tag{75}$$

From Equation (74), $pf_1 < GFAP$, the optimum global receiver operating characteristic using an OR combiner is

$$GDP(GFAP) = \operatorname*{Sup}_{GFAP < pf_1 < 1}\left(g(pf_1) + h\left(\frac{GFAP - pf_1}{1 - pf_1}\right) - g(pf_1)\, h\left(\frac{GFAP - pf_1}{1 - pf_1}\right)\right) \tag{76}$$

The result is plotted in Figure 67 in case of $d_1 = 0.1$ and $d_2 = 0.4$. Figure 67 shows the OR and the AND fusion rules (the optimal ROCs in case of two sensors [40]). The AND fusion rule is optimum at very low false alarm probabilities while the OR fusion rule is optimum at high false alarm probabilities The ROC achieved by the proposed algorithm is shown in Figure 68. From Figures 67 and 68 it is readily seen that the ROC achieved by the proposed algorithm is indeed the convex hull of the ROCs obtained with the OR and AND fusion rules; i.e. it is the optimal ROC [40].

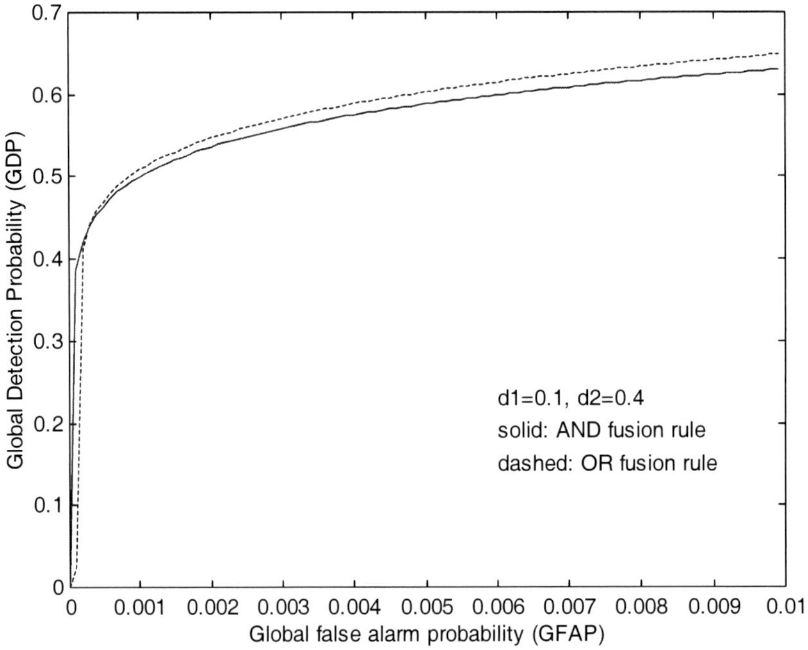

Figure 67. Comparison of ROCs of a two different sensors using "AND" and "OR" fusion rules.

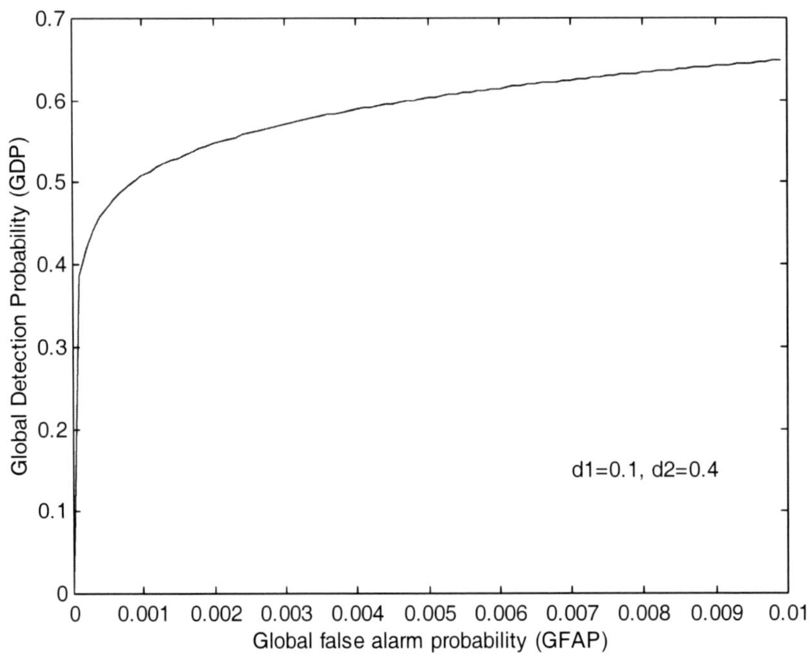

Figure 68. The global ROC of a two different sensors using the proposed approach.

The second selected special case is the case of identical sensor where the optimum fusion rule reduces to K -out-of- n fusion rule. For a specified K, we solve the equation

$$GFAP = \sum_{i=K}^{n} c_i^n \, pf^i \, (1 - p_f)^{n-i}, \tag{77}$$

for pf and then determine the corresponding pd from the common ROC of the identical sensors. The global detection probability, for the specified K, is calculated from

$$GDP = \sum_{i=k}^{n} c_i^n \, pd^i \, (1 - pd)^{n-i}. \tag{78}$$

For every desire value of GFAP, there is an optimum integer value of K that maximizes the GDP.

We consider the case of DDS with four identical sensors with Rayleigh distributed observations with $d = 0.02$. The ROCs corresponding to the K-out-of-n fusion rules, $K = 1,2,3,4,$ and to the proposed Neyman Pearson fusion of the decisions of the four identical sensors are given in Table 1. The points in the last row of Table 1 are determined using the proposed algorithm. It can easily be verified that the last ROC is the convex hull of the other ROCs. Moreover, when the initial operating points of the identical sensors are deliberately made non-identical, the algorithm always converged to identical optimum sensor operating points. Tables 2-4 present complete globally optimum solutions of DDSs with three, four, and five non-identical sensors at different GFAPs. The corresponding values of $\hat{GDP}_{\max}$ are precisely determined up to the sixth decimal digit.

Table 1. Comparison of ROC's of four-sensor distributed detection system with $K - \text{out - of} - n$ fusion rules, $K = 1,2,3,4,$ and the proposed Neyman-Pearson fusion rule (identical sensors, $d = 0.02$)

$GFAP$	GDP $(K=1)$	GDP $(K=2)$	GDP $(K=3)$	GDP $(K=4)$	w_{opt}	$\hat{GDP}_{\max}$
.00015	.99889	.9962	.9764	.8318	.56358	.99889
.00025	.99905	.9967	.9786	.8471	.56898	.99905
.00035	.99926	.9970	.9800	.8529	.57213	.99926
.00045	.99928	.9972	.9810	.8572	.57951	.99928
.00055	.99932	.9974	.9818	.8621	.58136	.99932
.00065	.99936	.9975	.9824	.8648	.58719	.99936
.00075	.99939	.9976	.9830	.8671	.59108	.99939
.00085	.99941	.99780	.9834	.8691	.61005	.99941

Table 2. Complete globally optimum solution of a three non identical sensors distributed detection system at different $GFAP$'s, $d_1 = 0.03,\ d_2 = 0.04,\ d_3 = 0.05$

$GFAP = 10^{-3}$, $w_{opt} = 0.6579110$ Required				
d_k	t_k	pf_k	pd_k	a_k
0.03	7.9577	$\times 10^{-4}$ 3.499607	0.787627	9.26802
0.04	8.0064	$\times 10^{-4}$ 3.333091	0.725962	8.98034
0.05	8.0564	$\times 10^{-4}$ 3.170633	0.668432	8.75719
$t_0 = 5.76146,\ G\hat{D}P_{max} = 0.980703$				
Required $GFAP = 10^{-6}$, $w_{opt} = 0.7989000$				
d_k	t_k	pf_k	pd_k	a_k
0.03	14.8282	$\times 10^{-7}$ 3.632433	0.640923	15.40757
0.04	14.9166	$\times 10^{-7}$ 3.325069	0.550645	15.11988
0.05	15.0082	$\times 10^{-7}$ 3.034163	0.472174	14.89674
$t_0 = 11.90101,\ G\hat{D}P_{max} = 0.914844$				
Required $GFAP = 10^{-9}$, $w_{opt} = 0.8591598$				
d_k	t_k	pf_k	pd_k	a_k
0.03	21.6948	$\times 10^{-10}$ 3.784973	0.521605	21.78128
0.04	21.8259	$\times 10^{-10}$ 3.319944	0.417681	21.49360
0.05	21.9629	$\times 10^{-10}$ 2.894878	0.333489	21.27046
$t_0 = 18.27473,\ G\hat{D}P_{max} = 0.814325$				

Table 5 presents typical total execution times of the proposed algorithm on a Siemens computer (Pentium 4, CPU 3.6 GHz, 512 MB of RAM). Different DDSs were considered. In every case a complete global optimization for a given GFAP with precise determination of $G\hat{D}P_{max}$ up to the 6th decimal digit was carried out. The entries of the 3rd row of Table 5 represent the total number of decision functions R_n which should be searched for the set of optimum thresholds. These data are extracted from [38] (if we don't use the proposed approach). The entries of the 2nd row are extracted using the proposed algorithm. It is clear from Table 5 that when the number of sensors increases the execution time increases as 2^n although R_n increases much faster. This demonstrates the power of the proposed algorithm particularly at large number of non-identical sensors.

Table 3. Complete globally optimum solution of a four non identical sensors distributed detection system at different $GFAP$'s, $d_1 = 0.03$, $d_2 = 0.04$, $d_3 = 0.05$, $d_4 = 0.06$

$GFAP = 10^{-3}$, $w_{opt} = 0.6802806$ Required				
d_k	t_k	pf_k	pd_k	a_k
0.03	8.2188	2.695460×10^{-4}	0.781482	9.49283
0.04	8.2690	2.563499×10^{-4}	0.718378	9.20514
0.05	8.3204	2.434892×10^{-4}	0.659666	9.98200
0.06	8.3732	2.309684×10^{-4}	0.605080	8.79968
$t_0 = 5.98627$, $G\hat{D}P_{max} = 0.991729$				
$GFAP = 10^{-6}$, $w_{opt} = 0.8116276$ Required				
d_k	t_k	pf_k	pd_k	a_k
0.03	15.0692	2.854565×10^{-7}	0.636306	15.62854
0.04	15.1590	2.609240×10^{-7}	0.545331	15.34068
0.05	15.2521	2.377316×10^{-7}	0.466449	15.11772
0.06	15.3486	2.158776×10^{-7}	0.398155	14.93540
$t_0 = 12.12199$, $G\hat{D}P_{max} = 0.946900$				
$GFAP = 10^{-9}$, $w_{opt} = 0.867891$ Required				
d_k	t_k	pf_k	pd_k	a_k
0.03	21.9166	3.032021×10^{-10}	0.518144	21.98932
0.04	22.0491	2.655715×10^{-10}	0.413967	21.70161
0.05	22.1877	2.312158×10^{-10}	0.329761	21.47845
0.06	22.3324	2.000590×10^{-10}	0.261860	21.29609
$t_0 = 18.48268$, $G\hat{D}P_{max} = 0.860297$				

The proposed algorithm can be used to optimize DDSs according to criteria other than the Neyman Pearson criterion. For example, in multiple sensor distributed IFF (Identification Friend or Foe) systems two types of decision errors may be committed: identifying a hostile target as a friendly one, or conversely identifying a friendly target as a hostile one. In heavy air combat situations both types of errors are almost equally significant. Hence it is more convenient to optimize the decision strategy of the fusion center according to the Ideal observer criterion rather than the Neyman Pearson criterion.

Table 4. Complete globally optimum solution of a five non identical sensors distributed detection system at different $GFAP$'s, $d_1 = 0.03$, $d_2 = 0.04$, $d_3 = 0.05$, $d_4 = 0.06$, $d_5 = 0.07$

Required $GFAP = 10^{-3}$, $w_{opt} = 0.6983120$				
d_k	t_k	pf_k	pd_k	a_k
0.03	8.4145	2.216384×10^{-4}	0.776908	9.66198
0.04	8.4657	2.105587×10^{-4}	0.712746	9.37430
0.05	8.5183	1.997695×10^{-4}	0.653170	9.15115
0.06	8.5723	1.892743×10^{-4}	0.597896	8.96883
0.07	8.6277	1.790768×10^{-4}	0.546654	8.81468
$t_0 = 6.15542$, $G\hat{D}P_{max} = 0.995948$				
Required $GFAP = 10^{-6}$, $w_{opt} = 0.8220415$				
d_k	t_k	pf_k	pd_k	a_k
0.03	15.2456	2.392884×10^{-7}	0.632947	15.79048
0.04	15.3365	2.184913×10^{-7}	0.541474	15.50280
0.05	15.4307	1.988468×10^{-7}	0.462302	15.27965
0.06	15.5283	1.803530×10^{-7}	0.393883	15.09733
0.07	15.6295	1.630027×10^{-7}	0.334853	14.94318
$t_0 = 12.28392$, $G\hat{D}P_{max} = 0.9628069$				
Required $GFAP = 10^{-9}$, $w_{opt} = 0.8750879$				
d_k	t_k	pf_k	pd_k	a_k
0.03	22.0735	2.591706×10^{-10}	0.515712	22.13640
0.04	22.2071	2.267752×10^{-10}	0.411362	21.84872
0.05	22.3467	1.972244×10^{-10}	0.327151	21.62558
0.06	22.4926	1.704503×10^{-10}	0.259356	21.44326
0.07	22.6449	1.463613×10^{-10}	0.204918	21.28911
$t_0 = 18.62985$, $G\hat{D}P_{max} = 0.8870487$				

Table 5. Comparison of typical measured values of the total execution time of the proposed algorithm with the total number of decision functions searched for the set of optimum thresholds

Number of sensors (n)	3	4	5	6
Total execution time (s)	38-50	75-100	130-180	225-330
The number of functions R_n	9	114	6894	7786338

The Ideal observer criterion implies the minimization of the global error probability GEP. Assuming equal prior probabilities of occurrence of a hostile or a friendly target, the global error probability can be expressed as

$$GEP = \frac{1}{2}(GFAP + 1 - G\hat{D}P_{\max}). \tag{79}$$

The proposed algorithm can be used to optimize such distributed IFF systems by varying GFAP and evaluating the corresponding $G\hat{E}P$. The value of GFAP giving minimum $G\hat{E}P$ is then easily determined. This procedure is applied to three non-identical sensor system with coefficients $d_1 = 0.007$, $d_2 = 0.010, d_3 = 0.014$. The results are presented in Figure 69 and Table 6.

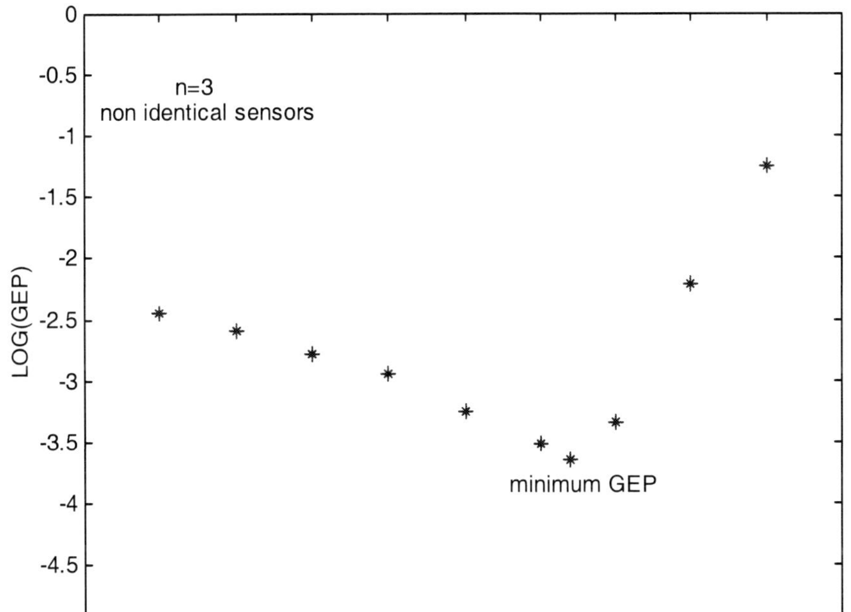

Figure 69. Dependence of global error probability on the global false alarm probability.

Table 6. Complete solution of a three non identical sensors distributed detection systems, optimized according to the Ideal Observer criterion,
$d_1 = 0.007,\ d_2 = 0.010,\ d_3 = 0.014$

$w_{opt} = 0.62872195,\ t_0 = 7.22861,\ G\hat{E}P_{min} = 4.75283\times10^{-4}$				
d_k	t_k	pf_k	pd_k	a_k
0.007	9.51505	7.37341×10^{-5}	0.93556	12.18821
0.010	9.53129	7.25453×10^{-5}	0.90909	11.83153
0.014	9.55317	7.09758×10^{-5}	0.87481	11.49508

Conclusion

In this paper, the problem of obtaining and comparing the global performance improvement in centralized and decentralized approaches has been considered, for any number of sensors, by adopting the more general optimum data fusion structures in case of identical sensors. Optimum structures and strategies have been considered in case of identical sensors with Gaussian and Rayleigh distributed observations. More complicated examples have been considered in case of large number of sensors to characterize the performance trade off between the centralized and decentralized approaches. Then the communication requirements can be compared to choose a preferred architecture. It is found that the increment in the global performance improvement is depending on the desired global false alarm probability, the number of sensors and the signal to noise ratio at each individual sensor. The performance loss of decentralized detection systems has been considered. It has been found that the performance loss increases as the number of sensor increases. It is found that for very small values of global false alarm probabilities it is impractical to use more than three or four sensors to avoid additional hardware costs without significant performance improvement. It has been found that letting the number of sensors be greater than or equal to $2n$ reliably ensures a performance of the optimum decentralized schemes non inferior to that of the optimum centralized schemes, with n sensors, for any value of n, SNR per sensor observations and $GFAP$.

A simple algorithm for optimizing multiple sensor detection system with data fusion, in terms of both the sensors and the fusion center, according to Neyman-Pearson strategy, has been proposed. The proposed algorithm determines, for a given false alarm probability, the corresponding optimum setting of the thresholds both at the fusion center and at the sensors. It scans all possible solutions and picks the solution that gives the largest detection probability. The proposed algorithm is found to be simple, accurate and fast. It is worth noting that the execution time of the proposed algorithm is mainly dependent on the number of considered sensors no matter they are identical or not, although from an analytic point of view finding optimum operating point of n sensors (2n unknowns) is far more difficult than finding a common optimum operating point (2 unknowns) in case of identical sensors. It is also shown that the proposed algorithm can be used to optimize distributed detection systems according to criteria other than the Neyman-Pearson one, such as the Ideal Observer criterion for example.

References

[1] Robert R Tenney and Nils R. Sandell, "Detection With Distributed Sensors", *IEEE Transactions on Aerospace and Electronic Systems,* vol. AES-17, No. 4, pp. 501-510, July 1981.

[2] R. Viswanathan and P. K. Varshney, "Distributed Detection With Multiple Sensors: Part I - Fundamentals", *Proceedings of the IEEE*, vol. 85, No. 1, pp. 54-63, Jan. 1997.

[3] Ashraf M. Aziz, *"New Data Fusion Algorithms for Distributed Multisensor Multitarget Environments,"* Ph.D. Dissertation, Dept. of Electrical and Computer Eng., Naval Postgraduate School, Monterey, California, Sept. 1999.

[4] Ashraf M. Aziz, "Fuzzy Track-to-Track Association and Track Fusion Approach in Distributed Multisensor-Multitarget Multiple-Attribute Environment", *Signal Processing,* Vol. 87, No. 6, pp. 1474-1492, June 2007.

[5] Ashraf M. Aziz, M. Tummala and R. Cristi, "Fuzzy Logic Data Correlation Approach in Multisensor-Multitarget Tracking Systems," *Signal Processing*, vol. 76, No. 2, pp 195-209, July 1999.

[6] E. Waltz and J. Llinas, Multisensor Data Fusion, Norwood, MA, Artech House, 1990.

[7] Ashraf M. Aziz, "Multiple-Sensor Distributed Detection Systems With Data Fusion," *Proceedings of the 13th IEEE SP Society and EURASIP International Conference on Digital Signal Processing*, Santorini, Greece, pp. 1031-1034, vol. 2, July 1997.

[8] Ashraf M. Aziz, Murali Tummala, and Roberto Cristi, "Optimal Data Fusion Strategies Using Multiple-Sensor Detection Systems," *Proceedings of the 31st IEEE Asilomar International Conference on Signals, Systems, and Computer*s, California, USA, pp. 941-945 vol. 1, Nov. 1997.

[9] H. L. Van Trees, *Detection, Estimation and Modulation Theory*, vol. 1, New York: John Wiley, 1969.

[10] F. Sadjadi, "Hypothesis Testing in a Distributed Environment," *IEEE Transactions on Aerospace and Electronic Systems*, vol. AES-22, pp. 134-137, March 1986.

[11] Z. Chair and P.K. Varshney, "Optimal Data Fusion in Multiple-Sensor Detection Systems", *IEEE Transactions on Aerospace and Electronic Systems*, vol. AES-22, No. 1, pp. 98-101, Jan. 1986.

[12] S. Thomopoulos, R. Viswanathan, and D. Bougoulias, "Optimal Decision Fusion in Multiple-Sensor Systems", *IEEE Transactions on Aerospace and Electronic Systems*, vol. AES-23, No-5, pp. 644-653, Sept. 1987.

[13] R. Srinivasan, "Distributed Radar Detection Theory," *IEE Proc.*, Part F, vol. 133, No. 1, pp. 55-60, Feb. 1986.

[14] J. N. Tsistsiklis, ''Distributed Detection," in *Advances in Statistical Signal Processing, Signal Detection,* vol. 2, H. V. Poor and J. B. Thomas, Eds. Greenwich, CT: JAI Press, 1993.

[15] Q. Zhang, P. Varshney, and R. Wesel, "Optimal Bi-Level Quantization of i.i.d Sensor Observations for Binary Hypothesis Testing," *IEEE Transactions on Information Theory,* vol. IT-48, No-7, pp. 2105-2111, July 2002.

[16] R. Blum and S. Kassam, "Optimum Distributed Detection of Weak Signals in Dependent Sensors," *IEEE Transactions on Information Theory*, vol. IT-38, No-3, pp. 1066-1079, May 1992.
[17] R. Blum, "Necessary Conditions for Optimum Distributed Sensor Detector Under the Neyman-Pearson Criterion ," *IEEE Transactions on Information Theory*, vol. IT-42, No-3, pp. 990-994, May 1996.
[18] J. N. Tsitsiklis and M. Athans, "On the Complexity of Decentralized Decision Making and Detection Problems," *IEEE Transactions on Automatic Control*, vol. AC-30, No-5, pp. 440-446, May 1985.
[19] D. L. Hall, *Mathematical Techniques in Multisensor Data Fusion*, Norwood, MA, Artech House, 1992.
[20] Ashraf M. Aziz, "A Simple and Efficient Suboptimal Multilevel Quantization Approach in Geographically Distributed Sensor Systems", *Signal Processing*, International Journal, 2008 (to appear).
[21] J. Chamberland and V. Veeravalli, "Decentralized Detection in Sensor Networks," *IEEE Transactions on Signal Processing*, vol. 51, No-2, pp. 407-416, Feb. 2003.
[22] M. Feder and N. Merhav, "Universal Composite Hypothesis Testing: A Competitive Minimax Approach," *IEEE Transactions on Information Theory*, vol. IT-48, No-6, pp. 1504-1517, June 2002.
[23] E. Levitan and N. Merhav, "A Competitive Neyman-Pearson Approach to Universal Hypothesis Testing With Applications," *IEEE Transactions on Information Theory*, vol. IT-48, No-8, pp. 2215-2229, Aug. 2002.
[24] S. Thomopoulos, "Theories in Distributed Decision Fusion: Comparison and Generalization," *SPIE* vol. 1383, Sensor Fusion III, pp. 623-634, 1990.
[25] E. Lehmann, Testing Statistical Hypotheses, John Wiley, New York, 1986.
[26] S. Thomopoulos, D. Bougoulias, and L. Zhang, "Optimal and Suboptimal Distributed Decision Fusion," *SPIE vol. 931*, *Sensor Fusion*, pp. 26-30, 1988.
[27] J. V. DiFranco and W. L. Rubin, *Radar Detection,* Prentice-Hall, Inc., Englewood Cliffs, N. J., 1968.
[28] R. Viswanthan and V. Aalo, "On Counting Rules in Distributed Detection", *IEEE Transactions on Acoustic, Speech, and Signal Processing*, vol. 37, No-5, pp. 772-775, May. 1989.
[29] E. Lehmann, *Testing Statistical Hypotheses*, John Wiley, New York, 1986.
[30] M. Longo, T. Lookabaugh, and R. Gray, "Quantization for Decentralized Hypothesis Testing Under Communication Constraints," *IEEE Transactions on Information Theory*, vol. IT-36, No-2, pp. 241-255, March. 1990.
[31] C. W. Helstrom, "Gradient Algorithm for Quantization Levels in Distributed Detection Systems", *IEEE Transactions on Aerospace and Electronic Systems*, vol. AES-31, No-1, pp. 390-398, Jan. 1995.
[32] J. N. Tsitsiklis, ''Distributed Detection," in *Advances in Statistical Signal Processing, Signal Detection*, vol. 2, H. V. Poor and J. B. Thomas, Eds. Greenwich, CT: JAI Press, 1993.
[33] D. Buede and E. Waltz, "Benefits of Soft Sensors and Probabilistic Fusion," SPIE vol. 1096, *Signal and Data Processing of Small Targets*, pp. 320-309, 1989.

[34] C. C. Lee and J. J. Chao, "Optimal Local Decision Space Partitioning for Distributed Detection," *IEEE Transactions on Aerospace and Electronic Systems*, vol. AES-25, No-4, pp. 536-543, July 1989.
[35] J. N. Tsitsiklis, ''Decentralized Detection by a large number of sensors," *Math. Contr. Signals Syst.*, Vol. 1, No.2, 1988, pp. 167-182.
[36] J. N. Tsitsiklis, ''External Properties of Likelihood Ratio Quantizers," *IEEE Trans. Inform. Theory,* Vol. 38, No.2, March 1992, pp. 335-362.
[37] R. Srinivasan, P. Sharma and V. Malik, *IEEE Trans. Aerospace Electronic Systems*, Vol.31, No.1, January 1995, pp. 390-398.
[38] S. Thomopoulos, R. Viswanathan, and D. Bougoulias, "Optimal Distributed Decision Fusion", *IEEE Transactions on Aerospace and Electronic Systems*, vol. AES-25, No-5, pp. 761-765, Sept. 1989.
[39] P. K. Willet and D. J. Warren, "The Suboptimality of Randomized Test in Distributed and Quantized Detection Systems", *IEEE Transactions on Information Theory*, vol. 38, No-2, pp. 335-362, March 1992.
[40] S. D. Stearns, "Optimum detection using multiple sensors", *Proc. Camahan Conf. on Security Technology*, Uxington, KY, USA, 1983.

In: New Signal Processing Research
Editor: Takumi Maeda, pp. 345-368
ISBN: 978-1-60456-479-2

Chapter 10

ON IMAGE ENHANCEMENT USING A GAUSSIAN TRANSFORMATION IN SPATIAL DOMAIN

Khoa N. Le*
Centre for Wireless Monitoring and Applications,
Griffith School of Engineering, Griffith University,
Gold Coast campus, Queensland, Australia

Abstract

This chapter proposes a gray-level Gaussian transformation for image enhancement in the spatial domain. Comparisons with a power-law gray-level transformation are made. Image histograms of both transformations are estimated. A new formula for the useful energy in images is proposed to automate the performance evaluation process of image enhancement. Relative merits of the new transformation are discussed.

Key words: Gaussian transformation, power-law transformation, point processing, image enhancement, image histogram

1. Introduction

Image enhancement is one of the most visually appealing and interesting areas in image processing (Gonzalez and Woods, 2002). To successfully process different images from different applications, unique methods or algorithms for each image type are required. The ultimate goal of image enhancement is to make images to be clearer and higher quality, making them easier to understand and thus more useful. Image enhancement is also important because of the high demand in processing practical images which are often blurry and noisy, making them difficult to interpret. Examples of practical images for enhancement are space images, X-ray images, mammogram and biomedical images.

* E-mail address: K.Le@griffith.edu.au, Tel. (+617) 5552 9175, Fax. (+617) 5552 8065

There are different methods to perform image enhancement in the spatial domain. The first method is to globally process pixels in an image using a transformation to improve its contrast level. This method lowers gray levels below a threshold (darkening) and increases those above the threshold (brightening), thus, improving the contrast level in images. The second method is to remodel images' histograms known as histogram equalization. Other histogram modeling methods include histogram matching and using histogram statistics can also be employed for image enhancement. A histogram $h(r_k) = n_k$ of an image shows the frequency of pixel occurrence n_k in a particular gray level r_k. The third method is to use spatial filtering for smoothing and sharpening in which the Laplacian operator has been successfully used. It should be noted that a combination of all of the above methods can also be used to yield effective image enhancement, depending on particular applications. Out of the three aforementioned methods, the first method is the most basic as it only involves direct processing of pixels or point processing in an image.

There has been significant progress in different context of image enhancement. Mukhopadhyay and Chanda (Mukhopadhyay and Chanda, 2001) used a multi-sensor data acquisition system to record image data which were shown to be clearer and thus effective for detection of soft tissues in the human brain. The technique was shown to enhance the image quality which directly improved patient clinical treatment. Chen et al. proposed an automatic gray-level grouping method for image enhancement in which satisfactory results were obtained (Chen et al., 2006). The basic method proposed in (Chen et al., 2006) was then further extended to improve its effectiveness (Chen et al., 2006). Bishnu et al. (Bishnu et al., 2005) used an Euler vector for image enhancement by modifying the four most significant bit planes of images. The vector was also used to store and to retrieve digital images. Raji (Raji et al., 1998) proposed a family of power-law transformations for image enhancement with a non-zero offset. In their approach, images were divided into a number of classes which were individually transformed, yielding satisfactory results. Munteanu and Rosa (Munteanu and Rosa, 2004) proposed an automatic image evaluation driven by an optimization process in which three criteria on assessing image enhancement were studied. Detailed comparisons were also made with histogram matching and contrast stretching methods. Ojala et al. (Ojala et al., 2002) employed a multi-resolution approach to gray scale and rotation invariant texture classification using local binary patterns. Histogram statistics were also used to make their method more effective. Chang and Chuang (Chang and Chuang, 2002) proposed an image protection method for intellectual property using a visual secret sharing scheme. The image quality in this scheme was also enhanced.

Even though different techniques of image enhancement have been proposed in the literature, there lack of techniques using gray-level transformations to process pixels, in which the most recent works were reported by Munteanu and Rosa (Munteanu and Rosa, 2004) and Raji et al. (Raji et al., 1998). A number of basic gray-level transformations such as power-law with a zero offset, log and inverse-log were extensively studied in (Gonzalez and Woods, 2002) in which relative merits of each method were discussed. A graphical form of a transformation $T(r)$ as a function of the gray level r is shown in Figure 1 from which gray levels of $r < m$ are darkened and those of $r > m$ are brightened by the transform respectively, thus improving the contrast level of images.

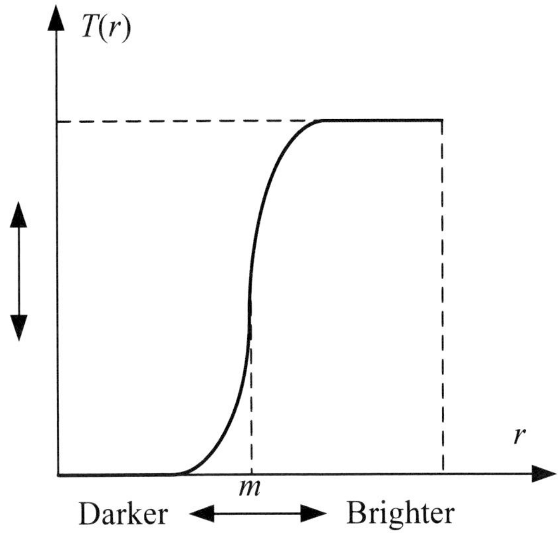

Figure 1. A graphical illustration of a transformation $T(r)$ as a function of r.

Figure 2 shows a schematic diagram of image enhancement using transformations in which pixel values s in the transformed image are computed using the relation: $s = T(r)$. The smaller the pixel value, the darker the image region and vice versa.

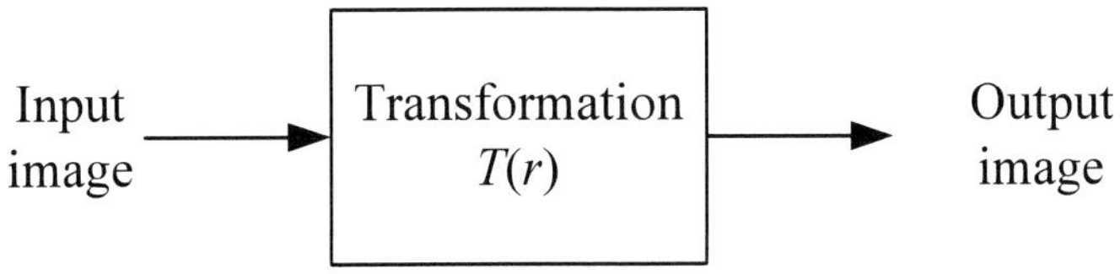

Figure 2. Image enhancement using a transformation $T(r)$.

From Figure 1, the transformation $T(r)$ can be obtained by using a Gaussian exponential function. The main motivation of using the Gaussian transformation is because of its rotational invariant nature which has been effectively employed to suppress cross-terms and supporting auto-terms in time-frequency planes (Choi and Williams, 1989) and many other areas of signal processing. The Gaussian function has also been successfully used as a spatial filter for image enhancement, edge detection and as a notch filter (Gonzalez and Woods, 2002). Thus it is appropriate to employ the Gaussian gray-level transformation for image enhancement, from that its performance can be assessed. In fact, the main aim of this paper is to show that the Gaussian transformation is effective and thus can be used for to improve image contrast by choosing appropriate values of its control parameter σ. Another aim of the paper is to propose a formula to estimate the useful energy in images to automate the image-enhancement performance process.

The paper is organized as follows. Section 2 gives mathematical background of the Gaussian and power-law transformations. A new formula to estimate the useful energy as a function of an image's entropic measurement is also given. Section 3 shows experimental results using both transformations for image enhancement in which six typical images are employed. Relative merits of each transformation are discussed. Section 4 concludes the main findings of the paper and outlines possible further work.

2. The Gaussian Transformation

Mathematically, the Gaussian and power-law transformations are given as

$$T_G(r) = L[1 - \exp(-\sigma r^2)], \tag{1}$$

$$T_p(r) = cr^{\gamma}, \tag{2}$$

where σ, c and γ are control parameters of the transformations respectively, and L the number of gray levels. It should be noted that the larger σ is, the smaller the exponential term and the larger the transformation which effectively brightens images. This is particularly useful for images with dark background such as magnetic resonant images and mammogram which are used for clinical treatment of patients. The Gaussian and power-law transformations are graphically given in Figure 3 from which it is clear that the Gaussian transformation transition is faster than that of the power-law transformation when σ increases which makes the former more useful for contrast enhancement than the latter.

However, as σ becomes large, say $\sigma > 10$, the Gaussian transformation may brighten the image and thus over enhances it as will be seen later. As σ becomes very large, the transformation becomes very sharp which can be used to create binary images. The power-law transformation is also an effective tool for image enhancement as was shown in (Gonzalez and Woods, 2002). By varying γ and c, satisfactory results can be obtained. In this paper, $\gamma = 1.5$ and $c = 1$ are chosen for comparison purposes.

To quantitatively measure the effectiveness of a transformation, an "entropic" measurement formula was used by Munteanu and Rosa (Munteanu and Rosa, 2004)

$$H = \sum_{i=1}^{L} q_i \log(q_i), \tag{3}$$

where q_i is the number of pixels in a gray level. It is assumed that there are $L = 256$ gray levels in images used in this paper. It should be noted that Eq. (3) resembles similar characteristics of an entropy formula with q_i in this case represents the number of pixels in different gray levels of an image. The amount of useful information in a gray level is therefore given by

$$H_g = \frac{\sum_{i=1}^{L} q_i \log(q_i)}{L}. \tag{4}$$

The energy of an image can be considered as the total energy sum of its pixels, which is given by

$$E = \sum_{j=1}^{MN} m_j^2, \tag{5}$$

where m_j is the value of pixel number j and there are MN pixels in an image of size $M \times N$.

Eq. (5) represents the total energy of an image, the energy per pixel in an image of size $M \times N$ is thus given by

$$E_p = \frac{\sum_{j=1}^{MN} m_j^2}{MN}. \tag{6}$$

The useful energy of images can then be defined as the product of the useful information as given in Eq. (4), and the energy per pixel as given in Eq. (6), and is therefore given by

$$E_u = \left(\frac{\sum_{j=1}^{MN} m_j^2}{MN} \right) \left(\frac{\sum_{i=1}^{L} q_i \log(q_i)}{L} \right). \tag{7}$$

The useful information in an image shows how effective its information can be represented and thus the effectiveness of the transformation used to enhance it. It is important to realize that E_u is dependent on a transformation used to enhance an image because the transformation effectively changes the number of pixels in gray levels. It should also be noted that Eq. (7) can be employed as a performance indicator and hence it can be used to automate the image-enhancement performance evaluation process. In this paper, the effectiveness of image enhancement employing the Gaussian and power-law transformations is assessed using two methods: (1) by human image experts, and (2) by quantitatively using Eq. (7). It will be shown later that the two methods yield consistent results which show that the Gaussian transformation is an effective gray-level transformation for image enhancement in the spatial domain.

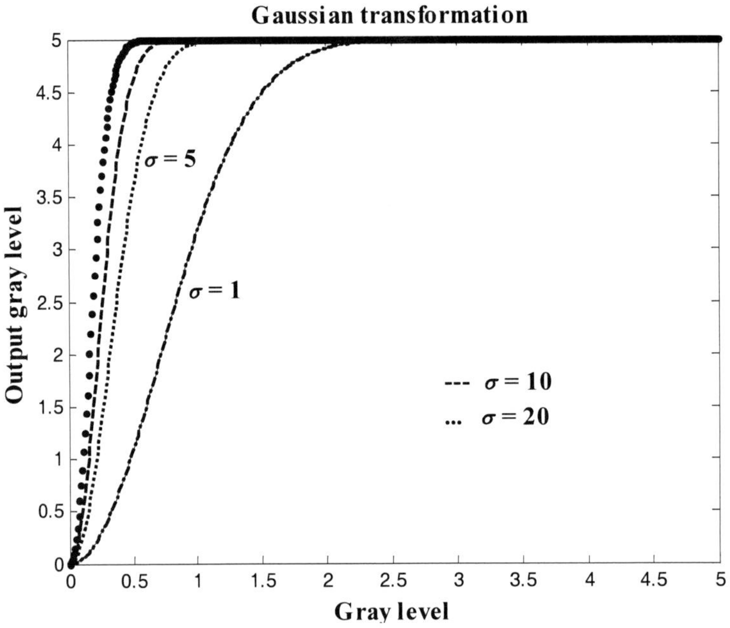

Figure 3. The Gaussian transformations for $\sigma = 1, 3, 10$ and 20.

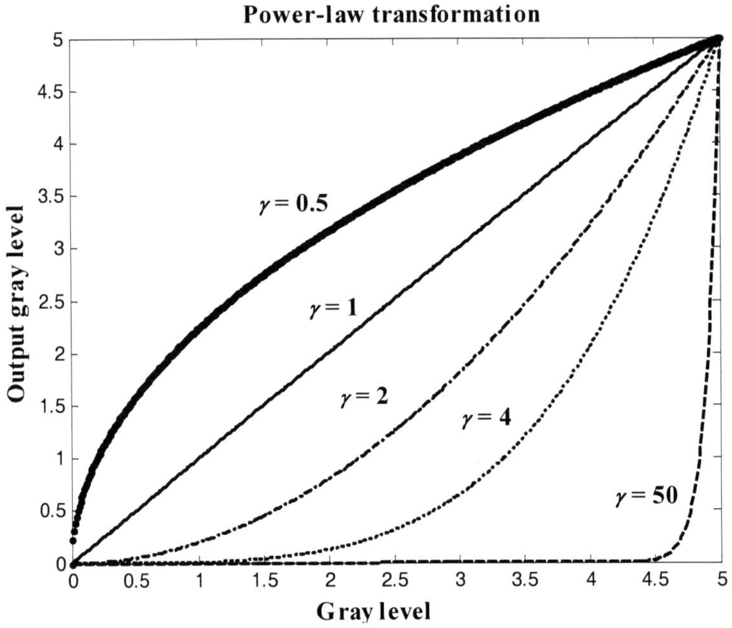

Figure 4. The power-law transformations for $\gamma = 0.5, 1, 2, 4$ and 50.

3. Experimental Results

This section presents experimental results using the Gaussian transformation with $\sigma = 1, 4, 5$ and 10, and the power-law transformation with $c = 1$ and $\gamma = 1.5$. The power-law transformation is employed in this paper as a performance comparison benchmark and it is not the main aim of this paper to thoroughly compare the performance of the Gaussian and power-law transformations. The histograms of the original image and transformed images using the two transformations are also given for comparison purposes. It should be noted that the more uniform the histogram, the higher the contrast level in images, and thus the more effective the transformation (Raji et al., 1998; Gonzalez and Woods, 2002; Munteanu and Rosa, 2004). Six different images are used in this section: (1) "pout" image, (2) "eight" image, (3) "mri" image, (4) "tire" image, (5) "cameraman" image and (6) "autumn" image. Relative performance merits of each transformation are discussed.

For $\sigma = 4$, image enhancement using both transformations are given in Figure 5 to Figure 10 for the six images. From Figure 5(a), it should be noted that the original image is blurry and some details in the background are darkened. The histogram in Figure 5(b) of the original image is non-uniform with the majority of pixels are in the middle region of the gray levels. The power-law transformation brightens the left-hand side area of the image, resulting in a more uniform histogram as seen in Figure 5(c) and Figure 5(e) respectively.

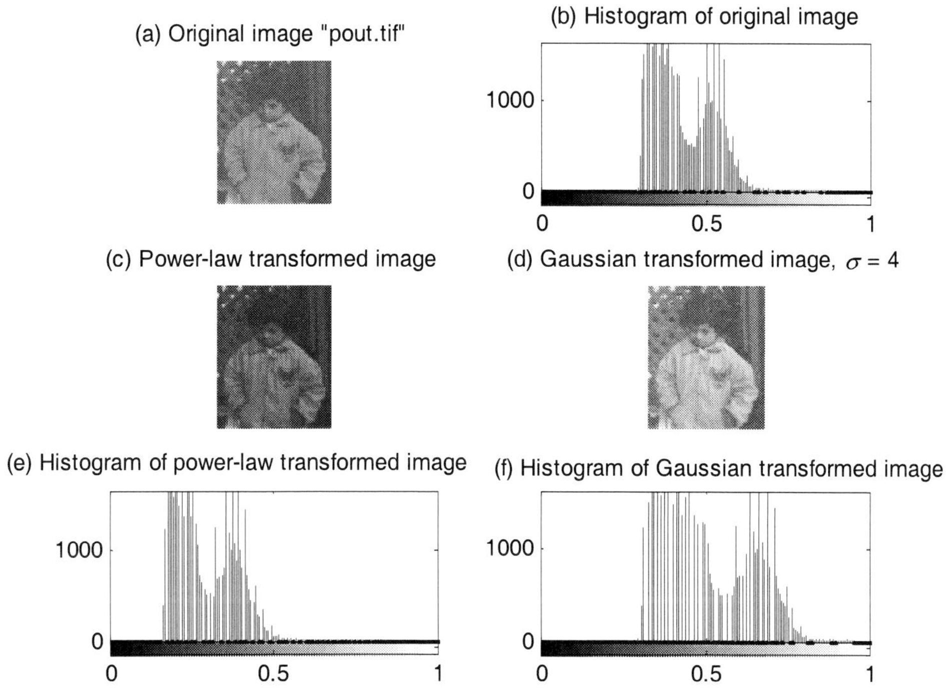

Figure 5. Image enhancement of Image 1 with $\sigma = 4$.

Even though slight improvement can be seen in Figure 5(c) using the power-law transformation, however, the image background is still dark, especially, the fine details of the fence behind the child cannot clearly be seen. The Gaussian transformed image in Figure 5(d) shows significant improvements by revealing fine details of the fence, the right-hand side wooden column and face expression of the child which were hidden in the background and could not be revealed by using the power-law transformation. The histogram of the Gaussian transformed image shows a wider pixel distribution than those of the power-law transformed image and the original image, which shows the effectiveness of the Gaussian transformation for image enhancement.

From Figure 6(a), the second original image consists of four circular objects whose details are not clearly displayed. Its histogram shows a large number of pixels located in the middle region of the input gray levels as seen in Figure 6(b). Figure 6(c) shows the power-law transformed image whose histogram (Figure 6(e)) is markedly similar to that of the original image even though it is slightly more uniform. From Figure 6(c), the power-law transformed image appears to be darker than the original image which means that this transformation is not very effective to enhance this particular image as also evidenced by Figure 6(e). The Gaussian-transformed image in Figure 6(d) shows improvement by revealing more fine details in the centers of the four circular objects. It is also clear that the Gaussian transformation can effectively enhance the image in this case as its histogram is wider than those of the original image and power-law transformed image as shown in Figure 6(b) and Figure 6(e) respectively.

A magnetic resonant scan of a human head is shown in Figure 7(a) as the original image which is severely darkened and cannot be used for clinical treatment purposes. The background of the image is dark and many of its fine details are possibly masked by noise and other disturbances.

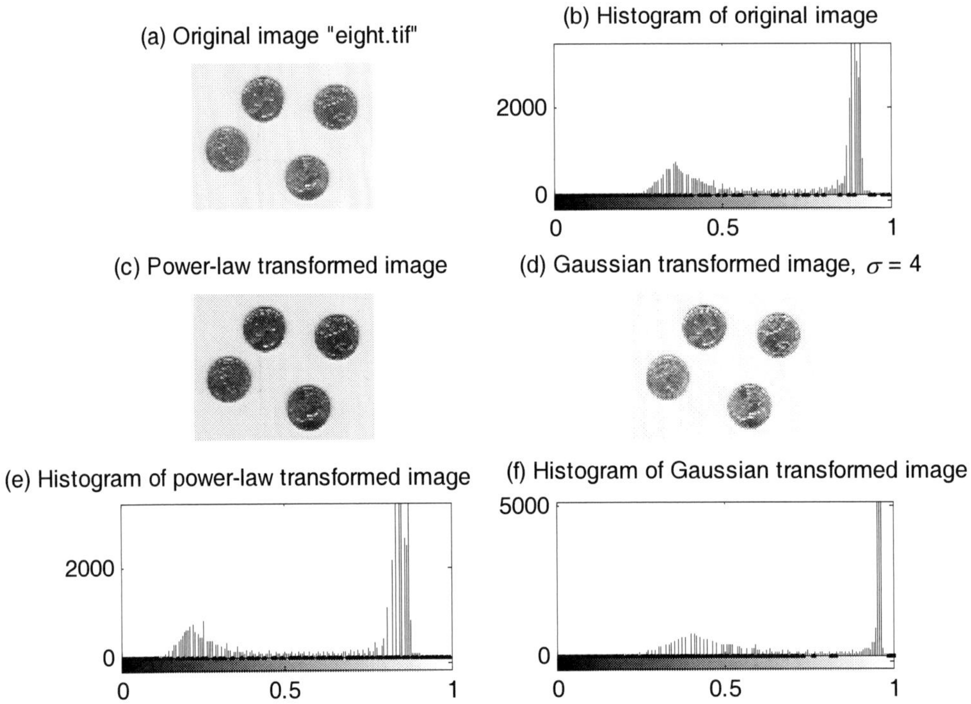

Figure 6. Image enhancement of Image 2 with $\sigma = 4$.

It should be noted that for this particular image, the main task is to reveal the fine details of the scan which is located in the middle of the image on a black background, and not to recover its dark background which is commonly present in magnetic resonant scans. The original image's histogram shows that a large number of pixels are narrowly distributed in the lower region of the gray levels, yielding an over-darkened image as seen in Figure 7(a). Similar to Figure 6(c), the power-law transformed image given in Figure 7(c) appears to be darker than the original image which suggests that the transformation is not very effective to enhance this particular image. The transformed image's histogram is also significantly narrower than that of the original image as seen in Figure 7(e).

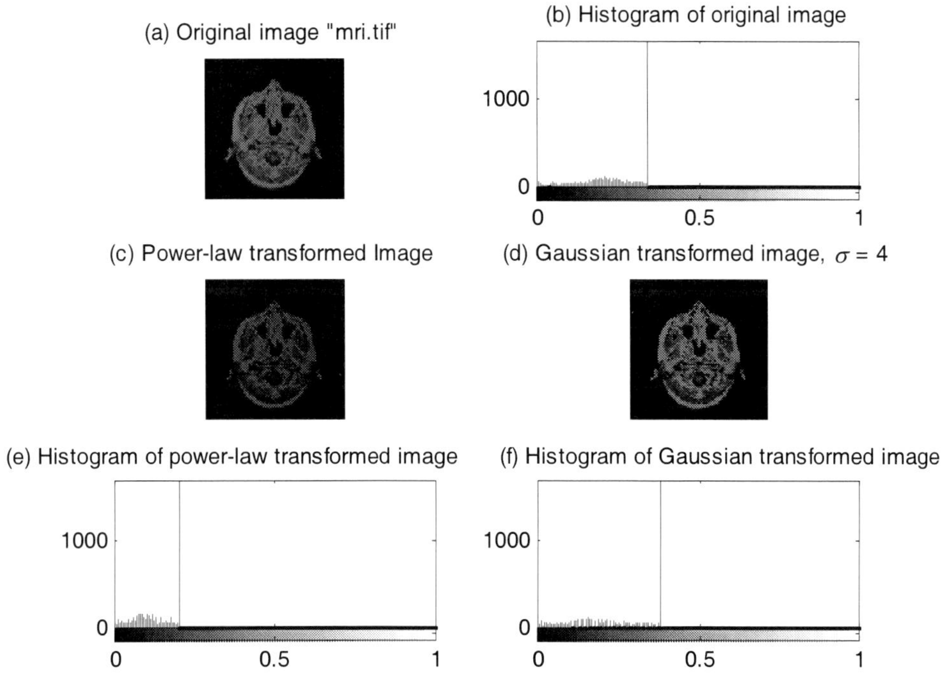

Figure 7. Image enhancement of Image 3 with σ = 4.

By comparing Figure 6 and Figure 7, it is clear that the power-law transformation performance is worse than when it is used to enhance the Image 2 which suggests that this transformation are not very effective for magnetic resonant scans when $c = 1$ and $\gamma = 1.5$. The Gaussian-transformed image is given in Figure 7(d) in which improvements from the original image can be clearly seen by revealing more fine details of the scan. It should be noted that the background in the transformed image is still very similar to that of the original image, suggesting that it is not possible for further enhancement as explained earlier. The histogram of the Gaussian-transformed image is also more widely distributed than those of the original image and power-law transformed image which consistently shows consistent performance of the Gaussian transformation for image enhancement.

For Image 4 as seen in Figure 8(a), the power-law transformed image (Figure 8(c)) is darker than the original image, whereas the Gaussian-transformed image (Figure 8(d)) is brighter with a more uniform histogram as seen in Figure 8(f). Since the original image's histogram is widely distributed, it is expected that both transformations may not yield very significant improvement over the original image.

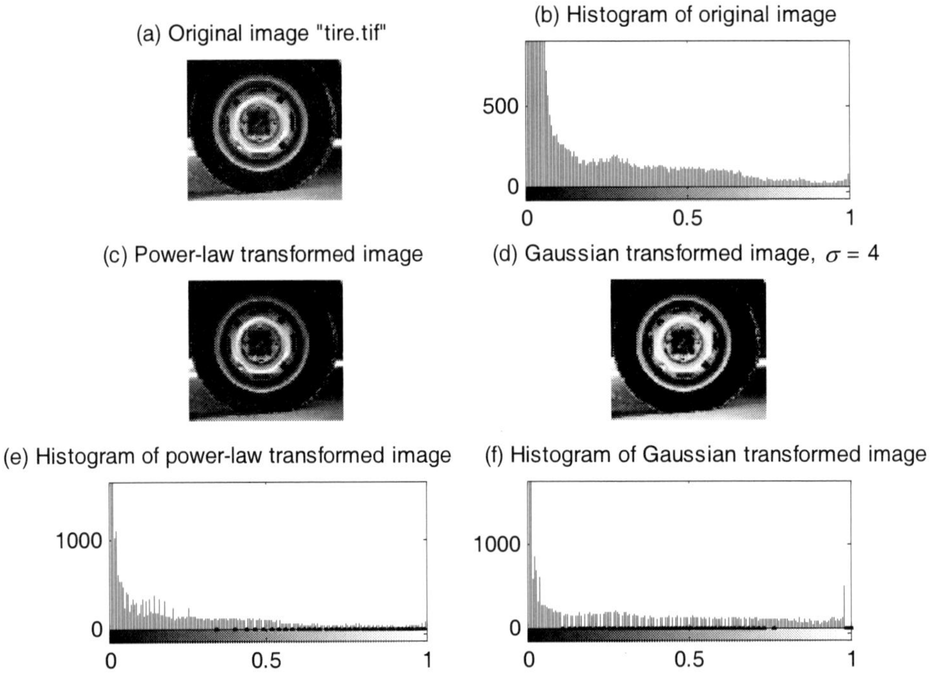

Figure 8. Image enhancement of Image 4 with $\sigma = 4$.

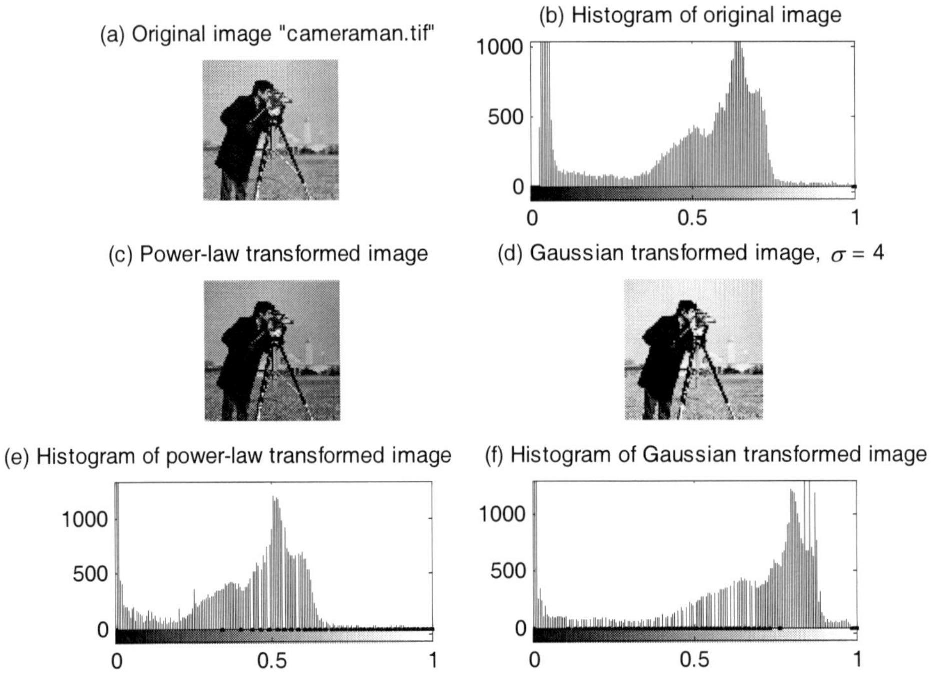

Figure 9. Image enhancement of Image 5 with $\sigma = 4$.

In fact, it should be noted that the original image has a slightly more uniform histogram than that of the Gaussian-transformed image, however, there are considerably more pixels locating in the lower region of the gray levels, yielding a darker original image. It is also clear that the Gaussian-transformed image can reveal more hidden details and better contrast in the top half of the image as evidently seen in Figure 8(d).

Figure 9(a) shows the original image along with two transformed images using the power-law and Gaussian transformations as shown respectively in Figure 9(c) and Figure 9(d). It is clear that the power-law transformed image appears slightly darker than the original image with a narrow histogram, whereas the Gaussian-transformed image is brighter with a broader and more uniform histogram. Both transformations in this case can reveal fine details in the background which could not be clearly seen in the original image. In addition, the Gaussian transformation yields a brighter background than the power-law transformation. Besides, the Gaussian transformation creates a better-contrast image than the original and power-law transformed images by more evenly and widely spreading pixels over the gray levels as clearly shown in Figure 9(b), Figure 9(e) and Figure 9(f).

Figure 10(a) shows the original gray-scale "autumn" image which was converted from its colorful version in MATLAB. In this case, the power-law transformed image is very much similar to the original image with a slightly more uniform histogram. The Gaussian-transformed image is brighter than both images shown in Figure 10(a) and Figure 10(c) in which finer details of the shrubs and bushes in the image background can be revealed. The Gaussian-transformed image's histogram is also more uniform than those of the original and power-law transformed images.

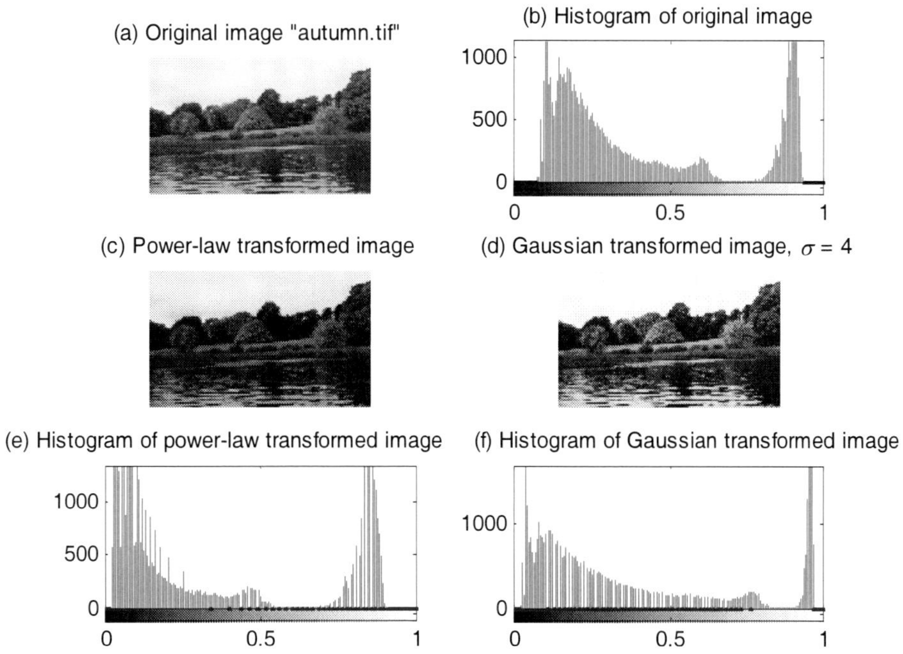

Figure 10. Image enhancement of Image 6 with $\sigma = 4$.

Figure 11 to Figure 16 show image enhancement using the Gaussian transformation with $\sigma = 5$ and power-law transformation in which similar results to those in the case of $\sigma = 4$ are obtained. The Gaussian-transformed images are brighter than those obtained when $\sigma = 4$.

Histograms of the transformed images are narrower than when $\sigma = 4$ as explained earlier in Section 2.

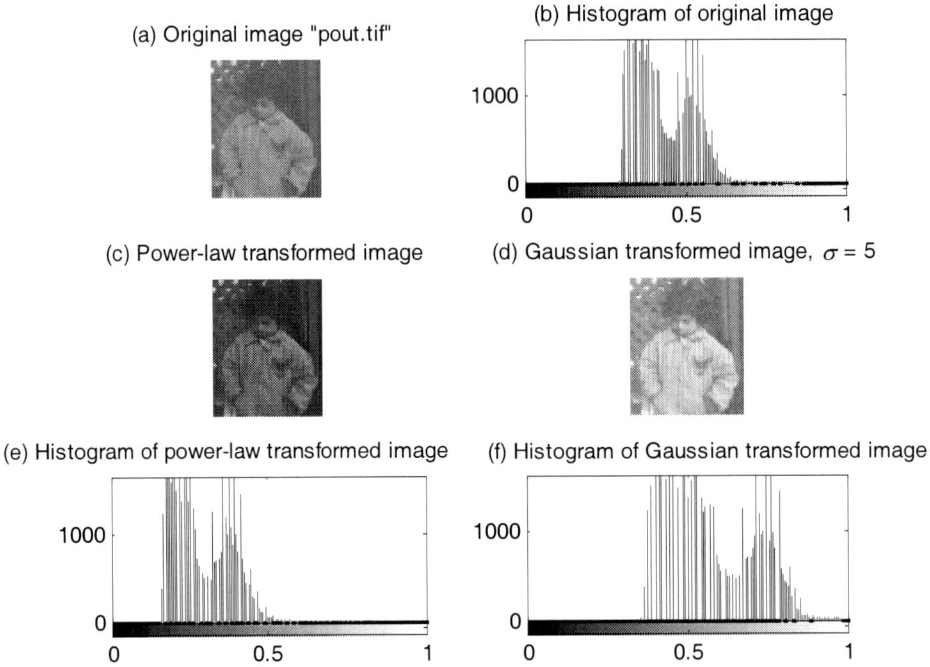

Figure 11. Image enhancement of Image 1 with $\sigma = 5$.

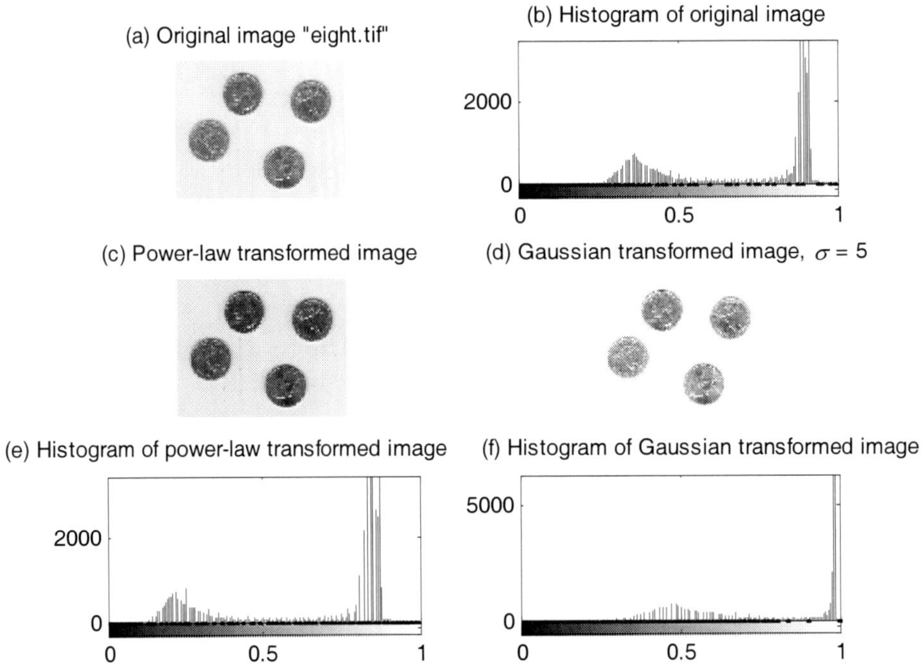

Figure 12. Image enhancement of Image 2 with $\sigma = 5$.

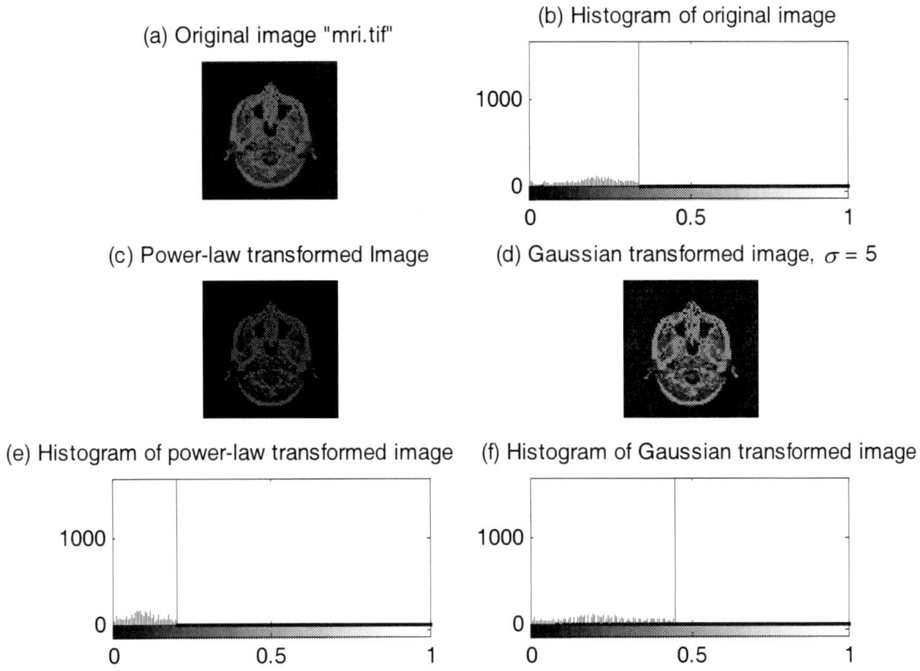

Figure 13. Image enhancement of Image 3 with $\sigma = 5$.

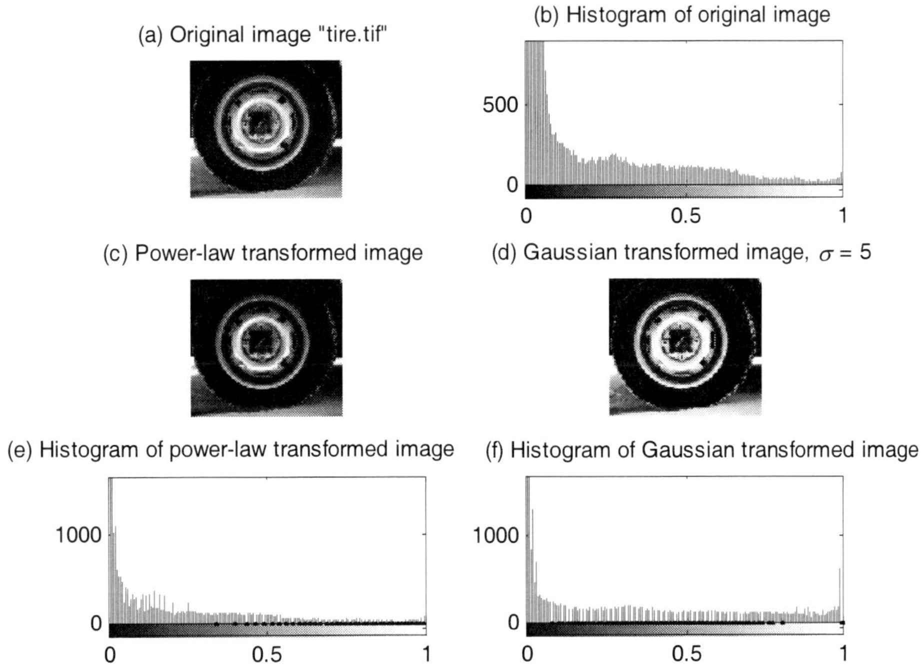

Figure 14. Image enhancement of Image 4 with $\sigma = 5$.

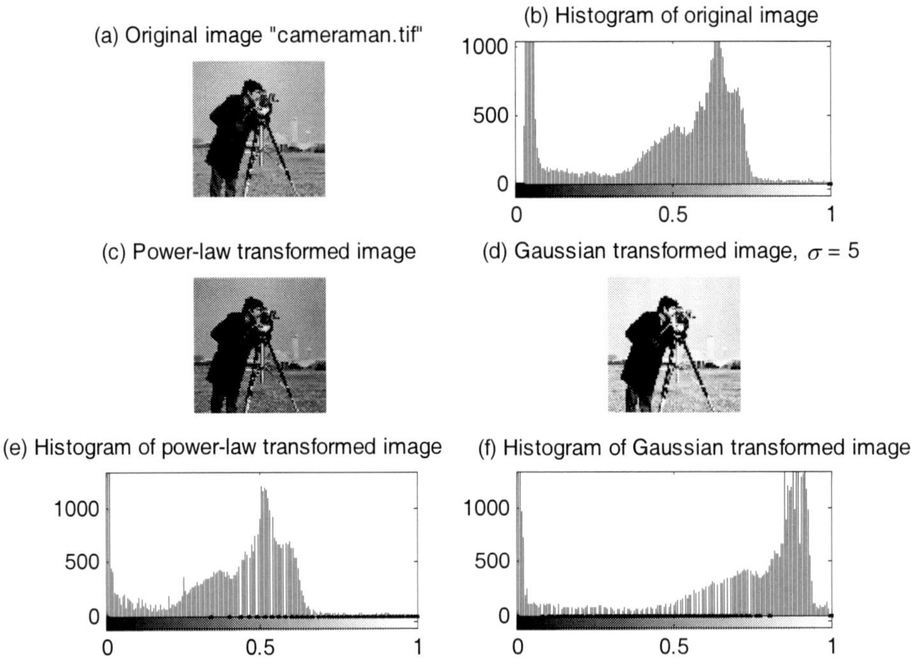

Figure 15. Image enhancement of Image 5 with $\sigma = 5$.

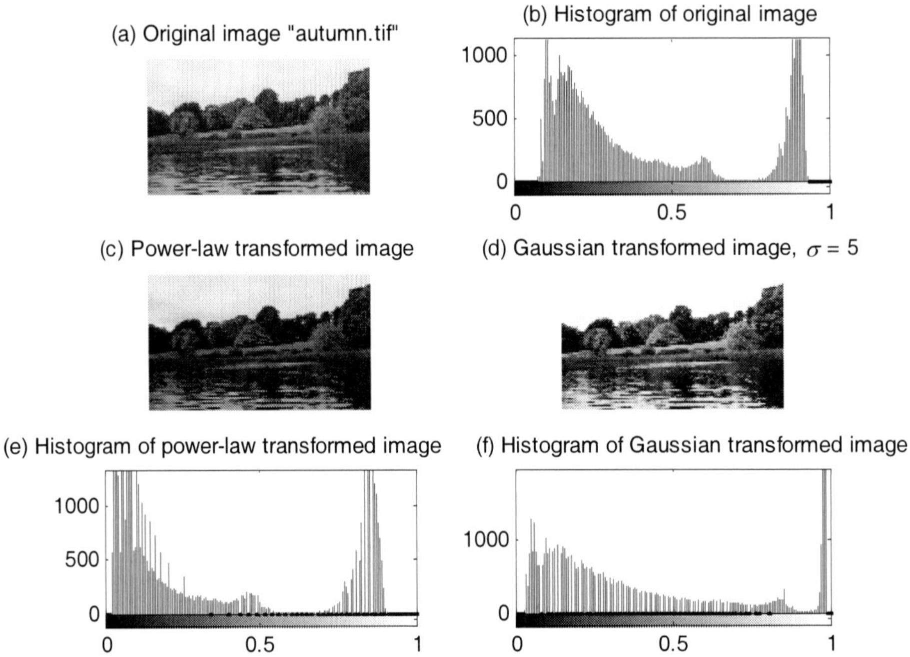

Figure 16. Image enhancement of Image 6 with $\sigma = 5$.

To further assess the performance of the Gaussian transformation for image enhancement, $\sigma = 1$ and 10 are chosen. For $\sigma = 1$, the transformation darkens the images and thus it is not effective as seen in Figure 17 to Figure 22.

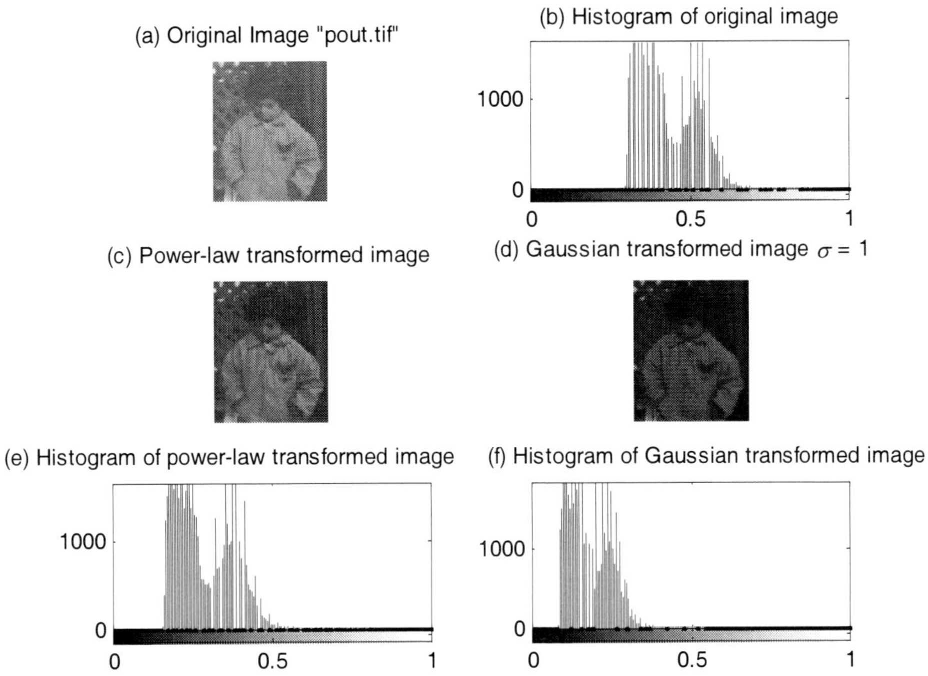

Figure 17. Image enhancement of Image 1 with $\sigma = 1$.

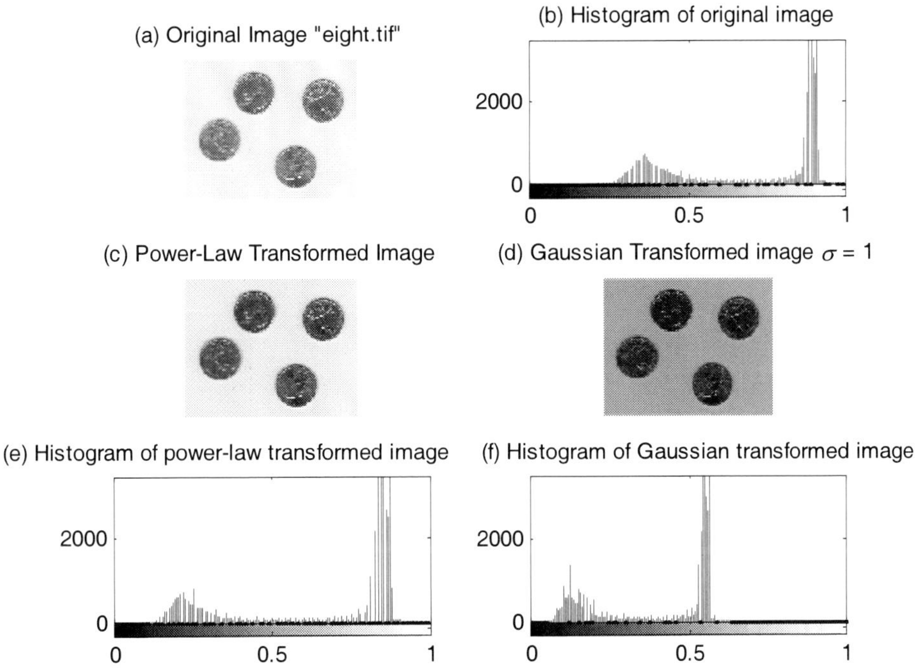

Figure 18. Image enhancement of Image 2 with $\sigma = 1$.

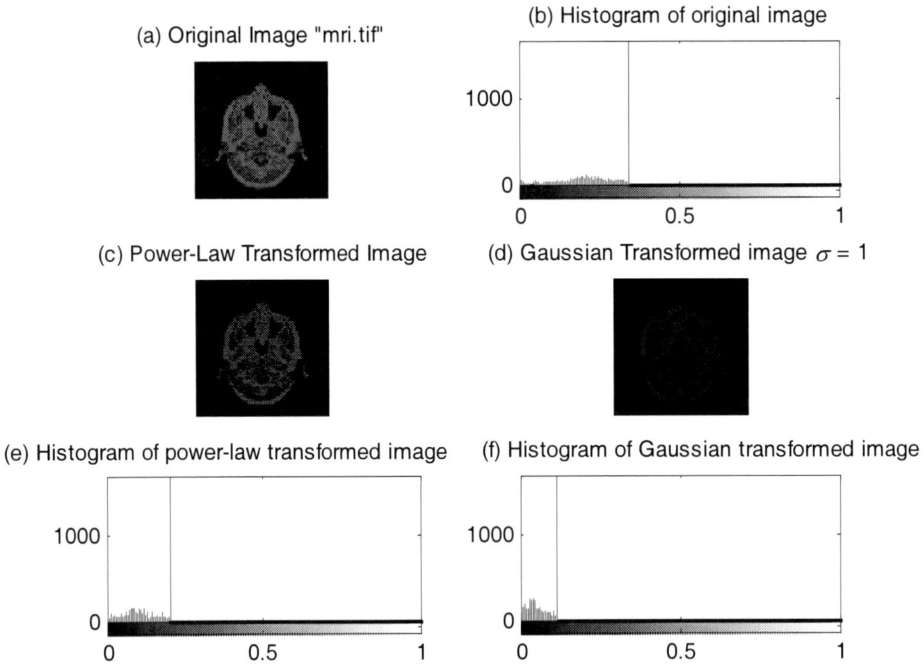

Figure 19. Image enhancement of Image 3 with $\sigma = 1$.

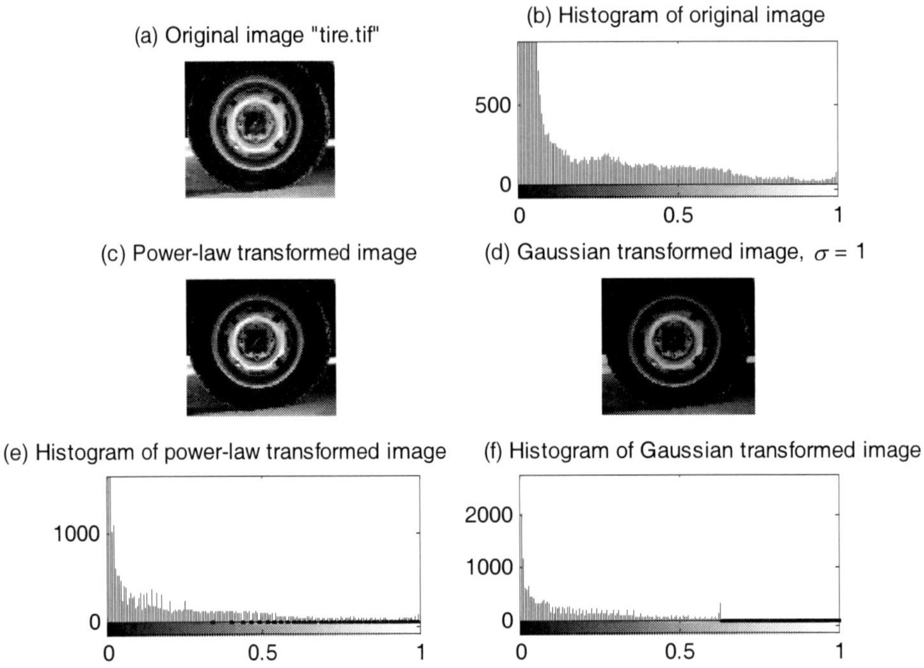

Figure 20. Image enhancement of Image 4 with $\sigma = 1$.

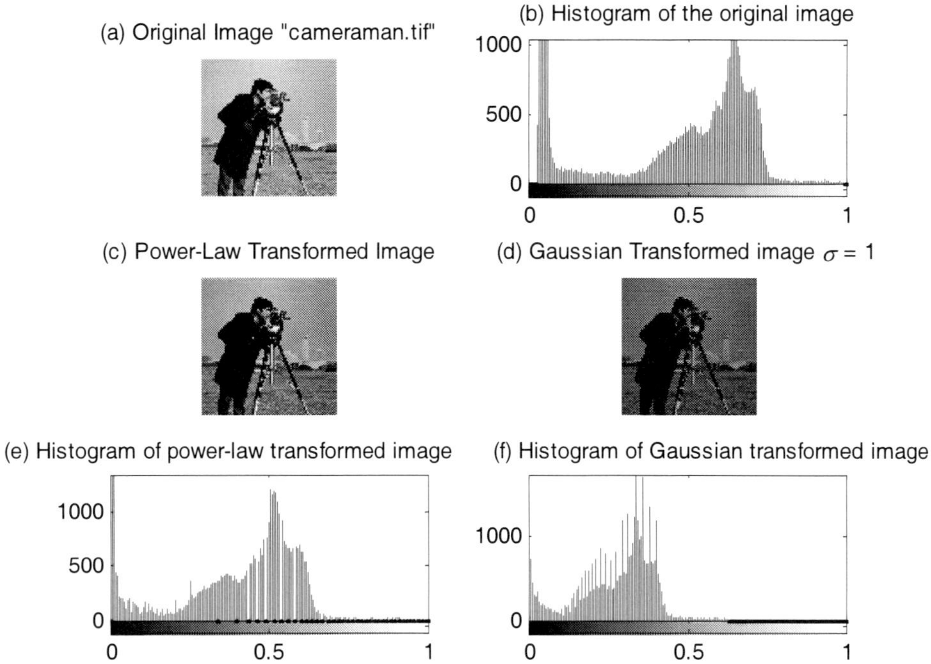

Figure 21. Image enhancement of Image 5 with $\sigma = 1$.

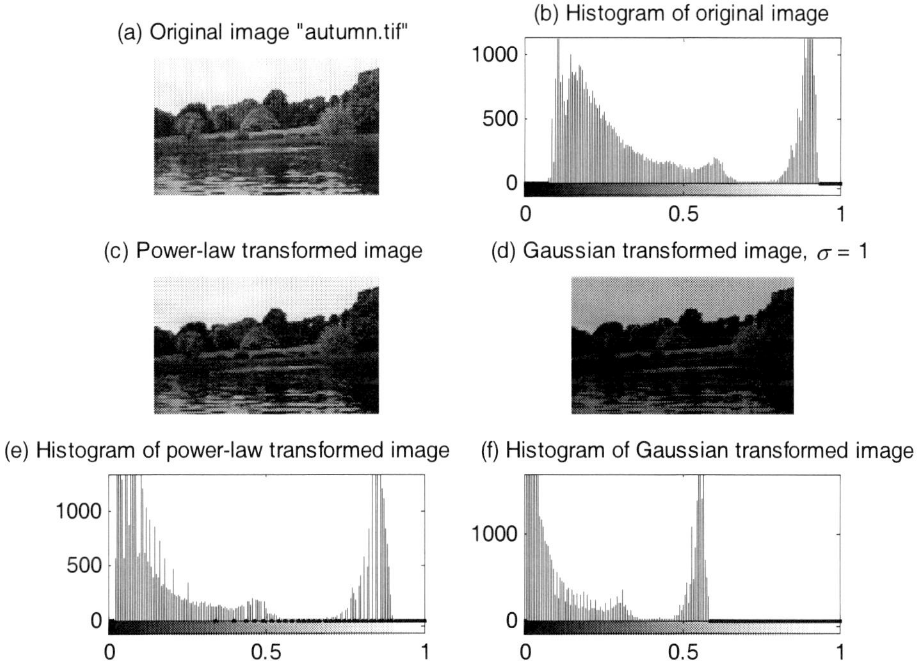

Figure 22. Image enhancement of Image 6 with $\sigma = 1$.

For $\sigma = 10$, the transformation over-brightens the images which are also not satisfactory as many fine details are masked as seen in Figure 23 to Figure 24. However, for the "mri" image as seen in Figure 25 and Figure 26, more details can be revealed by using $\sigma = 10$ which suggests that higher values of σ can be used to enhance magnetic resonant images.

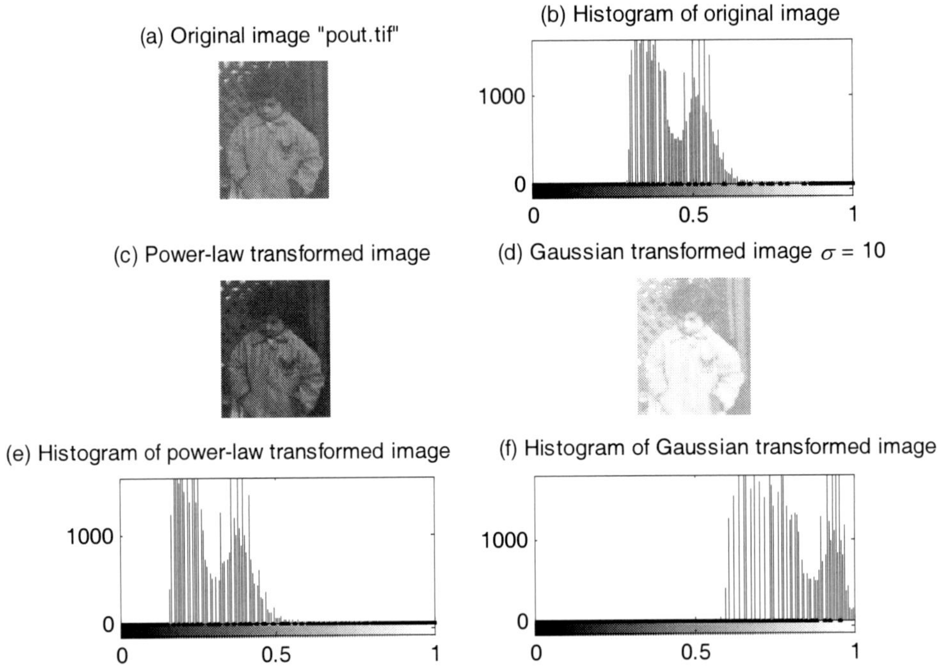

Figure 23. Image enhancement of Image 1 with $\sigma = 10$.

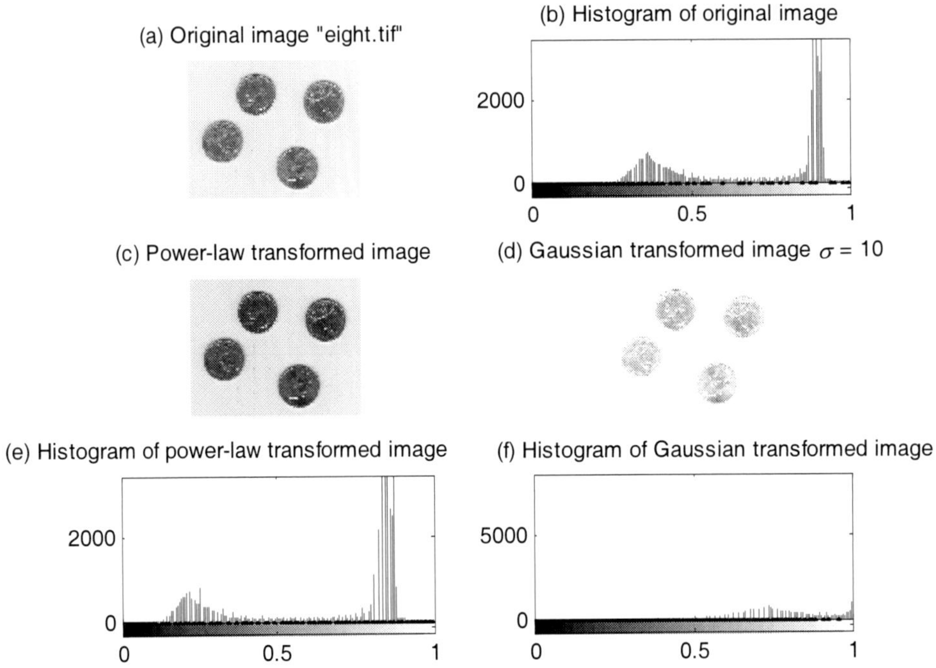

Figure 24. Image enhancement of Image 2 with $\sigma = 10$.

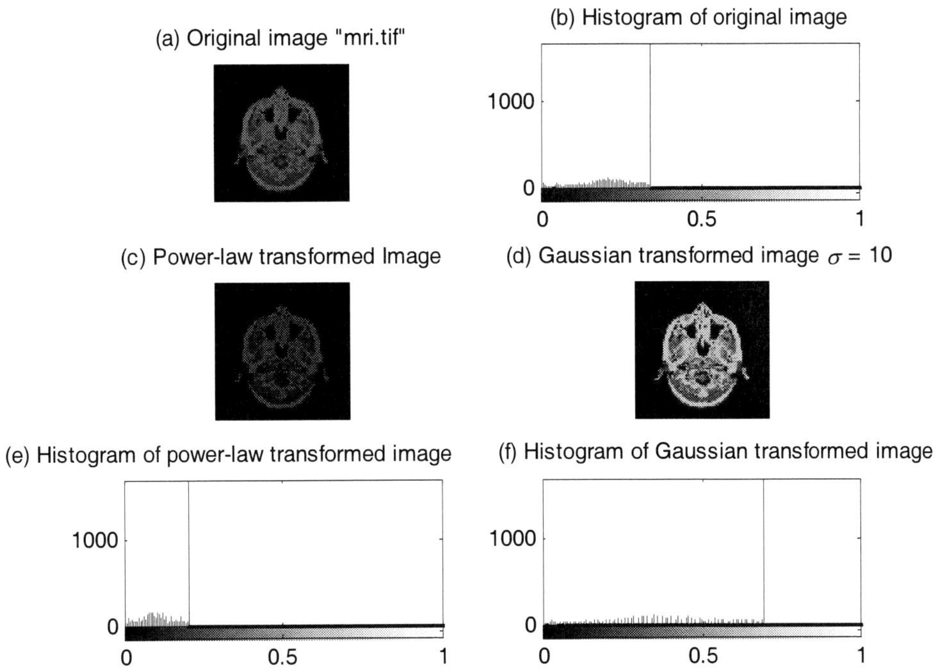

Figure 25. Image enhancement of Image 3 with σ = 10.

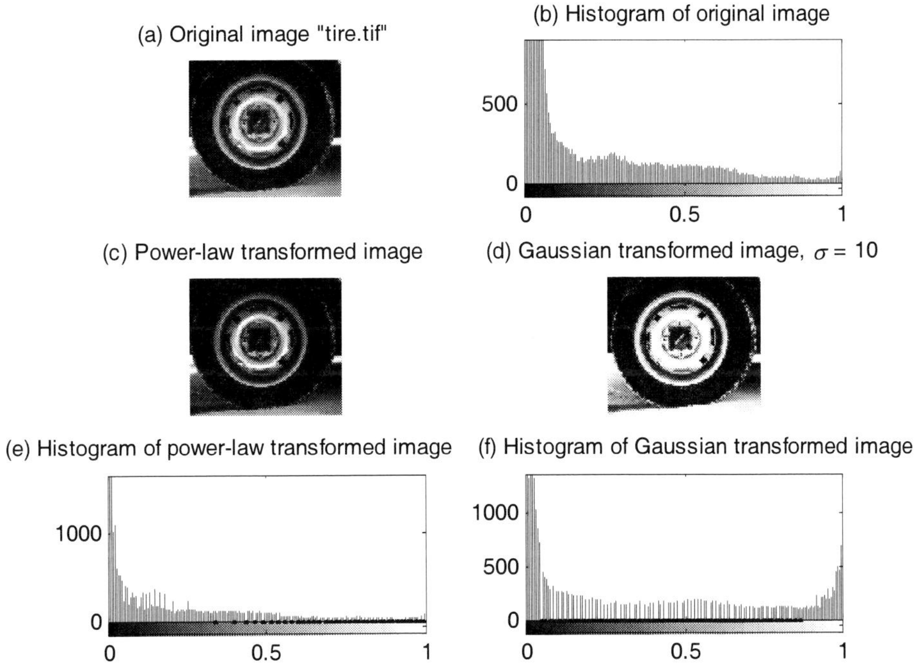

Figure 26. Image enhancement of Image 4 with $\sigma = 10$.

From Figure 27 and Figure 28, satisfactory performances are obtained based on assessment by image experts.

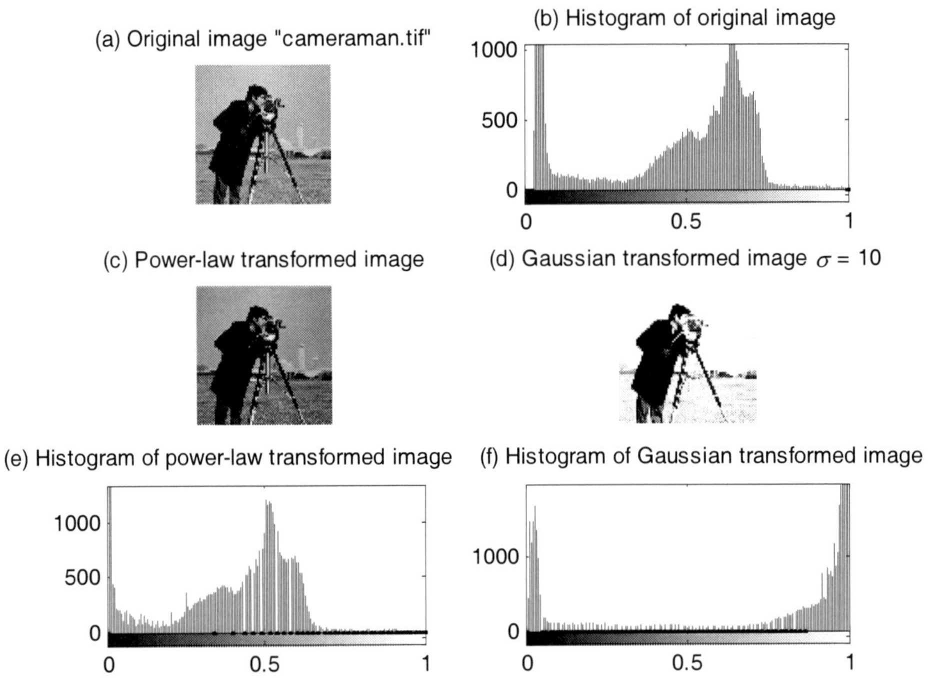

Figure 27. Image enhancement of Image 5 with $\sigma = 10$.

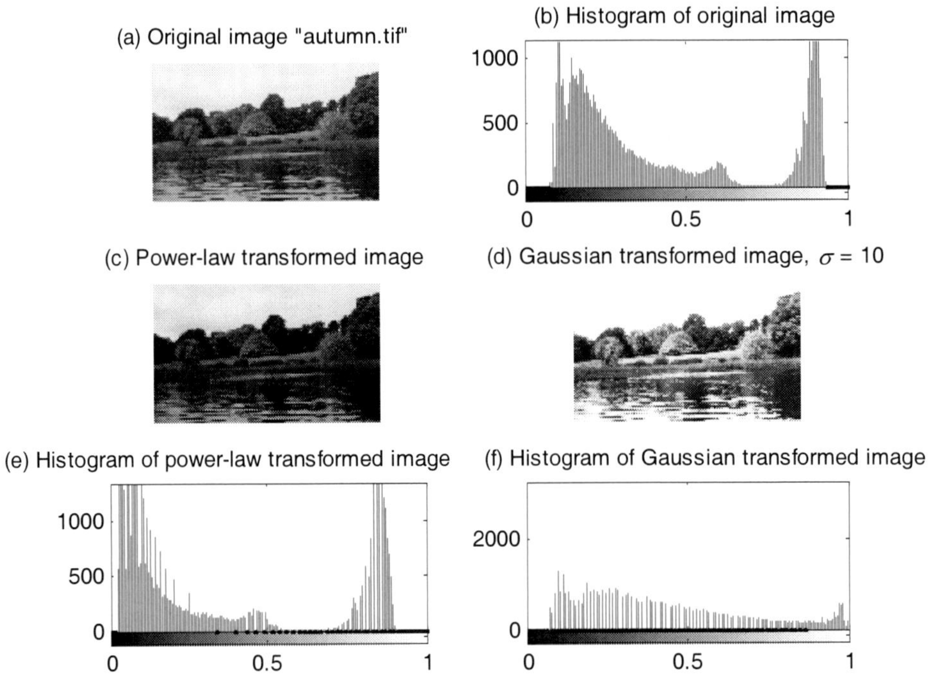

Figure 28. Image enhancement of Image 6 with $\sigma = 10$.

However, it is difficult to recognize fine differences between $\sigma = 4$ and $\sigma = 10$ which shows the necessity and importance of using an automatic performance indicator to assess image quality. In general, the range of $1 < \sigma \leq 10$ can be considered as the optimal range to use the Gaussian transformation for image enhancement. In particular, $\sigma = 4$ and 5 yield satisfactory results and should be typically employed. To quantitatively compare the performance of the two transformations, by using Eq. (7), the useful energy of original images, power-law transformed images and Gaussian-transformed images are plotted in Figure 29 to Figure 32 for $\sigma = 1$, 4, 5, and 10 respectively in which it is evident that the Gaussian transformation is the most effective because of its highest useful energy.

As expected, the useful energy of the original images is the lowest compared to that of the transformed images. The power-law transformation with $c = 1$ and $\gamma = 1.5$ is less effective than the Gaussian transformation except for the "pout" image and "mri" image in which the useful energy of both transformations are almost identical. It should also be noted that by noting the useful energy given in Figure 29 to Figure 32 for $\sigma = 1$, 4, 5, and 10 respectively, the effectiveness of the Gaussian and power-law transformations can be judged. It is important to realize that outcomes of the image enhancement performance evaluation process assessed by image experts and by using the useful energy given by Eq. (7) are consistent, which means that human interpretation may not be required (Munteanu and Rosa, 2004). This makes the elusive goal of assessing image quality an easier task.

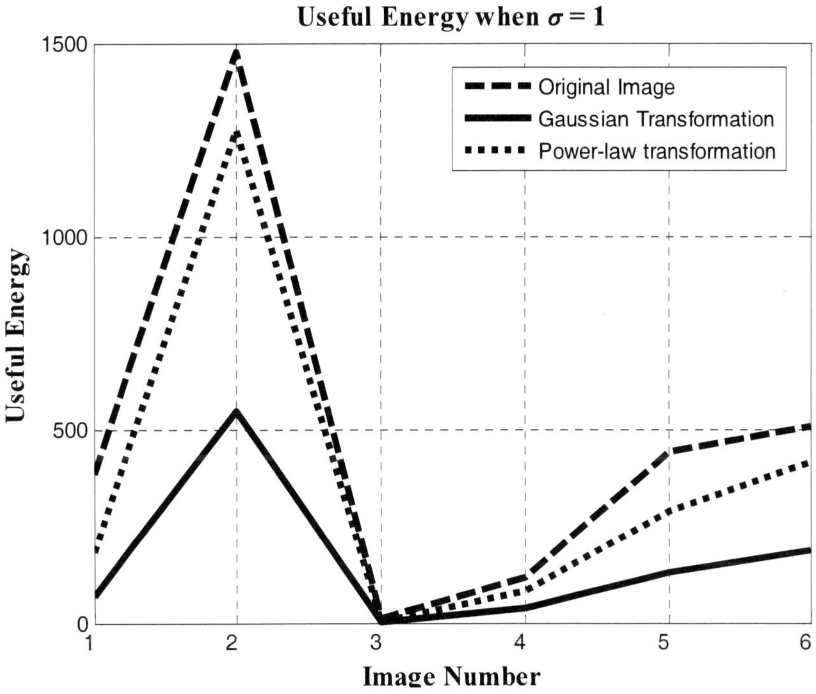

Figure 29. Useful energy of images using Eq. (7) when $\sigma = 1$.

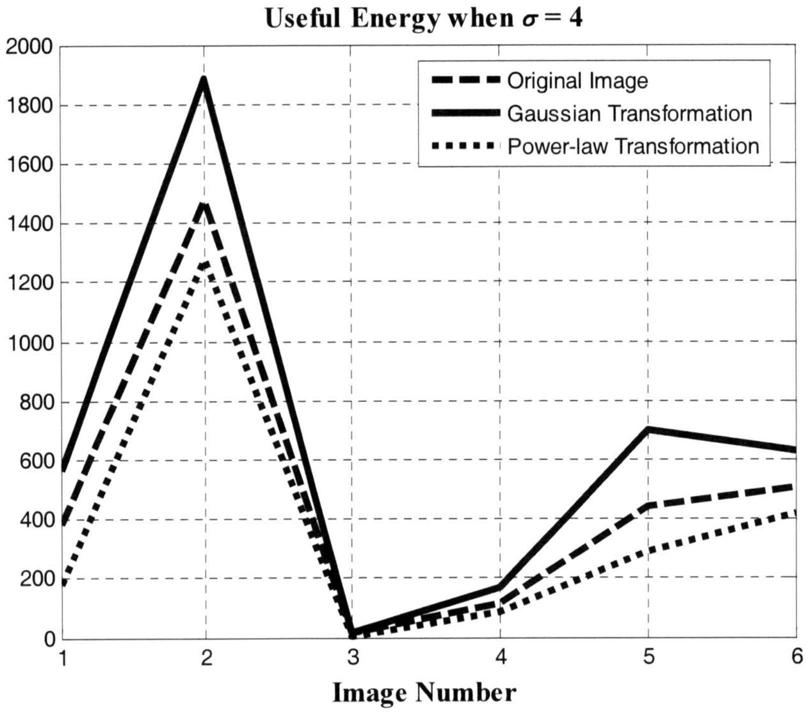

Figure 30. Useful energy of images using Eq. (7) when $\sigma = 4$.

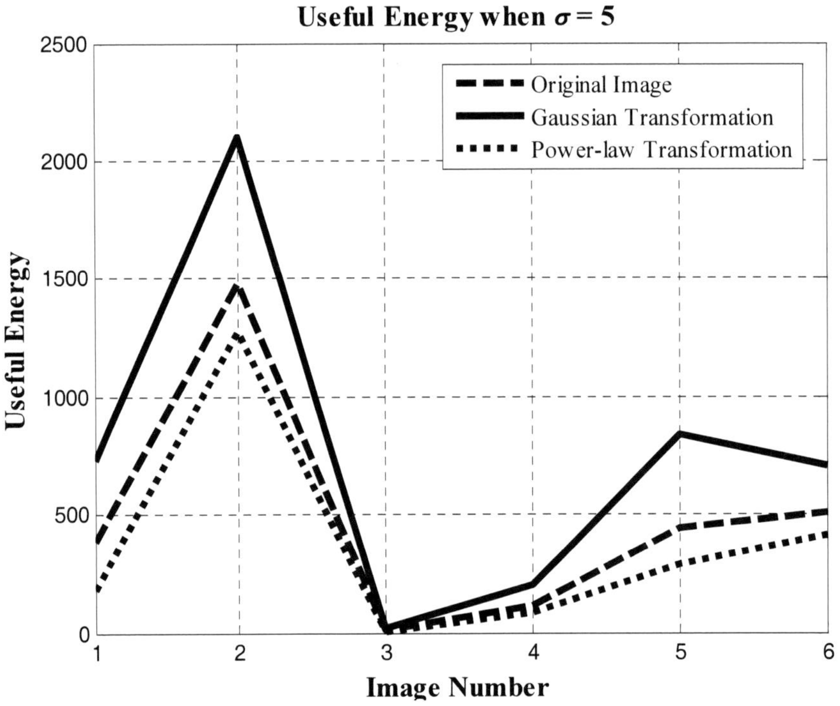

Figure 31. Useful energy of images using Eq. (7) when $\sigma = 5$.

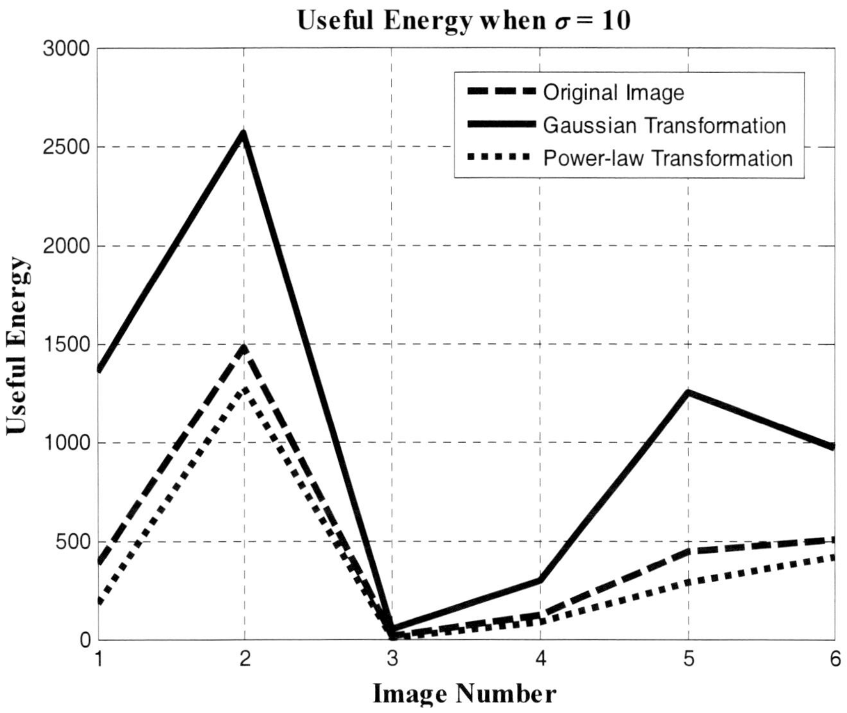

Figure 32. Useful energy of images using Eq. (7) when $\sigma = 10$.

4. Conclusion and Possible Future Work

The Gaussian transformation for image enhancement was proposed which has been shown to be more effective than the power-law transformation for $1 < \sigma \leq 10$. A formula to estimate the useful energy of images based on their useful entropic measurement and total energy has been proposed as a performance indicator for gray-level transformations in image enhancement in which it has been shown that the formula is effective. Performance of gray-level transformations has also been assessed by image experts in which consistent results have been obtained.

Possible further work can be carried out by employing the Gaussian probability density for histogram equalization and histogram matching. A family of Gaussian transformations can also be used for image enhancement, which is mathematically given as

$$T_{Gh}(r) = L[1 - \exp(-\sigma r^2)]^n, \tag{8}$$

where n is the order of the family.

Another closely-related transformation to the Gaussian transformation is a hyperbolic transformation, which has been shown to be more effective than the Gaussian distribution in tracking, edge detection, auto-term support and cross-term suppression in the time-frequency plane, and it can be employed for image enhancement. It should be noted that six typical images of different types have been enhanced in this paper. More images can also be tested to provide a more comprehensive conclusion on the effectiveness of both transformations by estimating images' useful energy using Eq. (7).

References

Bishnu, A., Bhattacharya, B. B., Kundu, M. K., Murthy, C. A. and Acharya, T., 2005. Euler Vector for Search and retrieval of Gray-Tone Images. *IEEE Transactions on Systems, Man, and Cybernetics--Part B: Cybernetics*, vol. 35, pp. 801-812.

Chang, C. C. and Chuang, J. C., 2002. An image intellectual property protection scheme for gray-level images using visual secret sharing strategy. *Pattern Recognition Letters*, vol. 23, pp. 931-941.

Chen, Z., Abidi, B. R., Page, D. L. and Abidi, M. A., 2006. Gray-Level Grouping (GLG): An Automatic Method for Optimized Image Constrast Enhancement -- Part I: The Basic Method. *IEEE Transactions on Image Processing*, vol. 15, pp. 2290-2302.

Chen, Z., Abidi, B. R., Page, D. L. and Abidi, M. A., 2006. Gray-Level Grouping (GLG): An Automatic Method for Optimized Image Contrast Enhancement -- Part II: The Variations. *IEEE Transactions on Image Processing*, vol. 15, pp. 2303-2314.

Choi, H. I. and Williams, W. J., 1989. Improved time-frequency representation of multi-component signals using exponential kernels. *IEEE Transactions on Signal Processing*, vol. 37, pp. 862-71.

Gonzalez, R. C. and Woods, R. E. (2002). *Digital Image Processing*. New Jersey, Prentice Hall, Upper Saddle River.

Mukhopadhyay, S. and Chanda, B., 2001. Fusion of 2D grayscale images using multiscale morphology. *Pattern Recognition*, vol. 34, pp. 1939-1949.

Munteanu, C. and Rosa, A., 2004. Gray-scale image enhancement as an automatic process driven by evolution. IEEE Transactions on Systems, Man, and Cybernetics---Part B: *Cybernetics*, vol. 34, pp. 1292-1298.

Ojala, T., Pietikainen, M. and Maenpaa, T., 2002. Multiresolution Gray-Scale and Rotation Invariant Texture Classification with Local Binary Patterns. *IEEE Transactions on Pattern Analysis and Machine Intelligence*, vol. 24, pp. 971-987.

Raji, A., Thaibaoui, A., Petit, E., Bunel, P. and Mimoun, G., 1998. A gray-level transformation-based method for image enhancement. *Pattern Recognition Letters*, vol. 19, pp. 1207-1212.

In: New Signal Processing Research
Editor: Takumi Maeda, pp. 369-387
ISBN: 978-1-60456-479-2

Chapter 11

A Hyperbolic Scatterer Distribution in Multi-Path Mobile Environments

Khoa N. Le** and Kishor P. Dabke**
Centre for Wireless Monitoring and Applications, Griffith School of Engineering,
Griffith University, Gold Coast, QLD 4215, Australia
Monash University, Department of Electrical and Computer Systems Engineering,
Melbourne, Clayton Campus, VIC. 3168, Australia

Abstract

In multi-path mobile environments, signal transmission between a base station (BS) and user equipment (UE) is usually complicated because the signals are blocked, reflected or refracted by random scatterers. These scatterers can be modelled by using a suitable distribution. An angle-of-arrival (AoA) probability density function (pdf) based on the scatterer distribution is used to describe features of signal transmission between the BS and the UE. The Gaussian AoA pdf was shown to be effective in modelling the scatterers [1]. The AoA pdf in which the scatterer distribution obeys a hyperbolic sech function is derived in this paper. Approximate hyperbolic and Gaussian AoA pdfs under small and large AoA conditions are also given. Power azimuth spectra (PAS) and entropy of the approximate pdfs as functions of the angular spread are computed. Relative merits of each pdf are discussed. Possible further work is also outlined.

Key words: Gaussian distribution, entropy, angle-of-arrival pdf, mobile antenna, hyperbolic distribution

1. Introduction

Consider the process of transmitting and receiving signals between the BS and the UE's antenna. This is a complicated problem as there are often random scatterers between them, effectively diverting the antenna signals from and to the BS and UE's antenna. It is thus clear

* E-mail address: K.Le@griffith.edu.au; Tel. (+617) 5552 9175, Fax. (+617) 5552 8065;
** E-mail address: Kishor.Dabke@eng.monash.edu.au

that signals transmitted via this medium travel in a multi-path fashion in which "ghost" signals may be detected. To rectify this problem, it can be assumed that the received signal at the BS is the sum of all arriving signals after being reflected and scattered by the random scatterers. It is clear that the scatterer actions are random and they obey a distribution which can be used to estimate their probability of occurrence.

Estimation of azimuthal dispersion is important in mobile environment for which a number of models have been proposed in the literature [2-4] under the assumption that the received signals at the BS transmitted from the UE's antenna are uniformly distributed. The common feature of these models is that they are only designed to suit specific applications and they are not bell-shaped. A more effective model was proposed by Janaswamy [5] and Bevan *et al.* [1] in which the probability of occurrence of the random scatterers was assumed to obey a Gaussian distribution to more effectively model the random scatterers than the previous models [2-4]. The Gaussian distribution also yields better fit to measured AoA data which was experimentally obtained in Aarhus, Denmark and Stockholm by Pedersen, Mogensen and Fleury [6]. It should be noted that because of its bell shape, the Gaussian distribution can yield better fit to most practical data. Janaswamy [5] gave fundamental derivations of the Gaussian AoA pdf, time-of-arrival (ToA) pdf, PAS and power delay spectrum (PDS). Detailed comparisons between the Gaussian scatterer model and the previous models [2-4] were made [5] in which it was shown that the Gaussian model possesses the smallest fitting error to the experimental data [6]. Thus, uniform and bell-shaped distributions are effective to model the random scatterers between the BS and the UE.

The Gaussian distribution was proposed to model the random scatterers [1], however, fitting to experimental data still possesses large errors [1-4]. This suggests that there is possible improvement by using an even more effective distribution than the Gaussian which motivates further research in antenna modelling. The work reported in this paper has been motivated by the following reasons: (1) Because the Gaussian distribution was used to model the random scatterers [1] to derive the Gaussian AoA pdf, and because of the greater effectiveness of the hyperbolic distribution compared to Gaussian distribution [7-11], there ought to be improvements if the random scatterers are modelled by the hyperbolic distribution; (2) The hyperbolic distribution yields better fit than the Gaussian distribution to the data given in [6] as will be shown in Figure 1 in Section 2 — this is the major motivation to propose the hyperbolic distribution to model random scatterers in this paper; (3) The hyperbolic distribution resembles the bell shape of the Gaussian distribution which makes the two distributions similar even though they are mathematically different; (4) To estimate the PAS of the hyperbolic AoA pdf [12]; (5) To estimate the entropy of the approximate pdfs to assess their uniformness and effectiveness, i.e. the larger the entropy, the more informative the pdf; and (6) To show that non-Gaussian distributions are also effective to model the random scatterers in multi-path mobile environment. It should also be noted that none of the models in [2-4] is bell-shaped which made them less effective than the Gaussian model proposed in [5] and [1].

To derive the AoA pdf, the following assumptions, which are similar to those in [1], are made with the exception of assumption (5): (1) The elevation angle is zero with scattered signals received by the BS in the horizontal plane; (2) Unity reflection coefficient and random phase, i.e. collisions between signals and scatterers are elastic; (3) It is impossible for signals to travel from the UE to BS without encountering random scatterers, i.e. infinite attenuation for a path that is a direct line-of-sight between the BS and UE's antenna; (4) There is no further coupling among the signals and the scatterers are omnidirectional, i.e. signals travelling from the UE's antenna to BS are scattered and reflected by scatterers only once and they are independent of each other; and (5) Random scatterers are assumed to obey the hyperbolic distribution which resembles the bell shape. In addition, the probability of occurrence of a scatterer is assumed to be inversely proportional to its distance from UE, i.e. there are more scatterers near the UE than further from it. It should be noted that assumption (5) differentiates the work reported in this paper from other works reported in the literature. It should also be clear that by modelling the scatterers in terms of their probability of occurrence, it is possible to effectively approximate the data histogram by varying the distribution's parameter/s.

The aims of this paper are thus four-fold: (1) To introduce a hyperbolic distribution (in place of the Gaussian distribution) to more accurately approximate the measured AoA data [6], especially toward the data's "tail" region; (2) To derive the AoA pdf when the random scatterers are governed by the hyperbolic distribution; (3) To approximate the Gaussian and hyperbolic AoA pdfs for small and large AoA; and (4) To estimate the PAS and entropy of the approximate Gaussian and hyperbolic AoA pdfs, from which the relative merits of each pdf can be assessed. The approximations of the Gaussian and hyperbolic AoA pdfs are necessary because of the complexity of their theoretical forms derived in steps (1) and (2) as will be shown later. The estimation of the PAS is also essential to determine the scattering power associated with specific azimuthal arrival angle θ.

The paper is organised as follows. Section 2 introduces and shows the Gaussian and hyperbolic approximations to the measured data histogram given in [6]. Section 3 outlines the general framework of the AoA in multi-path mobile environment. Section 4 mathematically derives the expression of the AoA pdf. Section 5 discusses the theoretical PAS. Sections 6 and 7 derive the approximate Gaussian and hyperbolic AoA pdfs along with their corresponding PAS and entropy. Relative merits of each pdf are also discussed. Section 8 concludes the findings of the paper and suggests further work.

2. Data Fitting

In this section, the hyperbolic and Gaussian distributions are discussed and their approximations to the real data histogram given in [6] are shown. The motivation for using the hyperbolic distribution to replace the Gaussian distribution is also explained. Mathematically, the Gaussian and hyperbolic distributions are given by Eqs. (1) and (2) respectively

$$p_G(x) = \frac{\exp\left(-\frac{(x-\mu)^2}{2\sigma^2}\right)}{\sigma\sqrt{2\pi}}, \text{ with variance of } \sigma \text{ and mean } \mu, \tag{1}$$

$$p_H(x) = \frac{\beta}{2.367}\operatorname{sech}[(\beta-\mu)^2 x^2], \tag{2}$$

with the distribution variance of about $1.5323/\beta^3$ and mean of μ.

It should be noted that the argument of the hyperbolic sech function used in Eq. (2) is different from that of a normal hyperbolic kernel used in time-frequency signal processing so that the distribution can be correctly matched with the measured AoA pdf.

From Figure 1, it is evident that the two distributions are closely related and yield satisfactory approximations to the data. It is also clear that the hyperbolic distribution provides a better fit than the Gaussian distribution toward the data's tail region corresponding to larger values of θ in the data used in [6] for specific values of β, which motivates the work presented in this paper.

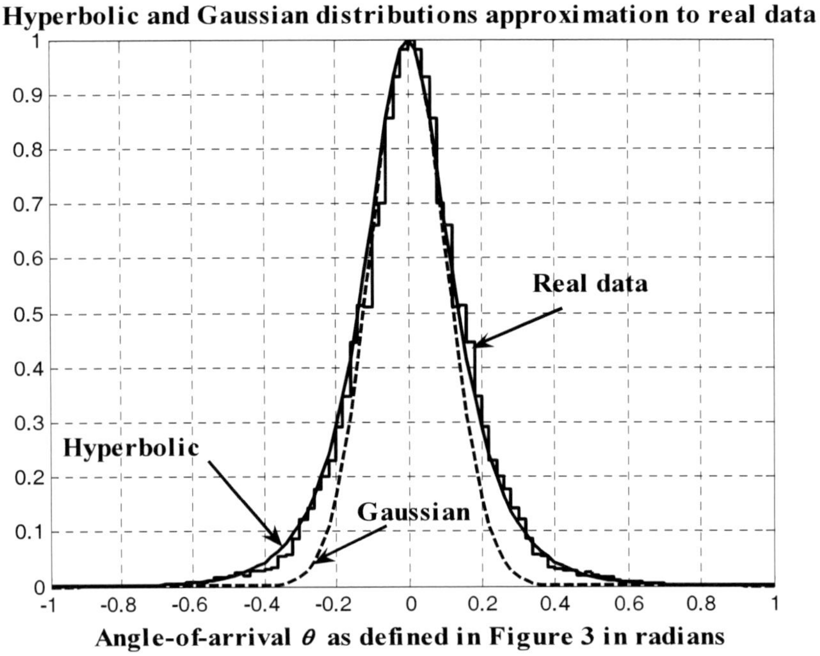

Figure 1. Approximation to the measured data histogram given in [6] using the hyperbolic and Gaussian distributions with $\beta \approx 9.5493$ and $\sigma \approx 0.1074$ respectively to yield the line of best fit to the data. The hyperbolic distribution clearly gives a better fit to the data than the Gaussian distribution toward the data's tail region.

It should also be noted that the hyperbolic distribution is more suitable for data with considerable energy toward its "tail" region which can be defined as that beyond six standard deviations (6σ). In the region within 6σ, the Gaussian distribution yields a slightly better match in the minimum squared error sense than the hyperbolic distribution. For this particular data set given in [6], the hyperbolic distribution gives a better approximation, hence better overall results. From Figure 1, it is evident that the hyperbolic distribution can be effectively used to model random scatterers, which extends antenna modelling in the mobile environment beyond the Gaussian distribution which has long been dominant in this field.

3. System Description

The Gaussian distribution used to model random scatterers was introduced in [1] to derive the AoA pdf in mobile environment. This section presents the AoA pdf using a hyperbolic scatterer distribution because the hyperbolic kernel function has been shown to be more effective [7-11] than the Gaussian distribution in the field of time-frequency signal processing. Consider the antenna system shown in Figure 2 in which a base station (BS) is located a distance D from the user equipment's (UE) antenna. The effective radius of the UE is R_{eff} which encloses a circular area as shown and within which random scatterers obey the hyperbolic distribution. Evidently, the largest value of the AoA at the BS is dependent on D and R_{eff}. More details on the system are given in [1].

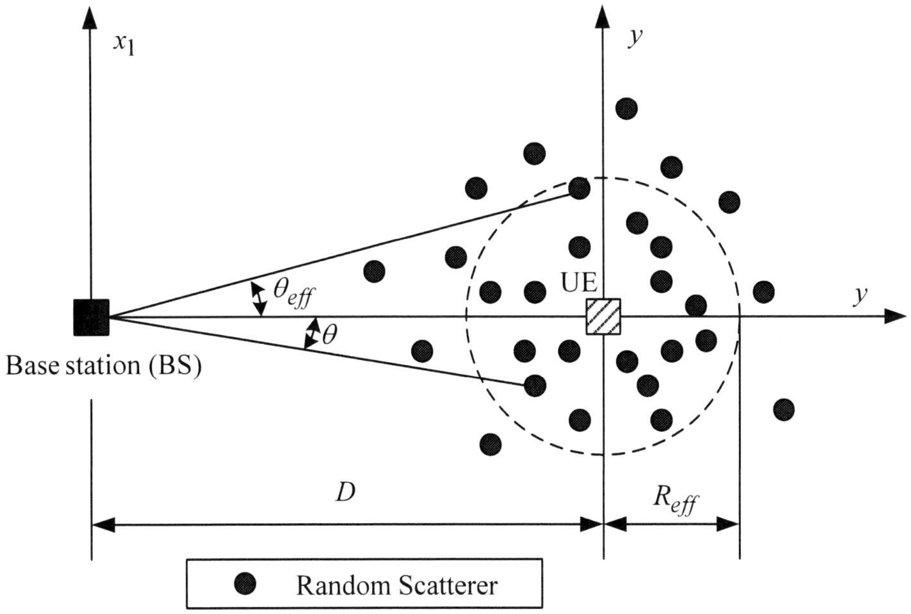

Figure 2. A schematic diagram of angle θ subtended by random scatterers at the BS. The angle θ seen at the BS is assumed to be small so that $\sin(\theta) \approx \theta$ [1].

It is clear that as D increases the power transmitted from the UE and received at the BS decreases, which reduces the effectiveness of the system. It should also be noted that the larger the effective radius R_{eff} the more signals are received from the random scatterers, hence the better is the performance of the UE. However, there are limitations to increasing R_{eff}. First, if R_{eff} becomes too large, the AoA seen at the BS becomes too large and does not meet certain constraints imposed on it. Second, in the case of general communication between the UE's antenna and the BS where full duplex traffic is employed, if R_{eff} becomes too small, then it will be much more difficult to receive signals at the UE as this lowers the system's efficiency. Therefore, it is clear that the effective radius of the antenna should meet certain conditions so that the constraints mentioned above are met.

4. Derivation of the Hyperbolic AoA PDF

The major difference between the approach in the present paper and the work reported in [1] which assumes a Gaussian distribution for the random scatterers is that the scatterers are assumed to obey the hyperbolic distribution. It should be noted that detailed derivations are included in this section, not in an appendix as it is believed that further insight can be gained from the details.

Mathematically, the AoA pdf is expressed as [1]

$$p(r,\varphi) = \frac{1}{\pi R_{eff}^2} \operatorname{sech}\left(\frac{r^2}{R_{eff}^2}\right) \tag{3}$$

where (r, φ) are the polar coordinates centred at the UE, r is the distance to a given scatterer from the UE, and R_{eff} is the distance at which the pdf decreases by a factor of e.

Converting the pdf given in Eq. (3) to the polar coordinates (R, θ) with its origin located at BS results in

$$x_1 = R\sin\theta \text{ and } y_1 = R\cos\theta. \tag{4}$$

Using the Jacobian transformation, one obtains [1]

$$p(R,\theta) = \frac{R}{\pi R_{eff}^2} \operatorname{sech}\left(\frac{R^2 - 2RD\cos(\theta) + D^2}{R_{eff}^2}\right), \tag{5}$$

where R is the new variable in the polar coordinates centered at BS.

To obtain the pdf of θ, one has to eliminate R, which means

$$p(\theta) = \int_0^{+\infty} \frac{R}{\pi R_{eff}^2} \operatorname{sech}\left(\frac{R^2 - 2RD\cos(\theta) + D^2}{R_{eff}^2}\right) dR \tag{6}$$

A series representation of the integral can be obtained as follows. Let

$$I = \int_0^{+\infty} \frac{x}{\pi R_{eff}^2} \operatorname{sech}\left(\frac{x^2 - 2xD\cos(\theta) + D^2}{R_{eff}^2}\right) dx \text{, and} \tag{7}$$

let

$$I = \frac{I_1}{\pi R_{eff}^2}, \tag{8}$$

where
$$I_1 = \int_0^{+\infty} x\text{sech}\left(\frac{x^2 - 2xD\cos(\theta) + D^2}{R_{eff}^2}\right)dx\,, \tag{9}$$

so that

$$I_1 = \int_0^{+\infty} x\text{sech}\left(ax^2 + bx + c\right)dx\,, \tag{10}$$

where

$$a = \frac{1}{R_{eff}^2}\,,\ x = R, \tag{11}$$

$$b = -2D\,a\cos(\theta) \tag{12}$$

and

$$c = D^2 a\,. \tag{13}$$

Using the result ([13], p.27, Sec. 1.232(2))

$\text{sech}(y) = 2\sum_{k=0}^{\infty}(-1)^k e^{-(2k+1)y}$; $y > 0$, and the substitution $y = ax^2 + bx + c$,

we have

$$I_1 = 2\int_0^{\infty} x\sum_{k=0}^{\infty}(-1)^k e^{-(2k+1)(ax^2+bx+c)}dx\,, \tag{14}$$

with

$$ax^2 + bx + c > 0\,. \tag{15}$$

Clearly, a sufficient condition for (15) to hold is

$$ax^2 + c > -bx\,,\ \text{or}\ R^2 + D^2 > 2RD\cos(\theta)\,,\ \text{i.e.}\ \cos\theta < \frac{1}{2}\left(\frac{R}{D} + \frac{D}{R}\right), \tag{16}$$

which is always true since $\cos(\theta) < 1$ except for $\theta = 0$. Another way of obtaining a sufficient condition for (15) is by forcing $b > 0$, implying $D\cos(\theta) < 0$. However, since $0 < \theta < \pi/2$,

$\cos(\theta) > 0$. Thus, (16) is used as the constraint for the derivation of the AoA pdf. With the sufficient condition (16) true, we have, after interchanging the order of summation and integration,

$$I_1 = 2\sum_{k=0}^{\infty}(-1)^k \int_0^{\infty} xe^{-(2k+1)ax^2-(2k+1)bx} e^{-(2k+1)c} dx \text{, or}$$

$$I_1 = 2\sum_{k=0}^{\infty}(-1)^k e^{-(2k+1)c} \int_0^{\infty} xe^{-\mu x^2 - 2\nu x} dx, \tag{17}$$

where

$$\mu = (2k+1)a \text{, and} \tag{18}$$

$$\nu = \frac{(2k+1)}{2} b. \tag{19}$$

Using the result ([13], p. 361, Sec. 3.462(5)) and after some algebraic manipulations, the theoretical AoA pdf using the hyperbolic distribution to model random scatterers can be written in terms of $\tau = D/R_{eff}$ and θ as

$$I = \frac{1}{\pi}\sum_{m=1,3,\ldots}^{+\infty}(-1)^{\frac{m-1}{2}}\exp(-m\tau^2)\left[\frac{1}{m} + \tau\sqrt{\frac{\pi}{m}}\cos(\theta)\exp[m\tau^2\cos^2(\theta)]\{1 + erf[\tau\sqrt{m}\cos(\theta)]\}\right]$$

$$= \frac{1}{\pi}\sum_{m=1,3,\ldots}^{+\infty}(-1)^{\frac{m-1}{2}}\exp(-m\tau^2)\frac{1}{m}$$

$$+ \frac{\tau\cos(\theta)}{\sqrt{\pi}}\sum_{m=1,3,\ldots}^{+\infty}(-1)^{\frac{m-1}{2}}\left(\frac{1}{\sqrt{m}}\right)\exp[-m\tau^2\sin^2(\theta)]\{1 + erf[\tau\sqrt{m}\cos(\theta)]\}, \tag{20}$$

which is plotted in Figure 3 in which the hyperbolic AoA pdf as a function of θ resembles the bell shape similar to that of the Gaussian AoA pdf shown in [5] and [1].

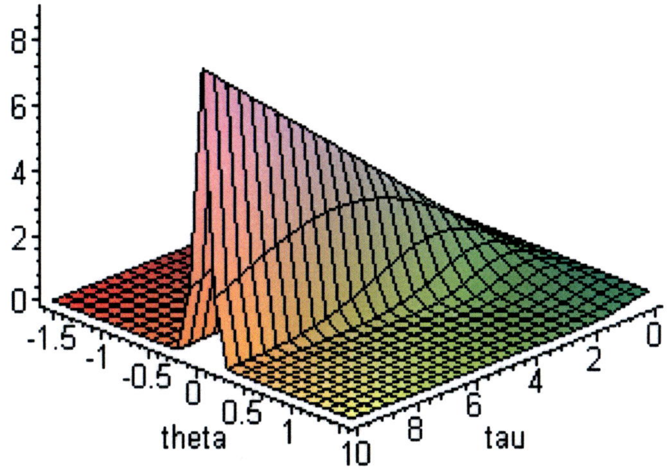

Figure 3. Theoretical hyperbolic AoA pdf [Eq. (20)] as a function of $\tau = D/R_{eff}$ and θ.

5. PAS

To measure the incoming power at the BS transmitted from the UE when the AoA is varied, the PAS at a fixed angle $\theta_{eff} = \eta$ is employed which is given as [5]

$$\mathrm{PAS}(\eta) = \frac{P_o \sigma_o l_e}{4\pi^2} \int_{\theta_{eff}=\eta}^{\pi} p(x_m, y_m)\big|_{\eta \text{ fixed}} \, d\theta_{eff} \tag{21}$$

where P_o is the total power radiated from the UE, σ_o the bistatic echo width of a scatterer and l_e the effective beam width. From Figure 2, we obtain

$$\theta_{eff} = \arctan\left(\frac{R_{eff}}{D}\right) = \arctan\left(\frac{1}{\tau}\right), \text{ hence } d\theta_{eff} = \frac{-d\tau}{\tau^2 + 1} \tag{22}$$

Using Eqs. (21) and (22), the absolute power distribution as a function of θ can be rewritten as a function of τ as

$$\mathrm{PAS}(\eta) = \frac{P_o \sigma_o l_e}{4\pi^2} \int_{1/\tan(\eta)}^{+\infty} p(x_m, y_m)\big|_{\eta \text{ fixed}} \, \frac{d\tau}{\tau^2 + 1} \tag{23}$$

It should be noted that by using the PAS, it is possible to determine the efficiency and the number of users a system can accommodate by taking ratio of the total transmitted power from the UE and the received power at the BS. Thus, PAS estimation also plays an important role in signal transmission in multi-path mobile environments. The Gaussian and hyperbolic PAS's are discussed for small and large θ in Sections 6 and 7 respectively.

As can be seen from Eq. (20), the theoretical expression of the hyperbolic AoA pdf is complicated and hence does not make interpretation easy. By evaluating the pdf for small and large θ, additional insight can be gained which reveals more salient features of the pdf and the effectiveness of the hyperbolic distribution in terms of modeling the random scatterers. The Gaussian AoA pdf is also derived under the same conditions for comparison purposes.

6. Approximations to the Hyperbolic AoA PDF

It is clear that when θ is small, there are two paths over which signals can be transmitted from the UE to the BS. The first path is a direct line-of-sight (LoS) between the UE and BS, i.e. angle between them is zero, and the received signals are in-phase with the transmitted signals from the UE. It is clear that most signals will arrive at the BS via the first path because it is the shortest between the UE and BS. In addition, the more signals arriving at the BS via the first path, the more efficient the transmission. Since the BS is omnidirectional, there exists a second critical signal path making an angle of 180° with the BS, yielding out-of-phase arrival signals to the transmitted signals from the UE. It should be noted that arrival signals at the BS via the second path usually consist of noise, reflected and/or refracted signals from the UE and other sources. Thus, they should be filtered out before reaching the BS to avoid undesirable distortion.

6.1. For Small AoA

From Eq. (20), the hyperbolic AoA pdf for small θ is theoretically given in Eq. (24) as

$$I = \frac{1}{\pi} \sum_{m=1,3,..}^{\infty} (-1)^{\frac{m-1}{2}} \exp(-m\tau^2) \left[\frac{1}{m} + \tau \sqrt{\frac{\pi}{m}} \cos(\theta) \exp[m\tau^2 \cos^2(\theta)] \Big(1 - erf[\tau\sqrt{m}\cos(\theta)] \Big) \right]. \tag{24}$$

Since θ is small, we can approximate

$$\cos\theta \approx 1;\ \sin(\theta) \approx \theta \text{ and note that } \sin(\theta_{\max}) = \frac{R_{eff}}{D} = \frac{1}{\tau}, \text{ with } \tau >> 1. \tag{25}$$

Using the result (5.11.53), p. 55 [Eldon R. Hansen "A Table of Series and Products", Prentice Hall, Englewood Cliffs, N.J., 1975], and after some algebraic manipulations, we obtain the approximate hyperbolic AoA pdf when θ is small as given in Eq. (26)

$$I \approx \frac{1}{\pi} \tan^{-1}(e^{-\tau^2}), \tag{26}$$

which is plotted in Figure 5, from which it can be noted that the hyperbolic AoA pdf is perfectly symmetrical which makes it practical and useful. Under the condition of small θ, from Eq. (23), the PAS of the hyperbolic AoA pdf is given by

$$\mathrm{PAS}_{\mathrm{small}}^{\mathrm{Hyperbolic}}(\eta) \approx \frac{P_o \sigma_o l_e}{4\pi^2} \int_{1/\tan(\eta)}^{+\infty} \frac{\tan^{-1}(e^{-\tau^2})}{\pi(1+\tau^2)} d\tau \tag{27}$$

which is plotted in Figure 4.

6.2. For Large AoA

Under this assumption, $\cos(\theta) \to 0$, thus Eq. (24) can be rewritten as

$$I = \frac{1}{\pi} \sum_{m=1,3,..}^{\infty} (-1)^{\frac{m-1}{2}} \left(\frac{\exp(-m\tau^2)}{m} \right) \tag{28}$$

which is plotted in Figure 8.

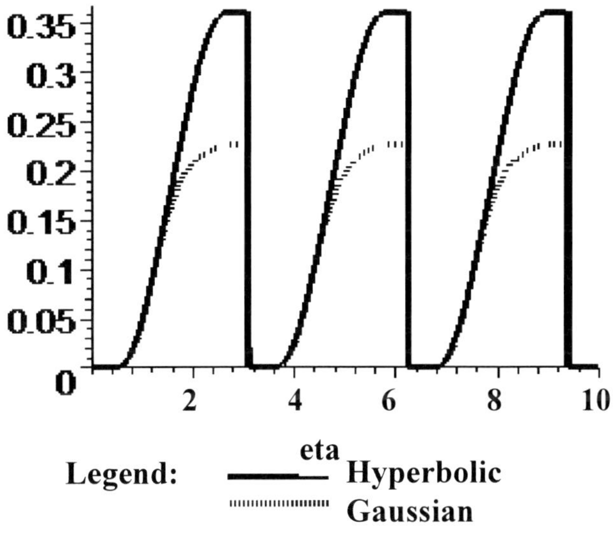

Figure 4. Normalised hyperbolic and Gaussian PAS as functions of η for small θ.

It should be noted that the conditions of large θ is critical in which arrival signals having a large AoA are usually undesirable. The hyperbolic AoA pdf under this condition still resembles an approximate bell shape as seen in Figure 8 with slight skewness toward the tail region of τ. The PAS under this condition is given by

$$\mathrm{PAS}_{\text{large}}^{\text{Hyperbolic}}(\eta) \approx \frac{P_o \sigma_o l_e}{4\pi^2} \int_{1/\tan(\eta)}^{+\infty} \frac{1}{\pi(\tau^2+1)} \sum_{m=1,3,..}^{\infty} (-1)^{\frac{m-1}{2}} \left(\frac{\exp(-m\tau^2)}{m} \right) d\tau. \quad (29)$$

which is plotted in Figure 9.

7. Approximations to the Gaussian AoA PDF

7.1. For Small AoA

The AoA pdf using the Gaussian distribution to model random scatterers is theoretically given by

$$p(\theta) = \frac{1}{2\pi} \exp\left(\frac{-D^2}{R_{eff}^2} \right) \left\{ 1 + \frac{\sqrt{\pi} D \cos(\theta)}{R_{eff}} \exp\left(\frac{-D^2 \cos^2(\theta)}{R_{eff}^2} \right) \left[1 + \frac{1}{2\sqrt{\pi}} \int_0^{\tau} e^{-t^2} dt \right] \right\} \quad (30)$$

where $\tau = D/R_{eff}$.

For small θ, Eq. (30) can be rewritten as

$$p(\tau) \approx \frac{1}{2\pi} \exp(-\tau^2) \left\{ 1 + \tau\sqrt{\pi} \exp(-\tau^2) \left[1 + \frac{erf(\tau)}{4} \right] \right\} \quad (31)$$

which is plotted in Figure 5.

The PAS of the Gaussian AoA pdf under this condition is given by

$$\mathrm{PAS}_{\text{small}}^{\text{Gaussian}}(\eta) \approx \frac{P_o \sigma_o l_e}{4\pi^2} \int_{1/\tan(\eta)}^{+\infty} \frac{\exp(-\tau^2)}{2\pi(1+\tau^2)} \left\{ 1 + \tau\sqrt{\pi} \exp(-\tau^2) \left[1 + \frac{erf(\tau)}{4} \right] \right\} d\tau. \quad (32)$$

The PAS of the hyperbolic and Gaussian AoA pdfs for small θ can be normalized with respect to $\frac{P_o \sigma_o l_e}{4\pi^2}$ and plotted in Figure 4 as a function of η from which it is clear that the hyperbolic AoA pdf possesses about one and a half times more the received power at the BS (about 0.35) than the Gaussian AoA pdf (about 0.23) when θ is small. From Figure 4, the received power is maximized at specific angles $\theta_{eff} = \eta$ of about 3, 6.2, 9.3 and so on, corresponding to τ of about 7, 12, 8 and so on for both the Gaussian and hyperbolic AoA pdfs under this condition. It should be noted that $\tau >> 1$ which is consistent with the assumption of small θ stated in Eq. (25).

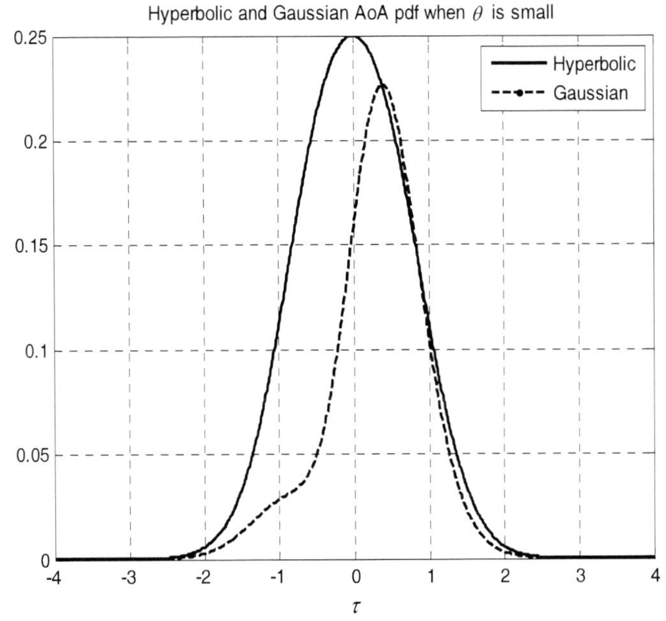

Figure 5. The hyperbolic and Gaussian AoA pdfs as functions of $\tau = D/R_{eff}$ for small θ.

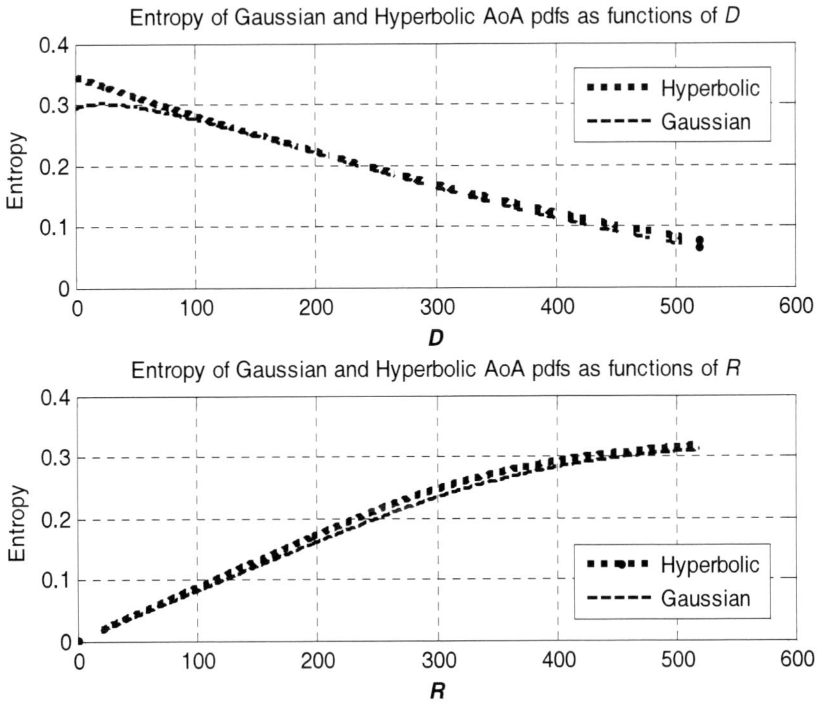

Figure 6. Entropy of the hyperbolic and Gaussian AoA pdfs as functions of R_{eff} and D for small θ.

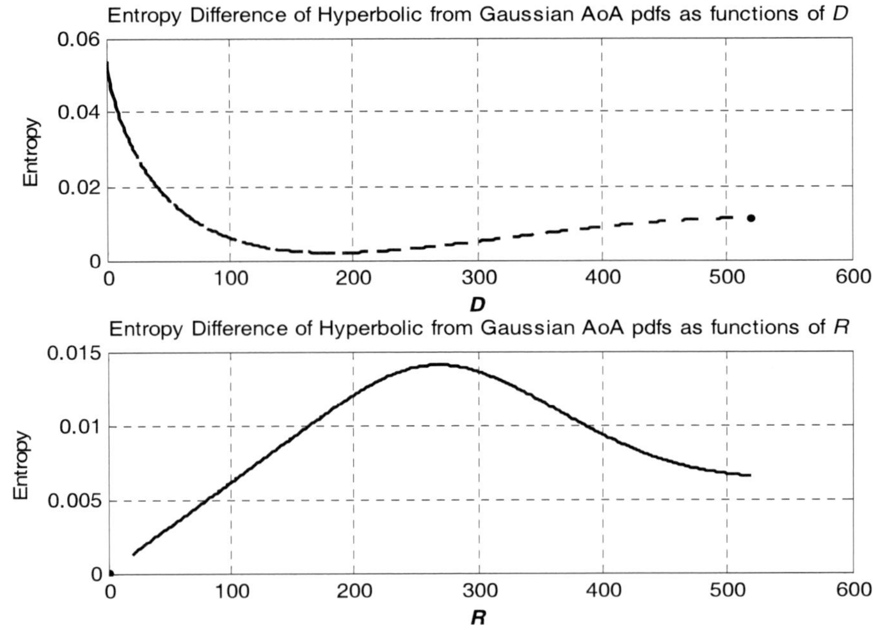

Figure 7. Entropy difference of the hyperbolic AoA pdf from the Gaussian AoA pdf as functions of R_{eff} and D for small θ.

From Figure 5, it is clear that the Gaussian AoA pdf for small θ is skewed to the right which means that it is not perfectly symmetrical with respect to τ. In addition, compared to the hyperbolic AoA pdf for small θ, it is clear that the hyperbolic AoA pdf is perfectly symmetrical with respect to τ, suggesting that it is more uniform than the Gaussian AoA pdf. The advantages of the hyperbolic distribution over the Gaussian distribution can be examined by estimating their corresponding entropy as shown in Figure 6. The entropy difference of the hyperbolic AoA pdf from the Gaussian AoA pdf, i.e. (hyperbolic – Gaussian) is plotted in Figure 7 as functions of R_{eff} and the coverage distance D.

From Figure 6 and Figure 7, it is clear that the hyperbolic AoA pdf possesses larger entropy than that of the Gaussian AoA pdf, indicating that the former distribution is more effective than the latter to model random scatterers. The maximum entropy difference is about 0.06 as a function of D. It is also clear that the hyperbolic distribution is more effective than the Gaussian distribution to model random scatterers in the range of $D \leq 200$. For $D > 200$, the entropy difference becomes smaller and reaches a saturated value of about 0.01 when $D \geq 500$. In addition, it appears that the entropy of both pdfs is identical for very large values of D.

With respect to R_{eff}, the entropy difference between the hyperbolic and Gaussian AoA pdfs attains its maximum value of about 0.015 at $R_{eff} \approx 250$. For $R_{eff} \geq 500$, the entropy difference reaches a saturated value of about 0.005. From Figure 7, it is clear that the hyperbolic distribution is robust over the entire range of R_{eff} and D, whereas the Gaussian distribution is limitedly effective for large D and R_{eff}. Table 1 summarizes the findings in this section.

Table 1. Summary of usable ranges of the hyperbolic and Gaussian distributions when θ is small

Distribution	$D \le 200$	$D > 200$	$R_{eff} \le 250$	$R_{eff} > 250$
Hyperbolic	suitable	suitable	suitable	suitable
Gaussian	unsuitable	suitable	unsuitable	suitable

7.2. For Large AoA

It should be noted that arrival signals to the BS at a large AoA should be filtered out to avoid distortion. For completeness, the Gaussian and hyperbolic AoA pdfs are approximated under this condition. Similar to the hyperbolic AoA pdf, under this condition, $\cos(\theta) \rightarrow 0$, from Eq. (30), the Gaussian pdf is approximately given by

$$p(\tau) \approx \frac{1}{2\pi}\exp(-\tau^2) \tag{33}$$

which is plotted in Figure 8.

The Gaussian AoA pdf for large θ resembles the perfect bell shape which indicates that it is more robust than the hyperbolic AoA pdf when θ is large. Thus, this can be considered as a trade-off between the two distributions. For comparison purposes, the Gaussian and hyperbolic AoA pdfs are plotted on the same scale as seen in Figure 8.

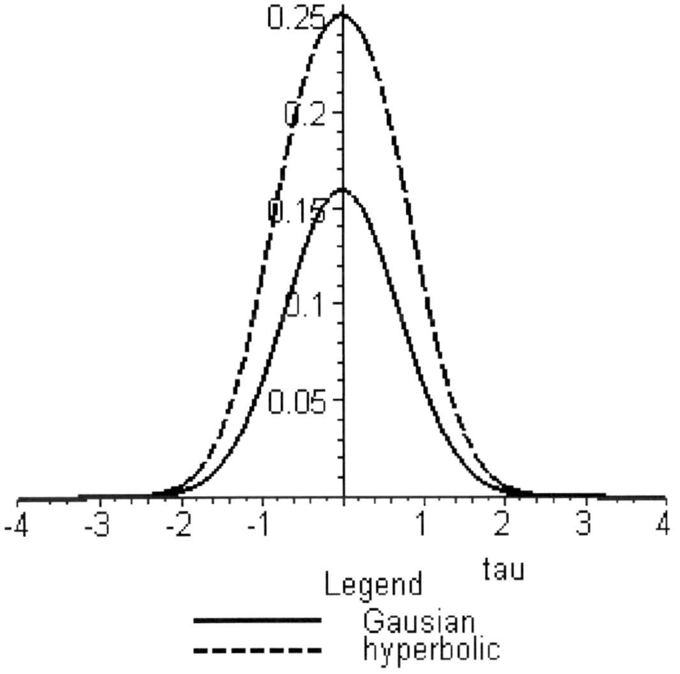

Figure 8. Approximate Gaussian and hyperbolic AoA pdfs as functions of $\tau = D/R_{eff}$ for large θ.

It is clear that the Gaussian AoA pdf is more uniform than the hyperbolic AoA pdf because of the imperfectness of the hyperbolic under this condition. However, since received signals with large θ are usually undesirable, the advantage of the Gaussian distribution over the hyperbolic distribution to model random scatterers is minimal.

The PAS of the Gaussian AoA pdf under this condition is given in Eq. (34) as

$$\mathrm{PAS}^{\mathrm{Gaussian}}_{\mathrm{large}}(\eta) \approx \frac{P_o \sigma_o l_e}{4\pi^2} \int_{1/\tan(\eta)}^{+\infty} \frac{\exp(-\tau^2)}{2\pi(\tau^2+1)} d\tau \tag{34}$$

which is plotted in Figure 9 from which it is clear that the hyperbolic AoA pdf possesses more received power from the UE to the BS for large θ which is similar to the case when θ is small. However, in this case the Gaussian AoA pdf exhibits some improvement at some specific values of $\eta = \theta_{eff} \approx 3, 6.2, 9.3, \ldots$ as reported in Section 7.1. In this case, the received power at the BS using the hyperbolic AoA pdf is about 3 times larger than that using the Gaussian AoA pdf.

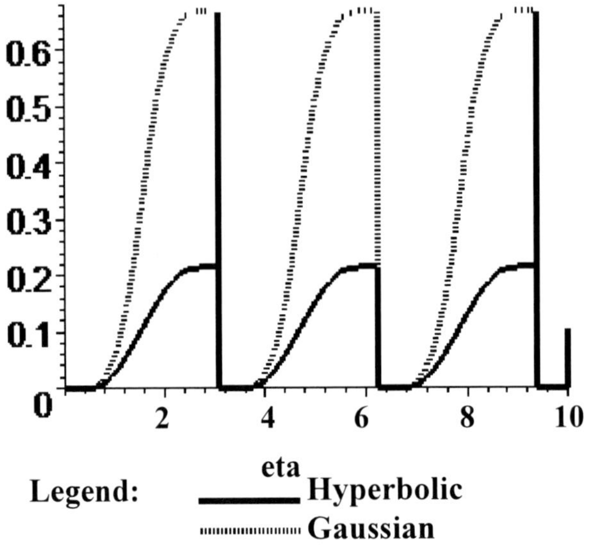

Figure 9. Normalised PAS of the hyperbolic and Gaussian AoA pdfs for large θ.

The entropy difference of the Gaussian AoA pdf from that of the hyperbolic AoA pdf is plotted in Figure 10 as functions of D and R_{eff}.

From Figure 10, as expected, the Gaussian AoA pdf possesses larger entropy than the hyperbolic AoA pdf as explained earlier. The entropy difference linearly decreases as D increases, and it increases as R_{eff} increases. The entropy difference appears to reach a saturated value of about 0.05 at $D > 200$ and of about 0.25 at $R_{eff} \approx 200$ which are consistent with the findings obtained from Figure 7 and Section 7.1. It should be noted that as R_{eff} increases beyond a limit, random scatterers located beyond the limit do not have significant effect on the quality of the received signals at the BS as the transmitted signals or antenna beams from the UE can not reach them. When the coverage distance D increases beyond a limit, the entropy difference under the condition of large θ decreases which is consistent with the findings reported in Section 7.1. Under the condition of small θ, the entropy decreases but reaches a saturated value with respect to D and R_{eff}. It should be noted that the entropy difference with respect to R_{eff} is about five times smaller than that with respect to D when θ is small. Thus, varying D can significantly affect the effectiveness of the Gaussian and hyperbolic distributions to model random scatterers and the robustness of received signals at the BS.

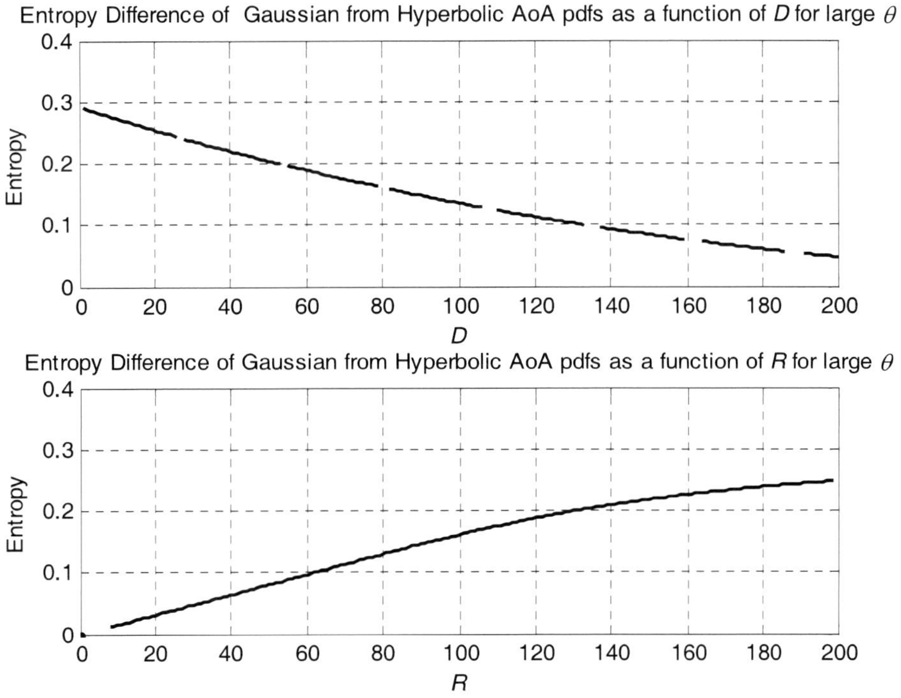

Figure 10. Entropy difference (hyperbolic – Gaussian) AoA pdf as functions of D and R_{eff} for large θ.

Conclusion and Further Work

The angle-of-arrival (AoA) probability density function (pdf) in multi-path mobile environment has been derived with the hyperbolic scatterer distribution between the BS and the UE. It has been shown that the hyperbolic AoA pdf is more effective than the Gaussian AoA pdf because the former possesses larger entropy than the latter. The power azimuth spectra of the hyperbolic and Gaussian pdfs have been derived to assess the received power transmitted from the UE to the BS. It was shown that about one and a half times more transmission power can be received at the BS by using the hyperbolic AoA pdf than by using the Gaussian Aoa pdf. It has been shown that the hyperbolic distribution can also be used to effectively model the random scatterers for large and small coverage distance in mobile macro cells for large and small AoA.

Since both the hyperbolic and Gaussian distributions can be used to model the random scatterers, it is thus possible that their joint distribution can be useful, employing the Gaussian distribution to model scatterers in the vicinity of the UE's antenna and the hyperbolic distribution for those further away from the device. A schematic diagram of this model is shown in Figure 11.

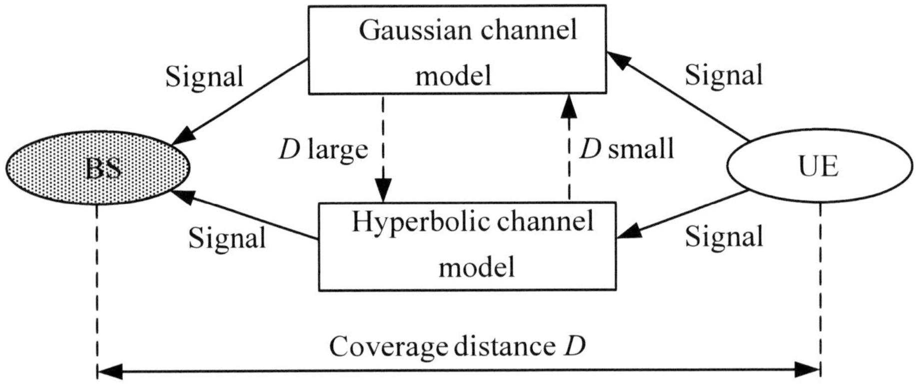

Figure 11. A communications model between BS and UE with the BS as the receiver and the UE as the transmitter employing the Gaussian and hyperbolic channels to model random scatterers.

Further, the marginal and joint AoA pdfs along with their PDSs can also be derived, yielding further insight on the effectiveness of the Gaussian and hyperbolic distributions for the random scatterer modelling.

References

[1] Bevan, D.D.N., et al., "Gaussian Channel Model for Mobile Multipath Environment," *EURASIP Journal on Applied Signal Processing*, 2004, 2004(9), pp. 1321-9.

[2] Petrus, P., Reed, J.H., and Rappaport, T.S., "Geometrical-based statistical macrocell channel model for mobile environments," *IEEE Transactions on Communications*, 2002, 50(3), pp. 495-502.

[3] Liberti, J.C. and Rappaport, T.S. *A geometrically based model for line-of-sight multipath radio channels*, in *Proceedings of the IEEE 46th Vehicular Technology Conference*, 1996, Atlanta, USA: IEEE.

[4] Clarke, R.H., "A statistical theory of mobile-radio reception," *Bell System Technical Journal*, 1968, 47(6), pp. 957-1000.

[5] Janaswamy, R., "Angle and time of arrival statistics for the Gaussian scatter density model," *IEEE Transactions on Wireless Communications*, 2002, 1(3), pp. 488-497.

[6] Pedersen, K.I., Mogensen, P.E., and Fleury, B.H., "A Stochastic Model of the Temporal and Azimuthal Dispersion Seen at the Base Station in Outdoor Propagation Environments," *IEEE Transactions on Vehicular Technology*, 2000, 49(2), pp. 437-47.

[7] Le, K.N., Dabke, K.P., and Egan, G.K., "Hyperbolic kernel for time-frequency power spectrum," *Optical Engineering*, 2003, 42(8), pp. 2400-15.

[8] Le, K.N., Dabke, K.P., and Egan, G.K., "Hyperbolic wavelet power spectra of non-stationary signals," *Optical Engineering*, 2003, 42(10), pp. 3017-37.
[9] Le, K.N., Dabke, K.P., and Egan, G.K., "On mathematical derivation of auto-term functions and signal-to-noise ratios of Choi-Williams, first- and *n*th-order hyperbolic kernels," *Digital Signal Processing: A Review Journal*, 2006, 16(1), pp. 84-104.
[10] Le, K.N., Dabke, K.P., and Egan, G.K. *The hyperbolic wavelet function,* in *SPIE Aerosense Proceedings: Wavelet Applications VIII*, April 2001, Orlando, Florida, USA: SPIE.
[11] Le, K.N., "Time-frequency distributions for crack detection in rotors -- A fundamental note," *Journal of Sound and Vibration*, 2006, 294(1-2), pp. 397-409.
[12] Le, K.N., "A new formula for the angle-of-arrival probability density function in mobile environment," *Signal Processing*, 2007, 87(6), pp. 1314-1325.
[13] Gradshteyn, I.S. and Ryzhik, I.M., *Table of Integrals, Series and Products*, 2000, New York: Academic Press, 23, 406, 480.

In: New Signal Processing Research
Editor: Takumi Maeda, pp. 389-410
ISBN: 978-1-60456-479-2

Chapter 12

EM Algorithm for Blind Channel Estimation and Data Detection with Noncoherent Receivers on Intersymbol Interference Communication Channels

Roman Lisnanski
Israel Aerospace Industries, Israel

Abstract

Noncoherent receivers are used in digital communication systems when carrier phase estimation is difficult or impossible. It is known that on the additive white Gaussian noise (AWGN) channel, optimal noncoherent receiver structure is usually implemented using matched filters followed by square law detectors. Square law detection results in non-Gaussian, signal dependent noise distribution at detector's output which significantly complicates channel estimation and data detection tasks especially if no training sequence is used (blind estimation). In this chapter, we present a new approach for optimum noncoherent receiver design. The proposed receiver performs blind channel estimation and data detection on channels characterized by intersymbol interference (ISI). We model the received signal as hidden Markov model (HMM) where the observation probability has a non-central chi-square distribution (NCCS). The proposed algorithm proceeds in 4 steps. The first step is based on a recently developed EM algorithm for the NCCS distribution and its finite mixtures. This novel algorithm involves only closed form expressions and can be easily implemented. The second step is a standard Baum-Welch algorithm. At the third step, the correspondence between the channel states and the transmitted bits is established. At the last step, the data is detected by symbol by symbol maximum a posteriori probability (MAP) algorithm or Viterbi algorithm.

Simulation results are presented for three communication channels: wireless communication with on-off keying (OOK) modulation on AWGN channel, wireless communication with OOK modulation on frequency selective fading channel and direct detection optical channel with ISI and OOK modulation. The proposed approach can also be used in noncoherent receivers for signal to noise ratio (SNR) estimation. Performance of EM based SNR estimation algorithm is presented.

1. Introduction

Joint channel estimation and data detection for linear or nonlinear unknown intersymbol interference (ISI) channels characterized by additive white Gaussian noise (AWGN) can be performed by applying the cluster-based blind equalization methods [4, 6, 9]. These equalizers utilize the clusters formed by the received observations at the receiver front end and treat equalization as a classification task. A major advantage of this approach is that no explicit channel modeling is required. However, when the channel model is known, the hidden Markov models (HMM) theory can achieve better performance. In this approach, the received signal is modeled as a HMM and the maximum likelihood (ML) solution is found via the Baum-Welch algorithm (BWA) which is a special case of Expectation-Maximization algorithm (EM) [14]. Such algorithms usually assume Gaussian observation densities in the HMM model. In [13], generalized EM algorithm was proposed for joint channel estimation and data detection on communication channels that can be modeled by HMM with non Gaussian observation densities. In this chapter, we describe a new method for joint channel estimation and data detection on communication channels that can be modeled by HMM with NCCS observation densities. Such communication channels are usually created when square law detectors are used at the receiver.

Most communication systems transmit digital data over a channel using carrier modulated signals.

Knowledge of carrier phase is required at the receiver if coherent demodulation is performed.

In many cases, the carrier phase is unknown at the receiver. Uncertainty in the carrier phase of the received signal may be due to one or more of the following reasons. First, the oscillators that are used at the transmitter and the receiver to generate the carrier signals are generally not phase synchronous. Second, the time delay in the propagation of the signal from the transmitter to the receiver is not generally known precisely. Large changes in carrier phase can occur due to relatively small changes in the propagation delay. Motion of the transmitter and receiver by only a fraction of a wavelength relative to each other also leads to large change in relative phase angle. In some channels, the time delay in the propagation of the signal from the transmitter to the receiver may change rapidly and in an apparently random manner, so that the carrier phase of the received signal varies in an apparently random fashion.

In many applications, it is not possible to perform coherent detection due to the inability of the receiver to produce carrier phase estimates, due, for example, to significant oscillator phase noise instability. In some scenarios, coherent systems are undesirable due to the additional complexity imposed by the phase estimation circuitry or phase estimation algorithms. Techniques for phase recovery typically require more complexity and cost in the receiver.

When carrier phase of the carrier modulated signal is unknown at the receiver and no attempt is made to estimate its value, the design of optimum noncoherent receiver should be considered. It is known that on the additive white Gaussian noise (AWGN) channel, optimal noncoherent receiver structure is usually implemented using matched filters followed by square law detectors. Square law detection results in NCCS noise distribution at detector's output which significantly complicates channel estimation and data detection tasks. In this chapter, we present a novel approach for optimum noncoherent receiver design. The proposed receiver performs blind (without training sequence) channel estimation and data detection and can operate under severe ISI and low SNR conditions. The receiver design is based on HMM model with continuous NCCS observation densities. HMM model parameters are estimated by a recently developed EM algorithm for NCCS distribution and its finite mixtures [5] and by standard BWA algorithm. Our chapter is organized as follows: in section 2, a brief overview of EM algorithm is presented. In section 3, the proposed communication system model is described. In section 4, the proposed four step algorithm for blind channel estimation and data detection is described. In section 5, the HMM model is defined and simulation results are presented for channel estimation performance of three different communications channels:

- Noncoherent receiver for binary OOK modulation on AWGN channel
- Noncoherent receiver for binary OOK on frequency selective fading channel
- Intensity modulated direct detection optical channel with chromatic and polarization-mode dispersion

In section 6, conclusions are drawn.

2. Brief Overview of EM Algorithm

Expectation-Maximization (EM) algorithm is a general-purpose algorithm for maximum likelihood estimation in a wide variety of situations [14] that can be formulated as incomplete data problems. The name EM algorithm was given by Dempster,Laird, and Rubin in a celebrated paper read before the Royal Statistical Society in 1976 and published in its journal in 1977 [3]. In this paper, a general formulation of the EM algorithm was presented, its basic properties established, and many examples and applications of it provided. Since then, the EM algorithm has become a widely used statistical method.

EM algorithm provides an iterative procedure for performing ML estimation in cases where some of the data is missing, or where one can treat as missing hypothetical additional data that, if known, would simplify the estimation. The EM algorithm consists of two major steps: an expectation step (E-step), followed by a maximization step (M-step). The expectation is with respect to the unknown underlying variables, using the current estimate of the parameters and conditioned upon the observations. The maximization step then provides a new estimate of the parameters. These two steps are iterated until convergence.

EM algorithm can be used for ML parameter estimation of finite mixture models[15]. Mixtures of distributions, especially Gaussian mixtures, have been used extensively as models in such problems where data can be viewed as arising from two or more populations mixed in varying proportions. Since the advent of the EM algorithm, it has become the most commonly used approach to the fitting of mixture distributions. When mixture components belong to the exponential family distribution, the application of EM algorithm becomes simpler because there is a closed form solution to the M-step of the EM algorithm. In [12], a generalized EM algorithm was proposed which is based on approximating by parabola the maximized function on the M-step of the generalized EM algorithm. This algorithm can be applied to mixtures that do not belong to the exponential family distributions. In [5], an EM algorithm for the NCCS distribution and its finite mixtures was developed which involves only closed form expressions.

Another important application of EM algorithm is for parameter estimation of HMM models. HMM model is a fusion of Markov chain and mixture model. EM algorithm for finding the ML estimate of the parameters of a hidden Markov model given a set of observations is known as Baum-Welch algorithm [16].

3. Communication System Model

We consider the transmission of a sequence of independent symbols $a_1, a_2, \ldots$ through a nonlinear finite memory channel characterized by non-Gaussian additive or signal dependent noise. The channel can cause distortion, fading, noise and interference of the transmitted symbols. We denote the sequence of received symbols by $r_1, r_2, \ldots$. The action of the channel is usually characterized by the conditional probabilities [17] $\Pr(r_1, r_2, \ldots | a_1 a_2, \ldots)$ of receiving the sequence $r_1, r_2, \ldots$ when the symbol sequence $a_1, a_2, \ldots$ was sent. We assume that the channel has a finite memory, that is, at any time instant t, the channel output r_t depends only on a finite number, say L, of past source symbols besides the one emitted at time t. Under this assumption, we can use the family of hidden Markov models (HMMs) for system description. A hidden Markov model (HMM) is a statistical model where the system being modelled is assumed to be a Markov process with unknown parameters, and the challenge is to determine the hidden parameters, from the observed data sequence [16]. The discrete HMM can be used in two different but mathematically equivalent ways: their outputs may be associated either with states [1] or with state transitions [8]. We associate HMM outputs with states. HMM model can be applied to modulation schemes with symbols drawn from an M-ary signal constellation in either one or two-dimensional Euclidean space. For simplicity, we will consider an on-off keying communication system so that the symbols are real and the correspondence between bits and symbols is one to one. We will use a transmitted bit sequence $I_1, I_2, \ldots$ in our equations instead of the symbol sequence $a_1, a_2, \ldots$. The channel model is shown in Figure 1.

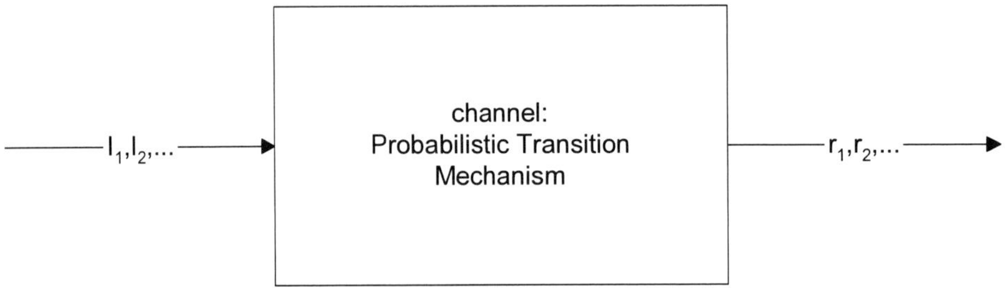

Figure 1. Finite memory channel.

A general finite memory channel can be modeled as a probabilistic transition mechanism characterized by the PDFs of the received samples $r_1, r_2, \ldots$ conditioned on channel state. We assume that the conditional PDF function of the received samples depends on the current channel state and has a NCCS probability density function (a comprehensive description of NCCS distribution can be found in [7]). We assume that the dependence on the channel state is through channel parameter which is a function of channel state. The channel state is determined by the current and L previously transmitted bits. This model can describe a finite memory channel characterized by NCCS noise which is non-Gaussian and signal dependent. The discrete HMM for such channel is defined as follows:

(1) The number of channel states is $N_s = 2^{L+1}$. It is the number of channel outputs in the absence of channel noise

(2) The probability density function (PDF) of the observation r_t conditioned on channel parameter c_t is characterized by NCCS distribution and the PDF value depends on the currently received sample r_t, on channel parameter c_t and on noise parameter σ :

$$f_{NCCS}(r_i \mid c_j;\sigma) = \frac{1}{2\sigma^2}\left(\frac{r_i}{c_j}\right)^{\frac{m}{2}} e^{\frac{c_j + r_i}{2\sigma^2}} I_m\left(\frac{\sqrt{r_i}\sqrt{c_j}}{\sigma^2}\right) \tag{1}$$

$$m = \frac{n-2}{2}$$

I_m - modified Bessel function of the first kind of order m

c – noncentrality parameter of the NCCS distribution

n – number of degrees of freedom of the distribution

The channel parameter is a function of the channel state $c_t = c(s_t)$. The channel state at time instant t is determined by the current and L previously transmitted bits: $s_t = (I_t, I_{t-1}, \ldots I_{t-L})$. For square law detection performed by noncoherent receiver on ISI communication channel, the channel parameter depends on the channel state as follows: $c_t = c(s_t) = \left(\sum_{k=0}^{k=L-1} h_k I_{t-k}\right)^2$ where the real coefficients $h_0, h_1, \ldots$ characterize the linear part of the channel prior to square law detector.

(3) The state transition probability matrix is

$$\begin{aligned} &\boldsymbol{A} = \{\boldsymbol{a}_{ij}\}, 1 \le \boldsymbol{i}, \boldsymbol{j} \le \boldsymbol{N}_s, \boldsymbol{N}_s = 2^{L+1} \\ &\boldsymbol{a}_{ij} = \boldsymbol{P}(\boldsymbol{s}[\boldsymbol{t}+1] = \boldsymbol{s}_j \mid \boldsymbol{s}[\boldsymbol{t}] = \boldsymbol{s}_i) \\ &\boldsymbol{a}_{ij} = 0.5 \text{ for allowable transition, 0 otherwise} \end{aligned} \tag{2}$$

(4) The initial state probability distribution: $\pi_0, \pi_1, ...\pi_{N_s-1}$. We assume that the initial state probability distribution is uniform: $\pi_0 = \pi_1 = ...\pi_{N_s-1} = \frac{1}{N_s}$.

Our goal is to perform blind channel estimation. In terms of HMM model, this means that based on the received samples $r_1, r_2, ...$, we should find the possible channel parameter values c_j for $j = 1, 2, ...N_s$, the state transition probability matrix A and to establish the correspondence between the channel parameters and the transmitted sequences of L+1 bits. After the channel estimation is completed, we can perform data detection by symbol by symbol MAP algorithm (SbS-MAP) algorithm or Viterbi algorithm. These detection algorithms are described in [11].

4. The Proposed Algorithm

4.1. General Overview of the Proposed Algorithm

The proposed algorithm proceeds in 4 stages as follows:

(1) Channel parameters estimation by EM algorithm for NCCS distribution and its finite mixtures
(2) Transition probability matrix estimation by BWA
(3) Establishing the correspondence between the channel states and the transmitted bits
(4) Data detection

At the first stage, we estimate the possible values of c_j for $j = 1, 2, ...N_s$. At this stage, the EM algorithm for NCCS distribution and its finite mixtures [5] is used. A sufficient number of EM iterations are performed to make sure that the estimated channel parameters converge close enough to their true values. At the second stage, a transition probability matrix is estimated by a BWA based on channel parameters estimated on the first stage. At the third stage, the correspondence between the channel parameters and the transmitted bits is established. The third stage completes the channel estimation. At the last stage, the data is detected by SbS-MAP algorithm or Viterbi algorithm.

4.2. Channel Parameters Estimation by Using EM Algorithm for NCCS Distribution and Its Mixtures

At this stage, we should estimate the channel parameters: c_j for $j = 1, 2, ...N_s$.

Let's denote the received sequence, corresponding to an input sequence of length N by $r_1 r_2 ....r_N$. If the transmitted sequence is long enough, then, with probability approaching 1, all possible channel states $s_j, j = 1,2,...N_s$ will be realized (the number of channel states $N_s = 2^{L+1}$).

The received sequence values will have a PDF which is a mixture of N_s NCCS component densities [15]. The PDF of the received samples will be given by the following formula:

$$\boldsymbol{f}_{mix}(\boldsymbol{r}_i;\boldsymbol{C}) = \frac{1}{\boldsymbol{N}_s}\sum_{j=1}^{N_s}\boldsymbol{f}_{NCCS}(\boldsymbol{r}_i \mid \boldsymbol{c}_j) \tag{3}$$

where $\boldsymbol{f}_{NCCS}(\boldsymbol{r}_i \mid \boldsymbol{c}_j)$ is defined by (1)

where $\mathrm{C}=(c_1, c_2, ...c_{N_s})$ is a vector of unknown channel parameters that we want to estimate. We omit the noise parameter σ from this derivation for convenience (it can be set to one without compromising the generality of the analysis). Each PDF function $\boldsymbol{f}_{NCCS}(\boldsymbol{r}_i \mid \boldsymbol{c}_j)$ is a NCCS PDF. Our objective is to estimate the parameters $c_1, c_2, ...c_{N_s}$ given the received sample sequence $r_1 r_2 ....r_N$. We do it by using a recently developed EM algorithm for the NCCS distribution and its finite mixtures [5]. This algorithm involves only closed form expressions. It was derived by using the infinite mixture representation of NCCS density. We briefly describe it below.

The E and M steps of this EM algorithm are defined as follows:

E-step:

$$w_{ij} = \frac{\pi_j f(r_i;c_j)}{\sum_{j=1}^{N_s}\pi_j f(r_i;c_j)} \tag{4}$$

$$s_{ij} = \sqrt{(c_j r_i)} \cdot R_{\frac{p}{2}-1}\left(\sqrt{(c_j r_i)}\right), i = 1,...N; j = 1,...N_s \tag{5}$$

where R is the Bessel quotient defined as follows:

$$R_v(\alpha) = \frac{I_{v+1}(\alpha)}{I_v(\alpha)} \tag{6}$$

$\boldsymbol{I}_v$ is the modified Bessel function of the first kind of order v

M-step:

$$\pi_j = \frac{\sum_{i=1}^{N} w_{ij}}{N}, j = 1,...N_s \tag{7}$$

$$c_j = \frac{\sum_{i=1}^{N} w_{ij} s_{ij}}{\sum_{i=1}^{N} w_{ij}} \tag{8}$$

4.3. Transition Probability Matrix Estimation by BWA

After estimating the channel parameters $c_1, c_2, \ldots c_{2^{L+1}}$, we use them to perform transition probability matrix estimation by a standard BWA [16]. We initialize the transition probability matrix $a_{ij} = \frac{1}{N_s};\ i, j = 1, 2, \ldots N_s$ and begin to iterate. We denote the length of the available data as T. T is equal to the number of received samples N. At each iteration of the BWA, the following steps are performed:

Forward variable calculation

(1) We can efficiently find $\alpha_t(i)$ inductively as follows:

Initialization:

$$\alpha_1(i) = \pi_i f_{NCCS}(r_1 \mid c_i), 1 \le i \le N_s \tag{9}$$

Induction:

$$\alpha_{t+1}(j) = \left[\sum_{i=1}^{N_s} \alpha_t(i) a_{ij} \right] f_{NCCS}(r_{t+1} \mid c_j), 1 \le t \le T-1, 1 \le j \le N_s \tag{10}$$

Backward variable calculation

(2) We can efficiently find $\beta_t(i)$ inductively as follows:

Initialization:

$$\beta_T(i) = 1, 1 \le i \le N_s \tag{11}$$

Induction:

$$\beta_t(i) = \sum_{j=1}^{N_s} a_{ij} f_{NCCS}(r_{t+1} \mid c_j)\beta_{t+1}(j), t = T-1, T-2, \ldots 1, 1 \leq i \leq N_s \tag{12}$$

(3) Transition probability matrix re-estimation is done according to the following formula:

$$\bar{a}_{ij} = \frac{\sum_{t=1}^{T-1} \alpha_t(i) a_{ij} f_{NCCS}(r_{t+1} \mid c_j)\beta_{t+1}(j)}{\sum_{t=1}^{T-1}\sum_{j=1}^{N_s} \alpha_t(i) a_{ij} f_{NCCS}(r_{t+1} \mid c_j)\beta_{t+1}(j)} \tag{13}$$

We continue to iterate through (1), (2), (3) until some stopping criteria is fulfilled (a certain amount of iterations is performed or the likelihood function increase is below some threshold).

4.4. Establishing the Correspondence between the Channel Parameters and the Transmitted Bits

In order to be able to perform data detection, we should establish the correspondence between the channel parameters $c_1, c_2, \ldots . c_{N_s}$ and the channel states defined by transmitted bit sequences$(I_n, I_{n-1}, \ldots, I_{n-L})$. Following [4], we call this 'cluster labeling'. Without performing the cluster labeling, the parameter vector $c_1, c_2, \ldots . c_{N_s}$ that we found may differ from the desired vector by an unknown permutation operation. This is explained in details in [6]. Some cluster labeling algorithms need the information of a starting point for their initialization. This starting point is usually a cluster that jumps to itself and corresponds to a label of same symbols, for example 00 or 11 when L=1. In our case of noncoherent detection communication channel with NCCS noise, the starting point is different from others and can be easily distinguished: it is a zero bit sequence point, that corresponds to central chi square PDF. Therefore we can always correctly identify it. We perform the cluster labeling by using the estimated transition probability matrix and a zero bit sequence starting point.

4.5. Data Detection

The data can be detected by either SbS-MAP algorithm or Viterbi algorithm.

In case of slow time-varying channel, the channel estimation needs not to be performed continuously but periodically, frequently enough to follow the channel changes. In a time interval between the channel estimations, a Viterbi algorithm can be used for data detection. In case of fast time-varying channel, when continuous channel estimation is needed, it is convenient to use a symbol by symbol MAP (SbS-MAP) algorithm because most calculations needed for SbS-MAP are already performed during the channel estimation stage. SbS-MAP algorithm also achieves better bit error rate (BER) performance than Viterbi algorithm because it is a minimum BER algorithm [11].

5. Application for 3 Different Channels

We present 3 different communication channels where our proposed method can be applied:

(1) Binary on-off signaling with noncoherent detection over AWGN channel
(2) Binary on-off signaling with noncoherent detection over frequency selective fading channel
(3) Intensity modulated direct detection optical channel

5.1. Binary on-off Signaling with Noncoherent Detection over AWGN Channel

A discrete HMM model for this channel can be defined as follows:

1) The number of states is $N_s = 2$. The state depends only on currently transmitted bit (0 or 1). When 0 is transmitted, the channel parameter is $\boldsymbol{c}_1 = 0$. When 1 is transmitted, the channel parameter is $\boldsymbol{c}_2 = \frac{4\boldsymbol{E}_b}{\boldsymbol{N}_0}$, where $\frac{E_b}{N_0}$ is average bit energy to thermal noise two sided spectral density ratio of on-off keying signal [10].

2) The probability density function of the observation r_i conditioned on state c_j has a NCCS PDF. It is given by the following formulas (formula (1) with two degrees of freedom and c1,c2):

$$f_{NCCS}(r_i \mid c = c_1; \sigma) = \frac{1}{2\sigma^2} e^{-\frac{r_i}{2\sigma^2}} \tag{14}$$

$$f_{NCCS}(r_i \mid c = c_2; \sigma) = \frac{1}{2\sigma^2} e^{-\frac{c_2 + r_i}{2\sigma^2}} I_m\left(\frac{\sqrt{r_i c_2}}{\sigma^2}\right) \tag{15}$$

3) The state transition probability matrix is

$$A = \begin{pmatrix} 0.5 & 0.5 \\ 0.5 & 0.5 \end{pmatrix}$$

4) The initial probability distribution: $\pi_0 = 0.5, \pi_1 = 0.5$

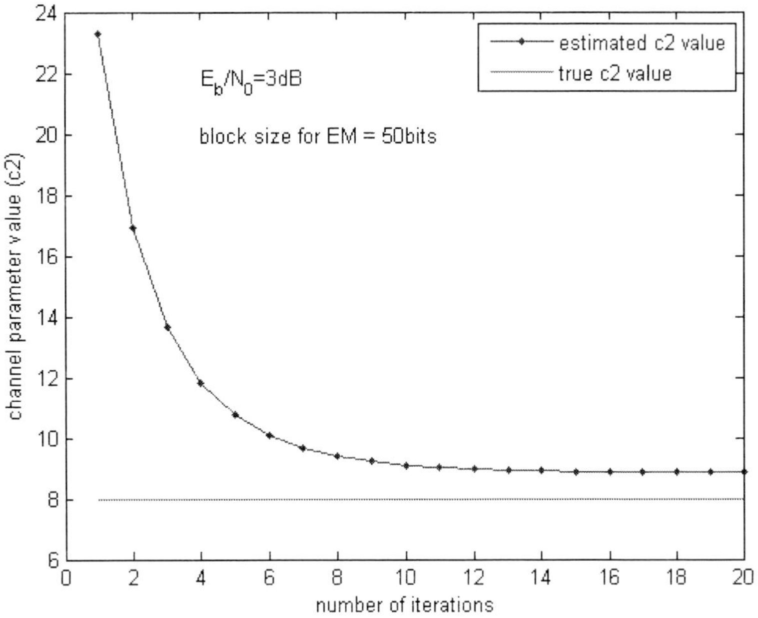

Figure 2. Evolution of channel parameter estimates with iterations.

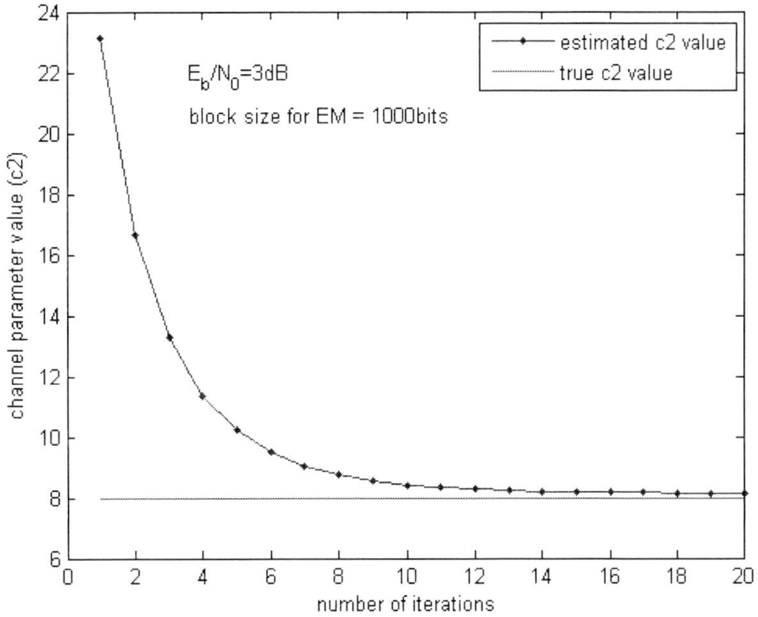

Figure 3. Evolution of channel parameter estimates with iterations.

In this case, the received signal samples have a PDF which is a mixture of 2 components: chi-square distribution resulting from channel parameter c_1 and NCCS resulting from channel parameter c_2. The results of channel parameter estimation are shown on Figure 2 for $\frac{E_b}{N_0} = 3dB$, 20 EM iterations and block of 50 bits.

We can see that after 20 EM iterations, there is still a noticeable channel parameter estimation error. We expect that channel parameter estimation error will be smaller if more bits are used for estimation. The results of channel parameter estimation are shown on Figure 3 for $\frac{E_b}{N_0} = 3dB$, block of 1000 bits.

We see the reduction of estimation error when more bits are used for the channel parameter estimation. The estimated channel parameter c_1 is directly connected to $\frac{E_b}{N_0}$. It means that the proposed algorithm can be used for optimum signal to noise ratio (SNR) estimation. The resulting SNR estimator can work at high as well as low SNR. The performance of the estimator is presented at Figure 4:

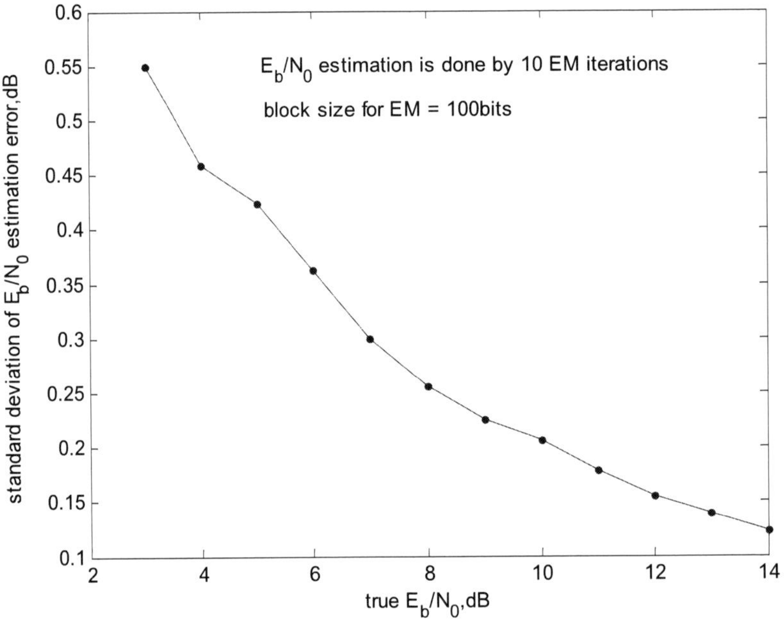

Figure 4. $\frac{E_b}{N_0}$ estimation error versus true values of $\frac{E_b}{N_0}$.

We see that the proposed SNR estimator can work at low SNR where many currently used estimators fail.

The optimal receiver for the noncoherent detection of binary OOK on AWGN channel is a simple threshold device that compares the received samples to the optimal threshold th_{opt} which is the solution of the following transcendental equation [2]:

$$I_0\left(\frac{2\sqrt{E}}{N_0}\sqrt{th_{opt}}\right) = \exp\left(\frac{E}{N_0}\right) \tag{16}$$

It can be seen from (16) that the optimum threshold depends on the values of E and N_0 (the estimate of which may not be an easy task).

Our method allows optimal detection without prior knowledge of $\frac{E_b}{N_0}$. In our method, the estimated value of $\frac{E_b}{N_0}$ naturally results from the detection algorithm. The performance of our detection algorithm that does not require any prior knowledge of $\frac{E_b}{N_0}$ is compared on Figure 5 to the performance of the well known optimum threshold detection algorithm which assumes perfect knowledge of $\frac{E_b}{N_0}$. Our algorithm performs channel estimation followed by ML detection. Channel estimation is performed on block of 100 bits. Ten EM iterations are performed.

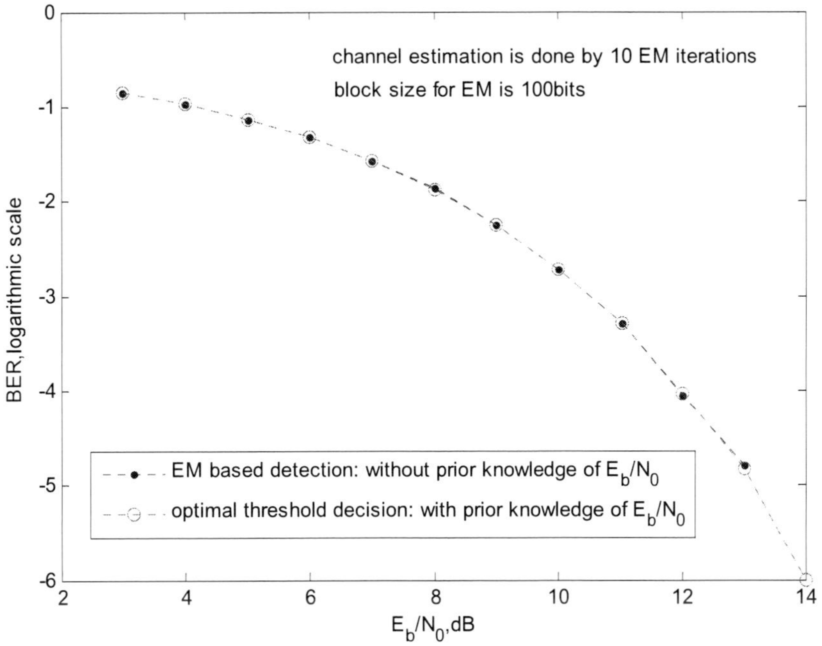

Figure 5. BER versus $\frac{E_b}{N_0}$ for classic versus EM based detection.

We can see from Figure 5 that the BER of the estimated channel is almost identical to the BER of the known channel.

5.2. Binary on-off Signaling with Noncoherent Detection over Frequency Selective Fading Channel

When OOK signals are passed through a frequency selective multipath fading channel, the multipath delays in the channel result in intersymbol interference between different symbols.

In this example, we select the channel of memory length L=2 and transfer function $H(z)=1+0.6z^{-1}+0.3z^{-2}$ ($h_0=1, h_1=0.6, h_2=0.3$). This is a transfer function of the linear part of the channel that precedes the square law detector. The channel parameter is a function of the channel state $c_t=c(s_t)$. In case of frequency selective fading channel with noncoherent square law detection, $f_{NCCS}(r_t \mid c_t;\sigma)$ has two degrees of freedom and channel parameter depends on the channel state as follows: $c_t=c(s_t)=\left(\sum_{k=0}^{k=L-1} h_k I_{t-k}\right)^2$

The HMM model of this channel can be defined as follows:

1) The number of states is $N_s=8$. The channel state at time instant t is determined by the current and two previously transmitted bits: $s_t=(I_t, I_{t-1}, I_{t-2})$. Channel parameter depends on the channel state as follows: $c_t=c(s_t)=\left(\sum_{k=0}^{k=2} h_k I_{t-k}\right)^2$

2) The probability density function of the observation r_i conditioned on state c_j has a NCCS PDF. It is given by the following formulas (insert two degrees of freedom in (1)):

$$f(r_i \mid c=c_j;\sigma)=\frac{1}{2\sigma^2}e^{-\frac{c_1+r_i}{2\sigma^2}}I_m\left(\frac{\sqrt{r_i c_j}}{\sigma^2}\right) \quad (17)$$

3) The state transition probability matrix is defined in Table1 (the correspondence between the transmitted bits and channel parameters is defined in Table 2):

Table 1. State Transition Probability Matrix

	c_1	c_2	c_3	c_4	c_5	c_6	c_7	c_8
c_1	0.5	0	0	0	0.5	0	0	0
c_2	0.5	0	0	0	0.5	0	0	0
c_3	0	0.5	0	0	0	0.5	0	0
c_4	0	0.5	0	0	0	0.5	0	0
c_5	0	0	0.5	0	0	0	0.5	0
c_6	0	0	0.5	0	0	0	0.5	0
c_7	0	0	0	0.5	0	0	0	0.5
c_8	0	0	0	0.5	0	0	0	0.5

4) The initial probability distribution: $\pi_i = \frac{1}{8}, i = 1,...8$

The resulting channel states (that are determined by transmitted bit sequences) and their corresponding channel parameters (around which the received samples are clustering) are shown in Table2 for $\frac{E_b}{N_0} = 9dB$.

Table 2. Channel parameters and their corresponding states

I_k	I_{k-1}	I_{k-2}	Corresponding channel parameter	Corresponding channel parameter value
0	0	0	C_1	0
0	0	1	C_2	2.85
0	1	0	C_3	11.41
0	1	1	C_4	25.68
1	0	0	C_5	31.7
1	0	1	C_6	53.57
1	1	0	C_7	81.15
1	1	1	C_8	114.43

The histogram of 100000 received samples for this channel is shown in Figure 6 for $\frac{E_b}{N_0} = 9dB$.

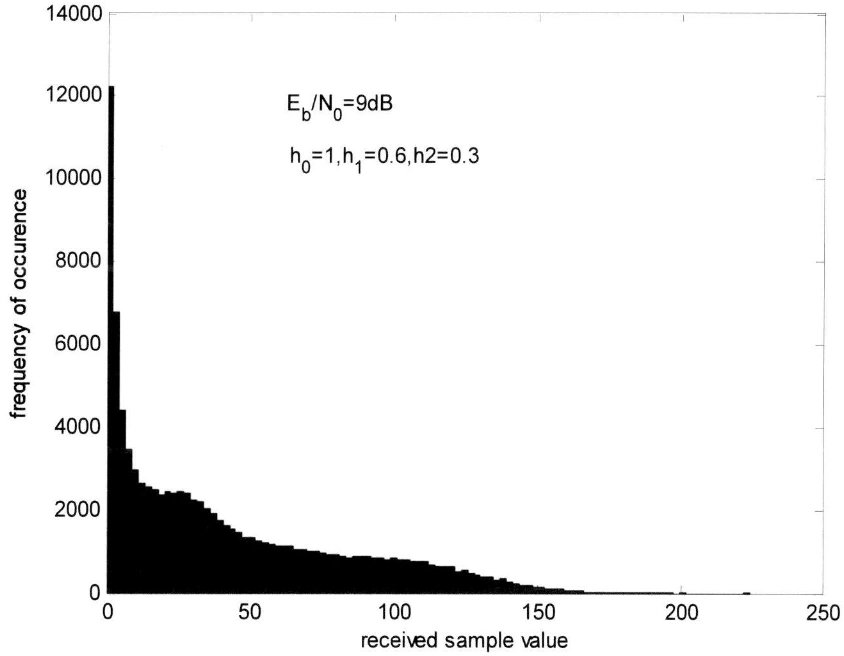

Figure 6. Histogram of received observations.

The results of channel parameter estimation are shown on Figure 7. We can see on Figure 7 the convergence of the channel estimates to their true values.

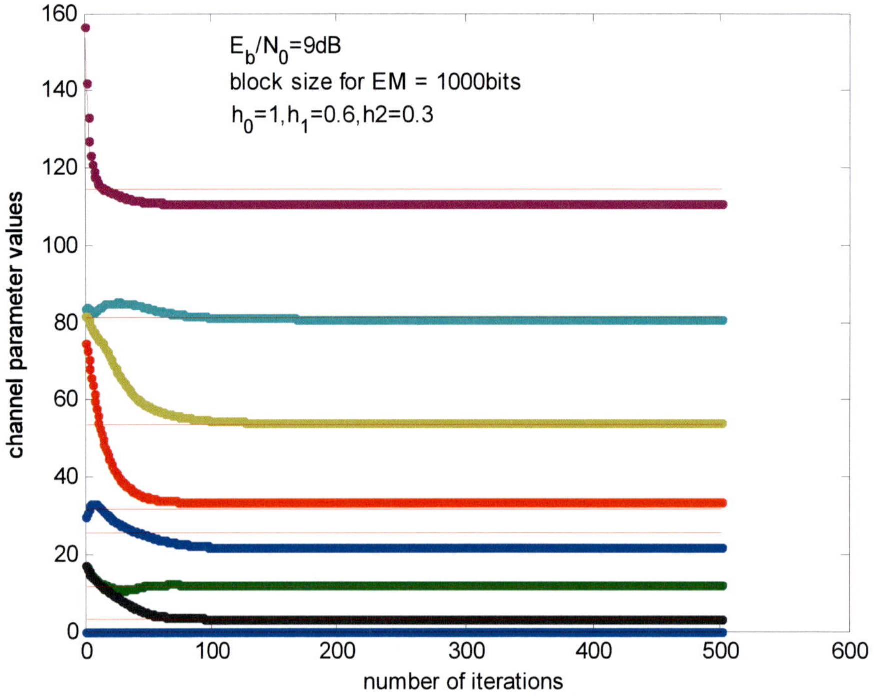

Figure 7. Evolution of channel parameter estimates with iterations.

State Transition Probability Matrix Estimation by BWA

Now when the channel states are estimated, BWA algorithm is applied in order to find the state transition probability matrix. We use 200 received samples for this purpose. 5 BWA iterations were performed. The initial matrix we took was uniform. All its elements were equal to 0.125.

The estimated transition probability matrix after the fifth BWA iteration is shown below:

$$\boldsymbol{A}_5 = \begin{pmatrix} 0.36 & 0.12 & 0.01 & 0.03 & 0.2 & 0.00 & 0.00 & 0.00 \\ 0.43 & 0.1 & 0.01 & 0.04 & 0.54 & 0.07 & 0.00 & 0.00 \\ 0.17 & 0.16 & 0.01 & 0.00 & 0.09 & 0.43 & 0.02 & 0.00 \\ 0.04 & 0.55 & 0.09 & 0.00 & 0.01 & 0.31 & 0.05 & 0.00 \\ 0.00 & 0.06 & 0.46 & 0.12 & 0.03 & 0.14 & 0.37 & 0.03 \\ 0.00 & 0.01 & 0.36 & 0.16 & 0.01 & 0.04 & 0.47 & 0.11 \\ 0.00 & 0.00 & 0.06 & 0.26 & 0.03 & 0.00 & 0.09 & 0.3 \\ 0.00 & 0.00 & 0.01 & 0.38 & 0.09 & 0.00 & 0.01 & 0.56 \end{pmatrix}$$

We know that all the elements of this matrix are either zero or 0.5. Looking at the matrix after the fifth BWA iteration we can say with a high level of confidence which element is zero and which is 0.5.

Establishing the Correspondence between the Channel States and the Transmitted Bits

We see that 5 iterations are enough and cluster labeling can be performed by comparing the estimated state transition probability matrix with the true state transition probability matrix (Table 1). Comparing between these matrices and using our knowledge that c_1 corresponds to (000) we complete the cluster labeling. This completes the channel estimation. Data detection can be performed by either SbS-MAP algorithm or Viterbi algorithm.

5.3. Intensity Modulated Direct Detection Optical Channel

The schematic diagram of an optical communication system is shown in Figure 8.

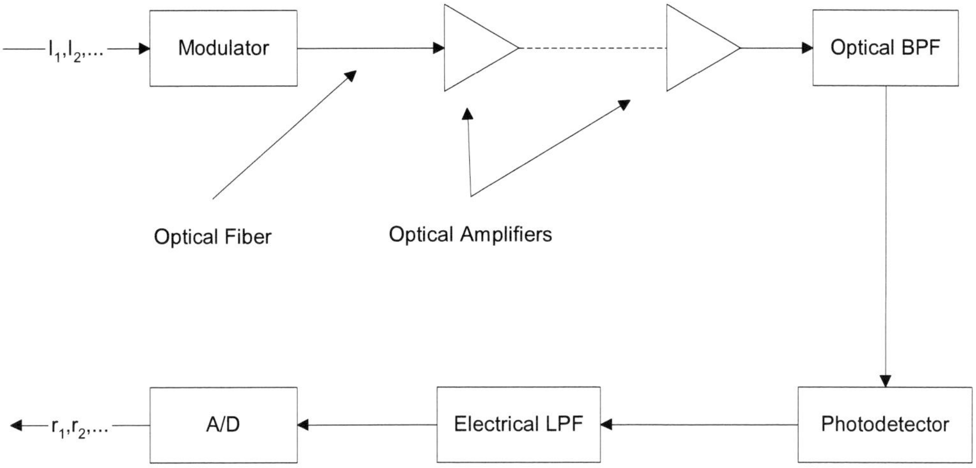

Figure 8. Schematic diagram of an optical communication system.

The transmitted bits $I_1, I_2,...$ are converted into the optical signal that is inserted into the fiber. It then passes through a transmission line that primarily consists of fiber spans and optical amplifiers. At the end of the transmission line, the signal is optically filtered and enters a receiver, where it is converted to electrical current by a high-speed photodiode. This current is low-pass filtered and sampled at one sample per bit by the A-D converter. The received samples are denoted by $r_1, r_2,...$. We consider optical communication system that is limited by Amplified Spontaneous Emission (ASE) noise and ISI and uses on-off keying (OOK) modulation. It can be shown, that the received samples have a NCCS PDF [18].

A discrete HMM model for this channel can be defined as follows:

1) The number of states is $N_s = 2^{L+1}$

2) The probability density function of the observation r_i conditioned on state c_j has a NCCS PDF given by formula (1). The channel parameter is given by the following formula:

$$c_t = c(s_t) = \left(\sum_{k=0}^{k=L-1} h_k I_{t-k} \right)^2 \tag{18}$$

$h_0, h_1, ..h_{L-1}$ are real constants that characterize the linear ISI in the optical channel before the square law detector. This linear optical ISI is a consequence of two major fiber impairments: chromatic dispersion and polarization-mode dispersion [18].

3) The state transition probability matrix is defined by formula (2)

4) The initial probability distribution: $\pi_0, \pi_1, ...\pi_{N_s-1}$

In our example, we modeled the channel of memory length L=2 and optical transfer function $H(z) = 1 + 0.6z^{-1} + 0.3z^{-2}$ ($\boldsymbol{h}_0 = 1, \boldsymbol{h}_1 = 0.6, \boldsymbol{h}_2 = 0.3$). This is a transfer function of the linear part of the channel that precedes the square law detector.

In this case we have eight NCCS PDF clusters corresponding to eight different bit combinations, where the corresponding noncentrality parameters are calculated by the following formula:

$$c = (A_{sc})^2 \left[(1 \cdot I_k + 0.6 \cdot I_{k-1} + 0.3 \cdot I_{k-2}) \right]^2 \tag{19}$$

Changing the scaling factor A_{sc} of channel parameters (while keeping $\sigma = 1$) enables us to simulate different optical signal to noise ratio (OSNR) conditions. The OSNR is defined by the following formula [18]:

$$OSNR = \frac{E\{c^2\}}{E_n} = \frac{E\{c^2\}}{4(BT_s + 1)\sigma} \tag{20}$$

where T_S is symbol duration, B is the optical filter bandwidth; σ is a noise parameter of NCCS PDF. We took a 10Gb/s transmission system using a 50-GHz optical filter which corresponds to NCCS number of degrees of freedom n = 24.

The resulting channel states (that are determined by transmitted bit sequences) and their corresponding channel parameters (around which the received samples are clustering) are shown in Table3 for $A_{sc} = 10$:

The histogram of 100000 received samples for this channel is shown on Figure 9 for $A_{sc} = 10$. Significant cluster overlap can be seen. We can see that the received observations form 8 different overlapping clusters for each of the channel parameters.

Table 3. Channel parameters and their corresponding states

I_k	I_{k-1}	I_{k-2}	Corresponding channel parameter	Corresponding channel parameter value
0	0	0	c_0	0
0	0	1	c_1	9
0	1	0	c_2	36
0	1	1	c_3	81
1	0	0	c_4	100
1	0	1	c_5	169
1	1	0	c_6	256
1	1	1	c_7	361

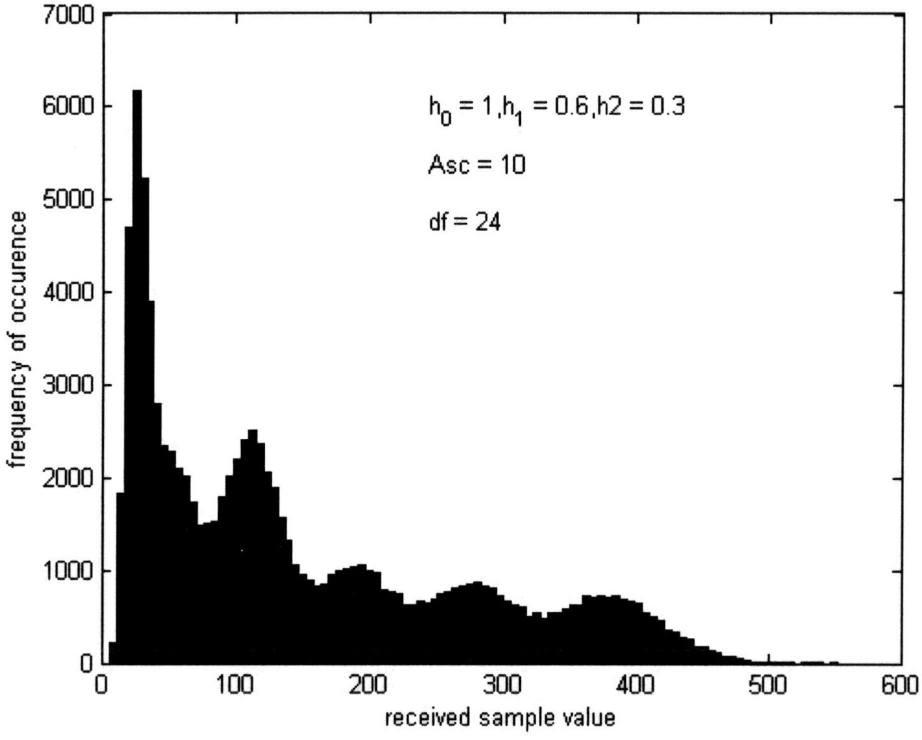

Figure 9. Histogram of received observations.

Channel States Estimation

The results of channel parameter estimation are shown in Figure 10 for A_{sc} = 10, where the horizontal lines indicate different channel parameters to be estimated.. Calculations were performed for N=1000 received samples.

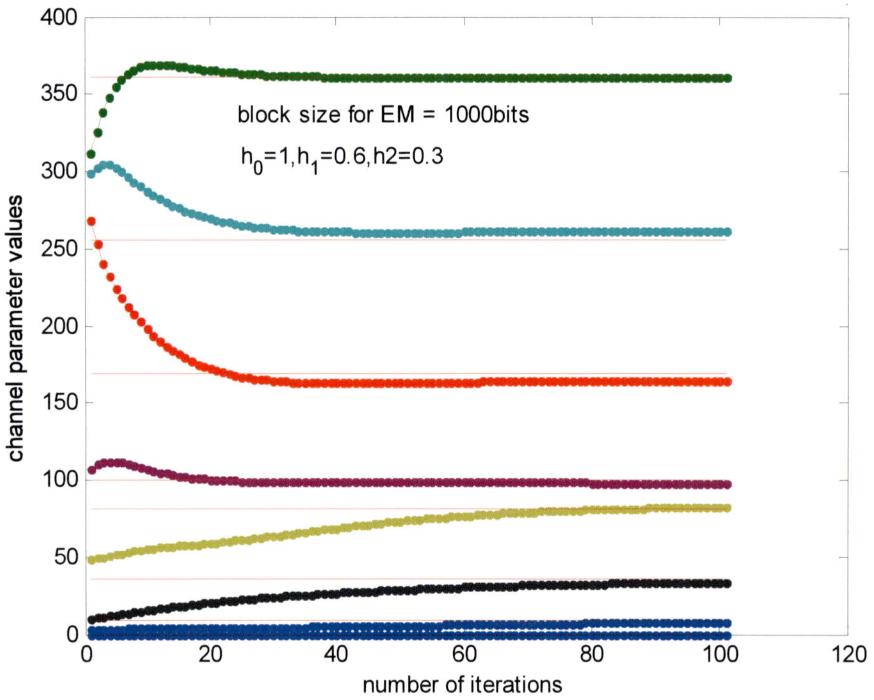

Figure 10. Evolution of channel parameter estimates with iterations.

State Transition Probability Matrix Estimation by BWA

Now when the channel states are estimated, BWA algorithm is applied in order to find the state transition probability matrix. We use 200 received samples for this purpose. 5 BWA iterations were performed. The initial matrix we took was uniform. All its elements were equal to 0.125.

The estimated transition probability matrix after the fifth BWA iteration allowed determining the entries of 0.5 and zero.

Establishing the Correspondence between the Channel States and the Transmitted Bits

Cluster labeling was performed by comparing the estimated state transition probability matrix with the true state transition probability matrix. Comparing between these matrices and using our knowledge that c_1 corresponds to (000) we complete the cluster labeling. This completes the channel estimation. Data detection can be performed by either SbS-MAP algorithm or Viterbi algorithm.

6. Conclusion

We have presented a new, EM based approach for optimum noncoherent receiver design. The proposed receiver can perform blind channel estimation and data detection over ISI channels.

The channel model is based on HMM model with NCCS observation densities. The estimation of HMM parameters is divided into two steps. On the first step, channel parameters are estimated by a recently developed algorithm for NCCS distribution and its finite mixtures. On the second step, transition probability matrix is estimated by a standard BWA algorithm. The proposed method is general and can be applied to different types of noncoherent receivers. Simulation results were presented for three communication channels: noncoherent receiver for binary OOK modulation on AWGN channel, noncoherent receiver for binary OOK on frequency selective fading channel and intensity modulated direct detection optical channel with chromatic and polarization-mode dispersion. Simulation results show that accurate channel estimation is possible. More research is required to find low complexity modifications of the proposed algorithms.

References

[1] Carles Anton-Haro, Jose A.R. Fonollosa, and Javier R. Fonollosa, "Blind Channel Estimation and Data Detection Using Hidden Markov Models", *IEEE Transactions on Signal Processing,* Vol.45, No.1, January 1997.

[2] Sergio Benedetto and Ezio Biglieri, *Principles of digital transmission*, Kluwer Academic/Plenum Publishers, 1999.

[3] A.P. Dempster, N.M. Laird, and D.B. Rubin, "Maximum likelihood from incomplete data via the EM algorithm", *J. Roy. Stat. Soc.*, **39**(1):1-38,1997.

[4] K.Georgoulakis, S. Theodoridis, "Blind and semi-blind equalization using hidden Markov models and clustering techniques", *Signal Processing* **80** (2000) 1795-1805.

[5] George Iliopoulos and Dimitris Karlis, ''EM algorithm for the non central chi square distribution and its finite mixtures'', *Technical Report, Department of Statistics*, Athens University of Economics, 2005.

[6] Yih-Jyi Jeng and Chien-Chung Yeh, "Cluster-Based Blind Nonlinear-Channel Estimation", *IEEE Transactions on Signal Processing*, Vol.45, No.5, May 1997.

[7] Norman L. Johnson, Samuel Kotz, N. Balakrishnan, Continuous Univariate Distributions, Volume 2, Second edition, *Wiley series in probability and mathematical statistics,* John Wiley and Sons, Inc, 1995.

[8] G.K.Kaleh and R.Vallet, "Joint Parameter Estimation and Symbol Detection for Linear or Nonlinear Unknown Channels", *IEEE Transactions on Communications*, Vol.42, No.7, July 1994.

[9] Yannis Kopsinis, Sergios Theodoridis, "A novel cluster based MLSE equalizer for M-PAM signaling schemes", *Signal Processing* **83** (2003) 1905-1918.

[10] Jhong Sam Lee and Leonard E. Miller, *CDMA Systems Engineering Handbook*, Artech House, 1998.

[11] Shu Lin and Daniel J. Costello, *Error Control Coding, Prentice Hall*, 2004, Second Edition.

[12] R. Lisnanski, "Generalized Expectation-Maximization algorithm for Maximum-Likelihood Fitting of Finite Mixture Models and its Applications in Signal Processing", *25th European Meeting of Statisticians*, Oslo – Norway, July 2005.

[13] Roman Lisnanski and Anthony J Weiss, ''Low complexity generalized EM algorithm for blind channel estimation and data detection in optical communication systems'', *Signal Processing* **86**(2006) 3393-3403.

[14] Geoffrey J.McLachlan and Thriyambakam Krishnan, *The EM Algorithm and Extensions*. John Wiley and Sons, 1997.

[15] Geoffrey McLachlan and David Peel, *Finite Mixture Models*, John Wiley and Sons, 2000.

[16] L.R. Rabiner, "A Tutorial on Hidden Markov Models and Selected Applications in Speech Recognition," *Proc. IEEE*, **77**(2): 257-286, February 1989.

[17] W. Turin, *Digital Transmission Systems: Performance Analysis and Modeling*, McGraw-Hill, 1998.

[18] A.J.Weiss, "On the Performance of Electrical Equalization in Optical Fiber Transmission Systems", *IEEE Photonics Technology Letters*, vol.15,NO.9, September 2003.

In: New Signal Processing Research
Editor: Takumi Maeda, pp. 411-422
ISBN: 978-1-60456-479-2

Chapter 13

A Solver Based on the Coding and Compressing of Solution Space with PSO for Sparse Solution of Linear Underdetermined System

Peng Xu and Dezhong Yao *

Center of Neuroinformatics, School of Life Science and Technology,
University of Electronic Science and Technology of China,
Chengdu, 610054, China

Abstract

In research and engineering applications, it is usually necessary to solve the underdetermined system for its sparse solution, whereas most of current methods cannot solve this problem effectively. In this paper, a novel algorithm is proposed to solve underdetermined system sparsely. In this algorithm, the solution space is coded with some particles firstly, and then the solution space is compressed in this particle-coded space by the evolution way of particle swarm optimization method to convert the underdetermined system to a non-underdetermined one as possible. Finally, the underdetermined system is solved in the compressed sub-space robustly after it is converted to a non-underdetermined one. During the solving procedure, a new l_0 constrained fitness function is introduced to select the global optimal sparse solution for the underdetermined system. The proposed method is compared with Minimum Norm (MN) and Focal Underdetermined system solver (FOCUSS) by solving two different underdetermined systems: one was a random system, the other was an underdetermined system derived from a 3-shell sphere head model, which is usually adopted in EEG source imaging. The current results confirmed a good sparse solution for the underdetermined system could be achieved with the proposed method in this paper.

Key words: underdetermined system, sparse solution, particle swarm, EEG source imaging

* E-mail address: dyao@uestc.edu.cn. Corresponding author: Dezhong Yao.

1. Introduction

The following underdetermined system will be usually encountered in current research and engineering application,

$$Y = AX \quad (1)$$

where Y is recording of $M \times 1$, M is the number of sensors. A is the lead field matrix of $M \times N$, where N is the dimension size of possible solution space. X is the solution vector to be estimated. For many problems, M is usually much smaller than N, which means that the system is underdetermined, thus the problem lacks a unique solution because there are an infinite number of possible solution configurations that could explain the measured recordings Y [1], [3], [5]-[7].

How to obtain a feasible solution consistent with the actual problem is still the focus in such research area as estimation of Direction of arrival (DOA), EEG or MEG inverse problem, etc. When solving such an underdetermined problem, some constraints are usually imposed to get the solution satisfying the actual problem. Some constraints such as minimum norm (MN) and various prior information of the solution space are currently adopted. Recently, much attention has been paid to the sparse constraint of an underdetermined system. For EEG inverse problem, usually the main neural electric activities are sparsely localized, thus a reasonable solution should not only explain the scalp recordings but also be localized sparsely [1], [4]-[5], [8], [22]. Compared to the sensor dimension, the number of DOA to be estimated is very small, so sparsity can be also considered as a constraint to solve the underdetermined system. The focal underdetermined system solver (FOCUSS) is the current technique to solve the underdetermined system sparsely and focally, but it is very common being trapped by the local minima and failed to converge to the actual solution configuration for some cases [5], [22].

The original and most effective way to measure the sparsity of a signal is the l_0 norm, which is to calculate the number of the non-null entries in the signal [11]. If the l_0 norm of the solution is taken as a constraint, we have a Lagrange multiplier expression of the inverse problem (1) as,

$$\min_{X} \| Y - AX \|_2 + \lambda \| X \|_0 \quad (2)$$

where λ is the punishment factor of sparsity. But formula (2) is not continuously differentiated and many effective optimization methods based on the gradient of object function can not be directly applied to solve this problem, so in practice, $l_p (p \le 1)$ norm is usually taken to replace l_0 norm, where l_1 norm is the mostly adopted in current practice, whereas the approximation will lower the solution sparsity to some degree [1], [4], [11].

In formula (2), the value of object function consists of two different components, one is the error between the estimated signal and the observed signal, and the other is the punishment weight for the sparsity. In the underdetermined system with sparse solution, the number of effective entries, i.e. the number of non-null entries in the solution X, is usually not larger than M, the dimension of the observed signal Y. Thus many entries in the solution X are null and have no contribution to the observed signal Y, and accordingly the corresponding columns in the lead field matrix A are invalid for the solution estimation. If the invalid columns in the lead field matrix can be totally or partly removed, formula (1) may be converted to a non-underdetermined system. As we know, the solving of a non-underdetermined system is much easier and more stable than solving of the underdetermined one [5], [8]. So formula (1) can be solved in a compressed solution space for the sparse solution in a way to solve the non-underdetermined system.

The particle swarm algorithm has been found to be robust and fast in solving problems featuring nonlinearity and non-differentiability, multiple optima [12]-[14]. In this work, we adopted the particle swarm to encode the solution space for underdetermined system firstly, and by the evolution way similar to that in the particle swarm optimization method, the solution space was iteratively compressed in the particle-coded space so that the underdetermined system could be converted to a non-underdetermined one and solved with the singular value decomposition (SVD) based method in this compressed sub-solution-space, and then a new l_0 constrained fitness function was introduced to guarantee the algorithm to select the global optimal sparse solution for the underdetermined system.

The method was introduced in Section 2. In Section 3, the algorithm was tested to solve the random linear underdetermined system and to localize sources on a 3-shell spherical head model, and the comparison with MN solution and FOCUSS was given in this section too. Discussions and conclusions concluded this paper in Section 4.

2. Methods

2.1. Standard Particle Swarm Algorithm

Particle swarm optimization (PSO), a novel evolutional algorithm, was originally designed and developed by Eberhart and Kennedy motivated from the simulation of social behavior [13], [15]-[17]. In recent years, PSO has been successfully adopted to solve many hard problems and it has attracted wide attentions [18]-[20]. In PSO, instead of using genetic operators, each particle (individual) adjusts its 'flying' according to its own flying experience and its companions' flying experience [12]-[13], [16]. Each particle is treated as a point in a D-dimensional space, where D is usually the dimension of solution space. The *ith* particle is represented as $X_i = (x_{i1}, x_{i2}, \ldots\ldots, x_{iD})$. The best previous position (the position giving the best fitness value) of the *i*th particle is recorded and represented as $P_i = (p_{i1,} p_{i2,} \ldots\ldots, p_{iD})$. And the best position ever achieved by the all particles is denoted by P_g. The velocity of the *i*th particle is represented as $V_i = (v_{i1,} v_{i2,} \ldots\ldots, v_{iD})$. In each offspring, the velocity and position information of the *i*th particle is updated as [15],

$$\begin{cases} v_{id} = wv_{id} + c_1 r_1 (p_{id} - x_{id}) + c_2 r_2 (p_{gd} - x_{id}) \\ x_{id} = x_{id} + v_{id} \end{cases} \tag{3}$$

where w is the inertia weight, c_1 and c_2 are two positive velocity constants, r_1 and r_2 are two random number in the range [0,1]. In literature [12], the authors proposed that using of the iteration decreasing inertia weight would avail for a good result. It is the updating way of formula (3) that gives PSO powerful ability to search for the global optima. The iteration procedure of the standard PSO with m particles can be briefly stated as the follows,

Step 1. Initialize the position and velocity information of m particles randomly; decide the inertia weight w, velocity constants c_1 and c_2; initialize the number of offspring, $G_{\max}$.

Step 2. Evaluate the fitness value of each particle: $f(i), 1 \le i \le m$.

Step 3. Update the best position information of each particle: Compare the fitness value $f(i)$ of the ith particle at current position X_i with the ever achieved best fitness value at position P_i, if $f(i)$ is better, P_i is replaced by the current position X_i, else P_i will be remained. Where $1 \le i \le m$.

Step 4. Update P_g, the best position information ever achieved by all the particles: Compare the ever achieved best fitness value of the ith particle at position P_i with the global best fitness value at P_g, if the value at P_i is better, P_g will be replaced by P_i, else P_g will be remained.

Step 5. Update the velocity and position information of each particle according to formula (3).

Step 6. Judge the stopping criteria: if the global fitness value satisfies the requirement or the number of offspring is larger than the predefined number $G_{\max}$, the iterations will be stopped, else return to Step 2 and go on.

2.2. The Solver Based on the Coding and Compressing of Solution Space with PSO for Sparse Solution of Underdetermined System

2.2.1. Coding and Compressing the Solution Space with Particles

There are an infinite number of solutions satisfying the underdetermined system denoted in formula (1), but for such problem as EEG inverse problem, DOA estimation, a sparse solution is reasonable and expected, and the current minimum norm based methods can't solve this problem effectively [1]-[2], [4]-[5], [8]. The sparsity of solution means that there are only a small number of non-null entries in the solution and most of the elements in the solution don't contribute to the observed signal [11]. Based on the sparsity of solution, the solution space can be adequately compressed by removing these sub–solution-spaces, which have no contribution to the observed signal. Let Θ represent one compressed solution space with dimension of *K*, and X_s represent the solution in space Θ, and Θ includes the whole source information. The compressed format of formula (1) in Θ can be denoted as,

$$Y = A_s X_s \tag{4}$$

If Θ is compressed enough to make *K* smaller than *M*, and the ideal case is that *K*, the number of columns in A_s, is equal to the number of sources. When *K* is smaller than *M*, the underdetermined system can be simplified to an over-determined one and for most of the over-determined systems, the solutions are certain [1], [5], [8], so in the compressed sub-solution-space Θ, the sources can be estimated as,

$$X_s = (A_s)^+ Y \tag{5}$$

where $(A_s)^+$ denotes the inverse operator for A_s. Finally, the estimated X_s in the compressed sub-space Θ will be mapped to the original solution space to get the final solution *X* for the underdetermined system. But for the real EEG inverse problem, the prior information about the solution spaces of sources is very poor and if the sub-space is not compressed suitably, the sources will not be correctly estimated [5], [8].

PSO has been proved to be powerful to search for the global optima [15]-[19]. In this work, particle swarm is adopted to divide and search for the optimal sub-solution space.

For the underdetermined system $Y = AX$, the solution space is encoded by particle as,

$$\begin{gathered} X = (x_1, x_{2,} ...., x_N) \\ \Downarrow \\ E = (e_1, e_2, ..., e_N) \end{gathered} \tag{6}$$

where *E* is the code-chain particle with the same dimension as that of the original solution space, and $e_j (1 \le j \le N)$ within range [-1,1], represents the activation probability of solution at the *j*th position in the original solution space. For a given source activation threshold ε within range [0,1], if $\| e_j \| \ge \varepsilon$, the *j*th position may be supposed to have source. By scanning each element in the code chain *E* from left to right, those possible source activation positions will be selected to construct the compressed sub-solution-space, and the corresponding columns in lead field matrix *A* are taken out to form the sub-lead-field matrix A_s. For convenience to describe the algorithm, the procedure of judging the possible source positions and compressing the sub-solution-space is represented by operator $Dec(\cdot,\cdot,\cdot)$ as,

$$(X_s, A_s) = Dec(A, X, E) \tag{7}$$

where X_s is the solution in the compressed sub-solution-space and A_s is the sub-lead-field matrix.

To get the good compressed sub-space, *m* particles are adopted to encode and compress the solution space. Let $(E_1, E_2, \ldots\ldots, E_m)$ represent the *m*-particle-encoded solution space, where $E_i (1 \le i \le m)$ is the encoded vector with the same dimension as the original solution space. The l_0 constrained fitness function, $f(X) = \| Y - AX \|_2 + \lambda \| X \|_0$, is adopted to evaluate the quality of the compressed sub-space in our algorithm.

2.2.2. Procedure of Particles Based Sparse Solution Algorithm

Suppose the dimension of the original solution space is *D*, the method based on the coding and compressing of solution space with PSO for sparse solution of underdetermined problem can be summarized as,

Step 1. Initialize the variables: Initialize the *m*-particle encoded solution space $(E_1, E_2, \ldots\ldots, E_m)$ with number within [0, 1] randomly; Initialize the velocity information of *m* particles, $(V_1, V_2, \ldots\ldots, V_m)$, with number within [0,1] randomly. Where $E_i, V_i (1 \le i \le m)$ are both vectors with *D*-dimension; Set the source activation threshold ε, the maximum offspring number $G_{\max}$, the sparsity punishment factor λ.

Step 2. Compress the sub-solution-space and lead field matrix denoted by each particle with formula (7) according to the code chain E_i as $(X_{si}, A_{si}) = Dec(A, X, E_i)$, where $1 \le i \le m$. Estimate the source information in the sub-space as $X_{si} = (A_{si})^{+} Y$. Finally, X_{si} is mapped to the original source space and the fitness value $f(i)$ of the *i*th particle is evaluated.

Step 3. Update the best position information of each particle: Compare the fitness value $f(i)$ of the *i*th particle at current position (code chain E_i) with the ever achieved best fitness value by this particle at position P_i, if $f(i)$ is better, P_i will be replaced by the current code chain E_i, else P_i will be remained. Where $1 \le i \le m$.

Step 4. Update P_g, the best position information ever achieved by all the particles: Compare the ever achieved best fitness value of the *i*th particle at position P_j with the global optimal fitness value at P_g, if the value at P_i is better, P_g will be replaced by P_i, else P_g will be remained.

Step 5. Update the velocity and position information of each particle as,

$$\begin{cases} V_i(n) = wV_i(n) + c_1 r_1 (P_i(n) - E_i(n)) + c_2 r_2 (P_g(n) - E_i(n)) \\ E_i(n) = E_i(n) + V_i(n) \end{cases}$$

where $1 \le n \le D$, $1 \le i \le m$. $E_i(n)$ and $V_i(n)$ are the nth code bit and the nth velocity element of the ith particle, respectively; w, c_1 and c_2 are same as those in the standard PSO. In our method, the particle velocity and element in code chain are constrained within range [-1,1] as: If $V_i(n) > 1, V_i(n) = 1$; if $V_i(n) < -1, V_i(n) = -1$; if $E_i(n) > 1, E_i(n) = 1$; if $E_i(n) < -1, E_i(n) = -1$.

Step 6. Judge the stopping criteria: If the number of offspring is larger than the predefined number $G_{\max}$, the iterations will be stopped and the solution corresponding to the global optimal position (code chain) P_g is the final solution for the underdetermined system, else return to Step 2 and go on.

3. Simulation Results

In this section, we will test the solving performance of MN, FOCUSS and the proposed method on a random underdetermined linear system and a 3-shell sphere head model, respectively.

3.1. Solving for the Random Underdetermined Linear System

In this simulation, the comparison is implemented by solving the random underdetermined linear system with the sparse solution. Similar to the dipoles in the EEG inverse problem, those sources with certain number are placed on some positions in the underdetermined system. For each case with certain number of sources, the lead field matrix A is firstly randomly generated, and then the positions and strengths of sources were randomly determined. With the randomly determined lead field matrix A and sources X, the recording Y could be achieved according to formula (1).

For certain source configuration with fixed number of sources, the simulation was repeated for 10 times, and then MN, FOCUSS and the proposed method were used to solve this system. Only when sources in the solution were consistent with the simulated one, the solution was considered to be successful.

We tested the solving for several random underdetermined system with different dimensions, we just listed the results for two tested cases with dimensions of 32×180 and 64×600 in the following Tables 1~2, and Figures 1~2. In this simulation, The parameters of the proposed method were initialized as: The swarm size (population size) was two times of the solution space dimension, i.e., 360 and 1200 for these two different dimensional system; the sparsity punishment factor was 10; the number of offspring was 100; the inertia weight was updated as $w(k) = 0.8 - \dfrac{0.8 - 0.4}{G_{\max}} \times k$, where k was the current offspring number; the velocity constants, c_1 and c_2 were both 2.0.

Table 1. The results for underdetermined with dimension of 32×180

Source Number	MN	FOCUSS	PSO
1	0/10 $^{\Delta}$	8/10	10/10
2	0/10	6/10	9/10
3	0/10	5/10	8/10
4	0/10	3/10	9/10
5	0/10	1/10	8/10
6	0/10	0/10	7/10

Δ: a/b denotes the ratio of correctly solving number,a, to the total simulated number, b.

Table 2. The results for underdetermined with dimension of 64×600

Source Number	MN	FOCUSS	PSO
3	0/10 $^{\Delta}$	3/10	9/10
5	0/10	0/10	8/10
8	0/10	1/10	8/10
10	0/10	0/10	6/10

Δ a/b denotes the ratio of correctly solving number, a, to the total simulated number, b.

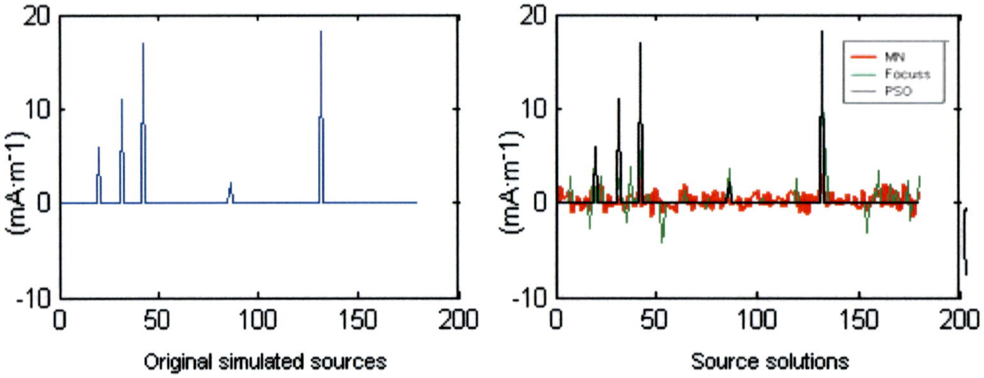

Figure 1. The sixth simulation test for the solving of 32×180 system with 5 sparse sources.

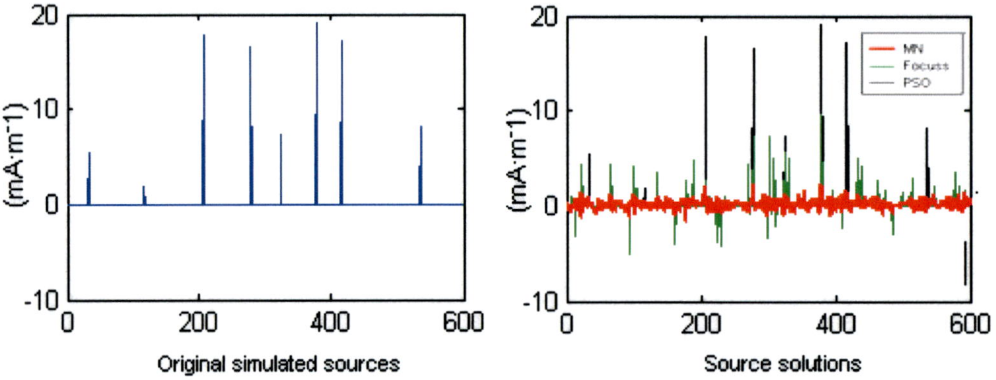

Figure 2. The first simulation test for the solving of 64×600 system with 8 sparse sources.

3.2. Source Localization on 3-Shell Spherical Head Model

In this simulation, the algorithm was tested on a normalized 3-shell spherical head model. The radiuses for the scalp, skull and brain were 1.0, 0.92 and 0.87, respectively. The sources were restricted on the cortex surface, consisting of 800 mesh nodes. The conductivities for cortex, skull and scalp were $1.0\Omega^{-1}m^{-1}$, $1/80\Omega^{-1}m^{-1}$ and $1.0\Omega^{-1}m^{-1}$, respectively. The standard 128-electrode system was adopted and the lead field matrix *A* was calculated by placing the unit radius dipole on nodes of the cortex surface, and *A* is a matrix with dimension of 128×800. Similar to the above underdetermined system, certain sources were placed on the solution meshes to simulate the scalp recording *Y* by formula (1). Three source configurations with 1, 3 and 5 sources were used to generate three different scalp mappings, and then MN, FOCUSS and the proposed method were used to localize the sources. The results for these three source configurations were shown in Figure 3. The parameters for the proposed method were as: The swarm size (population size) was 1200; the sparsity punishment factor was 10; the number of offspring was 100; the source activation threshold was 0.8; the inertia weight was updated as $w(k) = 0.8 - \frac{0.8 - 0.4}{G_{\max}} \times k$, where *k* was the current offspring number; the velocity constants, c_1 and c_2 were both 2.0, too.

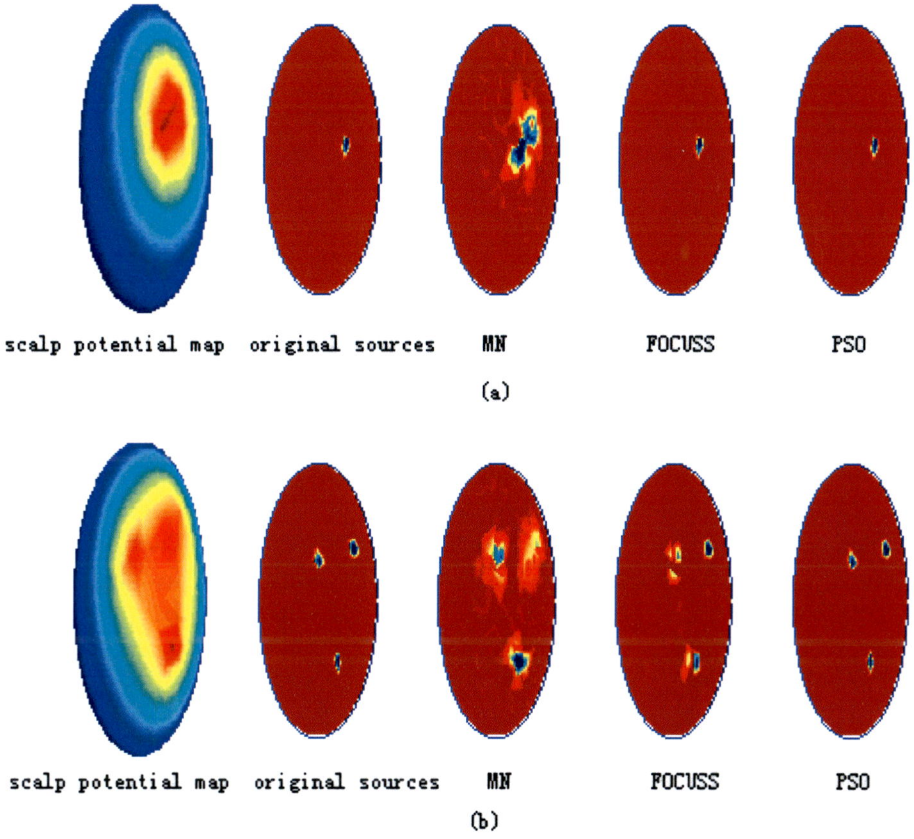

Figure 3. Continued on next page.

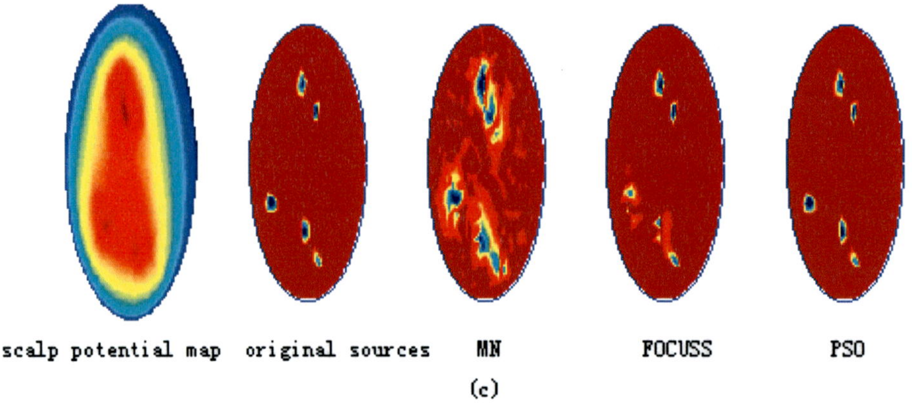

Figure 3. Source localization on 3-shell spherical head model for different source configurations. (a) 1 source; (2) 3 sources; (c) 5 sources.

4. Discussion and Conclusion

For the random underdetermined system, MN solution was power distributed and not suitable when solutions were sparsely distributed. Compared to MN solution, FOCUSS was more competitive for the sparse solution. Whereas, FOCUSS was greatly affected by the source configuration and characteristics of the underdeterminded system. As shown in Figs.1 and 2, MN solution was smooth without obvious source peak, and the solution configuration of MN was far from the original simulated configuration. With iteration strategy, FOCUSS made solution more energy localized and some obvious peaks were recovered on the due source position. Tables 1 and 2 showed that for the less complex source configurations, FOCUSS had good ability to get the desired sparse solution, however, with the increasing of source number, the FOCUSS performance was obviously lowered with fewer successful trials Compared to MN and FOCUSS, the PSO based solution space coding method showed good ability to solve the underdetermined system as those shown in those Figures and Tables in Section 3.1, where it solved most of the simulated cases robustly. For the EEG inverse problem on a 3-shell spherical head model, the source imaging with MN, FOCUSS and the proposed method on head model was similar to the above solution for the underdetermined system. The localization results of MN were scattered and blurring. Compared to MN, the solutions of FOCUSS were more focal and had relatively high spatial resolution. When source number was small, the sources estimated with MN and FOCUSS were acceptable, but when source configuration was more complex with more sources, some spurious sources were introduced. Especially in Fig. 3 (c), some sources in configurations with 5 sources were neighboring, and in the results localized with MN, and FOCUSS, the neighboring information of these sources was blurred and some sources were lost. For all the source configurations, the particles based method still localized them correctly and the imaging on head model was clear.

The above results showed that the particles based method was robust for different source configurations and different underdetermined systems. Compared to other localization methods, the improvement of localization ability of the particles based method is dependent on the following two techniques. First of all, the solution space is encoded and compressed by the swarm particles, and the powerful global searching ability of particle swarm can guarantee to find the optimal compressed sub-solution space with high probability, so the inverse problem can be converted to a non-underdetermined one and stably solved in the sub-solution-space; Secondly, the adopted new l_0 norm measured fitness function can make the solution sparsely. Moreover, in theory view, the l_0 norm can effectively eliminate the undesired source bias toward surface induced in current weighted minimum norm based methods for EEG inverse problem.

For PSO algorithm, the global searching ability is partly based on the population size [16], [17]. In the simulation for the random underdetermined system, we experimentally expanded the population size to be 4 times of the solution space size, and we found that the proposed method could solve all the simulated cases in Section 3.1 correctly, but the cost time is accordingly increased. The current study is implemented under the noise free conditions, and when noise is considered, we can use such techniques as singular value truncation and other regularization methods to eliminate the noise effect in the compressed solution space [21]. In the near future, we will use the proposed method to study the sources activated in such cognitive problems as inhibition or return, and Go/No-Go task, etc [22].

Acknowledgement

This work was supported by NSFC #60701015, #30570474 and #60571019.

References

[1] C. M. Michel, M. Murray, "EEG source imaging," *Clinical neurophysiology*, vol.115, pp. 2195-2222, 2004.

[2] R.D. Pascual-Marqui, C.M. Michel, D. Lehmann, "Low resolution electromagnetic tomography: a new method for localizing electrical activity in the brain," *Int. J. Psychophysiol*, vol.18, pp.49-65, 1994.

[3] J. Z. Wang, S. J. Williamson, and L. Kaufman, "Magnetic source images determined by a lead-field analysis: The unique minimum-norm least-squares estimation," *IEEE Trans. Biomed. Eng.*, vol.39, pp. 665-675, 1992.

[4] C. Silva, J.C. Maltez, E. Trindade, "Evaluation of L_1 and L_2 minimum norm performances on EEG localizations," *Clinical neurophysiology*, vol. 115, pp.1657-1668, 2004.

[5] H. S. Liu, X. R. Gao, " A recursive algorithm for the three-dimensional imaging of brain electric activity: shrinking LORETA-FOCUSS," *IEEE Trans. Biomed. Eng.*, vol.51, no.10, pp.1794-1802, 2004.

[6] D. Yao, "The equivalent source technique and cortical imaging," *Electroenceph and clinical neurophysiol*, vol.98, pp.478-483, 1996.

[7] I. F. Gorodnitsky , B. D. Rao, "Sparse signal reconstruction from limited data using FOCUSS: A re-weighted minimum norm algorithm," *IEEE Trans. S.P.*, vol. 45, pp. 600-616, 1997.

[8] D Yao, B. He, "A Self-Coherence Enhancement Algorithm and its Application to Enhancing 3D Source Estimation from EEGs," *Annals of Biomedical Engineering*, vol.29, no.11, pp.1019-1027, 2001.

[9] D. Yao, B. He, "The Laplacian weighted minimum norm estimated of three-dimensional equivalent charge distribution in the brain," *Proc. IEEE/EMBS,* vol.20, pp.2108-2111, 1998.

[10] B. He, D. Yao, J. Lian, D. Wu, "An equivalent current source model and Laplacian weighted minimum norm current estimates of Brain electrical activity," *IEEE Trans. Biomed. Eng*., 2002 vol.4, pp.277-288, 2002.

[11] D.M. Malioutov, M. Cetin, A. S. Willsky, "Optimal sparse representations in general overcomplete bases," *IEEE International Conference on Acoustics, Speech, and Signal Processing*, vol.2, pp.793-796, 2004.

[12] Y. Shi, R.C. Eberhart, "Parameter selection in particle swarm optimization," *Evolutionary Programming VII: Proc. EP,* New York: Springer-Verlag, pp. 591-600, 1998.

[13] R.C. Eberhart, Y. Shi, "Comparison between genetic algorithms and particle swarm optimization," *Proc 7th Annual Conf on Evolutionary Programming,* Berlin: Springer, pp. 611-615, 1998.

[14] I.C. Trelea, "The particle swarm optimization algorithm: convergence analysis and parameter selection," *Information Processing Letters, vol.* 85, pp.317-25, 2003.

[15] J. Kennedy, R. Eberhart, "Particle swarm optimization," *Proc IEEE Int Conf on Neural Networks,* Perth, pp. 1942-1948, 1995.

[16] J. Kennedy, "The behavior of particles," *Proc 7th Annual Conf on Evolutionary Programming,* Berlin: Springer, pp. 581-591, 1998.

[17] R.C. Eberhart, J. Kennedy, "A new optimizer using particle swarm theory," *IEEE Proceedings of the sixth international symposium on micro machine and human science.* Nagoya, Japan, pp. 39-43, 1995.

[18] M.P. Wachowiak, R. Smolíková, Y.F. Zheng, "An approach to multimodal biomedical image registration utilizing particle swarm optimization," *IEEE Transactions on Evolutionary Computation,* vol. 8, pp.289-301, 2004.

[19] K.E. Parsopoulos, M.N. Vrahatis, "Recent approaches to global optimization problems through Particle Swarm Optimization," *Natural Computing,* vol. 1, pp.235-306, 2002.

[20] X. Hu, Y. Shi, R.C. Eberhart. "Recent advances in particle swarm," *Proceedings of the IEEE International Conference on Evolutionary Computation,* pp.90-97, 2004.

[21] A.N. Tikhonov, V.Y. Arsenin. *"Solutions of ill-posed problems (trans. from Russian),"* New York: Wiley, 1977.

[22] X. Peng, T. Yin, C. HuaFu, Y. DeZhong. Lp norm iterative sparse solution for EEG source localization. *IEEE Trans Biomed Eng*, vol. 54, no.3, 400-409, 2007.

In: New Signal Processing Research
Editor: Takumi Maeda, pp. 423-438
ISBN 978-1-60456-479-2

Chapter 14

BISTATIC STAP FOR GROUND MOVING TARGET INDICATION

Chin-Heng Lim
Temasek Laboratories@NTU, Research Techno Plaza,
Singapore 637553, Singapore

Abstract

Space-time adaptive processing (STAP) for bistatic airborne radar offers several advantages, such as the higher possibility of detecting stealth targets. Moreover, the ability to passively provide surveillance data without alerting the target under observation has implications for peace keeping operations and counter-terrorism activities. One way to achieve these capabilities is to deploy reconfigurable bistatic sensor systems where passive receiver(s), such as unmanned aerial vehicles (UAVs), are not co-located with the active transmitter(s). These transmitters can be placed in secure and defensible locations remote from the surveillance area, thus decreasing the possibility of detection by other systems.

However, bistatic radar operation does have its disadvantages. The major problem with bistatic radar systems is the range dependent nature of the clutter echoes. Due to the motion of the two platforms, the ground clutter echoes will exhibit a Doppler shift that is proportional to the velocity of both platforms. The physical geometry of the environment causes the spatio-temporal clutter spectrum to change between different ranges. This results in training range gates that are statistically different from the test range gate, which broadens the clutter suppression filter notch and consequently degrades the processor performance. This region of research has not been emphasized enough in the current literature.

The major objective of this chapter is twofold: i) discuss the concept of STAP; ii) present training methods to address the range dependency problem for bistatic radar operation in a dense environment of ground-moving targets. The work is directed towards what may be called 'small STAP', where the number of spatial channels is small and the array is non-uniform. The work is motivated by a desire to minimize the amount of navigational data associated with both the transmitter and receiver. Furthermore, it is directed towards environments where all range gates may contain targets.

A novel technique is proposed to predict the range-dependent inverse covariance matrix, which is used to compute the STAP filter weights, by utilizing linear prediction theory. An extension of the technique to non-linear prediction is also presented. The two proposed techniques give an improvement in clutter suppression performance over conventional STAP techniques. They also provide mitigation against the additional clutter notches resulting from range and Doppler ambiguities. These techniques eliminate the requirement for a uniform linear array and can be applied to arrays of arbitrary configuration. No navigational data or parameters estimation are necessary as only the clutter data is required.

1. Introduction

Radars are designed to detect, locate and track targets. Space-time adaptive processing (STAP) is a well-established technique for detection of moving targets by an airborne radar system. Interest in bistatic STAP, where the transmitter and receiver are separated, has picked up in recent years. Bistatic radar offers several advantages over its monostatic counterpart, such as the higher possibility of detecting stealth targets and the lower probability of intercept by other radars.

Training and updating of the clutter covariance matrix is a key step in the implementation and effectiveness of any STAP system. In a bistatic or multi-static environment, the usual impediment and possible clutter in-homogeneity experienced in the linear monostatic side-looking case is further complicated by the range-dependent nature of the clutter ridge in the angle-Doppler plane induced by the physical geometry of the two (or more) aircraft [1]. Variations in the clutter covariance matrix with range in the bistatic case are thus due to a combination of these two factors, that is: (i) in-homogeneity of ground scatterers; (ii) the physical geometry of the aircraft. In this Chapter, the work considers only the range dependency of the homogeneous clutter echoes, resulting from the physical geometry of the bistatic radar system.

A number of compensation approaches exist to mitigate the impact of this range dependency on clutter suppression, such as Doppler compensation in the angle-Doppler domain [2] and angle-Doppler compensation [3]. However, these techniques effectively only manage to map the mainlobe clutter peak from one range gate to another and they take little account of sidelobe clutter. Other techniques have been proposed that attempt to map the sidelobe clutter, like angle and separate Doppler compensation [4]; registration-based methods [5]; estimation of clutter covariance by configuration system parameter estimation [6]; joint space-time interpolation [7].

The major draw-back with these techniques is that they require knowledge of the radar system navigational data, which has to be constantly updated. Although the data can be estimated, like in [6] and [8], an initial estimate or knowledge of the theoretical direction-Doppler curves are required respectively. To get around this requirement, a new technique was proposed in [9] to predict the range-dependent inverse covariance matrix using least squares estimation. A detailed analysis of this technique was provided in [10] and [11] while an extension to using non-linear prediction theory was proposed in [12]. No navigational data or parameters estimation has to be performed as only the clutter data is required. Moreover, the techniques are not restricted to uniform linear array (ULA) applications.

In this Chapter, a comprehensive analysis and performance evaluation of the linear and non-linear prediction techniques, with regards to their applications to relatively slow-moving, low-flying targets in bistatic airborne radar systems, are presented. Such targets tend to appear close to the clutter notch, which complicate the clutter suppression problem. In particular, it will be shown that the two techniques provide mitigation against the aliasing effects often experienced by practical bistatic airborne radar systems. Several more operational advantages of using these methods have been presented in [10] and [12].

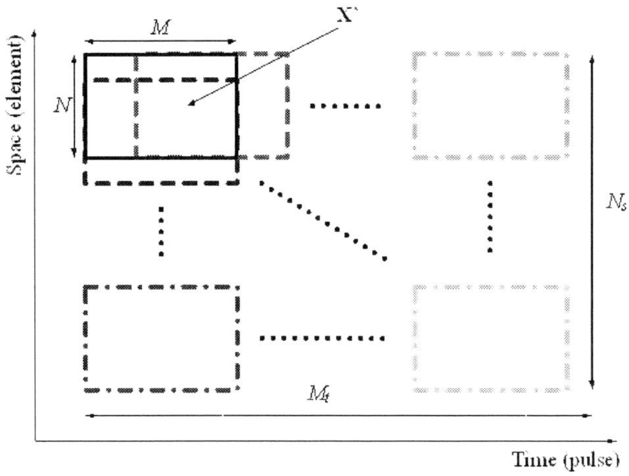

Figure 1. Illustration of data partitioning, by sliding the window across one pulse in the temporal domain and/or one element in the spatial domain, of the $(N_s \times M_t)$ received data snapshot for constructing the $(N \times M)$ data matrix $\mathbf{X}'$ in the context of STAP.

This Chapter is organized as follows. The data model of the STAP processor is formulated in Section 2.. Section 3. deal with the formulation and analysis of the two prediction techniques while a discussion on both techniques can be found in Section 4.. Simulation results are presented in Section 5.. A brief conclusion can be found in Section 6.. In what follows, bold lower case and upper case letters are used to denote column vectors and matrices respectively. The superscripts $(\cdot)^T$ and $(\cdot)^H$ denote the transpose and Hermitian operations respectively.

2. Problem Statement

Consider a radar system utilising an N_s-element antenna array with inter-element spacing d. The radar transmits an M_t-pulse (slow-time) waveform in its coherent processing interval (CPI). The received data can then be partitioned in both space and time by using a sliding window as shown in Figure 1, into an $(N \times M)$ space-time snapshot $\mathbf{X}'$, where N and M denote the spatial and temporal dimensions of the STAP filter system respectively. This partitioning will result in $K_T = (N_s - N + 1)(M_t - M + 1)$ snapshot matrices being generated for processing.

The columns of these space-time snapshots (spatial snapshots from each pulse) are then stacked into inter-leaved column vectors $\mathbf{x}_k$ of size $(NM \times 1)$. The K_T columns are then arranged as the columns of the $(NM \times K_T)$ matrix $\mathbf{X}$. The space-time interference (clutter + noise) covariance matrix is defined as $\mathbf{C}$, where $\mathbf{C} = \frac{\mathrm{E}[\mathbf{X}\mathbf{X}^H]}{K_T}$ and $\mathrm{E}[\cdot]$ is the expectation operator. The clutter model used in this work can be found in Chapter 2 of [1] and the

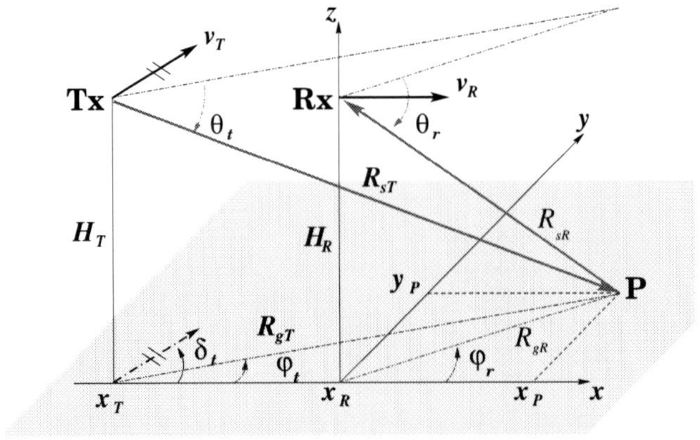

Figure 2. Geometry of an airborne bistatic radar.

noise is assumed to be additive white Gaussian noise with a variance of σ_n^2. Under the assumption of Gaussian interference, the optimum processor (Wiener filter) of the Neyman-Pearson's [13] likelihood ratio test is [14]:

$$\mathbf{w}_{opt} = \mathbf{C}^{-1}\mathbf{s}. \tag{1}$$

The $(NM \times 1)$ signal-steering vector $\mathbf{s}$ is given by $\mathbf{s} = \mathbf{s}_t \otimes \mathbf{s}_s$, where $\otimes$ is the Kronecker product. The temporal dimension $\mathbf{s}_t$ and spatial dimension $\mathbf{s}_s$ of the target steering vector are respectively given as:

$$\mathbf{s}_t = \left[1 \; e^{j2\pi f_D} \; \ldots \; e^{j2\pi(M-1)f_D}\right]^T \tag{2}$$

$$\mathbf{s}_s = \left[1 \; e^{j2\pi f_s} \; \ldots \; e^{j2\pi(N-1)f_s}\right]^T. \tag{3}$$

In the case of a bistatic forward-looking radar, as shown in Fig. 2, the normalised target Doppler frequency f_D and the normalised target spatial frequency f_s are:

$$f_D = \frac{v_{rad,rx} + v_{rad,tx}}{\lambda \text{PRF}} \tag{4}$$

$$f_s = \frac{d \sin\varphi_r \cos\theta_r}{\lambda}. \tag{5}$$

$v_{rad,rx}$ and $v_{rad,tx}$ are the relative velocities of the target to the receiver and transmitter respectively, λ is the wavelength of the radar signal, PRF refers to the pulse repetition frequency of the radar system, φ_r and θ_r are the azimuth and elevation angle of the receiver to the target respectively.

In practice, the true interference covariance matrix is usually unknown. Thus, in order to implement the detection test, an estimate of the interference covariance matrix, $\widehat{\mathbf{C}} = \frac{1}{K_t}\sum_{k=0}^{K_t} \mathbf{z}_k \mathbf{z}_k^H$, is obtained from the secondary/training data. The secondary data $\mathbf{z}_k$ of size K_t, which is drawn from surrounding range gates nearest to the range gate under test, is assumed to be target-free and homogeneous with the primary/test data. This gives the

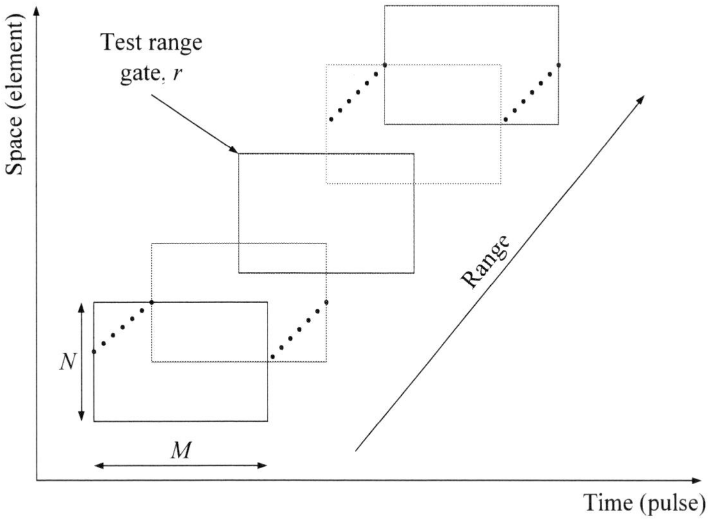

Figure 3. Illustration of training snapshots obtained in the range domain.

sample matrix inversion (SMI) algorithm [14]. Note that the size of the training data set K_t does not necessarily have to be the same as that of the test data set K_T.

For conventional STAP approaches, like the SMI algorithm, the K_t snapshots are usually obtained from other range gates (fast-time) [15], as shown in Figure 3, and only one snapshot is obtained from each range gate. These training range gates exclude the range gate under test and the two immediate surrounding range gates (guard-gates).

This approach makes the algorithms vulnerable to heterogeneity problems, especially for airborne bistatic radar, and degrades the detection performance [16]. However, in some scenarios, not all of the K_t snapshots may be obtained solely from the range domain. As a result, some of these training data may have to be obtained from the space and/or time domain as well, which have already been discussed by the sliding window approach in the beginning of this section.

The efficiency of any linear processors can be characterised by the improvement factor loss (IF_{loss}) metric [1], which is defined as:

$$IF_{loss} = \frac{IF_{opt}}{IF_{STAP}} = \frac{\frac{\mathbf{s}^H \mathbf{C}^{-1} \mathbf{s} \cdot \mathrm{tr}(\mathbf{C})}{\mathbf{s}^H \mathbf{s}}}{\frac{\mathbf{s} \widehat{\mathbf{C}}^{-1} \mathbf{s} \mathbf{s}^H \widehat{\mathbf{C}}^{-1} \mathbf{s} \cdot \mathrm{tr}(\mathbf{C})}{\mathbf{s}^H \widehat{\mathbf{C}}^{-1} \mathbf{C} \widehat{\mathbf{C}}^{-1} \mathbf{s} \cdot \mathbf{s}^H \mathbf{s}}} = \frac{\mathbf{s}^H \mathbf{C}^{-1} \mathbf{s} \mathbf{s}^H \widehat{\mathbf{C}}^{-1} \mathbf{C} \widehat{\mathbf{C}}^{-1} \mathbf{s}}{\mathbf{s}^H \widehat{\mathbf{C}}^{-1} \mathbf{s} \mathbf{s}^H \widehat{\mathbf{C}}^{-1} \mathbf{s}}, \tag{6}$$

where tr($\mathbf{C}$) is the trace of $\mathbf{C}$; IF_{opt} and IF_{STAP} are the improvement factor (IF) of the optimum processor and the STAP processor being evaluated respectively. The maximum attainable value of IF_{loss} is unity, indicating that the processor performance is not degraded by clutter. In practice, the processor performance is degraded by estimation losses and the bistatic clutter Doppler range-dependency problem.

3. Prediction Techniques for Bistatic STAP

3.1. Linear Prediction of Inverse Covariance Matrix (PICM)

Previous attempts to estimate the covariance matrix, which had been reported in the literature, are mostly based on the SMI algorithm [14]. The general formulation of this algorithm can be shown as follows:

$$\widehat{\mathbf{C}}_r = \frac{1}{2}\left[\mathbf{C}_{r-1} + \mathbf{C}_{r+1}\right] \tag{7}$$

where r refers to the range gate under test. The averaging of the covariance matrices is equivalent to averaging the power spectral densities (PSDs):

$$\hat{\mathbf{PS}}_r = \frac{1}{2}\left[\mathbf{PS}_{r-1} + \mathbf{PS}_{r+1}\right]. \tag{8}$$

However, this simply generates two clutter ridges, neither of which will be in the correct location for the range gate under test.

A new technique was proposed in [9], which uses linear prediction theory to obtain an estimate of the range-dependent inverse interference covariance matrix. Linear prediction [17] is widely used in coding and communications applications. It must also be noted that the linear prediction techniques are applied to the inverse covariance matrix sequence, as shown in Figure 4, and *not* the uncorrelated data snapshots. Since the interference covariance matrix is essentially made up of complex sinusoids, the covariance matrix changes smoothly between range gates. Therefore, the covariance matrix of the test range gate can be predicted by using the covariance matrices from surrounding training range gates.

However, the STAP filter weights from equation (1) are a function of the inverse covariance matrix $\mathbf{C}^{-1}$. By working with the inverse covariance matrix, there is no need to ensure that the positive-definite property of the covariance matrix exists for inverting the covariance matrix. Conventional STAP approaches, which deal with the covariance matrix, require this property (positive-definite eigenvalues) to ensure that the matrix is invertible. So naturally, the linear prediction should be done on the inverse covariance matrix.

For each kth range gate, a reasonable estimate of the covariance matrix $\mathbf{C}_k^{-1}$ is obtained by using $2NM$ snapshots [18]. These snapshots may be obtained by using a sliding window in the space and/or time domain, as mentioned in Section 2.. The interference covariance matrix so obtained is rank-deficient [1]. In this case, there is no unique solution for its inverse operation. The projection of the covariance matrix onto the inverse subspace results in errors, as inverting the clutter covariance matrix will amplify the high frequency noise components and introduce irregularities. As a result, the inverse covariance matrix does not change smoothly between range gates. Implementing linear prediction will further magnify the errors and result in a poor estimate. Therefore, diagonal loading [19], also known as regularisation, is added to the clutter covariance matrix before inversion, that is,

$$\ddot{\mathbf{C}} = \mathbf{C} + \sigma^2\mathbf{I}, \tag{9}$$

where σ^2 is the diagonal loading level. The diagonal-loaded inverse covariance matrix $\ddot{\mathbf{C}}^{-1}$ then changes smoothly between range gates.

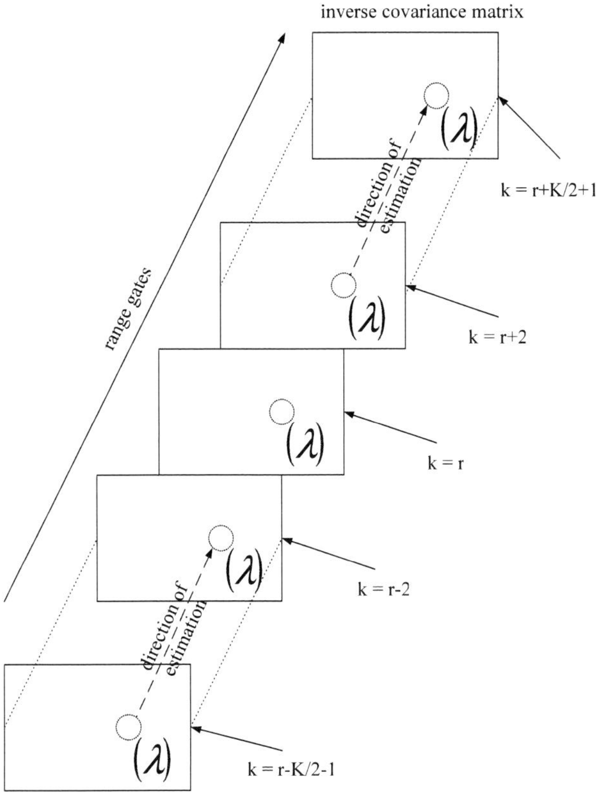

Figure 4. Inverse covariance matrix sequence, where λ denotes each individual index of the inverse covariance matrix.

Denote each index λ of the stacked inverse covariance matrix of the kth training range gate as $\bar{\mathbf{C}}_k^{-1}(\lambda)$, the two-taps linear prediction, shown in Figure 5, is (equation (10) of [10]):

$$\widehat{\bar{\mathbf{C}}}_k^{-1}(\lambda) = \alpha_{+1}(\lambda)\bar{\mathbf{C}}_{k+1}^{-1}(\lambda) + \alpha_{-1}(\lambda)\bar{\mathbf{C}}_{k-1}^{-1}(\lambda) \tag{10}$$

$$\text{for} \quad k = r - \frac{K}{2} - 1, \ldots, r-2, r+2, \ldots, r + \frac{K}{2} + 1,$$
$$\lambda = 1, 2, \ldots, (NM)^2,$$

where $\widehat{\bar{\mathbf{C}}}_k^{-1}$, $\bar{\mathbf{C}}_{k+1}^{-1}$ and $\bar{\mathbf{C}}_{k-1}^{-1}$ are the estimated stacked inverse covariance matrix for the kth range gate and the stacked inverse covariance matrices for the $(k+1)^{th}$ and $(k-1)^{th}$ range gate respectively; α_{+1} and α_{-1} are the prediction weights; K is the total number of training range gates required and $\frac{K}{2}$ training range gates are used on either side of the rth test range gate. The $(r-1)^{th}$ and $(r+1)^{th}$ range gates (guard-gates) are normally excluded to prevent any target signal attenuation [15]. Such target self-nulling occurs when the target energy leaks into the neighbouring range gates.

The inverse covariance matrices and prediction weights are a function of λ. The solution of $\alpha_{+1}(\lambda)$ and $\alpha_{-1}(\lambda)$ are given by the minimum mean squared error of the true and

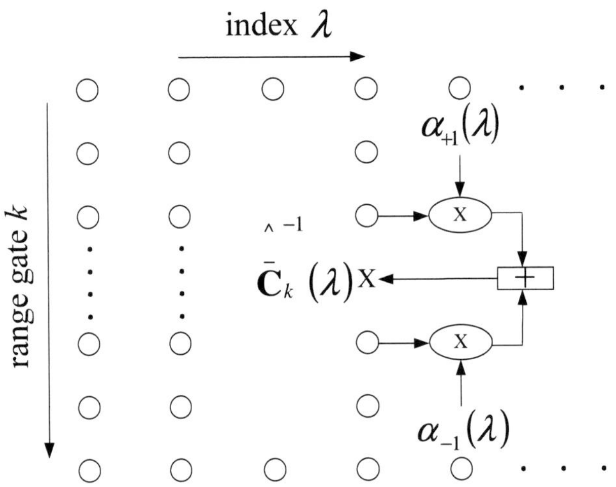

Figure 5. Two-taps linear prediction.

estimated stacked inverse covariance matrix

$$\sum_k \left| \bar{\mathbf{C}}_k^{-1}(\lambda) - \hat{\bar{\mathbf{C}}}_k^{-1}(\lambda) \right|^2 . \tag{11}$$

The series of resulting linear simultaneous equations from equation (10) can be re-arranged into a matrix form and a solution can be obtained by solving this system of linear simultaneous equations. From equation (10):

$$\widehat{\mathbf{C}}_k^{-1}(\lambda) = \left[\bar{\mathbf{C}}_{k+1}^{-1}(\lambda) \;\; \bar{\mathbf{C}}_{k-1}^{-1}(\lambda)\right] \left[\begin{array}{c} \alpha_{+1}(\lambda) \\ \alpha_{-1}(\lambda) \end{array} \right] \tag{12}$$

$$= \mathbf{\Theta}\, \mathbf{\Omega} \tag{13}$$

where

$$\mathbf{\Theta} = \left[\bar{\mathbf{C}}_{k+1}^{-1}(\lambda) \;\; \bar{\mathbf{C}}_{k-1}^{-1}(\lambda)\right], \tag{14}$$

$$\mathbf{\Omega} = \left[\alpha_{+1}(\lambda) \;\; \alpha_{-1}(\lambda)\right]^T . \tag{15}$$

The linear prediction is done on each λ for the inverse covariance matrix. The prediction weights $\widehat{\mathbf{\Omega}}$ for each λ of the inverse covariance matrix is obtained from the surrounding training data (regressors). From equation (12),

$$\mathbf{\Theta}^H \widehat{\mathbf{C}}_k^{-1}(\lambda) = \mathbf{\Theta}^H \mathbf{\Theta}\, \mathbf{\Omega} \tag{16}$$

$$\Longrightarrow \quad \left(\mathbf{\Theta}^H \mathbf{\Theta}\right)^{-1} \mathbf{\Theta}^H \widehat{\mathbf{C}}_k^{-1}(\lambda) = \widehat{\mathbf{\Omega}}. \tag{17}$$

There are more equations than unknowns required, thus equation (11) is a standard least squares (LS) problem with a known solution. Technically, this is smoothing [17] rather than forward or backward prediction. For bistatic STAP, simple forward or backward prediction will not give as accurate an estimate as smoothing does. It should also be noted here that a LS fit allows for the smoothing out of noise in inverse covariance matrices estimates.

3.2. Non-Linear PICM

The formulation of non-linear PICM (NL-PICM) has been discussed in [12]. This non-linear (polynomial/Volterra series) function, with linear parameters, provides good clutter suppression performance in the presence of aliasing (range and/or Doppler ambiguities) effects. The Volterra series (up to the 2nd order) is used to model the non-linear fit (other non-linear functions can be used). Using the same denotations as in Section 3.1., the non-linear prediction is (equation (8) in [12]):

$$\begin{aligned}\widehat{\bar{\mathbf{C}}}_k^{-1}(\lambda) &= \alpha_{+1}(\lambda)\bar{\mathbf{C}}_{k+1}^{-1}(\lambda) + \alpha_{-1}(\lambda)\bar{\mathbf{C}}_{k-1}^{-1}(\lambda)\\ &+\alpha_{+2}(\lambda)\bar{\mathbf{C}}_{k+1}^{-2}(\lambda) + \alpha_{-2}(\lambda)\bar{\mathbf{C}}_{k-1}^{-2}(\lambda)\\ &+\alpha_{12}(\lambda)\bar{\mathbf{C}}_{k+1}^{-1}(\lambda)\bar{\mathbf{C}}_{k-1}^{-1}(\lambda),\end{aligned} \tag{18}$$

where the range of k and λ follows that in equation (10); α_{+1}, α_{-1}, α_{+2}, α_{-2} and α_{12} are the NL prediction weights and K is the total number of training range gates required for the prediction sequence. Once again, the $(r-1)^{th}$ and $(r+1)^{th}$ range gates (guard-gates) are excluded for computation of the NL prediction weights.

As with PICM, the non-linear prediction is done on each λ for the inverse covariance matrix. The prediction weights $\widehat{\mathbf{\Gamma}}$ for each λ of the inverse covariance matrix is obtained from the surrounding training data (regressors) as:

$$\left(\mathbf{\Phi}^H\mathbf{\Phi}\right)^{-1}\mathbf{\Phi}^H\widehat{\bar{\mathbf{C}}}_k^{-1}(\lambda) = \widehat{\mathbf{\Gamma}}, \tag{19}$$

where

$$\mathbf{\Phi} = \left[\bar{\mathbf{C}}_{k+1}^{-1}(\lambda)\;\; \bar{\mathbf{C}}_{k-1}^{-1}(\lambda)\;\; \bar{\mathbf{C}}_{k+1}^{-2}(\lambda)\;\ldots\; \bar{\mathbf{C}}_{k-1}^{-2}(\lambda)\;\; \bar{\mathbf{C}}_{k+1}^{-1}(\lambda)\bar{\mathbf{C}}_{k-1}^{-1}(\lambda)\right] \tag{20}$$

$$\mathbf{\Gamma} = \left[\alpha_{+1}(\lambda)\;\; \alpha_{-1}(\lambda)\;\; \alpha_{+2}(\lambda)\;\; \alpha_{-2}(\lambda)\;\; \alpha_{12}(\lambda)\right]^T \tag{21}$$

4. Comments on Both Prediction Techniques

4.1. Computational Complexity

In practice, the exact covariance matrix $\mathbf{C}$ of each range gate is unknown and needs to be estimated from the data. For each kth range gate, a reasonable estimate of the covariance matrix $\mathbf{C}_k^{-1}$ is obtained by using $2NM$ snapshots [18]. For conventional STAP algorithms, the computational complexity of computing the STAP weights is given as $O(NM)^3$ for inverting the estimated covariance matrix $\widehat{\mathbf{C}}$. For PICM and NL-PICM, there is the additional computational complexity of $O(NM)^3$ for computing the inverse covariance matrix of each range gate. This procedure is repeated for all the range gates in the training sequence. The computational complexity of the two-taps linear prediction, of order $O(4)^3$, only involves the inversion of a (2×2) matrix. On the other hand, NL-PICM involves the inversion of a (5×5) matrix.

Since the covariance matrix is Hermitian, from [20], the inverse covariance matrix is also Hermitian. PICM and NL-PICM involves applying linear prediction techniques to the inverse covariance matrices in the training sequence, thus the estimated inverse covariance matrix $\widehat{\mathbf{C}}_k^{-1}$ retains the Hermitian property. By exploiting the Hermitian property of the inverse covariance matrix, the prediction only needs to be carried out for the upper triangular portion. The prediction weights for the lower triangular portion of the inverse covariance matrix are simply the complex conjugate of its upper triangular portion's counterparts. Therefore, the prediction weights are computed for only the $\sum_{\lambda=1}^{NM} \lambda$ components of $\bar{\mathbf{C}}_k^{-1}(\lambda)$ and the total computational complexity of the PICM and NL-PICM techniques are respectively $\left(K \times O(NM)^3 + \sum_{\lambda=1}^{NM} \lambda \times O(4)^3\right)$ and $\left(K \times O(NM)^3 + \sum_{\lambda=1}^{NM} \lambda \times O(25)^3\right)$.

4.2. No Toeplitz-block-Toeplitz (TBT) Structure Necessary

Since the prediction is done on the inverse covariance matrices sequence, there is no need to ensure that the matrix must be positive definite, which is a condition to ensure that any matrix is non-singular (pp. 125 of [20]). Therefore, by implementing PICM or NL-PICM, there is no need to exploit the TBT structure of the theoretical covariance matrix, thus eliminating the requirement for a uniform linear array (ULA), as required in [5] and [6]. Without any such restrictions, both techniques can be applied to arrays of arbitrary configuration. In addition, no navigational data of the radar system is required to implement these techniques nor is there a need to perform any radar parameters estimation.

4.3. Useful Applications of PICM and NL-PICM

4.3.1. Longer and Multi-dimensional Prediction

Longer prediction can be done by using more training range gates to obtain a better estimate of the inverse covariance matrix. In the case of PICM, the four-taps linear prediction is:

$$\widehat{\mathbf{C}}_k^{-1}(\lambda) = \alpha_{+2}(\lambda)\bar{\mathbf{C}}_{k+2}^{-1}(\lambda) + \alpha_{+1}(\lambda)\bar{\mathbf{C}}_{k+1}^{-1}(\lambda) + \alpha_{-1}(\lambda)\bar{\mathbf{C}}_{k-1}^{-1}(\lambda) + \alpha_{-2}(\lambda)\bar{\mathbf{C}}_{k-2}^{-1}(\lambda) \tag{22}$$

In this case, 12 training range gates are required to obtain 8 equations [18] in order to make a reasonable estimate of 4 parameters, α_{+2}, α_{+1}, α_{-1} and α_{-2}. The extension to even longer predictors is straight-forward. For multi-dimensional PICM, more prediction coefficients are obtained from within the same range gate. From equation (10), the two-taps, two-dimensional linear prediction is:

$$\begin{aligned}\widehat{\mathbf{C}}_k^{-1}(\lambda) &= \alpha_{+1}(\lambda)\bar{\mathbf{C}}_{k+1}^{-1}(\lambda) + \alpha_{-1}(\lambda)\bar{\mathbf{C}}_{k-1}^{-1}(\lambda)\\ &+\beta_{+1}(\lambda+1)\bar{\mathbf{C}}_{k+1}^{-1}(\lambda+1) + \beta_{-1}(\lambda+1)\bar{\mathbf{C}}_{k-1}^{-1}(\lambda+1)\\ &+\gamma_{+1}(\lambda-1)\bar{\mathbf{C}}_{k+1}^{-1}(\lambda-1) + \gamma_{-1}(\lambda-1)\bar{\mathbf{C}}_{k-1}^{-1}(\lambda-1)\end{aligned} \tag{23}$$

Increasing the number of prediction weights results in a corresponding increase in the number of training range gates required, as more prediction coefficients will require more training data. Therefore, there is a trade-off between more accurate estimates and computational complexity.

4.3.2. Mitigation against Aliasing

A pulse Doppler radar can be ambiguous in either range or Doppler frequency [1]. For conventional pulse Doppler radar, range ambiguities occur because of the transmission of repetitive pulses. The additional ambiguous clutter returns are also known as multiple-time-around clutter. Such clutter echoes may result in additional clutter notches, thus affecting the detection of targets.

Doppler ambiguities occur when the Nyquist sampling criterion is not followed. From equation (4), we can see that the bistatic Doppler frequency is associated with the relative velocities of the target to the receiver and transmitter. These velocities are functions of the azimuth and elevation angles of the receiver and transmitter to the target. Thus any ambiguity in azimuth causes ambiguity in Doppler frequency and vice versa.

PICM and NL-PICM are able to deal with the additional clutter notches resulting from range and Doppler ambiguities. In such situations, conventional compensation techniques such as angle-Doppler compensation (ADC) may not perform as well, since they were originally proposed to align only the mainlobe clutter between the training range gates and the range gate under test.

5. Simulation Results

This section will show the simulation results of both PICM and NL-PICM methods against conventional STAP algorithms. Different sets of simulations, bistatic forward-looking (BFL) and bistatic side-looking (BSL), are carried out and the common radar parameters used are shown in Table 1. Unless otherwise specified, the dimensions of the STAP filter and the radar system were taken to be $N = N_s = 8$ and $M = 24$, $M_t = 407$ respectively. A total of $K = 20$ training range gates (10 on either side of the range gate under test) were used to compute the STAP filter weights.

Table 1. Radar Parameters

pulse repetition frequency	PRF	20 kHz
operating frequency	f_o	10 GHz
receiver & transmitter height	H_R, H_T	1000 m
receiver & transmitter velocity	v_R, v_T	90 m/s
baseline separation (along flight direction)		2000 m
clutter-to-noise ratio (CNR)		30dB

For all the algorithms shown in this Section, $2NM$ snapshots (using a sliding window in the time domain only) were used to obtain the covariance matrix $\mathbf{C}_k$ for each kth training range gate. The simulation results were obtained from a Monte Carlo simulation comprising of 500 runs.

For the BFL radar (direction of travel normal to linear antenna array with half wavelength spacing), the transmitter's flight path is 90° from that of the receiver. The backlobe of the clutter scatterers' response was ignored and the receiver look angle is $\varphi_r = 45^\circ$.

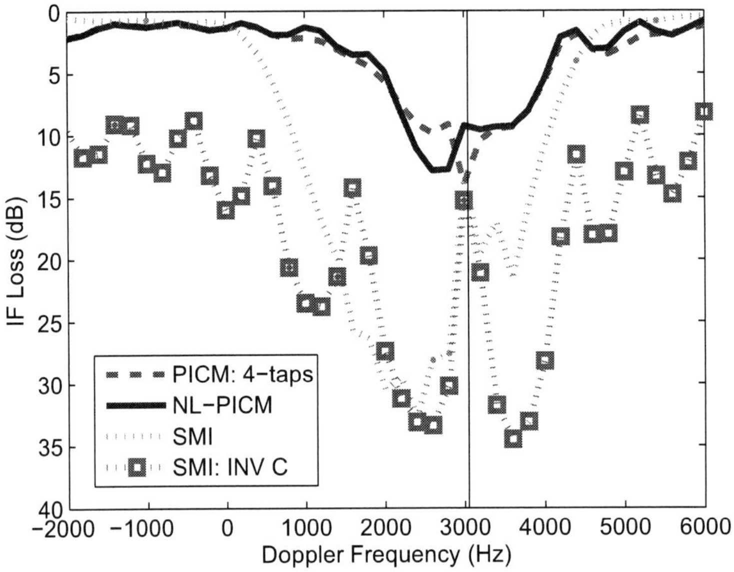

Figure 6. IF_{loss} plot for a bistatic forward-looking radar: PICM (4-taps), NL-PICM, SMI and SMI-(INV C)

Figure 6 shows the IF_{loss} plot of the NL-PICM technique (blue solid line) and the PICM technique with four-taps linear prediction (red dashed line). The magenta dotted line shows the sample matrix inversion (SMI) algorithm [14], which is obtained by straight averaging the covariance matrices $\mathbf{C}_k$ from the training sequence. Lastly, the green dotted line with squares shows the SMI algorithm obtained by straight averaging of the inverse covariance matrices $\mathbf{C}_k^{-1}$. This algorithm is included as a reference since the proposed non-linear prediction is carried out on the inverse covariance matrices.

At the critical clutter notch frequency of 3044Hz, it can be observed that the NL-PICM technique gives the best performance. There is an improvement of 4dB over the PICM technique. Another important point to note is the narrower clutter notch provided by both PICM and NL-PICM over the two SMI processors, hence enhancing the capability of detecting relatively slow-moving, low-flying targets. The importance of this region is indicated by the fact that such target signals tend to fall within this region and will be attenuated.

Next, the effects of aliasing (range and Doppler ambiguities) on the clutter suppression performance are presented. The results are shown in Figures 7-9, with the same plots used to denote the respective STAP algorithms as before. The ADC algorithm [3] (black dash-dotted line) is also included as another comparison.

For the first set of results, the BFL radar system with Doppler ambiguities includes both the frontlobe and backlobe of the clutter scatterers' response. The receiver look angle is changed to $\varphi_r = 0°$. The results are shown in Figures 7 and 8, with the latter being a zoomed-in version of the former.

For the second set of results, a BSL radar system with range ambiguities is used and the backlobe clutter response is excluded. To illustrate the robustness of the prediction techniques in aliasing mitigation for different bistatic airborne radar systems, the radar parameters are changed. In this case, the receiver look angle is taken to be $\varphi_r = 90^\circ$ and the results are shown in Figure 9.

The clutter suppression results differ for the range and Doppler ambiguities because the effects on the clutter spectrum due to each ambiguity are different. When range ambiguities are present, the clutter echoes of a certain range gate includes the clutter contributions from other range gates. Thus the different mainlobes in ambiguous range gates move in different directions and do not coincide, even when compensation is performed to mitigate for the bistatic clutter Doppler range dependency. Whereas for Doppler ambiguities, the clutter notches are repeated every PRF. In this case, the different mainlobes move in the same direction (constant shift).

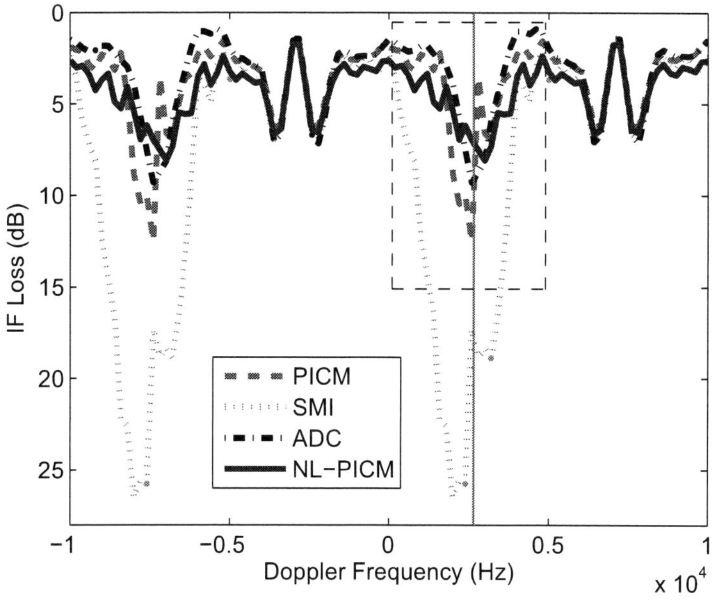

Figure 7. IF_{loss} plot: bistatic forward-looking radar with Doppler ambiguities.

From Figures 7 and 8, it can be seen that Doppler aliasing results in additional clutter notches being formed [1]. NL-PICM performs the best while the SMI technique fares the worst. It has been well-documented that the SMI technique degrades severely in bistatic airborne radar applications. At the critical clutter notch frequency of 2663Hz, NL-PICM gives an improvement of 2dB and 2.5dB over ADC and PICM respectively.

In Figure 9, the widened clutter notches are caused by the clutter contributions from other range gates folding into the test range gate. Although PICM performs slightly better than NL-PICM, the latter is still a good choice as PICM performs much poorer in the presence of Doppler ambiguities. Generally, NL-PICM gives good clutter suppression performance around the critical notch frequency of 2284Hz.

Another important point to note is the narrower clutter notch provided by NL-PICM

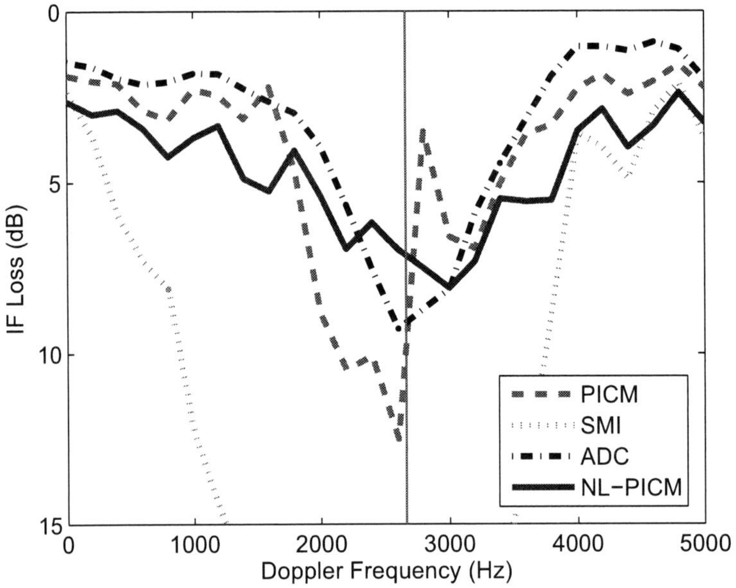

Figure 8. IF_{loss} plot: Close-up version of the indicated box in Fig. 7.

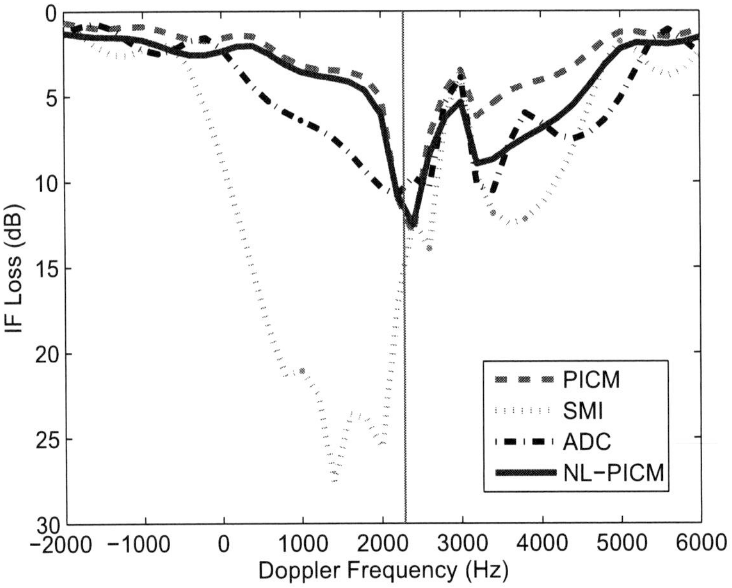

Figure 9. IF_{loss} plot: bistatic side-looking radar with range ambiguities.

over the other STAP methods, hence enhancing the capability of detecting relatively slow-moving, low-flying targets. This highlights the benefits of using NL-PICM in the presence of aliasing effects.

Specifically, from Figures 7-9, it can be noted that NL-PICM gives a better clutter suppression performance over ADC within the clutter notches, as NL-PICM deals well with the additional clutter notches caused by aliasing. The ADC technique mitigates for the bistatic clutter Doppler range dependency problem by compensating with the peak angle and Doppler frequencies. Therefore, it performs well within the mainlobe. However, aliasing results in additional clutter notches due to the overlapping of clutter spectrums and thus the ADC algorithm does not perform as well as NL-PICM.

6. Conclusion

In this Chapter, an analysis and evaluation of using PICM and NL-PICM to obtain an estimate of the inverse covariance matrix for bistatic STAP is carried out. Simulation results show the improvement in performance over conventional STAP techniques. Both methods are able to mitigate against additional/widened clutter notches caused by aliasing.

No navigational/system data or parameters estimation is necessary to implement these methods, as only the clutter data is required. The methods are not restricted to uniform linear array applications and can be directly applied to arrays of arbitrary configuration. Their robustness with respect to bistatic radar operations are shown by their useful applications to different bistatic radar systems.

References

[1] R. Klemm, *Principles of space-time adaptive processing*, IEE Radar, Sonar, Navigation and Avionics Series, second edition, 2002.

[2] C.H. Lim, E. Aboutanios and B. Mulgrew, Modified JDL with Doppler compensation for airborne bistatic radar, *Proc. IEEE Radar Conference*, May 2005, pp. 854-858.

[3] B. Himed, Y. Zhang and A. Haijari, STAP with angle-Doppler compensation for bistatic airborne radars, *Proc. IEEE Radar Conference*, 2002, pp. 311-317.

[4] C.H. Lim and B. Mulgrew, Filter banks based JDL with angle and separate Doppler compensation for airborne bistatic radar, *International Radar Symposium*, September 2005, pp. 583-587.

[5] F. Lapierre and J. Verley, Registration-based solutions to the range-dependence problem in STAP radars, *Proc. Adaptive Sensor Array Processing (ASAP) workshop*, March 2003.

[6] A. Jaffer, B. Himed and P. Ho, Estimation of range-dependent clutter covariance by configuration system parameter estimation, *Proc. IEEE Radar Conference*, May 2005, pp. 596-601.

[7] V. Varadarajan and J. Krolik, Joint space-time interpolation for bistatic STAP, *Proceedings of 32nd Asilomar Conference on Signales, Systems, Computers*, vol. 1, November 2003, pp. 60-65

[8] F. Lapierre and J. Verley, Registration-based range-dependence compensation for bistatic STAP radars, *EURASIP Journal on Applied Signal Processing*, 2005, vol. 1, pp. 85-98.

[9] C.H. Lim and B. Mulgrew, Prediction of inverse covariance matrix (PICM) sequences for STAP, *IEEE Signal Processing Letters*, April 2006, vol. 13, no. 4, pp. 236-239.

[10] C. H. Lim and B. Mulgrew, Linear prediction of range-dependent inverse covariance matrix (PICM) Sequences for STAP, *EURASIP Signal Processing Journal*, June 2007, vol. 87, no. 6, pp. 1412-1420.

[11] C. H. Lim, *Bistatic Space-Time Adaptive Processing for Ground Moving Target Indication*, Ph.D. Thesis, University of Edinburgh, United Kingdom, October 2006.

[12] C.H. Lim, C. M. See and B. Mulgrew, Non-linear prediction of inverse covariance matrix for STAP, *IEEE International Conference on Acoustics, Speech and Signal Processing (ICASSP)*, April, 2007, vol. 2, pp. 921-924.

[13] N. Goodman, *Statistical analysis based on a certain multivariate complex Gaussian distribution*, Annals Mathematical Statistics, March 1963, vol. 24, pp. 152-177.

[14] L. Brennan and I. Reed, Theory of adaptive radar, *IEEE Trans. Aerospace and Electronic Systems*, March 1973, vol. AES-9, no. 2, pp. 237-252.

[15] J. Ward, *Space-time adaptive processing for airborne radar*, Technical Report 1015, MIT Lincoln Laboratory, 1994.

[16] W. L. Melvin, Space-time adaptive radar performance in heterogeneous clutter, *IEEE Trans. Aerospace and Electronic Systems*, April 2000, vol. 36, pp. 621-633.

[17] S. Haykin, *Adaptive filter theory*, Prentice Hall, fourth edition, 2002.

[18] I. Reed, J. Mallet and L. Brennan, Rapid convergence rate in adaptive arrays, *IEEE Trans. Aerospace and Electronic Systems*, November 1974, vol. AES-10, no. 6, pp. 853-863.

[19] B.D. Carlson, Covariance matrix estimation errors and diagonal loading in adaptive arrays, *IEEE Trans. Aerospace and Electronic Systems*, July 1988, vol. 24, pp. 397-401.

[20] G.H. Golub and C.F. Van Loan, *Matrix Computations*, North Oxford Academic, 1983

In: New Signal Processing Research
Editor: Takumi Maeda, pp. 439-453
ISBN 978-1-60456-479-2

Chapter 15

ADVANCES IN NONNEGATIVE MATRIX FACTORIZATION: TOWARDS UNIQUENESS

Rafal Zdunek*
Institute of Telecommunications, Teleinformatics, and Acoustics
Wroclaw University of Technology
Wybrzeze Wyspianskiego 27, 50-370 Wroclaw, Poland

Abstract

Nonnegative Matrix Factorization (NMF) finds nonnegative and usually sparse lower-rank representations of the input data. Basically, the factorization is achieved by alternating minimization of a given cost function subject to nonnegativity constraints. Unfortunately, the underlying optimization problem in NMF is generally non-unique, and a variety of assumptions and computational strategies has been proposed to select suitable approximations to the true nonnegative factors. The problems of non-uniqueness in NMF have been studied by many researchers, e.g., D. Donoho and V. Stodden; F. Theis, K. Stadlthanner and T. Tanaka; and M. Heiler and C. Schnorr, who emphasize that additional constraints (aside from nonnegativity) considerably relax the non-uniqueness. A common approach is to use sparsity and/or orthogonality constraints. In our approach, we also point at the other constraints, e.g., smoothness, binarity or unimodality. Additionally, the optimization algorithm crucially affects the estimated factors since the cost function is non-convex, independent of the additional constraints. The gradient descent techniques that are often used for solving this problem do not guarantee a unique solution. Depending on initialization, NMF may even provide spurious or misleading factors. In this paper, we discuss a possibility of using other (morthological) constraints in NMF, and some optimization algorithms with iterative steering to the global solution.

1. Introduction

NMF [38, 39, 30, 31] is an emerging technique that has found a variety of applications, e.g. in data analysis (pattern recognition [30, 36, 50, 54], image retrieval and classification [22, 37], clustering [49, 34, 16, 37, 57, 2, 8, 9, 56], dimensionality reduction, language

*E-mail address: rafal.zdunek@pwr.wroc.pl

modeling, text mining [40, 3], music transcription [10]), signal and image processing (blind source separation [14, 42, 48, 64], spectra recovering [33, 32, 45, 46, 47]), and neurobiology (gene separation [6, 43], EEG signal analysis [44]).

The problem in NMF can be expressed as follows: given the observations in $\boldsymbol{Y} \in \mathbb{R}_+^{I \times K}$ and the lower-rank J, find such nonnegative matrices (factors) $\boldsymbol{A} \in \mathbb{R}_+^{I \times J}$ and $\boldsymbol{X} \in \mathbb{R}_+^{J \times K}$ that $\boldsymbol{Y} \cong \boldsymbol{A}\boldsymbol{X}$. The symbol $\mathbb{R}_+$ denotes the nonnegative orthant (subspace) of $\mathbb{R}$. Usually $K >> I \geq J$, and the lower-rank J is assumed to be known. If among the I rows of $\boldsymbol{Y}$, the $I - J$ rows is strongly correlated with the remaining I rows, the lower-rank J can be readily estimated from SVD of the covariance matrix $\boldsymbol{Y}\boldsymbol{Y}^T$ as the number of predominate singular values.

In many applications of NMF, the estimated factors have clear interpretations. For example, in blind source separation, the rows of $\boldsymbol{Y}$ represent the observed signal (or vectorized images) and the rows of $\boldsymbol{X}$ represent the hidden (lateral) components. In this case, $\boldsymbol{A}$ is the matrix of mixing coefficients. In data clustering, the columns of $\boldsymbol{Y}$ are the data points (in $\mathbb{R}^I$), J is the number of clusters, the columns of $\boldsymbol{X}$ represent the indicators to clusters, and the columns of $\boldsymbol{A}$ are centroids.

To solve the problem in NMF, the objective function $D(\boldsymbol{Y}||\boldsymbol{A}\boldsymbol{X})$ that measures misfitting between $\boldsymbol{Y}$ and $\boldsymbol{A}\boldsymbol{X}$ is assumed, and the following alternating minimization algorithm is used.

Algorithm 1. (NMF)

Set $\quad \boldsymbol{A}^{(0)}, \boldsymbol{X}^{(0)}$,

For $\quad s = 1, 2, \ldots$, until convergence **do**

$\qquad \boldsymbol{X}^{(s+1)} = \arg\min_{\mathbf{X} \geq 0} D(\boldsymbol{Y}||\boldsymbol{A}^{(s)}\boldsymbol{X})$

$\qquad \boldsymbol{A}^{(s+1)} = \arg\min_{\mathbf{A} \geq 0} D(\boldsymbol{Y}||\boldsymbol{A}\boldsymbol{X}^{(s+1)})$

End

The function $D(\boldsymbol{Y}||\boldsymbol{A}\boldsymbol{X})$ is non-convex with respect to both arguments $\boldsymbol{A}$ and $\boldsymbol{X}$, even though it might be convex with respect to one argument. Thus initialization for $\boldsymbol{A}^{(0)}$ and $\boldsymbol{X}^{(0)}$ is very important, and many initialization strategies for NMF have been proposed [1, 28, 29, 5, 63, 12].

Additionally, the function may have many local and even global minima, and hence any stationary point obtained with Algorithm 1 is not guarantied to be a global minimum. The problems of non-uniqueness in NMF have been studied by many researchers, e.g., D. Donoho, V. Stodden [17]; F. Theis, K. Stadlthanner, T. Tanaka [53]; and M. Heiler and C. Schnorr [24].

In the next Section, we briefly discuss some typical ambiguities in NMF with the stress on the constraints that help to select the desired solution. Section 3. presents the selected algorithms that steer the iterative updates to the desired solution. The exemplary numerical experiments are illustrated in Section 4. Finally, some conclusions are given in Section 5..

2. Anmbiguities in NMF

For any cost function $D(\boldsymbol{Y}||\boldsymbol{A}\boldsymbol{X})$ minimized with Algorithm 1, we have

$$D(\boldsymbol{Y}||\boldsymbol{A}\boldsymbol{X}) = D(\boldsymbol{Y}||\boldsymbol{A}\boldsymbol{P}^{-1}\boldsymbol{P}\boldsymbol{X}) = D(\boldsymbol{Y}||\tilde{\boldsymbol{A}}\tilde{\boldsymbol{X}}), \tag{1}$$

with $\tilde{A} = AP^{-1}$ and $\tilde{X} = PX$. There are many possibilities for selection of a nonnegative matrix P, for which P^{-1} is also a nonnegative matrix, and $A \neq \tilde{A}$ and $X \neq \tilde{X}$.

Obviously, if P is a permutation or nonnegative diagonal matrix, the relation in (1) holds. The corresponding permutation and scaling indeterminacies (ambiguities) are intrinsically related with NMF, and cannot be removed without adequate prior knowledge.

Unfortunately, such indeterminacies are not only the ones that may take place in NMF. For example, let us consider the following nonnegative mixing and source matrices

$$A = \begin{bmatrix} 2 & 5 \\ 4 & 3 \end{bmatrix}, \quad X = \begin{bmatrix} x_1(t) \\ x_2(t) \end{bmatrix}, \tag{2}$$

thus we have

$$Y = \begin{bmatrix} y_1(t) \\ y_2(t) \end{bmatrix} = AX = \begin{bmatrix} 2x_1(t) + 5x_2(t) \\ 4x_1(t) + 3x_2(t) \end{bmatrix}. \tag{3}$$

It is easy to show that there exists another nonnegative decomposition which gives us the following components:

$$Y = \begin{bmatrix} 2x_1(t) + 5x_2(t) \\ 4x_1(t) + 3x_2(t) \end{bmatrix} = \tilde{A}\tilde{X} = \begin{bmatrix} 0.2 & 0.8 \\ 1.8 & 0.2 \end{bmatrix} \begin{bmatrix} 2x_1(t) + x_2(t) \\ 2x_1(t) + 6x_2(t) \end{bmatrix}. \tag{4}$$

Thus

$$\tilde{A} = \begin{bmatrix} 0.2 & 0.8 \\ 1.8 & 0.2 \end{bmatrix}, \quad \tilde{X} = \begin{bmatrix} 2x_1(t) + x_2(t) \\ 2x_1(t) + 6x_2(t) \end{bmatrix} \tag{5}$$

are new nonnegative factors that do not come from the permutation or scaling indeterminacies. To remove this kind of ambiguity, the prior knowledge on the estimated factors is needed.

The prior knowledge can be incorporated to the optimization algorithm in many ways. For example, it can be modeled with additional penalty terms in the objective function. Thus

$$D_p(Y||AX) = D(Y||AX) + \alpha_A \Psi_A(A) + \alpha_X \Psi_X(X), \tag{6}$$

where $\Psi_A(A)$ and $\Psi_X(X)$ are the penalty terms, and $\alpha_A \geq 0$ and $\alpha_X \geq 0$ are penalty parameters for A and X, respectively.

The selected additional constraints (beside of nonnegativity) for NMF and the related penalty terms can be classified as follows:

2.1. Sparsity

This type of constraints is commonly explored in many algorithms for NMF. The uniqueness of NMF applied to a certain class of sparse signals was geometrically demonstrated by D. Donoho and V. Stodden [17]. They gave very strict sparsity conditions under which NMF is unique. F. Theis, K. Stadlthanner, T. Tanaka [53] formulated the analytic conditions under which the set of feasible solutions is very small. They demonstrated that the

NMF algorithm proposed by P. Hoyer [25] finds a unique solution if certain (less restrictive) sparsity constraints for the signals of interest are satisfied.

It was also experimentally demonstrated in [14, 11, 61] that if the signals of interest are sufficiently sparse, the penalty terms expressed in terms of the l_1 metric substantially improve the estimated factors. The experimental results given in [51] suggest that the l_0 metric better selects the true sparse factors than the sparsity measure proposed by P. Hoyer [25].

The penalty terms for sparse factors can also be defined with the l_p diversity measure that was used to derive the M-FOCUSS algorithm [15]. Enforcing sparsity in the column vectors of $\boldsymbol{X}$, we have

$$\Psi_X(\boldsymbol{X}) = \sum_k (||\boldsymbol{x}_k||_1)^2 = \mathrm{tr}\{\boldsymbol{X}^T \boldsymbol{E} \boldsymbol{X}\}, \tag{7}$$

where $\boldsymbol{x}_k$ is the k-th column of $\boldsymbol{X}$, and $\boldsymbol{E} \in \mathbb{R}^{J \times J}$ is a matrix of all ones entries. The penalized objective function with (7) can be easily minimized using a gradient descent approach but as shown in [61] the penalty parameter α_X in (6) should gradually decrease with an increase in the number of alternating steps.

2.2. Smoothness

Smoothness constraints give preference in selecting the solution of interest from a set of all admissible solutions. Obviously, such constraints may be used only if the desired solution is characterized by a spacial or temporal smooth structure. As demonstrated experimentally in [59], the temporal smooth structure in signals separated with NMF can be modeled with the Markov Random Field (MRF) that is commonly used for image restoration and reconstruction.

In this case, the penalty term for $\boldsymbol{X}$ in (6) has the following form

$$\Psi_X(\boldsymbol{X}) \quad = \quad \sum_{j=1}^{J} \sum_{k=1}^{K} \sum_{l \in S_k} \psi\left(x_{jk} - x_{jl}, \delta\right), \tag{8}$$

where $S_k = \{k-1, k+1\}$ determines the nearest neighborhood of the k-th column vector in $\boldsymbol{X}$, δ is a scaling factor, and $\psi\left(\xi, \delta\right)$ is some potential function of ξ, which can take different forms. Examples include the following functions: Besag (Laplacian) [4], Hebert and Leahy [23], Geman and McClure [19], Geman and Reynolds [20], Stevenson and Delp (Hubert) [52], or Green [21]. A typical Gaussian function may also be useful.

2.3. Orthogonality

Many researchers [16, 7, 33] observed that in many NMF problems one of the factors additionally satisfies the orthogonality constraints. In particular, this occurs in clustering, where the matrix of cluster indicators is orthogonal and nonnegative. The inclusion of orthogonality constraints to orthogonal NMF problems considerably reduces a degree of freedom in choosing the matrix $\boldsymbol{P}$ in (1). If both factors are orthogonal, the factorization problem usually boils down to SVD. However, in many applications only one factor is orthogonal but the other is nonnegative. In this case, we have a uni-orthogonal NMF [16]. The optimization problems in this version of NMF are defined as

$$\min_{\mathbf{A}\geq 0, \mathbf{X}\geq 0} D(\boldsymbol{Y}||\boldsymbol{AX}), \quad \text{s.t.} \quad \boldsymbol{A}^T\boldsymbol{A} = \boldsymbol{I}_J, \tag{9}$$

or

$$\min_{\mathbf{A}\geq 0, \mathbf{X}\geq 0} D(\boldsymbol{Y}||\boldsymbol{AX}), \quad \text{s.t.} \quad \boldsymbol{X}\boldsymbol{X}^T = \boldsymbol{I}_J, \tag{10}$$

where $\boldsymbol{I}_J \in \mathbb{R}^{J\times J}$ is an identity matrix. In (9), we assume the columns in $\boldsymbol{A}$ are orthogonal, whereas (10) enforces orthogonality in the rows of $\boldsymbol{X}$.

2.4. Binarity

The binary factors naturally appear in the problems of binary classification or partitional clustering.

Let us assume the following example in which NMF can be used to clustering:

$$\boldsymbol{Y} = \begin{bmatrix} 0.5847 & 0.5714 & 0.5867 & 0.7301 & 0.7309 & 0.7124 & 0.7246 \\ 0.1936 & 0.1757 & 0.1920 & 0.4193 & 0.3955 & 0.3954 & 0.4147 \end{bmatrix}. \tag{11}$$

It easy to notice that the first 3 columns should be in one cluster, and the last 4 columns in the other.

Assuming each column vector of $\boldsymbol{Y} = [\boldsymbol{y}_1, \ldots, \boldsymbol{y}_K]$ represents a single observation (a datum point in $\mathbb{R}^I$), the cluster are disjoint (non-overlapping), and J is *a priori* known number of clusters, we can interpret the column vectors of $\boldsymbol{A} = [\boldsymbol{a}_1, \ldots, \boldsymbol{a}_J]$ as the centroids (indicating the directions of central points of clusters in $\mathbb{R}^I$) and binary values in $\boldsymbol{X} = [x_{jk}]$ as indicators to the clusters. If $x_{jk} = 1$, $\boldsymbol{y}_k$ belongs to the j-th cluster; $x_{jk} = 0$, otherwise.

Thus for the data in (11), $\boldsymbol{X}$ should be

$$\boldsymbol{X} = \begin{bmatrix} 1 & 1 & 1 & 0 & 0 & 0 & 0 \\ 0 & 0 & 0 & 1 & 1 & 1 & 1 \end{bmatrix}. \tag{12}$$

The matrix $\boldsymbol{A}$ can be computed as $\boldsymbol{A} = \boldsymbol{Y}\boldsymbol{X}^+$, where $\boldsymbol{X}^+$ is a pseudo-inverse to $\boldsymbol{X}$. Thus

$$\boldsymbol{A} = \begin{bmatrix} 0.5809 & 0.7245 \\ 0.1871 & 0.4062 \end{bmatrix}. \tag{13}$$

The column vector $\boldsymbol{a}_1$ constitutes the algebraic mean over the first 3 columns of $\boldsymbol{Y}$, and $\boldsymbol{a}_2$ – the algebraic mean over the last 4 columns of $\boldsymbol{Y}$. Thus the column vectors in $\boldsymbol{A}$ can be interpreted as the centroids of the clusters.

It is easy to notice that $\boldsymbol{X}$ is an orthogonal matrix, i.e. $\boldsymbol{X}\boldsymbol{X}^T = \boldsymbol{I}_J$, thus the orthogonality constraints are met here intrinsically. Moreover, the binarity and orthogonality constraints are much stronger than only the nonnegative orthogonality constraints. In this case the NMF problem has a very limited set of admissible solution but the uniqueness (up to the scale and permutation ambiguities) is still an open question. The problem of finding the binary matrix $\boldsymbol{X}$ in NMF with the above-mentioned binary constraints can be expressed in terms of the l_0 sparsest problem (for selection of $\boldsymbol{X}$) that is NP-hard. A variety of simplified computational strategies for approximation of NP-hard problems exists, but since the alternating minimization is intrinsically non-convex, it is recommendable to use some globally convergent optimization algorithms for solving the binary NMF problem.

2.5. Unimodality

In many applications of NMF such as spectroscopy the lateral signals are expected to be unimodal, i.e. the curves with one extremum (minimum or maximum) in a given locality. This case was analyzed, e.g. in [18]. Unimodality constraints can be modeled with piecewise splines or other piecewise unimodal functions. The example of modeling the basis vectors in NMF with the exponential functions given by the Boltzmann distribution can be found in [55].

3. Algorithms

A suitably selected penalty term in the objective function undoubtedly restricts the area of feasible solutions, and hence the task of selecting a good approximation to the true solution is easier. Nevertheless, any objective function minimized with respect to both arguments $\boldsymbol{A}$ and $\boldsymbol{X}$ is non-convex, and the optimization process may get stuck to in some local minimum that may still exist in spite of the constraints. The choice of the optimization algorithm is therefore very important, and obviously the usage of global optimization algorithms gives a chance to pick up one of the global minima. However, the chance depends on the associated parameters that must be carefully tuned in global optimization algorithms to ensure a relatively fast convergence to a global minimum. Usually this process is very slow.

Maybe due to these reasons, the gradient-descent algorithms are more favored in the optimization to NMF. A typical and simple approach, which was proposed in [30, 31], is to use the multiplicative algorithms. Minimizing the squared Frobenius norm, the following iterative update rules can be obtained

$$\boldsymbol{A} \leftarrow \boldsymbol{A} \odot \left(\boldsymbol{Y}\boldsymbol{X}^T\right) \oslash \left(\boldsymbol{A}\boldsymbol{X}\boldsymbol{X}^T\right), \quad \boldsymbol{X} \leftarrow \boldsymbol{X} \odot \left(\boldsymbol{A}^T\boldsymbol{Y}\right) \oslash \left(\boldsymbol{A}^T\boldsymbol{A}\boldsymbol{X}\right), \tag{14}$$

and consequently for the generalized Kullback-Leibler divergence:

$$\boldsymbol{A} \leftarrow \left(\boldsymbol{A} \oslash \left(\boldsymbol{X}\boldsymbol{e}_K \otimes \boldsymbol{e}_I\right)\right) \odot \left(\boldsymbol{Y} \oslash \boldsymbol{A}\boldsymbol{X}\right)\boldsymbol{X}^T, \tag{15}$$

$$\boldsymbol{X} \leftarrow \left(\boldsymbol{X} \oslash \left(\boldsymbol{A}^T\boldsymbol{e}_I \otimes \boldsymbol{e}_K^T\right)\right) \odot \boldsymbol{A}^T\left(\boldsymbol{Y} \oslash \boldsymbol{A}\boldsymbol{X}\right), \tag{16}$$

where $\boldsymbol{e}_K = [1,\ldots,1] \in \mathbb{R}^K$ and $\boldsymbol{e}_I = [1,\ldots,1] \in \mathbb{R}^I$ are adequate vectors of all ones, and $\otimes$, $\odot$, $\oslash$ denote the Kronecker product, component-wise multiplication, component-wise division, respectively. It was proved in [31] that the algorithms (14)–(16) are convergent to stationary points of the respective objective functions, but it should be noted that the stationary points may be local minima, very different from the true solution. As demonstrated with the experiments [61, 62], this event takes place very often if the number of observations is comparable with the number of true (lateral) components, and the true components are not sufficiently sparse.

To solve the optimization problem in NMF many other algorithms were proposed (see e.g. the survey of the algorithms in [3]). However, in many cases the algorithms outperform the multiplicative algorithms only with respect to the increased convergence rate, i.e. they converge to a local minimum faster. In this respect, the Alternating Least Squares (ALS) [1, 3, 27, 11], projected gradient [35, 13, 25], or even second-order [26, 60, 61, 24, 62] methods are very profitable, but do not assure the convergence to a global minimum.

The issue of applying global optimization algorithms to NMF has been raised by many researchers, e.g. [51, 3, 24]. K. Stadlthanner *et al* proposed to use a genetic algorithm to solve the minimization problem in which the penalty term of the objective function is defined as the l_0-based sparsity measure. This approach to NMF seems to be very promising but it involves priorly setting of many parameters such as the number of individuals, offsprings, migrations in the population, and mutation probability. Moreover, genetic algorithms are usually very slow since the evolution process needs many generations to reach a global minimum.

Global optimization in NMF can be also performed with the simulated annealing technique. The promising results of applying semi-binary NMF with the simulated annealing to data clustering can be found in [58]. In general, if binary constraints can be imposed at least to one of the factors in NMF, the optimization problem can be solved by maximization of the Gibbs–Boltzmann distribution associated with the objective function. Assuming $\boldsymbol{X} = [\boldsymbol{x}_1, \ldots, \boldsymbol{x}_K]$ is a binary matrix, the Gibbs–Boltzmann distribution can be associated with the function $F(\boldsymbol{x}_k) = D(\boldsymbol{y}_k || \boldsymbol{A}\boldsymbol{x}_k)$ for $k = 1, \ldots, K$. Thus

$$P_F(\boldsymbol{x}_k) = \frac{\exp\left\{-\frac{1}{T}F(\boldsymbol{x}_k)\right\}}{\sum_{\boldsymbol{x}_k \in \{0,1\}^J} \exp\left\{-\frac{1}{T}F(\boldsymbol{x}_k)\right\}}, \tag{17}$$

where T is a temperature parameter controlling the ascent towards a global maximum of $P_F(\boldsymbol{x}_k)$.

To find a binary solution to the problem $\min_{\boldsymbol{x}_k \in \{0,1\}^J} F(\boldsymbol{x}_k)$, the following theorem can be applied:

Theorem 1. *[41] Let $\boldsymbol{x}_k^*$ be the global maximizer of the Gibbs–Boltzmann distribution (17), then*

$$\lim_{T \to 0} \langle \boldsymbol{x}_k \rangle_{P_F} \to \boldsymbol{x}_k^*,$$

where $< \cdot >_{P_F}$ means the expectation with respect to P_F.

Proof. Assuming $\forall \boldsymbol{x}_k \neq \boldsymbol{x}_k^* : \ F(\boldsymbol{x}_k) \geq F(\boldsymbol{x}_k^*)$:

$$\lim_{T \to 0} \langle \boldsymbol{x}_k \rangle_{P_F} = \lim_{T \to 0} \sum_{\boldsymbol{x}_k \in \{0,1\}^J} \boldsymbol{x}_k \frac{\exp\left\{-\frac{1}{T}F(\boldsymbol{x}_k)\right\}}{\sum_{\boldsymbol{x}_k \in \{0,1\}^J} \exp\left\{-\frac{1}{T}F(\boldsymbol{x}_k)\right\}} \tag{18}$$

$$= \lim_{T \to 0} \frac{\sum_{\boldsymbol{x}_k \in \{0,1\}^J} \boldsymbol{x}_k \exp\left\{\frac{1}{T}\left(F(\boldsymbol{x}_k^*) - F(\boldsymbol{x}_k)\right)\right\}}{\sum_{\boldsymbol{x}_k \in \{0,1\}^J} \exp\left\{\frac{1}{T}\left(F(\boldsymbol{x}_k^*) - F(\boldsymbol{x}_k)\right)\right\}} \tag{19}$$

$$= \lim_{T \to 0} \frac{\sum_{\boldsymbol{x}_k \neq \boldsymbol{x}_k^*} \boldsymbol{x} \exp\left\{\frac{1}{T}\left(F(\boldsymbol{x}_k^*) - F(\boldsymbol{x}_k)\right)\right\} + \boldsymbol{x}_k^*}{\sum_{\boldsymbol{x}_k \neq \boldsymbol{x}_k^*} \exp\left\{\frac{1}{T}\left(F(\boldsymbol{x}_k^*) - F(\boldsymbol{x}_k)\right)\right\} + 1} \to \boldsymbol{x}_k^* . \quad \square \tag{20}$$

In contrary to [58], here we assume one of the simplest approximation to (17) that involves replacing P_F with a Mean-Field (MF) reference distribution, i.e.

$$P_F(\boldsymbol{x}_k) \approx P_{MF}(\boldsymbol{x}_k) = \prod_{j=1}^{J} P_j(x_{jk}) \,. \tag{21}$$

Using the MF approximation, we have:

$$P_{MF}(\boldsymbol{x}_k) = \prod_{j=1}^{J} P_j(x_{jk}) = \prod_{j=1}^{J} \frac{\exp\{-\frac{1}{T}F(x_{jk})\}}{\sum_{x_{jk}\in\{0,1\}} \exp\{-\frac{1}{T}F(x_{jk})\}} \,, \tag{22}$$

where $F(x_{jk}) = F(x_{jk}| \langle x_{-jk} \rangle)$. Applying Theorem 1 to (22), the x_{jk} at temperature T can be approximated by the formula:

$$\begin{aligned} x_{jk}(T) &= \frac{\sum_{x_{jk}\in\{0,1\}} x_{jk} \exp\{-\frac{1}{T}F(x_{jk})\}}{\sum_{x_{jk}\in\{0,1\}} \exp\{-\frac{1}{T}F(x_{jk})\}} \\ &= \left(1 + \exp\left\{\frac{F(x_{jk}=1) - F(x_{jk}=0)}{T}\right\}\right)^{-1} . \end{aligned} \tag{23}$$

If $\boldsymbol{A}$ is a non-binary matrix, it can be updated with any above-mentioned algorithm for NMF, e.g. multiplicative update (14) or (15), quasi-Newton, projected gradient, or projected ALS.

The key issue in simulated annealing optimization is to carefully select the temperature schedule for T in (17). It cannot go to zero too fast since an iterative process may easily get trapped in a local minimum. Too slow annealing considerably slows down the overall NMF algorithm. In [58] the temperature schedule was defined with the following exponential rule:

$$T = \bar{T} + T_0 \exp\{-\lambda s\} \,, \tag{24}$$

where $\bar{T} > 0$ is the minimal temperature to avoid numerical instabilities, $T_0 > 0$ is an initial temperature, $\lambda > 0$ is a decay constant that determines how fast the temperature goes to $\bar{T}$.

The exponential temperature schedule given by (24) was also used for determining the regularization parameter in the Levenberg-Marquardt approach to the second-order methods in NMF [60, 61]. The parameter controls the nature of the Hessian inverses. For a high value of T, the iterative updates have gradient-descent nature with very a small steplength. A gradual decrease in T increases the step-length and steers the updates towards the Newton ones. This approach also tends to minimize the risk of getting stuck in local minima. This was experimentally demonstrated in [60, 61, 62].

4. Experiments

We present two examples that compare efficiency of the selected constraints. To analyze the immunity of the proposed methods to get stuck in local minima, the Monte Carlo (MC) analysis is applied in all the experiments. In each MC run the algorithm of interest was initialized with uniformly distributed random matrices $\boldsymbol{A}^{(0)}$ and $\boldsymbol{X}^{(0)}$, and the alternating updates are terminated after 1000 iterations.

In the first experiment we test the sparsity and smoothness constraints using the benchmark [61] of 5 statistically dependent nonnegative sparse signals with a smooth temporal structure. The signals were mixed with the uniformly distributed random matrix $\boldsymbol{A} \in \mathbb{R}^{5\times 5}$.

We used for the experiments the following NMF algorithms: (a) the standard Lee-Seung algorithm [30] for minimization of the Frobenious norm, (b) the quasi-Newton algorithm with sparsity constraints [61], (c) the multiplicative algorithm with smoothness constraints [59]. The results are evaluated in terms of the standard Signal-to-Interference (SIR) measure.

The histograms of mean-SIR values (averaged SIRs over all the estimated signals) from 100 MC runs are shown in Fig. 1.

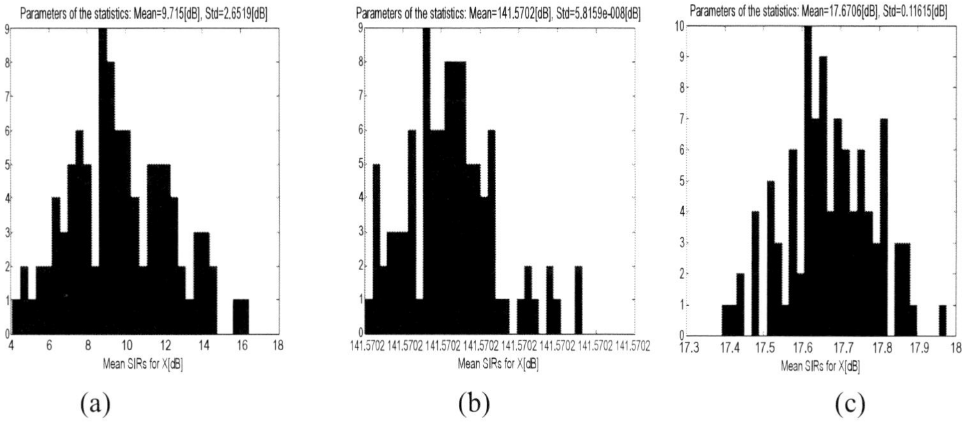

Figure 1. Histograms from 100 mean-SIR samples generated with the following algorithms: (a) multiplicative NMF (14); (b) quasi-Newton with sparsity constraints [61]; (c) multiplicative NMF with smoothness constraints [59]

In the second experiment, we test the binarity constraints on the mixed data with 3 weakly overlapped clusters that are illustrated in Fig. 2.

We compare the following algorithms: (a) the standard Lee-Seung algorithm [30] for minimization of the Frobenious norm, (b) the uni-orthogonal NMF proposed by Ding *et al* [16]; (c) the simulated annealing update for $\boldsymbol{X}$ with (23) and the quasi-Newton updates for $\boldsymbol{A}$ [61]. In this case, we evaluate the performance in terms of the number of misclassified data. The MC analysis was also used. The results are presented in Table 1.

5. Conclusion

The aim of this paper is to highlight the problem that NMF is non-unique, and some precautions should be taken in its applications. According to the current state-of-knowledge, there is no algorithm for NMF that gives unique factors (even up to the scale and permutation ambiguities), independently on a class of data. The nature of data to be decomposed with

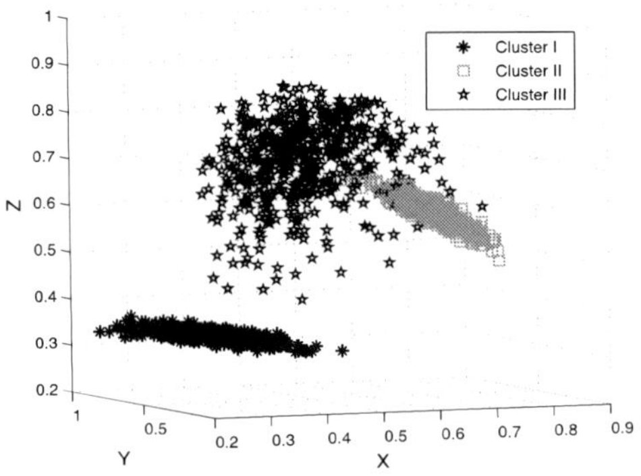

Figure 2. Distribution of clustered data in 3D space.

Table 1. Mean-values and standard deviations (in parenthesis) of the number of misclustered vectors versus the number of alternating steps (the number s) for the selected NMF algorithms

Algorithm	$s = 20$	$s = 50$	$s = 100$	$s = 200$
NMF (Frobenius)	172.9 (118.7)	138.3 (118.1)	151.1 (123.1)	161.5 (116.6)
Uni-orthogonal NMF	216.5 (96.6)	213.6 (121.1)	177 (134.3)	80.3 (70.1)
Algorithm (23)	96 (60.1)	82.1 (72.3)	56 (43.7)	56 (32.4)

NMF is very important, and knowing the nature a suitable algorithm can be selected. The suitable algorithm still rarely means the algorithm that provides unique factors even for a very specific class of data.

Despite NMF algorithms provide non-unique factors, the risk of getting spurious factors can be considerably minimized if the nature of the factors is known and can be well modeled with appropriate constraints. In the paper, we briefly discuss the selected constraints such as sparsity, smoothness, orthogonality, binarity, and unimodality. Closure or boundness could be also a subject of constraints. The constraints can be imposed to the estimated solution in many ways, e.g. by the penalty terms in the objective function or as a subject of equality or inequality constraints in optimization algorithms. Our experiments demonstrate that the algorithms with all kinds of constraints give more consistent results, i.e. standard deviations of the histograms of means-SIR samples (see Fig. 1) are substantially smaller than for the cases of unconstrained NMF algorithms. The same conclusion can be confirmed with the results in Table 1. Unfortunately, the problem of selecting the right penalty parameters is still open. Should it be constant or varying with the number of alternating steps? Our experiments with sparsity constraints showed that the penalty parameter should be very high at the beginning of the alternating process and it should exponentially decreases to a very small number. This is motivated with the simulated annealing technique.

Another problem in NMF is the choice of the optimization algorithm. Since an objective function minimized with the alternating minimization is non-convex, some global optimization algorithms should be here more appropriate than gradient descent techniques. From the literature [51] we know that the genetic algorithm gives very promising results for sparse data. Our numerical tests also confirm that the simulated annealing technique involving the Gibbs–Boltzmann distribution gives also very stable results. There are also many other methods for global optimization of non-convex problems that might find potential applications to NMF problems. Examples include branch and bound (D.C.), tabu search, swarm-based, and memetic algorithms. We believe that this kind of algorithms will find robust applications in NMF problems. Our future research will focuss on the above-mentioned issues.

References

[1] R. Albright, J. Cox, D. Duling, A. N. Langville, and C. D. Meyer. Algorithms, initializations, and convergence for the nonnegative matrix factorization. *Technical Report* **81706**, NCSU Math, 2006. submitted.

[2] A. Banerjee, S. Merugu, I. S. Dhillon, and J. Ghosh. Clustering with Bregman divergences. In *SIAM International Conf. on Data Mining*, Lake Buena Vista, Florida, April 2004. SIAM.

[3] M. Berry, M. Browne, A. Langville, P. Pauca, and R. Plemmons. Algorithms and applications for approximate nonnegative matrix factorization. *Computational Statistics and Data Analysis*, **52**(1):155–173, 2007.

[4] J. Besag. Toward Bayesian image analysis. *J. Appl. Stat.*, **16**:395–407, 1989.

[5] C. Boutsidis and E. Gallopoulos. SVD based initialization: A head start for nonnegative matrix factorization. *Pattern Recogn.*, **41**(4):1350–1362, 2008.

[6] J. P. Brunet, P. Tamayo, T. R. Golub, and J. P. Mesirov. Metagenes and molecular pattern discovery using matrix factorization. volume 101, pages 4164–4169. PNAS, 2000.

[7] Bin Cao, Dou Shen, Jian-Tao Sun, Xuanhui Wang, Qiang Yang, and Zheng Chen. Detect and track latent factors with online nonnegative matrix factorization. In *Proc. the 20th International Joint Conference on Artificial Intelligence (IJCAI)*, pages 2689–2694, Hyderabad, India, January 6–12 2007.

[8] P. Carmona-Saez, R. D. Pascual-Marqui, F. Tirado, J. M. Carazo, and A. Pascual-Montano. Biclustering of gene expression data by non-smooth non-negative matrix factorization. *BMC Bioinformatics*, **7**(78), 2006.

[9] H. Cho, I. S. Dhillon, Y. Guan, and S. Sra. Minimum sum squared residue based co-clustering of gene expression data. In *Proc. 4th SIAM International Conference on Data Mining (SDM)*, pages 114–125, Florida, 2004.

[10] Y. C. Cho and S. Choi. Nonnegative features of spectro-temporal sounds for classification. *Pattern Recognition Letters*, **26**:1327–1336, 2005.

[11] A. Cichocki and R. Zdunek. Regularized alternating least squares algorithms for nonnegative matrix/tensor factorizations. *Springer LNCS*, **4493**:793–802.

[12] A. Cichocki and R. Zdunek. Multilayer nonnegative matrix factorization. *Electronics Letters*, **42**(16):947–948, 2006.

[13] A. Cichocki and R. Zdunek. Multilayer nonnegative matrix factorization using projected gradient approaches. *International Journal of Neural Systems*, **17**:431–446, 2007.

[14] A. Cichocki, R. Zdunek, and S. Amari. New algorithms for non-negative matrix factorization in applications to blind source separation. In *Proc. IEEE International Conference on Acoustics, Speech, and Signal Processing, ICASSP2006*, volume 5, pages 621–624, Toulouse, France, May 14–19 2006.

[15] S. F. Cotter, B. D. Rao, K. Engan, and K. Kreutz-Delgado. Sparse solutions to linear inverse problems with multiple measurement vectors. *IEEE Trans. Signal Processing*, **53**(7):2477–2488, 2005.

[16] C. Ding, T. Li, W. Peng, and H. Park. Orthogonal nonnegative matrix tri-factorizations for clustering. In *KDD06: Proceedings of the 12th ACM SIGKDD international conference on Knowledge discovery and data mining*, pages 126–135, New York, NY, USA, 2006. ACM Press.

[17] D. Donoho and V. Stodden. When does nonnegative matrix factorization give a correct decomposition into parts? In *Neural Information Processing Systems*, volume 16. MIT press, 2003.

[18] Hong-Tao Gaoa, Tong-Hua Lia, Kai Chena, Wei-Guang Lia, and Xian Bia. Overlapping spectra resolution using non-negative matrix factorization. *Talanta*, **66**(1):65–73, 2005.

[19] S. Geman and D. McClure. Statistical methods for tomographic image reconstruction. *Bull. Int. Stat. Inst.*, LII-4:5–21, 1987.

[20] S. Geman and G. Reynolds. Constrained parameters and the recovery of discontinuities. *IEEE Trans. Pattern Anal. Machine Intell.*, **14**:367–383, 1992.

[21] P. J. Green. Bayesian reconstruction from emission tomography data using a modified EM algorithm. *IEEE Trans. Medical Imaging*, **9**:84–93, 1990.

[22] D. Guillamet, B. Schiele, and J. Vitrià. Analyzing non-negative matrix factorization for image classification. In *16th International Conference on Pattern Recognition (ICPR'02)*, volume 2, pages 116–119, Quebec City, Canada, August 2002.

[23] T. Hebert and R. Leahy. A generalized EM algorithm for 3-D Bayesian reconstruction from Poisson data using Gibbs priors. *IEEE Transactions on Medical Imaging*, **8**:194–202, 1989.

[24] M. Heiler and C. Schnörr. Learning sparse representations by non-negative matrix factorization and sequential cone programming. *J. Mach. Learn. Res.*, **7**:1385–1407, 2006.

[25] P. O. Hoyer. Non-negative matrix factorization with sparseness constraints. *Journal of Machine Learning Research*, **5**:1457–1469, 2004.

[26] D. Kim, S. Sra, and I. S. Dhillon. Fast Newton-type methods for the least squares non-negative matrix approximation problem. In *Proc. 6-th SIAM International Conference on Data Mining*, Minneapolis, Minnesota, USA, April 2007.

[27] Hyunsoo Kim and Haesun Park. Sparse non-negative matrix factorizations via alternating non-negativity-constrained least squares for microarray data analysis. *Bioinformatics*, **23**, 2007.

[28] Y.-D. Kim and S. Choi. A method of initialization for nonnegative matrix factorization. In *Proc. IEEE International Conference on Acoustics, Speech, and Signal Processing (ICASSP07)*, volume II, pages 537–540, Honolulu, Hawaii, USA, April 15–20 2007.

[29] A. N. Langville, C. D. Meyer, and R. Albright. Initializations for the nonnegative matrix factorization. In *Proc. of the Twelfth ACM SIGKDD International Conference on Knowledge Discovery and Data Mining*, Philadelphia, USA, August 20–23 2006.

[30] D. D. Lee and H. S. Seung. Learning the parts of objects by nonnegative matrix factorization. *Nature*, **401**:788–791, October 1999.

[31] D. D. Lee and H. S. Seung. Algorithms for nonnegative matrix factorization. In *NIPS*, pages 556–562, 2000.

[32] H. Li, T. Adali, and D. Emge W. Wang. Non-negative matrix factorization with orthogonality constraints for chemical agent detection in raman spectra. In *IEEE Workshop on Machine Learning for Signal Processing*. Mystic, USA, 2005.

[33] H. Li, T. Adali, W. Wang, D. Emge, and A. Cichocki. Non-negative matrix factorization with orthogonality constraints and its application to Raman spectroscopy. *Journal of VLSI Signal Processing*, **48**(1-2):83–97, August 2007.

[34] T. Li and Ch. Ding. The relationships among various nonnegative matrix factorization methods for clustering. In *Proceedings of the Sixth International Conference on Data Mining*, pages 362–371, Washington, DC, USA, 2006. IEEE Computer Society.

[35] C.-J. Lin. Projected gradient methods for non-negative matrix factorization. *Neural Computation*, **19**(10):2756–2779, October 2007.

[36] W. Liu and N. Zheng. Non-negative matrix factorization based methods for object recognition. *Pattern Recognition Letters*, **25**(8):893–897, 2004.

[37] O. G. Okun. Non-negative matrix factorization and classifiers: experimental study. pages 550–555, Marbella, Spain, 2004.

[38] P. Paatero. Least-squares formulation of robust nonnegative factor analysis. *Chemometrics and Intelligent Laboratory Systems*, **37**:23–35, 1997.

[39] P. Paatero and U. Tapper. Positive matrix factorization: A nonnegative factor model with optimal utilization of error estimates of data values. *Environmetrics*, **5**:111–126, 1994.

[40] V. P. Pauca, F. Shahnaz, M. W. Berry, and R. J. Plemmons. Text mining using nonnegative matrix factorizations. In *Proc. SIAM Inter. Conf. on Data Mining*, Orlando, FL, April 2004.

[41] J. W. Phillips, R. M. Leahy, and J. C. Mosher. MEG-based imaging of focal neuronal current sources. *IEEE Trans. Medical Imaging*, **16**:338–248, 1997.

[42] J. Piper, V. P. Pauca, R. J. Plemmons, and M. Giffin. Object characterization from spectral data using nonnegative matrix factorization and information theory. In *Proc. AMOS Tech. Conf.*, Maui, HI, September 2004.

[43] N. Rao, S. J. Shepherd, and D. Yao. Extracting characteristic patterns from genome – wide expression data by non-negative matrix factorization. In *Proc. of the 2004 IEEE Computational Systems Bioinformatics Conference (CSB 2004)*, Stanford, CA, August 2004.

[44] T. M. Rutkowski, R. Zdunek, and A. Cichocki. Multichannel EEG brain activity pattern analysis in time-frequency domain with nonnegative matrix factorization support. *International Congress Series*, **8611**:266–269, 2007.

[45] P. Sajda, S. Du, T. Brown, L. Parra, and R. Stoyanova. Recovery of constituent spectra in 3D chemical shift imaging using nonnegative matrix factorization. In *4th International Symposium on Independent Component Analysis and Blind Signal Separation*, pages 71–76, Nara, Japan, April 2003.

[46] P. Sajda, S. Du, T. R. Brown, R. Stoyanova D. C. Shungu, Xiangling Mao, and L. C. Parra. Nonnegative matrix factorization for rapid recovery of constituent spectra in magnetic resonance chemical shift imaging of the brain. *IEEE Trans. Medical Imaging*, **23**(12):1453–1465, 2004.

[47] P. Sajda, S. Du, and L. Parra. Recovery of constituent spectra using non-negative matrix factorization. In *Proceedings of SPIE*, volume 5207, pages 321–331. Wavelets: Applications in Signal and Image Processing, 2003.

[48] M. N. Schmidt and M. Mørup. Nonnegative matrix factor 2-D deconvolution for blind single channel source separation. *Springer LNCS*, **3889**:700–707, 2006.

[49] F. Shahnaz, M. Berry, P. Pauca, and R. Plemmons. Document clustering using nonnegative matrix factorization. *Journal on Information Processing and Management*, **42**:373–386, 2006.

[50] M. W. Spratling. Learning image components for object recognition. *Journal of Machine Learning Research*, **7**:793–815, 2006.

[51] K. Stadlthanner, F. J. Theis, E. W. Lang, A. M. Tome, C. G. Puntonet, and J. M. Gorriz. Hybridizing sparse component analysis with genetic algorithms for microarray analysis. *Neurocomputing*, 2008. in press.

[52] R. Stevenson and E. Delp. Fitting curves with discontinuities. In *Proc. 1-st Int. Workshop on Robust Computer Vision*, Seattle, Wash., USA, Oct. 1–3 1990.

[53] F. J. Theis, K. Stadlthanner, and T. Tanaka. First results on uniqueness of sparse non-negative matrix factorization. In *Proc. EUSIPCO 2005*, Antalya, Turkey, 2005.

[54] Y. Wang, Y. Jia, C. Hu, and M. Turk. Non-negative matrix factorization framework for face recognition. *International Journal of Pattern Recognition and Artificial Intelligence*, **19**(4):495–511, 2005.

[55] K. Watanabe, A. Hidaka, and T. Kurita. Automatic factorization of biological signals by using boltzmann non-negative matrix factorization. In *Proc. of World Congress on Computational Intelligence (WCCI 2008)*, Hong Kong, June 1–6 2008. to appear.

[56] S. Wild. *Seeding non-negative matrix factorization with the spherical k-means clustering*. M.Sc. Thesis, University of Colorado, 2000.

[57] R. Zass and A. Shashua. A unifying approach to hard and probabilistic clustering. In *International Conference on Computer Vision (ICCV)*, Beijing, China, Oct. 2005.

[58] R. Zdunek. Data clustering with semi-binary nonnegative matrix factorization. In *International Conference on Artificial Intelligence and Soft Computing (ICAISC 2008)*, Zakopane, Poland, June 22–26 2008. to appear.

[59] R. Zdunek and A.Cichocki. Gibbs regularized nonnegative matrix factorization for blind separation of locally smooth signals. In *15th IEEE International Workshop on Nonlinear Dynamics of Electronic Systems (NDES 2007)*, pages 317–320, Tokushima, Japan, July 23–26 2007.

[60] R. Zdunek and A. Cichocki. Non-negative matrix factorization with quasi-Newton optimization. *Springer LNAI*, **4029**:870–879, 2006.

[61] R. Zdunek and A. Cichocki. Nonnegative matrix factorization with constrained second-order optimization. *Signal Processing*, **87**:1904–1916, 2007.

[62] R. Zdunek and A. Cichocki. Nonnegative matrix factorization with quadratic programming. *Neurocomputing*, 2008. in press.

[63] Zhonglong Zheng, Jie Yang, and Yitan Zhu. Initialization enhancer for non-negative matrix factorization. *Eng. Appl. Artif. Intell.*, **20**(1):101–110, 2007.

[64] A. Ziehe, P. Laskov, K. Pawelzik, and K.-R. Mueller. Non-negative sparse coding for general blind source separation. In *NIPS 2003*, Vancouver, Canada, 2003.

In: New Signal Processing Research
Editor: Takumi Maeda, pp. 455-475
ISBN: 978-1-60456-479-2

Chapter 16

THE EMPIRICAL MODE DECOMPOSITION: A USEFUL TOOL FOR SIGNAL ANALYSIS

R.T. Rato*, M.D. Ortigueira and A.G. Batista
UNINOVA
Campus da FCT da UNL, Quinta da Torre,
2825 – 114 Monte da Caparica, Portugal

Abstract

The Empirical Mode Decomposition is a technique to decompose a given signal into a set of elemental signals called Intrinsic Mode Functions. The Empirical Mode Decomposition is the base of the so-called "Hilbert-Huang Transform" that comprises also a Hilbert Spectral Analysis and an instantaneous frequency computation. A modified improved algorithm for the Empirical Mode Decomposition is described. Its interpretation in terms of AM/FM modulation is done. Numerical simulations are carried out to empirically evaluate the algorithm.

Key words: Empirical mode decomposition, intrinsic mode function, parabolic interpolation, extrema detection, amplitude-frequency modulation

1. Introduction[1]

The Empirical Mode Decomposition (EMD as a technique to decompose a given signal into a set of elemental signals has been gaining importance and being used in an increasing number of scientific and engineering fields. The EMD decomposes a given signal into several elemental signals called "intrinsic mode functions" (IMF). This decomposition is the base of the so-called "Hilbert-Huang Transform (HHT)" that is essentially a time-frequency analysis

* E-mail addresses: rtr@uninova.pt, mdo@fct.unl.pt, and agb@fct.unl.pt.
Phone. +351 1 2948520, Fax. +351 1 2957786

[1] This text is mostly based of the article "On the HHT, its problems, and some solutions", Mechanical Systems and Signal Processing, to appear soon.

and, besides EMD, it comprises also the Hilbert Spectral Analysis. This performs a spectral analysis using the Hilbert Transform (HT) to compute the analytic signal. From this we obtain an amplitude and an instantaneous frequency.

The EMD, as proposed by Huang et al., is a signal decomposition algorithm based on a successive removal of the Intrinsic Mode Functions. In the following we are going to describe it.

Given any signal, x(t), the IMFs are found by an iterative procedure called *sifting algorithm,* composed of the following steps:

1. Find all the local maxima, M_i, i=1, 2 … and minima, m_k, k=1,2,… in x(t)
2. Compute the corresponding interpolating signals $M(t) := f_M(M_i, t)$, and $m(t) := f_m(m_k, t)$. These signals are the upper and lower envelopes of the signal.
3. Let $e(t) := \dfrac{M(t)+m(t)}{2}$
4. Subtract e(t) from the signal: $x(t) := x(t) - e(t)$
5. Return to step 1. – stop when x(t) remains nearly unchanged.
6. Once we obtain an IMF, $\varphi(t)$, we remove it from the signal $x(t) := x(t) - \varphi(t)$ and return to 1. if x(t) has more than one extremum (not a constant nor a trend).

The interpolating function is a cubic spline. By construction, the number of extrema should decrease when going from one IMF to the next, and the whole decomposition is expected to be completed with a finite number of IMFs. We must remark that, at least conceptually, the algorithm:

- Is simple
- Appears naturally
- Does not assume anything about the signal, mainly stationarity
- Can be applied to a wide class of signals.

Besides this, the algorithm gives good results in situations where other methods fail. However, it has some drawbacks, tied with some of the assumptions needed to implement the algorithm, leading to unexpected results.

Here, we make a global appreciation to the different aspects of the algorithm, consider its drawbacks and present modifications that alleviate such problems. The most important is the extrema determination by using a parabolic approximation and the instantaneous amplitude and frequency computation that is done through an amplitude demodulation and first order autoregressive (AR) approximation.

2. The Analysis of EMD

2.1. The Intrinsic Mode Functions

The Intrinsic Mode Functions are very important in EMD, since they perform the role of basis functions, although they are not orthogonal. In the following we will do a brief study of its characteristics.

For the continuous case, an abstract IMF is defined as a signal that satisfies two conditions:

(a) In the whole signal segment, the number of extrema and the number of zero crossings must be either equal or differ at most by one;
(b) At any point, the mean value of the envelope defined by the local maxima and the envelope defined by the local minima is zero.

This is the original definition of IMF presented in Huang et al. However the first condition is redundant unless the function at hand is discontinuous. In fact, if a function verifies the second condition, it verifies the first also, because between a peak and a valley there is always a zero, since the envelopes are symmetric.

It is not difficult to see that sinusoidal signals $\sin(2\pi ft)$ or $\cos(2\pi ft)$ for any real f and t are IMF. It is not difficult to recognize that if $\Phi(t)$ is a continuous function, the same happens with $\sin[\Phi(t)]$ and $\cos[\Phi(t)]$. These are well known functions in Telecommunications where they are studied under the theme "Angle Modulation". The instantaneous frequency is defined, aside a constant, by the derivative of $\Phi(t)$. Now, let us consider a function $g(t) = A(t).\sin[\Phi(t)]$. This is what is called "double side band" modulated sinusoid. In general, it is not an IMF, even if A(t) is also a sinusoid. But, if A(t) changes slowly when compared with the changes of $\sin[\varphi(t)]$, we have really an IMF. So, we can say that a function of the type $A(t).\sin[\Phi(t)]$ represents an IMF, provided that A(t) is a slowly varying function. This is the so-called envelope. But we must remark that the envelope assumed in EMD may not be a true envelope. A simple example shows this. The function $x(t) = e^{-|t|}.\sin(2\pi ft)$ has $e^{-|t|}$ as envelope, but almost all the extrema points of x(t) do not belong to $e^{-|t|}$. In some signals this fact has as consequence that the function may have segments above (below) the envelope defined by the maxima (minima). As a conjecture, we can say that the extrema envelope coincides with the "true" envelope when this has extrema locations equal to those of the signal. However, there is no simple way of defining the "true" envelope and the extrema defining envelope seems to be suitable to our objectives.

We can conclude that the definition of IMF is tied with the definition of envelope that depends also on the interpolating function used to estimate it. So, the Huang et al IMFs correspond to the cubic spline interpolator. With other interpolator we may obtain a different set of IMFs. In our simulations we used also the Akima interpolator[2]. The results were very similar.

We may wonder if the use of the term "mode" is acceptable here and if there is any relation between this "mode" concept and the usual mode term used when dealing with linear systems. We must remember that these correspond to the impulse response of a second order linear system with two complex conjugate poles; such impulse response is given by: $e^{-at}.\sin(2\pi ft+\varphi).u(t)$, where u(t) is the Heaviside unit step function, $a\in R^+$ and $\varphi\in[-\pi,\pi)$. It is not hard to conclude that this function is an IMF and must be let indecomposable by an EMD algorithm. However, the sum of at least two modes may lead to a function that is not an IMF and its decomposition may not lead to a recover of the original modes.

[2] We used the routine available at the MatLab site.

2.2. The EMD as an AM/FM Decomposition

Consider the sifting process. The first step finds two sets of points that constitute samples of two discrete-time signals, the maxima and minima, obtained by means of an irregular sampling. The interpolated signals give estimates of the upper and lower envelopes. If the envelopes were symmetric we would say that x(t) was an amplitude (AM) modulated signal. Therefore the sifting procedure is an iterative way of removing the dissymmetry between the upper and lower envelopes in order to transform the original signal into an AM signal. At least conceptually, this goal would be achieved in a few steps if:

- The extrema were correctly determined.
- We had no problems with the interpolation at the extremities.
- We had no computational errors.

The above procedure allows us to conclude that each IMF is an amplitude modulated signal. Moreover, as the instantaneous frequency can change from instant to instant, we can say that each IMF is an amplitude-frequency (AM/FM) modulated signal. So, the EMD is nothing else than a decomposition into a set of AM/FM modulated signals. As it is not hard to understand, the envelopes cannot vary as fast as the signal, x(t). In spectral terms, we can say that the bandwidth of the envelopes must be a fraction of the central frequency of x(t) (normally called carrier). This means that when performing the sifting we are removing low frequency components. So, we are leaving a high frequency signal. This explains why

- The IMF appear in a high to low frequency order
- EMD is essentially a time-frequency decomposition.
- EMD behaves like a bank of filters.

This explains also an interesting phenomenon: if we add two IMFs with non intercepting bands, they will be decomposed without great distortion, but if the bands intercept, they will be decomposed into a set of several IMFs. We will show this later.

2.3. Main Difficulties of the EMD

The algorithm has some implicit difficulties and procedures that can originate "strange" decompositions. For example, when trying to decompose a sinusoidal segment, it is expected to obtain only one IMF and no residual. This may not happen with some available algorithm implementations.

We are going to look into each step of the algorithm and identify the difficulties and their causes. Later we will present some solutions. The referred steps are:

1. *Extrema locations* – the extrema locations computation does not have an obvious solution. Although most data is originated by continuous-time processes, in practice, the algorithm operates on quantized discrete-time signals. When working with this kind of signals, some special attention is necessary because the extrema may not be correctly identified. Most, if not all, of the continuous waveform actual extrema will

fall in between sampling instants, and will not be correctly localized. Following Rilling *et al* we can alleviate this difficulty with a fair amount of oversampling.

2. *Extrema interpolation* – We have considered before the interpolation problem and the interpolation function choice. However, even if we made a reasonable choice, the effectiveness of the approximation is highly dependent on the extrema computation and may lead to some undesirable results. One of them is the "overshoots": the signal crosses the envelope. This affects the IMF estimation
3. *End effects* – The end effects appear when we have to decide what to do with the first and last samples. The solution will affect the final decomposition:
   - To consider them as maxima and minima simultaneously (this forces all the IMFs to be zero at that points)
   - To consider them as maxima or minima according to the nearest extremum in order to guarantee the alternation between maxima and minima – this is the option we adopted.
   - To leave them free.
4. *Sifting stopping criterion* – The stopping criterion is another source of problems due to its degree of arbitrariness, since it may not guarantee a total signal removal to obtain a "true" IMF.
5. *IMF removal* – According to the above description the final step is the removal of the IMF from the signal. However, if the IMF is not well computed, we may be "adding" to the remaining signal a component that will appear in a following IMF. This explains partially why we do not obtain only one IMF in the case of a pure sinusoid.

2.4. A Framework for Improved EMD Algorithms

The EMD does not have an analytical formulation: it is based on a computational algorithm. So it performs accordingly to its implementation details. When the same signal is fed into different EMD implementations, different results are obtained. This is confusing and inadequate. To achieve consistency and to be able to compare results, different implementations should be equivalent. Here we will establish a framework for EMD implementation to avoid the referred drawbacks.

Based on this rationale, we state the following guiding principles to be followed in the design, implementation and checking of EMD programs:

(a) The IMF set obtained by multiplying a constant value to all the samples in the signal, should be the IMF set of the original signal, multiplied by the same constant.
(b) Changing the mean of the signal, it should only change the trend related IMF (the last one), leaving all the others unchanged.
(c) The EMD of an IMF should be the IMF itself.
(d) The IMF set obtained from a time reversed signal should be the time reversed IMF set of the original signal[3].

[3] This condition is not fundamental and can be released.

These rules suggest us to

- Remove the mean.
- Normalize the signal to a unit power.

This last procedure is important when dealing with signals with very low amplitudes as it is the case of biomedical signals.

2.5. Some Solutions for the Difficulties

2.5.1. The Extrema Locations

In practice, the algorithm operates on discrete-time signals. This deserves some special attention because the extrema definition cannot be based on a continuous neighborhood, as in calculus. It is necessary to define a classificatory function to assert if a given sample is, or is not, an extremum. The classification of a sample as an extremum must be based on the relation of the actual sample and its left and right neighbors. Except for the first and last samples, all samples vicinity will have just those two values. Larger vicinities, with more than two values, will distress the algorithm local feature. So, the classificatory function must receive as input, just these three values. A classification based solely on ">"and ">=" type relations will adhere naturally to principles b) and c) seen in the last section. For a given sample, x[n], the classificatory function is:

If ((x[n] > x[n-1]) AND (x[n] >= x[n+1]))

OR ((x[n] >= x[n-1]) AND (x[n] > x[n+1]))Then Return ("Is a Maximum")

Elseif((x[n] < x[n-1]) AND (x[n] <= x[n+1]))

OR ((x[n] <= x[n-1]) AND (x[n] < x[n+1]))Then Return ("Is a Minimum")

Else Return("Not an extrema");

However the envelope fitting depends on the extrema accuracy (position and value). As referred above, dealing with a discrete-time signal produces an error in the location of an extremum that can be equal to half the sampling interval. To alleviate this problem, we interpolate the signal in the interval defined by the two neighbours to obtain a better extremum location and individualization. Classic interpolation for band limited signals relies on the computationally demanding *sinc* function as the interpolation kernel. This has the disadvantage of needing many samples on the right and on the left sides.

In agreement with our comments in 2.1, we are expecting that the IMFs have some degree of smoothness, mainly near the extrema. In particular they may have also a null derivative at the extrema. This means that, near the extrema, it can be approximated by a parabola. So, parabolic interpolation comes into sight as a practical compromise between no interpolations at all and *sinc* based interpolations, for better location of the extrema values.

Using parabolic interpolation, each extremum is estimated from only the above defined three samples. A new interpolated extremum sample is obtained, usually with no integer abscissa, and this new sample is fed into the EMD algorithm, replacing the integer abscissa extremum sample as an envelope defining point. To see what happens, assume the situation where we have an extremum near time t = n, and let y(1) = x[n-1], y(2)= x[n], and y(3)= x[n+1]. The interpolating parabola is defined by $y(k) = a.k^2 + b.k + c$, (k=1,2,3). We have:

$$[y(1);y(2);y(3)] = [1\ \ 1\ \ 1;4\ \ 2\ \ 1;9\ \ 3\ \ 1]\,[a;b;c] \text{ and}$$
$$[a;b;c] = [1/2\ \ -1\ \ 1/2;-5/2\ \ 4\ \ -3/2;3\ \ -3\ \ 1]\,[y(1);y(2);y(3)] \tag{1}$$

The sign of "a" will allow us to decide which kind of extremum we have:

- a > 0, it is a minimum,
- a < 0, it is a maximum.
- a = 0 it is not an extremum.

From simple geometric considerations, the coordinates (t_p,x_p) for the extremum obtained from the interpolating parabola will be:

$$t_p = -\frac{b}{2a};\ y_p = -\frac{b}{2}t_p+c \tag{2}$$

If $1.5 < t_p < 2.5$, we will have an extremum with a true location point at $t = n-2+t_p$; otherwise, there will not be an extremum.

A simple reasoning shows that this procedure gets round the use of the above classificatory function. In fact we can run the interpolation algorithm just described for all the triplets x[n-1], x[n], and x[n+1] formed with the samples of the signal. The first and last samples are not considered, leaving them as non extrema, by now.

2.5.2. The End Effects

In section 2.3 we referred the end effects. These appear due to the fact that a given interpolator may not be a good extrapolator. In fact, unless we consider the initial and final points as maximum and minimum simultaneously, we have, at least one extremum in each side that must be free. This means that the envelope near that point will be obtained by extrapolation. This extrapolation is poorer according to the distance to the first (last) extremum. To be more concrete let us assume that the first extremum calculated as described in the last section is a maximum. Then, the first point must be a minimum. So, we are not constraining the upper envelope estimator that near the first point will be an extrapolator. This gives a poor result. To avoid this problem, several attempts were done such as repetition or reflection of the signal. The results were not encouraging, because we were really extrapolating the signal. Instead we decided to extrapolate the maxima and the minima. This has an important effect: the envelope estimator behaves always as an interpolator. As we will see later the end effects are almost invisible. Let us consider the beginning of the signal that we assume to be at time, t=0. The extrapolation of the extrema was done by the following procedure:

1. Find the first maximum, M_1, and minimum, m_1, and their time locations, T_1 and t_1. Assume, for example, that $T_1 > t_1$.
2. Insert a new maximum, $M_0 = M_1$, located at $T_0 = - t_1$, and a new minimum, $m_0 = m_1$, located a $t_0 = - T_1$.

To the end of the signal the extrapolation is similar. This procedure is correct in face of our comments at section 2.2: the envelope is a low pass signal in face of the original signal. Alternatively, we can:

1. Consider the first and last samples as maxima or minima according to the nearest extremum in order to guarantee the alternation between maxima and minima.
2. Extrapolate only one of the maxima or minima using the above procedure.

In our implementation we used this procedure with very satisfactory results.

2.5.3. The Mean Envelope Removal

Once we determined the upper and lower envelopes we have to remove its mean from the signal. This is an important aspect of the algorithm, since we may be adding a non existing component that can distort the actual IMF and will appear in, at least, one of the following IMFs. To attenuate this we modified step 4. in the sifting by introducing a step size $0<\alpha<1$: $x(t):= x(t)-\alpha.e(t)$. This increases the iteration time duration, but the algorithm becomes more reliable. In principle, the parameter α can be arbitrary, but we found better to choose it accordingly to an acceptable criterion: *minimize the energy of the resulting signal.* This criterion leads to α equal to the correlation coefficient between x(t) and e(t).

2.5.4. The Stopping Criterion

To state a stopping criterion in the sifting procedure, we will define a resolution factor by:

the ratio between energy of the signal at the beginning of the sifting, x(t), and the energy of the average of the envelopes, e(t).

If this ratio grows above the allowed resolution, then the IMF computation must stop. This criterion is guideline (a) complying, as opposed to criteria based on iteration count. In practice, a resolution between 40 and 50 dB is enough.

An useful property of the resolution factor is that it enables the researcher to set the number IMFs. Reducing the resolution factor reduces the number of obtained IMFs. This has important practical applications, in situations where different signals should be decomposed into the same IMF number in order to perform a side by side analysis. There is interdependence between the number of IMFs and resolution. When conducting signal analysis, if the resolution is fixed, the number of IMFs is allowed to vary from signal to signal. If the number of IMFs is to be fixed, then the resolution must be adjusted on a per signal basis.

2.6. Flowchart of the Algorithm

The EMD algorithm proposed here is summarized in the following flowchart that can be used as a guide for the implementation procedure.

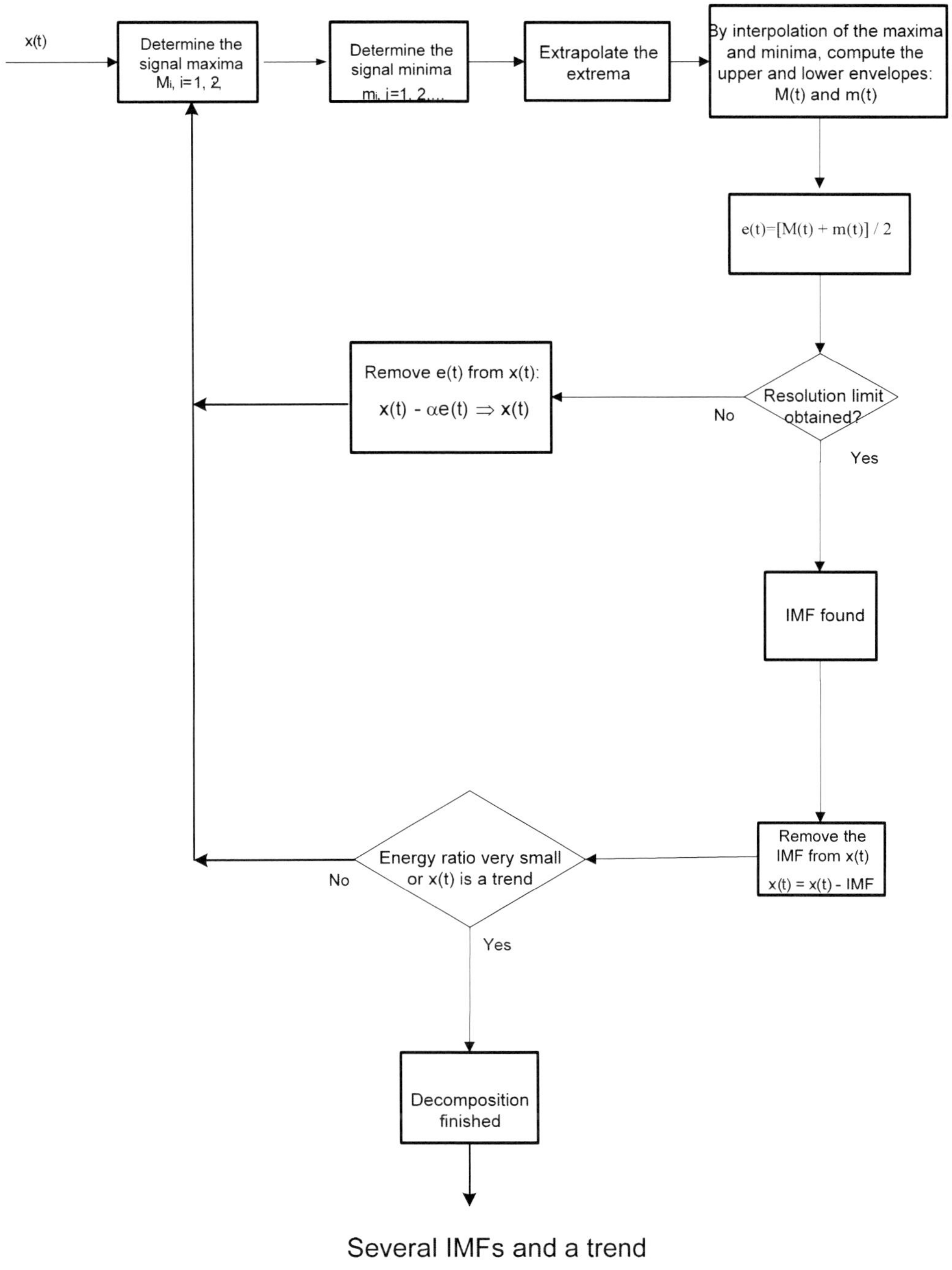

2.7. Illustrating Results

2.7.1. Adding Two Intrinsic Mode Functions

To illustrate the behaviour of the proposed algorithm and justify some of the affirmations we did before, we computed the decompositions for a signal that is a sum of two intrinsic mode functions:

$$x_1(n) = \sin(2\pi f_0 n)$$

$$x_2(n) = 2|\sin(2\pi f_1 n)| - 1$$

with n=0,1, …, 299, f_1=0.23 Hz. We made two trials with f_0 = 0.1 Hz and f_0 =0.3 Hz. In the following two figures, the signals in the upper two strips are the original signals. In figures 3, we present the results corresponding to the case of f_0 = 0.1 Hz.

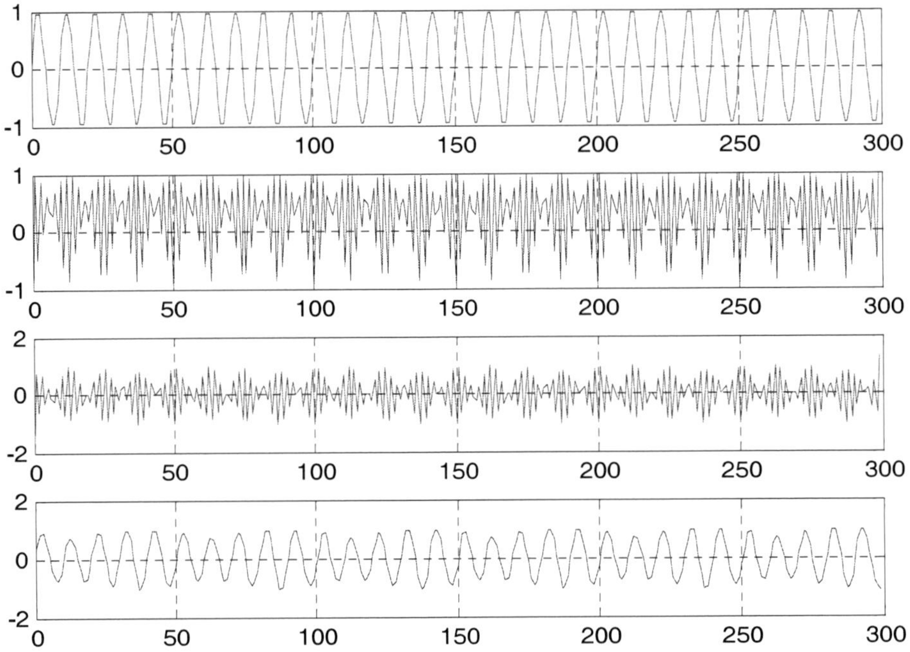

Figure 2. Decomposition of the signal $\sin(2\pi f_0 n) + 2|\sin(2\pi f_1 n)|-1$ with f_0 = 0.1 Hz and f_1 = 0.23 Hz. The uppermost two strips represent the original signals.

In the following pictures we show the results obtained with the same algorithms, but now with f_0 = 0.3Hz.

As we can see, if the bands of the IMFs do not intercept, they are obtained, but in the reverse situation we have a decomposition with a larger number of IMFs.

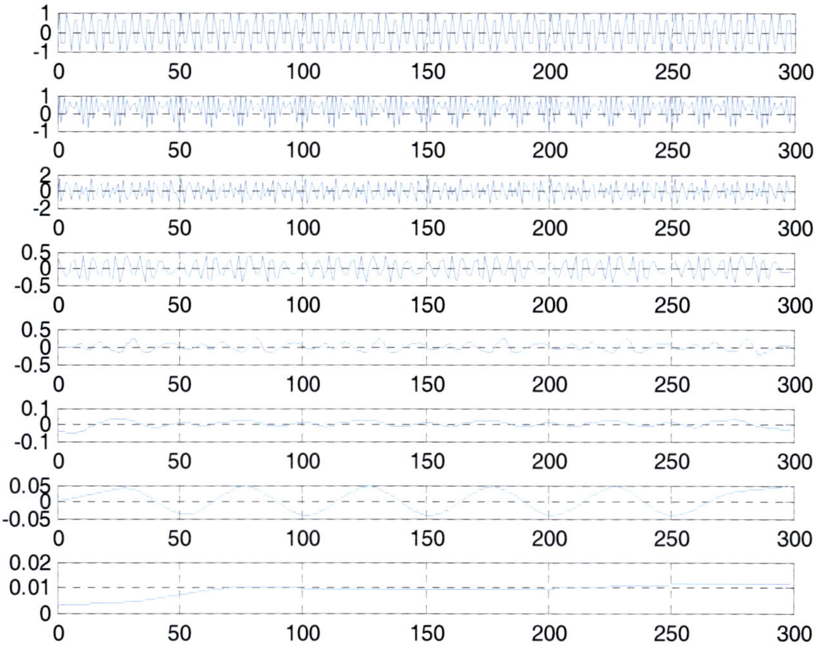

Figure 3. Decomposition of the signal $\sin(2\pi f_0 n) + 2|\sin(2\pi f_1 n)|-1$ with $f_0 = 0.3$ Hz and $f_1 = 0.23$ Hz. The uppermost two strips represent the original signals.

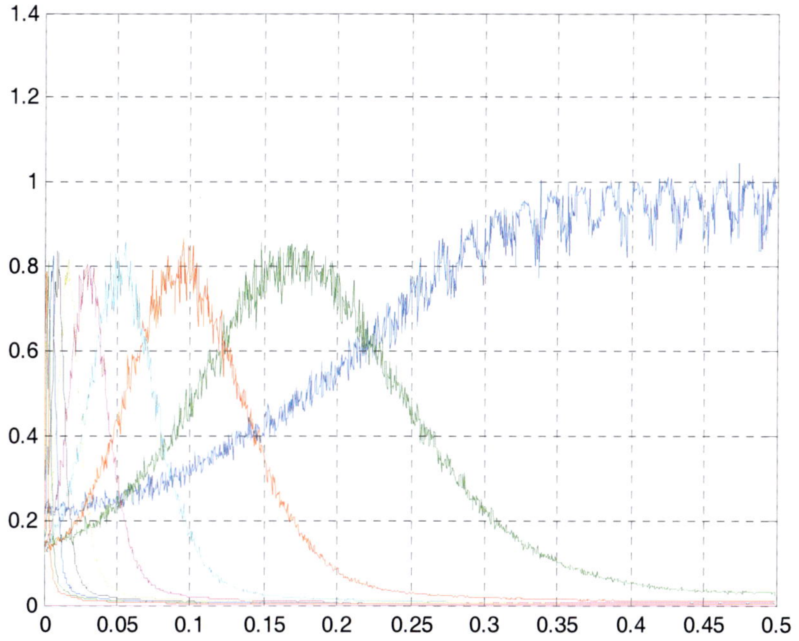

Figure 4. Illustration of the bank of filters behaviour of the EMD.

We represented the results obtained with 10 IMFs. We used L=2000. It is interesting to observe that the different curves cross near 0.6.

2.7.2. The Bank of Filters Behaviour

As we said before, the EMD behaves as a bank of filters. To illustrate this we generated 100 realizations of a white noise like signal. For each realization, we made the EMD and computed the Fourier transforms of all the IMFs retaining their absolute values. In the previous picture we show their averages. The white noise like signal was generated in such a way to have absolute value of its Fourier transform equal to 1. This is done with the following procedure. Assume that we want a signal with L points. We

1. Generate (L-1)/2 uniformly random points in the interval (-1;1), $\varphi(n)$;
2. Compute the complex vector p with elements given by $p(n)=e^{j\pi\varphi(n)}$, n=1, 2 ...(L-1)/2;
3. Obtain the Fourier transform given by: X=[1 p -1 p']; the vector p' is the conjugate reverse of p.
4. Compute the inverse Fourier transform of X to obtain a vector with L values of a noise signal.

3. The EMD Based Time-Frequency Analysis

3.1. The Hilbert Spectral Analysis

Spectral estimation is the second step of the HHT. This consists in computing the instantaneous amplitude and frequency for each IMF by using the Hilbert transform and the analytic signal. This is another drawback of the HHT, because the Hilbert transform uses the whole signal (theoretically from -∞ to +∞). As we have a finite segment of a signal, the window effect will distort its spectrum and as consequence its Hilbert transform. This can give poor frequency estimation. Besides, it is not easy to accept a global operator as base for a local estimation.

The Hilbert spectral analysis allows us to obtain estimations of instantaneous amplitude and frequency of a given modulated signal, $\varphi(t)$. It consists of a five step procedure:

1. Computation of the Hilbert transform, $\hat{\varphi}(t)$, of the signal, $\varphi(t)$.

2. Form the so-called analytic signal given by:

$$z(t) = \varphi(t) + i\cdot\hat{\varphi}(t) \quad (3)$$

3. Compute the instantaneous amplitude by:

$$A(t) = \sqrt{\varphi^2(t) + \hat{\varphi}^2(t)} \quad (4)$$

4. Obtain the instantaneous phase from:

$$\theta(t) = \text{arctg}\frac{\hat{\varphi}(t)}{\varphi(t)} \quad (5)$$

5. The instantaneous frequency is, aside a constant, the derivative of the phase

$$f(t) = \frac{d\theta(t)}{dt} \tag{6}$$

Having the triplet (t,f(t),A(t)) we can obtain a time-frequency representation.

3.2. Analysis by Demodulating the IMF

Let φ(t) an IMF and y(t) the corresponding analytic signal. So,

$$\varphi(t) = \mathrm{Re}\{|y(t)|.e^{j\arg(y(t))}\} = |y(t)|.\cos[\theta(t)], \tag{7}$$

where θ(t) = arg[y(t)]. So, we obtained the instantaneous amplitude and an oscillating function that is a constant amplitude frequency modulated signal (not necessarily a sinusoid).

If |y(t)| is known, we can perform an amplitude demodulation and obtain

$$s(t) = \cos[\theta(t)], \tag{8}$$

such that

$$|\, s(t)\,| \le 1 \tag{9}$$

s(t) can be considered as a FM signal. Its demodulation leads us to the instantaneous frequency. This will be considered later in this section. By now we will consider the amplitude demodulation.

At the end of the sifting procedure leading to the referred IMF, φ(t), we have also its envelopes, M(t) and m(t). If these were the "true" envelopes, they would be symmetric and its difference would be the estimate of the amplitude modulating signal

$$\varphi(t) = A(t).s(t) \tag{10}$$

and

$$A(t) = |y(t)| = M(t) - m(t) \tag{11}$$

As M(t) and m(t) are not truly symmetric, we must look for a more reliable estimate of A(t). This can be achieved by the following procedure:

- Make g(t) = | φ(t) |.
- Compute the maxima of g(t) and extrapolate them as prescribed in 2.5.2.
- Interpolate those maxima, to obtain an estimate of A(t).

Now, its enough to divide φ(t) by A(t) to obtain a FM signal, $s_{FM}(t)$.

Assume now that the instantaneous frequency of $s_{FM}(t)$ is a slowly time varying signal, so that we may consider to be constant over small time intervals. Moreover, sample it to obtain a discrete-time signal that we can express in the format:

$$s_{FM}(n) \approx \cos[2\pi f(n_0).n] \quad (12)$$

valid for n_0-N≤n≤ n_0+N. So, we assume that the frequency is constant in a window with length 2N+1, where N is a positive integer. As shown in appendix A, in this situation the instantaneous frequency can be obtained from

$$\cos[2\pi f(n_0)] = \frac{\sum_{2}^{L-1} s_{FM}(n)[s_{FM}(n-1) + s_{FM}(n+1)]}{2\sum_{2}^{L-1} s_{FM}^2(n)} \quad (13)$$

where L is the number of available samples. In the case of a pure sinusoid this formula gives the correct value, provided that we have, at least, 3 samples. For a FM signal we put L=2N+1, as referred above. It defines the window of validity of the approximations. The choice of N depends on the practical application. In the case of the chirp signal presented later we used N=10. We have used values from 20 to 40 in EEG applications. Therefore and as in the above case we obtain also a triplet that allows us to obtain a time-frequency representation.

3.3. Illustrating Examples

3.3.1. Frequency Estimation

To illustrate the performances of the above methods for frequency estimation we used them with some signals.

In the following figures, we present illustrations of the frequency estimation for one chirp noisy signal.

As it can be seen, the AR method outperforms the Hilbert transform estimation.

3.3.2. An ECG Signal

In the following we will present the results concerning the analysis of a segment of an ECG signal with 5 beats. In the following figure, we present the original signal and the corresponding IMFs obtained with the algorithm we propose here.

To compare we are going to present the results of the time-frequency analysis, using 2 algorithms:

- Ours, that consists in the IMF amplitude demodulation followed by a frequency estimation using the AR modelling.
- Huang's approach based on the Hilbert transform.

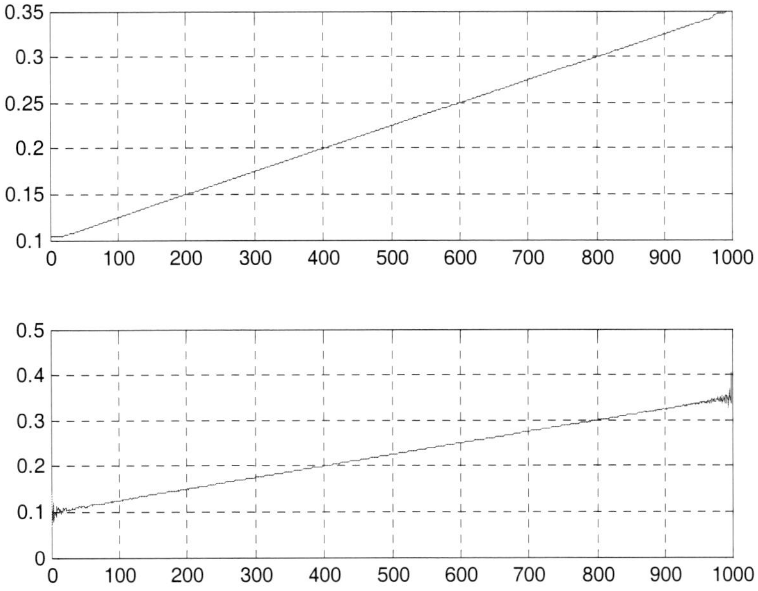

Figure 5. Estimation of the instantaneous frequency with AR, above, and Hilbert transform methods.

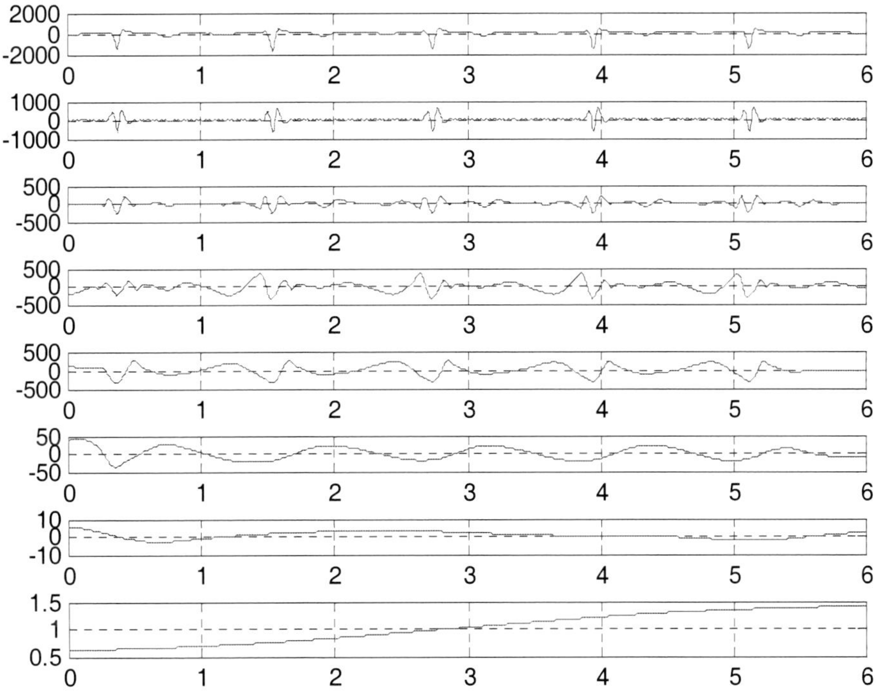

Figure 6. EMD of a segment of an ECG. The original signal is in the upper strip.

The energy operator demodulation proposed by Junsheng et al is not suitable to be used with short length signals since it cannot assure that the energy operator assumes only non negative values.

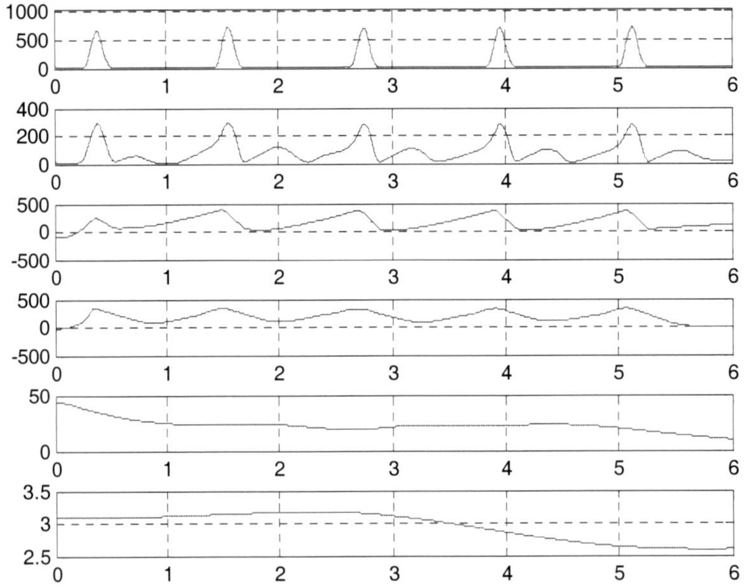

a)

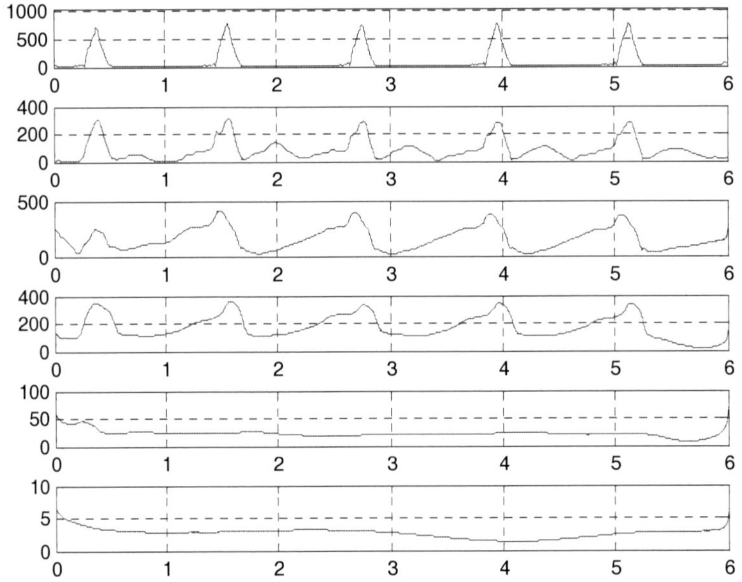

b)

Figure 7. Estimation of the amplitudes by demodulation, a), and Hilbert transform, b).

As expected from our considerations in 3.1 and 3.2 they are not very different. This does not happen with the frequency estimates. It is clear that our approach gives more realistic and convincing results, mainly in the frequency computation.

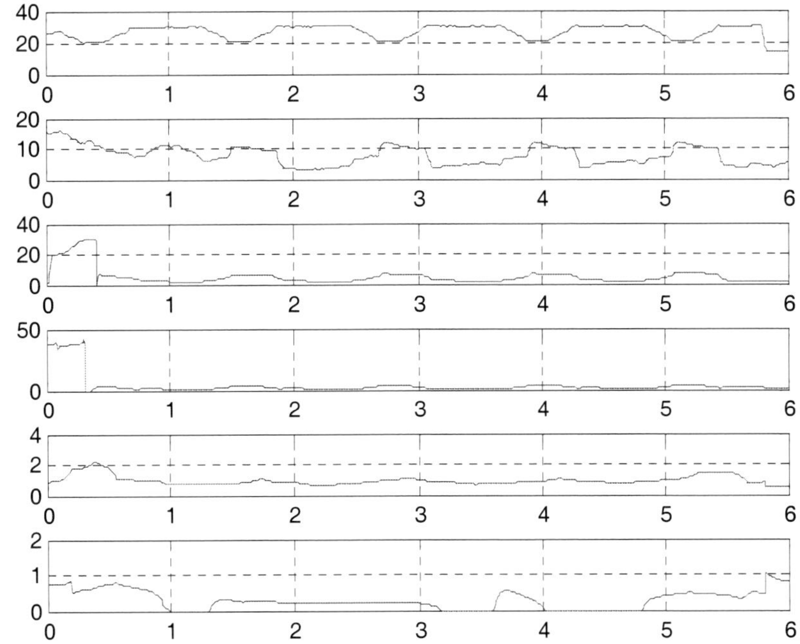

a)

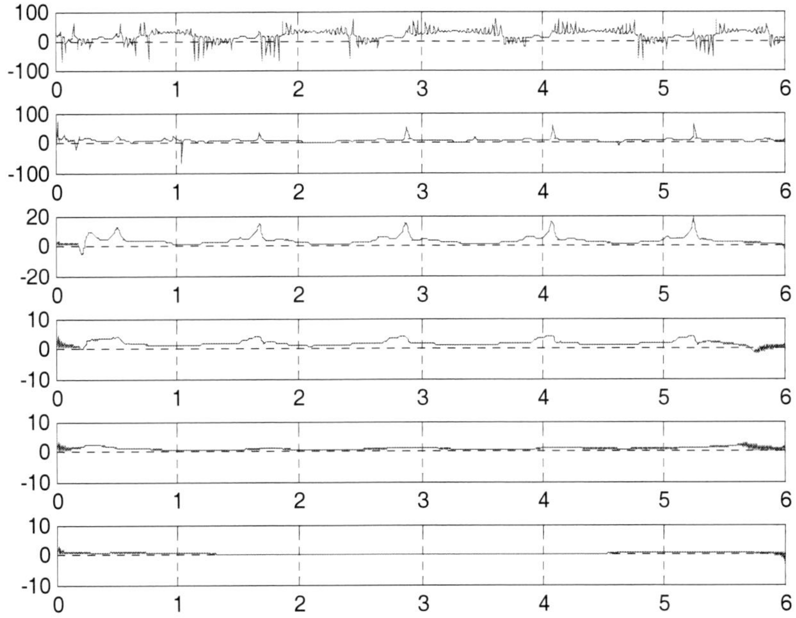

b)

Figure 8. Estimation of the frequencies by AR modelling, a), and Hilbert transform, b).

Conclusions

We started from the original Empirical Mode Decomposition to present a more reliable algorithm. As a data driven alternative approach to the analysis of non stationary signals it is very useful in many practical applications. We studied it and revealed its features and some drawbacks. We studied the intrinsic mode functions and made an interpretation of the EMD in terms of the AM/FM decomposition. We also criticised the use of the Hilbert transform for spectral estimation, presenting an alternative based on an AR local approximation. We presented also some simulation results to show the features of the algorithm..

Appendix A

Consider the signal $s_{FM}(t)$ introduced in 3.2. Assume that its instantaneous frequency f(t) is a slowly time varying signal, so that we may consider to be constant in small intervals. We can write:

$$s_{FM}(t) \approx \cos[2\pi f(t_0).t] \tag{a.1}$$

for t in a small interval around t_0. Sampling this signal we can express it in the format:

$$s_{FM}(n) \approx \cos[2\pi f(n_0).n] \tag{a.2}$$

valid for $n_0\text{-}N \le n_0 \le n_0+N$. So, we assume that the frequency is constant in a window with length 2N+1. In practice, we used N=15.

Here, we propose an instantaneous frequency estimator based on a local AR approximation. It is known that, if $x_n = \cos(2\pi fn + \varphi)$, with $f\in(0,1/2]$ and $\varphi\in(-\pi,\pi]$, it verifies an AR equation. In fact, we have:

$$\cos[2\pi f\, n+ \varphi] = \cos[2\pi f(n-1) + \varphi]\cos(2\pi f) - \sin[2\pi f(n-1) + \varphi]\sin(2\pi f)$$

and

$$\cos[2\pi f\,(n-2)+ \varphi] = \cos[2\pi f(n-1) + \varphi]\cos(2\pi f) + \sin[2\pi f(n-1) + \varphi]\sin(2\pi f)$$

Adding both equations, we obtain:

$$\cos[2\pi f\, n+ \varphi] + \cos[2\pi f\,(n-2)+ \varphi] = 2\cos[2\pi f(n-1) + \varphi]\cos(2\pi f)$$

and

$$\cos[2\pi f\, n + \varphi] - 2\cos[2\pi f(n-1) + \varphi]\cos(2\pi f) + \cos[2\pi f\,(n-2)+ \varphi] = 0$$

giving:

$$x(n) - 2\cos(2\pi f)\, x(n-1) + x(n-2) = 0 \tag{a.3}$$

that leads to an AR model with denominator polynomial

$$1 - 2\cos(2\pi f)z^{-1} + z^{-2}. \tag{a.4}$$

This polynomial is obtained using the Levinson recursion {see below} with the reflection coefficients:

$$C_1 = -\cos(2\pi f) \qquad \text{(a.5)}$$

and

$$C_2 = 1.$$

So, computing the first reflection coefficient in a window centred in the reference sample, we can obtain an estimate of the frequency.

Equation (a.3) is a special case of the Autoregressive (AR) model

$$\sum_{i=0}^{N_0} a_i x(n-i) = v(n) \qquad \text{(a.6)}$$

where v(n) is white noise and a_i, i=0, 1, …, N_0, with a_0=1 are the AR parameters. These can be computed from a set of reflection coefficients by means of the Levinson recursion {see below}.

To compute the reflection coefficient estimates we use a modified Burg method. We are going to introduce it in the following steps. Let x(n) be a discrete time signal. For N=1, …, N_0

1 – Define the forward and backward prediction errors by

$$f_n^N = \sum_{i=0}^{N} a_i^N x(n-i) \qquad \text{(a.7)}$$

and

$$b_n^N = \sum_{i=0}^{N} a_i^N x(n-i) \qquad \text{(a.8)}$$

and the error prediction power by

$$P_N = \frac{1}{2(L-N)} \sum_{N+1}^{L} \left[(f_n^N)^2 + (b_n^N)^2 \right] \qquad \text{(a.9)}$$

2 – Use the Levinson recursion [10]

$$a^{N;i} = a^{N-1;i} + C_N\, a^{N-1;N-i} \qquad i = 0, \dots, N \qquad \text{(a.10)}$$

to obtain

$$f^{N;n} = f^{N-1;n} + C_N\, b^{N-1;n} \qquad \text{(a.11)}$$

and

$$b^{N;n} = b^{N-1;n-1} + C_N f^{N-1;n-1} \quad \text{(a.12)}$$

with $f^{0;n} = x(n)$ and $b^{0;n} = x(n-1)$.

3 – With the above recursions transform P_N into a function of a unique variable – the reflection coefficient C_N:

$$P_N = \frac{1}{2(L-N)} \sum_{N+1}^{L} \left[(f_n^{N-1} + C_N \cdot b_{n-1}^{N-1})^2 + (b_{n-1}^{N-1} + C_N \cdot f_n^{N-1})^2 \right] \quad \text{(a.13)}$$

4 – Derive P_N relatively to C_N and equate to zero, to obtain:

$$C_N = -\frac{\sum_{N+2}^{L} \left[f_n^{N-1} b_n^{N-1} + f_{n-1}^{N-1} \cdot b_{n-1}^{N-1} \right]}{\sum_{N+2}^{L} \left[\left(f_{n-1}^{N-1}\right)^2 + \left(b_n^{N-1}\right)^2 \right]} \quad \text{(a.14)}$$

For N=1, we obtain immediately equation (13).

References

de Boor, C. ,"*A Practical Guide to Splines,*" Springer-Verlag, New York, 1978.

Carlson, A. B., "*Communication Systems*", 3rd edition, McGraw-Hill Book Company, 1986.

Chen, Q.H., Huang, N.E., Riemenschneider, S.: "A B-spline approach for empirical mode decompositions," *Advances in Computational Mathematics*, Vol. 24. (2006) 171–195

Cohen, L., "*Time-Frequency Analysis,*" Prentice-Hall, Englewood Cliffs, NJ, 1995.

Deering, R. Kaiser, J.F., "The use of a masking signal to improve empirical mode decomposition,"*ICASSP2005*, Vol. 4. (2005) 485-488.

Flandrin, P., Rilling, G. and Gonçalves, P., "On Empirical Mode Decomposition as a Filter Bank," *IEEE Signal Processing Letters*, vol.11, No.2, Feb. 2004.

Huang, N.E. , Long, S.R. and. Shen, Z, "The mechanism for frequency downshift in nonlinear wave evolution," *Adv. Appl. Mech*. **32** (1996) 59–111.

Huang, N.E. Shen, Z. Long, S.R. Wu, M.L. Shih, H.H. Zheng, Q. Yen, N.C. Tung C.C. and Liu, H.H. "The empirical mode decomposition and Hilbert spectrum for nonlinear and non-stationary time series analysis," 1998, *Proc. Roy. Soc*. London A, Vol. 454, pp. 903–995.

Huang, N.E. Chern, C.C. Huang, K. Salvino, L. Long ,S.R. and Fan, K.L., "Spectral analysis of the Chi-Chi earthquake data: Station TUC129, Taiwan, September 21, 1999," *Bull. Seismol. Soc. Amer.* **91** 1,310-1,338.

Huang,N.E. Tung, C.C. and Long, S.R. ,"The probability structure of the ocean surface," *The Sea* **9** (1990) 335–366.

Huang,N.E. Tung, C.C. and Long, S.R. , "A new view of nonlinear water waves: The Hilbert spectrum," *Ann. Rev. Fluid Mech.* **31** (1999) 417–457.

Huang, N.E. Wu, M.-L.C. Long, S.R. Shen, S.S.P. Qu, W. Gloersen, P. and Fan, K.L., "A confidence limit for the empirical mode decomposition and Hilbert spectral analysis," *Proc. Roy. Soc. London A* **459** (2003) 2317–2346.

Junsheng, C., Dejie, Y., and Yu, Y., "Research on Intrinsic Mode Function (IMF) criterion in EMD method," *Mechanical Systems and Signal Processing*, **20** (2006), 817-824.

Junsheng, C., Dejie, Y., and Yu, Y., "The application of the energy operator demodulation approach based on EMD in machinery fault diagnosis," *Mechanical Systems and Signal Processing,* **21** (2007), 668-677.

Maragos, P., Kaiser, J.F., and Quatieri, T.F., "On Amplitude and Frequency Demodulation Using Energy Operators," *IEEE transactions on Signal Processing*, Vol. 41, No, 4, April 1993.

Ortigueira, M.D. and Tribolet, J. M.,"Global versus local minimization in least-squares AR Spectral Estimation", *Signal Processing*, Vol.7, N° 3, December, 1984, pp 267-281.

Qin, S.R. and Zhong, Y.M., "A new envelope algorithm of Hilbert-Huang Transform", *Mechanical Systems and Signal Processing*, **20** (2006), 1941-1952.

Rato, R.T. and Ortigueira, M. D., "A Modified EMD Algorithm for Application in Biomedical Signal Processing", *Proceedings of the International Conference on Computational Intelligence in Medicine and Healthcare, CIMED 2005*, Costa da Caparica, June 29 – July 1, 2005 (CD edition).

G. Rilling, P. Flandrin and P. Gonçalves., "On Empirical Mode Decomposition and its Algorithms," 2003 *IEEE-EURASIP Workshop on Nonlinear Signal and Image Processing NSIP-03*, Grado (I).

G. Rilling and P. Flandrin, "On the Influence of Sampling on the Empirical Mode Decomposition," *Proceedings of the International Conference on Acoustics, Speech and Signal Processing, ICASSP 2006*, pp III-444 – III-447.

http://perso.ens-lyon.fr/patrick.flandrin/emd.html

In: New Signal Processing Research
Editor: Takumi Maeda, pp. 477-490
ISBN 978-1-60456-479-2

Chapter 17

A DCT-BASED IMAGE RESTORATION FROM A BLURRED AND DOWNSAMPLED IMAGE

Yoshinori Abe and Youji Iiguni
Department of Systems Innovation, Graduate School of Engineering Science,
Osaka University, Toyonaka 560-8531, Japan

Abstract

A fast computation method of acquiring a high resolution (HR) image from its blurred and downsampled low resolution (LR) image by using the discrete cosine transform (DCT) is presented. The blurring and downsampling process is mathematically modeled in matrix form and then the inverse problem of the process is solved by using the regularization approach. Since the blurring and downsampling matrix is very large, a large computation time is needed to directly solve the problem. The DCT makes the blurring and downsampling matrix be a sparse matrix, which leads to the analytical solution in scalar form in the DCT domain. Therefore, the HR image can be quickly generated from the LR image by the DCT fast computation method. Computer simulation has shown that the proposed method is superior to the cubic spline interpolation in the HR image restoration performance.

1. Introduction

In recent years, high resolution (HR) digital images are a requirement in many applications. Since there is a difficulty in manufacturing the HR image capturing device, signal processing techniques to enhance the resolution of digital images have been widely studied [1]. HR image restoration is the problem of acquiring a HR image from its blurred and downsampled low resolution (LR) image. The charge coupled device (CCD) image capturing device is commonly used in the standard digital imaging system. The image capturing device for the HR image has a smaller size CCD and the device for the LR image has a larger size CCD. When the same scene is captured by using the HR and LR CCD devices, the LR image is expressed as a blurred and downsampled version of the HR image. Therefore, the HR image restoration problem in the CCD device is considered as an inverse problem of the blurring and downsampling process.

The regularization approach is widely used to solve a general restoration problem. The problem is often solved by using the iterative method, which requires long computation time. If the downsampling process is not included in the restoration problem, the FFT based computation method proposed by Hunt [2] is straightforwardly applicable. However, the fast computation method is not applicable to the HR image restoration problem, because the downsampling is performed in the current problem. The blurring and downsampling matrix is not square and thus it cannot be diagonalized by using the FFT.

We therefore clarify that the blurring and downsampling matrix becomes sparse by a similar transformation with the DCT matrix. The sparse matrix representation allows us to derive the analytical solution of the HR restoration problem. Since the analytical solution can be expressed in scalar form in the DCT domain, the HR image can be quickly generated from the LR image by the properties of the Kronecker product and the DCT fast computation method.

The proposed method requires only two DCT operations, which is much reduced from the computation cost of the conventional iterative restoration method [3]. Unlike the fast computation method derived by El-Khamy [4], the proposed method does not degrade the HR image restoration performance since we do not use the block by block operation, which segments the LR image into overlapping segments and restores each of segments. Computer simulation has shown that the proposed method is superior to the cubic spline interpolation in the HR image restoration performance.

2. Mathematical Formulation of the Image Restoration

2.1. Observation Model

Let the $(aM \times aM)$ HR image be $\{x_{ij}\}$ and the $(M \times M)$ LR image be $\{y_{kl}\}$. The HR and LR images are obtained from the same scene captured by the different resolution cameras at the same position. The pixel density of the HR image is a^2 times higher than that of the LR image. We assume that $\{y_{kl}\}$ is a blurred and downsampled version of $\{x_{ij}\}$. Then we can formulate y_{kl} as a weighted sum of $x_{ij}s$. For the ease of analysis, we here consider the observation noise is negligibly small.

When a is even, the blurring and downsampling process is formulated as

$$y_{kl} = \sum_{m=-P}^{P-1} \sum_{n=-P}^{P-1} h_m h_n x_{ak+m+a/2, al+n+a/2}, \tag{1}$$

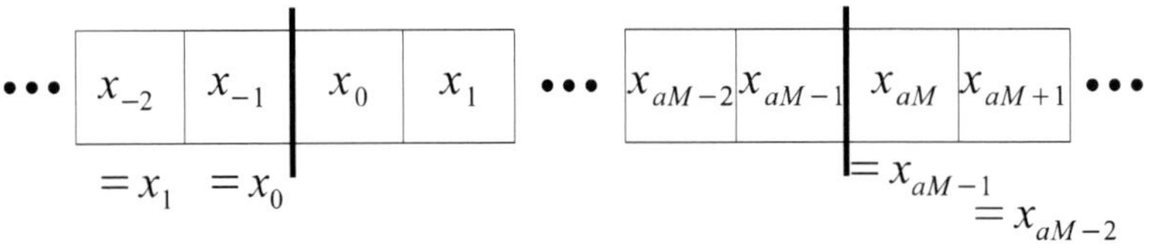

Figure 1. Symmetric expansion of x_i

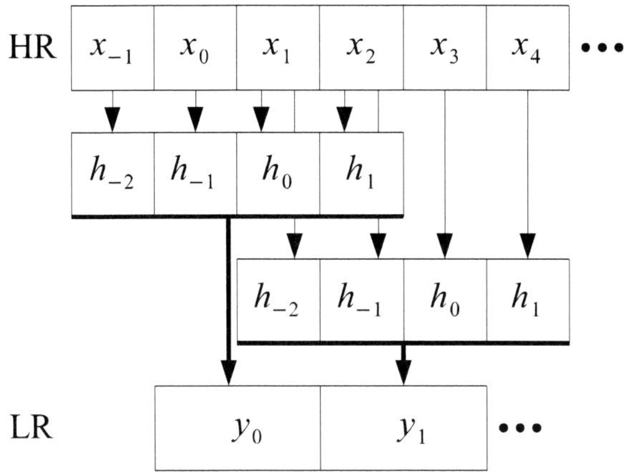

Figure 2. Relation between the HR and LR images with $a = 2$ and $P = 2$

where $2P$ blurring and downsampling coefficients $\{h_m\}_{m=-P}^{P-1}$ are assumed to be symmetric such as

$$h_m = h_{-m-1}. \tag{2}$$

We also assume that x_{ij} is symmetrically expanded outside the region $0 \leq i, j \leq aM - 1$, where x_{ij} is symmetric about $i = -0.5, aM - 0.5$ and $j = -0.5, aM - 0.5$. Figure 1 shows the symmetric expansion of the one dimensional (1-D) data $\{x_i\}_{i=0}^{aM-1}$ about $i = -0.5, aM - 0.5$. Figure 2 illustrates the relation between the HR and LR images in 1-D case with $a = 2$ and $P = 2$. When a is odd, we similarly have

$$y_{kl} = \sum_{m=-P}^{P} \sum_{n=-P}^{P} h_m h_n x_{ak+m+(a-1)/2, al+n+(a-1)/2} + v_{kl}, \tag{3}$$

where $(2P + 1)$ coefficients $\{h_m\}_{m=-P}^{P}$ are symmetric such as

$$h_m = h_{-m}. \tag{4}$$

Figure 3 illustrates the relation between the HR and LR images in 1-D case with $a = 3$ and $P = 2$. Both Equations (1) and (3) can be rewritten in matrix form as

$$\boldsymbol{y} = H\boldsymbol{x}, \tag{5}$$

where $\boldsymbol{x}$ and $\boldsymbol{y}$ are lexicographically ordered vectors consisting of the HR image $\{x_{ij}\}$ and the LR image $\{y_{kl}\}$. Since the coefficients are separable, the $(M^2 \times a^2M^2)$ matrix H can be written as

$$H = H_s \otimes H_s, \tag{6}$$

where $\otimes$ means the Kronecker product.

A typical example of $\{h_m\}$ is the averaging filter defined by

$$h_m = \begin{cases} 1/a & 0 \leq m \leq \lfloor (a-1)/2 \rfloor \\ 0 & \text{otherwise}, \end{cases} \tag{7}$$

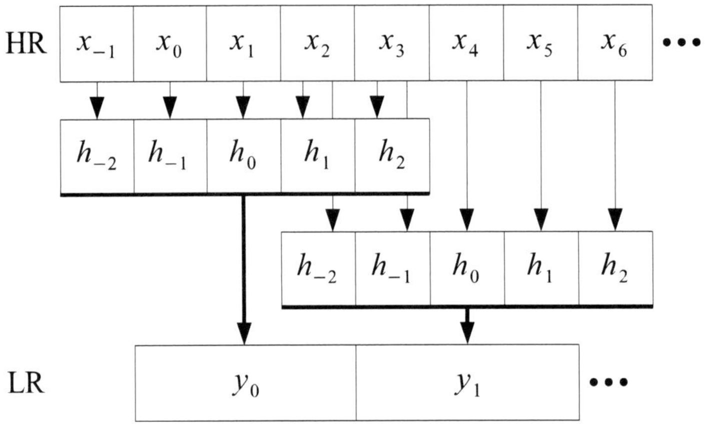

Figure 3. Relation between the HR and LR images with $a = 3$ and $P = 2$

where $\lfloor x \rfloor$ gives the largest integer less than or equal to x. Then H_s is written as

$$H_s = \begin{pmatrix} 1/a & \cdots & 1/a & & & & & & \mathbf{0} \\ & & & 1/a & \cdots & 1/a & & & \\ & & & & & & \ddots & & \\ \mathbf{0} & & & & & & 1/a & \cdots & 1/a \end{pmatrix}. \tag{8}$$

Another example of $\{h_m\}$ is the Gaussian filter. Capel experimentally pointed out that the camera point spread function is well approximated by the Gaussian filter [5]. In the following simulation, we use the Gaussian filter defined by

$$h_m = \begin{cases} e^{-\frac{(m+0.5)^2}{2\sigma^2}} & (m = -P, -P+1, \ldots, P-1) \quad a \text{ is even} \\ e^{-\frac{m^2}{2\sigma^2}} & (m = -P, -P+1, \ldots, P) \quad a \text{ is odd.} \end{cases} \tag{9}$$

2.2. HR Image Restoration Problem

We shall consider the problem of restoring the HR image $\boldsymbol{x}$ from the LR image $\boldsymbol{y}$. Since the problem is ill-defined, we impose a smoothness constraint on $\boldsymbol{x}$ to uniquely determine $\boldsymbol{x}$. The constraint we imposed is the square sum of the discrete Laplacian of $\boldsymbol{x}$.

The discrete Laplacian of each x_{ij} is defined by $x_{i-1,j} + x_{i,j-1} - 4x_{ij} + x_{i,j+1} + x_{i+1,j}$. The lexicographically ordered vector consisting of the discrete Laplacians of $\{x_{ij}\}$ is then expressed as [6]

$$P\boldsymbol{x} = (I \otimes P_s + P_s \otimes I)\boldsymbol{x}, \tag{10}$$

where I is the $(aM \times aM)$ identity matrix, and P_s is the $(aM \times aM)$ matrix defined by

$$P_s = \begin{pmatrix} -1 & 1 & & & \mathbf{0} \\ 1 & -2 & 1 & & \\ & \ddots & \ddots & \ddots & \\ & & 1 & -2 & 1 \\ \mathbf{0} & & & 1 & -1 \end{pmatrix}. \tag{11}$$

Then the square sum of the discrete Laplacians of $\boldsymbol{x}$ can be represented by the quadratic form $\boldsymbol{x}^T Q \boldsymbol{x}$, where the $(a^2M^2 \times a^2M^2)$ matrix Q is defined by

$$Q = P^T P. \tag{12}$$

Under the condition (5) we shall estimate the HR image $\boldsymbol{x}$ from the LR image $\boldsymbol{y}$ by minimizing the square sum of the discrete Laplacian $\boldsymbol{x}^T Q \boldsymbol{x}$. The estimation problem is formulated as

$$\arg\min_{\boldsymbol{x}} \boldsymbol{x}^T Q \boldsymbol{x} \quad \text{subject to } \boldsymbol{y} = H\boldsymbol{x}. \tag{13}$$

We shall solve the problem (13) by using the Lagrange multiplier method.

Let the Lagrangian L be

$$L(\boldsymbol{x}, \boldsymbol{\lambda}) = \boldsymbol{x}^T Q \boldsymbol{x} + \boldsymbol{\lambda}^T (\boldsymbol{y} - H\boldsymbol{x}), \tag{14}$$

where $\boldsymbol{\lambda} = (\lambda_{00}, \lambda_{01}, \ldots, \lambda_{0,M-1}, \lambda_{10}, \ldots, \lambda_{M-1,M-1})^T$ is the M^2 dimensional vector consisting of the Lagrange multipliers. Then the following equations must hold:

$$\frac{\partial}{\partial \boldsymbol{x}} L(\boldsymbol{x}, \boldsymbol{\lambda}) = 2Q\boldsymbol{x} - H^T \boldsymbol{\lambda} = \boldsymbol{0}, \tag{15}$$

$$\frac{\partial}{\partial \boldsymbol{\lambda}} L(\boldsymbol{x}, \boldsymbol{\lambda}) = \boldsymbol{y} - H\boldsymbol{x} = \boldsymbol{0}. \tag{16}$$

Then we perform the similar transformations of Q and H by the DCT matrix, and we show that they become sparse matrices.

2.3. Similar Transformations of Q and H

We put the 1-D DCT matrix of size M as W_M. The ij element of W_M is given by

$$(W_M)_{ij} = \frac{\sqrt{2}}{\sqrt{M}} c(i) \cos\left(\frac{\pi i (j + 0.5)}{M}\right) \tag{17}$$

with

$$c(i) = \begin{cases} \frac{1}{\sqrt{2}}, & i = 0 \\ 1, & \text{otherwise} \end{cases}. \tag{18}$$

We should note that W_M is a unitary matrix. We transform the variables $\boldsymbol{x}$, $\boldsymbol{y}$, and $\boldsymbol{\lambda}$ into $\tilde{\boldsymbol{x}}$, $\tilde{\boldsymbol{y}}$, and $\tilde{\boldsymbol{\lambda}}$ by the two dimensional (2-D) DCT matrix as follows:

$$\tilde{\boldsymbol{x}} = (W_{aM} \otimes W_{aM})\boldsymbol{x}, \tag{19}$$

$$\tilde{\boldsymbol{y}} = (W_M \otimes W_M)\boldsymbol{y}, \tag{20}$$

$$\tilde{\boldsymbol{\lambda}} = (W_M \otimes W_M)\boldsymbol{\lambda}. \tag{21}$$

Then the equations (15) and (16) can be rewritten as

$$2\widetilde{Q}\tilde{\boldsymbol{x}} - \widetilde{H}^T \tilde{\boldsymbol{\lambda}} = \boldsymbol{0}, \tag{22}$$

$$\tilde{\boldsymbol{y}} - \widetilde{H}\tilde{\boldsymbol{x}} = \boldsymbol{0}, \tag{23}$$

where the matrices $\widetilde{Q}$ and $\widetilde{H}$ are the similar transformations of Q and H defined by

$$\widetilde{Q} = (W_{aM} \otimes W_{aM})Q(W_{aM} \otimes W_{aM})^T, \tag{24}$$
$$\widetilde{H} = (W_M \otimes W_M)H(W_{aM} \otimes W_{aM})^T. \tag{25}$$

In our previous study, we have shown that the matrix $\widetilde{Q}$ can be transformed into a diagonal matrix as follows [7]:

$$\widetilde{Q} = \mathrm{diag}\Big\{ \tilde{q}_{00}, \tilde{q}_{01}, \ldots, \tilde{q}_{0,aM-1}, \tilde{q}_{10}, \ldots, \tilde{q}_{aM-1,aM-1} \Big\} \tag{26}$$

with

$$\tilde{q}_{ij} = 4\left(\tilde{p}_i + \tilde{p}_j\right)^2, \tag{27}$$
$$\tilde{p}_i = \left(1 - \cos\left(\frac{\pi i}{aM}\right)\right). \tag{28}$$

We here show that $\widetilde{H}$ is a sparse matrix. We can rewrite $\widetilde{H}$ as

$$\begin{aligned}
\widetilde{H} &= (W_M \otimes W_M)H(W_{aM} \otimes W_{aM})^T \\
&= (W_M \otimes W_M)(H_s \otimes H_s)(W_{aM} \otimes W_{aM})^T \\
&= (W_M H_s W_{aM}^T) \otimes (W_M H_s W_{aM}^T) \\
&= \widetilde{H}_s \otimes \widetilde{H}_s,
\end{aligned} \tag{29}$$

where we put

$$\begin{aligned}
\left(\widetilde{H}_s\right)_{ij} &= \left(W_M H_s W_{aM}^T\right)_{ij} \\
&= \begin{cases} \tilde{h}_j, & j = i + 2mM,\ 2mM - i \\ 0, & \text{otherwise.} \end{cases}
\end{aligned} \tag{30}$$

Equation (30) can be written in matrix form as

$$\widetilde{H}_s = \left(\begin{array}{ccccccc|c} \tilde{h}_0 & & & & & & 0 & \\ & \tilde{h}_1 & & & & & \tilde{h}_{2M-1} & \\ & & \ddots & & & \cdot^{\cdot^{\cdot}} & & \cdots \\ & & & \tilde{h}_{M-1} & 0 & \tilde{h}_{M+1} & & \end{array} \right.$$
$$\left. \begin{array}{|ccccccc} \tilde{h}_{(a-2)M} & & & & & & 0 \\ & \tilde{h}_{(a-2)M+1} & & & & & \tilde{h}_{aM-1} \\ & & \ddots & & & \cdot^{\cdot^{\cdot}} & \\ & & & \tilde{h}_{(a-1)M-1} & 0 & \tilde{h}_{(a-1)M+1} & \end{array} \right). \tag{31}$$

When a is even, $\tilde{h}_i$ is given by

$$\tilde{h}_i = \begin{cases} \frac{2}{\sqrt{a}c_{2M}(i)} \sum_{n=0}^{P-1} h_n \cos\left(\frac{\pi i(n+0.5)}{aM}\right), & (4m-1)M < i < (4m+1)M \\ -\frac{2}{\sqrt{a}c_{2M}(i)} \sum_{n=0}^{P-1} h_n \cos\left(\frac{\pi i(n+0.5)}{aM}\right), & (4m+1)M < i < (4m+3)M \\ 0, & \text{otherwise} \end{cases}, \tag{32}$$

with

$$c_M(i) = \begin{cases} \frac{1}{\sqrt{2}}, & i \neq 0,\ i = mM \\ 1, & \text{otherwise} \end{cases}. \tag{33}$$

When a is odd, $\tilde{h}_i$ is given by

$$\tilde{h}_i = \begin{cases} \frac{2}{\sqrt{a}c_{2M}(i)} \left(\frac{h_0}{2} + \sum_{n=1}^{P} h_n \cos\left(\frac{\pi i n}{aM}\right)\right), & (4m-1)M < i < (4m+1)M \\ -\frac{2}{\sqrt{a}c_{2M}(i)} \left(\frac{h_0}{2} + \sum_{n=1}^{P} h_n \cos\left(\frac{\pi i n}{aM}\right)\right), & (4m+1)M < i < (4m+3)M \\ 0, & \text{otherwise} \end{cases}. \tag{34}$$

The derivation of Equation (30) is shown in Appendix. We verify from Equation (30) that $\widetilde{H}$ is a sparse matrix.

2.4. Fast Computation by the DCT

Here we show that the solution of the system of Equations (22) and (23) can be written in scalar form by using the sparse representation of G and H, i.e. (26) and (29). For simplicity, we here discuss the case of even a. Using (29), we can rewrite Equation (23) in scalar form as

$$\begin{aligned} \tilde{y}_{00} &= \sum_{m=0}^{a/2-1} \sum_{n=0}^{a/2-1} \tilde{h}_{2mM} \tilde{h}_{2nM} \tilde{x}_{2mM,2nM} \\ \tilde{y}_{kl} &= \sum_{m=0}^{a/2-1} \sum_{n=0}^{a/2-1} \tilde{h}_{k+2mM} \tilde{h}_{l+2nM} \tilde{x}_{k+2mM,l+2nM} \\ &+ \sum_{m=0}^{a/2-1} \sum_{n=0}^{a/2-1} \tilde{h}_{k+2mM} \tilde{h}_{2M-l+2nM} \tilde{x}_{k+2mM,2M-l+2nM} \\ &+ \sum_{m=0}^{a/2-1} \sum_{n=0}^{a/2-1} \tilde{h}_{2M-k+2mM} \tilde{h}_{l+2nM} \tilde{x}_{2M-k+2mM,l+2nM} \\ &+ \sum_{m=0}^{a/2-1} \sum_{n=0}^{a/2-1} \tilde{h}_{2M-k+2mM} \tilde{h}_{2M-l+2nM} \tilde{x}_{2M-k+2mM,2M-l+2nM}. \end{aligned} \tag{35}$$

Substituting (26) and (29) into (22) gives

$$2\tilde{q}_{ij}\tilde{x}_{ij} = \begin{cases} 0, & i = (2m+1)M \text{ or } j = (2m+1)M \\ \tilde{h}_i \tilde{h}_j \tilde{\lambda}_{\kappa(i)\kappa(j)}, & \text{otherwise} \end{cases}, \tag{36}$$

with the definition

$$\kappa(i) = \begin{cases} \text{mod}(i, 2M), & 2mM \leq i \leq (2m+1)M \\ 2M - \text{mod}(i, 2M), & (2m+1)M + 1 \leq i \leq (2m+2)M - 1 \end{cases}, \tag{37}$$

where $\text{mod}(i, 2M)$ stands for the common residue of i (modulo $2M$). Here we assume that $\tilde{h}_{2mM} \neq 0$, which is valid in many situations. Since $\tilde{q}_{00} = 0$, putting $i = j = 0$ in Equation (36) gives

$$\tilde{\lambda}_{00} = 0. \tag{38}$$

Since $\kappa(2mM) = 0$ and $\tilde{q}_{2mM,2nM} \neq 0$, putting $i = 2mM$ and $j = 2nM$ in (36) gives

$$\tilde{x}_{2mM,2nM} = 0 \quad m > 0 \text{ or } n > 0. \tag{39}$$

Substituting Equation (39) into the top equation of (35), we can straightly determine $\tilde{x}_{00}$ as

$$\tilde{x}_{00} = \frac{\tilde{y}_{00}}{\tilde{h}_0^2}. \tag{40}$$

Therefore, we shall consider the case of $i \neq 0$ or $j \neq 0$ in the following. Since $\tilde{q}_{ij} \neq 0$, Equation (36) and (39) are combined into

$$\tilde{x}_{ij} = \begin{cases} 0, & i = mM \text{ or } j = mM \\ \dfrac{\tilde{h}_i \tilde{h}_j \tilde{\lambda}_{\kappa(i)\kappa(j)}}{2\tilde{q}_{ij}}, & \text{otherwise} \end{cases}. \tag{41}$$

Substituting (41) into the bottom equation of (35), we have

$$\tilde{\lambda}_{kl} = \frac{2\tilde{y}_{kl}}{D_{kl}}, \tag{42}$$

with the definition

$$D_{kl} = \sum_{m=0}^{a/2-1} \sum_{n=0}^{a/2-1} \frac{\tilde{h}^2_{k+2mM} \tilde{h}^2_{l+2mM}}{\tilde{q}_{k+2mM,l+2mM}} + \sum_{m=0}^{a/2-1} \sum_{n=0}^{a/2-1} \frac{\tilde{h}^2_{k+2mN} \tilde{h}^2_{2M-l+2nM}}{\tilde{q}_{k+2mM,2M-l+2nM}}$$
$$+ \sum_{m=0}^{a/2-1} \sum_{n=0}^{a/2-1} \frac{\tilde{h}^2_{2M-k+2mM} \tilde{h}^2_{l+2nM}}{\tilde{q}_{2M-k+2mM,l+2nM}} + \sum_{m=0}^{a/2-1} \sum_{n=0}^{a/2-1} \frac{\tilde{h}^2_{2M-k+2mM} \tilde{h}^2_{2M-l+2nM}}{\tilde{q}_{2M-k+2mM,2M-l+2nM}}. \tag{43}$$

Substituting (42) into (41), and summarizing it together with the result of (40), we have

$$\tilde{x}_{ij} = \begin{cases} \dfrac{\tilde{y}_{00}}{\tilde{h}_0 \tilde{h}_0}, & i = j = 0 \\ 0, & i = mM \text{ or } j = nM \quad (mn \neq 0) \\ \dfrac{\tilde{h}_i \tilde{h}_j}{\tilde{q}_{ij} D_{\kappa(i)\kappa(j)}} \tilde{y}_{\kappa(i)\kappa(j)}, & \text{otherwise} \end{cases}. \tag{44}$$

This equation shows that the DCT of the HR image $\tilde{x}_{ij}$ can be computed from the DCT of the LR image $\tilde{y}_{kl}$ by scalar computation.

The algorithm of the proposed method is summarized in Figure 4. We can compute $\tilde{\boldsymbol{x}}$ from $\tilde{\boldsymbol{y}}$ by using (44) in $O(M^2)$. The DCT of $\boldsymbol{y}$ and the inverse DCT of $\tilde{\boldsymbol{x}}$ can be computed in $O(M^2 \log M)$. Therefore, the proposed method can compute the HR image $\boldsymbol{x}$ from the LR image $\boldsymbol{y}$ in $O(M^2 \log M)$ processing time.

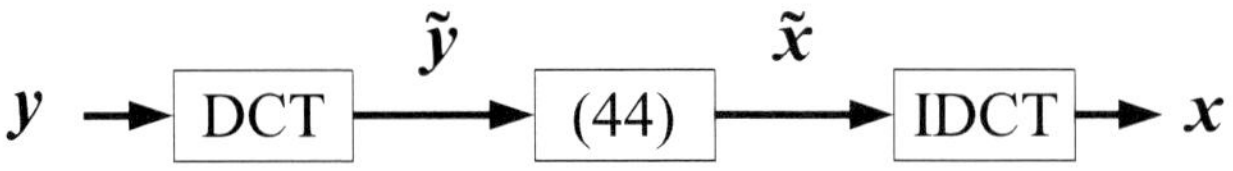

Figure 4. Flow diagram of the proposed method

3. Simulation Results

In this section, we compared the restoration performances of the conventional and proposed methods. All simulations were done on an IBM PC/AT compatible computer with an Intel Pentium 4 2.4 GHz and 512 Mbyte DRAM's. We used eight images of size (256×256) with 8 bit grayscale: "barbara", "boat", "bridge", "camera", "girl", "lena", "mandrill", and "peppers".

First, we generated the (128×128) LR image $\boldsymbol{y}$ from the original (256×256) HR image $\boldsymbol{x}$ according to Equation (5). We used the Gaussian filter (9) with $\sigma = 1$ and $P = aM$. Then we restored the (256×256) HR image from $\boldsymbol{y}$ by the linear interpolation, the cubic spline interpolation [8], and the proposed methods. Figure 5 shows the original "barbara" and the restored images by the three methods. The restored image by the proposed method seems to be more "high-passed" than the other two images.

We next quantitatively measured the restoration performance of each method by the peak signal to noise ratio (PSNR) defined by

$$\mathrm{PSNR} = 10 \log_{10} \left(\frac{255^2}{\frac{1}{4M^2} \sum_{i=0}^{2M-1} \sum_{j=0}^{2M-1} e_{ij}^2} \right), \tag{45}$$

where e_{ij} is the difference between the pixel values of the original and restored images. Table 1 summarizes the PSNRs of 8 images restored by the three methods. This shows that the proposed method is better than the linear interpolation method by 3.1 dB, and is better than the cubic interpolation methods by 2.0 dB. The computation times of the proposed, linear, and cubic spline interpolation methods were 0.0097 [s], 0.0014 [s], and 0.0052 [s], respectively. The proposed method achieves better restoration performance at the expense of a small increase of computation time.

4. Conclusion

We derived a fast computation method of the HR image restoration problem by using the DCT. When the blurring and downsampling process is symmetric, a sparse transformation

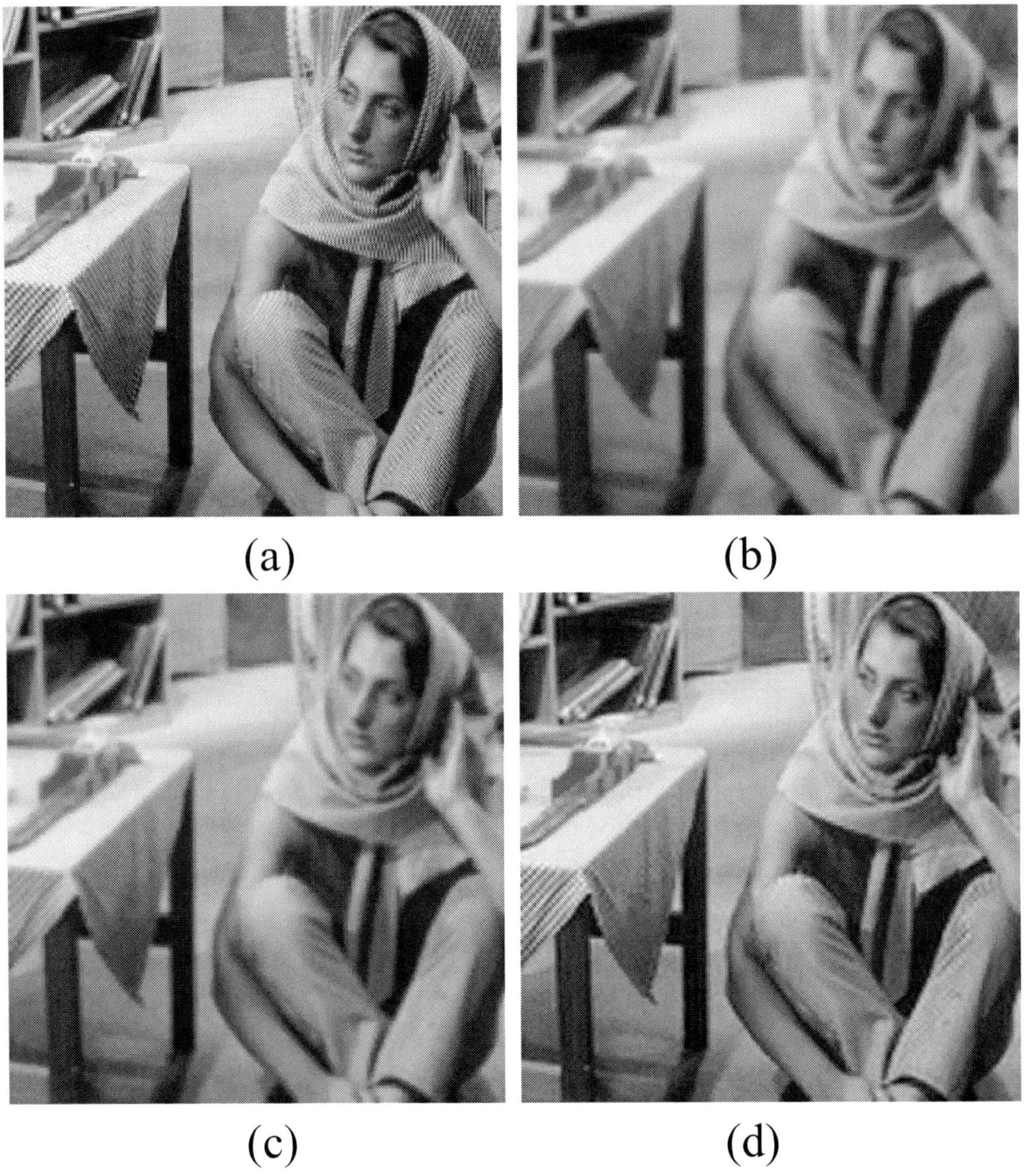

Figure 5. Original and restored images: (a) Original "barbara", (b) Linear interpolation, (c) Cubic spline interpolation, (d) Proposed method

with the DCT is applicable to the blurring and downsampling matrix. Then the analytical solution of the HR image restoration problem is expressed in scalar form in the DCT domain. Using the fast DCT computation method, HR image is quickly acquired from the blurred and downsampled LR image. Computer simulation showed that the restoration performance of the proposed method is superior to that of the cubic spline interpolation at the expense of an increase of computation time.

Table 1. PSNRs for the two interpolation methods and the proposed method

	linear	cubic spline	proposed
barbara	26.03	26.89	28.31
boat	27.76	29.03	31.13
bridge	22.15	22.81	23.96
camera	24.40	25.35	27.26
girl	30.07	31.55	34.41
lena	27.18	28.26	30.32
mandrill	22.74	23.54	25.37
peppers	24.54	26.20	29.08
average	25.61	26.70	28.73

A. Appendix

Let the similar transformation of H_s by using the DCT matrix W_M be

$$\widetilde{H}_s = W_M H_s W_{aM}^T. \tag{46}$$

First, we consider the case of even a. From the symmetry of the cosine function, we have

$$\cos\left(\frac{\pi j(l+0.5)}{aM}\right) = \cos\left(\frac{\pi j(-l-1+0.5)}{aM}\right). \tag{47}$$

Then the ij element of $H_s W_{aM}^T$ can be written as

$$\begin{aligned}(H_s W_{aM}^T)_{ij} &= \frac{\sqrt{2}c(j)}{\sqrt{aM}} \sum_{l=0}^{aM-1} (H_s)_{il} \cos\left(\frac{\pi j(l+0.5)}{aM}\right) \\ &= \frac{\sqrt{2}c(j)}{\sqrt{aM}} \sum_{n=-P}^{P-1} h_n \cos\left(\frac{\pi j(ai+0.5a+n+0.5)}{aM}\right).\end{aligned} \tag{48}$$

Therefore, the ij element of $\widetilde{H}_s$ is

$$\begin{aligned}(\widetilde{H}_s)_{ij} &= (W_M H_s W_{aM}^T)_{ij} \\ &= \frac{\sqrt{2}c(i)}{\sqrt{M}} \sum_{l=0}^{M-1} (H_s W_{aM}^T)_{lj} \cos\left(\frac{\pi i(l+0.5)}{M}\right) \\ &= \frac{2c(i)c(j)}{\sqrt{a}M} \sum_{l=0}^{M-1} \sum_{n=-P}^{P-1} h_n \cos\left(\frac{\pi j(al+0.5a+n+0.5)}{aM}\right) \cos\left(\frac{\pi i(l+0.5)}{M}\right).\end{aligned} \tag{49}$$

From the property of cosine function, we have

$$\begin{aligned}
&\cos\left(\frac{\pi j(al+0.5a+n+0.5)}{aM}\right)\cos\left(\frac{\pi i(l+0.5)}{M}\right)\\
&=\frac{1}{2}\left(\cos\left(\frac{\pi((i+j)a(l+0.5)+j(n+0.5))}{aM}\right)\right.\\
&\quad\left.+\cos\left(\frac{\pi((i-j)a(l+0.5)-j(n+0.5))}{aM}\right)\right)\\
&=\frac{1}{2}\left(\cos\left(\frac{\pi(i+j)(l+0.5)}{M}\right)\cos\left(\frac{\pi j(n+0.5)}{aM}\right)\right.\\
&\quad-\sin\left(\frac{\pi(i+j)(l+0.5)}{M}\right)\sin\left(\frac{\pi j(n+0.5)}{aM}\right)\\
&\quad+\cos\left(\frac{\pi(i-j)(l+0.5)}{M}\right)\cos\left(\frac{\pi j(n+0.5)}{aM}\right)\\
&\quad\left.-\sin\left(\frac{\pi(i-j)(l+0.5)}{M}\right)\sin\left(\frac{\pi j(n+0.5)}{aM}\right)\right). \qquad (50)
\end{aligned}$$

Since h_n is assumed to be symmetric about $n=-0.5$ as (2), we have

$$\sum_{n=-P}^{P-1} h_n\cos\left(\frac{\pi j(n+0.5)}{aM}\right)=2\sum_{n=0}^{P-1} h_n\cos\left(\frac{\pi j(n+0.5)}{aM}\right), \qquad (51)$$

$$\sum_{n=-P}^{P-1} h_n\sin\left(\frac{\pi j(n+0.5)}{aM}\right)=0. \qquad (52)$$

Using (50), (51), and (52), Equation (49) can be rewritten as

$$\begin{aligned}
(\widetilde{H}_s)_{ij}=&\frac{2c(i)c(j)}{\sqrt{a}M}\sum_{l=0}^{M-1}\left(\cos\left(\frac{\pi(i+j)(l+0.5)}{M}\right)+\cos\left(\frac{\pi(i-j)(l+0.5)}{M}\right)\right)\\
&\times\sum_{n=0}^{P-1} h_n\cos\left(\frac{\pi j(n+0.5)}{aM}\right). \qquad (53)
\end{aligned}$$

Here we use the property

$$\sum_{l=0}^{M-1}\cos\left(\frac{\pi k(l+0.5)}{M}\right)=\begin{cases}M, & k=4mM\\ -M, & k=(4m+2)M\\ 0, & \text{otherwise}\end{cases}. \qquad (54)$$

Substituting (54) into (53), we have Equation (30) with the definition (32).

We can similarly treat the case of odd a. The ij element of $H_sW_{aM}^T$ can be written as

$$\begin{aligned}
(H_sW_{aM}^T)_{ij}&=\frac{\sqrt{2}c(j)}{\sqrt{aM}}\sum_{l=0}^{aM-1}(H_s)_{il}\cos\left(\frac{\pi j(l+0.5)}{aM}\right)\\
&=\frac{\sqrt{2}c(j)}{\sqrt{aM}}\sum_{n=-P}^{P} h_n\cos\left(\frac{\pi j(ai+0.5a+n)}{aM}\right). \qquad (55)
\end{aligned}$$

Therefore, the ij element of $\widetilde{H}_s$ is

$$
\begin{aligned}
(\widetilde{H}_s)_{ij} &= (W_M H_s W_{aM}^T)_{ij} \\
&= \frac{\sqrt{2}c(i)}{\sqrt{M}} \sum_{l=0}^{M-1} (H_s W_{aM}^T)_{lj} \cos\left(\frac{\pi i(l+0.5)}{M}\right) \\
&= \frac{2c(i)c(j)}{\sqrt{a}M} \sum_{l=0}^{M-1} \sum_{n=-P}^{P} h_n \cos\left(\frac{\pi j(al+0.5a+n)}{aM}\right) \cos\left(\frac{\pi i(l+0.5)}{M}\right). \qquad (56)
\end{aligned}
$$

From the property of cosine function, we have

$$
\begin{aligned}
&\cos\left(\frac{\pi j(al+0.5a+n)}{aM}\right) \cos\left(\frac{\pi i(l+0.5)}{M}\right) \\
&= \frac{1}{2}\left(\cos\left(\frac{\pi((i+j)a(l+0.5)+jn)}{aM}\right) + \cos\left(\frac{\pi((i-j)a(l+0.5)-jn)}{aM}\right)\right) \\
&= \frac{1}{2}\left(\cos\left(\frac{\pi(i+j)(l+0.5)}{M}\right)\cos\left(\frac{\pi jn}{aM}\right) - \sin\left(\frac{\pi(i+j)(l+0.5)}{M}\right)\sin\left(\frac{\pi jn}{aM}\right)\right. \\
&\quad \left. + \cos\left(\frac{\pi(i-j)(l+0.5)}{M}\right)\cos\left(\frac{\pi jn}{aM}\right) - \sin\left(\frac{\pi(i-j)(l+0.5)}{M}\right)\sin\left(\frac{\pi jn}{aM}\right)\right).
\end{aligned}
\qquad (57)
$$

Since h_n is assumed to be symmetric about $n = 0$ as (4), we have

$$
\sum_{n=-P}^{P} h_n \cos\left(\frac{\pi jn}{aM}\right) = h_0 + 2\sum_{n=1}^{P} h_n \cos\left(\frac{\pi jn}{aM}\right), \qquad (58)
$$

$$
\sum_{n=-P}^{P} h_n \sin\left(\frac{\pi jn}{aM}\right) = 0. \qquad (59)
$$

Using (57), (58), and (59), Equation (56) can be rewritten as

$$
\begin{aligned}
(\widetilde{H}_s)_{ij} = \frac{2c(i)c(j)}{\sqrt{a}M} \sum_{l=0}^{M-1} &\left(\cos\left(\frac{\pi(i+j)(l+0.5)}{M}\right) + \cos\left(\frac{\pi(i-j)(l+0.5)}{M}\right)\right) \\
&\times \left(h_0 + 2\sum_{n=1}^{P} h_n \cos\left(\frac{\pi jn}{aM}\right)\right).
\end{aligned}
\qquad (60)
$$

Substituting (54) into (60), $(\widetilde{H}_s)_{ij}$ can be written as Equation (30) with the definition (34).

References

[1] S. C. Park and M. K. Park and M. G. Kang, Figure Super-Resolution Image Reconstruction -A Technical Overview-. *Signal Processing Magazine*, vol. 20, no. 3, pp. 21–36, May 2003.

[2] B. R. Hunt, Figure The application of constrained least squares estimation to image restoration by digital computer. *IEEE Trans. Computers*, **C-22**(9), pp. 805–812, 1973.

[3] J. H. Shin, J. H. Jung, and J. K. Paik, Figure Regularized iterative image interpolation and its application to spatially scalable coding. *IEEE Trans. Consumer Electronics*, vol. 44, no. 3, pp. 1042–1047, August 1998.

[4] S. E. El-Khamy, M. M. Hadhoud, M. I. Dessouky, B. M. Salam and F. E. Abd El-Samie, Figure Sectioned implementation of regularized image interpolation. *Proceedings of the 46th IEEE International Midwest Symposium on Circuits and Systems* (MWSCAS '03), vol. 2, pp. 656–659, 2003.

[5] D. Capel, Figure Image Mosaicing and Super-Resolution. Springer Verlag, 2004.

[6] F. O'Sullivan, Figure Discretized Laplacian smoothing by Fourier methods. *Journal of the American Statistical Association* , vol. 86, September, 1991, pp. 634–642.

[7] Y.Abe and Y.Iiguni, Figure Image restoration from a downsampled image by using the DCT. Signal Processing, vol. 87, no. 10, pp. 2370–2380, October 2007.

[8] P. Thévenaz, T. Blu, and M. Unser, Figure Image Interpolation and Resampling. in Handbook of Medical Imaging, *Processing and Analysis*, I.N. Bankman, Ed., Academic Press, San Diego Ca, USA, pp. 393–420, 2000.

In: New Signal Processing Research
Editor: Takumi Maeda, pp. 491-503

ISBN 978-1-60456-479-2

Chapter 18

ROBUST ESTIMATION OF SPATIALLY DISTRIBUTED SOURCES

Ahmed Zoubir and Yide Wang
Institut de Recherche en Électrotechnique et Électronique de
Nantes Atlantique, Nantes, France

Abstract

This chapter describes some robust techniques for the localization of distributed sources. For a distributed source, the system receives a spatially dispersed signal consisting of several arriving rays spread around some nominal direction. In this study we distinguish two types of distributed sources: coherently distributed (CD) and incoherently distributed (ID) sources. We present, for these two types of sources, some robust methods for decoupled estimation of their angular parameters (DOA and angular spread). Compared with existing methods, the proposed methods improve the robustness to the mismodeling of the spatial distribution of source, and can work even in the case where the different sources have different angular distribution shapes.

1. Introduction

In most applications of array signal processing, source localization techniques are generally based on point source assumption, where the energy of each source is considered to be concentrated at discrete direction. However, in many pratical applications, the transmitted signal is often obstructed by buildings, vehicles, *etc.* The multipath propagation in the vicinity of each transmitter may cause angular spreading of its energy. In this situation, a *distributed source* model is more appropriate than the point source one [1].

In the literature, we can find two types of distributed sources, namely, coherently distributed (CD) and incoherently distributed (ID) sources [2]. More precisely, for a CD source the observation period is much smaller than the channel coherency time, in this case the signal components arriving from different directions can be modeled as the delayed and attenuated replicas of the same signal. On the other hand, if all signals coming from different directions are assumed to be uncorrelated, the source is said ID source.

Many methods have been proposed for estimating these two types of distributed sources. In CD source case, the rank of the noise-free covariance matrix is equal to the number of sources. Consequently, most classical direction-finding methods, which are based on the point source assumption, can be easily extended to this situation (see [2, 3, 4, 5] and references therein). For example, in [2], the MUSIC algorithm has been generalized from the point source case to the distributed source situation, the resulting method is named Distributed Sources Parameter Estimator (DSPE). Note that, the main drawback of all estimators, in CD source case, is that they assume that all sources have an identical and known angular distribution shape.

On the other hand, for ID sources, the rank of the noise-free covariance matrix increases as the angular spread of the source increases. Therefore the classical subspace based techniques cannot be easily generalized to the case of ID sources, because the choice of the effective dimension of the pseudo-signal subspace depends on the unknown parameters [6]. Due to the complexity of the subspace based methods in the ID sources case, some interesting ideas are proposed: a beamforming method [7], the covariance fitting methods [8, 9] and recently a new subspace method [10]. However, some of these methods are limited to the single source case whereas the others assume that all sources have an identical and known angular distribution shape.

In this chapter, we propose some robust techniques that allow a decoupled estimation of the angular parameters of CD and ID sources. Compared with the original estimators, the proposed estimators impove the robustness to the mismodeling of the spatial distribution of the distributed CD and ID sources.

This chapter is organized as follows. In Section 2, we present the adopted CD and ID source models. In Section 3, we show how the estimation of the DOAs can be decoupled from that of the angular spreads. Based on this idea, some new methods are presented. Computer simulation results are presented in Section 4, and conclusions are drawn in Section 5.

2. Problem Formulation

Consider a uniform linear array (ULA) of m sensors. The distance between two adjacent sensors is denoted by d. Assume q distributed narrowband far-field sources are impinging on the array from nominal DOAs $\theta_k,\ k = 1, \ldots, q$. As in [2], the signals received at the antenna array can be modeled as:

$$\boldsymbol{x}(t) = \sum_{i=1}^{q} \int_{\theta \in \Theta} \boldsymbol{a}(\theta) s_i(\theta, t; \boldsymbol{\eta}_i) d\theta + \boldsymbol{n}(t) \tag{1}$$

In the above equation, $s_i(\theta, t; \boldsymbol{\eta}_i)$ is the angular signal distribution of the ith source, $\boldsymbol{\eta}_i = [\theta_i\ \sigma_i]^T$ is the unknown parameter vector of the ith source, with θ_i its nominal DOA and σ_i the corresponding angular spread, Θ is the angular field of view, $\boldsymbol{n}(t)$ is an additive white Gaussian noise vector and $\boldsymbol{a}(\theta)$ is the steering vector for a point source at DOA θ, denoted by

$$\boldsymbol{a}(\theta) = \left[1, e^{j\frac{2\pi d}{\lambda}\sin\theta}, \ldots, e^{j\frac{2\pi d}{\lambda}(m-1)\sin\theta}\right]^T \tag{2}$$

where $(.)^T$ denotes the transpose operator and λ is the wavelength of the impinging signals.

Assuming that all sources are mutually uncorrelated, the noise and signals are uncorrelated and that the noise is spatially and temporally white, the covariance matrix of the array measurements (1) can be written as:

$$\mathbf{R} = \sum_{i=1}^{q} \int_{\theta \in \Theta} p_i(\theta, \boldsymbol{\eta}_i)\boldsymbol{a}(\theta)\boldsymbol{a}^H(\theta)d\theta + \sigma_n^2 \mathbf{I} \tag{3}$$

where $p_i(\theta, \boldsymbol{\eta}_i) = \mathbb{E}\left[s_i(\theta, t; \boldsymbol{\eta}_i)s_i^*(\theta, t; \boldsymbol{\eta}_i)\right]$ is the power distribution of the ith source, $\mathbb{E}\left[.\right]$ denotes the expectation and σ_n^2 is the noise power.

2.1. CD Source Model

For a CD source, the shape of the angular distribution does not change temporally, and the received signal components from that source at different directions are fully correlated. Thus, the angular signal distribution, associated with the ith CD source can be expressed as: $s_i(\theta, t; \boldsymbol{\eta}_i) = s_i(t)g_i(\theta; \boldsymbol{\eta}_i)$, where $s_i(t)$ is the ith complex, random signal source and $g_i(\theta; \boldsymbol{\eta}_i)$ is the corresponding deterministic angular weighting function [2]. The function $g_i(\theta; \boldsymbol{\eta}_i)$ is assumed to be non-vanishing only around the actual DOA θ_i. Therefore, the covariance matrix (3) can be rewritten for the CD case as:

$$\mathbf{R}_{\text{CD}} = \mathbf{C}\,\boldsymbol{\Gamma}_s \mathbf{C}^H + \sigma_n^2 \mathbf{I} \tag{4}$$

where $\boldsymbol{\Gamma}_s = \mathbb{E}[\boldsymbol{s}(t)\boldsymbol{s}^H(t)]$ is the emitted signal covariance matrix. The matrix $\mathbf{C} = [\boldsymbol{c}_1(\boldsymbol{\eta}_1), \ldots, \boldsymbol{c}_q(\boldsymbol{\eta}_q)]$ contains the generalized steering vectors. For small angular spreads, each generalized steering vector is given approximately by [4, 3]:

$$\boldsymbol{c}_i(\boldsymbol{\eta}_i) \simeq \boldsymbol{\Phi}(\theta_i)\boldsymbol{b}_i(\boldsymbol{\eta}_i) \tag{5}$$

where $\boldsymbol{\Phi}(\theta_i) = \text{diag}\big(\boldsymbol{a}(\theta_i)\big)$ and $\boldsymbol{b}_i(\boldsymbol{\eta}_i)$ is the real-valued vector that depends on the shape of the angular weighting function $g_i(\theta; \boldsymbol{\eta}_i)$. Note that, in the case of Gaussian shape, the $(k)^{\text{th}}_{1 \le k \le m}$ element of the vector $\boldsymbol{b}_i(\boldsymbol{\eta}_i)$ can be expressed as:

$$\left[\boldsymbol{b}_i(\boldsymbol{\eta}_i)\right]_k^{\text{Gauss}} = \exp\left[-\frac{1}{2}\Big(\frac{2\pi d}{\lambda}(k-1)\sigma_i \cos\theta_i\Big)^2\right]. \tag{6}$$

In the case of uniform shape, we have:

$$\left[\boldsymbol{b}_i(\boldsymbol{\eta}_i)\right]_k^{\text{unif}} = \text{sinc}\left[\sqrt{3}\frac{2\pi d}{\lambda}(k-1)\sigma_i \cos\theta_i\right]. \tag{7}$$

2.2. ID Source Model

A distributed source is said to be ID if its components arriving from different directions are uncorrelated. Thus, for the ith ID source, we have: $p_i(\theta, \boldsymbol{\eta}_i) = \sigma_i^2 \rho_i(\theta; \boldsymbol{\eta}_i)$ where $\sigma_{s_i}^2$ is the power of the ith source and $\rho_i(\theta; \boldsymbol{\eta}_i)$ is the corresponding normalized angular power density, it is assumed non-vanishing only around the nominal DOA θ_i. Using this last formula and under the assumption of small angular spreads, the covariance matrix can be rewritten for the ID case as [2, 6]:

$$\mathbf{R}_{\mathrm{ID}} = \sum_{i=1}^{q} \sigma_{s_i}^2 \mathbf{\Psi}_i + \sigma_n^2 \mathbf{I} \tag{8}$$

where $\mathbf{\Psi}_i = \mathbf{\Phi}_i \mathbf{B}_i \mathbf{\Phi}_i^H$ is the noise-free covariance matrix of ith ID source, with the diagonal matrix $\mathbf{\Phi}(\theta_i)$ defined before and $\mathbf{B}_i$ a real-valued symmetric Toeplitz matrix. Its elements are given, for the Gaussian shape, by:

$$\left[\mathbf{B}_i\right]_{k\ell}^{\text{Gauss}} = \exp\left[-\frac{1}{2}\left(\frac{2\pi d}{\lambda}(k-\ell)\sigma_i \cos\theta_i\right)^2\right] \tag{9}$$

and for the uniform shape by:

$$\left[\mathbf{B}_i\right]_{k\ell}^{\text{unif}} = \text{sinc}\left[\sqrt{3}\,\frac{2\pi d}{\lambda}(k-\ell)\sigma_i \cos\theta_i\right]. \tag{10}$$

3. Parameters Estimation Techniques

In this section, we shortly describe some existing estimators for CD and ID sources and then we present some possible improvements.

3.1. CD Sources Methods

In [2], the MUSIC algorithm is generalized for CD source case, the proposed technique, called DSPE method, minimizes the following cost function for estimating the angle parameters:

$$\hat{\boldsymbol{\eta}} = \arg\min_{\boldsymbol{\eta}} \frac{\boldsymbol{c}^H(\boldsymbol{\eta})\hat{\mathbf{E}}_n\hat{\mathbf{E}}_n^H\boldsymbol{c}(\boldsymbol{\eta})}{\boldsymbol{c}^H(\boldsymbol{\eta})\boldsymbol{c}(\boldsymbol{\eta})} \tag{11}$$

where $\hat{\mathbf{E}}_n$ is the estimated noise subspace spanned by the $(m-q)$ orthonormal eigenvectors associated with the $(m-q)$ smallest eigenvalues of the sample covariance matrix $\hat{\mathbf{R}} = (1/T)\sum_{t=1}^{T}\boldsymbol{x}(t)\boldsymbol{x}^H(t)$.

As we can observe, the DSPE estimator has two drawbacks:

- Shape of the angular distribution of different sources has to be identical,
- Distribution shape has to be known.

To overcome theses problems, we propose to use the fact that $\boldsymbol{b}(\boldsymbol{\eta})$ in (5) is real valued and the matrix $\mathbf{\Phi}^*\hat{\mathbf{E}}_n\hat{\mathbf{E}}_n^H\mathbf{\Phi}$ is Hermitian. In addition, the cost function $\frac{\boldsymbol{c}^H\hat{\mathbf{E}}_n\hat{\mathbf{E}}_n^H\boldsymbol{c}}{\boldsymbol{c}^H\boldsymbol{c}} = \frac{\boldsymbol{b}^T\mathbf{\Phi}^*\hat{\mathbf{E}}_n\hat{\mathbf{E}}_n^H\mathbf{\Phi}\boldsymbol{b}}{\boldsymbol{b}^T\mathbf{\Phi}^*\mathbf{\Phi}\boldsymbol{b}}$ is real positive valued. Therefore, (11) can be rewritten as:

$$\hat{\boldsymbol{\eta}} = \arg\min_{\boldsymbol{\eta}} \frac{\boldsymbol{b}^T\text{Re}\left(\mathbf{\Phi}^*\hat{\mathbf{E}}_n\hat{\mathbf{E}}_n^H\mathbf{\Phi}\right)\boldsymbol{b}}{\boldsymbol{b}^T\boldsymbol{b}} \tag{12}$$

where Re(.) denotes the real part. We have then an objective function $\frac{\boldsymbol{b}^T \text{Re}\left(\boldsymbol{\Phi}^* \hat{\mathbf{E}}_n \hat{\mathbf{E}}_n^H \boldsymbol{\Phi}\right) \boldsymbol{b}}{\boldsymbol{b}^T \boldsymbol{b}}$ to minimize, which can be done in an efficient manner by taking into account some properties of the vector $\boldsymbol{b} = [b_1, b_2, \ldots, b_m]^T$. Indeed, according to formula (5), the elements $\left[b_k\right]_{k=1,\ldots,m}$ of the vector $\boldsymbol{b}(\boldsymbol{\eta})$ for a Gaussian distribution have to be positive and satisfy the constraints: $1 = b_1 \geq b_2 \geq \cdots \geq b_m \geq 0$. Note that, the same properties are also true in the case of small angular spread for a uniform distribution.

By exploiting these results, the estimation of the DOA in (11) can be replaced by the following constrained problem:

$$\hat{\theta} \;=\; \arg\min_{\theta} \left[\min_{\boldsymbol{x}} \frac{[1 \;\; \boldsymbol{x}^T] \text{Re}\left(\boldsymbol{\Phi}^* \hat{\mathbf{E}}_n \hat{\mathbf{E}}_n^H \boldsymbol{\Phi}\right) [1 \;\; \boldsymbol{x}^T]^T}{\boldsymbol{x}^T \boldsymbol{x} + 1} \quad \text{subject to} \quad \mathbf{J}\boldsymbol{x} \leq \boldsymbol{e} \right] \tag{13}$$

where $\boldsymbol{x} = [x_1, \ldots, x_{m-1}]^T$, $\boldsymbol{e} = [1, 0, \ldots, 0]^T$ and $\mathbf{J}$ is a $m \times (m-1)$ matrix such as:

$$\mathbf{J} = \begin{bmatrix} 1 & 0 & \cdots & 0 \\ -1 & 1 & \ddots & \vdots \\ 0 & -1 & \ddots & 0 \\ \vdots & \ddots & \ddots & 1 \\ 0 & \cdots & 0 & -1 \end{bmatrix}$$

The minimization problem under brackets (13) can be solved using a Sequential Quadratic Programming method (SQP) [11, 12, 13, 14]. In our simulations, the Matlab function `fmincon`, based on the SQP method is used to solve this problem. A full description of this algorithm can be found in `www.mathworks.com`.

After the DOA search procedure in (13), the angular spread of the ith source can be estimated by using the original DSPE estimator:

$$\hat{\sigma}_i = \arg\min_{\sigma} \frac{\boldsymbol{b}_i^T(\hat{\theta}_i, \sigma) \text{Re}\left(\boldsymbol{\Phi}^*(\hat{\theta}_i) \hat{\mathbf{E}}_n \hat{\mathbf{E}}_n^H \boldsymbol{\Phi}(\hat{\theta}_i)\right) \boldsymbol{b}_i(\hat{\theta}_i, \sigma)}{\boldsymbol{b}_i^T(\hat{\theta}_i, \sigma) \boldsymbol{b}_i(\hat{\theta}_i, \sigma)} \tag{14}$$

where $\hat{\theta}_i$ is the estimated DOA of the ith source given by (13).

3.2. ID Sources Methods

In [7] the conventional Capon estimator is generalized for ID source case as:

$$\hat{\boldsymbol{\eta}} = \arg\min_{\boldsymbol{\eta}} \mu_{\max}\left(\hat{\mathbf{R}}^{-1} \boldsymbol{\Psi}\right) \tag{15}$$

where $\mu_{\max}(.)$ stands for the maximal eigenvalue of a matrix and $\hat{\mathbf{R}}$ is the sample covariance matrix.

As we can observe, the above generalized Capon estimator (15) has the same limitations as for the DSPE estimator presented in section 3.1.

To solve these problems, we propose to take into account some properties of the matrix $\mathbf{B}$ in (8). Indeed, the elements $[\gamma_n]_{n=1,\ldots,m}$ of the first column vector γ of the matrix $\mathbf{B}$

for a Gaussian distribution have to be positive and satisfy the constraints: $1 = \gamma_1 \geq \gamma_2 \geq \cdots \geq \gamma_m \geq 0$, where $\gamma_n = [\mathbf{B}]_{k\ell}$ with $| k - \ell | = n - 1$. Note that, the same properties are also true in the case of small angular spread for a uniform distribution. From these results, the estimation of the DOA in (15) can be replaced by the following constrained problem:

$$\hat{\theta} = \arg\min_{\theta} \left[\min_{\boldsymbol{y}} \mu_{\max} \left[\hat{\mathbf{R}}^{-1} \mathbf{\Phi}(\theta) \mathbf{Y} \mathbf{\Phi}^H(\theta) \right] \quad \text{subject to} \quad \mathbf{J}\boldsymbol{y} \leq \boldsymbol{e} \right] \tag{16}$$

where $\boldsymbol{y} = [y_1, \ldots, y_{m-1}]^T$, $\mathbf{Y}$ is a real-valued symmetric Toeplitz matrix of size $(m \times m)$, given by:

$$\mathbf{Y} = \begin{bmatrix} 1 & y_1 & y_2 & \cdots & y_{m-1} \\ y_1 & 1 & y_1 & \ddots & \vdots \\ y_2 & y_1 & 1 & \ddots & y_2 \\ \vdots & \ddots & \ddots & \ddots & y_1 \\ y_{m-1} & \cdots & y_2 & y_1 & 1 \end{bmatrix} \tag{17}$$

and $\mathbf{J}$ is a $m \times (m-1)$ matrix defined as before.

As in (13), the minimization problem under brackets (16) can be solved using a SQP method.

Similarly, after the DOA search procedure in (16), the angular spread of the ith source can be estimated by replacing θ in (15) by the estimated DOA $\hat{\theta}_i$ in (16).

4. Simulation Results

In this section we present some simulation results that illustrate the performances of the proposed methods: Decoupled Distributed Signal Parameter Estimator (13) named DDSPE (CD case) and Robust generalized Capon (16) named RGC (ID case). Consider a uniform linear array of 8 sensors separated by a half wavelength of incoming signals. In these simulations the proposed DDSPE estimator is compared with DSPE and MUSIC estimators, while the proposed RGC estimator is compared with GC estimator.

The performances of all estimators are obtained from 200 Monte-Carlo simulations by calculating the root-mean-square error (RMS Error) of DOA and spread estimates. The proposed methods are initialized with $\boldsymbol{x}^{(0)} = \boldsymbol{y}^{(0)} = \frac{1}{m}[m-1, \ldots, 1]^T$.

4.1. CD case

In the first example, the influence of the signal to noise ratio (SNR) is examined. The parameter vectors of the two CD Gaussian sources are $\boldsymbol{\eta}_1^{\text{Gauss}} = [-10^\circ \;\; 7^\circ]^T$ and $\boldsymbol{\eta}_2^{\text{Gauss}} = [10^\circ \;\; 6^\circ]^T$. In this simulation, the actual distribution shape is assumed to be known for the DSPE estimator. It can be observed in Fig.1, that the proposed DDSPE estimator has performance similar to that of the original DSPE method. In fact, the difference between the results provided by these two estimators increases as the SNR decreases but this increase remains small even for small values of SNR. For example, for SNR=-10 dB, this difference

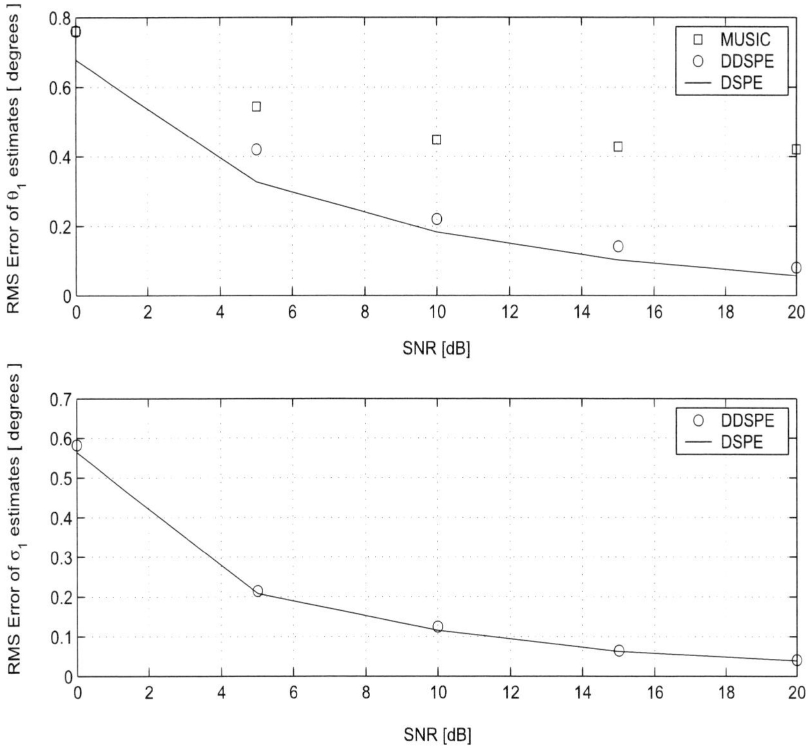

Figure 1. RMS Error of θ_1 and σ_1 estimate versus the signal to noise ratio (SNR), $\boldsymbol{\eta}_1 = [-10^\circ\ 7^\circ]^T$, $\boldsymbol{\eta}_2 = [10^\circ\ 6^\circ]^T$, $T = 300$ snapshots, $m = 6$ sensors (Gaussian distribution).

equal 0.1° for the DOA estimation and 0.05° for the spread estimation. In the second example, we illustrate the behaviors of the estimators [MUSIC, DSPE and DDSPE] when the spread angle is varied between 1° and 10°. Two Gaussian CD sources are simulated with DOAs $\theta_1 = -10^\circ$ and $\theta_2 = 10^\circ$, the SNR is equal to 10 dB. Here also the DSPE estimator assumes that the distribution shape is known. We can observe in Fig.2, that the proposed DDSPE estimator presents a good alternative for estimating the angular parameters of CD sources. Note that, the particular behavior (with a minimum at about 5°) of the RMS error of angular spread can be explained by the fact that the two estimators are trying to fit an angular dispersion to cope with the noise. In fact, the position of the minimum decreases as the SNR increases. In the third example, we compare the estimators DSPE and DDSPE when two CD sources have a same but unknown distribution shape. For simplicity, we assume that the angular spread is known for the DSPE estimator. Actually, the sources have a Gaussian shape, but we consider in the search procedure that they have a uniform shape. In this simulation, the theoretical covariance matrix is used and the angular parameters of the two equipower CD sources are $\boldsymbol{\eta}_1^{\text{Gauss}} = [-10^\circ\ \ 7^\circ]^T$ and $\boldsymbol{\eta}_2^{\text{Gauss}} = [10^\circ\ \ 7^\circ]^T$. We can show in Fig.3, that the proposed DDSPE estimator is robust to the mismodeling of the angular distribution, whereas the DSPE estimator fails to detect the two sources. It is interesting to examine the behaviors of the two estimators (DSPE and DDSPE) when different CD sources do not necessarily have the same angular distribution shape. In our

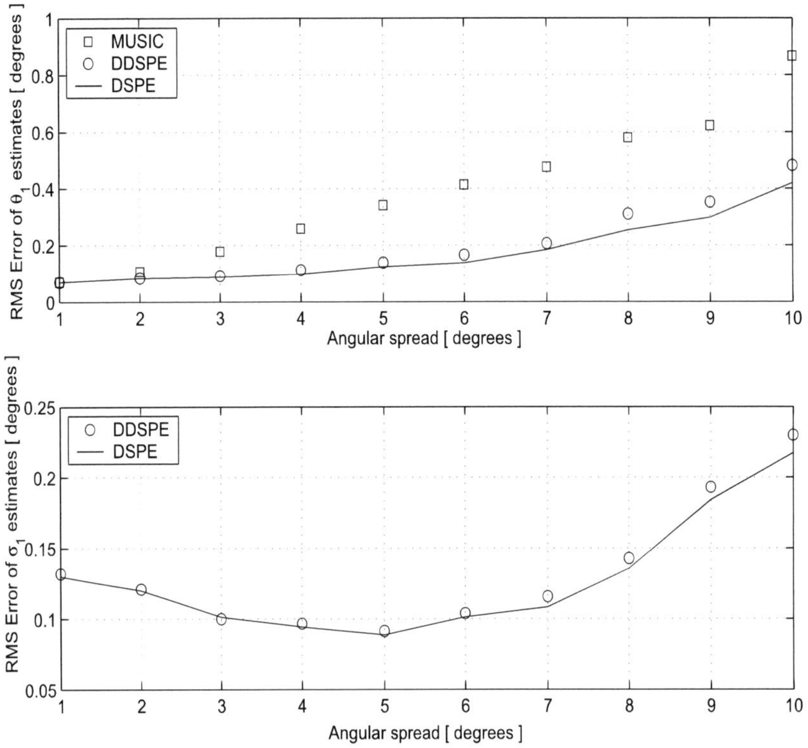

Figure 2. RMS Error of θ_1 and σ_1 estimate versus the true angular spread, $\theta_1 = -10°$, $\theta_2 = 10°$, SNR $= 10$ dB, $T = 300$ snapshots, $m = 6$ sensors (Gaussian distribution).

fourth example, we consider two equipower CD sources with different angular distribution shapes. The shape of the first source is assumed to be Gaussian with $\boldsymbol{\eta}_1^{\text{Gauss}} = [-10° \;\; 6°]^T$. The shape of the second source is assumed to be uniform with $\boldsymbol{\eta}_2^{\text{unif}} = [10° \;\; 6°]^T$. In the search procedure, it is assumed that the two sources have uniform shape. As it can be seen in Fig.4, that the proposed DDSPE estimator is robust to the mismodeling of the spatial distribution of the sources and may be used when different sources have different angular distribution shapes.

4.2. ID Case

We propose here to examine the performances of the estimators GC (15) and RGC (16) when the various ID sources have an identical distribution shape. In the fifth example, we examine the influence of the signal to noise ratio (SNR) when the two ID sources have the Gaussian shape. The parameter vectors of the two equipower sources are $\boldsymbol{\eta}_1^{\text{Gauss}} = [-10° \;\; 5°]^T$ and $\boldsymbol{\eta}_2^{\text{Gauss}} = [10° \;\; 4°]^T$. For the GC estimator (16), we consider two situations. In the first case, we consider that the actual distribution shape is known (or correctly estimated). Whereas, in the second case, we suppose that this shape is wrongly estimated (for this second case, the distribution shape of all sources is assumed to be uniform). Figures 5 (a) and (b) show the RMSE of DOA and angular spread estimates, respectively. As

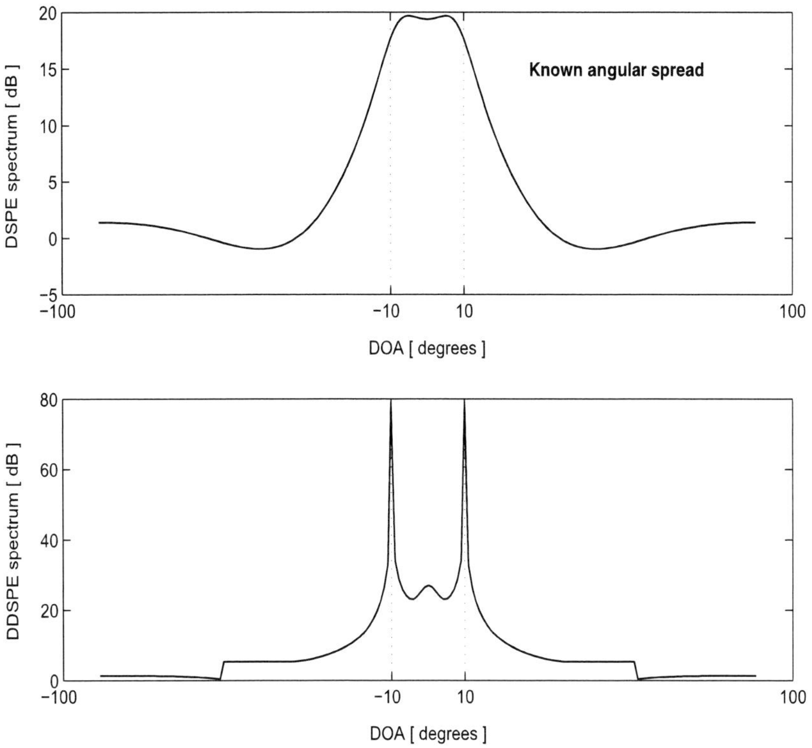

Figure 3. Spatial spectra of the DSPE (11) and DDSPE (13) [2 sources case with same but unknown angular distribution shape]

it can be seen in these two figures, that the GC method is very sensitive to the mismodeling of the angular distribution shape of the sources, while the proposed method gives the best estimation performance because it doesn't need the *a priori* knowledge of this distribution shape.

We examine now the behaviors of the two estimators (GC and RGC) when different ID sources do not necessarily have the same angular distribution shape. In the last example, we consider two equipower ID sources with different angular distribution shapes. The shape of the first source is assumed to be Gaussian with $\eta_1^{\mathrm{Gauss}} = [-10^\circ \;\; 5^\circ]^T$. The shape of the second one is assumed to be uniform with $\eta_2^{\mathrm{unif}} = [10^\circ \;\; 4^\circ]^T$. As in the fifth example, two cases are discussed for the GC method. In the search procedure, it is assumed first that the two sources have Gaussian shape, while in the second case it is assumed the two sources have uniform shape. As we can observe in figures 6 (a) and (b), that compared with the original GC estimator, the proposed RGC estimator is robust to the mismodeling of the spatial distribution of the sources and may be used when different sources have different angular distribution shapes.

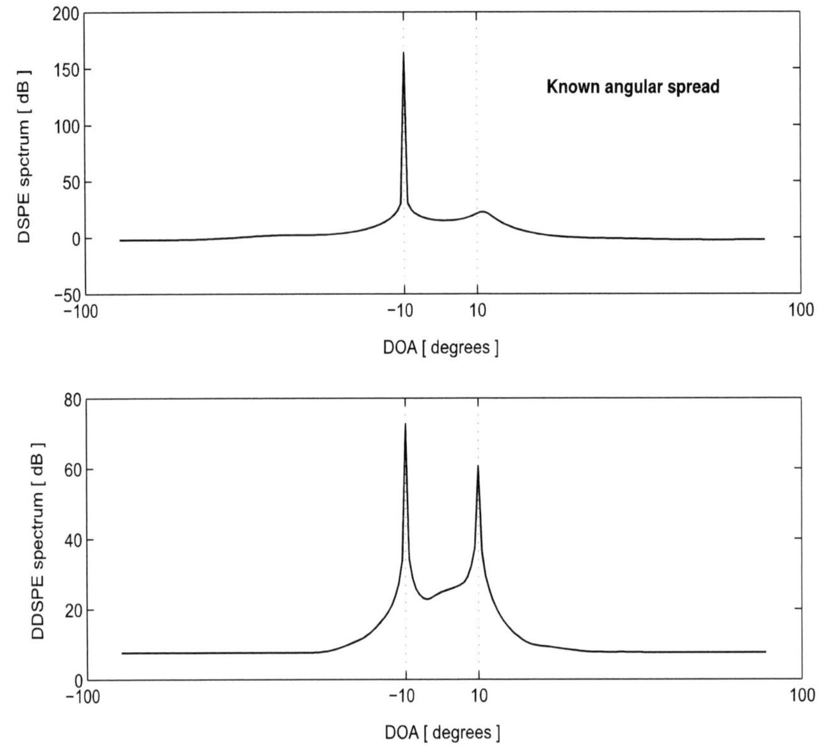

Figure 4. Spatial spectra of the DSPE (11) and DDSPE (13) [2 sources case with different angular distribution shapes]

5. Conclusion

In this chapter, we have presented some new techniques for estimating the DOAs and angular spreads of CD and ID sources. Using a ULA, our approach tries to exploit some properties of the generalized steering vector in the case of CD sources and some properties of the covariance matrix in the case of ID sources to enable a decoupled estimation of the DOAs from that of the angular spreads of sources. The proposed estimators impove the robustness to the mismodeling of the spatial distribution of the distributed CD and ID sources. Numerical examples show that the proposed estimators exhibit better estimation performances compared with the original estimators.

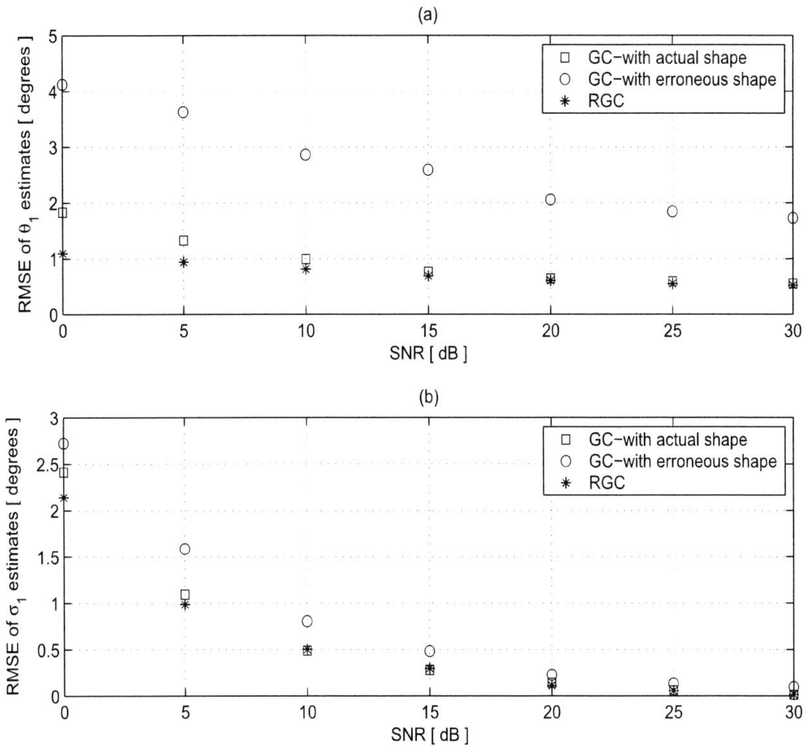

Figure 5. RMS Error of θ_1 and σ_1 estimate versus the signal to noise ratio (SNR) when 2 sources have the identical Gaussian shape.

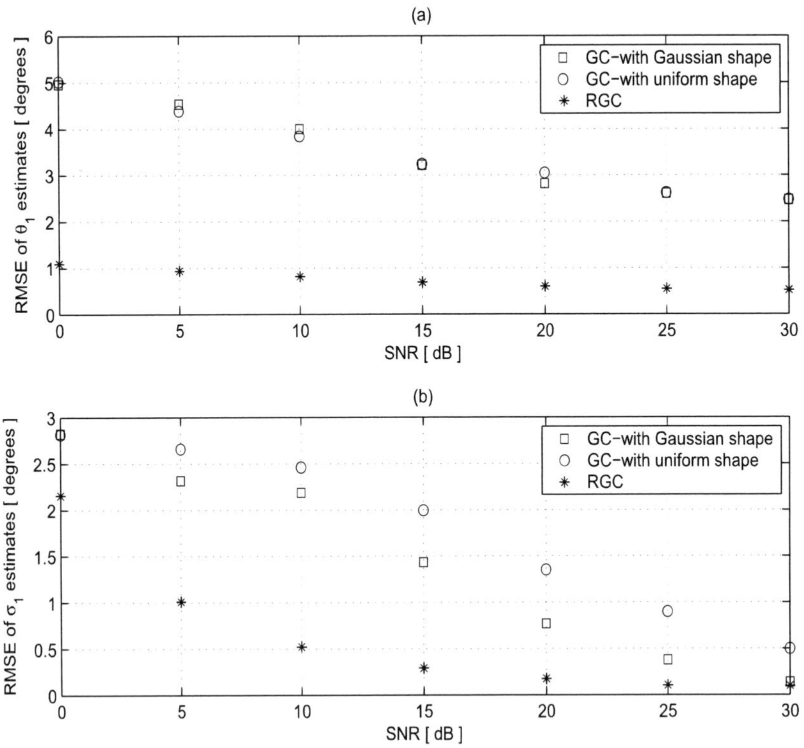

Figure 6. RMS Error of θ_1 and σ_1 estimate versus the signal to noise ratio (SNR) when 2 sources have different shapes.

References

[1] Zetterberg P. Mobile cellular communications with base station antenna arrays: Spectrum, algorithms and propagation models. *PhD thesis, Royal Institute of technology, S3, KTH, SE-10044 Stockholm*, Sweden, 1997.

[2] Valaee S., Champagne B. and Kabal P. Parametric localization of distributed sources. *IEEE Trans. on Signal Processing*, **43**, pp. 2144–2153, Sept. 1995.

[3] Zoubir A. and Wang Y. Performance analysis of the generalized beamforming estimators in the case of coherently distributed sources. *Signal Processing*, **88**, pp. 428–435, 2008.

[4] Zoubir A., Wang Y. Ould Haiballa M. M. and Chargé P. Decoupled estimation of angular parameters for coherently distributed sources. *Annals of Telecommunications*, **62**, no. 9-10, pp. 1171–1184, 2007.

[5] Zoubir A. and Wang Y. Efficient DSPE algorithm for estimating the angular parameters of coherently distributed sources. *Signal Processing*, **88**, pp. 1071–1071, 2008.

[6] Bengtsson M. Antenna array signal processing for high rank models. *PhD thesis, Royal Institute of technology, S3, KTH, SE-10044 Stockholm*, Sweden, 1999.

[7] Hassanien A., Shahbazpanahi S. and Gershman A.B. A generalized Capon estimator for localization of multiple spread sources. *IEEE Transaction on Signal Processing*, **52**, pp. 280–283, January 2004.

[8] Besson O. and Stoica P. Decoupled estimation of DOA and angular spread for a spatially distributed source. *IEEE Trans. on Signal Processing*, **48**(7), pp. 1872–1882, July 2000.

[9] Zoubir A., Wang Y. and Chargé P. On the ambiguity of COMET-EXIP algorithm for estimating a scattered source. *Signal Processing*, **86**, pp. 733–743, February 2006.

[10] Zoubir A., Wang Y. and Chargé P. Efficient subspace based estimator for localization of multiple spatially distributed sources. *IEEE trans. on Signal Processing*, **52**(2), pp. 532–542, February 2008.

[11] Fletcher R. and Powell M.J.D. A Rapidly Convergent Descent Method for Minimization. *Computer Journal*, **6**, 163–168, 1963.

[12] Han S.P. A Globally Convergent Method for Nonlinear Programming. *Journal of Optimization Theory and Applications*, **22**, 1977.

[13] Powell M.J.D. A Fast Algorithm for Nonlinearly Constrained Optimization Calculations. *Numerical Analysis, ed. G.A. Watson, Lecture Notes in Mathematics*, Springer Verlag., 630, 1978.

[14] Powell M.J.D. The Convergence of Variable Metric Methods For Nonlinearly Constrained Optimization Calculations. *Nonlinear Programming*, **3**, 1978.

In: New Signal Processing Research
Editor: Takumi Maeda, pp. 505-527

ISBN: 978-1-60456-479-2

Chapter 19

Maximum Entropy Fuzzy Clustering for Maneuvering Target Tracking in a Cluttered Environment

Li Liang-qun, Xie Wei-xi and Huang Jing-xiong
School of Information Engineering, Shenzhen University,
Shenzhen, Guangdong 518060, China

Abstract

For tracking maneuvering target in a cluttered environment, a novel tracking algorithm based on maximum entropy fuzzy clustering is proposed. First, the interacting multiple model (IMM) approach is used to solve the maneuver problem, and the false alarms generated by clutter are accommodated through a maximum entropy fuzzy probabilistic data association filter (MEF-PDAF). Second, to avoid the unobservability problem of passive target tracking, a nonlinear measurement model of multiple passive sensors is formulated. Moreover, to reduce the computational load, the characteristic of the discrimination factor is analyzed, and the influence of the clutter density on it is considered, which enables the algorithm eliminate invalid measurements. Finally, simulation results show that the proposed algorithm has advantages over the conventional IMM-PDAF algorithm in terms of simplicity and efficiency.

Key words: Tracking Maneuvering Target, Interacting Multiple Models (IMM), MEF-PDAF.

1. Introduction

The key to successful maneuvering target tracking in a cluttered environment lies in the effective extraction of useful information about the target's state from observations and deal with the problem arising from uncertainty about the maneuvers target and the uncertain origin of the measurements. The underlying statistical model for our problem is Jump Markov Linear System (JMLS) [8, 9], which is a linear system whose parameters evolve with time according to a finite state Markov chain. JMLS are widely used in target tracking. It is well known that the exact computation of conditional mean or state estimates involves a prohibitive computational cost exponential in the number of observations [10].

Several algorithms have been presented in the literature to address this problem, including the Interacting Multiple Model Probabilistic Data Association Filter (IMM-PDAF) algorithm, the Probabilistic Multi-Hypothesis Tracking (PMHT) algorithm, batch Expectation Maximization (EM) , Markov chain Monte Carlo (MCMC) methods, and particle filtering [5]. A promising approach is the interacting multiple model (IMM) algorithm originally developed by Blom [6], which is based on a hybrid system description of the maneuver scenarios; the occurrence of target maneuvers is explicitly included in the kinematics equations through regime jumps. In the presence of clutter, the integration of the IMM and PDAF [7] is an efficient solution to the uncertainty of measurement origins [1]. In the tracking benchmark problem [8] designed to compare the performance of different algorithms for tracking highly maneuvering targets in the presence of electronic countermeasures, the PDA-based estimator, in conjunction with the interacting multiple model (IMM) estimator, yielded one of the best solutions. Its performance was comparable to that of the MHT algorithm [9].

Passive tracking of moving targets using only bearing or line of sight (LOS) angle measurements is a known nonlinear estimation problem. Since the problem is nonlinear, the usual approach for recursive estimation is to employ an extended Kalman filter (EKF). Because the LOS is an incomplete position observation, it cannot be converted into Cartesian coordinates to allow for linear filtering as in [10]. In recursive bearings-only tracking, the use of the Cartesian coordinate EKF exhibits erratic estimation results and unstable behavior [l], even without the detrimental effects caused by the presence of false detections or clutter. Furthermore, the azimuth and elevation measurements of a passive sensor do not allow an instantaneous range determination. To solve this problem, two solutions have been considered. First, the passive sensor platform is allowed to move freely, it is possible to recover range observability by selecting an appropriate path for the platform. However, in some applications, the sensor platforms have very slow mobility compared with the target dynamics, so this solution is not feasible. A possible solution is to use several passive sensors and fuse their information in some way to estimate the range.

Dufour and Michel [12] proposed an extended version of IMMPDAF to track a maneuvering target in three dimensions with two passive sensors in a cluttered environment. The contribution of their work is twofold. First, a novel application of the IMM algorithm is studied, where passive-only sensors are fused for tracking a target maneuvering in three dimensions. Second, several accurate models of target motion are proposed to improve the performance. The main shortcoming of their method is its heavy computational load, particularly in heavily cluttered environment. Recently, we proposed a target tracking method based on maximum entropy fuzzy clustering, which has the same performance as PDAF with a lower computational load [13]. Inspired by the literatures [12,13], this paper proposes a novel algorithm based on Maximum entropy fuzzy clustering for real time maneuvering target tracking.

The remainder of the paper is organized as follows. The problem formulation is presented in Section 2. The maximum entropy clustering algorithm for single target tracking is described and a novel maneuvering target tracking algorithm is introduced in Section 3. Simulation results and performance comparison with the existing algorithms are presented in Section 4. Finally, some conclusions are provided in Section 5.

2. Problem Formulation

2.1. System Setup

Given a system

$$x_k = F(c_k)x_{k-1} + v_k \tag{1}$$

$$y_k = h(x_k) + w_k \tag{2}$$

where $x_k \in R^n$ is the dynamical state of the system under model c_k, $c_k \in \{1,2,\cdots,M\}$ is the model state of the system (e.g, $c_k = 1$ when the target moves at constant speed and $c_k = 2$ when the target accelerates). The process noise v_k and the measurement noise w_k are independent, zero mean noise with known covariance $Q(k)$ and $R(k)$, respectively. The mode transition of the system is modeled by a Markov chain with

$$\pi_{ij} = P(c_{k+1} = j | c_k = i), \qquad \forall i, j \in M \tag{3}$$

2.2. Probabilistic Data Association Filter (Pdaf)

Suppose the set of validate measurements at time k is $Z_k = \{z(k,j)\}_{j=1}^{m_k}$, m_k is the number of validated measurements at time k. The cumulative set of validated measurements up to time k is denoted as $Z^k = \{Z_l\}_{l=1}^{k}$.

At each scan, a validation gate, centered around the predicted measurement of the target, is setup to select the measurements associated probabilistically with the target. The validation region is

$$[z(k) - \hat{z}(k|k-1)]' [S(k)]^{-1} [z(k) - \hat{z}(k|k-1)] < g^2 \tag{4}$$

where $S(k)$ is the covariance of the innovation corresponding to the true measurement. If the Kalman filter is used to estimate the state of a target, the standard PDAF equations are described as follows.

The predictions of the state and the measurement at time k are defined as

$$\hat{x}(k|k-1) = F(k)\hat{x}(k-1|k-1) \tag{5}$$

$$\hat{z}(k|k-1) = H(k)\hat{x}(k|k-1) \tag{6}$$

Similarly the covariance of the prediction state is

$$P(k \mid k-1) = F(k)P(k-1 \mid k-1)F(k)^T + G(k)QG(k)^T \tag{7}$$

The innovation covariance (for the correct measurement)

$$S(k) = H(k)P(k \mid k-1)H(k)^T + R(k) \tag{8}$$

The Kalman gain is

$$W(k) = P(k \mid k-1)H(k)^T S(k)^{-1} \tag{9}$$

The state update equation of PDAF is

$$\hat{x}(k \mid k) = \hat{x}(k \mid k-1) + W(k)v(k) \tag{10}$$

where the combined innovation is

$$v(k) = \sum_{j=1}^{m_k} \beta_j(k)v_j(k) \tag{11}$$

$$v_j(k) = z_j(k) - \hat{z}(k \mid k-1) \tag{12}$$

The overall covariance associated with the updated state is

$$P(k \mid k) = P^0(k \mid k) + W(k)\left\{\beta_0(k)S(k) + \sum_{j=1}^{m_k}\left[\beta_j(k)v_j(k)v_j(k)^T\right] - v(k)v(k)^T\right\}W^T(k) \tag{13}$$

where $P^0(k|k)$ is defined as

$$P^0(k \mid k) = P(k \mid k-1) - W(k)S(k)W(k)^T \tag{14}$$

$\beta_j(k)$ denotes the probability that the jth measurement comes from the target in track at time k, $\beta_0(k)$ denotes the probability that none of the measurements is originated from the target, or equivalently, the probability that the current measurement is a false alarm(or clutter).The derivation of the probability $\beta_j(k)$ is briefly described here, and the details can be found in [1,7].

To evaluate the association probabilities, the conditioning is broken down into the past data Z^{k-1} and the latest data $Z(k)$. A probabilistic inference can be made on both the number of measurements in the validation region (from the clutter density, if know) and on their locations, expressed as

$$\beta_j(k) = P\{\varepsilon_i(k) \mid Z^k\}$$
$$= \frac{1}{c_k} p\left[Z(k) \mid \varepsilon_i(k), m_k, Z^{k-1}\right] \times P\{\varepsilon_i(k) \mid m_k, Z^{k-1}\} \tag{15}$$
$$j = 0, 1, \ldots, m_k$$

where c_k is a normalization constant, $\varepsilon_i(k)$ denotes the event that the measurement $z_i(k)$ is originated from the target, $\varepsilon_0(k)$ denotes the event that none of the measurements is originated from the target.

Using the parametric PDAF [7], the association probabilities are

$$\beta_0(k) = \frac{b}{b + \sum_{i=1}^{m_k} e_i} \quad j = 0$$
$$\beta_j(k) = \frac{e_j}{b + \sum_{i=1}^{m_k} e_i} \quad j = 1, \cdots, m_k \tag{16}$$

where

$$b = \left[\frac{2\pi}{g}\right]^{m/2} \lambda V(k) \frac{1 - P_D P_G}{P_D},$$

$$e_j = \exp\left[-v_i(k)'(k) S^{-1}(k) v_i(k)/2\right]$$

$$V_k = \frac{\pi^{m/2}}{\Gamma(m/2 + 1)} g^{m/2} \left|S(k)\right|^{1/2}$$

$V(k)$ is the volume of the ellipsoidal validation region, P_D is the probability of detection, P_G is the probability that the target measurement falls in the m-dimensional validation region, λ is the spatial density of false measurements.

At the same time, the weights $\beta_j(k)$ fulfill the constraint

$$\sum_{j=0}^{m_k} \beta_j(k) = 1, \quad 0 \le \beta_j(k) \le 1 \tag{17}$$

3. Proposed Maneuvering Target Tracking Method

3.1. Maximum Entropy Fuzzy Clustering

To be specific, suppose that an N-point data set $\{x_i, i = 1, 2, \cdots, N\}$ is related to one of the clusters $\{c_j, j = 1, 2, \cdots, c\}$. The fuzzy clustering process can be formulated as an optimization problem with the corresponding cost function to be minimized defined as

$$E = \sum_{i=1}^{N} \sum_{j=1}^{c} u_{ij} \cdot d(x_i, c_j) \tag{18}$$

where $d(x_i, c_j)$ is the squared Euclidean distance between the given data point x_i and the cluster center c_i. The probabilistic constraint is that the summation of all $u_{ij} (j = 1, 2, \cdots c)$ must be equal to one, i.e.

$$\sum_{j=1}^{c} u_{ij} = 1 \qquad \forall u_{ij} \in [0,1] \tag{19}$$

The maximum entropy principle (MEP) of information theory suggests choosing the u_{ij} for the Shannon entropy

$$H = H(u_{ij}) = -\sum_{i=1}^{N} \sum_{j=1}^{c} u_{ij} \ln u_{ij} \tag{20}$$

is maximized under the constraints Eq.(18) and Eq.(19).Using the Lagrange multiplier method, the objective function can be defined as

$$J(U,C) = -\sum_{i=1}^{N} \sum_{j=1}^{c} u_{ij} \ln u_{ij} - \sum_{i=1}^{N} \alpha_i \sum_{j=1}^{c} u_{ij} \cdot d(x_i, c_j) + \sum_{i=1}^{N} \lambda_i (\sum_{j=1}^{c} u_{ij} - 1) \tag{21}$$

By maximizing Eq (20), the membership function of data x_i belonging to cluster center c_j is derived as

$$u_{ij} = \frac{e^{-\alpha_i d(x_i, c_j)}}{\sum_{k=1}^{c} e^{-\alpha_i d(x_i, c_k)}} \tag{22}$$

where α_i and λ_i are Lagrange multipliers. Varying α_i, adjusts the value of membership of data point x_i with its nearest cluster center c_j as compared to that of other cluster centers. Thus, we call α_i as "discriminating factor". Ref[14,15]discuss choosing a proper value of α_i.

3.2. Maximum Entropy Fuzzy Clustering for Target Tracking

Eq. (19) shows that the summation of the u_{ij} values is equal to one, and the membership value is also nonnegative. This makes u_{ij} a latent candidate and an alternative to the association probability in PDAF. Suppose there is a data set with m_k an n-dimensional measurements

$\left(x_1, x_2, \cdots x_{m_k}\right)$, and a given target T. To apply maximum-entropy fuzzy clustering to target tracking, Eq. (19) can be modified as

$$\sum_{j=1}^{m_k} u_j = 1, \qquad \forall u_j \in [0,1] \tag{23}$$

where m_k denotes the number of validated measurements, u_j can be viewed as the probability that the measurement z_j belongs to the cluster center c. The corresponding cost function is defined as

$$E = \sum_{i=1}^{m_k} u_i \cdot d(x_i, c) \tag{24}$$

According to the maximum-entropy fuzzy clustering principle described above, the corresponding main objective function $J(U,C)$ to be maximized by the Lagrange method is defined as

$$J(U,C) = -\sum_{j=1}^{m_k} u_j \ln u_j - \alpha \sum_{j=1}^{m_k} u_j \cdot d(x_j, c) + \lambda\left(\sum_{j=1}^{m_k} u_j - 1\right) \tag{25}$$

where α and λ are Lagrange multipliers, and $d(z_j, c)$ may be the squared Euclidean distance (or the Mahalanobis distance) between the measurement z_j and the cluster center c. The final probabilities are

$$u_j = \frac{e^{-\alpha d(x_j, c)}}{\sum_{i=1}^{m_k} e^{-\alpha d(x_i, c)}} \qquad j = 1, 2, \cdots m_k \tag{26}$$

The probabilities given by Eq. (26) are the least biased estimate possible in the information theoretic sense for $u_j, i = 1, 2, \cdots m_k$. This claim is justified in [14]. It is noted that the membership degree u_j is very approximate to the association probability in PDAF [7]. The only difference between them is the exponential terms. So if a proper value of α is selected, the membership degree can completely replace the association probability with the same or better performance. The association probability can be defined as

$$\beta_j = u_j \tag{27}$$

Furthermore, if there is no measurement within the validation gate of target T, the updated state estimation will be the prediction estimate.

$$\hat{x}(k \mid k) = \hat{x}(k \mid k-1) \tag{28}$$

Another important question we have not answered yet is how to determine the cluster center c. To avoid being trapped into a local maximum, a good initial guess is useful. For this purpose, a good initial for the center c is provided by the predicted measurement, that is,

$$c = H(k)F(k)\hat{x}(k-1 \mid k-1) \tag{29}$$

3.3. IMM-MEFPDAF

According to the descriptions above, the MEFPDAF developed in this paper consists of modifying the PDAF algorithm so that the association probabilities are substituted with membership grades provided by maximum entropy fuzzy clustering. This section proposes the IMM-MEFPDAF, which integrates interacting multiple models with the maximum entropy fuzzy probabilistic data association. The basic structure of the algorithm is the same as the IMM-PDAF. A detailed derivation of the IMM algorithm structure can be founded in [6], so only our new algorithm is described here.

Step 1: Interaction

Mixing probabilities:

$$\mu_{j|i}(k-1 \mid k-1) = \frac{1}{c_i} p_{ji}\mu_j(k-1) \tag{30}$$

Normalizing factors:

$$c_i = \sum_{j \in M} p_{ji}\mu_j(k-1) \tag{31}$$

Mixing initial state estimated in mode j

$$\hat{X}^{0i}(k-1 \mid k-1) = \sum_{j=1}^{M} X_j(k-1 \mid k-1)\mu_{j|i}(k-1 \mid k-1) \quad i = 1,2,\cdots,M \tag{32}$$

The corresponding state covariance is

$$\begin{aligned} P^{0i}(k-1 \mid k-1) = \sum_{j=1}^{M} \mu_{j|i}(k-1 \mid k-1)\{P^j(k-1 \mid k-1) + \left[X^j(k-1 \mid k-1)\right. \\ \left. - \hat{X}^{0i}(k-1 \mid k-1)\right]\left[X^j(k-1 \mid k-1) - \hat{X}^{0i}(k-1 \mid k-1)\right]\} \end{aligned} \tag{33}$$

Step 2: Prediction

Using $\hat{X}^i(k-1|k-1)$ and its covariance $P^i(k-1|k-1)$, one computes the predicted state $\hat{X}^i(k|k-1)$ and its covariance $P^i(k|k-1)$, and the predicted measurement $\hat{z}^i(k)$ and the corresponding covariance $S^i(k)$

$$\hat{X}^i(k|k-1) = F(k-1)\hat{X}^{0i}(k-1|k-1) \tag{34}$$

$$P^i(k|k-1) = F(k-1)P^{0i}(k-1|k-1)F(k-1)' + Q^j(k-1) \tag{35}$$

$$\hat{z}^i(k) = h(\hat{X}^i(k|k-1)) \tag{36}$$

$$S^i(k) = H_j(k)P^i(k|k-1)H_j(k)' + R^j(k) \tag{37}$$

where $F(k-1)$, $Q^j(k-1)$ and $R^j(k)$ are from (1) ~ (2). In a passive sensor system, the measurement equation is nonlinear, so the nonlinearity $y_k = h(x_k) + w_k$ must be linearized. $H_j(k)$ is the Jacobian matrix of $h(x_k)$

$$H_j(k) = \frac{\partial h(x_k)}{\partial x_k}\Big|_{\hat{X}^i(k|k-1)} \tag{38}$$

Step 3: Measurement Validation

The measurement $z_l^i(k)$ is validated if and only if

$$\left(z_l^i(k) - \hat{z}^i(k)\right)S(k)\left(z_l^i(k) - \hat{z}^i(k)\right)' < g \tag{39}$$

where $S(k)$ denotes the largest one of the model conditional covariance.

Step 4: Model Probabilities Update

The model probabilities are updated by

$$\mu_j = \frac{1}{c}\Lambda_j(k)\sum_{i\in M} p_{ij}\mu_i(k-1) = \frac{1}{c}\Lambda_j(k)\bar{c}_j \tag{40}$$

where $\bar{c}_j$ is the expression from Eq.(20) and c denotes the normalization constant. $\Lambda_j(k)$ is the likelihood function, written as [1]:

$$\begin{aligned}\Lambda_j(k) &= p\left(Z(k) \mid M_j, Z(k-1)\right) \\ &= p\left(v_1^j(k), \cdots, v_m^j(k) \mid m(k), Z(k-1)\right) \\ &= \left(\sum_{i=1}^{m(k)} e_i(k)\right) P_d P_g m(k)^{-1} V_k^{-m(k)+1}\end{aligned} \tag{41}$$

where P_d is the detection probability, $m(k)$ is the number of the measurements at time k, and V_k is the volume of the validation region at time k for model j.

Step 5: Filtering and Fusion

With the maximum-entropy fuzzy probabilistic data association filter, the update of model i is computed as follows:

$$\hat{X}^i(k \mid k) = \hat{X}^i(k \mid k-1) + \sum_{j=1}^{m_k} \beta_j^i (z_j^i(k) - \hat{z}^i(k)) \tag{42}$$

$$P^i(k \mid k) = P^i(k \mid k-1) - W^i(k \mid k) S^i(k) W^i(k \mid k)' \tag{43}$$

where β_j^i denotes the association probability computed via (15)-(16).

Finally, the model-conditioned estimates are combined:

$$\hat{X}(k \mid k) = \sum_{i=1}^{M} \mu_i \hat{X}^i(k \mid k) \tag{44}$$

$$P(k \mid k) = \sum_{i=1}^{M} \mu_i \left\{ P^i(k \mid k) + \left[\hat{X}^i(k \mid k) - \hat{X}(k \mid k)\right]\left[\hat{X}^i(k \mid k) - \hat{X}(k \mid k)\right] \right\} \tag{45}$$

3.4. Discrimination Factor $\alpha_{(opt)}$

A key issue remains unanswered, namely how to choose an optimal value for α. The optimal value α, denoted as $\alpha_{(opt)}$, is derived subject to the minimization of the expected distortion given in Equation (7). Suppose that the number of validation measurements, m_k, is given. for any given data sample set $\{x_j\}$ [14,15], the cost function E is always monotonically and exponentially decreasing as α increase. Due to this monotonicity, the optimal membership value $u_i\left(i = 1, 2, \cdots, m_i\right)$ is reached only when $\alpha \to \infty$. But since the cost function E decreases exponentially with the increase in α, when the discriminating factor α reaches some value, the value of E reaches a saturation state, and further increase in α only causes a slight change in E. At this point, we suppose that α has reached the optimal value.

A detailed derivation of the optimal discrimination factor $\alpha_{(opt)}$ can be founded in [15] which give:

$$\alpha_{(opt)} = -\frac{\ln(\varepsilon)}{d_{(\min)}} \tag{46}$$

In practice, the choice of an appropriate value for α depends on the specific applications. To a specified problem, it is interesting to note that the larger the value of α is the larger the probability of the measurements closed to the cluster center. So it is natural to require that when the density of clutter is low, α should be large, and vice versa. That is, the value of α is inversely proportional to the density of clutter λ, thus we define

$$\alpha = \frac{\eta}{\lambda \cdot d_{\min}} \tag{47}$$

where η is a positive constant, $\eta \in [0,1]$.

On the other hand, given a fixed α, the probability u_i approximate decreases exponentially with the increase in the value of d. So when the distance between the measurement and the cluster center is sufficiently large, the measurement can be viewed as invalidated even though it may be fall in the target validation region.

Substituting α in Eq (26) with Eq (47), it follows that

$$u_i = \frac{e^{-\frac{\eta}{\lambda \cdot d_{\min}} d(x_i, C)}}{\sum_{j=1}^{m_k} e^{-\frac{\eta}{\lambda \cdot d_{\min}} d(x_j, C)}} \tag{48}$$

Thus,

$$\begin{aligned} u_{i(\max)} &= \frac{e^{-\frac{\eta}{\lambda \cdot d_{\min}} d(x_i, C)}}{\sum_{j=1}^{m_k} e^{-\frac{\eta}{\lambda \cdot d_{\min}} d(x_j, C)}} \\ &= \frac{e^{-\frac{\eta}{\lambda \cdot d_{\min}} d_{\min}}}{\sum_{\substack{j=1 \\ j \neq i}}^{m_k} e^{-\frac{\eta}{\lambda \cdot d_{\min}} d(x_j, C)} + e^{-\frac{\eta}{\lambda \cdot d_{\min}} d_{\min}}} \\ &= \frac{e^{-\frac{\eta}{\lambda}}}{\sum_{\substack{j=1 \\ j \neq i}}^{m_k} e^{-\frac{\eta}{\lambda \cdot d_{\min}} d(x_j, C)} + e^{-\frac{\eta}{\lambda}}} \end{aligned} \tag{49}$$

Eq (49) shows that the numerator is an independent constant with respect to $d(x_i, C)$, while the dominator can be approximately considered a normalization constant. So, the maximum association probability value $u_{i(\max)}$ can also be considered to approximately constant. Therefore, when $u_{j(j \neq i)} / u_{i(\max)} \leq \xi$ (when ξ is a sufficiently small positive constant, such as $\xi = 10^{-3}$ or less), the j th measurement has only a slight effect on the update of state estimate of the target.To reduce the computational load of the algorithm, the j th measurement can be regarded as an invalid measurement. Thus, we suppose the threshold for selecting the set of measurements associated with the target as follows

$$\begin{aligned}
&\frac{u_{j(j \neq i)}}{u_{i(\max)}} = \frac{e^{-\frac{\eta}{\lambda d_{\min}} d_j}}{e^{-\frac{\eta}{\lambda}}} \geq \xi \\
&\Leftrightarrow d_{j(j \neq i)} \leq -\frac{\lambda \ln(\xi)}{\eta} d_{\min} + d_{\min} = (1 - \frac{\lambda \ln(\xi)}{\eta}) d_{\min} \\
&\Leftrightarrow d_{j(j \neq i)} \leq (1 - \frac{\lambda \ln(\xi)}{\eta}) d_{\min} = D_{\max}
\end{aligned} \tag{50}$$

In this scheme, the distance between the j th measurement and the cluster center c must be less than $D_{\max}$ for the j th measurement to be considered a validate measurement. Thus, we call $D_{\max}$ the "maximum validation distance". Furthermore, because the minimum distance $d_{\min}$ is different at each time step, the maximum validation distance $D_{\max}$ can be adjusted automatically according to the distribution of the measurements. There is a tradeoff, since the reduction of computational cost results in a slight degradation in estimation performance of the target estimated. To make this tradeoff, the size of threshold can be adjusted according to the application.

4. Simulation Results

This section evaluates and compares the performance of the proposed algorithm with the conventional IMMPDAF algorithm [12].

4.1. Observation Model

With a single passive sensor, when the sensor is allowed to move freely, it is possible to recover range observability by selecting an appropriate path. For a stationary passive sensor system, to avoid observability problems, the range must be acquired through multiple passive sensors. This paper considers multiple passive sensors, the observation equation is defined as

$$Z(k) = H(X(k)) + W(k), \tag{51}$$

where

$$\begin{aligned} Z(k) &= [z_1(k)^T \; z_2(k)^T \cdots z_n(k)^T]^T \\ &= [(\alpha_1, \beta_1)^T \; (\alpha_2, \beta_2)^T \cdots (\alpha_n, \beta_n)^T]^T, \end{aligned} \tag{52}$$

$$H(X(k)) = [H_1(X(k))^T \; H_2(X(k))^T \cdots H_n(X(k))^T]^T, \tag{53}$$

$$W(k) = [w_1(k)^T \; w_2(k)^T \cdots w_n(k)^T]^T. \tag{54}$$

where $H_i(X(k))$ is the nonlinear equation, and $W(k)$ denotes the measurement noise vector. We assume that the angle errors $w_i(k)$ $(i = 1, 2, \cdots, n)$ are mutually independent and the measurement noise $W(k)$ has a diagonal covariance $R(k)$.

$$\begin{aligned} R(k) &= E\left[W(k) W(k)^T \right] \\ &= diag[\sigma_{\alpha_1}^2 \; \sigma_{\beta_1}^2, \ldots, \sigma_{\alpha_n}^2 \; \sigma_{\beta_n}^2] \end{aligned} \tag{55}$$

where $\sigma_{\alpha_i}^2$ $(i = 1, 2, \ldots n)$ is the azimuth error variance of sensor *i*, and $\sigma_{\beta_i}^2$ $(i = 1, 2, \ldots n)$ is the elevation error variance of sensor *i*.

4.2. Target Motion Models

Suppose that the state vector of the target is $\left(x(k), y(k), z(k), \dot{x}(k), \dot{y}(k), \dot{z}(k)\right)$.In this paper, the interacting multiple model approach is used to address the maneuver problem. Three target motion models are used:

Model 1: Constant Velocity Motion

The state transition matrix and the process noise covariance matrix are defined by

$$A = \begin{pmatrix} 1 & 0 & 0 & T & 0 & 0 \\ 0 & 1 & 0 & 0 & T & 0 \\ 0 & 0 & 1 & 0 & 0 & T \\ 0 & 0 & 0 & 1 & 0 & 0 \\ 0 & 0 & 0 & 0 & 1 & 0 \\ 0 & 0 & 0 & 0 & 0 & 1 \end{pmatrix} \tag{56}$$

$$Q_1 = \begin{pmatrix} \frac{1}{4}T^4 & 0 & 0 & \frac{1}{2}T^3 & 0 & 0 \\ 0 & \frac{1}{4}T^4 & 0 & 0 & \frac{1}{2}T^3 & 0 \\ 0 & 0 & \frac{1}{4}T^4 & 0 & 0 & \frac{1}{2}T^3 \\ \frac{1}{2}T^3 & 0 & 0 & T^2 & 0 & 0 \\ 0 & \frac{1}{2}T^3 & 0 & 0 & T^2 & 0 \\ 0 & 0 & \frac{1}{2}T^3 & 0 & 0 & T^2 \end{pmatrix} \quad (57)$$

Model 2: Constant Turn Motion

The state transition matrix is

$$A = \begin{pmatrix} 1 & 0 & 0 & \frac{\sin(w)}{w} & \frac{\cos(w)-1}{w} & 0 \\ 0 & 1 & 0 & \frac{1-\cos(w)}{w} & \frac{\sin(w)}{w} & 0 \\ 0 & 0 & 1 & 0 & 0 & 1 \\ 0 & 0 & 0 & \cos(w) & -\sin(w) & 0 \\ 0 & 0 & 0 & \sin(w) & \cos(w) & 0 \\ 0 & 0 & 0 & 0 & 0 & 1 \end{pmatrix}, \quad (58)$$

where w is a constant angular rate. The process noise covariance matrix is the same as in Model 1.

Model 3: For $w < 0$ describes a clockwise turn, and Model 3 is its natural counterpart for a counterclockwise turn $w > 0$.

4.3. Scenario Design and Performance Analysis

In this example, simulation results are given for the cases of two sensors and three sensors, respectively.

Two sensors are located at $(5km,0,0)$ and $(-5km,0,0)$ for both case. The "three sensor" case has an additional sensor at $(0,8.66km,0)$. The false alarms are generated from a Gaussian distribution with centered on the true target position. The number of false alarms is uniformly distributed between 0 and λ. The probability via matrix of the Markovian transition of the system mode is defined

$$\Pi = \begin{bmatrix} 0.9 & 0.05 & 0.05 \\ 0.1 & 0.8 & 0.1 \\ 0.1 & 0.1 & 0.8 \end{bmatrix},$$

Two different algorithms were tested using the same trajectory. The target makes two circular turns with rectilinear segments connecting them. The speed modulus is kept constant throughout (=300m/s). The initial position is $(2km, 8km, 1km)$, and the initial velocity is $(150m/s, 259.8m/s, 0)$. The segments are defined as follows.

1st segment. Rectilinear flight until the plane is at $(6.35km, 15.53km, 1km)$ space (from t=0s to t=30s).

2nd segment. Circular turn with turn rate $6^{\circ}/s$ (from t=31s to t=50s).

3rd segment. Rectilinear flight until the plane is at $(14.31km, 10.33km, 1km)$ space (from t=51s to t=70s).

4th segment. Circular turn with turn rate $4.8^{\circ}/s$ (from t=71s to t=95s).

5th segment. Rectilinear flight until the plane is at $(21.26km, 11.63km, 1km)$ space (from t=96s to t=100s).

4.3.1. Effect of Measurement Noise

Here we fix the clutter density and the process noise variances, and vary the measurement noise. The clutter density is set to $\lambda = 2$. The process noise variance is set to $Q_{ii} = 0.0004$. The simulation results are shown in Figures 1 and 2 for two sensors and three sensors, respectively.

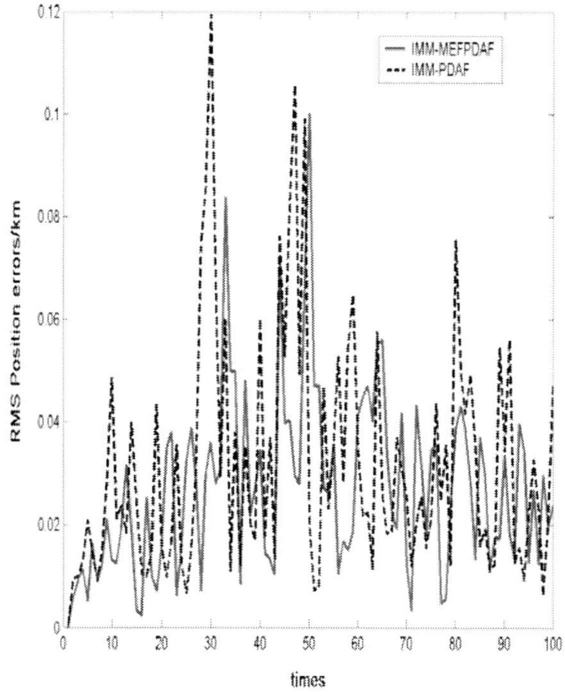

Figure 1. Continued on next page.

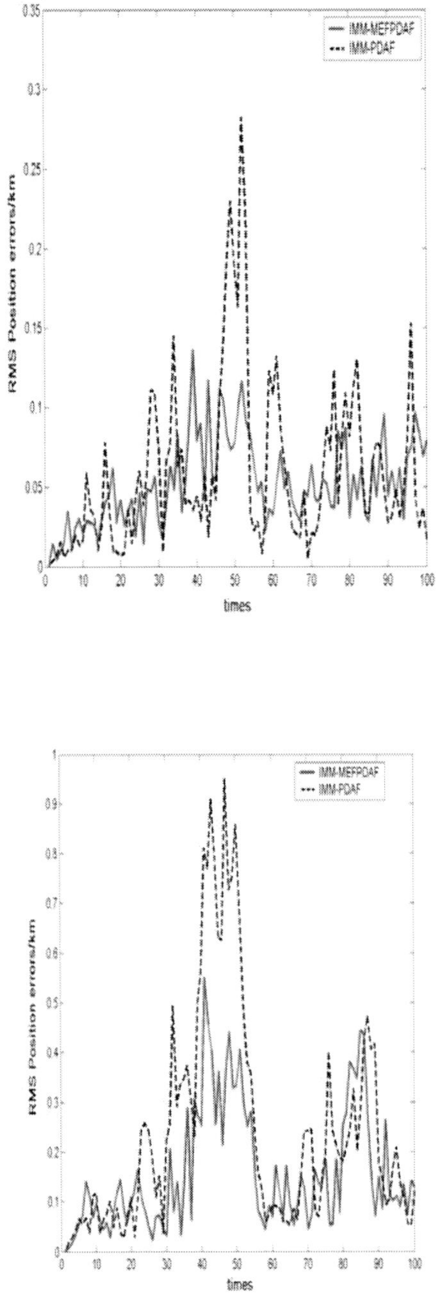

Figure 1. RMS error statistics for two sensors. RMS Position error (in *km*) for IMM-MEFPDAF (solid line (-)) and IMM-PDAF (dashed line (--)). Left is for measurement noise variance (σ_w = 0.0005 *mrad*), middle for measurement noise variance (σ_w = 0.001 *mrad*), and right for measurement noise variance (σ_w = 0.005 *mrad*).

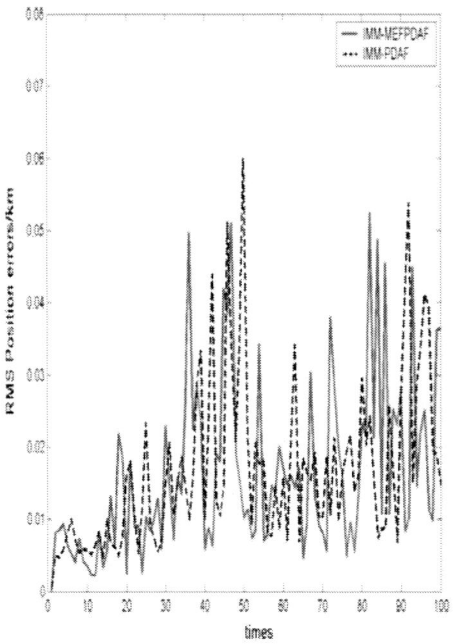

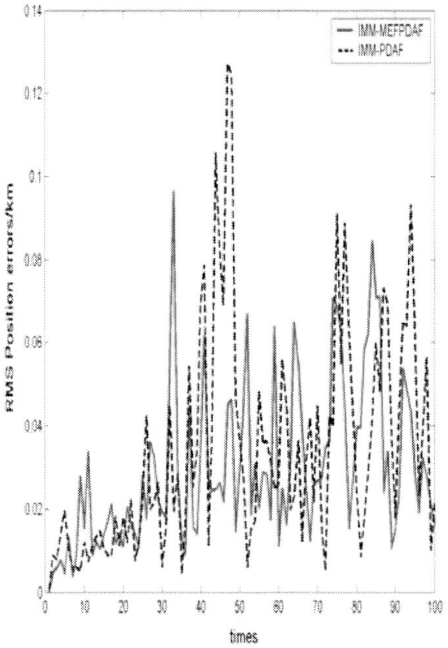

Figure 2. Continued on next page.

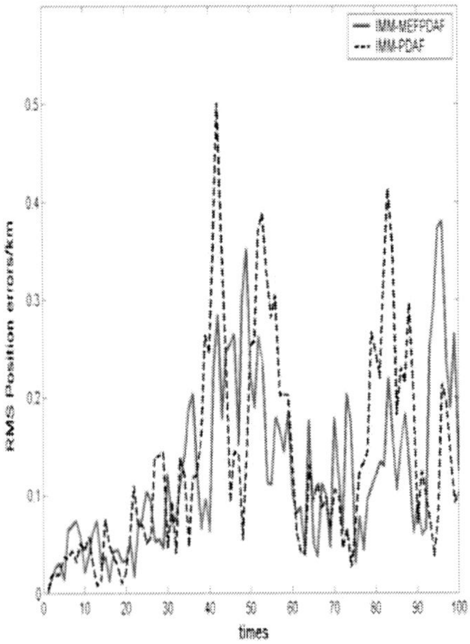

Figure 2. RMS error statistics for three sensors. RMS Position error (in *km*) for IMM-MEFPDAF (solid line (-)) , IMM-PDAF (dashed line (--)). Left is for measurement noise variance (σ_w = 0.0005 *mrad*), middle for measurement noise variance (σ_w =0.001 *mrad*), and right for measurement noise variance (σ_w = 0.005 mrad).

All algorithms exhibit a monotonic decrease in performance as σ_w increases and a monotonic increase in performances as the number of sensors increases. The figures also suggests that the IMM-MEFPDAF has performance similar to the IMM-PDAF for low measurement noise($\sigma_w = 0.0005$ *mrad*), but significantly superior to the IMM-PDAF for large measurement noise($\sigma_w = 0.001$ or $\sigma_w = 0.005$)when the target maneuvers.

4.3.2. Effect of Clutter Density

We now fix the measurement noise and the process noise variances, and vary clutter density. The measurement noise is set to $\sigma_w = 0.001 mrad$. The process noise variance is set to $Q_{ii} = 0.0004$. The simulation results are shown in Figure 3 and Figure 4 for the cases of two sensors and three sensors respectively.

All algorithms exhibit a monotonic decrease in performance as clutter density λ increases. However, the degradation in performance is less for IMM-MEFPDAF than for IMM-PDAF. The performance difference between the PDAF and the MEF-PDAF appears to increase with increasing clutter density. At the same time, we find that IMM-MEFPDAF is significantly superior to the IMM-PDAF when the target maneuvers. As expected, the performance of all algorithms is increases as the number of sensors increases.

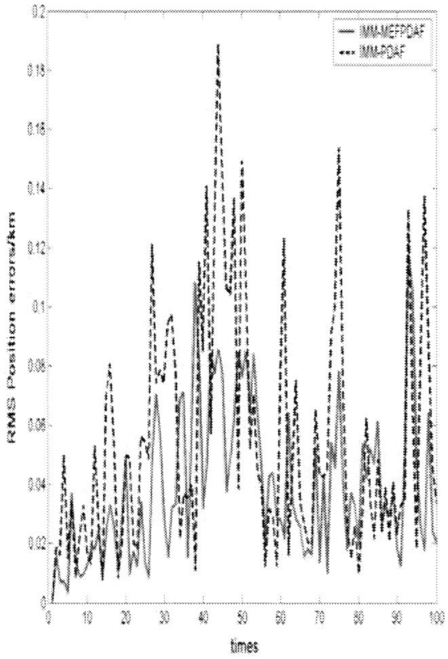

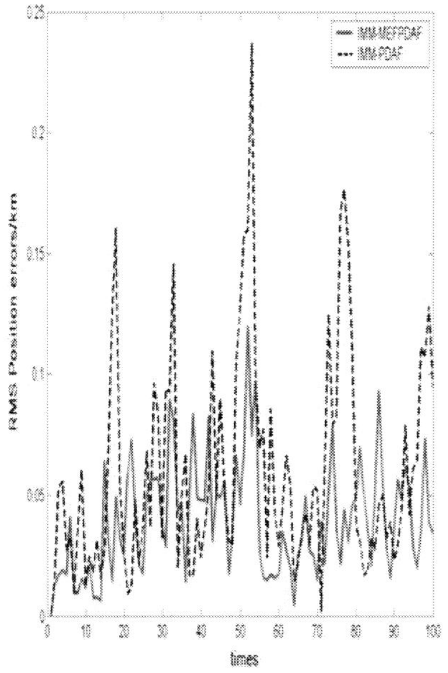

Figure 3. Continued on next page.

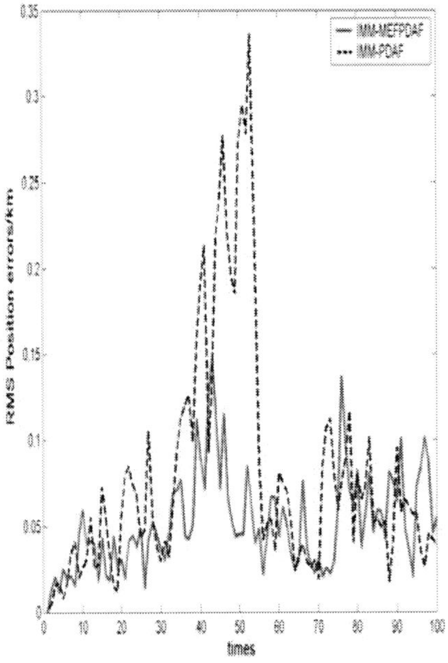

Figure 3. RMS error statistics for two sensors. RMS Position error (in *km*) for IMM-MEFPDAF (solid line (-)) , IMM-PDAF (dashed line (--)). Left is for clutter density (λ=0.5), middle for clutter density (λ=1.5), and right for clutter density (λ=2.5).

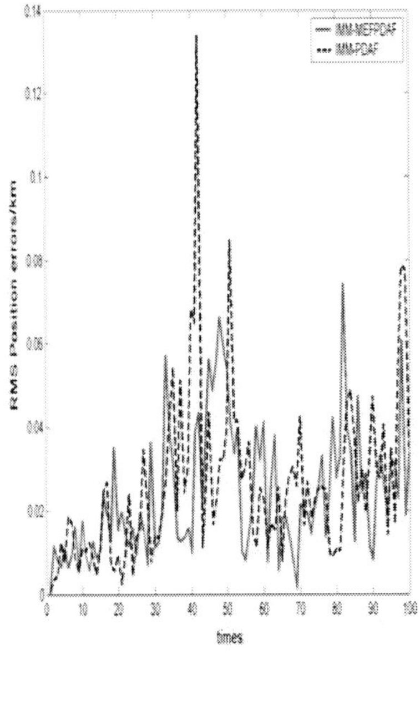

Figure 4. Continued on next page.

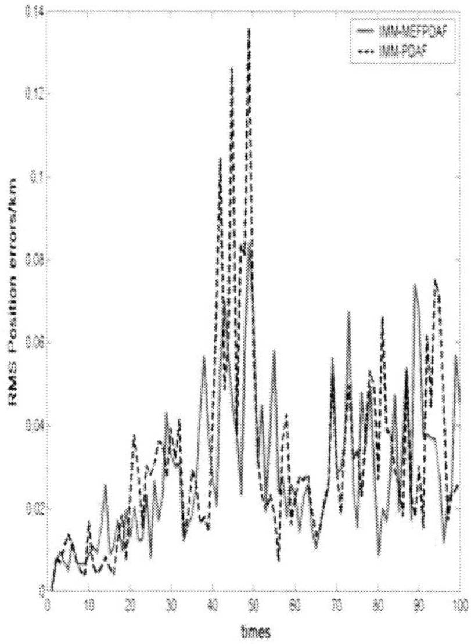

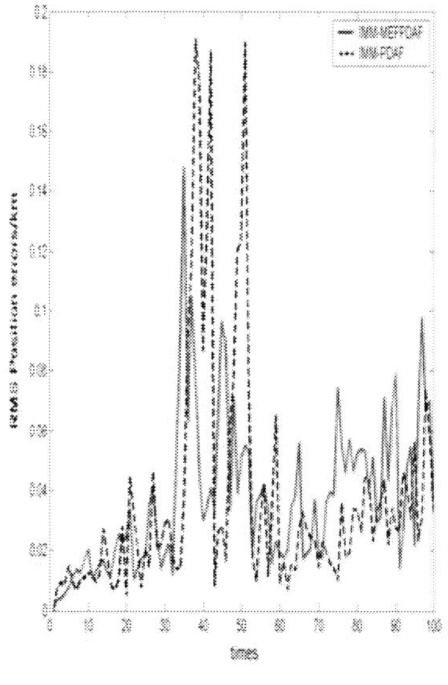

Figure 4. RMS error statistics for three sensors. RMS Position error (in *km*) for IMM-MEFPDAF (solid line (-)) , IMM-PDAF (dashed line (--)). Left is for clutter density (λ=0.5), middle for clutter density (λ=1.5), and right for clutter density (λ=2.5).

Finally, the runtime statistics of two algorithms for the case of two sensors are illustrated in Table 1. In each case, the program is run on a Pentium 4 3.0GHZ computer with 1.GB RAM and 50 Monte Carlos runs. Table 1 is shown that the runtime of the proposed algorithms is less than that of the conventional IMM-PDAF. One major reason for this is that in the proposed algorithms if the distance between the measurement and the cluster center exceeded the maximum validation distance, the measurement is considered invalid and eliminated from the validated measurement set; however, the measurement was still considered to be a validated measurement by the IMM-PDAF. Moreover, the difference of the runtime of two algorithms appears to increase with increasing clutter density. Hence, the proposed algorithm is much faster than the IMM-PDAF, and can be applied to real time maneuvering target tracking.

Table 1. Comparison of the run time of CPU(s)

	IMM-PDAF	IMM-MEFPDAF
$\lambda = 1$	10.700	9.1400
$\lambda = 2$	12.1050	10.2350
$\lambda = 3$	14.5350	10.9400

Conclusion

This paper described an efficient and novel approach for maneuvering target tracking based on maximum entropy fuzzy probabilistic data association. First, the derivation of maximum entropy fuzzy clustering to single target tracking was introduced. Second, the integration of interacting multiple models and maximum entropy fuzzy probabilistic data association was described, and applied to the maneuvering target tracking with multiple passive sensors in a cluttered environment. Simulation results show the performance of this approach is better than the conventional IMMPDAF approach, and achieve the real-time requirement of maneuvering target tracking. Moreover, the performance of the proposed algorithm did not degrade significantly with an increase in the difficulty of the association problem or the amount of clutter.

References

[1] Bar_shalom Y, and LI X. *Multitarget_Multisensor tracking: Principles and Techniques* [M]. Storrs, CT: YBS Publishing, 1995.

[2] *Y.* Bar-Shalom and X. R. Li. *Multi-target Mult-sensor tracking: Principles and Techniques.* Artech 1995.

[3] E. Mazor, A. Averbuck, Y. Bar-Shalom, and J. Dayan. Interacting multiple model methods in target tracking: a survey. *IEEE Trans. Aero. Elec. Sys.*, **34**:103-123, 1998.

[4] J. K. Tugnait. Adaptive estimation and identification for discrete systems with Markov jump parameters. *IEEE Trans. Automatic Control*, **25**:1054-1065, 1982.

[5] M. R. MORELANDE, S.CHALLA. Manoeuvring Target Tracking in Clutter using Particle Filters. *IEEE Transactions on aerospace and electronic systems* Vol. 41, No. 1 JANUARY 2005.

[6] Blom H A, Bar-Shalom. Y. The Interacting Multiple Model Algorithm for Systems with Markovian Switching Coefficient. *IEEE Trans on AC*-**33**(8), 1988: 780~783.

[7] Kirubarajan.T; Bar-Shalom.Y, "Probabilistic Data Association Techniques for Target Tracking in Clutter", *Proceedings of the IEEE*, Vol. 92, Issue: 3, March 2004, pp. 536 – 557.

[8] W. D. Blair, G. A. Watson, T. Kirubarajan, and Y. Bar-Shalom, "Benchmark for radar resource allocation and tracking in the presence of ECM," *IEEE Trans. Aerosp. Electron. Syst.*, vol. 34, pp.1015–1022, Oct. 1998.

[9] S. S. Blackman, R. J. Dempster, M. T. Busch, and R. F. Popoli, "IMM/MHT solution to radar benchmark tracking problem," *IEEE Trans. Aerosp. Electron. Syst.*, vol. 35, pp. 730–738, Apr. 1999.

[10] Blom,H.A.P, A sophisticated tracking algorithm for ATC surveillance data , In *Proceedings of in ternational radar conference*, Paris, France, may, 1984.

[11] Boers Y., Driessen J.N. interacting multiple model particle filter [J]. IEE Proc Radar Sonar Navig, 2003, 150(5): 344~349.

[12] Francois Dufour, Michel Mariton. Tracking a 3D Maneuvering Target with Passive Sensors, *IEEE Transactions on aerospace and electronic systems*, JULY 1991,27(4):725- 739.

[13] Liangqun Li,Hongbing Ji,Xinbo Gao. Maximum entropy fuzzy clustering with application to Real-Time Target Tracking, *International Journal of Signal Processing*, Nov.2006, 86(11): 3432-3447.

[14] P.X.Liu, M.Q.-H.Meng, "Online Data-Driven Fuzzy Clustering With Applications to Real-Time Robotic Tracking", *IEEE Transactions on Fuzzy Systems,*, Volume: 12, Issue: 4 , Aug. 2004 ,pp.516 – 523.

[15] K .Rose, "Deterministic annealing for clustering, compression, classification, regressions, and related optimization problems", *Proceedings of the IEEE*, Vol.86, No.11, November 1998, pp.2210-2239.

INDEX

D

H

I

J

K

L

M

O

P

Q

R

S

T

U

V